T0255362

Differential Equations

An Introduction to Basic Concepts, Results and Applications

Third Edition

Differential Equations

An Introduction to Basic Concepts, Results and Applications

Third Edition

Ioan I Vrabie

"Al. I. Cuza" University of Iaşi and
"O. Mayer" Mathematics Institute of the
Romanian Academy, Iaşi, Romania

World Scientific

NEW JERSEY • LONDON • SINGAPORE • BEIJING • SHANGHAI • HONG KONG • TAIPEI • CHENNAI

Published by

World Scientific Publishing Co. Pte. Ltd.

5 Toh Tuck Link, Singapore 596224

USA office: 27 Warren Street, Suite 401-402, Hackensack, NJ 07601

UK office: 57 Shelton Street, Covent Garden, London WC2H 9HE

Library of Congress Cataloging-in-Publication Data

Names: Vrabie, I. I. (Ioan I.), 1951–

Title: Differential equations : an introduction to basic concepts, results and applications / by Ioan I. Vrabie ("Al I. Cuza" University of Iaşi, Iaşi, Romania & "O. Mayer" Mathematics Institute of the Romanian Academy, Iaşi, Romania).

Description: 3rd edition. | New Jersey : World Scientific, 2016. | Includes bibliographical references and index.

Identifiers: LCCN 2016009661 | ISBN 9789814749787 (hardcover : alk. paper) | ISBN 9789814759205 (pbk. : alk. paper)

Subjects: LCSH: Differential equations--Textbooks.

Classification: LCC QA371 .V7313 2016 | DDC 515/.35--dc22

LC record available at https://lccn.loc.gov/2016009661

British Library Cataloguing-in-Publication Data

A catalogue record for this book is available from the British Library.

Printed in Singapore

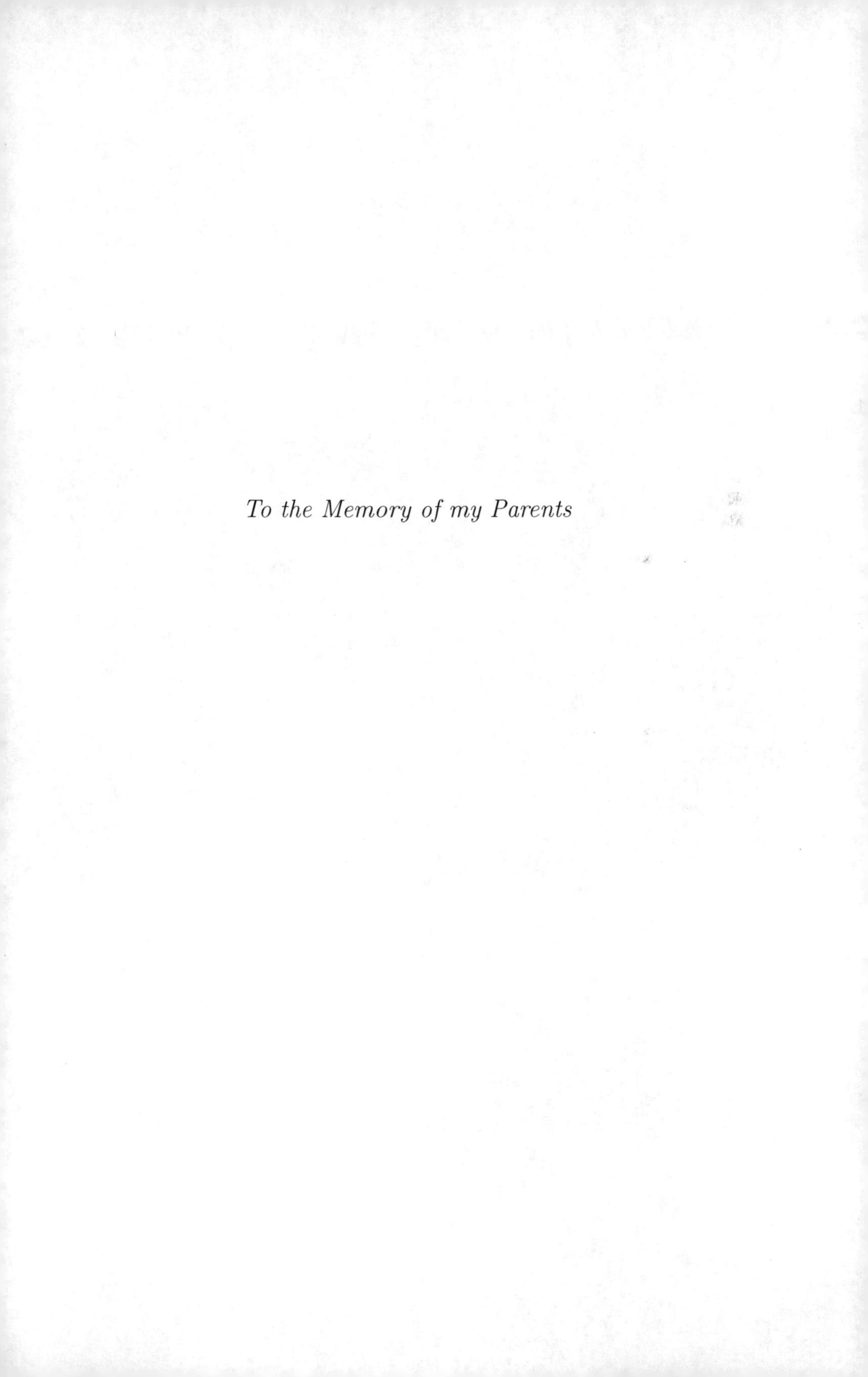

To the Memory of my Parents

Preface to the Third Edition

In this Third Edition some new additions and changes have been made. First, a few misprints have been corrected, several English inaccuracies fixed and some proofs simplified. Second, two new (and somewhat non-standard for a textbook) chapters have been added with the intention to offer the reader a perspective on some flourishing topics of great importance in applications. We refer here to Chapter 10 on Nonlocal Problems and Chapter 11 on Delay Functional Differential Equations. Of course, each of the newly introduced chapters contains a section of Exercises and Problems, whose complete solutions are included in the Solutions section. We have also added a new Section 12.3 introducing very briefly some basic facts on Banach spaces. However, keeping in mind that this third edition, like the preceding ones, should be accessible to a large and rather heterogeneous audience, I have tried as much as possible to give elementary proofs, avoiding the use of advanced results in Algebra, Topology and Functional Analysis. As a consequence, the rather complicated method for finding $e^{t\mathcal{A}}$ based on the Jordan normal form of a matrix, presented in the previous editions, has been substituted by Putzer's algorithm, which is both easy to prove and convenient to use in concrete applications.

Several new entries have been added to the Reference list, some of them containing information on both nonlocal problems and delay functional differential equations, as for instance [Burlică *et al.* (2016)], [Byszewski (1991)] and [McKibben (2011)]. Both [Kelley and Peterson (2010)] and [Putzer (1966)] concern a method of finding the exponential of a matrix which avoids the use of the Jordan normal form of the matrix. Moreover, for several references in Romanian, which are not accessible to a large numbers of readers, we have provided some English alternatives, e.g. [Gantmacher

(1959)], [Rudin (1976)] and [Wrede and Spiegel (2002)].

Acknowledgements. During the preparation of this new edition, I have benefited from the invaluable help of several colleagues, collaborators and friends. Their comments and remarks have been important in improving the presentation. I would like to mention here the very careful reading of both the newly added parts of the book and of the older ones, done by some of my former students, graduate students and collaborators: Professors Monica-Dana Burlică and Daniela Roşu from the "G. Asachi" Technical University of Iaşi, Mihai Necula and Eugen Vărvăruca from the "Al. I. Cuza" University of Iaşi.

My warmest thanks and sincere gratitude are addressed to all of them, and especially to Professor Eugen Vărvăruca, whose suggestions were essential in the improvement of the main statements in Sections 2.4–2.6, as well as in the simplification of the corresponding proofs.

I would also like to thank the whole staff of World Scientific for the excellent job of proofreading the manuscript. Finally, I thank Ms E. H. Chionh, the editor who took care of all the three editions, for her extremely effective and pleasant cooperation.

Iaşi, February 23rd, 2016 *Ioan I. Vrabie*

Preface to the Second Edition

The Second Edition of the book differs from the First Edition in that several additions and changes have been made. First, in order to offer the reader some basic information concerning two important topics which, in one way or another, are related to differential equations, the following insertions have been operated. So, keeping in mind that Volterra Equations are not only nontrivial extensions of differential equations but one of the starting points of research in Operator Theory, we have included Chapter 8, entirely devoted to this old but still modern and of great interest topic. Once this being done, taking into consideration that a very efficient tool in solving linear integral equations of Volterra type is the Laplace Transform, we have decided to present the fundamentals of this theory in Section 12.4. Further, we have added Chapter 9 dealing with the Calculus of Variations, a discipline having its roots in Mechanics and which is essentially based on nonlinear differential equations. It is almost needless to say that the Calculus of Variations is indispensable not only for mathematicians but for physicists as well. In addition, this branch of Mathematics was – and still is – the starting point of many deep researches in both Differential Equations and Differential Geometry.

As in the case of all other chapters, each of Chapters 8 and 9 ends with an exercises and problems section. The complete solutions of the exercises and problems newly added have been included in the last section of the book.

Second, trying to make the book as self-contained as possible, a simple form of the Implicit Functions Theorem, i.e. Theorem 2.8.1, whose proof was derived from Peano's Theorem 2.2.1, has been included. In addition, an extension to the case of implicitly defined vector-valued functions has been proposed in Problem 2.28. It should be noticed that the proof here presented, in spite of its relative simplicity – due, of course, to the very deep result on which is based, i.e. Peano's Theorem 2.2.1 –, is, with no doubt, far from being elementary. However, the reader which is familiar with Peano's Theorem 2.2.1 would have no problem in following the arguments there involved.

It should be emphasized that both newly added chapters and sections are intended as an incentive for further study rather than a main source of information.

Several new entries on the reference list have been included. These are [Arnold (1978)], [Barbu (2010)], [Bliss (1946)], [Cârjă *et al.* (2007)], [Gelfand and Fomin (1963)], [Giusti (2003)], [Havârneanu (2007)], [Krasnov *et al.* (1977)], [Lavrentiev and Sabat (1959)], [Lefter (2006a)], [Lefter (2006b)], [Sneddon (1971)] [Şabac (1981)], [Widder (1946)] and two of them, i.e. [Cârjă *et al.* (2004)] and [Cârjă and Vrabie (2005)], which were just in print at the time when the First Edition appeared, were updated. Some small corrections have been operated and a direct proof of the implication (iii) implies (i) in Theorem 4.1.4 has been included, substituting the indirect proof given in the First Edition.

Acknowledgements. I gratefully acknowledge the help of Professors Teodor Havârneanu, Cătălin-George Lefter and Mihai Necula, from the Department of Mathematics of "Al. I. Cuza" University of Iaşi, and that of Professor Marius Popescu, from the Department of Mathematics of the University of Galaţi, for the very careful reading of the newly added parts to the Second Edition of the book, and for the useful suggestions and remarks they have made. All their comments, which I took into consideration, have led to a substantial improvement of the presentation. I would also like to mention that Example 9.4.1 is due to Professor Eugen Popa, from the Department of Mathematics of "Al. I. Cuza" University of Iaşi.

I take this opportunity to express my warmest thanks to all of them as well as to the editors of World Scientific for the excellent job of proofreading the whole manuscript.

Iaşi, January 31st, 2011 *Ioan I. Vrabie*

Preface to the First Edition

The book is an entirely rewritten English version of the lecture notes of a course on Differential Equations I taught during the last twelve years at the Faculty of Mathematics of "Al. I. Cuza" University of Iaşi. These lecture notes were written in 1999 in Romanian. Their goal was to present in a unitary frame and from a new perspective the main concepts and results belonging to a discipline which, due to the continuous interplay between theory and applications, is by far one of the most fascinating branches of modern mathematics, i.e. *differential equations*. It was my intention to give the reader the opportunity to know a point of view — rather different from the traditional one — offering a possible way to learn differential equations with main emphasis on the Cauchy problem. So, I decided to treat separately the problems of: existence, uniqueness, approximation, continuation of the solutions and, at the same time, to give the simplest possible but complete proofs to some fundamental results which are at the core of the discipline: Peano's local existence theorem, the classification of non-continuable solutions from the viewpoint of their behavior at the end of the existence interval, the continuous dependence of the solution on the data and parameters, *etc.* This goal was by far very hard to accomplish due to the existence of a long list of very good, or even exceptional, textbooks and monographs on this subject covering all levels of difficulty: [Arnold (1974)], [Arrowsmith and Place (1982)], [Barbu (1985)], [Braun (1983)], [Coddington and Levinson (1955)], [Corduneanu (1977)], [Cronin (1980)], [Elsgolts (1980)], [Halanay (1972)], [Hale (1969)], [Hartman (1964)], [Hirsch and Smale (1974)], [Hubbard and West (1995)], [Ionescu (1972)], [Piccinini *et al.* (1984)], [Pontriaghin (1969)], to cite only a few. However, in spite of this challenging competition, I hope that the reader will find this text attractive enough from both the viewpoint of the chosen topics and the presentation.

The book contains a preface, a list of symbols, seven main chapters, a short chapter on auxiliary results, a rather long section including detailed solutions to all exercises and problems, a bibliography and ends with an index. With the sole exception of Chapters 6 and 7, which require some basic results on Lebesgue integral and Measure Theory, it is completely accessible to any reader having satisfactory knowledge of Linear Algebra and Mathematical Analysis. The 36 figures included illustrate the concepts introduced and smooth the way towards a complete understanding of the arguments used in the proofs.

The first chapter starts with a very brief presentation of the main steps made along the last four centuries toward the modern theory of differential equations. It continues with some preliminary notions and results referring to: the concept of solution, some methods of solving elementary equations, various mathematical models described either by differential equations or systems of differential equations, and some basic integral inequalities.

The second chapter contains several fundamental results concerning the Cauchy Problem: the local existence, the continuation of the solutions, the existence of global solutions, the relationship between the local and the global uniqueness, the continuous dependence and the differentiability of the solutions with respect to the data and to the parameters.

The third chapter is merely concerned with some classical facts about the approximation of the solutions: the method of power series, the method of successive approximations, the method of polygonal lines, the implicit Euler method and a particular, and therefore simplified, instance of Crandall–Liggett exponential formula.

In the fourth chapter we apply the previously developed theory to a systematic study of one of the most important class of systems: first-order linear differential systems. Here we present the main results concerning the global existence and uniqueness, the structure of the space of solutions, the fundamental matrix and the Wronskian, the variation of constants formula, the properties of the mapping $t \mapsto e^{t\mathcal{A}}$ and the basic results referring to n^{th}-order linear differential equations.

The fifth chapter is mainly concerned with the study of an extremely important problem of the discipline: the stability of solutions. We introduce four concepts of stability and we successively study the stability of the null solution of linear systems, perturbed systems and fully nonlinear systems respectively, in the last case by means of the Lyapunov's function method. We also include some facts about instability which is responsible for the so-called *unpredictability* and *chaos*.

In the sixth chapter, we start with the study of the concept of prime integral, first for autonomous, and thereafter for non-autonomous systems. Next, with this background material at hand, we present the basic results concerning linear and quasi-linear first-order partial differential equations. Some examples of conservation laws are also included.

The seventh chapter, rather heterogeneous, has a very special character being conceived to help the reader to go deeper within this discipline. So, here, we discuss some concepts and results concerning distributions and solutions in the sense of distributions, Carathéodory solutions, differential inclusions, variational inequalities, viability and invariance and gradient systems.

In the last chapter we include some auxiliary concepts and results needed for a good understanding of some parts of the book: the operator norm of a matrix, compact sets in $C([\,a,b\,];\mathbb{R}^n)$, the projection of a point on a convex set.

Each chapter, except the one on Auxiliary Results, ends with a special section containing exercises and problems ranging from extremely simple to challenging ones. The complete proofs of all these are included into a rather developed final section (more than 60 pages).

Acknowledgements. The writing of this book was facilitated by a very careful reading of some parts of the manuscript, by several remarks and suggestions made by Professors Ovidiu Cârjă, Mihai Necula from "Al. I. Cuza" University of Iaşi, by Professors Silvia-Otilia Corduneanu and Silviu Nistor from "Gh. Asachi" Technical University of Iaşi, remarks and suggestions which I took into account. The simplified version of the Frobenius theorem was called to my attention by Dr. Constantin Vârsan, Senior Researcher at The Mathematical Institute of the Romanian Academy in Bucharest. Some of the examples in Physics and Chemistry have been reformulated taking into account the remarks made by Professors Dumitru Luca and Gelu Bourceanu. Professor Constantin Onică had a substantial contribution in solving and correcting most part of the exercises and the problems proposed. A special mention deserves the very careful — and thus critical — reading of the English version by Professor Mircea Bârsan.

It is a great pleasure to express my appreciation to all of them.

Iaşi, November 30th, 2003 *Ioan I. Vrabie*

List of Symbols

$\mathcal{A}^{\tau}$	—	the transpose of the matrix $\mathcal{A} \in \mathcal{M}_{n\times n}(\mathbb{R})$
$B(\xi, r)$	—	the closed ball centered at ξ of radius $r > 0$
$\overset{\circ}{B}$	—	the interior of the set B
$\mathcal{B}^*$	—	the adjoint of the matrix $\mathcal{B} \in \mathcal{M}_{n\times n}(\mathbb{R})$
$\operatorname{conv} F$	—	the convex hull of F, i.e. the set of all convex combinations of elements in F
$\overline{\operatorname{conv} F}$	—	the closed convex hull of F, i.e. the closure of $\operatorname{conv} F$
$C_b([\,a, +\infty); \mathbb{R}^n)$	—	the space of all bounded continuous functions from $[\,a, +\infty)$ to $\mathbb{R}^n$
$\mathcal{CP}(\mathbb{I}, \Omega, f, a, \xi)$	—	the Cauchy problem $x' = f(t, x)$, $x(a) = \xi$, where $f : \mathbb{I} \times \Omega \to \mathbb{R}^n$, $a \in \mathbb{I}$ and $\xi \in \Omega$
$C([\,a, b\,]; \mathbb{R}^n)$	—	the space of continuous functions from $[\,a, b\,]$ to $\mathbb{R}^n$
$\mathcal{D}$	—	the data $(\mathbb{I}, \Omega, f, a, \xi)$. So, $\mathcal{CP}(\mathcal{D})$ denotes $\mathcal{CP}(\mathbb{I}, \Omega, f, a, \xi)$
$\operatorname{dist}(\mathcal{K}, \mathcal{F})$	—	the distance between the sets $\mathcal{K}$ and $\mathcal{F}$, i.e. $\operatorname{dist}(\mathcal{K}, \mathcal{F}) = \inf\{\lVert x - y \rVert\,;\ x \in \mathcal{K}, y \in \mathcal{F}\}$
$\operatorname{dist}(\eta, \mathcal{K})$	—	the distance between the point η and the set $\mathcal{K}$, i.e. $\operatorname{dist}(\eta, \mathcal{K}) = \inf\{\lVert \eta - \xi \rVert\,;\ \xi \in \mathcal{K}\}$
$\mathcal{D}(\mathbb{R})$	—	the space of indefinite differentiable functions from $\mathbb{R}$ to $\mathbb{R}$, with compact support
Δ	—	the compact set $[\,a, a + h\,] \times B(\xi, r)$
$\delta J[q][\eta]$	—	the first variation of J calculated at q along the direction η, i.e. $\frac{d}{d\varepsilon} J[q + \varepsilon\eta]\rvert_{\varepsilon=0}$
$\delta^2 J[q][\eta]$	—	the second variation of J calculated at q along the direction η, i.e. $\frac{d^2}{d\varepsilon^2} J[q + \varepsilon\eta]\rvert_{\varepsilon=0}$
$\mathcal{D}'(\mathbb{R})$	—	the set of linear continuous functionals from $\mathcal{D}(\mathbb{R})$ to $\mathbb{R}$
$\operatorname{graph}(x)$	—	the graph of $x : \mathbb{I} \to \mathbb{R}^n$, i.e. $\operatorname{graph}(x) = \{(t, x(t))\,;\ t \in \mathbb{I}\}$
$\mathbb{I}$, $\mathbb{J}$, $\mathbb{K}$	—	nonempty intervals in $\mathbb{R}$
$\mathcal{L}[f(t)](p)$	—	the Laplace Transform of the function f
$\mathbb{N}$	—	the set of nonnegative integers, i.e. $0, 1, 2, \ldots$
$\mathbb{N}^*$	—	the set of positive integers, i.e. $1, 2, 3, \ldots$
$\nabla_x z$, or ∇z	—	the gradient of the function z with respect to $x_1, x_2, \ldots, x_n$, i.e. $\nabla_x z = \operatorname{grad}_x z = \left(\frac{\partial z}{\partial x_1}, \frac{\partial z}{\partial x_2}, \ldots, \frac{\partial z}{\partial x_n}\right)$

$\|\mathcal{A}\|_{\mathcal{M}}$	—	the norm of the matrix $\mathcal{A} \in \mathcal{M}_{n\times m}(\mathbb{R})$, i.e. $\|\mathcal{A}\|_{\mathcal{M}} = \sup\{\|\mathcal{A}x\|_n\,;\ x \in \mathbb{R}^m,\ \|x\|_m \leq 1\}$
$\|u\|_{C_b([\,a,+\infty);\mathbb{R}^n)}$	—	is defined by $\|u\|_{C_b([\,a,+\infty);\mathbb{R}^n)} = \sup\{\|u(t)\|;\ t \in [\,a,+\infty)\}$ for each $u \in C_b([\,a,+\infty);\mathbb{R}^n)$
Ω	—	a nonempty and open subset of $\mathbb{R}^n$
ω_f	—	the growth index of the function f, i.e. $\omega_f = \inf\{\omega \in \mathbb{R};\ \exists\ M > 0,\ \|f(t)\| \leq Me^{\omega t},\ \forall\, t > 0\}$
$\mathcal{P}_K(x)$	—	the projection of the vector $x \in \mathbb{R}^n$ on the subset $K \subset \mathbb{R}^n$
$\mathbb{R}$	—	the set of real numbers
$\mathbb{R}^*$	—	the set of real numbers excluding 0
$\mathbb{R}_+$	—	the set of nonnegative real numbers
$\mathbb{R}_+^*$	—	the set of positive real numbers
$\operatorname{supp}\phi$	—	the set $\overline{\{t \in \mathbb{R}\,;\ \phi(t) \neq 0\}}$
Σ	—	a nonempty and locally closed subset of Ω
x'	—	the derivative of the function x
$\dot{x}$	—	the derivative in the sense of distributions of the distribution x
$\mathcal{X}^{-1}(s)$	—	the inverse of the matrix $\mathcal{X}(s)$, i.e. $\mathcal{X}^{-1}(s) = [\mathcal{X}(s)]^{-1}$

Contents

Preface to the Third Edition vii

Preface to the Second Edition ix

Preface to the First Edition xi

List of Symbols xv

1. **Generalities** 1

1.1 Brief History 1
1.1.1 The Birth of the Discipline 1
1.1.2 Major Themes 3
1.2 Introduction 10
1.3 Elementary Equations 18
1.3.1 Equations with Separable Variables 18
1.3.2 Linear Equations 20
1.3.3 Homogeneous Equations 21
1.3.4 Bernoulli Equations 22
1.3.5 Riccati Equations 23
1.3.6 Exact Differential Equations 23
1.3.7 Equations Reducible to Exact Differential Equations 24
1.3.8 Lagrange Equations 25
1.3.9 Clairaut Equations 26
1.3.10 Higher-Order Differential Equations 27
1.4 Some Mathematical Models 28
1.4.1 Radioactive Disintegration 29
1.4.2 The Carbon Dating Method 30
1.4.3 Equations of Motion 31

1.4.4 The Harmonic Oscillator 31
1.4.5 The Mathematical Pendulum 32
1.4.6 Two Demographic Models 33
1.4.7 A Spatial Model in Ecology 35
1.4.8 The Prey-Predator Model 35
1.4.9 The Spreading of a Disease 37
1.4.10 Lotka Model 39
1.4.11 An Autocatalytic Generation Model 40
1.4.12 An *RLC* Circuit Model 41
1.5 Integral Inequalities 43
1.6 Exercises and Problems 45

2. The Cauchy Problem 49

2.1 General Presentation 49
2.2 The Local Existence Problem 55
2.3 The Uniqueness Problem 59
2.3.1 The Locally Lipschitz Case 60
2.3.2 The Dissipative Case 62
2.4 Saturated Solutions 63
2.4.1 Characterization of Continuable Solutions 64
2.4.2 The Existence of Saturated Solutions 65
2.4.3 Types of Saturated Solutions 66
2.4.4 The Existence of Global Solutions 71
2.5 Continuous Dependence on Data and Parameters 74
2.5.1 The Dissipative Case 75
2.5.2 The Locally Lipschitz Case 77
2.5.3 Continuous Dependence on Parameters 79
2.6 Problems of Differentiability 80
2.6.1 Differentiability with Respect to the Data 81
2.6.2 Differentiability with Respect to the Parameters . 84
2.7 The Cauchy Problem for the nth-Order Differential Equation 88
2.8 The Implicit Function Theorem 91
2.9 Exercises and Problems 93

3. Approximation Methods 97

3.1 The Power Series Method 97
3.1.1 An Example 97
3.1.2 The Existence of Analytic Solutions 99

3.2 The Successive Approximations Method 103
3.3 The Method of Polygonal Lines 106
3.4 The Euler Implicit Method. The Exponential Formula . . 110
3.4.1 The Semigroup Generated by $\mathcal{A}$ 110
3.4.2 Two Auxiliary Lemmas 112
3.4.3 The Exponential Formula 114
3.5 Exercises and Problems 118

4. **Systems of Linear Differential Equations** 121

4.1 Homogeneous Systems. The Space of Solutions 121
4.2 Non-homogeneous Systems. The Variation of Constants Formula 130
4.3 The Exponential of a Matrix 132
4.4 A Method of Finding $e^{t\mathcal{A}}$ 137
4.5 The nth-Order Linear Differential Equation 141
4.6 The nth-Order Linear Differential Equation with Constant Coefficients 145
4.7 Exercises and Problems 149

5. **Elements of Stability** 155

5.1 Types of Stability 155
5.2 Stability of Linear Systems 161
5.3 The Case of Perturbed Systems 168
5.4 The Lyapunov Function Method 173
5.5 The Case of Dissipative Systems 181
5.6 The Case of Controlled Systems 187
5.7 Unpredictability and Chaos 190
5.8 Exercises and Problems 197

6. **Prime Integrals** 201

6.1 Prime Integrals for Autonomous Systems 201
6.2 Prime Integrals for Non-Autonomous Systems 211
6.3 First Order Partial Differential Equations 212
6.4 The Cauchy Problem for Quasi-Linear Equations 216
6.5 Conservation Laws 221
6.5.1 Some Examples 221
6.5.2 A Local Existence and Uniqueness Result 223
6.5.3 Weak Solutions 224

6.6 Exercises and Problems 232

7. **Extensions and Generalizations** 239

7.1 Distributions of One Variable 239
7.2 The Convolution Product 248
7.3 Generalized Solutions 251
7.4 Carathéodory Solutions 258
7.5 Differential Inclusions 264
7.6 Variational Inequalities 272
7.7 Problems of Viability 279
7.8 Proof of the Nagumo's Viability Theorem 281
7.9 Sufficient Conditions for Invariance 287
7.10 Necessary Conditions for Invariance 290
7.11 Gradient Systems. Frobenius Theorem 294
7.12 Exercises and Problems 300

8. **Volterra Equations** 305

8.1 Volterra Equations of the Second Kind 305
8.2 The Resolvent Kernel 308
8.3 Volterra Equations of the First Kind 310
8.4 The Nonlinear Case 311
8.5 Exercises and Problems 314

9. **Calculus of Variations** 319

9.1 Some Examples 319
9.2 Necessary Conditions for Extremum 323
9.2.1 The Scalar-Valued Case 325
9.2.2 The Vector-Valued Case 328
9.3 Some Particular Cases 330
9.4 Regularity of the Extremals 331
9.5 Higher-Order Necessary Conditions 335
9.6 Sufficient Conditions for Extremum 341
9.7 The Canonical Euler–Lagrange System 343
9.8 Prime Integrals of the Euler–Lagrange System 345
9.9 Exercises and Problems 346

10. **Nonlocal Problems** 349

10.1 Nonlocal Initial Conditions 349

10.2 An Existence and Uniqueness Result 350
10.3 A Simple Existence Result 350
10.4 Proof of Theorem 10.2.1 353
10.5 The Dissipative Case 354
10.6 A Stability Result 357
10.7 Particular Cases 359
10.8 Exercises and Problems 362

11. **Delay Functional Differential Equations** 365

11.1 Preliminaries 365
11.2 Local Existence Under Lipschitz Condition 366
11.3 Some Auxiliary Results 370
11.4 Local Existence. The Continuous Case 372
11.5 Global Existence 375
11.6 A Spring Mass System with Delay 379
11.7 A Delay Glucose Level-Dependent Dosage 381
11.8 Existence of Global Bounded Solutions 382
11.9 Exercises and Problems 386

12. **Auxiliary Results** 389

12.1 Elements of Vector Analysis 389
12.2 Compactness in $C([a,b];\mathbb{R}^n)$ 394
12.3 Banach Spaces 399
12.4 The Projection of a Point on a Convex Set 401
12.5 The Laplace Transform 403
12.5.1 The Transforms of Some Elementary Functions . 411

Solutions 413

Chapter 1 413
Chapter 2 423
Chapter 3 434
Chapter 4 442
Chapter 5 450
Chapter 6 459
Chapter 7 472
Chapter 8 477
Chapter 9 485
Chapter 10 489

Chapter 11 494

Bibliography 497

Index 501

Chapter 1

Generalities

The present chapter serves as an introduction. The first section contains several historical comments, while the second one is dedicated to a general presentation of the discipline. The third section reviews the most representative differential equations which can be solved by elementary methods. In the fourth section we gathered several mathematical models which illustrate the applicative power of the discipline. The fifth section is dedicated to some integral inequalities which will prove useful later, while the last sixth section contains several exercises and problems (whose proofs can be found at the end of the book).

1.1 Brief History

1.1.1 *The Birth of the Discipline*

The name "*equatio differentialis*" was used for the first time in 1676 by Gottfried Wilhelm von Leibniz in order to designate a relation of functional dependence to be satisfied by an a priori unknown function and some of its derivatives. This concept provides a unified abstract framework for a wide variety of problems in Mathematical Analysis and Mathematical Modelling formulated (and some of them even solved) by the middle of the XVII century. One of the first such problems is the so-called *problem of inverse tangents* consisting in the determination of a plane curve by knowing the properties of its tangent at any point. The first who tried to reduce this problem to quadratures[1] was Isaac Barrow[2] (1630–1677) who, using a geometric procedure invented by himself (in fact a substitute of the method of separation of variables), solved several problems of this type. In 1687 Sir Isaac Newton integrated a linear differential equation and, in 1694, Jean Bernoulli (1667–1748) used the *method of integrating factor* in order to solve some n^{th}-order linear differential equations. In 1693 Leibniz employed the substitution $y = tx$ in order to solve homogeneous equations, and, in 1697, Jean Bernoulli succeeded to integrate the homonymous equation in the particular case of constant coefficients. Eighteen years later, Jacopo Riccati (1676–1754) presented a procedure of reduction of the order for a class of second-order differential equations and began a systematic study of the equation which inherited his name. In 1760 Leonhard Euler (1707–1783) observed that, whenever a particular solution of the Riccati equation is known, the latter can be reduced, by means of a substitution, to a linear equation. Moreover, he remarked that, if one knows two particular solutions of the same equation, then its general solution can be found by a quadrature. By the systematic study of this type of equation, Euler was one of the first important forerunners of this discipline. It is the merit of Jean le Rond D'Alembert (1717–1783) to have had observed that an nth-order differential equation is equivalent to a system of n first-order differential equations. In 1775 Joseph Louis de Lagrange (1736–1813) introduced the *variation of constants method*, which, as one can deduce from a letter to Daniel Bernoulli (1700–1782) in 1739, had already been known to Euler. The equations of the form $Pdx + Qdy + Rdz = 0$ were for a long time considered absurd whenever the left-hand side was not an exact differential, although they had been studied by Newton. It was Gaspard Monge (1746–1816) who, in 1787, gave their geometric interpretation and rehabilitated them in the mathematical world. The notion of *singular solution* was introduced in 1715 by Brook Taylor (1685–1731) and was studied in 1736 by Alexis Clairaut (1713–1765). However, it is the merit of Lagrange who, in 1801, defined the concept of singular solution in its current acceptation,

---

[1] By quadrature we mean the method of reducing a given problem to the computation of an integral. The name comes from the homonymous procedure, known from the early times of Greek Geometry, which consists in finding the area of a plane figure by constructing, only by means of the ruler and the compass, of a square with the same area.

[2] Isaac Barrow, a professor at Cambridge who was a teacher of Sir Isaac Newton (1642–1727), is considered one of the forerunners of the Differential Calculus independently invented by two brilliant mathematicians: his former student and Gottfried Wilhelm von Leibniz (1646–1716).

making a net distinction between this type of solution and that of particular solution. The scientists realized soon that many classes of differential equations cannot be solved explicitly, a fact which led to the development of a wide variety of approximating methods. Newton's statement, in the treatise on *fluxional equations*, written in 1671 but published in 1736, that: *all differential equations can be solved by using power series with undetermined coefficients*, had a deep influence on the mathematical thinking of the XVIII century. So, in 1768, Euler devised this type of approximation methods based on the development of the solution in power series. It is interesting to notice that, during this research process, Euler defined the *cylindrical functions* which, however, later came to be known by the name of the astronomer Friedrich Wilhelm Bessel (1784–1846), who had used them very efficiently in his work. We emphasize that, at this stage, mathematicians did not question the convergence of the power series used, nor the existence of the "solution to be approximated".

1.1.2 *Major Themes*

In what follows we present briefly some of the most important achievements in the study of the *initial-value problem*, also called *Cauchy problem*. This consists in determining a solution x, of a differential equation $x' = f(t, x)$, which for a preassigned value a of the argument takes a preassigned value ξ, i.e. $x(a) = \xi$. We deliberately do not touch here upon some other problems, as for instance the boundary-value problems, very important in fact, but which do not belong to the proposed topic of this book.

As we have already mentioned, the mathematicians realized pretty early on that many differential equations cannot be solved explicitly. This situation presented them with several major, but quite difficult, problems to be solved. One such problem consists in finding general sufficient conditions on the data of an initial-value problem in order that this problem has at least one solution. The first who established a notable result in this respect was Baron Augustin Cauchy[3] who, in 1820, employed the *polygonal lines method* in order to prove the local existence for the initial-value problem associated to a differential equation whose right-hand side is of class C^1. The method, improved in 1876 by Rudolf Otto Sigismund Lipschitz (1832–1903), was placed in its most general and natural framework in 1890 by

[3]French mathematician (1789–1857). He is the founder of Complex Analysis and the author of the first modern course in Mathematical Analysis (1821). He observed the link between convergent and fundamental sequences of real numbers.

Giuseppe Peano.[4] This explains why, in many monographs, this is referred to as the *Cauchy–Lipschitz–Peano's method.*

As we have already mentioned, the method of power series, one of the most popular among the mathematicians of both XVII and XVIII centuries, was the favorite approach in approximating the solutions of initial-value problems. Its class of applicability, i.e. the class of differential equations whose right-hand sides are analytic, became very clear only at the middle of the XIX century, at the same time with the development of the modern Complex Function Theory. This might explain why, the first rigorous existence result concerning analytic solutions for an initial-value problem dealt with a class of differential equations in the complex field $\mathbb{C}$ and not, as one might have expected, in the real field $\mathbb{R}$. More precisely, in 1842, Cauchy, re-examining in a critical manner Newton's statement referring to the possibility of solving all differential equations in $\mathbb{R}$ by means of power series, placed this problem within its most natural framework: the Theory of Complex Functions of Several Complex Variables. In this context, in order to prove the convergence of the power series whose partial sum defines the approximate solution for an initial-value problem, he was led to invent the so-called *method of majorant series.* This method consists in the construction of a convergent series with positive terms, with the property that its general term is a majorant for the absolute value of the general term of the approximate solution's series. Such a series is called a majorant for the original one. The method was refined by Ernst Lindelöff who, in 1896, proposed a majorant series, better than that used by Cauchy, and showed that the very subtle arguments of Cauchy, based on the Theory of Complex Functions of Several Complex Variables, are also available in the real field, and moreover, *one can use simpler arguments.*

Another important step concerning the approximation of the solutions of an initial-value problem is due to Emile Picard (1856–1941) who, in 1890, in a paper mainly dedicated to partial differential equations, introduced *the method of successive approximations.* This method, who became well-known very soon, has its roots in Newton's *method of tangents*, and was the starting point for several fundamental results in Functional Analysis such as Banach's fixed point theorem.

During the same period, the so-called *Qualitative Theory of Differential*

[4]Italian mathematician (1858–1932) with notable contributions in Mathematical Logic. He formulated an axiomatic system for the natural numbers and the Axiom of Choice. However, his excessive formalism was very often a real impediment for his contributions to be understood by others.

Equations led the foundations for the fundamental contributions of Henri Poincaré.[5] As we have already noticed, the main concern of the equationists of the XVII and XVIII centuries was to find efficient methods, either to solve explicitly a given initial-value problem, or at least to approximate its solutions as accurately as possible. Unfortunately, neither of these objectives were achievable, and for that reason, they were soon abandoned. Without any doubt, it is the great merit of Poincaré for being the first who made the remarkable observation that, in all the cases in which quantitative arguments are not efficient, one can however obtain crucial information on a solution which can be neither expressed explicitly, nor approximated accurately.[6] More precisely, he put the problem of finding, in the first instance, of "allure" of the curve, associated with the solution in question, leaving aside any continuous transformation which could modify it. For instance, in Poincaré's acceptation, the two curves in $\mathbb{R}^3$ illustrated in Figure 1.1.1 (a) and (b) can be identified modulo "allure", while the other two, i.e. (c) and (d) in the same Figure 1.1.1, cannot. At around the same time, the modern *Theory of Stability* was initiated. The fundamental contributions of Poincaré, of James Clerk Maxwell[7] to the study of planetary motion, and especially those of Alexsandr Mihailovici Lyapunov,[8] emerged into a tremendous stream of a new theory of great practical interest. A similar moment, from the viewpoint of its importance for Stability Theory, will come only after seven decades, with the first results of Vasile M. Popov concerning the stability of automatic controlled systems.

[5]French mathematician (1854–1912), the initiator of the Theory of Dynamical Systems (an abstract version of the Theory of Differential Equations which is mainly concerned with the qualitative aspects of solutions) and of Algebraic Topology. In *Les méthodes nouvelles de la mécanique celéste*, Volumes I, II, III, Gauthier-Villars, 1892–1893–1899, he proved several stability results and applied them to the study of planetary motion.

[6]In his address to the International Congress of Mathematicians in 1908, Poincaré said: "In the past an equation was only considered to be solved when one had expressed the solutions with the aid of a finite number of known functions; but this is hardly possible one time in one hundred. What we can always do, or rather what we should always try to do, is to solve the *qualitative* problem so to speak, that is to try to find the general form of the curve representing the unknown function." (M. W. Hirsch's translation.)

[7]British physicist and mathematician (1831–1879) who succeeded in unifying the general theories of electricity and magnetism by establishing the general laws of electromagnetism on whose basis he predicted the existence of the electromagnetic field. This prediction was later confirmed by the experiments of Heinrich Hertz (1857–1894). At the same time, he was the first who applied the general concepts and results of stability in the study of the evolution of the rings of Saturn.

[8]Russian mathematician (1857–1918) who, in his doctoral thesis defended in 1892, defined the main concepts of stability as known nowadays. He also introduced two fundamental methods of studying the stability problems.

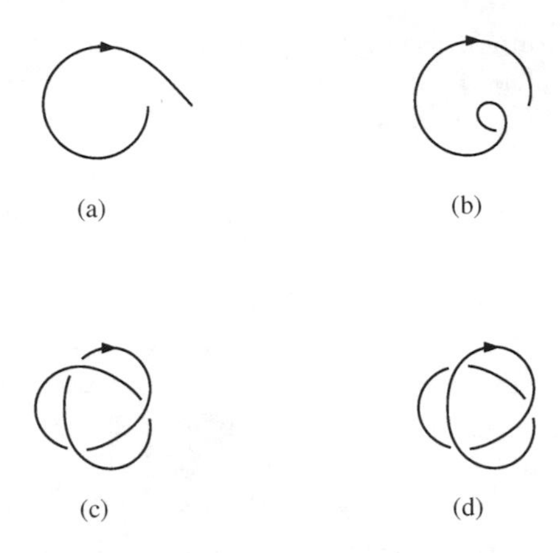

Fig. 1.1.1

The last years of the XIX century were, for sure, the most prolific from the viewpoint of Differential Equations. Some fundamental results were proved during those golden times, among which we can mention: the local existence of at least one solution (Peano 1890), the approximation of the solutions (Picard 1890), the analyticity of the solutions as functions of parameters (Poincaré 1890), the simple or asymptotic stability of solutions (Lyapunov 1892), (Poincaré 1892), the uniqueness of the solution of a given initial-value problem (William Fogg Osgood 1898). Also during the last two decades of the XIX century, Poincaré outlined the concept of *dynamical system* in its current acceptation and began a systematic study of one of the most important and, at the same time, most fascinating problems of the Qualitative Theory of Differential Equations: the classification of solutions according to their intrinsic topological properties. These referential moments were the starting points of two new mathematical disciplines: the *Theory of Dynamical Systems* and *Algebraic Topology*, which developed on their own even from the first years of the XX century. It should also be mentioned that, motivated by an astrophysical problem he was investigated in 1885, Poincaré founded another new discipline: *Bifurcation Theory*, to which, among the most notable contributors, one can mention: Lyapunov, Erhald Schmidt, Mark Alexsandrovici Krasnoselski, David H. Sattinger and Paul Rabinowitz. Also during the last decade of the XIX century, another fundamental result referring to the differentiability of the solution with respect to the initial data was discovered. Namely, in 1896, Ivar Bendixon proved the above mentioned result for the scalar differential equation, in

1897 Peano extended it to the case of a system of differential equations; it was, however, the merit of Thomas H. Gronwall who, in 1919, using the homonymous integral inequality, gave the most elegant proof and, therefore, the most popular nowadays.

The beginning of the XX century was deeply influenced by Poincaré's innovating ideas. Namely, in 1920, Garret David Birkhoff rigorously founded the *Theory of Dynamical Systems*. For a concise presentation of the Theory of Dynamical Systems see [Lefter (2006b)]. At this point, one should mention that subsequent fundamental contributions to this field are due mainly to Andrej Nikolaevich Kolmogorov[9], Vladimir Igorevich Arnold, Jürgen Kurt Moser, Joseph Pierre LaSalle (1916–1983), Morris W. Hirsch, Stephen Smale and George Sell. A special mention in this respect deserves the so-called KAM Theory, i.e. Kolmogorov-Arnold-Moser Theory. During the third decade of the XX century, a very important step was made toward a functional approach for problems of the Theory of Dynamical Systems. Birkhoff, together with Oliver Dimon Kellogg, were the first who, in 1922, used fixed point topological arguments in order to prove some existence and uniqueness results for certain classes of differential equations. These topological methods were initiated by Luitzen Egbertus Jan Brouwer,[10] extended and generalized subsequently by Solomon Lefschetz (1984–1972), and refined in 1934 by Jean Leray and Juliusz Schauder who expressed them in a very general abstract and elegant form, known nowadays under the name of *Leray–Schauder Topological Degree*. Renato Cacciopoli was the first who, in 1930, employed the *Contraction Principle* as a method of proof for an existence and uniqueness theorem. However, it is the merit of Stefan Banach who, even earlier, in 1922, gave its general abstract form, known as a result under the name of *Banach's fixed point theorem*, and as a method of proof under the name of *the method of successive approximations*.

Concerning the qualitative properties of solutions, the mathematicians focused their attention on the study of the so-called *ergodic behavior* beginning with Birkhoff (1931) and continuing with John von Neumann[11]

[9]Russian mathematician (1903–1987). He is the founder of the modern Probability Theory. He has remarkable contributions in the Theory of Dynamical Systems with application to Hamiltonian systems.

[10]Dutch mathematician and philosopher (1881–1966). He is one of the founders of the Intuitionists School. His famous fixed point theorem states that *every continuous function f, from a nonempty convex compact set $K \subset \mathbb{R}^n$ into K, has at least one fixed point $x \in K$, i.e. $f(x) = x$.*

[11]American mathematician born in Budapest (1903–1957). He is the creator of the Game Theory and made notable contributions to Functional Analysis and to Information Theory.

(1932), Kôsaku Yosida (1938), Yosida and Shizuo Kakutani (1938), *etc.* Due mainly to their applications in Chemistry, Electricity and Biology, the existence and properties of the so-called *limit cycles*, whose study was also initiated by Poincaré (1881), became another subject of great interest. Motivated by the study of self-sustained oscillations in nonlinear electric circuits, the theory of limit cycles grew up rapidly since the 1920s and 1930s with the contributions of G. Duffing, M. H. Dulac, B. Van der Pol and A. A. Andronov. Notable contributions to this topic (especially to the study of some specific classes of quadratic systems) are mostly due to Chinese, Russian and Ukrainian mathematicians such as N. N. Bautin, A. N. Sharkovskij, S.-L. Shi, S. I. Yashenko, Y. C. Ye, and others.

It is also worth mentioning the contribution of Erich Kamke, who established the theorem on the continuous dependence of the solution of an initial-value problem on the data and on the parameters, theorem extended in 1957 by Jaroslav Kurzweil. Also Kamke, following Paul Montel, Enrico Bompiani, Leonida Tonelli and Oscar Perron, introduced the so-called *comparison method* in order to obtain sharp uniqueness results. This method proved useful in the study of some stability problems and, surprisingly, as subsequently observed by Felix E. Browder, even in the proof of existence theorems.

Concerning the concept of solution, the new type of integral defined in 1904 by Henri Lebesgue, offered the possibility to extend the classical theory of differential equations based on the Riemann (in fact Cauchy) integral to another theory relying on the *Lebesgue integral.* This major step was made in 1918 by Constantin Carathéodory. Subsequent extensions, based on another type of integral, more general than that of Lebesgue, and known as the *Kurzweil–Henstock integral,* were initiated in 1957 by Kurzweil.

With the same idea in mind, namely to enlarge the class of admissible solutions, but from a completely different perspective, a new discipline was born: the *Theory of Distributions*, initiated in 1936 by Serghei Sobolev and given a definitive form in 1950–1951 by Laurent Schwartz. Initially thought of as a theory exclusively useful in the linear case, the Theory of Distributions proved its efficiency in the study of various nonlinear problems as well.

Other types of generalized solutions on which to construct an effective theory, especially in the nonlinear case, the so-called *viscosity solutions*, were introduced in 1950 by Eberhard Hopf and subsequently studied by Olga Oleinik and Paul Lax (1957), Stanislav Kružkov (1970), Michael G. Crandall and Pierre-Louis Lions (1983), and Daniel Tătaru (1990), among others. Notable results on the uniqueness problem, very important but at the same time extremely difficult in this context, were obtained in 1987 by Michael G. Crandall, Hitoshi Ishii and Pierre-Louis Lions.

Since 1950, with the publication of a famous counter-example by Jean Dieudonné, it was realized that, on some infinite dimensional spaces, as for instance c_0,[12] only the continuity of the right-hand side is not enough to ensure the local existence for an initial-value problem. This strange, but not unexpected situation, was completely elucidated in 1975 by Alexsandr Nicolaevici Godunov, who proved that, *for every infinite-dimensional Banach space X there exist a continuous function $f : X \to X$ and $\xi \in X$ such that the Cauchy problem $x' = f(x)$, $x(0) = \xi$ has no local solution.* Perhaps due to these reasons, starting with the end of the fifties, there has been a growing interest in the study of the local existence problem in infinite-dimensional Banach spaces and in some qualitative problems in the same setting. In this respect we mention the results of Constantin Corduneanu and Aristide Halanay.

The development of a functional calculus based on the Theory of Functions of a Complex Variable taking values into a Banach algebra was accomplished in parallel with the study of the "Abstract Theory of Differential Equations". So, in 1935, Nelson Dunford introduced the curvilinear integral of an analytic function with values in a Banach algebra and proved a Cauchy type representation formula for the exponential of an operator. In 1948, Einar Hille and Kôsaku Yosida, starting from the study of some partial differential equations, introduced and studied independently an abstract class of linear differential equations, with possibly discontinuous right-hand side, and proved the famous generation theorem concerning C_0-semigroups, known as *the Hille–Yosida Theorem.* The necessary and sufficient condition expressed in this theorem was extended in 1967 to the fully nonlinear case, but only in a Hilbert space setting, by Yukio Kõmura, while the sufficiency part, by far the most interesting, was proved in the general Banach space framework in 1971 by Michael G. Crandall and Thomas M. Liggett. This result[13] is known as the *Crandall–Liggett Generation Theorem*, while the formula established in the proof as the *Exponential Formula.*

In parallel with the extension of the differential equations' framework to infinite-dimensional spaces via the already mentioned contributions, but also through those of Philippe Bénilan (1940–2000), Haïm Brezis, Toshio

[12] We recall that c_0 is the space of all real sequences approaching 0 as n tends to ∞. Endowed with the supremum norm this is an infinite dimensional real Banach space.
[13] A simplified version of this fundamental result is presented in Section 3.4 of this book.

Kato, Jaques-Louis Lions (1928–2001), Amnon Pazy, the study of some problems of major interest in this new and fairly general context was reconsidered. So, in 1979, Ciprian Foiaş and Roger Temam obtained one of the first deep result concerning the existence of inertial manifolds and estimated the dimension of such manifolds in the case of the Navier–Stokes system in fluid dynamics. Results of this type essentially state that some infinite-dimensional systems have, for large values of the time variable, a "finite-dimensional-type" behavior.

The systematic study of optimal control problems in $\mathbb{R}^n$, initiated in the fifties by Lev Pontriaghin (1908–1988), Revaz Valerianovici Gamkrélidze and Vladimir Grigorievici Boltianski, was continued in the sixties and seventies by: Lamberto Cesari, Richard Bellman, Rudolf Emil Kalman, Wendell Helms Fleming, Jaques-Louis Lions, Hector O. Fattorini, among others. We notice that Lions was the first who extended this theory to the framework of linear differential equations in infinite-dimensional spaces in order to handle control problems governed by partial differential equations as well. Notable results in this direction, more difficult in the fully nonlinear case, were obtained subsequently by Viorel Barbu.

We conclude these brief historical considerations which reflect a rather subjective viewpoint of the author and which are far from being complete,[14] by emphasizing that the Theory of Differential Equations is a continuously growing discipline, whose by now classical results are very often extended and generalized in order to handle new cases suggested by practical applications and which is also enriched by completely new results having no direct correspondence within its classical counterpart. For this reason, all readers interested in mathematical research may find in this field a wealth of various open problems awaiting to be solved, and moreover, may be able to formulate and solve themselves some new and interesting problems.

1.2 Introduction

Differential Equations and Systems. *Differential Equations* have their roots as an "on its own" discipline in the natural interest of scientists to predict, as accurately as possible, the future evolution of a certain physical, biological, chemical, sociological, *etc.* system. It is easy to realize that, in

[14]The interested reader willing to get additional information concerning the evolution of this discipline is referred to [Wieleitner (1964)], [Hirsch (1984)] and [Piccinini *et al.* (1984)].

order to get a fairly acceptable prediction close enough to reality, we need fairly precise data on the present state of the system, as well as, sound knowledge on the law(s) according to which the instantaneous state of the system affects its instantaneous rate of change. *Mathematical Modelling* is that discipline which comes into play at this point, offering the scientist the description of such laws in a mathematical language, laws which, in many specific situations, take the form of differential equations, or even of systems of differential equations.

The goal of the present section is to define the concept of a differential equation, as well as that of a system of differential equations, and to give a brief overview of the main problems to be studied in this book.

Roughly speaking, a *scalar differential equation* represents a relationship of functional dependence between the values of a real valued function, called *unknown function*, at least one of its ordinary (partial) derivatives up to a given order n, and the independent variable(s).

The highest order of differentiation of the unknown function involved in the equation is called the *order of the equation.*

A differential equation whose unknown function depends on only one real variable is called *ordinary differential equation*, while a differential equation whose unknown function depends on at least two real independent variables is called a *partial differential equation.* For instance the equation

$$x'' + x = \sin t,$$

whose unknown function x depends on one real variable t, is an ordinary differential equation of second-order, while the equation

$$\frac{\partial^3 u}{\partial x^2 \partial y} + \frac{\partial u}{\partial y} = 0,$$

whose unknown function u depends on two independent real variables x and y, is a third-order partial differential equation.

In the present book we will focus our attention mainly on the study of ordinary differential equations, which from now on, whenever no confusion may occur, we simply refer to as differential equations. However, we will occasionally touch upon some problems referring to a special class of partial differential equations for which the most appropriate and natural approach is offered by the framework of ordinary differential equations.

The general form of an nth*-order scalar differential equation* with the unknown function x is

$$F(t, x, x', \ldots, x^{(n)}) = 0, \tag{$\mathcal{E}$}$$

where F is a function defined on a subset $D(F)$ in $\mathbb{R}^{n+2}$ and taking values in $\mathbb{R}$, which is not constant with respect to the last variable.

Under the standard regularity assumptions on the function F required by the applicability of the *Implicit Functions Theorem*, $(\mathcal{E})$ may be rewritten as

$$x^{(n)} = f(t, x, x', \ldots, x^{(n-1)}), \tag{$\mathcal{N}$}$$

where f is a function defined on a subset $D(f)$ in $\mathbb{R}^{n+1}$ with values in $\mathbb{R}$, which explicitly defines $x^{(n)}$ (at least locally) as a function of $t, x, x', \ldots, x^{(n-1)}$, by means of the relation $F(t, x, x', \ldots, x^{(n)}) = 0$. An equation of the form $(\mathcal{N})$ is called *n^{th}-order scalar differential equation in normal form.* With few exceptions, in all what follows, we will focus our attention on the study of *first-order differential equations in normal form*, i.e. on the study of differential equations of the form

$$x' = f(t, x), \tag{$\mathcal{O}$}$$

where f is a function defined on $D(f) \subseteq \mathbb{R}^2$ taking values in $\mathbb{R}$.

By analogy, if $g : D(g) \to \mathbb{R}^n$ is a given function, $g = (g_1, g_2, \ldots, g_n)$, where $D(g)$ is included in $\mathbb{R} \times \mathbb{R}^n$, we may define a *system of n first-order differential equations* with n unknown functions: $y_1, y_2, \ldots, y_n$, as a system of the form

$$\begin{cases} y_i' = g_i(t, y_1, y_2, \ldots, y_n) \\ i = 1, 2, \ldots, n, \end{cases} \tag{$\mathcal{S}$}$$

which, in its turn, represents the componentwise expression of a *first-order vector differential equation*

$$y' = g(t, y). \tag{$\mathcal{V}$}$$

By means of the transformations[15]

$$\begin{cases} y = (y_1, y_2, \ldots, y_n) = (x, x', \ldots, x^{(n-1)}) \\ g(t, y) = (y_2, y_3, \ldots, y_n,\ f(t, y_1, y_2, \ldots, y_n)), \end{cases} \tag{$\mathcal{T}$}$$

$(\mathcal{N})$ can be equivalently rewritten as system of n scalar differential equations with n unknown functions:

$$\begin{cases} y_1' = y_2 \\ y_2' = y_3 \\ \vdots \\ y_{n-1}' = y_n \\ y_n' = f(t, y_1, y_2, \ldots, y_n), \end{cases}$$

[15]Transformations proposed by Jean Le Rond D'Alembert. It should be noted that here $x, x', \ldots, x^{(n-1)}$ denote the components of a vector and, at this stage, $x', \ldots, x^{(n-1)}$ do not designate the derivatives of a given function x.

or, in other words, as a first-order vector differential equation $(\mathcal{V})$, with g defined by $(\mathcal{T})$. This way, the study of the equation $(\mathcal{N})$ reduces to the study of an equation of the type $(\mathcal{V})$ or, equivalently, to the study of a first-order differential system. This explains why, in all what follows, we will merely study the equation $(\mathcal{V})$, noticing only, whenever necessary, how to transcribe the results referring to $(\mathcal{V})$ in terms of $(\mathcal{N})$ by means of the transformations $(\mathcal{T})$.

If the function g in $(\mathcal{V})$ does not depend explicitly on t, equation $(\mathcal{V})$ is called *autonomous*. Under similar circumstances, the system $(\mathcal{S})$ is called *autonomous*. For instance, the equation

$$y' = 2y$$

is autonomous, while the equation

$$y' = 2y + t$$

is not. We emphasize, however, that every non-autonomous equation of the form $(\mathcal{V})$ may be equivalently rewritten as an autonomous one:

$$z' = h(z), \qquad (\mathcal{V}')$$

where the unknown function z has an extra-component (than y). More precisely, setting $z = (z_1, z_2, \ldots, z_{n+1}) = (t, y_1, y_2, \ldots, y_n)$ and defining $h : D(g) \subset \mathbb{R}^{n+1} \to \mathbb{R}^{n+1}$ by

$$h(z) = (1, g_1(z_1, z_2, \ldots, z_{n+1}), \ldots, g_n(z_1, z_2, \ldots, z_{n+1}))$$

for each $z \in D(g)$, we observe that $(\mathcal{V}')$ represents an equivalent writing of $(\mathcal{V})$. So, the first-order scalar differential equation $y' = 2y + t$ may be rewritten as a first-order vector differential equation in $\mathbb{R}^2$, of the form $z' = h(z)$, where $z = (z_1, z_2) = (t, y)$ and $h(z) = (1, 2z_2 + z_1)$. Similar considerations are in effect for the differential system $(\mathcal{S})$ too.

Type of Solutions. As defined so far, somehow descriptive and far from being rigorous, the concept of differential equation is ambiguous because we have not specified what is the sense in which the equality $(\mathcal{E})$ should be understood.[16] Namely, let us observe from the very beginning that any of the two formal equalities $(\mathcal{E})$, or $(\mathcal{N})$ may be thought as being satisfied at least in one of the three particular ways:

(i) for every t in the domain $\mathbb{I}_x$ of the unknown function x;

[16]In fact, we indicated only a formal relation which could define a predicate (the differential equation) but we did not specify the domain on which it acts (it is defined).

(ii) for every t in $\mathbb{I}_x \setminus \mathbb{E}$, with $\mathbb{E}$ an exceptional set (finite, countable, negligible, *etc.*);

(iii) in a generalized sense which might have nothing to do with the usual point-wise equality.

It becomes now clear that a crucial problem arising at this stage is that of how to define the concept of solution for $(\mathcal{E})$ by specifying what is the precise meaning of the equality $(\mathcal{E})$. It should be noted that any construction of a rigorous theory of Differential Equations is very sensitive to the manner in which we solve this preliminary problem. The following examples are of some help in order to understand the importance, and to evaluate the exact scale of this challenge.

Example 1.2.1. Let us consider the so-called *eikonal equation*

$$|x'| = 1. \tag{1.2.1}$$

It is easy to see that the only C^1 functions, $x : \mathbb{R} \to \mathbb{R}$, satisfying (1.2.1) for each $t \in \mathbb{R}$ are of the form $x(t) = t + c$, or $x(t) = -t + c$, with $c \in \mathbb{R}$. On the other hand, if we ask that (1.2.1) be satisfied for each $t \in \mathbb{R}$, with the possible exception of those points in a finite subset, besides the functions specified above, we may easily see that any function having the graph as in Figure 1.2.1 is a solution of (1.2.1) in this new acceptation.

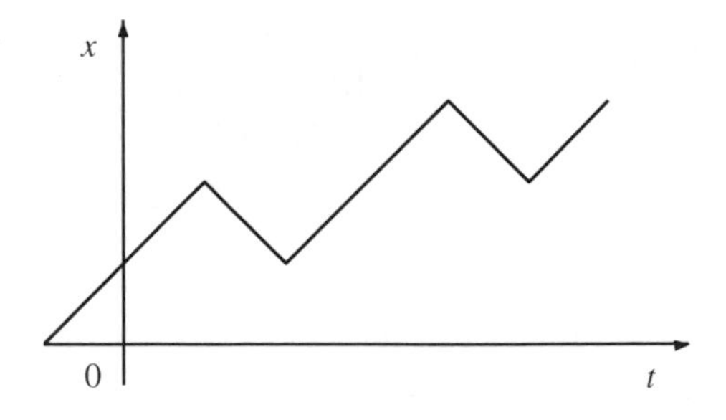

Fig. 1.2.1

Example 1.2.2. Now, let us consider the differential equation

$$x' = h,$$

where $h : \mathbb{I} \to \mathbb{R}$ is a given function. It is obvious that if h is continuous, then x is of class C^1, while if h is discontinuous, then the equation above cannot have any C^1 solutions defined on the whole interval $\mathbb{I}$.

These examples emphasize the importance of the class of functions in which we agree to accept the candidates to the title of solution. So, if this class is too narrow, the chance to be able to prove the existence of at least one solution is very small, while if this class is too broad, then this chance, which is obviously greater, is drastically counterbalanced by the price paid by the possible lack of some desirable regularity properties of solutions. Therefore, the concept of solution for a differential equation has to be defined having in mind a compromise, namely that on one hand it should have at least one solution and, on the other hand, that each solution should have sufficient regularity properties in order to be of some use in practice. From the examples previously analyzed, it is easy to see that the definition of this concept should take into account firstly the regularity properties of the function F. Throughout, we shall say that an interval is *nontrivial* if it has nonempty interior. So, assuming that F is of class C^n, it is natural to adopt:

Definition 1.2.1. A *solution* of the nth-order scalar differential equation $(\mathcal{E})$ is a function $x : \mathbb{I}_x \to \mathbb{R}$ of class C^n on the nontrivial interval $\mathbb{I}_x$, which satisfies $(t, x(t), x'(t), \ldots, x^{(n)}(t)) \in D(F)$ and

$$F(t, x(t), x'(t), \ldots x^{(n)}(t)) = 0$$

for each $t \in \mathbb{I}_x$.

Definition 1.2.2. A *solution* of the nth-order scalar differential equation in the normal form $(\mathcal{N})$ is a function $x : \mathbb{I}_x \to \mathbb{R}$ of class C^n on the nontrivial interval $\mathbb{I}_x$, which satisfies $(t, x(t), x'(t), \ldots, x^{(n-1)}(t)) \in D(f)$ and

$$x^{(n)}(t) = f(t, x(t), x'(t), \ldots x^{(n-1)}(t))$$

for each $t \in \mathbb{I}_x$.

Definition 1.2.3. A *solution* of the system of first-order differential equations $(\mathcal{S})$ is an n-tuple of functions $(y_1, y_2, \ldots, y_n) : \mathbb{I}_y \to \mathbb{R}^n$ of class C^1 on the nontrivial interval $\mathbb{I}_y$, which satisfies $(t, y_1(t), y_2(t), \ldots, y_n(t)) \in D(g)$ and $y_i'(t) = g_i(t, y_1(t), y_2(t), \ldots, y_n(t))$, $i = 1, 2, \ldots, n$, for each $t \in \mathbb{I}_y$. *The trajectory* corresponding to the solution y is the set $\tau(y) = \{y(t);\ t \in \mathbb{I}_y\}$.

The trajectory corresponding to a given solution $y = (y_1, y_2)$ of a differential system in $\mathbb{R}^2$ is illustrated in Figure 1.2.2 (a), while the graph of the solution in Figure 1.2.2 (b).

Definition 1.2.4. A *solution* of the first-order vector differential equation $(\mathcal{V})$ is a function $y : \mathbb{I}_y \to \mathbb{R}^n$ of class C^1 on the nontrivial interval $\mathbb{I}_y$, which satisfies $(t, y(t)) \in D(g)$ and $y'(t) = g(t, y(t))$ for each $t \in \mathbb{I}_y$. *The trajectory* corresponding to the solution y is the set $\tau(y) = \{y(t);\ t \in \mathbb{I}_y\}$.

Let us observe that the problem of finding the antiderivatives of a continuous function h on a given interval $\mathbb{I}$ may be embedded into a first-order differential equation of the form $x' = h$ for which, from the set of solutions given by Definition 1.2.1, we keep only those defined on $\mathbb{I}$, the "maximal domain" of the function h.

Definition 1.2.5. A family $\{x(\cdot, c) : \mathbb{I}_{x,c} \to \mathbb{R};\ c = (c_1, c_2, \ldots, c_n) \in \mathbb{R}^n\}$ of functions, implicitly defined by a relation of the form

$$G(t, x, c_1, c_2, \ldots, c_n) = 0, \tag{$\mathcal{G}$}$$

where $G : D(G) \subseteq \mathbb{R}^{n+2} \to \mathbb{R}$, is a function of class C^n with respect to the first two variables, with the property that, by eliminating the constants $c_1, c_2, \ldots, c_n$ from the system

$$\begin{cases} \dfrac{d}{dt}\left[G(\cdot, x(\cdot), c_1, c_2, \ldots, c_n)\right](t) = 0 \\ \dfrac{d^2}{dt^2}\left[G(\cdot, x(\cdot), c_1, c_2, \ldots, c_n)\right](t) = 0 \\ \vdots \\ \dfrac{d^n}{dt^n}\left[G(\cdot, x(\cdot), c_1, c_2, \ldots, c_n)\right](t) = 0 \end{cases}$$

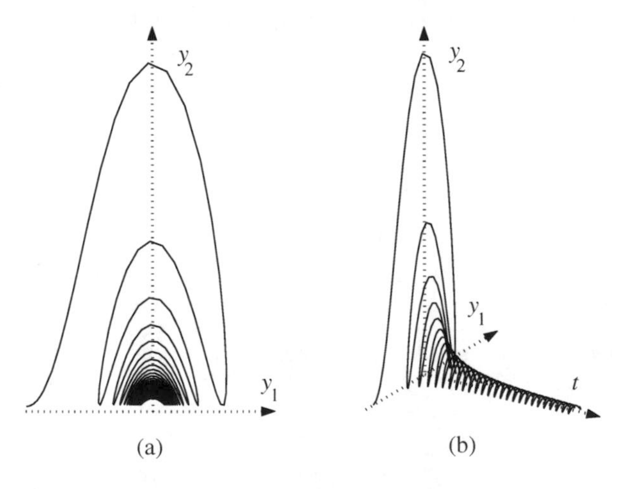

Fig. 1.2.2

and substituting these in $(\mathcal{G})$ one gets exactly $(\mathcal{E})$, is called *the general integral*, or *the general solution* of $(\mathcal{E})$.

Usually, we identify the general solution by its relation of definition saying that $(\mathcal{G})$ is *the general solution*, or *the general integral* of $(\mathcal{E})$.

Example 1.2.3. The general integral of the second-order differential equation

$$x'' + a^2 x = 0,$$

with $a > 0$, is $\{x(\cdot, c_1, c_2);\ (c_1, c_2) \in \mathbb{R}^2\}$, where

$$x(t, c_1, c_2) = c_1 \sin at + c_2 \cos at$$

for $t \in \mathbb{I}_{x,c}$[17]. Indeed, it is easy to see that the equation is obtained by eliminating the constants c_1, c_2 from the system

$$\begin{cases} (x - c_1 \sin at - c_2 \cos at)' = 0 \\ (x - c_1 \sin at - c_2 \cos at)'' = 0. \end{cases}$$

In this case, $G : \mathbb{R}^4 \to \mathbb{R}$ is defined by

$$G(t, x, c_1, c_2) = x - c_1 \sin at - c_2 \cos at$$

for each $(t, x, c_1, c_2) \in \mathbb{R}^4$ and $(\mathcal{G})$ may be equivalently rewritten as

$$x = c_1 \sin at + c_2 \cos at,$$

relation which defines explicitly the general integral. As we shall see later, in many other specific cases too, in which from $(\mathcal{G})$ one can get the explicit form of x as a function of $t, c_1, c_2, \ldots, c_n$, the general integral of $(\mathcal{E})$ can be expressed in an explicit form as $x(t, c_1, c_2, \ldots, c_n) = H(t, c_1, c_2, \ldots, c_n)$, with $H : D(H) \subseteq \mathbb{R}^{n+1} \to \mathbb{R}$ a function of class C^n.

Problems to be Studied. Next, we shall list several problems which we shall address in the study of equation $(\mathcal{V})$. We begin by noticing that the main problem we are going to treat is the so-called *Cauchy problem*, or *initial value problem* associated to $(\mathcal{V})$. More precisely, given $(a, \xi) \in D(g)$, the *Cauchy problem* for $(\mathcal{V})$ with data a and ξ consists in finding a particular solution $y : \mathbb{I}_y \to \mathbb{R}^n$ of $(\mathcal{V})$, with $a \in \mathbb{I}_y$ and satisfying the *initial condition* $y(a) = \xi$. Customarily a is called the *initial time*, while ξ the *initial state*.

In the study of this problem we shall encounter the following subproblems of obvious importance: (1) the existence problem, which consists in

[17] We mention that, in this case, the general integral also contains functions defined on the whole set $\mathbb{R}$, i.e. for which $\mathbb{I}_{x,c} = \mathbb{R}$.

finding reasonable sufficient conditions on the function g so that, for each $(a, \xi) \in D(g)$, the Cauchy problem for the equation $(\mathcal{V})$, with a and ξ as data, has at least one solution[18]; (2) the uniqueness problem, which consists in finding sufficient conditions on the function g so that, for each $(a, \xi) \in D(g)$, the Cauchy problem for the equation $(\mathcal{V})$, with a and ξ as data, has at most one solution defined on a given interval containing a; (3) the problem of continuation of the solutions; (4) the problem of the behavior of the non-continuable solution at the endpoint(s) of the maximal interval of definition; (5) the problem of approximation of a given solution; (6) the problem of continuous dependence of the solution on both the initial datum ξ and the right-hand side g; (7) the problem of differentiability of the solution with respect to the initial datum ξ; (8) the problem of getting additional information in the particular case in which $g : \mathbb{I} \times \mathbb{R}^n \to \mathbb{R}^n$ and, for each $t \in \mathbb{I}$, $g(t, \cdot)$ is a linear function; (9) the study of the behavior of the solutions as t approaches $+\infty$.

1.3 Elementary Equations

The goal of this section is to collect several types of differential equations whose general solutions can be found by means of a finite number of integration procedures. Since the integration of real functions of one real variable is also called *quadrature*, these equations are known under the name of *equations solvable by quadratures*.

1.3.1 *Equations with Separable Variables*

An equation of the form

$$x' = f(t)g(x), \tag{1.3.1}$$

where $f : \mathbb{I} \to \mathbb{R}$ and $g : \mathbb{J} \to \mathbb{R}$ are two continuous functions with $g(y) \neq 0$ for each $y \in \mathbb{J}$, is called *with separable variables*.

Theorem 1.3.1. *Let $\mathbb{I}$ and $\mathbb{J}$ be two nontrivial intervals in $\mathbb{R}$ and let f :*

[18] In many circumstances, in the process of establishing a mathematical model, one deliberately ignores the contribution of certain "parameters" whose influence on the evolution of the system in question is considered irrelevant. For this reason, almost all mathematical models are not exact representations of the reality and, accordingly, a first problem of great importance we face in this context (problem which is superfluous in the case of the real phenomenon) is that of the consistency of the model. But this consists in showing that the model in question has at least one solution.

$\mathbb{I} \to \mathbb{R}$ and $g : \mathbb{J} \to \mathbb{R}$ be two continuous functions with $g(y) \neq 0$ for each $y \in \mathbb{J}$. Then, the general solution of equation (1.3.1) is given by

$$x(t) = G^{-1}\left(\int_{t_0}^{t} f(s)\,ds\right) \tag{1.3.2}$$

for each $t \in \mathrm{Dom}(x)$, where t_0 is a fixed point in $\mathbb{I}$, and $G : \mathbb{J} \to \mathbb{R}$ is defined by

$$G(y) = \int_{\xi}^{y} \frac{d\tau}{g(\tau)}$$

for each $y \in \mathbb{J}$, with $\xi \in \mathbb{J}$.

Proof. Since g does not vanish on $\mathbb{J}$ and is continuous, it preserves constant sign on $\mathbb{J}$. Changing the sign of the function f if necessary, we may assume that $g(y) > 0$ for each $y \in \mathbb{J}$. Then, the function G is well-defined and strictly increasing on $\mathbb{J}$.

We begin by observing that the function x defined by means of the relation (1.3.2) is a solution of equation (1.3.1) which satisfies $x(t_0) = \xi$. Namely,

$$x'(t) = \left[G^{-1}\left(\int_{t_0}^{t} f(s)\,ds\right)\right]' = \frac{1}{G'\left(G^{-1}\left(\int_{t_0}^{t} f(s)\,ds\right)\right)} f(t) = g(x(t))f(t)$$

for each t in the domain of the function x. In addition, from the definition of G, it follows that $x(t_0) = \xi$.

To complete the proof it suffices to show that every solution of equation (1.3.1) is of the form (1.3.2). To this aim, let $x : \mathrm{Dom}(x) \to \mathbb{J}$ be a solution of equation (1.3.1) and let us observe that this may be equivalently rewritten as

$$\frac{x'(t)}{g(x(t))} = f(t)$$

for each $t \in \mathrm{Dom}(x)$. Integrating this equality on both sides over $[\,t_0, t\,]$, we get

$$\int_{t_0}^{t} \frac{x'(s)\,ds}{g(x(s))} = \int_{t_0}^{t} f(s)\,ds$$

for each $t \in \mathrm{Dom}(x)$. Consequently we have

$$G(x(t)) = \int_{t_0}^{t} f(s)\,ds,$$

where G is defined as above with $\xi = x(t_0)$. Recalling that G is strictly increasing on $\mathbb{J}$, we conclude that it is invertible from its range $G(\mathbb{J})$ onto $\mathbb{J}$. From this remark and the last equality we deduce (1.3.2). □

1.3.2 *Linear Equations*

A *linear equation* is an equation of the form

$$x' = a(t)x + b(t), \tag{1.3.3}$$

where $a, b : \mathbb{I} \to \mathbb{R}$ are continuous functions on $\mathbb{I}$. If $b \equiv 0$ on $\mathbb{I}$, the equation is called *linear homogeneous*, otherwise, it is called *linear non-homogeneous.*

Theorem 1.3.2. *If a and b are continuous on $\mathbb{I}$, then the general solution of equation* (1.3.3) *is given by the so-called variation of constants formula*

$$x(t) = \exp\left(\int_{t_0}^{t} a(s)\,ds\right)\xi + \int_{t_0}^{t} \exp\left(\int_{s}^{t} a(\tau)\,d\tau\right) b(s)\,ds \tag{1.3.4}$$

for each $t \in \mathrm{Dom}(x)$, where $t_0 \in \mathrm{Dom}(x)$ is fixed, $\xi \in \mathbb{R}$ and $\exp(y) = e^y$ for each $y \in \mathbb{R}$.

Proof. A simple computational argument shows that x defined by (1.3.4) is a solution of (1.3.3) which satisfies $x(t_0) = \xi$. So, we merely have to show that each solution of (1.3.3) is of the form (1.3.4) on its interval of definition. To this aim, let $x : \mathbb{I}_0 \to \mathbb{R}$ be a solution of equation (1.3.3), where $\mathbb{I}_0$ is a nontrivial interval included in $\mathbb{I}$. Fix $t_0 \in \mathbb{I}_0$ and multiply both sides of (1.3.3) (with t substituted by s) by

$$\exp\left(-\int_{t_0}^{s} a(\tau)\,d\tau\right)$$

where $s \in \mathbb{I}_0$. After some obvious rearrangements, we obtain

$$\frac{d}{ds}\left(x(s)\exp\left(-\int_{t_0}^{s} a(\tau)\,d\tau\right)\right) = b(s)\exp\left(-\int_{t_0}^{s} a(\tau)\,d\tau\right)$$

for each $s \in \mathbb{I}_0$. Integrating this equality on both sides between t_0 and $t \in \mathbb{I}_0$, and then multiplying the equality thus obtained by

$$\exp\left(\int_{t_0}^{t} a(\tau)\,d\tau\right),$$

we deduce (1.3.4), which completes the proof. □

Remark 1.3.1. From (1.3.4) it follows that every solution of (1.3.3) may be continued as a solution of the same equation to the whole interval $\mathbb{I}$.

1.3.3 *Homogeneous Equations*

A *homogeneous equation* is an equation of the form

$$x' = h\left(\frac{x}{t}\right), \tag{1.3.5}$$

where $h : \mathbb{I} \to \mathbb{R}$ is continuous and $h(r) \neq r$ for each $r \in \mathbb{I}$.

Theorem 1.3.3. *If $h : \mathbb{I} \to \mathbb{R}$ is continuous and $h(r) \neq r$ for each $r \in \mathbb{I}$, then the general solution of* (1.3.5) *is given by*

$$x(t) = tu(t)$$

for $t \neq 0$, where u is the general solution of the equation with separable variables

$$u' = \frac{1}{t}\left(h(u) - u\right).$$

Proof. We merely have to express x' by means of u and to impose the condition that x be a solution of equation (1.3.5). $\square$

An important class of differential equations which can be reduced to homogeneous equations is

$$x' = \frac{a_{11}x + a_{12}t + b_1}{a_{21}x + a_{22}t + b_2}, \tag{1.3.6}$$

where a_{ij} and b_i, $i, j = 1, 2$ are constants and

$$\begin{cases} a_{11}^2 + a_{12}^2 + b_1^2 > 0 \\ a_{21}^2 + a_{22}^2 + b_2^2 > 0. \end{cases}$$

According to the compatibility of the linear algebraic system

$$\begin{cases} a_{11}x + a_{12}t + b_1 = 0 \\ a_{21}x + a_{22}t + b_2 = 0, \end{cases} \tag{$\mathcal{AS}$}$$

we distinguish between three different cases. More precisely we have:
Case I. If the system $(\mathcal{AS})$ has a unique solution (ξ, η) then, by means of the change of variables

$$\begin{cases} x = y + \xi \\ t = s + \eta, \end{cases}$$

(1.3.6) can be equivalently rewritten under the form of the homogeneous equation below

$$y' = \frac{a_{11}\dfrac{y}{s} + a_{12}}{a_{21}\dfrac{y}{s} + a_{22}};$$

Case II. If the system $(\mathcal{AS})$ has infinitely many solutions, then there exists $\lambda \neq 0$ such that

$$(a_{11}, a_{12}, b_1) = \lambda\,(a_{21}, a_{22}, b_2)$$

and therefore (1.3.6) reduces to $x' = \lambda$;
Case III. If the system $(\mathcal{AS})$ is incompatible then there exists $\lambda \neq 0$ such that

$$\begin{cases} (a_{11}, a_{12}) = \lambda\,(a_{21}, a_{22}) \\ (a_{11}, a_{12}, b_1) \neq \lambda\,(a_{21}, a_{22}, b_2) \end{cases}$$

and, by means of the substitution $y = a_{21}x + a_{22}t$, the equation reduces to an equation with separable variables.

1.3.4 *Bernoulli Equations*

An equation of the form

$$x' = a(t)x + b(t)x^{\alpha}, \tag{1.3.7}$$

where $a, b : \mathbb{I} \to \mathbb{R}$ are non-identically zero continuous functions which are not proportional on $\mathbb{I}$, and $\alpha \in \mathbb{R} \setminus \{0, 1\}$, is called *Bernoulli equation.*

Remark 1.3.2. The restrictions imposed on the data a, b and α can be explained by the simple observations that: if $a \equiv 0$ then (1.3.7) is with separable variables; if there exists $\lambda \in \mathbb{R}$ such that $a(t) = \lambda b(t)$ for each $t \in \mathbb{I}$, (1.3.7) is with separable variables too; if $b \equiv 0$ then (1.3.7) is linear homogeneous; if $\alpha = 0$ then (1.3.7) is linear; if $\alpha = 1$ then (1.3.7) is linear homogeneous.

Theorem 1.3.4. *If $a, b : \mathbb{I} \to \mathbb{R}$ are continuous and non-identically zero on $\mathbb{I}$ and $\alpha \in \mathbb{R} \setminus \{0, 1\}$, then x is a positive solution of equation* (1.3.7) *if and only if the function y, defined by*

$$y(t) = x^{1-\alpha}(t) \tag{1.3.8}$$

for each $t \in \mathrm{Dom}(x)$, is a positive solution of the linear non-homogeneous equation

$$y' = (1 - \alpha)a(t)y + (1 - \alpha)b(t). \tag{1.3.9}$$

Proof. Let x be a positive solution of equation (1.3.7). Expressing x' as a function of y and y' and using the fact that x is a solution of (1.3.7) we deduce that y is a positive solution of (1.3.9). A similar argument shows that if y is a positive solution of equation (1.3.9), then x given by (1.3.8) is, in its turn, a positive solution of (1.3.7), and the proof is complete. □

1.3.5 *Riccati Equations*

An equation of the form

$$x' = a(t)x + b(t)x^2 + c(t), \tag{1.3.10}$$

where $a, b, c : \mathbb{I} \to \mathbb{R}$ are continuous, with b and c non-identically zero on $\mathbb{I}$, is called *Riccati Equation.*

By definition we have excluded the cases $b \equiv 0$ when (1.3.10) is a linear equation and $c \equiv 0$ when (1.3.10) is a Bernoulli equation with $\alpha = 2$.

Remark 1.3.3. In general, there are no effective methods of solving a given Riccati equation, except the fortunate case when we dispose of an *a priori* given particular solution. The next theorem refers exactly to this particular but important case.

Theorem 1.3.5. *Let $a, b, c : \mathbb{I} \to \mathbb{R}$ be continuous with b and c non-identically zero on $\mathbb{I}$. If $\varphi : \mathbb{J} \to \mathbb{R}$ is a solution of* (1.3.10), *then the general solution of* (1.3.10) *on $\mathbb{J}$ is given by*

$$x(t) = y(t) + \varphi(t),$$

where y is the general solution of the Bernoulli equation

$$y' = (a(t) + 2b(t)\varphi(t))y + b(t)y^2.$$

Proof. One verifies by direct computation that $x = y + \varphi$ is a solution of equation (1.3.10) if and only if $y = x - \varphi$ is a solution of the Bernoulli equation above. □

1.3.6 *Exact Differential Equations*

Let D be a nonempty and open subset of $\mathbb{R}^2$ and let $g, h : D \to \mathbb{R}$ be two functions of class C^1 on D, with $h(t, x) \neq 0$ on D. An equation of the form

$$x' = \frac{g(t, x)}{h(t, x)} \tag{1.3.11}$$

is called *exact* if there exists a function of class C^2, $F : D \to \mathbb{R}$, such that

$$\begin{cases} \dfrac{\partial F}{\partial t}(t, x) = -g(t, x) \\[2ex] \dfrac{\partial F}{\partial x}(t, x) = h(t, x). \end{cases} \tag{1.3.12}$$

The condition above shows that $-g(t, x)\, dt + h(t, x)\, dx$ is the differential dF of the function F calculated at $(t, x) \in D$.

Theorem 1.3.6. *If* (1.3.11) *is an exact equation, then its general solution is implicitly given by*

$$F(t, x) = c, \tag{1.3.13}$$

where $F : D \to \mathbb{R}$ *satisfies* (1.3.12), *and* c *ranges over* $F(D)$.

Proof. If (1.3.11) is an exact differential equation, then x is one of its solutions if and only if

$$-g(t, x(t)) + h(t, x(t))x'(t) = 0$$

for $t \in \mathrm{Dom}(x)$, equality which, by virtue of the fact that F satisfies (1.3.12), is equivalent to

$$\frac{d}{dt}[F(\cdot, x(\cdot))](t) = 0$$

for each $t \in \mathrm{Dom}(x)$. Since this last equality is, in its turn, equivalent to (1.3.13), the proof is complete. □

Theorem 1.3.7. *If* D *is a simply connected domain, and both* g *and* h *are* C^1 *functions from* D *to* $\mathbb{R}$, *then* (1.3.11) *is exact if and only if*

$$\frac{\partial h}{\partial t}(t, x) = -\frac{\partial g}{\partial x}(t, x),$$

for each $(t, x) \in D$.

For the proof see Teorema 3 in [Nicolescu *et al.* (1971b)], p. 180, or 10.13 (a) in [Wrede and Spiegel (2002)], p. 244.

1.3.7 *Equations Reducible to Exact Differential Equations*

In general, if the system (1.3.12) has no solutions, the method of finding the general solution of (1.3.11) described above is no longer applicable. There are, however, some specific cases in which, even though (1.3.12) has no solutions, (1.3.11) can be reduced to an exact equation. We describe in what follows such a method of reduction known under the name of *the method of integrating factor*. More precisely, if (1.3.11) is not exact, one looks for a function $\rho : D \to \mathbb{R}$ of class C^1 with $\rho(t, x) \neq 0$ for each $(t, x) \in D$ such that

$$-\rho(t, x)g(t, x)\, dt + \rho(t, x)h(t, x)\, dx$$

is the differential of a function $F : D \to \mathbb{R}$. Assuming that D is simply connected, from Theorem 1.3.7, we know that a necessary and sufficient condition for this to happen is that

$$h(t,x)\frac{\partial \rho}{\partial t}(t,x) + g(t,x)\frac{\partial \rho}{\partial x}(t,x) + \left(\frac{\partial g}{\partial x}(t,x) + \frac{\partial h}{\partial t}(t,x)\right)\rho(t,x) = 0$$

for each $(t,x) \in D$. This is a first-order partial differential equation for the unknown function ρ. We shall study the possibility of solving this type of equations, later on, in Chapter 6. For the moment, let us observe that, if

$$\frac{1}{h(t,x)}\left(\frac{\partial g}{\partial x}(t,x) + \frac{\partial h}{\partial t}(t,x)\right) = f(t)$$

does not depend on x, we can look for a solution ρ of the equation above which also does not depend on x. This function ρ is a solution of the linear homogeneous equation

$$\rho'(t) = -f(t)\rho(t).$$

Analogously, if $g(t,x) \neq 0$ for $(t,x) \in D$ and

$$\frac{1}{g(t,x)}\left(\frac{\partial g}{\partial x}(t,x) + \frac{\partial h}{\partial t}(t,x)\right) = k(x),$$

does not depend on t, we can look for a solution ρ of the equation above which does not depend on t either.

1.3.8 *Lagrange Equations*

A differential equation of the non-normal form

$$x = t\varphi(x') + \psi(x')$$

in which φ and ψ are functions of class C^1 from $\mathbb{R}$ to $\mathbb{R}$ and $\varphi(r) \neq r$ for each $r \in \mathbb{R}$, is called *Lagrange Equation*. This type of differential equation can be integrated by using the so-called *parameter method*. By this method we can find only the solutions of class C^2 of the parametric form

$$\begin{cases} t = t(p) \\ x = x(p), \quad p \in \mathbb{R}. \end{cases}$$

More precisely, let x be a solution of class C^2 of the Lagrange equation. Differentiating both sides of the equation, we get

$$x' = \varphi(x') + t\varphi'(x')\,x'' + \psi'(x')\,x''.$$

Denoting by $x' = p$, we have $x'' = p'$ and consequently

$$\frac{dp}{dt} = -\frac{\varphi(p) - p}{t\varphi'(p) + \psi'(p)}.$$

Assuming now that p is invertible and denoting its inverse by $t = t(p)$, the above equation may be equivalently rewritten as

$$\frac{dt}{dp} = -\frac{\varphi'(p)}{\varphi(p) - p}t - \frac{\psi'(p)}{\varphi(p) - p}.$$

This is a linear differential equation which can be solved by the variation of constants method. We will find then $t = \theta(p, c)$ for $p \in \mathbb{R}$, with c constant, from where, using the original equation, we deduce *the parametric equations* of the general C^2 solution of the Lagrange Equation, i.e.

$$\begin{cases} t = \theta(p, c) \\ x = \theta(p, c)\varphi(p) + \psi(p), \quad p \in \mathbb{R}. \end{cases}$$

1.3.9 *Clairaut Equations*

An equation of the form

$$x = tx' + \psi(x'),$$

where $\psi : \mathbb{R} \to \mathbb{R}$ is of class C^1, is called *Clairaut equation.* This can also be solved by the parameter method. More precisely, let x be a solution of class C^2 of the equation. Differentiating both sides of the equation, we get

$$x''[t + \psi'(x')] = 0.$$

Denoting by $x' = p$, the equation above is equivalent to $p'[t + \psi'(p)] = 0$. If $p' = 0$, it follows that $x(t) = ct + d$, with $c, d \in \mathbb{R}$, from where, imposing the condition on x to satisfy the equation, we deduce the so-called *general solution of the Clairaut equation*

$$x(t) = ct + \psi(c)$$

for $t \in \mathbb{R}$, where $c \in \mathbb{R}$. Obviously, these functions represent a family of straight lines. If $t + \psi'(p) = 0$, we deduce

$$\begin{cases} t = -\psi'(p) \\ x = -p\psi'(p) + \psi(p), \quad p \in \mathbb{R}, \end{cases}$$

a system that defines a plane curve called the *singular solution of the Clairaut equation* and which, is nothing but the envelope of the family of straight lines in the general solution. (We recall that the *envelope* of a family of straight lines is a curve with the property that the family of straight lines coincides with the family of all tangents to the curve.)

Remark 1.3.4. In general, Clairaut equation admits certain solutions which are merely of class C^1. Such a solution can be obtained by continuing a particular arc of curve of the singular solution with those half-tangents at the endpoints of the arc in such a way as to get a C^1 curve. See the solutions to Problems 1.11 and 1.12.

1.3.10 *Higher-Order Differential Equations*

In what follows we shall present two classes of nth-order scalar differential equations which, even though they cannot be solved by quadratures, can be reduced to equations of order strictly less than n. Let us consider first the *incomplete* n^{th}*-order scalar differential equation*

$$F(t, x^{(k)}, x^{(k+1)}, \dots, x^{(n)}) = 0, \tag{1.3.14}$$

where $0 < k < n$ and $F : D(F) \subset \mathbb{R}^{n-k+2} \to \mathbb{R}$. By means of the substitution $y = x^{(k)}$ this equation reduces to an $(n-k)^{\text{th}}$-order scalar differential equation with the unknown function y

$$F(t, y, y', \dots, y^{(n-k)}) = 0.$$

Let us assume for the moment that we are able to obtain the general solution $y = y(t, c_1, c_2, \dots, c_{n-k})$ of the latter equation. In these circumstances, we can obtain the general solution $x(t, c_1, c_2, \dots, c_n)$ of equation (1.3.14) by integrating k-times the identity $x^{(k)} = y$. Namely, for $a \in \mathbb{R}$ suitably chosen, we have

$$\begin{aligned} x(t, c_1, c_2, \dots, c_n) = \frac{1}{(k-1)!} \int_a^t (t-s)^{k-1} y(s, c_1, c_2, \dots, c_{n-k})\, ds \\ + \sum_{i=1}^{k} c_{n-k+i} t^{i-1}, \end{aligned}$$

where $c_{n-k+1}, c_{n-k+2}, \dots, c_n \in \mathbb{R}$ are constants that occur in the iterated integration process.

Example 1.3.1. Find the general solution of the third-order scalar differential equation

$$x''' = -\frac{1}{t}x'' + 3t, \quad t > 0.$$

The substitution $x'' = y$ leads to the non-homogeneous linear equation

$$y' = -\frac{1}{t}y + 3t, \quad t > 0$$

whose general solution is $y(t, c_1) = t^2 + c_1/t$ for $t > 0$. Integrating two times the identity $x'' = y$ we get $x(t, c_1, c_2, c_3) = t^4/12 + c_1(t \ln t - t) + c_2 t + c_3$, where $c_1, c_2, c_3 \in \mathbb{R}$.

A second class of higher-order differential equations which can be reduced to equations whose order is strictly less than the original one is the class of autonomous higher-order differential equations. So, let us consider the autonomous n^{th}-order differential equation

$$F(x, x', \dots, x^{(n)}) = 0, \tag{1.3.15}$$

where $F : D(F) \subset \mathbb{R}^{n+1} \to \mathbb{R}$. Let us denote by $p = x'$, and let us express p as a function of x. To this aim, let us observe that

$$\begin{cases} x'' = \dfrac{dp}{dt} = \dfrac{dp}{dx}\dfrac{dx}{dt} = \dfrac{dp}{dx}p, \\ x''' = \dfrac{d}{dt}\left(\dfrac{dp}{dx}p\right) = \dfrac{d}{dx}\left(\dfrac{dp}{dx}p\right)p, \\ \vdots \\ x^{(n)} = \dots \end{cases}.$$

In this way, for each $k = 1, 2, \dots, n$, $x^{(k)}$ can be expressed as a function of p, $\frac{dp}{dx}, \dots, \frac{dp^{k-1}}{dx^{k-1}}$. Substituting in (1.3.15) the derivatives of x as functions of p, $\frac{dp}{dx}, \dots, \frac{dp^{n-1}}{dx^{n-1}}$ we get an $(n-1)^{\text{th}}$-order differential equation.

Example 1.3.2. The second-order differential equation $x'' + \frac{g}{\ell}\sin x = 0$, i.e. *the pendulum equation*, reduces by the method described above to the first-order differential equation (with separable variables) $p\frac{dp}{dx} = -\frac{g}{\ell}\sin x$ whose unknown function is $p = p(x)$.

1.4 Some Mathematical Models

In this section we shall present several phenomena in Physics, Biology, Chemistry, Demography whose evolution can be described accurately by means of some differential equations, or even systems of differential equations. We begin with an example from Physics, which is well-known due to its use in archeology as a tool used for dating old objects. We emphasize that, in this example, as in many others that will follow, we shall substitute the discrete mathematical model, which is the most realistic one, by a continuously differentiable one, and this for purely mathematical reasons. More precisely, in order to take advantage of the concepts and results of Mathematical Analysis, we shall assume that every function which describes the evolution in time of the state of the system: the number of individuals in a given species, the number of molecules in a given substance, *etc.*, is of class C^1 on its interval of definition, even though, in reality, this function takes values in a very large but finite set. From a mathematical point of view, this means substituting of the discontinuous function x_r, whose graph is illustrated in Figure 1.4.1 as a union of segments which are parallel to the Ot axis, by the function x whose graph is a curve of class C^1. See Figure 1.4.1.

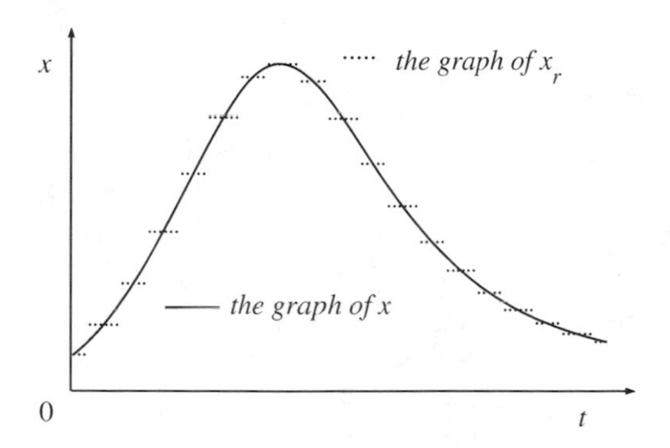

Fig. 1.4.1

1.4.1 *Radioactive Disintegration*

In 1902 Ernest Rutherford Lord of Nelson[19] and Sir Frederick Soddy[20] formulated *the law of radioactive disintegration* which states that *the instantaneous rate of disintegration of a given radioactive element is proportional to the number of radioactive atoms existing at the time considered, and does not depend on any other external factors.* Therefore, denoting by $x(t)$ the

[19]British chemist and physicist born in New Zealand (1871–1937). Laureate of the Nobel Prize for Chemistry in 1908, he successfully performed the first artificially induced transmutation of one chemical element into another: the Nitrogen into Oxygen by means of the alpha radiations (1919). He also proposed the atomic model which inherited his name.

[20]British chemist (1877–1956). Laureate of the Nobel Prize for Chemistry in 1921.

number of non-disintegrated atoms at the time t and assuming that x is a function of class C^1 on $[0, +\infty)$, by virtue of the above mentioned law, we deduce that

$$-x' = ax$$

for every $t \geq 0$, where $a > 0$ is a constant, specific to the radioactive element, called *disintegration constant* and which can be determined experimentally with a sufficient degree of accuracy. This is a first-order linear homogeneous differential equation, whose general solution is given by

$$x(t) = ce^{-at} = x(0)e^{-at}$$

for $t \geq 0$, with $c \in \mathbb{R}_+$.

1.4.2 *The Carbon Dating Method*

This method[21] is essentially based on radioactive disintegration. So, following [Hubbard and West (1995)], Example 2.5.4, p. 85, we recall that living organisms, besides the stable isotope C^{12}, contain a small amount of radioactive isotope C^{14} arising from cosmic ray bombardment. We notice that C^{14} enters the living bodies and due to some specific exchange processes, such as respiration, the ratio C^{14}/C^{12} is kept constant. If an organism dies, these exchange processes stop, and the radioactive C^{14} begins to decrease at a constant rate, whose approximate value (determined experimentally) is $1/8000$, i.e. one part in 8000 per year. Consequently, if $x(t)$ represents the ratio C^{14}/C^{12}, after t years from the death, we conclude that the function $t \mapsto x(t)$ satisfies

$$x' = -\frac{1}{8000}x.$$

Consequently, if we know $x(T)$, we can find the number T, of years after death, by means of

$$T = 8000 \ln \frac{r_0}{x(T)},$$

where r_0 is the constant ratio C^{14}/C^{12} in the living matter. For more details on similar methods of dating see [Braun (1983)].

[21] The carbon-14 method was proposed around 1949 by Willard Libby.

1.4.3 *Equations of Motion*

The equations of motion of n point particles in the three-dimensional Euclidean space are described by means of *Newton's second law* which states that "*Force equals mass times acceleration*". Indeed, in this case, this fundamental law takes the following mathematical expression

$$m_i x_i''(t) = F_i(x_1(t), x_2(t), \ldots, x_n(t)), \quad i = 1, 2, \ldots, n,$$

where x_i is the position vector in three-dimensional space of the ith-particle of mass m_i and F_i is the force acting on that particle. According to what kind of forces are involved: strong, weak, gravitational, or electromagnetic, we get various equations of motion. The last two forces, i.e. occurring in gravitation and electromagnetism, can be expressed in a rather simple manner in the case when the velocities of the particles are considerably less than the speed of light. In that case, the F_i's are the gradients of Newtonian and Coulombic potentials, i.e.

$$F_i(x_1, x_2, \ldots, x_n) = \sum_{j \neq i} \frac{k m_i m_j - e_i e_j}{\|x_j - x_i\|^3}(x_j - x_i),$$

where k is the gravitational constant and e_i is the charge of the ith-particle. For a more detailed discussion on this subject see [Thirring (1978)].

In the case of only one particle moving in the one-dimensional space, i.e. in a straight line, we mention:

1.4.4 *The Harmonic Oscillator*

Let us consider a particle of mass m that moves on a straight line under the action of an elastic force. We denote by $x(t)$ the abscissa of the particle at the time t and by $F(x)$ the force exerted upon the particle in motion situated at the point of abscissa x. Since the force is elastic, $F(x) = -kx$ for each $x \in \mathbb{R}$, where $k > 0$. On the other hand, the motion of the particle should obey Newton's Second Law which, in this specific case, takes the form $F(x(t)) = ma(t)$, where $a(t)$ is the acceleration of the particle at the time t. But $a(t) = x''(t)$ and denoting by $\omega^2 = k/m$, from the considerations above, it follows that x has to verify the second-order scalar linear differential equation:

$$x'' + \omega^2 x = 0,$$

called the *equation of the harmonic oscillator*. As we have already seen in Example 1.2.3, the general solution of this equation is

$$x(t, c_1, c_2) = c_1 \sin \omega t + c_2 \cos \omega t$$

for $t \in \mathbb{R}$, where $c_1, c_2 \in \mathbb{R}$.

1.4.5 *The Mathematical Pendulum*

Let us consider a pendulum of length ℓ and let us denote by $s(t)$ the arc-length of the curve described by the free extremity of the pendulum by the time t. We have $s(t) = \ell x(t)$, where $x(t)$ is the measure expressed in radians of the angle between the pendulum at the time t and the vertical axis Oy. See Figure 1.4.2.

The force which acts upon the pendulum is $F = mg$, where g is the gravitational acceleration. This force can be decomposed along two components, one having the direction of the thread, and the other having the direction of the tangent to the arc of circle described by the free end of the pendulum. See Figure 1.4.2. The component having the direction of the thread is counterbalanced by the resistance of the latter, so that the motion takes place only under the action of the component $-mg \sin x(t)$.

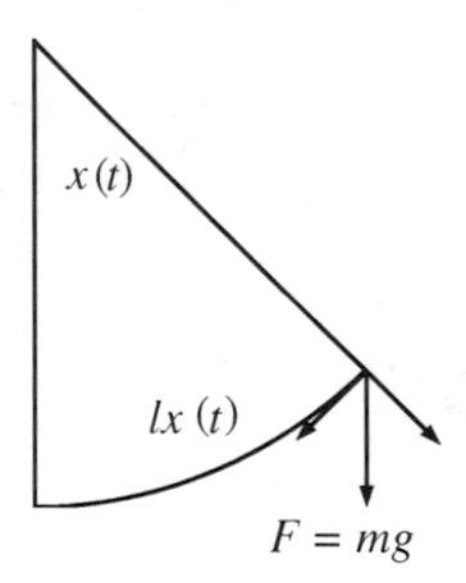

Fig. 1.4.2

But x should obey Newton's Second Law, which in this case takes the form of the second-order scalar differential equation $m\ell x'' = -mg \sin x$, or equivalently

$$x'' + \frac{g}{\ell} \sin x = 0,$$

a nonlinear equation which is called the *equation of the mathematical pendulum*, or *the equation of the gravitational pendulum*.

If we intend to study only the small oscillations around the vertical equilibrium position of the pendulum, we can approximate $\sin x$ by x and we obtain the *equation of the small oscillations of the pendulum*

$$x'' + \frac{g}{\ell} x = 0,$$

a second-order scalar linear differential equation. For this equation, which is formally the same as that of the harmonic oscillator, we know the general solution, i.e.

$$x(t, c_1, c_2) = c_1 \sin \sqrt{\frac{g}{\ell}} t + c_2 \cos \sqrt{\frac{g}{\ell}} t$$

for $t \in \mathbb{R}$, where $c_1, c_2 \in \mathbb{R}$.

1.4.6 *Two Demographic Models*

A first demographic model describing the growth of the human population was proposed in 1798 by Thomas Robert Malthus.[22] We shall present here a continuous variant of the model proposed by Malthus. More precisely, if we denote by $x(t)$ the population, i.e. the number of individuals of a given species at the time t, and by $y(t)$ the amount of food supplies available, according to *Malthus' Law*: *the instantaneous rate of change of x at the time t is proportional with $x(t)$, while the instantaneous rate of change of the amount of food supplies available is constant at any time.* Then we have the following mathematical model, expressed by means of a system of first-order differential equations of the form

$$\begin{cases} x' = cx \\ y' = k, \end{cases}$$

where c and k are strictly positive constants. This system of uncoupled equations (in the sense that each equation contains only one unknown function) can be solved explicitly. Its general solution is given by

$$\begin{cases} x(t, \xi) = \xi e^{ct} \\ y(t, \eta) = \eta + kt \end{cases}$$

for $t \geq 0$, where ξ and η represent the population and respectively the amount of food supplies available, at the time $t = 0$. One may see that this model describes rather well the real phenomenon only on very short intervals of time. For this reason, some more refined and more realistic models have been proposed. The aim was to take into consideration that, at any time, the number of individuals of a given species cannot exceed a certain critical value which depends on the amount of food at that time. So, if we denote by $h > 0$ the quantity of resources necessary to one individual

[22] British economist (1766–1834). In his *An essay on the principle of population as it affects the future improvement of society* (1798) he enunciated the principle stipulating that *a population, which evolves freely increases in a geometric ratio, while the amount of food follows an arithmetic ratio growth.* This principle, expressed as a discrete mathematical model, had a deep influence on the economical thinking even up to the middle of the XX century.

to remain alive after the time t, we may assume that x and y satisfy a system of the form

$$\begin{cases} x' = cx\left(\dfrac{y}{h} - x\right) \\ y' = k. \end{cases}$$

This system describes a more natural relationship between the amount of food and the growth, or decay, of a given population. In certain models, as for instance in the one proposed in 1835 by Verhulst, for simplicity, one considers $k = 0$, which means that the amount of food is constant ($y(t) = \eta$ for each $t \in \mathbb{R}$). Thus, one obtains a first-order nonlinear differential equation of the form

$$x' = cx(b - x),$$

for $t \geq 0$, where $b = \eta/h > 0$. This equation, i.e. the *Verhulst model*, known under the name of *logistic equation*, is with separable variables and can be integrated. More precisely, the general solution is

$$x = \frac{b\mu e^{cbt}}{1 + \mu e^{cbt}}$$

for $t \geq 0$, where $\mu \geq 0$ is a constant. To this solution we have to add the singular solution $x = b$ which was eliminated during the integration process. In order to individualize a certain solution x from the general one, we have to determine the corresponding constant μ. Usually this is done by imposing the initial condition

$$x(0) = \frac{b\mu}{1 + \mu} = \xi,$$

where ξ represents the number of individuals at the time $t = 0$, number which is assumed to be known. We deduce that the solution $x(\cdot, \xi)$ of the logistic equation that satisfies the initial condition $x(0, \xi) = \xi$ is given by

$$x(t, \xi) = \frac{b\xi e^{cbt}}{b + \xi(e^{cbt} - 1)}$$

for each $t \geq 0$.

We note that there are some other demographic models of the form

$$x' = d(t, x),$$

where $d(t, x)$ represents the difference between the rate of birth and the rate of mortality corresponding to the time t and to a population x.

1.4.7 *A Spatial Model in Ecology*

Following [Neuhauser (2001)], we consider an infinite number of sites which are linked by migration and we assume that all sites are equally accessible, while spatial distances between the sites are not taken into consideration. We denote by $x(t)$ the number of occupied sites and we assume that the time is scaled so that the rate at which the sites become vacant equals 1. Then, assuming that the colonization rate x' is proportional to the product of the number of occupied sites and the vacant sites, we get the so-called *Levins Model*

$$x' = \lambda x(1 - x) - x$$

which is formally equivalent to the logistic equation.

1.4.8 *The Prey-Predator Model*

Immediately after the First World War, in the Adriatic Sea area, a significant decay of the fish population was observed. This decay, at first glance in contradiction with the fact that almost all fishermen in the area, enrolled in the army, were in the impossibility to practice their usual job, was a big surprise. Under these circumstances, it seemed natural to expect growth, rather than decay, of the fish population. In his attempt to explain this strange phenomenon, Vito Volterra[23] proposed a mathematical model describing the evolution of two species both living within the same area, but which compete for surviving. Namely, in [Volterra (1926)], he considered two species of animals living in the same region, the first one having at disposal unlimited food supplies, species called *prey*, and the second one, called *predator*, having as unique source of amount of food the members of the first species. Think of the case of herbivores versus carnivores. Denoting by $x(t)$ and by $y(t)$ the population of the prey species, and respectively of the predator species at the time t, and assuming that both x and y are functions of class C^1, he proposed that x and y satisfy the system of first-order nonlinear differential equations

$$\begin{cases} x' = (a - ky)x \\ y' = -(b - hx)y, \end{cases} \tag{1.4.1}$$

where a, b, k, h are positive constants. The first equation is the mathematical expression of the fact that the instantaneous rate of growth of x at time

[23] Italian mathematician (1860–1940) with notable contributions in Functional Analysis and in Applied Mathematics (especially in Physics and in Biology).

t is proportional with the population of the prey species at the time considered ($x' = ax - \dots$), while the instantaneous rate of decay of x at the same time t is proportional with the number of all possible contacts between prey and predators at the same time t ($x' = \dots - kyx$). Analogously, the second equation expresses the fact that the instantaneous rate of decay of y at the time t is proportional with the population of the predator species at time t ($y' = -by \dots$), while the instantaneous rate of growth of y at the same time t is proportional with the number of all possible contacts between prey and predators. It should be mentioned that the very same model had been proposed earlier by [Lotka (1925)], and therefore the system (1.4.1) is known under the name of *Lotka–Volterra System.*

As we shall see later on,[24] each solution of the Lotka–Volterra System (1.4.1) with nonnegative initial data has nonnegative components as long as it exists, while each solution with positive initial data is periodic (with the principal period depending on the initial data). The trajectory of such a solution is illustrated in Figure 1.4.3 (a), while its graph in Figure 1.4.3 (b).

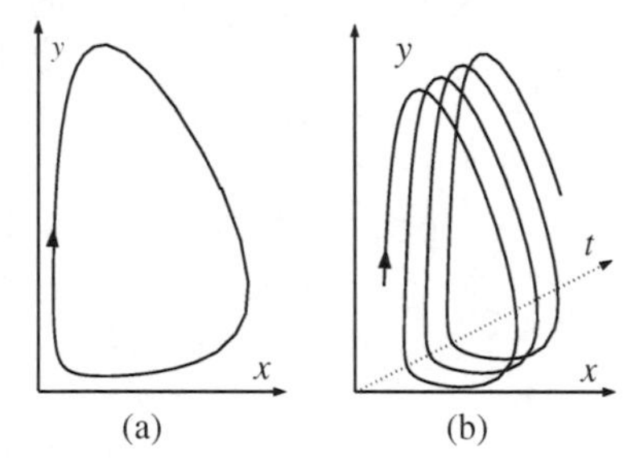

Fig. 1.4.3

For this reason the function $t \mapsto x(t) + y(t)$, which represents the total number of animals in both species at time t, is periodic too, and thus it has infinitely many local minima. Under these circumstances, it is not difficult to realize that the seemingly non-understandable decay of the fish population in the Adriatic Sea was a simple consequence of the fact that the moment in question (the end of First World War) was (presumably) quite close to a local minimum of the function above.

Finally, let us observe that the system above has two constant solutions called (for obvious reasons) *stationary solutions, or equilibria*: $(0, 0)$ and

[24]See Problems 6.1, 6.3, 6.4.

$(b/h, a/k)$. The first one has the property that, there exist solutions of the system, which start from initial points as close as we wish to $(0,0)$, but which do not remain close to $(0,0)$ as t tends to infinity. Indeed, if at a certain moment the predator population is absent, it remains absent for all t, while the prey population evolves obeying Malthus' law. More precisely, the solution starting from the initial point $(\xi, 0)$, with $\xi > 0$, is $(x(t), y(t)) = (\xi e^{at}, 0)$ for $t \geq 0$, which moves away from $(0,0)$ as t tends to infinity. For this reason we say that $(0,0)$ is *unstable* with respect to small perturbations in the initial data. We shall see later on that the second stationary solution is stable with respect to small perturbations of the initial data. Roughly speaking, this means that, all solutions having the initial data close enough to $(b/h, a/k)$ are defined on the whole half-axis and remain close to the solution $(b/h, a/k)$ on their whole domain of definition. The precise definition of this concept will be formulated in Section 5.1. See Definition 5.1.1.

1.4.9 *The Spreading of a Disease*

In 1976 A. Lajmanovich and J. Yorke proposed a model of the spread of a disease which confers no immunity. Following [Hirsch (1984)], we present a slight generalization of this model. We start with the description of a very specific variant and then we present the model in its full generality. More precisely, let us consider a disease which could affect a given population and that confers no immunity. This means that anyone who does not have the disease at a given time is susceptible to get infected, even though he or she had already been infected, but meanwhile recovered. Let us denote by p the population which is assumed to be constant (assumption which is plausible if, for instance, during the spreading of the disease there are neither births, nor deaths) and by x the number of infected people in the population. As we have already mentioned at the beginning of this section, we may assume that x is a positive continuously differentiable function of the time variable t. Consequently, $p - x$ is a nonnegative continuously differentiable function too. Obviously, for each $t \geq 0$, $p - x(t)$ represents the number of people susceptible to be infected at the time t. Then, if we assume that, at any time t, the instantaneous rate of change of the number of infected members is proportional to the number of all possible contacts between infected and non-infected members, number which obviously equals $x(t)(p - x(t))$, we deduce that x must obey the following nonlinear differential equation

$$x' = ax(p - x),$$

where $a > 0$ is a constant. This is an equation with separable variables, of the very same form as that described in the Verhulst's model, and whose general solution is given by

$$x = \frac{p\mu e^{apt}}{1 + \mu e^{apt}},$$

where μ is a positive constant. Apart from this general solution, there exists also the singular stationary solution $x = p$, eliminated during the integration process. As in the case of the logistic equation, the solution $x(\cdot, \xi)$ of the equation above, which satisfies the initial condition $x(0, \xi) = \xi$, is

$$x(t, \xi) = \frac{p\xi e^{apt}}{p + \xi(e^{apt} - 1)}$$

for each $t \geq 0$. It is of interest to note that, for each $\xi > 0$, we have

$$\lim_{t \to +\infty} x(t, \xi) = p,$$

relation which shows that, *in the absence of any external intervention (cure), a population which has at the initial moment a positive number* $\xi > 0$ *of infected members, approaches asymptotically the state of being entirely infected.* The graph of $x(\cdot, \xi)$ is illustrated in Figure 1.4.4.

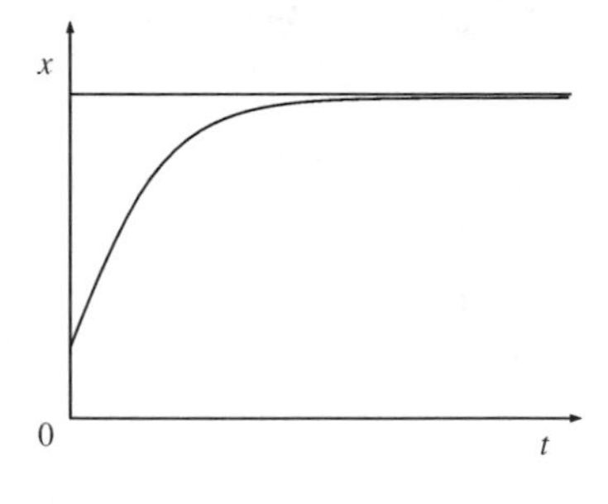

Fig. 1.4.4

We may now proceed to a more general case. More precisely, let us consider that the population in question is divided into n disjoint classes (on social criteria, for instance), each one having a constant number of members. We denote by p_i the number of elements of the class i and by x_i the number of infected members in the class i, $i = 1, 2, \ldots, n$. Then, the number of susceptible members in the class i is $p_i - x_i$. As above, from purely mathematical reasons, we shall consider that x_i is a positive continuously differentiable function of the time variable t. We denote by R_i the rate of infection corresponding to the class i and by C_i the rate of recovery corresponding to the same class i. For the sake of simplicity, we shall assume that R_i depends only on $x = (x_1, x_2, \ldots, x_n)$, while C_i depends only on x_i, $i = 1, 2, \ldots, n$. Finally, it is fairly realistic to consider that $\frac{\partial R_i}{\partial x_j} \geq 0$ for $i, j = 1, 2, \ldots, n$, relations which express the fact that the rate of infection R_i is increasing with respect to each of its arguments x_j, that represents the number of people infected in the class j.

Let us observe that all these assumptions lead to the mathematical model described by the system of first-order nonlinear differential equations

$$x_i' = R_i(x) - C_i(x_i) \quad (i = 1, 2, \ldots, n).$$

We mention that the model proposed by A. Lajmanovich and J. Yorke has the specific form

$$x_i' = \sum_{j=0}^{n} a_{ij} x_j (p_i - x_i) - k_i x_i \quad (i = 1, 2, \ldots, n),$$

where $a_{ij} \geq 0$ and $k_i \geq 0$, for $i, j = 1, 2, \ldots, n$ and was obtained via analogous considerations as those used for the simplified model, i.e. to that one corresponding to a single class.

For more details on models in both population dynamics and ecology see [Neuhauser (2001)].

1.4.10 *Lotka Model*

In 1920 A. J. Lotka considered a chemical reaction mechanism described by

$$\begin{cases} A + X \xrightarrow{k_1} 2X \\ X + Y \xrightarrow{k_2} 2Y \\ Y \xrightarrow{k_3} B, \end{cases} \tag{1.4.2}$$

where X and Y are intermediary chemical by-products, k_1, k_2 and k_3 are the reaction rate constants, and the concentrations of both the reactant A and the product B are kept constant. See [Lotka (1920a)] and [Lotka (1920b)]. Noticing that the meaning of the first relation is that one molecule of A combines with one molecule of X giving two molecules of X, the meaning of the next two relations becomes obvious.

Before obtaining the corresponding mathematical model of these reactions, we recall for easy reference a fundamental law which governs chemical reactions, i.e. *the law of mass action.* Namely this law states that: *the rate of a chemical reaction is proportional to the active concentrations of the reactants, i.e. only to the amounts of reactants that are taking part in the reaction.* For instance, for the irreversible reaction $X + Y \longrightarrow A$, if x and y denote the active concentrations of X and Y respectively, the law of mass action says that $x' = -kxy$, where $k > 0$ is the so-called *rate constant of the reaction.* If one assumes that the reaction is reversible with rate constants of reaction k_1 and k_{-1}, i.e. $X + Y \underset{k_{-1}}{\overset{k_1}{\rightleftarrows}} A$, then the active concentrations x and y must satisfy $x' = -k_1xy + k_{-1}a$. Finally, for the simplest irreversible reaction $X \longrightarrow C$, the law of mass action implies that $x' = -kx$, while for the reversible one $X \underset{k_{-1}}{\overset{k_1}{\rightleftarrows}} C$, it says that $x' = -k_1x + k_{-1}c$.

Now, coming back to (1.4.2), let us denote the concentrations of A, B, X and Y by a, b, x and y respectively, and let us observe that, by virtue of the law of mass action just mentioned, x and y must obey the kinetic equations

$$\begin{cases} x' = k_1ax - k_2xy \\ y' = -k_3y + k_2xy. \end{cases} \tag{1.4.3}$$

We emphasize that the system (1.4.3) is formally equivalent to the Lotka–Volterra system (1.4.1), and thus all the considerations made for the latter apply here as well. For this reason, in all that follows, we will refer to either system (1.4.1) or (1.4.3) as the Lotka–Volterra system, or the *prey-predator system.* For more details on this subject see [Murray (1977)], pp. 136–141.

1.4.11 *An Autocatalytic Generation Model*

Following [Nicolis (1995)], let us consider a tank containing a substance X whose concentration at the time t is denoted by $x(t)$, and another substance A whose concentration $a > 0$ is kept constant, and let us assume that the following reversible chemical reactions:

$$\begin{cases} A + X \underset{k_{-1}}{\overset{k_1}{\rightleftarrows}} 2X \\ X \underset{k_{-2}}{\overset{k_2}{\rightleftarrows}} B, \end{cases}$$

take place in the tank, in which B is a residual product whose concentration at the time t is $b(t)$.[25]

Here $k_i \geq 0$, $i = \pm 1, \pm 2$ are the reaction rate constants of the four reactions in question. The mathematical model describing the evolution of this chemical system, obtained by means of the law of mass action, is

$$\begin{cases} x' = k_1 a x - k_{-1} x^2 - k_2 x + k_{-2} b \\ b' = k_2 x - k_{-2} b. \end{cases}$$

If the second reaction does not take place, a situation described mathematically by $k_2 = k_{-2} = 0$, then the system above reduces to

$$x' = k_1 a x - k_{-1} x^2.$$

Let us notice the remarkable similarity of the equation above with the logistic equation in Verhulst's model, as well as with the equation describing the spread of a disease.

1.4.12 *An RLC Circuit Model*

Following [Hirsch and Smale (1974)], pp. 211–214, let us consider an electric circuit consisting of a resistance R, a coil L, and a capacitor C in which the orientation of currents on each of the three portions of the circuit is illustrated in Figure 1.4.5.

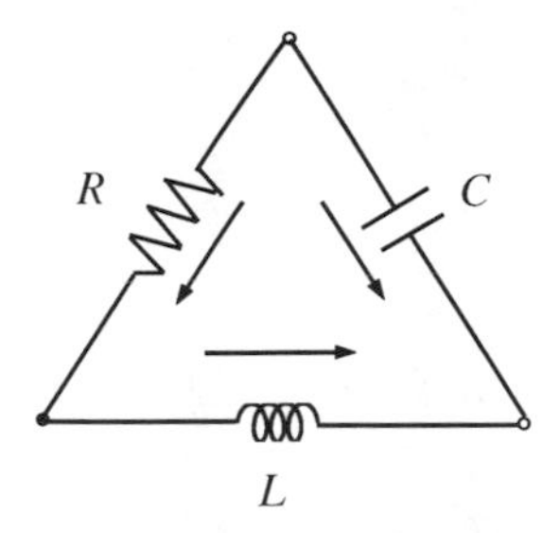

Fig. 1.4.5

Let us denote by $i(t) = (i_R(t), i_L(t), i_C(t))$ the state of the current in the circuit at the time t. Here i_R, i_L, i_C represent the currents on the portions of the circuit containing the resistance R, the coil L and respectively the

[25] This model was proposed in 1971 by Schlögl in order to describe some isothermal autocatalytic chemical reactions. For more details on this type of models the interested reader is referred to [Nicolis (1995)].

capacitor C. Analogously, let $v(t) = (v_R(t), v_L(t), v_C(t))$ be the state of the voltages in the circuit at the time t. Following Kirchhoff's Laws, we deduce

$$\begin{cases} i_R(t) = i_L(t) = -i_C(t) \\ v_R(t) + v_L(t) - v_C(t) = 0, \end{cases}$$

while from the generalized Ohm's Law $g(i_R(t)) = v_R(t)$ for each $t \geq 0$. Finally, from Faraday's Law, we obtain

$$\begin{cases} \mathcal{L}\dfrac{di_L}{dt} = v_L \\ \\ \mathcal{C}\dfrac{dv_C}{dt} = i_C \end{cases}$$

for each $t \geq 0$, where $\mathcal{L} > 0$ and $\mathcal{C} > 0$ are the *inductance* of L and respectively the *capacity* of C. From these relations we observe that i_L and v_C satisfy the system of first-order nonlinear differential equations

$$\begin{cases} \mathcal{L}\dfrac{di_L}{dt} = v_C - g(i_L) \\ \\ \mathcal{C}\dfrac{dv_C}{dt} = -i_L \end{cases}$$

for $t \geq 0$.

For simplicity, let us assume now that $\mathcal{L} = 1$ and $\mathcal{C} = 1$, and let us denote by $x = i_L$ and $y = v_C$. Then the above system can be rewritten under the form

$$\begin{cases} \dfrac{dx}{dt} = y - g(x) \\ \\ \dfrac{dy}{dt} = -x \end{cases}$$

for $t \geq 0$. Assuming in addition that g is of class C^1, differentiating the first equation on both sides and using the second one in order to eliminate y, we finally get

$$x'' + g'(x)x' + x = 0$$

for $t \geq 0$. This is the *Liénard Equation*. In the case in which $g(x) = x^3 - x$ for each $x \in \mathbb{R}$, the equation above takes the form

$$x'' + (3x^2 - 1)x' + x = 0$$

for $t \geq 0$, and it is known as the *Van der Pol Equation*. For a detailed study of mathematical models describing the evolution of both current and voltage in electrical circuits see also [Hirsch and Smale (1974)], Chapter 10. For many other interesting mathematical models see [Braun (1983)].

1.5 Integral Inequalities

In this section we include several inequalities that are very useful in proving the boundedness of solutions of certain differential equations or systems. We start with the following nonlinear integral inequality.

Lemma 1.5.1. (Bihari) *Let* $x : [a,b] \to \mathbb{R}_+$, $k : [a,b] \to \mathbb{R}_+$ *and* $\omega : \mathbb{R}_+ \to \mathbb{R}_+^*$ *be three continuous functions with* ω *nondecreasing on* $\mathbb{R}_+$ *and let* $m \geq 0$. *If*

$$x(t) \leq m + \int_a^t k(s)\,\omega(x(s))\,ds$$

for each $t \in [a,b]$, *then*

$$x(t) \leq \Phi^{-1}\left(\int_a^t k(s)\,ds\right)$$

for each $t \in [a,b]$, *where* $\Phi : \mathbb{R}_+ \to \mathbb{R}$ *is defined by*

$$\Phi(u) = \int_m^u \frac{d\eta}{\omega(\eta)}$$

for each $u \in \mathbb{R}_+$.

Proof. Let us observe that it suffices to prove the lemma in the case in which $m > 0$, since the case $m = 0$ can then be obtained by passing to the limit for m tending to 0. So, let $m > 0$, and let us consider the function $y : [a,b] \to \mathbb{R}_+^*$ defined by

$$y(t) = m + \int_a^t k(s)\,\omega(x(s))\,ds$$

for each $t \in [a,b]$. Obviously y is of class C^1 on $[a,b]$. In addition, since, $x(t) \leq y(t)$ for $t \in [a,b]$ and ω is nondecreasing, it follows that

$$y'(t) = k(t)\,\omega(x(t)) \leq k(t)\,\omega(y(t))$$

for each $t \in [a,b]$. This relation can be rewritten under the form

$$\frac{y'(s)}{\omega(y(s))} \leq k(s)$$

for each $s \in [a,b]$. Integrating both sides of the last inequality from a to t, we obtain

$$\Phi(y(t)) \leq \int_a^t k(s)\,ds$$

for each $t \in [a, b]$. Since Φ is strictly increasing, it is invertible on its range, which includes $[0, +\infty)$, and has strictly increasing inverse. From the last inequality we get

$$y(t) \leq \Phi^{-1}\left(\int_a^t k(s)\, ds\right),$$

relation which, in combination with $x(t) \leq y(t)$ for $t \in [a, b]$, completes the proof. □

The next two consequences of Lemma 1.5.1 are useful in applications.

Lemma 1.5.2. (Gronwall) *Let $x : [a, b] \to \mathbb{R}_+$ and $k : [a, b] \to \mathbb{R}_+$ be two continuous functions and let $m \geq 0$. If*

$$x(t) \leq m + \int_a^t k(s)\, x(s)\, ds$$

for each $t \in [a, b]$, then

$$x(t) \leq m \exp\left(\int_a^t k(s)\, ds\right)$$

for each $t \in [a, b]$.

Proof. Let us remark that, for each $\varepsilon > 0$, we have

$$x(t) \leq m + \int_a^t k(s)(x(s) + \varepsilon)\, ds$$

for each $t \in [a, b]$. Taking $\omega : \mathbb{R}_+ \to \mathbb{R}_+^*$, defined by $\omega(r) = r + \varepsilon$ for each $r \in \mathbb{R}_+$, from Lemma 1.5.1, we obtain

$$x(t) \leq (m + \varepsilon) \exp\left(\int_a^t k(s)\, ds\right) - \varepsilon$$

for each $\varepsilon > 0$ and $t \in [a, b]$. Passing to the limit for ε tending to 0 in this inequality, we get the conclusion of the lemma. The proof is complete. □

Some generalizations of Gronwall's Lemma 1.5.2 are stated in Section 1.6. See Problems 1.16 and 1.17.

Lemma 1.5.3. (Brezis) *Let $x : [a, b] \to \mathbb{R}_+$ and $k : [a, b] \to \mathbb{R}_+$ be two continuous functions and let $m \geq 0$. If*

$$x^2(t) \leq m^2 + 2\int_a^t k(s)\, x(s)\, ds$$

for each $t \in [a, b]$, then

$$x(t) \leq m + \int_a^t k(s)\, ds$$

for each $t \in [a, b]$.

Proof. As in the proof of Lemma 1.5.2, let us observe that, for each $\varepsilon > 0$, we have $x^2(t) \leq m^2 + 2\int_a^t k(s)\sqrt{x^2(s)+\varepsilon}\,ds$ for each $t \in [a,b]$. This inequality and Lemma 1.5.1 with $\omega : \mathbb{R}_+ \to \mathbb{R}_+^*$, defined by $\omega(r) = 2\sqrt{r+\varepsilon}$ for each $r \in \mathbb{R}_+$, yield $x^2 \leq \left(\sqrt{m^2+\varepsilon} + \int_a^t k(s)\,ds\right)^2 - \varepsilon$ for each $\varepsilon > 0$ and $t \in [a,b]$. We complete the proof by passing to the limit for ε tending to 0 in this inequality and then taking the square root on both sides of the inequality thus obtained. □

For a generalization of Lemma 1.5.3, see Problem 1.18.

1.6 Exercises and Problems

Problem 1.1. *Find a plane curve for which the ratio of the ordinate by the subtangent*[26] *equals the ratio of a given positive number k by the difference of the ordinate by the abscissa.*[27] ([Halanay (1972)], p. 7)

Problem 1.2. *Find a plane curve passing through the point* $(3,2)$ *for which the segment of any tangent line contained between the coordinate axes is divided in half by the point of tangency.* ([Demidovich (1973)], p. 329)

Exercise 1.1. *Solve the following differential equations.*

(1) $x'\cos^2 t \cot x + \tan t \sin^2 x = 0$.	(2) $tx' = x + x^2$.
(3) $tx'x = 1 - t^2$.	(4) $x' = (t+x)^2$.
(5) $x' = (8t + 2x + 1)^2$.	(6) $x'(4t + 6x - 5) = -(2t + 3x + 1)$.
(7) $x'(4t - 2x + 3) = -(2t - x)$.	(8) $x'(t^2x - x) + tx^2 + t = 0$.

Problem 1.3. *Find a plane curve passing through the point* $(1,2)$ *whose segment of the normal at any point of the curve lying between the coordinate axes is divided in half by the current point.* ([Demidovich (1973)], 2758, p. 330)

Problem 1.4. *Find a plane curve whose subtangent is of constant length a.* ([Demidovich (1973)], 2759, p. 330)

Problem 1.5. *Find a plane curve in the first quadrant whose subtangent is twice the abscissa of the point of tangency.* ([Demidovich (1973)], 2760, p. 330)

[26] We recall that the subtangent to a given curve of equation $x = x(t)$, $t \in [a,b]$ at a point $(t, x(t))$ equals $x(t)/x'(t)$.

[27] This problem, considered to be the first in the field of Differential Equations, was formulated by Debeaune and conveyed, in 1638, by Mersenne to Descartes. The latter realized not only the importance of the problem but also the impossibility to solve it the methods available at that time.

Exercise 1.2. *Solve the following differential equations.*

(1) $tx' = x - t$. (2) $tx' = -(t + x)$.
(3) $t^2x' = x(t - x)$. (4) $2txx' = t^2 + x^2$.
(5) $(2\sqrt{tx} - t)x' = -x$. (6) $tx' = x + \sqrt{t^2 + x^2}$.
(7) $(4x^2 + 3tx + t^2)x' = -(x^2 + 3tx + 4t^2)$. (8) $2txx' = 3x^2 - t^2$.

Problem 1.6. *Find the equation of a curve that passes through the point* $(1, 0)$ *and having the property that the segment cut off by the tangent line at any current point P on the t-axis equals the length of the segment OP.* ([Demidovich (1973)], 2779, p. 331)

Problem 1.7. *Let* $f : \mathbb{R}_+ \times \mathbb{R}_+ \to \mathbb{R}$ *be a continuous function for which there exists a real number m such that* $f(\lambda t, \lambda^m x) = \lambda^{m-1} f(t, x)$ *for each* $(t, x) \in \mathbb{R}_+ \times \mathbb{R}_+$ *and each* $\lambda \in \mathbb{R}_+$*. Show that, by the substitution* $x(t) = t^m y(t)$*, the differential equation* $x' = f(t, x)$*, called quasi-homogeneous, reduces to an equation with separable variables. Prove that the differential equation* $x' = x^2 - \frac{2}{t^2}$ *is quasi-homogeneous and then solve it.* ([Glăvan *et al.* (1993)], p. 34)

Exercise 1.3. *Solve the following differential equations.*

(1) $tx' = x + tx$. (2) $tx' = -2x + t^4$.
(3) $tx' = -x + e^t$. (4) $(x^2 - 3t^2)x' + 2tx = 0$.
(5) $tx' = -x - tx^2$. (6) $2txx' = x^2 - t$.
(7) $(2t - t^2x)x' = -x$. (8) $tx' = -2x(1 - tx)$.

Problem 1.8. *Let x, x_1, x_2 be solutions of the linear equation* $x' = a(t)x + b(t)$*, where a, b are continuous functions on* $\mathbb{I}$*. Prove that the ratio*

$$R(t) = \frac{x_2(t) - x(t)}{x(t) - x_1(t)}$$

is constant on $\mathbb{I}$*. What is the geometrical meaning of this result?*

Problem 1.9. *Let x_1, x_2 be solutions of the Bernoulli equation* $x' = a(t)x + b(t)x^2$*, where a, b are continuous functions on* $\mathbb{I}$*. Prove that, if* $x_1(t) \neq 0$ *and* $x_2(t) \neq 0$ *on* $\mathbb{J} \subset \mathbb{I}$*, then the function y, defined by* $y(t) = \frac{x_1(t)}{x_2(t)}$ *for each* $t \in \mathbb{J}$*, satisfies the linear equation* $y' = b(t)[x_1(t) - x_2(t)]y$.

Problem 1.10. *Let x, x_1, x_2, x_3 be solutions of the Riccati equation*

$$x' = a(t)x + b(t)x^2 + c(t),$$

where a, b, c are continuous functions on $\mathbb{I}$*. Prove that the ratio*

$$B(t) = \frac{x_2(t) - x(t)}{x_2(t) - x_1(t)} : \frac{x_3(t) - x(t)}{x_3(t) - x_1(t)}$$

is constant on $\mathbb{I}$.

Exercise 1.4. *Solve the following differential equations.*

(1) $(t + 2x)x' + t + x = 0$. (2) $2tx' + t^2 + 2x + 2t = 0$.
(3) $(3t^2x - x^2)x' - t^2 + 3tx^2 - 2 = 0$. (4) $(t^2x + x^3 + t)x' - t^3 + tx^2 + x = 0$.
(5) $(x^2 - 3t^2)x' + 2tx = 0$. (6) $2txx' - (t + x^2) = 0$.
(7) $tx' - x(1 + tx) = 0$. (8) $t(x^3 + \ln t)x' + x = 0$.

Exercise 1.5. *Solve the following differential equations.*

(1) $x = \frac{1}{2}tx' + x'^3$. (2) $x = x' + \sqrt{1 - x'^2}$.
(3) $x = (1 + x')t + x'^2$. (4) $x = -\frac{1}{2}x'(2t + x')$.
(5) $x = tx' + x'^2$. (6) $x = tx' + x'$.
(7) $x = tx' + \sqrt{1 + x'^2}$. (8) $x = tx' + \frac{1}{x'}$.

Problem 1.11. *Find a plane curve for which the distance from a given point to any line tangent to this curve is constant.* ([Demidovich (1973)], 2831, p. 340)

Problem 1.12. *Find the curve for which the area of the triangle formed by a tangent line at any point and by the coordinate axes is constant.* ([Demidovich (1973)], 2830, p. 340)

Problem 1.13. *Prove that, for a heavy liquid rotating about the vertical symmetry axis in a cylindric tank, the free surface is situated on a paraboloid of revolution.* ([Demidovich (1973)], 2898, p. 344)

Problem 1.14. *Find the relationship between the air pressure and the altitude if it is known that the pressure is of* $1kgf$ *per* $1cm^2$ *at the sea level and of* $0.92kgf$ *per* $1cm^2$ *at an altitude of* $500m$. ([Demidovich (1973)], 2899, p. 344)

Problem 1.15. *According to Hooke's law an elastic band of length* l *increases in length* klF *(k=constant) due to a tensile force* F. *By how much will the band increase in length due to its weight* W *if it is suspended at one end? (The initial length of the band is* l.) ([Demidovich (1973)], 2900, p. 344)

Problem 1.16. (Bellman's Inequality) *Let* $x : [\,a, b\,] \to \mathbb{R}_+$, $h : [\,a, b\,] \to \mathbb{R}$ *and* $k : [\,a, b\,] \to \mathbb{R}_+$ *be three continuous functions. If*

$$x(t) \le h(t) + \int_a^t k(s)\,x(s)\,ds \quad \textit{for each} \quad t \in [\,a, b\,],$$

then

$$x(t) \le h(t) + \int_a^t k(s)\,h(s)\exp\left(\int_s^t k(\tau)\,d\tau\right) ds \quad \textit{for each} \quad t \in [\,a, b\,].$$

Problem 1.17. *Let* $x : [a,b] \to \mathbb{R}_+$, $v : [a,b] \to \mathbb{R}$ *and* $k : [a,b] \to \mathbb{R}_+$ *be three continuous functions and* $\xi \in \mathbb{R}$. *If*

$$x(t) \le \xi + \int_a^t [k(s)\,x(s) + v(s)]\,ds \quad \text{for each} \quad t \in [a,b],$$

then

$$x(t) \le \xi \exp\left(\int_a^t k(s)\,ds\right) + \int_a^t v(s) \exp\left(\int_s^\tau k(\tau)\,d\tau\right) ds$$

for each $t \in [a,b]$. ([Halanay (1972)], p. 196)

Problem 1.18. *If* $x : [a,b] \to \mathbb{R}_+$ *and* $k : [a,b] \to \mathbb{R}_+$ *are continuous,* $m \ge 0$, $p > 1$, *and*

$$x^p(t) \le m^p + p\int_a^t k(s)\,x^{p-1}(s)\,ds \quad \text{for each} \quad t \in [a,b],$$

then

$$x(t) \le m + \int_a^t k(s)\,ds \quad \text{for each} \quad t \in [a,b].$$

Problem 1.19. *Let* $f : \mathbb{R} \to \mathbb{R}$ *be non-increasing and let* $x, y : [0,T] \to \mathbb{R}$ *be two functions of class* C^1. *If* $x'(t) + f(x(t)) \le y'(t) + f(y(t))$ *for each* $t \in [0,T]$ *and* $x(0) \le y(0)$ *then* $x(t) \le y(t)$ *for each* $t \in [0,T]$.

Chapter 2

The Cauchy Problem

This chapter is devoted to the study of the fundamental concepts and results concerning the main topic of this book: the so-called Cauchy problem, or the initial-value problem. In the first section we define the Cauchy problem for a given first order differential system and the basic concepts referring to: local solution, saturated solution, global solution, *etc.* In the second section we prove that a sufficient condition for the Cauchy problem to have at least one local solution is the continuity of the function f. In the third section we present several specific situations in which every two solutions of a Cauchy problem coincide on the common part of their domains. The existence of saturated solutions and that of global solutions are studied in the fourth section. In the fifth section we prove several results concerning the continuous dependence of the saturated solutions on the initial data and on the parameters, while in the sixth section we discuss the differentiability of saturated solutions with respect to the data and to the parameters. The seventh section reconsiders, in the setting of the nth-order scalar differential equation, all the problems previously studied for first order differential systems. In the eighth section we prove a simple variant of the implicit function theorem, while the last section contains several exercises and problems illustrating the most delicate aspects of the abstract theory.

2.1 General Presentation

Let $\mathbb{I}$ be a nontrivial interval in $\mathbb{R}$, Ω a nonempty and open subset of $\mathbb{R}^n$, $f : \mathbb{I} \times \Omega \to \mathbb{R}^n$ a given function, $a \in \mathbb{I}$ and $\xi \in \Omega$.

The Cauchy problem, or *the initial-value problem*, for a first-order differential system with data $\mathcal{D} = (\mathbb{I}, \Omega, f, a, \xi)$ consists in finding a C^1-function $x : \mathbb{J} \to \Omega$, where $\mathbb{J} \subseteq \mathbb{I}$ is a nontrivial interval, $a \in \mathbb{J}$, satisfying both "*the equation*" $x'(t) = f(t, x(t))$ for each $t \in \mathbb{J}$ and "*the initial condition*" $x(a) = \xi$. We denote such a problem by

$$\begin{cases} x' = f(t,x) \\ x(a) = \xi. \end{cases} \qquad \mathcal{CP}(\mathcal{D})$$

A function $x : \mathbb{J} \to \Omega$ with the properties mentioned above is called a *solution* of $\mathcal{CP}(\mathcal{D})$. We distinguish between several types of solutions of $\mathcal{CP}(\mathcal{D})$. Thus, if $\mathbb{J} = \mathbb{I}$, the solution x is called *global*; otherwise the solution is called *local.* If $\mathbb{J} = [\,a, b)$, or $\mathbb{J} = [\,a, b\,]$, then x is called a *right solution.* Analogously, if $\mathbb{J} = (c, a\,]$, or $\mathbb{J} = [\,c, a\,]$, x is called a *left solution*, while if $\inf\ \mathbb{J} < a < \sup\ \mathbb{J}$, x is called a *bilateral solution.* A right (left) solution $x : \mathbb{J} \to \Omega$ is called a *global at the right* (*left*) *solution* if $\mathbb{J} = \{t \in \mathbb{I};\ t \geq a\}$ ($\mathbb{J} = \{t \in \mathbb{I};\ t \leq a\}$). The solution $x : \mathbb{J} \to \Omega$ is called *continuable at the right* (*left*) if there exists a right (left) solution $y : \mathbb{K} \to \Omega$ with $\sup\ \mathbb{J} < \sup\ \mathbb{K}$ ($\inf\ \mathbb{J} > \inf\ \mathbb{K}$) and such that $x(t) = y(t)$ for each $t \in \mathbb{J} \cap \mathbb{K}$. A solution $x : \mathbb{J} \to \Omega$ is called *saturated at the right* (*left*) if it is not continuable at the right (left). Obviously each global at the right (left) solution is saturated at the right (left) but not conversely, as we can easily see from the example below.

Example 2.1.1. Take $\mathbb{I} = \mathbb{R}$, $\Omega = \mathbb{R}$, $f : \mathbb{R} \times \mathbb{R} \to \mathbb{R}$, $f(t, x) = -x^2$ for each $(t, x) \in \mathbb{R} \times \mathbb{R}$, $a = 0$ and $\xi = 1$. Obviously, $x : (-\infty, 1) \to \mathbb{R}$, defined by $x(t) = (t-1)^{-1}$, for each $t \in (-\infty, 1)$, is a solution of $\mathcal{CP}(\mathcal{D})$ which is saturated at the right, but which is not a global right solution. However, x is a global left solution of $\mathcal{CP}(\mathcal{D})$. Furthermore, $x|_{(-\infty,1/2)}$ is continuable at the right but not at the left, while $x|_{(-1,1)}$ is continuable at the left, but not at the right.

This example is instructive since it shows that even though f does not depend on t, no matter how regular it is with respect to x in Ω, $\mathcal{CP}(\mathcal{D})$ may have no global solutions.

Remark 2.1.1. Inasmuch as all the considerations involving left solutions for $\mathcal{CP}(\mathcal{D})$ are quite similar to those concerning right solutions, and since the study of bilateral solutions reduces to the study of the previous two types of solutions, our further discussions will refer only to right solutions. In addition, whenever no confusion may occur, we shall omit the word "right", and speak about "solutions" instead of "right solutions".

In a similar manner as for first-order differential systems, we formulate the Cauchy problem for an nth-order differential equation in the normal form. Namely, let $\mathbb{I}$ be a nontrivial interval in $\mathbb{R}$, Ω a nonempty and open subset of $\mathbb{R}^n$, $g : \mathbb{I} \times \Omega \to \mathbb{R}$ a function, $a \in \mathbb{I}$ and $\xi = (\xi_1, \xi_2, \dots, \xi_n) \in \Omega$.

The Cauchy Problem for an nth-order differential equation in the normal form with data $\mathcal{D}' = (\mathbb{I}, \Omega, g, a, \xi)$ consists in finding a function $y : \mathbb{J} \to \mathbb{R}$, of class C^n, where $\mathbb{J} \subset \mathbb{I}$ is a nontrivial interval with $a \in \mathbb{J}$ and $(y(t), y'(t), \dots, y^{(n-1)}(t)) \in \Omega$ for each $t \in \mathbb{J}$, which satisfies "*the equation*" $y^{(n)}(t) = g(t, y(t), y'(t), \dots, y^{(n-1)}(t))$ for each $t \in \mathbb{J}$ and "*the initial conditions*" $y(a) = \xi_1,\ y'(a) = \xi_2, \dots, y^{(n-1)}(a) = \xi_n$. We denote this problem by

$$\begin{cases} y^{(n)} = g(t, y, y', \dots, y^{(n-1)}) \\ y(a) = \xi_1,\ y'(a) = \xi_2, \dots, y^{(n-1)}(a) = \xi_n. \end{cases} \qquad \mathcal{CP}(\mathcal{D}')$$

By means of the transformations

$$\begin{cases} x = (x_1, x_2, \dots, x_n) = (y, y', \dots, y^{(n-1)}) \\ f(t, x) = (x_2, x_3, \dots, x_n,\ g(t, x_1, x_2, \dots, x_n)), \end{cases} \qquad (\mathcal{T})$$

we see that $\mathcal{CP}(\mathcal{D}')$ may be reformulated as a Cauchy problem for a first-order system of differential equations

$$\begin{cases} x_1' = x_2 \\ x_2' = x_3 \\ \vdots \\ x_{n-1}' = x_n \\ x_n' = g(t, x_1, x_2, \dots, x_n) \\ x_1(a) = \xi_1,\ x_2(a) = \xi_2, \dots, x_n(a) = \xi_n, \end{cases} \qquad \mathcal{CP}(\mathcal{D}'')$$

which, in turn, may be rewritten as a problem of the form $\mathcal{CP}(\mathcal{D})$, where $\mathcal{D} = (\mathbb{I}, \Omega, f, a, \xi)$, with f defined as above.

This way, the two Cauchy problems are equivalent, in the sense that all the concepts and results referring to $\mathcal{CP}(\mathcal{D})$ may be transferred to $\mathcal{CP}(\mathcal{D}')$ and conversely via the transformations $(\mathcal{T})$. However, we emphasize that, in order to minimize the possibility of confusion, we have to take special care in understanding this equivalence. More precisely, there are properties of the solutions of the two Cauchy problems (such as, for instance: the boundedness, the behavior at $+\infty$, *etc.*) which are not invariant with respect to these transformations. This happens because the solution of $\mathcal{CP}(\mathcal{D}')$ is the first component of the solution of $\mathcal{CP}(\mathcal{D}'')$. To illustrate this point more convincingly, let us analyze the following two examples.

Example 2.1.2. Consider the Cauchy problem

$$\begin{cases} y'' = -4t^2 y + 2\cos t^2 \\ y(0) = 0 \\ y'(0) = 0, \end{cases} \tag{$\mathcal{P}$}$$

which, by means of the transformations ($\mathcal{T}$), can be rewritten as a Cauchy problem for a first-order system of differential equations of the form

$$\begin{cases} x_1' = x_2 \\ x_2' = -4t^2 x_1 + 2\cos t^2 \\ x_1(0) = 0 \\ x_2(0) = 0. \end{cases} \tag{$\mathcal{P}'$}$$

It is easy to see that the function $y : \mathbb{R} \to \mathbb{R}$, defined by $y(t) = \sin t^2$ for $t \in \mathbb{R}$, is a solution of the Cauchy problem ($\mathcal{P}$). In addition, this solution is bounded on $\mathbb{R}$. By means of ($\mathcal{T}$), y corresponds to the solution $(x_1, x_2) : \mathbb{R} \to \mathbb{R} \times \mathbb{R}$ of ($\mathcal{P}'$), $(x_1(t), x_2(t)) = (\sin t^2, 2t\cos t^2)$ for each $t \in \mathbb{R}$, solution which is unbounded on $\mathbb{R}$. So, the boundedness of solutions is not invariant with respect to ($\mathcal{T}$).

Example 2.1.3. Consider the Cauchy problem

$$\begin{cases} y'' = y - \cos e^t - e^t \sin e^t \\ y(0) = \sin 1 \\ y'(0) = \cos 1 - \sin 1. \end{cases} \tag{$\mathcal{Q}$}$$

One verifies by direct computation that the function $y : \mathbb{R} \to \mathbb{R}$ defined by

$$y(t) = e^{-t} \sin e^t$$

for $t \in \mathbb{R}$ is a solution of the problem ($\mathcal{Q}$). Obviously

$$\lim_{t \to +\infty} y(t) = 0.$$

However, the solution

$$(x_1(t), x_2(t)) = (e^{-t} \sin e^t, -e^{-t} \sin e^t + \cos e^t),$$

for $t \in \mathbb{R}$, of the Cauchy problem

$$\begin{cases} x_1' = x_2 \\ x_2' = x_1 - \cos e^t - e^t \sin e^t \\ x_1(0) = \sin 1 \\ x_2(0) = \cos 1 - \sin 1, \end{cases} \tag{$\mathcal{Q}'$}$$

corresponding to ($\mathcal{Q}$) via the transformations ($\mathcal{T}$), has no limit as t approaches $+\infty$.

This example shows that the property of a certain solution of the Cauchy problem $\mathcal{CP}(\mathcal{D}')$ to have finite limit as t tends to $+\infty$ is not inherited by the corresponding solution of the Cauchy problem $\mathcal{CP}(\mathcal{D})$.

Remark 2.1.2. We also notice that we may use very similar arguments to reduce the study of the Cauchy problem for a higher-order system of differential equations in the normal form to that of a suitable Cauchy problem for a first-order differential system. We leave it to the reader to formulate the Cauchy problem for a higher-order system of differential equations.

In view of the preceding observations it is clear why, in all what follows, we restrict our attention only to the study of the Cauchy problem for first-order differential systems.

Next, we will prove two simple but useful results that will used frequently in the sequel.

Proposition 2.1.1. *Let* $f : \mathbb{I} \times \Omega \to \mathbb{R}^n$ *be a continuous function and* $\mathbb{J} \subset \mathbb{I}$ *a nontrivial interval such that* $a \in \mathbb{J}$. *Then, a function* $x : \mathbb{J} \to \Omega$ *is a solution of* $\mathcal{CP}(\mathcal{D})$ *if and only if* x *is continuous on* $\mathbb{J}$ *and satisfies the integral equation*

$$x(t) = \xi + \int_a^t f(\tau, x(\tau))d\tau \tag{$\mathcal{IE}$}$$

for each $t \in \mathbb{J}$.

Proof. If x is a solution of $\mathcal{CP}(\mathcal{D})$, then it is a continuous function on $\mathbb{J}$, being a function of class C^1. Thus, $\tau \mapsto f(\tau, x(\tau))$ is continuous on $\mathbb{J}$ too. Consequently, we may integrate both sides in

$$x'(\tau) = f(\tau, x(\tau))$$

from a to t. Taking into account that $x(a) = \xi$, we get $(\mathcal{IE})$.

Conversely, if x is continuous on $\mathbb{J}$ and satisfies $(\mathcal{IE})$, then $\tau \mapsto f(\tau, x(\tau))$ is continuous on $\mathbb{J}$ too. Hence, in view of $(\mathcal{IE})$, x is C^1 on $\mathbb{J}$. Differentiating both sides in $(\mathcal{IE})$ we obtain $x'(t) = f(t, x(t))$, for each $t \in \mathbb{J}$, while setting $t = a$ in $(\mathcal{IE})$ we get $x(a) = \xi$. Thus x is a solution of $\mathcal{CP}(\mathcal{D})$ and this completes the proof. □

Proposition 2.1.2. (Concatenation Principle). *Let* $f : \mathbb{I} \times \Omega \to \mathbb{R}^n$ *be a continuous function, let* $[a, b] \subset \mathbb{I}$, $[b, c] \subset \mathbb{I}$ *and let* $\xi \in \Omega$. *Let* $x : [a, b] \to \Omega$ *be a solution of* $\mathcal{CP}(\mathbb{I}, \Omega, f, a, \xi)$ *and* $y : [b, c] \to \Omega$ *a solution of* $\mathcal{CP}(\mathbb{I}, \Omega, f, b, x(b))$. *Then, the concatenate function* $z : [a, c] \to \Omega$, *i.e. the function defined by*

$$z(t) = \begin{cases} x(t), & \text{for } t \in [a, b] \\ y(t), & \text{for } t \in (b, c], \end{cases}$$

is a solution of $\mathcal{CP}(\mathbb{I}, f, \Omega, a, \xi)$.

Proof. Clearly z is continuous on $[\,a,c\,]$. In view of Proposition 2.1.1, it suffices to show that z satisfies

$$z(t) = \xi + \int_a^t f(\tau, z(\tau))d\tau, \tag{2.1.1}$$

for each $t \in [\,a,c\,]$. If $t \in [\,a,b\,]$, this is certainly the case since $z(t) = x(t)$ and x satisfies $(\mathcal{IE})$. Thus, let $t \in (b,c\,]$, and let us observe that, again by Proposition 2.1.1, we have

$$z(t) = y(t) = x(b) + \int_b^t f(\tau, y(\tau))d\tau = x(b) + \int_b^t f(\tau, z(\tau))d\tau.$$

Since

$$x(b) = \xi + \int_a^b f(\tau, x(\tau))d\tau = \xi + \int_a^b f(\tau, z(\tau))d\tau,$$

substituting in the last equality we get (2.1.1). The proof is complete. □

Remark 2.1.3. It is not difficult to observe that the regularity properties of the solutions of $\mathcal{CP}(\mathcal{D})$ depend on the regularity properties of f. More precisely, one may easily check that, if f is of class C^{k-1} on $\mathbb{I} \times \Omega$ $(k \geq 1)$, then each solution of $\mathcal{CP}(\mathcal{D})$ is of class C^k on its domain. Therefore, if f is a C^∞-function on $\mathbb{I} \times \Omega$, then so is each solution of $\mathcal{CP}(\mathcal{D})$ on its domain. Furthermore, it is also true that if f is analytic on $\mathbb{I} \times \Omega$, then each solution of $\mathcal{CP}(\mathcal{D})$ is analytic on its domain – this important and nontrivial result due to Cauchy will be proved in Section 3.1.

We conclude this section with some simple but useful considerations concerning the autonomous case. We recall that a differential equation is called *autonomous* if it has the form $x' = f(x)$, where $f : \Omega \to \mathbb{R}^n$. In other words, a differential equation is autonomous if its right-hand side f does not depend explicitly on t. So, let us consider the *autonomous Cauchy problem*

$$\begin{cases} x' = f(x) \\ x(a) = \xi, \end{cases} \tag{$\mathcal{ACP}(\mathcal{D})$}$$

where $\mathcal{D} = (\Omega, f, a, \xi)$.

Proposition 2.1.3. *A function $x : \mathbb{I}_x \to \Omega$ is a solution of $\mathcal{ACP}(\Omega, f, a, \xi)$ if and only if the function $x_a : \mathbb{I}_{x_a} \to \Omega$ defined by $x_a(t) = x(t+a)$ for each $t \in \mathbb{I}_{x_a}$, where $\mathbb{I}_{x_a} = \{t \in \mathbb{R};\ t + a \in \mathbb{I}_x\}$, is a solution of $\mathcal{ACP}(\Omega, f, 0, \xi)$.*

Proof. Clearly x is of class C^1 if and only if x_a is. Moreover, $x(a) = \xi$ if and only if $x_a(0) = \xi$ and $x'_a(t) = x'(t+a) = f(x(t+a)) = f(x_a(t))$ for each $t \in \mathbb{I}_{x_a}$ if and only if $x'(s) = f(x(s))$ for each $s \in \mathbb{I}_x$, which completes the proof. □

Proposition 2.1.3 explains why, in the case of autonomous systems, we will consider only the Cauchy problem with the initial datum given at $a = 0$, i.e. $x(0) = \xi$.

2.2 The Local Existence Problem

In general not every Cauchy problem admits a solution, as we can see from the example below.

Example 2.2.1. Let $f : \mathbb{R} \to \mathbb{R}$ be given by

$$f(x) = \begin{cases} -1 & \text{if } x \geq 0 \\ 1 & \text{if } x < 0. \end{cases}$$

Then the autonomous Cauchy problem

$$\begin{cases} x' = f(x) \\ x(0) = 0 \end{cases}$$

has no local right solution. Indeed, if we assume that $x : [0, \delta) \to \mathbb{R}$ is such a solution, then x is of class C^1 and $x'(0) = -1$. Therefore, on a certain right neighborhood of 0, x' has the same sign as -1. Without any loss of generality (by choosing a smaller value for δ if necessary), we may assume that $x'(t) < 0$ for each $t \in [0, \delta)$. It then follows that x is strictly decreasing on $[0, \delta)$ and consequently, $x(t) < x(0) = 0$ for each $t \in (0, \delta)$. So we have $x'(t) = f(x(t)) = 1$ for each $t \in (0, \delta)$, and $x'(0) = -1$, relations which show that x', which is continuous on $[0, \delta)$, is discontinuous at $t = 0$. This contradiction can be eliminated only if the Cauchy problem considered has no local right solution. As we shall see later, this phenomenon of nonexistence is due to the discontinuity of the function f.

The purpose of this section is to prove that, if $f : \mathbb{I} \times \Omega \to \mathbb{R}^n$ is continuous, then for each $(a, \xi) \in \mathbb{I} \times \Omega$, $\mathcal{CP}(\mathcal{D})$ has at least one local solution. This fundamental result was first proved in 1890 by the Italian mathematician Giuseppe Peano.

We analyze first the case in which $\Omega = \mathbb{R}^n$ and f is continuous and bounded on $\mathbb{I} \times \mathbb{R}^n$, and then, we will show how the result can be proved without these additional assumptions on f.

Let $f : \mathbb{I} \times \mathbb{R}^n \to \mathbb{R}^n$, let $a \in \mathbb{I}$, $\xi \in \mathbb{R}^n$, $\mathcal{D} = (\mathbb{I}, \mathbb{R}^n, f, a, \xi)$, and let us consider the Cauchy problem

$$\begin{cases} x' = f(t,x) \\ x(a) = \xi. \end{cases} \tag{$\mathcal{CP}(\mathcal{D})$}$$

Let us also consider the integral equation with the delay $\lambda \geq 0$

$$x_\lambda(t) = \begin{cases} \xi, & \textit{for } t \in [\,a-\lambda, a\,] \\ \xi + \displaystyle\int_a^t f(\tau, x_\lambda(\tau - \lambda))\, d\tau, & \textit{for } t \in (a, a+\delta\,], \end{cases} \tag{$(\mathcal{IE})_\lambda$}$$

where $\delta > 0$ is such that $[\,a, a+\delta\,] \subset \mathbb{I}$. Note that, for $\lambda = 0$, $(\mathcal{IE})_\lambda$ reduces to $(\mathcal{IE})$ in Proposition 2.1.1, which by virtue of Proposition 2.1.1 is equivalent to $\mathcal{CP}(\mathcal{D})$. Roughly speaking, this remark suggests that, in order to prove that $\mathcal{CP}(\mathcal{D})$ has at least one local solution, it suffices to show that, for each $\lambda > 0$, $(\mathcal{IE})_\lambda$ has at least one local solution and then, if possible, to pass to the limit for λ tending to 0 in $(\mathcal{IE})_\lambda$. This is exactly what we are going to do in the sequel in a rigorous manner. We begin with the following result which will prove useful later.

Lemma 2.2.1. *If $f : \mathbb{I} \times \mathbb{R}^n \to \mathbb{R}^n$ is continuous and $\lambda > 0$ is such that $[\,a, a+\lambda\,] \subseteq \mathbb{I}$ then, for each $(a, \xi) \in \mathbb{I} \times \mathbb{R}^n$ and each $\delta > 0$ such that $[\,a, a+\delta\,] \subseteq \mathbb{I}$, there exists a unique continuous function x_λ, defined on $[\,a-\lambda, a+\delta\,]$ and satisfying $(\mathcal{IE})_\lambda$.*

Proof. Obviously x_λ is uniquely determined and continuous on $[\,a-\lambda, a\,]$. Let then $t \in [\,a, a+\lambda\,]$. Note that, for each $\tau \in [\,a, t\,]$, we have $\tau - \lambda \in [\,a-\lambda, a\,]$ and consequently $x_\lambda(\tau - \lambda) = \xi$. Therefore

$$x_\lambda(t) = \xi + \int_a^t f(\tau, \xi)\, d\tau$$

and accordingly, x_λ is uniquely determined and continuous on $[\,a, a+\lambda\,]$. In a similar manner, we find a unique continuous function x_λ, defined successively on $[\,a+\lambda, a+2\lambda\,]$, $[\,a+2\lambda, a+3\lambda\,]$, and so on, up to a complete cover of $[\,a, a+\delta\,]$. Namely, after a finite number m of steps, with $m\lambda \geq a+\delta$, we can define the function x_λ on the whole interval $[\,a, a+\delta\,]$. The proof is complete. □

As we have already mentioned, we will prove first the following existence result which, although auxiliary, is interesting in itself.

Lemma 2.2.2. *If $f : \mathbb{I} \times \mathbb{R}^n \to \mathbb{R}^n$ is continuous and bounded on $\mathbb{I} \times \mathbb{R}^n$, then, for each $(a, \xi) \in \mathbb{I} \times \mathbb{R}^n$ and each $\delta > 0$ such that $[\,a, a+\delta\,] \subset \mathbb{I}$, $\mathcal{CP}(\mathcal{D})$ has at least one solution defined on $[\,a, a+\delta\,]$.*

Proof. Let $(a,\xi) \in \mathbb{I} \times \mathbb{R}^n$ and $\delta > 0$ be such that $[a, a+\delta] \subset \mathbb{I}$, let $m \in \mathbb{N}^*$ and let us consider the integral equation with the delay $\delta_m = \delta/m$

$$x_m(t) = \begin{cases} \xi, & \text{for } t \in [a-\delta_m, a] \\ \xi + \displaystyle\int_a^t f(\tau, x_m(\tau - \delta_m))\, d\tau, & \text{for } t \in (a, a+\delta]. \end{cases} \qquad (\mathcal{IE})_m$$

Let us remark that, by virtue of Lemma 2.2.1, for each $m \in \mathbb{N}^*$, $(\mathcal{IE})_m$ has a unique continuous solution $x_m : [a-\delta_m, a+\delta] \to \mathbb{R}^n$.

The idea is to prove that the family of functions $\{x_m;\ m \in \mathbb{N}^*\}$ satisfies the hypotheses of Theorem 12.2.1 which implies that the sequence $(x_m)_{m\in\mathbb{N}^*}$ has at least one convergent subsequence. To this aim, we shall show that $\{x_m;\ m \in \mathbb{N}^*\}$ is uniformly bounded – see Definition 12.2.3 – and equicontinuous on $[a, a+\delta]$ – see Definition 12.2.2.

First, let us recall that f is bounded on $\mathbb{I} \times \mathbb{R}^n$ and accordingly, there exists $M > 0$ such that

$$\|f(\tau, y)\| \leq M$$

for each $(\tau, y) \in \mathbb{I} \times \mathbb{R}^n$. From $(\mathcal{IE})_m$, we deduce that

$$\|x_m(t)\| \leq \|\xi\| + (t-a)M \leq \|\xi\| + \delta M$$

for each $m \in \mathbb{N}^*$ and $t \in [a, a+\delta]$. Hence $\{x_m;\ m \in \mathbb{N}^*\}$ is uniformly bounded on $[a, a+\delta]$.

Let us observe next that, again, from $(\mathcal{IE})_m$, we have

$$\|x_m(t) - x_m(s)\| \leq \left| \int_s^t \|f(\tau, x_m(\tau - \delta_m))\|\, d\tau \right| \leq M|t-s|,$$

for each $m \in \mathbb{N}^*$ and $t, s \in [a, a+\delta]$. Consequently, $\{x_m;\ m \in \mathbb{N}^*\}$ is equicontinuous on $[a, a+\delta]$. By virtue of Theorem 12.2.1 it follows that $(x_m)_{m\in\mathbb{N}^*}$ has at least one uniformly convergent subsequence, denoted for simplicity also by $(x_m)_{m\in\mathbb{N}^*}$. Let us denote by x its uniform limit which is a continuous function. Clearly we have,

$$\lim_{m\to\infty} x_m(\tau - \delta_m) = x(\tau),$$

uniformly for $\tau \in [a, a+\delta]$. Since f is continuous on $\mathbb{I} \times \mathbb{R}^n$, the relation above and Corollary 12.2.2 show that we may legitimately pass to the limit for $m \to \infty$ under the integral sign in $(\mathcal{IE})_m$. Moreover, using again Corollary 12.2.2, we deduce that x satisfies

$$x(t) = \xi + \int_a^t f(\tau, x(\tau))\, d\tau$$

for each $t \in [a, a+\delta]$. Now, Proposition 2.1.1 shows that $x : [a, a+\delta] \to \mathbb{R}^n$ is a solution of $\mathcal{CP}(\mathcal{D})$, and this completes the proof. $\square$

Remark 2.2.1. Under the hypotheses of Lemma 2.2.2, we may prove that for each $(a, \xi) \in \mathbb{I} \times \mathbb{R}^n$, $\mathcal{CP}(\mathcal{D})$ has at least one global solution.

We may now proceed to the statement of the main result of this section. To this aim, let $\mathbb{I}$ be a nonempty and open interval in $\mathbb{R}$, Ω a nonempty and open subset of $\mathbb{R}^n$, and $f : \mathbb{I} \times \Omega \to \mathbb{R}^n$ a given function.

Theorem 2.2.1. (Peano) *If $f : \mathbb{I} \times \Omega \to \mathbb{R}^n$ is continuous on $\mathbb{I} \times \Omega$, then for each $(a, \xi) \in \mathbb{I} \times \Omega$, $\mathcal{CP}(\mathbb{I}, \Omega, f, a, \xi)$ has at least one local solution.*

Proof. Let $(a, \xi) \in \mathbb{I} \times \Omega$. Since both $\mathbb{I}$ and Ω are open, there exist $d > 0$ and $r > 0$ such that $[\,a - d, a + d\,] \subset \mathbb{I}$, and

$$B(\xi, r) = \{\eta \in \mathbb{R}^n;\ \|\eta - \xi\| \le r\} \subset \Omega.$$

We define $\rho : \mathbb{R}^n \to \mathbb{R}^n$ by

$$\rho(y) = \begin{cases} y & \text{for } y \in B(\xi, r) \\ \dfrac{r}{\|y - \xi\|}(y - \xi) + \xi & \text{for } y \in \mathbb{R}^n \setminus B(\xi, r). \end{cases}$$

For $n = 1$, the graph of ρ is illustrated in Figure 2.2.1. We may easily verify that ρ maps $\mathbb{R}^n$ into $B(\xi, r)$ and is continuous on $\mathbb{R}^n$.

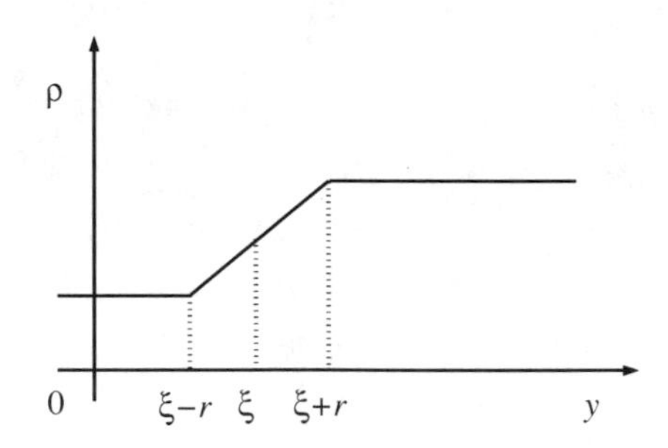

Fig. 2.2.1

Now, let us define $g : (a - d, a + d) \times \mathbb{R}^n \to \mathbb{R}^n$ by

$$g(t, y) = f(t, \rho(y)),$$

for each $(t, y) \in (a - d, a + d) \times \mathbb{R}^n$.

Since f is continuous, by Weierstrass' Theorem, its restriction to the compact set $[\,a - d, a + d\,] \times B(\xi, r)$ is bounded. Therefore, g is continuous and bounded on $(a - d, a + d) \times \mathbb{R}^n$.

By virtue of Lemma 2.2.1, we know that, for each $d' \in (0, d)$, the Cauchy problem

$$\begin{cases} x' = g(t, x) \\ x(a) = \xi \end{cases}$$

has at least one solution $x : [a, a + d'] \to \mathbb{R}^n$. Since $x(a) = \xi$ and x is continuous at $t = a$, for $r > 0$, there exists $\delta \in (0, d']$ such that for each $t \in [a, a + \delta]$, $\|x(t) - \xi\| \leq r$. But in this case $g(t, x(t)) = f(t, x(t))$, and therefore $x : [a, a + \delta] \to \Omega$ is a solution of $\mathcal{CP}(\mathcal{D})$, thereby completing the proof. □

2.3 The Uniqueness Problem

As we have already seen, the continuity of the right-hand side in $\mathcal{CP}(\mathcal{D})$ is sufficient to ensure the existence of at least one local solution. It is worth mentioning that, if f is merely continuous then, for certain choices of $(a, \xi) \in \mathbb{I} \times \Omega$, it is possible that $\mathcal{CP}(\mathcal{D})$ have more than one local solution, as we can see from the next example due to Peano (1890).

Example 2.3.1. Take $\mathbb{I} = \mathbb{R}$, $\Omega = \mathbb{R}$, $f : \mathbb{R} \times \mathbb{R} \to \mathbb{R}$, $f(t, x) = 3\sqrt[3]{x^2}$ for each $(t, x) \in \mathbb{R} \times \mathbb{R}$, $a = 0$ and $\xi = 0$. Then, we may easily verify that both $x(t) = 0$ and $y(t) = t^3$, for $t \in [0, \infty)$ are solutions of $\mathcal{CP}(\mathbb{R}, \mathbb{R}, f, 0, 0)$.

We note that there are examples of functions $f : \mathbb{I} \times \Omega \to \mathbb{R}^n$ such that for each $(a, \xi) \in \mathbb{I} \times \Omega$, $\mathcal{CP}(\mathcal{D})$ has at least two solutions. For such an example, the interested reader is referred to [Hartman (1964)], p. 18.

Definition 2.3.1. We say that $\mathcal{CP}(\mathcal{D})$ has the *local uniqueness property* if for any $(a, \xi) \in \mathbb{I} \times \Omega$ and any two solutions x and y of $\mathcal{CP}(\mathcal{D})$, there exists $\delta > 0$ such that $[a, a + \delta) \subset \mathbb{I}$ and $x(t) = y(t)$ for each $t \in [a, a + \delta)$.

We say that $\mathcal{CP}(\mathcal{D})$ has the *global uniqueness property* if for any fixed data $(a, \xi) \in \mathbb{I} \times \Omega$, any two solutions of $\mathcal{CP}(\mathcal{D})$ coincide on the common part of their domains.

We begin with the following useful result.

Proposition 2.3.1. *The problem $\mathcal{CP}(\mathcal{D})$ has the local uniqueness property if and only if it has the global uniqueness property.*

Proof. The "if" part is obvious. So, let us assume that $\mathcal{CP}(\mathcal{D})$ has the local uniqueness property, let $(a, \xi) \in \mathbb{I} \times \Omega$ and let $x : \mathbb{J} \to \Omega$ and $y : \mathbb{K} \to \Omega$

be two solutions of $\mathcal{CP}(\mathcal{D})$. Since x and y are continuous, the set

$$\mathcal{C}(x,y) = \{t \in \mathbb{J} \cap \mathbb{K};\ x(s) = y(s) \text{ for each } s \in [\,a,t\,]\}$$

is nonempty and closed. To complete the proof it suffices to show that

$$\sup \mathcal{C}(x,y) = \sup(\mathbb{J} \cap \mathbb{K}).$$

To this aim, let us assume the contrary. Since $\sup \mathcal{C}(x,y) \leq \sup(\mathbb{J} \cap \mathbb{K})$, it follows that $\sup \mathcal{C}(x,y) < \sup(\mathbb{J} \cap \mathbb{K})$. But in this case both x and y are defined at the right of $b = \sup \mathcal{C}(x,y)$, are solutions of $\mathcal{CP}(\mathbb{I}, \Omega, f, b, x(b))$, and then, by hypothesis, they must coincide on an interval of the form $[\,b, b+\delta)$, with $\delta > 0$ sufficiently small. Since this statement contradicts the definition of b, it follows that the supposition $\sup \mathcal{C}(x,y) < \sup(\mathbb{J} \cap \mathbb{K})$ is false. Then $\sup \mathcal{C}(x,y) = \sup(\mathbb{J} \cap \mathbb{K})$ and this completes the proof. □

By virtue of Proposition 2.3.1 "local uniqueness" and "global uniqueness" describe one and the same property of $\mathcal{CP}(\mathcal{D})$. Therefore, in all what follows, we shall say that $\mathcal{CP}(\mathcal{D})$ *has the uniqueness property* instead of saying that $\mathcal{CP}(\mathcal{D})$ *has the local, or global uniqueness property.*

2.3.1 *The Locally Lipschitz Case*

In order to state the first main result of this subsection, some definitions and remarks are needed.

Definition 2.3.2. A function $f : \mathbb{I} \times \Omega \to \mathbb{R}^n$ is called *locally Lipschitz* on Ω if for each compact subset $\mathcal{K}$ in $\mathbb{I} \times \Omega$, there exists $L = L(\mathcal{K}) > 0$ such that, for each $(t,u), (t,v) \in \mathcal{K}$ we have

$$\|f(t,u) - f(t,v)\| \leq L\|u - v\|. \tag{2.3.1}$$

Remark 2.3.1. The use of the term "locally" in Definition 2.3.2 is somewhat improper, but is justified by the fact that $f : \mathbb{I} \times \Omega \to \mathbb{R}^n$ is locally Lipschitz on Ω if and only if for each $(a, \xi) \in \mathbb{I} \times \Omega$, there exists a neighborhood $\mathcal{V}$ of (a, ξ), $\mathcal{V} \subset \mathbb{I} \times \Omega$ and $L = L(\mathcal{V}) > 0$ such that, for each $(t,u), (t,v) \in \mathcal{V}$, (2.3.1) holds. We leave it to the reader to prove this nice result in Real Analysis.

Remark 2.3.2. If $f : \mathbb{I} \times \Omega \to \mathbb{R}^n$ has first-order partial derivatives with respect to the last n arguments, and for each $i, j \in \{1, 2, \ldots, n\}$, $\partial f_i / \partial x_j$ is continuous on $\mathbb{I} \times \Omega$, then f is locally Lipschitz on Ω.

A first sufficient condition for uniqueness is:

Theorem 2.3.1. *If $f : \mathbb{I} \times \Omega \to \mathbb{R}^n$ is locally Lipschitz on Ω, then $\mathcal{CP}(\mathcal{D})$ has the uniqueness property.*

Proof. In view of Proposition 2.3.1, it suffices to show that if f is locally Lipschitz on Ω, then $\mathcal{CP}(\mathcal{D})$ has the local uniqueness property. Thus, let $(a,\xi) \in \mathbb{I} \times \Omega$ and let $x : \mathbb{J} \to \Omega$ and $y : \mathbb{K} \to \Omega$ be two solutions of $\mathcal{CP}(\mathcal{D})$. Since $(a,\xi) \in \mathbb{I} \times \Omega$, which is open, there exist $r > 0$ and $\delta > 0$ such that $a + \delta < \sup(\mathbb{J} \cap \mathbb{K})$ and $B(\xi, r) = \{\eta \in \mathbb{R}^n;\ \|\eta - \xi\| \le r\} \subset \Omega$.

Furthermore, inasmuch as both x and y are continuous at $t = a$, and $x(a) = y(a) = \xi$, by choosing a smaller value of δ if necessary, we may assume that

$$x(t) \in B(\xi, r) \text{ and } y(t) \in B(\xi, r) \tag{2.3.2}$$

for each $t \in [\,a, a+\delta\,]$. As f is locally Lipschitz on Ω, and $[\,a, a+\delta\,] \times B(\xi, r)$ is compact, there exists $L > 0$ such that

$$\|f(t,u) - f(t,v)\| \le L\|u - v\| \tag{2.3.3}$$

for each $t \in [\,a, a+\delta\,]$ and $u, v \in B(\xi, r)$. At this point, let us observe that we may assume with no loss of generality that

$$\delta L < 1. \tag{2.3.4}$$

(If not, we can take a smaller δ satisfying all the conditions above including the preceding one.)

Now, as both x and y satisfy $(\mathcal{IE})$ in Proposition 2.1.1, we have

$$\|x(t) - y(t)\| \le \int_a^t \|f(\tau, x(\tau)) - f(\tau, y(\tau))\| d\tau$$

for each $t \in [\,a, a+\delta\,]$. From (2.3.2) and (2.3.3), it follows

$$\sup_{t \in [\,a, a+\delta\,]} \|x(t) - y(t)\| \le \delta L \cdot \sup_{t \in [\,a, a+\delta\,]} \|x(t) - y(t)\|.$$

Since by (2.3.4) $\delta L < 1$, the last inequality holds only if

$$\sup_{t \in [\,a, a+\delta\,]} \|x(t) - y(t)\| = 0,$$

i.e. $x(t) = y(t)$ for each $t \in [\,a, a+\delta\,]$. Thus $\mathcal{CP}(\mathcal{D})$ has the local uniqueness property and, by virtue of Proposition 2.3.1, this completes the proof. □

A simple, but important consequence of Theorems 2.2.1 and 2.3.1 is stated below.

Theorem 2.3.2. *If $f : \mathbb{I} \times \Omega \to \mathbb{R}^n$ is continuous on $\mathbb{I} \times \Omega$ and locally Lipschitz on Ω, then for each $(a, \xi) \in \mathbb{I} \times \Omega$ there exists $\delta > 0$ such that $[\,a, a+\delta\,] \subset \mathbb{I}$ and $\mathcal{CP}(\mathcal{D})$ has a unique solution defined on $[\,a, a+\delta\,]$.*

2.3.2 *The Dissipative Case*

Another important class of functions f for which $\mathcal{CP}(\mathcal{D})$ has the uniqueness property is defined below.

Definition 2.3.3. A function $f : \mathbb{I} \times \Omega \to \mathbb{R}^n$ is called *dissipative on* Ω if for each $t \in \mathbb{I}$ and $u, v \in \Omega$, we have

$$\langle f(t,u) - f(t,v), u - v \rangle \leq 0,$$

where $\langle \cdot, \cdot \rangle$ stands for the usual inner product in $\mathbb{R}^n$, i.e. for each $u, v \in \mathbb{R}^n$, $u = (u_1, u_2, \ldots, u_n)$ and $v = (v_1, v_2, \ldots, v_n)$,

$$\langle u, v \rangle = \sum_{i=1}^{n} u_i v_i.$$

Remark 2.3.3. If $n = 1$, i.e. if $\Omega \subset \mathbb{R}$ and $f : \mathbb{I} \times \Omega \to \mathbb{R}$, f is dissipative on Ω if and only if for each $t \in \mathbb{I}$, $f(t, \cdot)$ is non-increasing on Ω.

Theorem 2.3.3. *If $f : \mathbb{I} \times \Omega \to \mathbb{R}^n$ is dissipative on Ω, then $\mathcal{CP}(\mathcal{D})$ has the uniqueness property.*

The conclusion of Theorem 2.3.3 follows from the next lemma which will prove useful in the sequel.

Lemma 2.3.1. *Let $f : \mathbb{I} \times \Omega \to \mathbb{R}^n$ be dissipative on Ω, let $a \in \mathbb{I}$ and $\xi, \eta \in \Omega$. If $x : \mathbb{I}_x \to \Omega$ and $y : \mathbb{I}_y \to \Omega$ are two solutions of the Cauchy problems $\mathcal{CP}(\mathbb{I}, \Omega, f, a, \xi)$ and respectively $\mathcal{CP}(\mathbb{I}, \Omega, f, a, \eta)$ then*

$$\|x(t) - y(t)\| \leq \|\xi - \eta\|$$

for each $t \in \mathbb{I}_x \cap \mathbb{I}_y$, $t \geq a$.

Proof. Let $a \in \mathbb{I}$, $\xi, \eta \in \Omega$ and let x, y be two solutions of $\mathcal{CP}(\mathbb{I}, \Omega, f, a, \xi)$ and $\mathcal{CP}(\mathbb{I}, \Omega, f, a, \eta)$, respectively. Then, for each $t \in \mathbb{I}_x \cap \mathbb{I}_y$, we have

$$x'(t) - y'(t) = f(t, x(t)) - f(t, y(t)).$$

Taking the inner product both sides in the preceding relation by $x(t) - y(t)$ and recalling that, by (i) in Lemma 12.1.2,

$$\langle x'(t) - y'(t), x(t) - y(t) \rangle = \frac{1}{2} \frac{d}{dt} \|x(t) - y(t)\|^2,$$

we get

$$\frac{1}{2} \frac{d}{dt} \|x(t) - y(t)\|^2 = \langle f(t, x(t)) - f(t, y(t)), x(t) - y(t) \rangle.$$

Here the fact that f is dissipative comes into play and shows that

$$\frac{1}{2}\frac{d}{dt}\|x(t)-y(t)\|^2 \leq 0.$$

Thus, $t \mapsto \frac{1}{2}\|x(t)-y(t)\|^2$ and, by consequence, $t \mapsto \|x(t)-y(t)\|$ are non-increasing on the common part of the domains of x and y. From this remark it follows that

$$\|x(t)-y(t)\| \leq \|x(a)-y(a)\| = \|\xi-\eta\|$$

for each $t \in \mathbb{I}_x \cap \mathbb{I}_y$, $t \geq a$, and this completes the proof. □

Remark 2.3.4. In contrast with the Lipschitz condition which ensures the bilateral uniqueness, in the dissipative case we can only deduce the uniqueness at the right but not the one at the left, as we can see from the next example.

Example 2.3.2. Let $\Omega = \mathbb{R}$ and let $f : \mathbb{R} \to \mathbb{R}$ be defined by

$$f(x) = \begin{cases} 3\sqrt[3]{x^2} & \text{if } x < 0 \\ 0 & \text{if } x \geq 0. \end{cases}$$

Then $\mathcal{CP}(f, \mathbb{R}, 0, 0)$ has only one right saturated solution, but $x(t) = 0$ and $y(t) = t^3$ for $t < 0$, are saturated left solutions of $\mathcal{CP}(\mathbb{R}, f, 0, 0)$.

We conclude this section with a useful consequence of Theorems 2.2.1 and 2.3.3.

Theorem 2.3.4. *If $f : \mathbb{I} \times \Omega \to \mathbb{R}^n$ is continuous on $\mathbb{I} \times \Omega$ and dissipative on Ω, then for each $(a, \xi) \in \mathbb{I} \times \Omega$ there exists $\delta > 0$ such that $[\,a, a+\delta\,] \subset \mathbb{I}$, and $\mathcal{CP}(\mathcal{D})$ has a unique solution defined on $[\,a, a+\delta\,]$.*

2.4 Saturated Solutions

Let $\mathbb{I}$ be a nontrivial interval in $\mathbb{R}$, Ω a nonempty and open subset of $\mathbb{R}^n$, let $f : \mathbb{I} \times \Omega \to \mathbb{R}^n$ be a given function, $a \in \mathbb{I}$ and $\xi \in \Omega$. Let $\mathcal{D} = (\mathbb{I}, \Omega, f, a, \xi)$ and let us consider the Cauchy problem

$$\begin{cases} x' = f(t, x) \\ x(a) = \xi. \end{cases} \qquad \mathcal{CP}(\mathcal{D})$$

We recall that a solution $x : \mathbb{J} \to \Omega$ of $\mathcal{CP}(\mathcal{D})$ is called *continuable at the right* (*left*) if there exists a right (left) solution $y : \mathbb{K} \to \Omega$ with $\sup \mathbb{J} < \sup \mathbb{K}$ ($\inf \mathbb{J} > \inf \mathbb{K}$) and such that $x(t) = y(t)$ for each $t \in \mathbb{J} \cap \mathbb{K}$.

We also recall that a solution $x : \mathbb{J} \to \Omega$ is called *saturated at the right* (*left*) if it is not continuable at the right (left). A right (left) solution $x : \mathbb{J} \to \Omega$ is called a *global at the right* (*left*) *solution* if $\mathbb{J} = \{t \in \mathbb{I};\ t \geq a\}$ ($\mathbb{J} = \{t \in \mathbb{I};\ t \leq a\}$). Since we merely consider right solutions, in all that follows, by a "continuable", respectively "saturated" solution we shall mean a "continuable at the right", respectively "saturated at the right" solution.

2.4.1 *Characterization of Continuable Solutions*

We begin with a very simple but useful lemma.

Lemma 2.4.1. *Let $f : \mathbb{I}\times\Omega \to \mathbb{R}^n$ be continuous on $\mathbb{I}\times\Omega$. Then, a solution $x : [\,a, b) \to \Omega$ of $\mathcal{CP}(\mathcal{D})$ is continuable if and only if*

(i) $b < \sup\ \mathbb{I}$
and there exists
(ii) $x^* = \lim_{t\uparrow b} x(t)$ and $x^* \in \Omega$.

Proof. The necessity is obvious, while the sufficiency is a consequence of both Theorem 2.2.1 and Proposition 2.1.2. Indeed, if both (i) and (ii) hold, a simple argument shows that x can be extended to $[\,a, b\,]$ as a solution of $\mathcal{CP}(\mathbb{I}, \Omega, f, a, \xi)$, denoted again by x, by setting $x(b) = x^*$. Then, by virtue of Theorem 2.2.1, $\mathcal{CP}(\mathbb{I}, \Omega, f, b, x^*)$ has at least one local solution $y : [\,b, b+\delta\,] \to \Omega$, where $\delta > 0$. By Proposition 2.1.2, we know that the concatenate function z is a solution of $\mathcal{CP}(\mathbb{I}, \Omega, f, a, \xi)$ defined on $[\,a, b+\delta)$ which coincides with x on $[\,a, b)$. Hence x is continuable, and this completes the proof. □

Remark 2.4.1. From Lemma 2.4.1 it readily follows that, whenever $\mathbb{I}$ and Ω are open and f is continuous, then each saturated solution of $\mathcal{CP}(\mathcal{D})$ is necessarily defined on an interval of the form $[\,a, b)$, i.e. on an interval which is open at the right. This is no longer true if either $\mathbb{I}$, or Ω is not open. A specific situation of this type will be analyzed in Sections 3.2 and 3.3. See also Section 7.7.

A sufficient condition for the existence of the finite limit in (ii) is stated below.

Proposition 2.4.1. *Let $x : [\,a, b) \to \Omega$ be a solution of $\mathcal{CP}(\mathcal{D})$ and let us assume that $b < +\infty$, and there exists $M > 0$ such that*

$$\|f(\tau, x(\tau))\| \leq M,$$

for each $\tau \in [a, b)$. *Then, there exists* $x^* \in \overline{\Omega}$,

$$x^* = \lim_{t \uparrow b} x(t).$$

Proof. In view of Proposition 2.1.1, for each $t, s \in [a, b)$ we have

$$\|x(t) - x(s)\| \le \left| \int_s^t \|f(\tau, x(\tau))\| d\tau \right| \le M|t - s|,$$

and thus x satisfies the hypothesis of the Cauchy test on the existence of a finite limit at b. □

We may now proceed to the statement of a useful characterization of continuability of the solutions of $\mathcal{CP}(\mathcal{D})$.

Theorem 2.4.1. *Let* $f : \mathbb{I} \times \Omega \to \mathbb{R}^n$ *be continuous on* $\mathbb{I} \times \Omega$. *Then, a solution* $x : [a, b) \to \Omega$ *of* $\mathcal{CP}(\mathcal{D})$ *is continuable if and only if the graph of* x, *i.e.*

$$\text{graph}\, x = \{(t, x(t)) \in \mathbb{R} \times \mathbb{R}^n;\ t \in [a, b)\},$$

is included in a compact subset of $\mathbb{I} \times \Omega$.

Proof. The "if" part. Assume that graph x is included in a compact subset of $\mathbb{I} \times \Omega$. Since $\mathbb{I}$ is open it follows that $b < \sup \mathbb{I}$ and f is bounded on graph x, i.e. there exists $M > 0$ such that

$$\|f(\tau, x(\tau))\| \le M,$$

for each $\tau \in [a, b)$. The claimed result is a consequence of Lemma 2.4.1 and Proposition 2.4.1.

The "only if" part. Assume that x is continuable. Then, x may be extended by continuity to $[a, b] \subset \mathbb{I}$. Denote this extension by y, and let us observe that the mapping $t \mapsto (t, y(t))$ is continuous from $[a, b]$ to $\mathbb{I} \times \Omega$. Therefore, its range which coincides with graph y is compact and included in $\mathbb{I} \times \Omega$. Since graph $x \subset$ graph y, the proof is complete. □

2.4.2 *The Existence of Saturated Solutions*

We continue with a fundamental result concerning saturated solutions for $\mathcal{CP}(\mathcal{D})$. We notice that in the next theorem f is completely arbitrary.

Theorem 2.4.2. *If* $x : \mathbb{J} \to \Omega$ *is a solution of* $\mathcal{CP}(\mathcal{D})$, *then either* x *is saturated, or* x *can be continued up to a saturated one.*

Proof. If x is saturated then there is nothing to prove. Thus, let us assume that x is continuable and let us define $\mathcal{S}$ to be the set of all solutions of $\mathcal{CP}(\mathcal{D})$ which extend x. Obviously, $x \in \mathcal{S}$ and thus $\mathcal{S}$ is nonempty. Moreover, since x is continuable, $\mathcal{S}$ contains at least two elements. On $\mathcal{S}$ let us define the relation "$\preceq$" by $y \preceq z$ if z extends y. It is a simple exercise to show that $(\mathcal{S}, \preceq)$ is an inductively ordered set. So, from Zorn's Lemma, there exists at least one maximal element $y \in \mathcal{S}$ such that $x \preceq y$. From the definition of " $\preceq$ ", and from the maximality of y it follows that y is a saturated solution of $\mathcal{CP}(\mathcal{D})$ which coincides with x on $\mathbb{J}$, thereby completing the proof. □

Remark 2.4.2. Under the hypothesis of Theorem 2.4.2, if $x : \mathbb{J} \to \Omega$ is a continuable solution of $\mathcal{CP}(\mathcal{D})$, two or more saturated solutions extending x may exist. See Problem 2.16. It is not difficult to see that this phenomenon occurs as a consequence of non-uniqueness. On the other hand, if $\mathcal{CP}(\mathcal{D})$ has the uniqueness property, we may easily conclude that for each continuable solution $x : \mathbb{J} \to \Omega$ of $\mathcal{CP}(\mathcal{D})$ there exists exactly one saturated solution of $\mathcal{CP}(\mathcal{D})$ extending x.

From Theorems 2.2.1 and 2.4.2 it follows:

Corollary 2.4.1. *If $f : \mathbb{I} \times \Omega \to \mathbb{R}^n$ is continuous, then for each data $(a, \xi) \in \mathbb{I} \times \Omega$, $\mathcal{CP}(\mathbb{I}, \Omega, f, a, \xi)$ has at least one saturated solution.*

2.4.3 *Types of Saturated Solutions*

We recall that a *limit point* of a function $x : [a, b) \to \mathbb{R}^n$ as t tends to b is an element x^* of $\mathbb{R}^n$ for which there exists a sequence $(t_k)_{k \in \mathbb{N}}$ in $[a, b)$ tending to b and such that $\lim_{k \to \infty} x(t_k) = x^*$.

Concerning the behavior of saturated solutions at the right end point of their interval of definition, we have the following fundamental result.

Theorem 2.4.3. *Let $f : \mathbb{I} \times \Omega \to \mathbb{R}^n$ be continuous on $\mathbb{I} \times \Omega$, and let $x : [a, b) \to \Omega$ be a saturated solution of $\mathcal{CP}(\mathcal{D})$. If $b <$ sup $\mathbb{I}$ then, either*

(i) $\lim_{t \uparrow b} \|x(t)\| = +\infty$, *or*
(ii) $\liminf_{t \uparrow b} \|x(t)\| < +\infty$, *in which case every limit point of x as $t \uparrow b$ belongs to the boundary of Ω.*

Proof. If $\lim_{t \uparrow b} \|x(t)\| = +\infty$ there is nothing to prove. Next, let us analyze the case in which $\liminf_{t \uparrow b} \|x(t)\| < +\infty$. To prove (ii) let us assume for a contradiction that there exists at least one sequence $(t_k)_{k \in \mathbb{N}}$

in $[a, b)$ tending to b and such that $(x(t_k))_{k\in\mathbb{N}}$ is convergent to some x^* in $\mathbb{R}^n$, but x^* does not belong to the boundary $\partial\Omega$ of Ω. Since x^* lies in the closure of Ω and $x^* \notin \partial\Omega$, we necessarily have $x^* \in \Omega$. In view of Lemma 2.4.1, to get a contradiction, it suffices to show that there exists $\lim_{t\uparrow b} x(t)$ which, of course, must coincide with x^*. To this aim, let us observe that, since Ω is open and $x^* \in \Omega$, there exists $r_0 > 0$ such that $B(x^*, r_0) = \{\eta \in \mathbb{R}^n;\ \|\eta - x^*\| \leq r_0\} \subset \Omega$. Furthermore, inasmuch as $b < \sup \mathbb{I}$, f is continuous on $[a, b] \times B(x^*, r_0)$, and the latter is compact, from Weierstrass' theorem, it follows that there exists $M > 0$ such that

$$\|f(\tau, y)\| \leq M, \tag{2.4.1}$$

for each $(\tau, y) \in [a, b] \times B(x^*, r_0)$.

Let $r \in (0, r_0]$ be arbitrary. Taking into account that $\lim_{k\to\infty} t_k = b$ and $\lim_{k\to\infty} x(t_k) = x^*$, we may choose $k \in \mathbb{N}$ such that

$$\begin{cases} b - t_k < \dfrac{r}{2M} \\[2ex] \|x(t_k) - x^*\| < \dfrac{r}{2}. \end{cases} \tag{2.4.2}$$

We shall show that for each $t \in [t_k, b)$, we have $x(t) \in B(x^*, r)$. Let

$$t^* = \sup\{t \in [t_k, b);\ x(s) \in B(x^*, r),\ \text{for } s \in [t_k, t]\}.$$

If $t^* = b$, the preceding statement is obviously true. Next, let us assume for a contradiction that $t^* < b$. This means that $x(t) \in B(x^*, r)$ for each $t \in [t_k, t^*]$, $\|x(t^*) - x^*\| = r$ and there exist points $t > t^*$, as close to t^* as we wish, satisfying $\|x(t) - x^*\| > r$. In other words, t^* is the "first moment in (t_k, b) after which x leaves the set $B(x^*, r)$". See Figure 2.4.1.

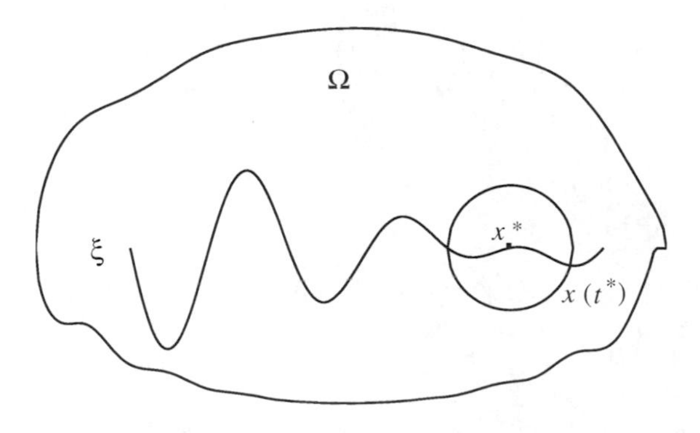

Fig. 2.4.1

The condition $\|x(t^*) - x^*\| = r$ signifies that, at t^*, x must cross the boundary of $B(x^*, r)$. Next, from the remark above combined with (2.4.1), and (2.4.2), we get

$$r = \|x(t^*) - x^*\| \leq \|x(t^*) - x(t_k)\| + \|x(t_k) - x^*\|$$

$$\leq \int_{t_k}^{t^*} \|f(\tau, x(\tau))\| d\tau + \|x(t_k) - x^*\| \leq (t^* - t_k)M + \|x(t_k) - x^*\|$$

$$\leq (b - t_k)M + \|x(t_k) - x^*\| < \frac{r}{2} + \frac{r}{2} = r.$$

This contradiction ($r < r$) comes from our supposition that for at least one $t \in [t_k, b)$, $x(t) \notin B(x^*, r)$. Then, for each $t \in [t_k, b)$, $x(t) \in B(x^*, r)$. Since $r > 0$ can be taken as small as we wish, we conclude that $\lim_{t \uparrow b} x(t) = x^*$. But $x^* \in \Omega$ and thus, in view of Lemma 2.4.1, it follows that x is continuable thereby contradicting the hypothesis. This contradiction can be eliminated only if $x^* \in \partial\Omega$. The proof is complete. □

Corollary 2.4.2. *Let $f : \mathbb{I} \times \mathbb{R}^n \to \mathbb{R}^n$ be continuous on $\mathbb{R} \times \mathbb{R}^n$ and let $x : [\,a, b) \to \mathbb{R}^n$ be a saturated solution of $\mathcal{CP}(\mathcal{D})$. If x is not global, i.e. $b < \sup \mathbb{I}$, then there exists*

$$\lim_{t \uparrow b} \|x(t)\| = +\infty.$$

Proof. Since the boundary of $\mathbb{R}^n$ is the empty set, alternative (ii) in Theorem 2.4.3 cannot hold. Therefore, alternative (i) holds. □

In the hypotheses of Corollary 2.4.2, if $b < +\infty$ and $\lim_{t \uparrow b} \|x(t)\| = +\infty$, we say that x *blows up in finite time.*

Corollary 2.4.3. *Let $f : \mathbb{I} \times \mathbb{R}^n \to \mathbb{R}^n$ be continuous on $\mathbb{R} \times \mathbb{R}^n$ and let $x : [\,a, b) \to \mathbb{R}^n$ be a solution of $\mathcal{CP}(\mathcal{D})$. Then x is continuable if and only if $b < \sup \mathbb{I}$ and x is bounded on $[\,a, b)$.*

Proof. The "only if" part is trivial, while the "if" part is an immediate consequence of Corollary 2.4.2. □

Lemma 2.4.2. *Let Ω be a nonempty and open subset of $\mathbb{R}^n$ and let $x : [\,a, b) \to \Omega$ be a continuous function. Then the following conditions are equivalent:*

(h) *for each compact subset K in Ω there exists $c \in (a, b)$ such that $x(t) \notin K$ for each $t \in (c, b)$;*

(hh) *either* $\lim_{t\uparrow b}\|x(t)\| = +\infty$, *or* $\liminf_{t\uparrow b}\|x(t)\| < +\infty$, *in which case every limit point of* x *as* $t\uparrow b$ *belongs to the boundary of* Ω.

Proof. We begin by proving that (h) implies (hh). So, let us assume that $x : [a,b) \to \Omega$ is a continuous function and that (h) holds true. If $\lim_{t\uparrow b}\|x(t)\| = +\infty$ there is nothing to prove. Alternatively, if $\liminf_{t\uparrow b}\|x(t)\| < +\infty$, we conclude that every limit point x^* of x, as t tends to b, belongs to the boundary of Ω. Indeed, if we assume for contradiction that there exists a limit point $x^* \in \Omega$, as Ω is open, there exists $r > 0$ such that $B(x^*, r) \subseteq \Omega$. Since $B(x^*, r)$ is compact, by virtue of (h), there exists $c \in (a,b)$ such that $x(t)$ lies outside $B(x^*, r)$ for each $t \in (c,b)$, which is absurd. This contradiction can be eliminated only if $x^* \in \partial\Omega$.

To prove that (hh) implies (h) we proceed by contraposition. So, let us assume that (h) does not hold. Then there exists a compact set $K \subseteq \Omega$ such that for each $c \in (a,b)$ there exists $t_c \in (c,b)$ with $x(t_c) \in K$. Let $(b_m)_{m\in\mathbb{N}^*}$ be a sequence in (a,b) with $b_m \uparrow b$, and let us fix a $t_m \in (b_m, b)$ such that $x(t_m) \in K$. Since $(x(t_m))_{m\in\mathbb{N}^*}$ is in K which is compact, we may assume without loss of generality, by extracting a subsequence if necessary, that $\lim_{m\to\infty} x(t_m) = x^*$. But $x^* \in K$. Since $K \subseteq \Omega$, clearly $x^* \notin \partial\Omega$, assertion which shows that (hh) does not hold true. So (hh) implies (h) and this completes the proof. □

From Lemma 2.4.2 and Theorem 2.4.3, we deduce:

Theorem 2.4.4. *Let* $f : \mathbb{I}\times\Omega \to \mathbb{R}^n$ *be continuous on* $\mathbb{I}\times\Omega$, *and let* $x : [a,b) \to \Omega$ *be a saturated solution of* $\mathcal{CP}(\mathcal{D})$. *If* $b <$ *sup* $\mathbb{I}$ *then* x *satisfies* (h) *in Lemma* 2.4.2.

Under an additional hypothesis on f, we have:

Theorem 2.4.5. *Let* $f : \mathbb{I}\times\Omega \to \mathbb{R}^n$ *be continuous on* $\mathbb{I}\times\Omega$ *and let us assume that it maps bounded subsets in* $\mathbb{I}\times\Omega$ *into bounded subsets in* $\mathbb{R}^n$. *Let* $x : [a,b) \to \Omega$ *be a saturated solution of* $\mathcal{CP}(\mathcal{D})$ *with* $b < \sup\mathbb{I}$. *Then, either*

(i) $\lim_{t\uparrow b}\|x(t)\| = +\infty$, *or*
(jj) $\liminf_{t\uparrow b}\|x(t)\| < +\infty$, *in which case there exists* $\lim_{t\uparrow b} x(t) = x^*$ *and* x^* *belongs to the boundary of* Ω.

Proof. In view of Theorem 2.4.3, the only fact we have to prove here is that, if $\liminf_{t\uparrow b}\|x(t)\| < +\infty$, then there exists $\lim_{t\uparrow b} x(t) = x^*$. So, let us assume that $\liminf_{t\uparrow b}\|x(t)\| < +\infty$. From Theorem 2.4.3, it follows that there exists at least one limit point x^* of x as $t \uparrow b$ and $x^* \in \partial\Omega$. This means that there exists an increasing sequence $(t_k)_{k\in\mathbb{N}}$ in (a, b) with

$$\lim_{k\to\infty} t_k = b$$

and

$$\lim_{k\to\infty} x(t_k) = x^*.$$

We will show that

$$\lim_{t\uparrow b} x(t) = x^*. \tag{2.4.3}$$

To this aim, let us assume for a contradiction that there exist $r > 0$ and an increasing sequence $(s_k)_{k\in\mathbb{N}}$ such that

$$\lim_{k\to\infty} s_k = b$$

and

$$\|x(s_k) - x^*\| > r \tag{2.4.4}$$

for each $k \in \mathbb{N}$. At this point, let us observe that we may assume with no loss of generality that

$$t_k < s_k$$

and

$$\|x(t_k) - x^*\| < r \tag{2.4.5}$$

for each $k \in \mathbb{N}$. Next, for each $k \in \mathbb{N}$, let us consider $\theta_k \in (t_k, s_k)$ satisfying

$$\|x(t) - x^*\| < r \tag{2.4.6}$$

for each $t \in [\, t_k, \theta_k)$ and

$$\|x(\theta_k) - x^*\| = r.$$

The existence of $\theta_k \in (t_k, s_k)$ is ensured by (2.4.4), (2.4.5) and Darboux intermediate value theorem applied to the function $t \mapsto \|x(t) - x^*\|$, which is continuous on the interval $[\, t_k, s_k\,]$. In other words, θ_k is the first point in (t_k, s_k) at which x crosses the boundary of $B(x^*, r)$.

Finally, since $[\, a, b) \times [B(x^*, r) \cap \Omega]$ is bounded and f maps bounded subsets in $\mathbb{I} \times \Omega$ into bounded subsets in $\mathbb{R}^n$, it follows that there exists $M > 0$ such that

$$\|f(t, u)\| \leq M$$

for each $(t, u) \in [a, b) \times [B(x^*, r) \cap \Omega]$. From (2.4.6), we deduce that

$$r = \|x(\theta_k) - x^*\| = \left\| x(t_k) + \int_{t_k}^{\theta_k} f(s, x(s))\, ds - x^* \right\|$$
$$\leq \|x(t_k) - x^*\| + M(\theta_k - t_k)$$

for each $k \in \mathbb{N}$. Since $\lim_{k\to\infty} [\|x(t_k) - x^*\| + M(\theta_k - t_k)] = 0$, taking a sufficiently large k in the preceding inequality, we get a contradiction. This contradiction can be eliminated only if (2.4.3) holds true, which completes the proof. □

Remark 2.4.3. If $\mathbb{I} = \mathbb{R}$ and $\Omega = \mathbb{R}^n$, then, each continuous function $f : \mathbb{I} \times \Omega \to \mathbb{R}^n$ maps bounded subsets of $\mathbb{I} \times \Omega$ into bounded subsets of $\mathbb{R}^n$. Indeed, if B is a bounded subset of $\mathbb{R} \times \mathbb{R}^n$, its closure $\bar{B}$ is included in $\mathbb{R} \times \mathbb{R}^n$ too. But $\bar{B}$ is compact, and f is continuous and thus $f(\bar{B})$ is bounded. Since $f(B) \subset f(\bar{B})$, $f(B)$ is bounded. On the other hand, if $\mathbb{I} \neq \mathbb{R}$, or $\Omega \neq \mathbb{R}^n$, then there are examples of continuous functions $f : \mathbb{I} \times \Omega \to \mathbb{R}^n$ which do not map bounded subsets of $\mathbb{I} \times \Omega$ into bounded subsets of $\mathbb{R}^n$.

2.4.4 *The Existence of Global Solutions*

We conclude this section with two sufficient conditions on f ensuring the existence of global solutions of $\mathcal{CP}(\mathcal{D})$.

Theorem 2.4.6. *Let $f : \mathbb{I} \times \mathbb{R}^n \to \mathbb{R}^n$ be continuous on $\mathbb{I} \times \mathbb{R}^n$, and let us assume that there exist two continuous functions $h, k : \mathbb{I} \to \mathbb{R}_+$ such that*

$$\|f(\tau, y)\| \leq k(\tau)\|y\| + h(\tau), \tag{2.4.7}$$

for each $(\tau, y) \in \mathbb{I} \times \mathbb{R}^n$. Then, for each $(a, \xi) \in \mathbb{I} \times \mathbb{R}^n$, $\mathcal{CP}(\mathcal{D})$ has at least one global solution.

Proof. In view of Corollary 2.4.1 it suffices to show that each saturated solution of $\mathcal{CP}(\mathcal{D})$ is global. To this aim, let $x : [a, b) \to \mathbb{R}^n$ be a saturated solution of $\mathcal{CP}(\mathcal{D})$.

Since x satisfies $(\mathcal{IE})$ in Proposition 2.1.1, from (2.4.7), we get

$$\|x(t)\| \leq \|\xi\| + \int_a^t h(\tau)d\tau + \int_a^t k(\tau)\|x(\tau)\|d\tau,$$

for each $t \in [a, b)$. We will show next that $b = \sup \mathbb{I}$. Indeed, if we assume the contrary, inasmuch as $[a, b]$ is compact and h, k are continuous on $[a, b] \subset \mathbb{I}$, there exists $M > 0$ such that

$$h(t) \leq M \quad \text{and} \quad k(t) \leq M,$$

for each $t \in [a, b]$.

The preceding inequalities along with Gronwall's Lemma 1.5.2 show that

$$\|x(t)\| \leq [\|\xi\| + M(b-a)]e^{M(b-a)},$$

for each $t \in (a, b)$. Hence x is bounded on $[a, b)$. Then $\{(t, x(t));\ t \in [a, b)\}$ is included in the compact set $\overline{\{(t, x(t));\ t \in [a, b)\}}$, which is a subset of $\mathbb{I} \times \Omega$. By Theorem 2.4.1, it follows that x is continuable, which is absurd. This contradiction can be eliminated only if $b = \sup \mathbb{I}$, and this completes the proof. □

A significant consequence of Theorem 2.4.6 refers to linear systems of first-order differential equations, whose thorough study is the subject of Chapter 4. Let $\mathbb{I}$ be a nontrivial interval in $\mathbb{R}$, and let $\mathcal{A} : \mathbb{I} \to \mathcal{M}_{m\times m}(\mathbb{R})$ and $\mathcal{B} : \mathbb{I} \to \mathcal{M}_{m\times p}(\mathbb{R})$ be two continuous *matrix-valued functions*, i.e. two matrices whose elements are continuous functions from $\mathbb{I}$ to $\mathbb{R}$. Let $a \in \mathbb{I}$, $\mathcal{X}_a \in \mathcal{M}_{m\times p}(\mathbb{R})$ and let us consider the Cauchy problem

$$\begin{cases} \mathcal{X}' = \mathcal{A}(t)\mathcal{X} + \mathcal{B}(t) \\ \mathcal{X}(a) = \mathcal{X}_a, \end{cases} \tag{2.4.8}$$

for the unknown function $\mathcal{X} : \mathbb{J} \to \mathcal{M}_{m\times p}(\mathbb{R})$.

Corollary 2.4.4. *If $\mathcal{A} : \mathbb{I} \to \mathcal{M}_{m\times m}(\mathbb{R})$ and $\mathcal{B} : \mathbb{I} \to \mathcal{M}_{m\times p}(\mathbb{R})$ are continuous, then for each $a \in \mathbb{I}$ and $\mathcal{X}_a \in \mathcal{M}_{m\times p}(\mathbb{R})$ the Cauchy problem* (2.4.8) *has a unique global solution.*

Proof. We recall that, as stated in Section 12.1, $\mathcal{M}_{m\times p}(\mathbb{R})$ is an $m \times p$-dimensional linear space over $\mathbb{R}$ and therefore it can be identified with $\mathbb{R}^{m\times p}$. Moreover, the norm $\|\cdot\|_{\mathcal{M}}$, defined on $\mathcal{M}_{m\times p}(\mathbb{R})$ by

$$\|\mathcal{X}\|_{\mathcal{M}} = \sup\{\|\mathcal{X}\xi\|_m;\ \xi \in \mathbb{R}^p,\ \|\xi\|_p \leq 1\},$$

is equivalent to the Euclidean norm. See Remark 12.1.1. Let us define the function $f : \mathbb{I} \times \mathbb{R}^{m\times p} \to \mathbb{R}^{m\times p}$ by

$$f(t, \mathcal{X}) = \mathcal{A}(t)\mathcal{X} + \mathcal{B}(t),$$

for each $(t, \mathcal{X}) \in \mathbb{I} \times \mathbb{R}^{m\times p}$. Clearly, f is continuous and

$$\|f(t, \mathcal{X})\|_{\mathcal{M}} \leq \|\mathcal{A}(t)\|_{\mathcal{M}}\|\mathcal{X}\|_{\mathcal{M}} + \|\mathcal{B}(t)\|_{\mathcal{M}},$$

for each $(t, \mathcal{X}) \in \mathbb{I}\times\mathbb{R}^{m\times p}$. Thus f satisfies the hypotheses of Theorem 2.4.6. Accordingly, for each $a \in \mathbb{I}$ and $\mathcal{X}_a \in \mathcal{M}_{m\times p}(\mathbb{R})$, the Cauchy problem (2.4.8) has at least one global solution.

Finally, since

$$\|f(t,\mathcal{X}) - f(t,\mathcal{Y})\|_{\mathcal{M}} \leq \|\mathcal{A}(t)\|_{\mathcal{M}}\|\mathcal{X} - \mathcal{Y}\|_{\mathcal{M}}$$

for each $(t,\mathcal{X}),(t,\mathcal{Y}) \in \mathbb{I} \times R^{m\times p}$, it follows that f is locally Lipschitz on $\mathbb{R}^{m\times p}$. Thus, in view of Theorem 2.3.1, (2.4.8) has the uniqueness property and this completes the proof. □

Theorem 2.4.7. *Let $f : \mathbb{I} \times \mathbb{R}^n \to \mathbb{R}^n$ be continuous on $\mathbb{I} \times \mathbb{R}^n$ and dissipative on $\mathbb{R}^n$. Then, for each $(a,\xi) \in \mathbb{I} \times \mathbb{R}^n$, $\mathcal{CP}(\mathcal{D})$ has a unique global solution.*

Proof. Let $(a,\xi) \in \mathbb{I} \times \mathbb{R}^n$. By Theorems 2.3.4 and 2.4.2, $\mathcal{CP}(\mathcal{D})$ has a unique saturated solution $x : [\,a,b) \to \mathbb{R}^n$. To complete the proof it suffices to show that $b = \sup\mathbb{I}$. To this aim, let us observe that $\mathcal{CP}(\mathcal{D})$ may be equivalently rewritten as

$$\begin{cases} x'(\tau) = f(\tau,x(\tau)) - f(\tau,0) + f(\tau,0), \\ x(a) = \xi, \end{cases}$$

for each $\tau \in [\,a,b)$. Taking the inner product with $x(\tau)$ in the above relation, recalling that $\langle x'(\tau),x(\tau)\rangle = \frac{1}{2}\frac{d}{d\tau}\|x(\tau)\|^2$, using the fact that f is dissipative, and integrating over $[\,a,t\,]$, we get

$$\frac{1}{2}\|x(t)\|^2 \leq \frac{1}{2}\|\xi\|^2 + \int_a^t \langle f(\tau,0),x(\tau)\rangle d\tau,$$

for each $t \in [\,a,b)$. Using the Cauchy-Schwarz Inequality, we obtain

$$\frac{1}{2}\|x(t)\|^2 \leq \frac{1}{2}\|\xi\|^2 + \int_a^t \|f(\tau,0)\| \cdot \|x(\tau)\| d\tau,$$

for each $t \in [\,a,b)$. Thus, Brezis Lemma 1.5.3 is applicable and leads to

$$\|x(t)\| \leq \|\xi\| + \int_a^t \|f(\tau,0)\| d\tau,$$

for each $t \in [\,a,b)$. Consequently, if $b < \sup\mathbb{I}$, then x is bounded on $[\,a,b)$. Thus the graph of x is included in a compact set of $\mathbb{I} \times \Omega$. By Theorem 2.4.1, x must be continuable which is absurd. This contradiction can be eliminated only if $b = \sup\mathbb{I}$, and this completes the proof. □

2.5 Continuous Dependence on Data and Parameters

Let us consider the Cauchy problem

$$\begin{cases} x'(t) = a(t)x(t) + b(t) \\ x(t_0) = \xi, \end{cases}$$

where $a, b : \mathbb{I} \to \mathbb{R}$ are continuous functions on $\mathbb{I}$, $t_0 \in \mathbb{I}$ and $\xi \in \mathbb{R}$. As we have already seen in Section 1.3, this problem has a unique global solution given by the so-called variation of constants formula

$$x(t, t_0, \xi) = \exp\left(\int_{t_0}^{t} a(s)\,ds\right)\xi + \int_{t_0}^{t} \exp\left(\int_{s}^{t} a(\tau)\,d\tau\right) b(s)\,ds.$$

See Theorem 1.3.2 and Remark 1.3.1. From the specific form of the solution, we may easily see that it is continuous with respect to the initial datum ξ, from $\mathbb{R}$ into $C(\mathbb{I};\mathbb{R})$, the latter being endowed with the uniform convergence topology on compact intervals in $\mathbb{I}$. This means that, for each $\xi \in \mathbb{R}$ and each $(\xi_m)_{m\in\mathbb{N}^*}$ with $\lim_m \xi_m = \xi$, we have $\lim_m x(t, t_0, \xi_m) = x(t, t_0, \xi)$ uniformly for t in each compact interval $\mathbb{J} \subseteq \mathbb{I}$. Indeed, if $\mathbb{J}$ is a compact interval in $\mathbb{I}$ and $\xi, \eta \in \mathbb{R}$, then

$$\sup_{t\in\mathbb{J}} |x(t, t_0, \xi) - x(t, t_0, \eta)| \le e^{\ell M}|\xi - \eta|,$$

where ℓ is the length of the interval $\mathbb{J}$, and $M = \sup_{\tau\in\mathbb{J}}\{|a(\tau)|\}$.

Starting from this simple observation, we intend to show that, under some natural hypotheses on the function f, the solution of the Cauchy problem $\mathcal{CP}(\mathcal{D})$ depends continuously not only on the data but also on the parameters in the problem (if any).

We begin with the study of the continuous dependence of the solution on the data and next, we will show how the continuous dependence on the parameters can be inferred from there. Let $\mathbb{I}$ be a nontrivial interval in $\mathbb{R}$, Ω a nonempty and open subset of $\mathbb{R}^n$ and let $f : \mathbb{I} \times \Omega \to \mathbb{R}^n$ be a continuous function. Let $a \in \mathbb{I}$ and $\xi \in \Omega$ and let us consider the Cauchy problem

$$\begin{cases} x' = f(t, x) \\ x(a) = \xi. \end{cases} \qquad \mathcal{CP}(\mathbb{I}, \Omega, f, a, \xi)$$

The next lemma will prove useful in what follows.

Lemma 2.5.1. *Let $\mathcal{K}$ be a compact subset of $\mathbb{R}^n$ and $\mathcal{F}$ a closed subset of $\mathbb{R}^n$ with $\mathcal{K} \cap \mathcal{F} = \emptyset$. Then* $\operatorname{dist}(\mathcal{K}, \mathcal{F}) > 0$[1].

[1] Here and in what follows, $\operatorname{dist}(\mathcal{K},\mathcal{F}) = \inf\{\|x - y\| : x \in \mathcal{K},\ y \in \mathcal{F}\}$ represents the distance between the two subsets $\mathcal{K}$ and $\mathcal{F}$.

Proof. Let us assume for a contradiction that there exist two subsets $\mathcal{K}$ and $\mathcal{F}$ in $\mathbb{R}^n$, the former compact and the latter closed, with $\mathcal{K} \cap \mathcal{F} = \emptyset$, but for which $\text{dist}(\mathcal{K}, \mathcal{F}) = 0$. From the characterization of the infimum of a set of reals by means of sequences, we deduce that there exist $(x_k)_{k\in\mathbb{N}} \subset \mathcal{K}$ and $(y_k)_{k\in\mathbb{N}} \subset \mathcal{F}$ such that

$$\lim_{k\to\infty} \|x_k - y_k\| = 0.$$

Since $\mathcal{K}$ is compact, by virtue of Cesàro Lemma, we know that there exists $x \in \mathcal{K}$ such that, at least on a subsequence, we have

$$\lim_{k\to\infty} x_k = x.$$

From the preceding equality it follows that $\lim_{k\to\infty} y_k = x$ too. But $y_k \in \mathcal{F}$ for each $k \in \mathbb{N}$ and inasmuch as $\mathcal{F}$ is closed, we conclude that $x \in \mathcal{F}$, relation which is in contradiction with the initial supposition that $\mathcal{F}$ and $\mathcal{K}$ are disjoint. This contradiction can be eliminated only if the distance between $\mathcal{K}$ and $\mathcal{F}$ is strictly positive. The proof is complete. □

Definition 2.5.1. A function $\varphi : \Omega \to \mathbb{R} \cup \{+\infty\}$ is called *lower semi-continuous at* $\xi \in \Omega$ (l.s.c.) if for each $\varepsilon > 0$ there exists $r > 0$ such that, for each $\eta \in B(\xi, r) \cap \Omega$ we have

$$\varphi(\xi) - \varepsilon \leq \varphi(\eta).$$

A function $\varphi : \Omega \to \mathbb{R} \cup \{+\infty\}$ is called *lower semi-continuous on* Ω if it is lower semi-continuous at any $\xi \in \Omega$.

Definition 2.5.2. A mapping between two metric spaces is *nonexpansive* if it is Lipschitz continuous with constant 1.

2.5.1 *The Dissipative Case*

The first main result of this section is stated below.

Theorem 2.5.1. *Let $f : \mathbb{I} \times \Omega \to \mathbb{R}^n$ be continuous on $\mathbb{I} \times \Omega$ and dissipative on Ω, let $a \in \mathbb{I}$ be fixed, $\xi \in \Omega$ and let $x(\cdot, \xi) : [a, b_\xi) \to \Omega$ be the unique saturated solution of $\mathcal{CP}(\mathbb{I}, \Omega, f, a, \xi)$. Then:*

(i) *the function $\xi \mapsto b_\xi$ is lower semi-continuous on Ω, i.e. for each $\xi \in \Omega$ and each $b \in (a, b_\xi)$, there exists $r > 0$ with the property that $B(\xi, r) \subset \Omega$ and for each $\eta \in B(\xi, r)$, $x(\cdot, \eta)$ is defined on $[a, b]$ at least, i.e. $b < b_\eta$;*

(ii) *for* $\xi \in \Omega$, *each* $b \in (a, b_\xi)$ *and* $r > 0$ *as in* (i), *the mapping* $\eta \mapsto x(\cdot, \eta)$ *is nonexpansive from* $B(\xi, r)$ *to* $C([\,a, b\,]; \mathbb{R}^n)$, *the latter being endowed with its usual supremum norm.*

Proof. Let $b \in (a, b_\xi)$ and let us define $\mathcal{K}_0 = \{x(t, \xi);\ t \in [\,a, b\,]\}$. Since x is continuous and $[\,a, b\,]$ is compact, $\mathcal{K}_0$ is compact too and is included in Ω. Inasmuch as Ω is open, it follows that $\mathcal{K}_0 \cap \partial\Omega = \emptyset$ and accordingly, by virtue of Lemma 2.5.1, the distance between $\mathcal{K}_0$ and $\partial\Omega$ is strictly positive. For this reason there exists $\rho > 0$ such that the set

$$\mathcal{K} = \{y \in \mathbb{R}^n;\ \text{dist}\,(y, \mathcal{K}_0) \le \rho\}$$

(which is obviously compact) is included in Ω. See Figure 2.5.1.

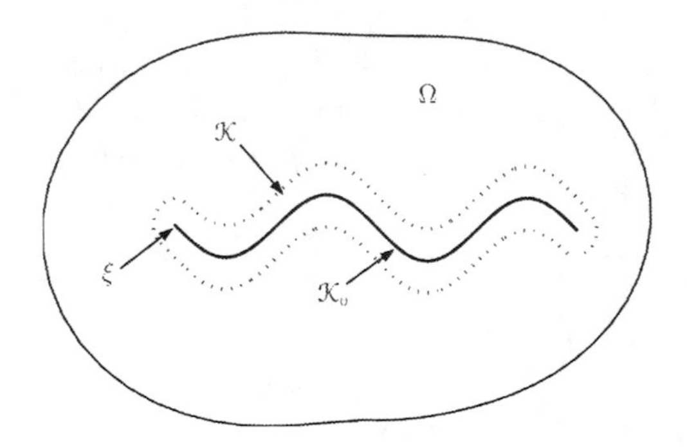

Fig. 2.5.1

Let $r \in (0, \rho)$. We will show first that for each $\eta \in B(\xi, r)$, $x(\cdot, \eta)$ is defined on $[\,a, b\,]$ at least. As for each $\eta \in B(\xi, r)$ we have

$$\text{dist}\,(\eta, \mathcal{K}_0) \le \text{dist}(\eta, \xi) = \|\eta - \xi\| \le r < \rho,$$

it follows that $\eta \in \mathcal{K} \subseteq \Omega$. Let $x(\cdot, \eta) : [\,a, b_\eta) \to \Omega$ be the unique saturated solution of the problem $\mathcal{CP}(\mathbb{I}, \Omega, f, a, \eta)$ whose existence and uniqueness are ensured by Theorems 2.3.4 and 2.4.2. We will show that $b < b_\eta$. To this aim, let us assume for contradiction that $b_\eta \le b$. Let us observe that, from the dissipativity condition and Lemma 2.3.1, we get

$$\|x(t, \xi) - x(t, \eta)\| \le \|\xi - \eta\| \tag{2.5.1}$$

for each $t \in [\,a, b_\eta)$. From (2.5.1), we deduce that

$$\|x(t, \xi) - x(t, \eta)\| \le r,$$

and thus $\operatorname{dist}(x(t,\eta),\mathcal{K}_0) \leq r$ for each $t \in [a, b_\eta)$. Clearly, this shows that $x(t,\eta) \in \mathcal{K}$ for each $t \in [a, b_\eta)$. Since $b_\eta \leq b < b_\xi$, it follows that $[a, b_\eta] \times \mathcal{K}$ is compact. In addition, $[a, b_\eta] \times \mathcal{K} \subseteq \mathbb{I} \times \Omega$, and therefore, by Theorem 2.4.1, it follows that $t \mapsto x(t,\eta)$ can be continued at the right of b_η, which is absurd. This contradiction can be eliminated only if $b < b_\eta$. Consequently, for each $\eta \in B(\xi, r)$, $x(\cdot,\eta)$ is defined on $[a,b]$ at least, which proves that (i) holds true.

Finally, from (2.5.1), we deduce that

$$\sup_{t\in[a,b]} \|x(t,\eta) - x(t,\mu)\| \leq \|\eta - \mu\|$$

for each $\eta, \mu \in B(\xi, r)$. So, the mapping $\eta \mapsto x(\cdot,\eta)$ is nonexpansive from $B(\xi,r)$ to $C([a,b];\mathbb{R}^n)$, the latter being endowed with the supremum norm, and this completes the proof of Theorem 2.5.1. □

2.5.2 *The Locally Lipschitz Case*

Unlike the case considered in Theorem 2.5.1, which is rather simple due to the fact that f satisfies a global condition with respect to the second variable, i.e. the dissipativity, the proof of Theorem 2.5.2 below, although similar to that of Theorem 2.5.1, is a little bit more complicated. The explanation of this fact rest in that the Lipschitz condition on Ω, which is assumed to hold only locally, deserves a more delicate analysis.

Theorem 2.5.2. *Let $f : \mathbb{I} \times \Omega \to \mathbb{R}^n$ be continuous on $\mathbb{I} \times \Omega$ and locally Lipschitz on Ω, let $a \in \mathbb{I}$ be fixed, $\xi \in \Omega$ and let $x(\cdot,\xi) : [a, b_\xi) \to \Omega$ be the unique saturated solution of $\mathcal{CP}(\mathbb{I},\Omega,f,a,\xi)$. Then:*

(i) *the function $\xi \mapsto b_\xi$ is lower semi-continuous on Ω, i.e. for each $\xi \in \Omega$ and each $b \in (a, b_\xi)$, there exists $r > 0$ with the property that $B(\xi,r) \subset \Omega$ and for each $\eta \in B(\xi,r)$, $x(\cdot,\eta)$ is defined on $[a,b]$ at least, i.e. $b < b_\eta$;*

(ii) *for each $\xi \in \Omega$, each $b \in (a, b_\xi)$ and $r > 0$ as in* (i), *the mapping $\eta \mapsto x(\cdot,\eta)$ is Lipschitz continuous from $B(\xi,r)$ to $C([a,b];\mathbb{R}^n)$, the latter being endowed with the supremum norm.*

Proof. Let $b \in (a, b_\xi)$ and let us define $\mathcal{K}_0 = \{x(t,\xi);\ t \in [a,b]\}$. Since x is continuous and $[a,b]$ is compact, $\mathcal{K}_0$ is compact too and is included in Ω. Then, by Lemma 2.5.1, it follows that there exists $\rho > 0$ such that the set

$$\mathcal{K} = \{y \in \mathbb{R}^n;\ \operatorname{dist}(y,\mathcal{K}_0) \leq \rho\} \tag{2.5.2}$$

(which is obviously compact) is included in Ω. In addition, as f is locally Lipschitz on Ω and $[a,b] \times \mathcal{K}$ is compact, there exists $L > 0$ such that

$$\|f(t,u) - f(t,v)\| \leq L\|u - v\|$$

for each $t \in [a,b]$ and each $u, v \in \mathcal{K}$.

Let $r \in (0, \rho e^{-L(b-a)})$, where $L > 0$ is the Lipschitz constant of f on $[a,b] \times \mathcal{K}$. We will show first that for each $\eta \in B(\xi, r)$, the unique saturated solution of the Cauchy problem $\mathcal{CP}(\mathbb{I}, \Omega, f, a, \eta)$, $x(\cdot, \eta) : [a, b_\eta) \to \Omega$, is defined on $[a,b]$ at least, i.e. that $b < b_\eta$. Let

$$c^* = \sup\{c;\ c \in (a, b_\eta),\ x(t,\eta) \in \mathcal{K},\ t \in [a,c]\}.$$

Since

$$\|x(a,\xi) - x(a,\eta)\| = \|\xi - \eta\| \leq r < \rho,$$

from the continuity of both $t \mapsto x(t,\xi)$ and $t \mapsto x(t,\eta)$, and the fact that $r < \rho$, it follows that c^* is well-defined and $c^* \in (a, b_\eta]$.

Next, since both $x(t,\xi) \in \mathcal{K}$ and $x(t,\eta) \in \mathcal{K}$ for each $t \in [a, c^*)$ – see the definition of c^* – we can use the Lipschitz condition to get

$$\|x(t,\xi) - x(t,\eta)\| \leq \|\xi - \eta\| + L \int_a^t \|x(s,\xi) - x(s,\eta)\|\, ds$$

for each $t \in [a, c^*) \cap [a,b]$. Thus, from Gronwall's Lemma 1.5.1, we obtain

$$\|x(t,\xi) - x(t,\eta)\| \leq e^{L(b-a)}\|\xi - \eta\|$$

for each $t \in [a, c^*) \cap [a,b]$.

Recalling that $r \in (0, \rho e^{-L(b-a)})$, and $\eta \in B(\xi, r)$, we get

$$\|x(t,\xi) - x(t,\eta)\| \leq e^{L(b-a)}\|\xi - \eta\| \leq e^{L(b-a)} r < \rho \tag{2.5.3}$$

for each $t \in [a, c^*) \cap [a,b]$.

The crucial point of the proof is to show that $c^* \geq b$, which would imply $b < b_\eta$. To this aim, let us assume for contradiction that $c^* < b$. Since $[a, c^*] \subseteq [a,b]$, $b < b_\xi \leq \sup I$ and the compact set $[a, c^*] \times \mathcal{K}$ is included in $\mathbb{I} \times \Omega$, it follows from Theorem 2.4.1 that $c^* < b_\eta$.

Next, from the continuity of both $t \mapsto x(t,\xi)$ and $t \mapsto x(t,\eta)$ at c^*, we deduce first that (2.5.3) holds for $t = c^*$ also, and then that there exists $d \in (c^*, b)$, sufficiently close to c^*, such that

$$\|x(t,\xi) - x(t,\eta)\| < \rho$$

for each $t \in [a,d]$. But $x(t,\xi) \in \mathcal{K}_0$ for each $t \in [a,b]$, and therefore $x(t,\eta) \in \mathcal{K}$ for each $t \in [a,d]$, relation which contradicts the definition of

c^*. This contradiction can be eliminated only if $c^* \geq b$ and thus $b < b_\eta$. As a consequence, for each $\eta \in B(\xi, r)$, $x(\cdot, \eta)$ is defined on $[a, b]$ at least, which proves (i).

In order to prove (ii), let us observe that, by (2.5.3), we have that

$$\sup_{t \in [a,b]} \|x(t, \eta) - x(t, \mu)\| \leq e^{L(b-a)} \|\eta - \mu\|$$

for each $\eta, \mu \in B(\xi, r)$. Hence the mapping $\eta \mapsto x(\cdot, \eta)$ is Lipschitz continuous, with Lipschitz constant $e^{L(b-a)}$, from $B(\xi, r)$ to $C([a, b]; \mathbb{R}^n)$, the latter being endowed with the supremum norm, and this completes the proof. □

2.5.3 *Continuous Dependence on Parameters*

We may now go on with the study of the continuous dependence on the parameters. To this aim, let $\mathbb{I}$ be a nonempty and open interval in $\mathbb{R}$, Ω a nonempty and open subset in $\mathbb{R}^n$, and $\mathbb{P}$ a nonempty and open subset of $\mathbb{R}^m$, the latter being called *the set of parameters*. If $u \in \Omega$ and $p \in \mathbb{P}$, we denote by

$$(u, p) = (u_1, u_2, \dots u_n, p_1, p_2, \dots, p_m).$$

Let $f : \mathbb{I} \times \Omega \times \mathbb{P} \to \mathbb{R}^n$ be a function which is continuous on $\mathbb{I} \times \Omega \times \mathbb{P}$ and locally Lipschitz on $\Omega \times \mathbb{P}$, i.e. for each compact subset $\mathcal{C} \subset \mathbb{I} \times \Omega \times \mathbb{P}$ there exists $L = L(\mathcal{C}) > 0$, such that for each $(t, u, p), (t, v, q) \in \mathcal{C}$, we have

$$\|f(t, u, p) - f(t, v, q)\|_n \leq L\|(u, p) - (v, q)\|_{n+m},$$

where, for $k \in \mathbb{N}^*$, $\|\cdot\|_k$ stands for the Euclidean norm of $\mathbb{R}^k$.

Let $a \in \mathbb{I}$, $\xi \in \Omega$ and $p \in \mathbb{P}$ and let us consider the Cauchy problem

$$\begin{cases} x' = f(t, x, p) \\ x(a) = \xi. \end{cases} \qquad \mathcal{CP}(\mathbb{I}, \Omega, f, a, \xi)_p$$

The third main result of this section is:

Theorem 2.5.3. *Let $f : \mathbb{I} \times \Omega \times \mathbb{P} \to \mathbb{R}^n$ be continuous on $\mathbb{I} \times \Omega \times \mathbb{P}$, locally Lipschitz on $\Omega \times \mathbb{P}$, let $(a, \xi, p) \in \mathbb{I} \times \Omega \times \mathbb{P}$ and let $x(\cdot, \xi, p) : [a, b_{\xi,p}) \to \mathbb{R}^n$ be the unique saturated solution of $\mathcal{CP}(\mathbb{I}, \Omega, f, a, \xi)_p$. Then, for each $b \in (a, b_{\xi,p})$ there exists $r > 0$, which depends on ξ, p, b, such that $B((\xi, p), r) \subset \Omega \times \mathbb{P}$, and for each $(\eta, q) \in B((\xi, p), r)$, the unique saturated solution $x(\cdot, \eta, q)$ of $\mathcal{CP}(\mathbb{I}, \Omega, f, a, \eta)_q$ is defined at least on $[a, b]$, i.e. $b < b_{\eta,q}$. Moreover, the mapping $(\eta, q) \mapsto x(\cdot, \eta, q)$ is Lipschitz continuous from $B((\xi, p), r)$ to $C([a, b]; \mathbb{R}^n)$, the latter being endowed with the usual supremum norm.*

Proof. For each $x \in \Omega$ and $p \in \mathbb{P}$, we denote by

$$z = (z_1, z_2, \ldots, z_{n+m}) = (x, p) = (x_1, x_2, \ldots, x_n, p_1, p_2, \ldots, p_m),$$

and we define $F : \mathbb{I} \times \Omega \times \mathbb{P} \to \mathbb{R}^{n+m}$ by

$$F(t, z) = (f_1(t, z), f_2(t, z), \ldots, f_n(t, z), 0, 0, \ldots, 0).$$

Then $\mathcal{CP}(\mathbb{I}, \Omega, f, a, \xi)_p$ may be rewritten as

$$\begin{cases} z' = F(t, z) \\ z(a) = (\xi, p). \end{cases} \qquad \mathcal{CP}(\mathbb{I}, \Omega \times \mathbb{P}, F, a, (\xi, p))$$

Hence, the continuous dependence of x on p reduces to the continuous dependence of z on (ξ, p), and thus Theorem 2.5.2 applies. □

2.6 Problems of Differentiability

As seen in Section 2.5, if $f : \mathbb{I} \times \Omega \times \mathbb{P} \to \mathbb{R}^n$ is continuous on $\mathbb{I} \times \Omega \times \mathbb{P}$ and locally Lipschitz on $\Omega \times \mathbb{P}$, the unique saturated solution $x(\cdot, \xi, p)$ of $\mathcal{CP}(\mathbb{I}, \Omega, f, a, \xi)_p$ depends continuously on $(\xi, p) \in \Omega \times \mathbb{P}$. Our aim here is to show that, whenever f is differentiable on $\Omega \times \mathbb{P}$, the saturated solution $x(\cdot, \xi, p)$ is differentiable too as a function of $(\xi, p) \in \Omega \times \mathbb{P}$.

In order to motivate what we are going to do next, let us consider the Cauchy problem governed by the first-order scalar differential equation

$$\begin{cases} x' = f(t, x) \\ x(a) = \xi \end{cases}$$

and let us assume that $f : \mathbb{I} \times \mathbb{R} \to \mathbb{R}$ is of class C^1. Let us assume also that, for each $\xi \in \mathbb{R}$, the Cauchy problem above has a unique saturated solution $x(\cdot, \xi)$ defined on $[\,a, a + \delta\,]$ at least, where $\delta > 0$ is independent of $\xi \in \mathbb{R}$. Finally, let us assume that, for each fixed $t \in \mathbb{R}$, the mapping $\xi \mapsto x(t, \xi)$ is differentiable on $\mathbb{R}$. Then, differentiating both sides in

$$x(t, \xi) = \xi + \int_a^t f(s, x(s, \xi))\, ds,$$

with respect to ξ, we deduce

$$x_\xi(t, \xi) = 1 + \int_a^t f_x(s, x(s, \xi)) x_\xi(s, \xi)\, ds,$$

where x_ξ, f_x represent the partial derivative of the function x with respect to ξ, and respectively of the function f with respect to x. So, the function

$y : [a, a+\delta] \to \mathbb{R}$, defined by $y(t) = x_\xi(t,\xi)$ for each $t \in [a, a+\delta]$, is a solution of the linear Cauchy problem

$$\begin{cases} y' = f_x(t, x(t,\xi))y \\ y(a) = 1. \end{cases}$$

We shall see in what follows that a completely analogous result holds in the general case too. Of course, in this case, f_x has to be substituted by the Jacobian matrix $(\partial f_i / \partial x_j)_{n \times n}$ and the constant 1 by the $n \times n$-unit matrix $\mathcal{I}_n$. We emphasize however, that the true difficulty of the problem consists in proving that $\xi \mapsto x(t,\xi)$ is differentiable and not that it satisfies a linear Cauchy problem similar to the scalar one.

2.6.1 *Differentiability with Respect to the Data*

As in the preceding section, we analyze first the simpler case in which f does not depend on any parameters, and then we will show how the result in the general case can be obtained as a consequence. So, let $\mathbb{I}$ be a nontrivial interval in $\mathbb{R}$, Ω a nonempty and open subset of $\mathbb{R}^n$, $f : \mathbb{I} \times \Omega \to \mathbb{R}^n$ a continuous function, $a \in \mathbb{I}$ and $\xi \in \Omega$. Let us consider the Cauchy problem

$$\begin{cases} x' = f(t,x) \\ x(a) = \xi. \end{cases} \qquad \mathcal{CP}(\mathbb{I}, \Omega, f, a, \xi)$$

If for each $t \in \mathbb{I}$, $f(t,\cdot)$ is a function of class C^1 on Ω, we denote by $f_x(t,x)$ its derivative at $x \in \Omega$, i.e. the $n \times n$-matrix whose generic element on the ith row and j^{th} column is $\frac{\partial f_i}{\partial x_j}(t,x)$, $i,j = 1,2,\ldots,n$.

Remark 2.6.1. If $f : \mathbb{I} \times \Omega \to \mathbb{R}^n$ is continuous and for each $t \in \mathbb{I}$, $f(t,\cdot)$ is of class C^1 on Ω and f_x is continuous on $\mathbb{I} \times \Omega$ then f is locally Lipschitz on Ω. In this case, for each $(a,\xi) \in \mathbb{I} \times \Omega$, $\mathcal{CP}(\mathbb{I}, \Omega, f, a, \xi)$ has a unique saturated solution $x(\cdot,\xi) : [a, b_\xi) \to \Omega$. See Remark 2.3.2, Theorems 2.3.2 and 2.4.2.

Remark 2.6.2. Let us assume that $f : \mathbb{I} \times \Omega \to \mathbb{R}^n$ is continuous and for each $t \in \mathbb{I}$, $f(t,\cdot) : \Omega \to \mathbb{R}^n$ is of class C^1. Let us assume further that f_x is continuous on $\mathbb{I} \times \Omega$. Then, for each $\tau \in \mathbb{I}$ and $x, y \in \Omega$ for which the line segment

$$[y,x] = \{\theta x + (1-\theta)y;\ \theta \in [0,1]\}$$

is contained in Ω, we have

$$f(\tau,x) - f(\tau,y) = \int_0^1 f_x(\tau, \theta x + (1-\theta)y)(x-y)\,d\theta. \qquad (2.6.1)$$

This is a consequence of the fact that $f(\tau, \cdot)$ is a function of class C^1 on Ω and of the obvious equality

$$\frac{d}{d\theta}\left(f(\tau, \theta x + (1-\theta)y)\right) = f_x(\tau, \theta x + (1-\theta)y)(x-y)$$

for each $\theta \in [0,1]$.

We may now state the first fundamental differentiability result.

Theorem 2.6.1. *Let $f : \mathbb{I} \times \Omega \to \mathbb{R}^n$ be continuous and such that, for each $t \in \mathbb{I}$, $f(t,\cdot) : \Omega \to \mathbb{R}^n$ is of class C^1 and $f_x : \mathbb{I} \times \Omega \to \mathcal{M}_{n\times n}(\mathbb{R})$ is continuous on $\mathbb{I} \times \Omega$. Let $(a,\xi) \in \mathbb{I} \times \Omega$ and let $x(\cdot,\xi) : [a, b_\xi) \to \Omega$ be the unique saturated solution of $\mathcal{CP}(\mathbb{I}, \Omega, f, a, \xi)$. Then, for each $b \in (a, b_\xi)$, there exists $r > 0$ (depending on ξ and on b) such that $B(\xi, r) \subset \Omega$ and, for each $\eta \in B(\xi, r)$, $x(\cdot, \eta)$ is defined on $[a, b]$ at least, i.e. $b < b_\eta$. In addition, for each $t \in [a,b]$, the mapping $\eta \mapsto x(t,\eta)$ is differentiable on $B(\xi,r)$ and, for each $\eta \in B(\xi, r)$, its derivative $x_\eta(t,\eta)$ satisfies $x_\eta(\cdot,\eta) : [a,b] \to \mathcal{M}_{n\times n}(\mathbb{R})$ and $t \mapsto x_\eta(t,\eta)$ is the unique solution of the Cauchy problem*

$$\begin{cases} \mathcal{X}' = f_x(t, x(t,\eta))\mathcal{X} \\ \mathcal{X}(a) = \mathcal{I}_n, \end{cases} \tag{2.6.2}$$

where $\mathcal{I}_n = \text{diag}(1,1,\ldots,1)$.

Proof. Since, by virtue of Remark 2.6.1, f satisfies the hypotheses of Theorem 2.5.2, it follows that, for each $b \in (a, b_\xi)$, there exist $r > 0$ and a compact subset $\mathcal{K}$ in Ω such that $B(\xi, r) \subseteq \Omega$ and for each $\eta \in B(\xi, r)$, $x(\cdot,\eta)$ is defined on $[a,b]$ at least and, for each $t \in [a,b]$, $x(t,\eta) \in \mathcal{K}$. From the definition of $\mathcal{K}$ – see (2.5.2) – and the validity of (2.5.3) for each $t \in [a,b]$ (since, as we have proved, $b \leq c^*$), we deduce that, for each $\mu, \eta \in B(\xi, r)$ and each $\tau \in [a,b]$, we have $x(\tau,\mu), x(\tau,\eta) \in B(x(\tau,\xi),\rho)$. Since $B(x(\tau,\xi),\rho) \subseteq \mathcal{K}$, it follows that

$$\theta x(\tau,\mu) + (1-\theta)x(\tau,\eta) \in \mathcal{K} \tag{2.6.3}$$

for each $\theta \in [0,1]$. See Figure 2.6.1.

Fix $\eta \in B(\xi, r)$ and let us observe that $f_x(\cdot, x(\cdot,\eta)) : [a,b] \to \mathcal{M}_{n\times n}(\mathbb{R})$ is continuous. So, the hypotheses of Corollary 2.4.4 hold, from where it follows that (2.6.2) has a unique solution $\mathcal{X} : [a,b] \to \mathcal{M}_{n\times n}(\mathbb{R})$.

In order to show that, for each $t \in [a,b]$, $x(t,\cdot)$ is differentiable at $\eta \in B(\xi,r)$ and its derivative $x_\eta(t,\eta)$ coincides with $\mathcal{X}(t)$, it suffices to prove that, for each $t \in [a,b]$ and $\eta \in B(\xi,r)$, we have

$$\lim_{\mu\to\eta} \frac{1}{\|\mu-\eta\|}\|x(t,\mu) - x(t,\eta) - \mathcal{X}(t)(\mu-\eta)\| = 0. \tag{2.6.4}$$

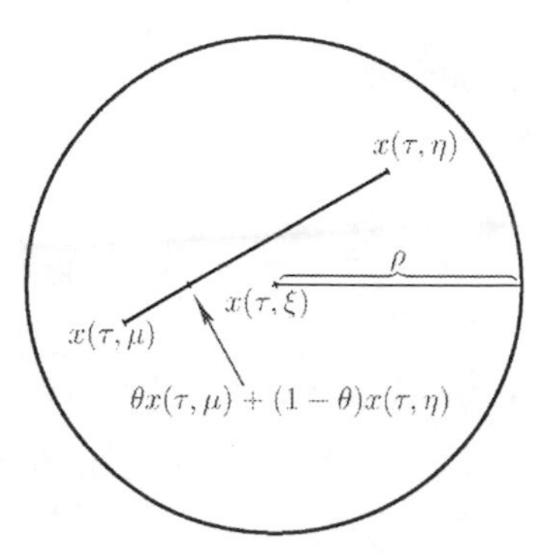

Fig. 2.6.1

Let $t \in [a,b]$, $\mu, \eta \in B(\xi, r)$. By Proposition 2.1.1, it follows that

$$x(t,\mu) - x(t,\eta) = \mu - \eta + \int_a^t [f(\tau, x(\tau,\mu)) - f(\tau, x(\tau,\eta))]d\tau$$

and

$$\mathcal{X}(t)(\mu - \eta) = \mathcal{I}_n(\mu - \eta) + \int_a^t f_x(\tau, x(\tau,\eta))\mathcal{X}(\tau)(\mu - \eta)\, d\tau.$$

Then, for each $t \in [a,b]$ and $\mu, \eta \in B(\xi, r)$, we have

$$x(t,\mu) - x(t,\eta) - \mathcal{X}(t)(\mu - \eta)$$

$$= \int_a^t [f(\tau, x(\tau,\mu)) - f(\tau, x(\tau,\eta))]d\tau - \int_a^t f_x(\tau, x(\tau,\eta))\mathcal{X}(\tau)(\mu - \eta)\, d\tau.$$

By virtue of (2.6.1) and (2.6.3), we successively obtain

$$x(t,\mu) - x(t,\eta) - \mathcal{X}(t)(\mu - \eta)$$

$$= \int_a^t \int_0^1 f_x(\tau, \theta x(\tau,\mu) + (1-\theta)x(\tau,\eta))[x(\tau,\mu) - x(\tau,\eta)]d\theta\, d\tau$$

$$- \int_a^t f_x(\tau, x(\tau,\eta))\mathcal{X}(\tau)(\mu - \eta)\, d\tau$$

and

$$x(t,\mu) - x(t,\eta) - \mathcal{X}(t)(\mu - \eta)$$

$$= \int_a^t \int_0^1 f_x(\tau, \theta x(\tau,\mu) + (1-\theta)x(\tau,\eta))[x(\tau,\mu) - x(\tau,\eta) - \mathcal{X}(\tau)(\mu - \eta)]d\theta\, d\tau$$

$$+\int_a^t \left(\int_0^1 f_x(\tau, \theta x(\tau,\mu) + (1-\theta)x(\tau,\eta))d\theta - f_x(\tau, x(\tau,\eta)) \right) \mathcal{X}(\tau)(\mu-\eta)\, d\tau \tag{2.6.5}$$

for each $t \in [\,a, b\,]$ and $\mu, \eta \in B(\xi, r)$.

On the other hand, since f_x and $\mathcal{X}$ are continuous, there exists $M > 0$ such that, for each $t \in [\,a, b\,]$ and $y \in \mathcal{K}$, we have

$$\|f_x(t, y)\|_{\mathcal{M}} \leq M \textit{ and } \|\mathcal{X}(t)\|_{\mathcal{M}} \leq M, \tag{2.6.6}$$

where $\|\cdot\|_{\mathcal{M}}$ is the norm on $\mathcal{M}_{n\times n}(\mathbb{R})$ defined in Section 12.1.

Next, let us define $g : B(\xi, r) \times B(\xi, r) \to \mathbb{R}_+$ by

$$g(\mu,\eta) = \int_a^b \left\| \int_0^1 f_x(\tau, \theta x(\tau,\mu) + (1-\theta)x(\tau,\eta))d\theta - f_x(\tau, x(\tau,\eta)) \right\|_{\mathcal{M}} d\tau,$$

for each $(\mu, \eta) \in B(\xi, r) \times B(\xi, r)$. Since f_x is continuous and, by virtue of Theorem 2.5.2,

$$\lim_{\mu\to\eta} x(\tau, \mu) = x(\tau, \eta)$$

uniformly for $\tau \in [\,a, b\,]$, in view of Corollary 12.2.2, it follows that

$$\lim_{\mu\to\eta} g(\mu, \eta) = 0, \tag{2.6.7}$$

for each $\eta \in B(\xi, r)$.

From (2.6.5) and (2.6.6), we obtain

$$\|x(t,\mu) - x(t,\eta) - \mathcal{X}(t)(\mu - \eta)\| \leq \|\mu - \eta\| g(\mu,\eta) M$$

$$+M \int_a^t \|x(\tau,\mu) - x(\tau,\eta) - \mathcal{X}(\tau)(\mu-\eta)\|\, d\tau,$$

for each $t \in [\,a, b\,]$ and $\mu, \eta \in B(\xi, r)$. Then, from Gronwall's Lemma 1.5.2, we deduce

$$\|x(t,\mu) - x(t,\eta) - \mathcal{X}(t)(\mu - \eta)\| \leq \|\mu - \eta\| g(\mu,\eta) M e^{M(b-a)},$$

for each $\mu, \eta \in B(\xi, r)$ and $t \in [\,a, b\,]$. As this inequality together with (2.6.7) implies (2.6.4), the proof is complete. □

2.6.2 *Differentiability with Respect to the Parameters*

We may now proceed to the study of the differentiability of the solution with respect to the parameters. To this aim, as in Section 2.5, let us consider a nontrivial interval $\mathbb{I}$ in $\mathbb{R}$, a nonempty and open subset Ω in $\mathbb{R}^n$ and a nonempty and open set of parameters $\mathbb{P}$ in $\mathbb{R}^m$. Let $f : \mathbb{I} \times \Omega \times \mathbb{P} \to \mathbb{R}^n$

be a continuous function, let $a \in \mathbb{I}$, $\xi \in \Omega$, $p \in \mathbb{P}$, and let us consider the Cauchy problem

$$\begin{cases} x' = f(t, x, p) \\ x(a) = \xi. \end{cases} \qquad \mathcal{CP}(\mathbb{I}, \Omega, f, a, \xi)_p$$

If, for each $t \in \mathbb{I}$, $f(t, \cdot, \cdot) : \Omega \times \mathbb{P} \to \mathbb{R}^n$ is of class C^1, we denote by $f_x(t, x, p)$ the derivative of the partial function $f(t, \cdot, p)$ calculated at $x \in \Omega$, i.e. the $n \times n$ matrix whose generic element on the ith row and j^{th} column is $\frac{\partial f_i}{\partial x_j}(t, x, p)$, $i, j = 1, 2, \ldots, n$, and by $f_p(t, x, p)$ the derivative of the partial function $f(t, x, \cdot)$ calculated at $p \in \mathbb{P}$, i.e. the $n \times m$ matrix whose generic element on the ith row and j^{th} column is $\frac{\partial f_i}{\partial p_j}(t, x, p)$, $i = 1, 2, \ldots, n$, $j = 1, 2, \ldots, m$.

The second fundamental differentiability result is stated below.

Theorem 2.6.2. *Let $f : \mathbb{I} \times \Omega \times \mathbb{P} \to \mathbb{R}^n$ be continuous and let us assume that, for each $t \in \mathbb{I}$, $f(t, \cdot, \cdot) : \Omega \times \mathbb{P} \to \mathbb{R}^n$ is of class C^1 and both functions $f_x(\cdot, \cdot, \cdot) : \mathbb{I} \times \Omega \times \mathbb{P} \to \mathcal{M}_{n \times n}(\mathbb{R})$ and $f_p(\cdot, \cdot, \cdot) : \mathbb{I} \times \Omega \times \mathbb{P} \to \mathcal{M}_{n \times m}(\mathbb{R})$ are continuous on $\mathbb{I} \times \Omega \times \mathbb{P}$. Let $(a, \xi, p) \in \mathbb{I} \times \Omega \times \mathbb{P}$ and let $x(\cdot, \xi, p) : [a, b_{\xi,p}) \to \Omega$ be the unique saturated solution of $\mathcal{CP}(\mathbb{I}, \Omega, f, a, \xi)_p$. Then, for each $b \in (a, b_{\xi,p})$, there exists $r > 0$ (depending on ξ, p, b) such that $B(p, r) \subset \mathbb{P}$ and for each $q \in B(p, r)$, $x(\cdot, \xi, q)$ is defined on $[a, b]$ at least, i.e. $b < b_{\xi,q}$. In addition, for each $t \in \mathbb{I}$, the mapping $q \mapsto x(t, \xi, q)$ is differentiable on $B(p, r)$, and for each $q \in B(p, r)$, its derivative $x_q(t, \xi, q)$ satisfies $x_q(\cdot, \xi, q) : [a, b] \to \mathcal{M}_{n \times m}(\mathbb{R})$ and $t \mapsto x_q(t, \xi, q)$ is the unique solution of the Cauchy problem*

$$\begin{cases} \mathcal{Y}' = f_x(t, x(t, \xi, q), q)\mathcal{Y} + f_q(t, x(t, \xi, q), q) \\ \mathcal{Y}(a) = \mathcal{O}_{n \times m} \end{cases} \tag{2.6.8}$$

where $\mathcal{O}_{n \times m}$ is the $n \times m$ null matrix.

The linear differential system (2.6.8) is known under the name of *system in variations* associated to the problem $\mathcal{CP}(\mathcal{D})_p$.

Proof. Let $x \in \Omega$, $p \in \mathbb{P}$ and let us denote by $z = (x, p)$, i.e. the vector in $\mathbb{R}^{n+m}$ whose first n components are the components of x, and whose last m components are those of p. Let us define the function $F : \mathbb{I} \times \Omega \times \mathbb{P} \to \mathbb{R}^{n+m}$ by

$$F(t, z) = (f_1(t, z), \ldots, f_n(t, z), 0, \ldots, 0)$$

for each $(t, z) \in \mathbb{I} \times \Omega \times \mathbb{P}$, i.e. the vector in $\mathbb{R}^{n+m}$ whose first n components are those of $f(t, z)$ and whose last m components are zero. Let us denote by $\zeta = (\xi, p)$ and let us observe that $\mathcal{CP}(\mathbb{I}, \Omega, f, a, \xi)_p$ may be rewritten as

$$\begin{cases} z' = F(t, z) \\ z(a) = \zeta. \end{cases} \qquad \mathcal{CP}(\mathbb{I}, \Omega \times \mathbb{P}, F, a, \zeta)$$

Then, we may apply Theorem 2.6.1 from where, it follows that, for each $b \in (a, b_{\xi,p})$, there exists $r = r(\zeta, b) > 0$, such that $B(\zeta, r) \subset \Omega \times \mathbb{P}$ and for each $\varsigma \in B(\zeta, r)$, $z(\cdot, \varsigma)$ is defined on $[\,a, b\,]$ at least. In addition, for each $t \in \mathbb{I}$, the partial function $\varsigma \mapsto z(t, \varsigma)$ is differentiable on $B(\zeta, r)$ and its derivative z_ς is the unique solution of the Cauchy problem

$$\begin{cases} \mathcal{Z}' = F_z(t, z(t, \varsigma))\mathcal{Z} \\ \mathcal{Z}(a) = \mathcal{I}_{n+m}. \end{cases} \tag{2.6.9}$$

Observing that

$$\mathcal{Z}(t) = z_\varsigma(t) = \begin{pmatrix} x_\eta(t, \eta, q) & x_q(t, \eta, q) \\ \mathcal{O}_{m\times n} & \mathcal{I}_{m\times m} \end{pmatrix}$$

and

$$F_z(t, z) = \begin{pmatrix} f_x(t, x, q) & f_q(t, x, q) \\ \mathcal{O}_{m\times n} & \mathcal{O}_{m\times m} \end{pmatrix}$$

and identifying the corresponding block matrices, from (2.6.9), we obtain (2.6.8), and this completes the proof. □

Remark 2.6.3. In practice, the parameters $p \in \mathbb{P}$ in the system

$$x' = f(t, x, p) \tag{$\mathcal{S}$}$$

may have various meanings. They may be, either the mathematical expression of an external command by means of which we can modify x according to some *a priori* performances or criteria, or the mathematical expression of a random perturbation which may modify the evolution of the system, or even a set of coefficients which are specific to some physical object, *etc.* In all these cases, it is very important to know "how fast" does $x(t, \xi, p)$ vary when p varies in some neighborhood of a certain $p_0 \in \mathbb{P}$. Due to the differentiability of the mapping $p \mapsto x(t, \xi, p)$ in p_0, the "magnitude" of this variation may be estimated in terms of the "magnitude" of the norm of the derivative $x_p(t, \xi, p)$ at $p = p_0$. This explains why this matrix is called the *sensitivity matrix* of $(\mathcal{S})$ and its elements *sensitivity functions*, while, whenever the norm of the sensitivity matrix is "sufficiently small", the system $(\mathcal{S})$ is called *robust.*

Remark 2.6.4. Theorem 2.6.2 is the main tool in the so-called *small parameter method* used for the first time in Mechanics. This consists in approximating $x(t,\xi,p)$ by $x(t,\xi,p_0) + x_p(t,\xi,p_0)(p-p_0)$ for p sufficiently close to p_0.

In order to understand the spirit of this method, let us analyze the following example.

Example 2.6.1. Find the solution of the scalar Cauchy problem

$$\begin{cases} x' = x + p(x^2+t) \\ x(0) = 0, \end{cases} \tag{$\mathcal{CP}$)$_p$}$$

for p close enough to 0.

Since, for $p \neq 0$, the differential equation in $(\mathcal{CP})_p$ is nonlinear (in fact a Riccati equation) there exists very few chances to solve it explicitly. Hence, we will try to find the solution of the problem $(\mathcal{CP})_p$ approximately.

Take $\mathbb{I} = \mathbb{R}$, $\Omega = \mathbb{R}$, $\mathbb{P} = \mathbb{R}$ and $f : \mathbb{R}\times\mathbb{R}\times\mathbb{R} \to \mathbb{R}$, defined by $f(t,x,p) = x + p(x^2+t)$ for each $(t,x,p) \in \mathbb{R}\times\mathbb{R}\times\mathbb{R}$. For $p=0$, $(\mathcal{CP})_p$ reduces to

$$\begin{cases} x'(t) = x(t) \\ x(0) = 0, \end{cases} \tag{$\mathcal{CP}$)$_0$}$$

whose unique saturated solution $x(\cdot,0,0) : [0,\infty) \to \mathbb{R}$ is $x(t,0,0) = 0$ for each $t \in \mathbb{R}_+$.

Let us observe that

$$\frac{\partial f}{\partial x}(t, x(t,0,0), 0) = 1$$

and

$$\frac{\partial f}{\partial p}(t, x(t,0,0), 0) = t$$

for each $(t,x,p) \in \mathbb{R}\times\mathbb{R}\times\mathbb{R}$. Accordingly, the system in variations (2.6.8) associated to $\mathcal{CP}(\mathbb{I},\Omega,f,0,0)_p$ has the specific form

$$\begin{cases} y'(t) = y(t) + t \\ y(0) = 0. \end{cases}$$

The solution of this Cauchy problem is $y(t) = e^t - t - 1$ for each $t \in \mathbb{R}_+$. Let $T > 0$. Then, if p is "sufficiently close to" 0, by virtue of Remark 2.6.4, we may take

$$x(t,0,p) \cong (e^t - t - 1)p$$

for each $t \in [0,T]$.

For the sake of completeness we include here the following result of differentiability with respect to the “initial time” a. The reader interested in the proof of this result is referred to [Halanay (1972)], p. 100.

Theorem 2.6.3. *Let $f : \mathbb{I} \times \Omega \to \mathbb{R}^n$ be of class C^1, let $(a, \xi) \in \mathbb{I} \times \Omega$ and let $x(\cdot, a) : (c_a, b_a) \to \Omega$ be the unique saturated bilateral solution of the $\mathcal{CP}(\mathbb{I}, \Omega, f, a, \xi)$. Then, for each $c \in (c_a, a)$ and $b \in (a, b_a)$, there exists $\delta = \delta(a, b, c) > 0$, such that, for each $\tilde{a} \in (a - \delta, a + \delta)$, the unique saturated bilateral solution $x(\cdot, \tilde{a})$ of $\mathcal{CP}(\mathbb{I}, \Omega, f, \tilde{a}, \xi)$ is defined at least on $[\,c, b\,]$. In addition, for each $t \in [\,c, b\,]$, the partial function $\tilde{a} \mapsto x(t, \tilde{a})$ is differentiable on $(a - \delta, a + \delta)$, and its derivative at $\tilde{a}$ satisfies $x_{\tilde{a}}(\cdot, \tilde{a}) : [\,c, b\,] \to \mathbb{R}^n$, and $t \mapsto x_{\tilde{a}}(\cdot, \tilde{a})$ is the solution of the linear Cauchy problem*

$$\begin{cases} \mathcal{Z}' = f_x(t, x(t, \tilde{a}))\mathcal{Z} \\ \mathcal{Z}(\tilde{a}) = -f(\tilde{a}, \xi). \end{cases} \tag{2.6.10}$$

Remark 2.6.5. Under the hypotheses of Theorem 2.6.3, let us denote by $x(\cdot, a, \xi) : (c_a, b_a) \to \Omega$ the unique saturated bilateral solution of $\mathcal{CP}(\mathbb{I}, \Omega, f, a, \xi)$. Then x satisfies

$$\frac{\partial x}{\partial a}(a, a, \xi) + \sum_{i=1}^{n} f_i(a, \xi) \frac{\partial x}{\partial \xi_i}(a, a, \xi) = 0. \tag{2.6.11}$$

Indeed, from (2.6.10) we have

$$\frac{\partial x}{\partial a}(a, a, \xi) = -f(a, \xi),$$

while, from (2.6.2) it follows

$$x_\xi(a, a, \xi) = \mathcal{I}_n.$$

Accordingly

$$f(a, \xi) = \mathcal{I}_n f(a, \xi) = x_\xi(a, a, \xi) f(a, \xi) = \sum_{i=1}^{n} f_i(a, \xi) \frac{\partial x}{\partial \xi_i}(a, a, \xi)$$

and thus (2.6.11) holds.

2.7 The Cauchy Problem for the nth-Order Differential Equation

The goal of this section is to present several results referring to the Cauchy problem for an nth-order scalar differential equation, which, as we have seen in Section 2.1, may be reduced to a Cauchy problem for a suitably defined first-order system of differential equations.

More precisely, let $\mathbb{I}$ be a nontrivial interval in $\mathbb{R}$, Ω a nonempty and open subset of $\mathbb{R}^n$, $g : \mathbb{I} \times \Omega \to \mathbb{R}$ a given function, $a \in \mathbb{I}$, $\xi = (\xi_1, \xi_2, \ldots, \xi_n) \in \Omega$ and let us consider the Cauchy problem for the nth-order scalar differential equation in the normal form with data $\mathcal{D}' = (\mathbb{I}, \Omega, g, a, \xi)$

$$\begin{cases} y^{(n)} = g(t, y, y', \ldots, y^{(n-1)}) \\ y(a) = \xi_1,\ y'(a) = \xi_2, \ldots, y^{(n-1)}(a) = \xi_n. \end{cases} \qquad \mathcal{CP}(\mathcal{D}')$$

We recall that, by means of the transformations,

$$\begin{cases} x = (x_1, x_2, \ldots, x_n) = (y, y', \ldots, y^{(n-1)}) \\ f(t, x) = (x_2, x_3, \ldots, x_n,\ g(t, x_1, x_2, \ldots, x_n)), \end{cases} \qquad (\mathcal{T})$$

$\mathcal{CP}(\mathcal{D}')$ can be reformulated as a Cauchy problem for the system of first-order differential equations

$$\begin{cases} x_1' = x_2 \\ x_2' = x_3 \\ \vdots \\ x_{n-1}' = x_n \\ x_n' = g(t, x_1, x_2, \ldots, x_n) \\ x_1(a) = \xi_1,\ x_2(a) = \xi_2, \ldots, x_n(a) = \xi_n. \end{cases}$$

In turn, this Cauchy problem can be reformulated as a Cauchy problem for a first-order vector differential equation in the normal form

$$\begin{cases} x' = f(t, x) \\ x(a) = \xi, \end{cases} \qquad \mathcal{CP}(\mathcal{D})$$

where $\mathcal{D} = (\mathbb{I}, \Omega, f, a, \xi)$, while f is defined as above.

We may now proceed to state the fundamental results referring to $\mathcal{CP}(\mathcal{D}')$. We begin with the following local existence theorem.

Theorem 2.7.1. (Peano). *If $g : \mathbb{I} \times \Omega \to \mathbb{R}$ is continuous on $\mathbb{I} \times \Omega$, then, for each $(a, \xi) \in \mathbb{I} \times \Omega$, $\mathcal{CP}(\mathbb{I}, \Omega, g, a, \xi)$ has at least one local solution.*

Proof. Clearly g is continuous if and only if f, defined by means of the transformations $(\mathcal{T})$, has the same property. In view of Theorem 2.2.1, for each $(a, \xi) \in \mathbb{I} \times \Omega$, there exists at least one local solution $x : [\,a, a+\delta\,] \to \Omega$ of $\mathcal{CP}(\mathbb{I}, \Omega, f, a, \xi)$. Taking into account $(\mathcal{T})$, it is easy to see that the function $y : [\,a, a+\delta\,] \to \mathbb{R}$, defined by $y(t) = x_1(t)$ for each $t \in [\,a, a+\delta\,]$, is a local solution of $\mathcal{CP}(\mathbb{I}, \Omega, g, a, \xi)$. The proof is complete. □

Definition 2.7.1. We say that $\mathcal{CP}(\mathcal{D}')$ has the *uniqueness property* if, for each $(a,\xi) \in \mathbb{I} \times \Omega$, any two solutions y and z of $\mathcal{CP}(\mathbb{I},\Omega,g,a,\xi)$ coincide on their common interval of definition.

As in the case of the Cauchy problem for first-order differential systems, the continuity of the right-hand side of $\mathcal{CP}(\mathcal{D}')$ alone ensures only the local existence of at least one solution, but not its uniqueness on the interval of existence. In order to get the uniqueness as well, we have to impose additional conditions on the function g. One of the most frequently used hypotheses is the locally Lipschitz condition.

Definition 2.7.2. A function $g : \mathbb{I} \times \Omega \to \mathbb{R}$ is called *locally Lipschitz* on Ω if for each compact subset $\mathcal{K}$ in $\mathbb{I} \times \Omega$, there exists $L = L(\mathcal{K}) > 0$ such that, for each $(t,u),(t,v) \in \mathcal{K}$ we have $|g(t,u) - g(t,v)| \leq L\|u - v\|$.

A sufficient condition for uniqueness is:

Theorem 2.7.2. *If $g : \mathbb{I} \times \Omega \to \mathbb{R}$ is locally Lipschitz on Ω, then $\mathcal{CP}(\mathcal{D}')$ has the uniqueness property.*

Proof. Let us observe that, if g is locally Lipschitz on Ω, then f defined by means of $(\mathcal{T})$ has the same property. Indeed, let $\mathcal{K}$ be a compact subset of $\mathbb{I} \times \Omega$ and let $L = L(\mathcal{K}) > 0$ as in Definition 2.7.2. Then

$$\|f(t,u) - f(t,v)\| = \left(\sum_{i=2}^{n} |u_i - v_i|^2 + |g(t,u) - g(t,v)|^2\right)^{1/2}$$

$$\leq \left(\sum_{i=1}^{n} |u_i - v_i|^2 + L^2\|u - v\|^2\right)^{1/2} = \sqrt{1 + L^2}\|u - v\|$$

for each $(t,u),(t,v) \in \mathcal{K}$ and thus f is locally Lipschitz in the sense of Definition 2.3.2. The conclusion follows from Theorem 2.3.1. $\square$

A simple but important consequence of Theorems 2.7.1 and 2.7.2 is:

Theorem 2.7.3. *If $g : \mathbb{I} \times \Omega \to \mathbb{R}$ is continuous on $\mathbb{I} \times \Omega$ and locally Lipschitz on Ω then, for each $(a,\xi) \in \mathbb{I} \times \Omega$, there exists $\delta > 0$ such that $[\,a, a+\delta\,] \subset \mathbb{I}$ and $\mathcal{CP}(\mathbb{I},\Omega,g,a,\xi)$ has a unique solution defined on $[\,a, a+\delta\,]$.*

2.8 The Implicit Function Theorem

The aim of this section is to give a proof of a deep result in Mathematical Analysis, The implicit function theorem, by using the local existence Theorem 2.2.1 of Peano. More precisely, let $D \subseteq \mathbb{R}^2$ be a nonempty and open set, let $F : D \to \mathbb{R}$ be a continuous function, and let $(a, \xi) \in D$.

Definition 2.8.1. We say that a continuous function $x : [\,a, b\,] \to \mathbb{R}$ is *implicitly defined* by

$$F(t, x) = F(a, \xi) \tag{2.8.1}$$

if $(t, x(t)) \in D$ for each $t \in [\,a, b\,]$ and

$$\begin{cases} x(a) = \xi \\ F(t, x(t)) = F(a, \xi) \ \text{ for } t \in [\,a, b\,]. \end{cases} \tag{2.8.2}$$

Our goal is to prove the following sufficient condition which ensures the existence of a unique function implicitly defined by (2.8.1).

Theorem 2.8.1. (The implicit function theorem) *Let $F : D \to \mathbb{R}$ be a C^1 function and let us assume that*

$$\frac{\partial F}{\partial x}(a, \xi) \neq 0.$$

Then, there exist $b > a$ and a unique C^1-function, $x : [\,a, b\,] \to \mathbb{R}$, implicitly defined by (2.8.1).

Proof. To begin, let us assume for the moment that such a C^1-function $x : [\,a, b\,] \to \mathbb{R}$ exists. Since $t \mapsto F(t, x(t))$ is constant on $[\,a, b\,]$ it follows that its derivative on $[\,a, b\,]$ is 0. This means that

$$0 = \frac{d}{dt}[F(t, x(t))] = \frac{\partial F}{\partial t}(t, x(t)) + \frac{\partial F}{\partial x}(t, x(t))x'(t),$$

for each $t \in [\,a, b\,]$. From this identity, by assuming that

$$\frac{\partial F}{\partial x}(t, x(t)) \neq 0 \tag{2.8.3}$$

for each $t \in [\,a, b\,]$, we conclude that x is a solution of the Cauchy problem

$$\begin{cases} x'(t) = -\dfrac{\dfrac{\partial F}{\partial t}(t, x(t))}{\dfrac{\partial F}{\partial x}(t, x(t))} \\ x(a) = \xi, \end{cases} \tag{2.8.4}$$

for each $t \in [a, b]$.

This remark suggests that the stated result may be proved by reversing the above argument. More precisely, we will show that, under the hypotheses of Theorem 2.8.1, the Cauchy problem (2.8.4), has at least one solution x, and then we will conclude that x is in fact implicitly defined by (2.8.1). So, thanks to the continuity of $\frac{\partial F}{\partial x}$ combined with the fact that $\frac{\partial F}{\partial x}(a, \xi) \neq 0$, we conclude that there exists an open neighborhood $\mathbb{I} \times \Omega \subseteq D$ of (a, ξ) such that

$$\frac{\partial F}{\partial x}(t, x) \neq 0,$$

for each $(t, x) \in \mathbb{I} \times \Omega$. Thus, we can define the function $f : \mathbb{I} \times \Omega \to \mathbb{R}$, by

$$f(t, x) = -\frac{\dfrac{\partial F}{\partial t}(t, x)}{\dfrac{\partial F}{\partial x}(t, x)},$$

for $(t, x) \in \mathbb{I} \times \Omega$. Clearly f is continuous on $\mathbb{I} \times \Omega$ and since $a \in \mathbb{I}$ and $\xi \in \Omega$, where $\mathbb{I}$ and Ω are open, by virtue of Theorem 2.2.1 of Peano, we conclude that there exists $b > a$ such that the Cauchy problem

$$\begin{cases} x'(t) = f(t, x(t)) \\ x(a) = \xi, \end{cases}$$

which is nothing but (2.8.4), has at least one solution $x : [a, b] \to \mathbb{R}$. From (2.8.4) we easily deduce that

$$\frac{d}{dt}[F(t, x(t))] = 0$$

for each $t \in [a, b]$, which means that $F(t, x(t)) = F(a, x(a)) = F(a, \xi)$. Thus x is a C^1 function which is implicitly defined by (2.8.1). To show the uniqueness, we begin by observing that, for each $t \in [a, b]$, the partial function $x \mapsto F(t, x)$ is strictly monotone on a sufficiently small neighborhood of ξ, and thus it is injective. Consequently, by choosing a smaller value b if necessary, we conclude that, for each $t \in [a, b]$, the equation $F(t, x) = F(a, \xi)$ has at most one solution. So, the implicitly defined function is unique and this completes the proof. $\square$

2.9 Exercises and Problems

Exercise 2.1. *Solve the following Cauchy problems*

$$(a)\begin{cases} tx' = x + x^2 \\ x(1) = 1. \end{cases} \qquad (b)\begin{cases} tx' = (1-t^2)x \\ x(2) = 1. \end{cases}$$

$$(c)\begin{cases} x' = (8t + 2x + 1)^2 \\ x(0) = -\frac{1}{2}. \end{cases} \qquad (d)\begin{cases} x'(t^2x - x) + tx^2 + t = 0 \\ x(0) = 2. \end{cases}$$

$$(e)\begin{cases} tx' = x - t \\ x(1) = 2. \end{cases} \qquad (f)\begin{cases} tx' = -(t + x) \\ x(1) = 0. \end{cases}$$

$$(g)\begin{cases} t^2x' = x(t - x) \\ x(1) = 1. \end{cases} \qquad (h)\begin{cases} 2txx' = 3x^2 - t^2 \\ x(1) = 2. \end{cases}$$

$$(i)\begin{cases} tx' = x + tx \\ x(1) = e. \end{cases} \qquad (j)\begin{cases} tx' = -2x + t^4 \\ x(1) = 2. \end{cases}$$

$$(k)\begin{cases} tx' = -x + e^t \\ x(1) = 0. \end{cases} \qquad (l)\begin{cases} tx' = -x - tx^2 \\ x(1) = -1. \end{cases}$$

$$(m)\begin{cases} 2txx' = x^2 - t \\ x(1) = 2. \end{cases} \qquad (n)\begin{cases} (2t - t^2x)x' = -x \\ x(1) = 1. \end{cases}$$

$$(o)\begin{cases} tx' = -2x(1 - tx) \\ x(1) = 1. \end{cases} \qquad (p)\begin{cases} (x^2 - 3t^2)x' = -2tx \\ x(0) = 1. \end{cases}$$

Problem 2.1. *Give another proof to Proposition 2.1.2 avoiding the use of the equivalence between* $\mathcal{CP}(\mathcal{D})$ *and* $(\mathcal{IE})$.

Problem 2.2. *Show that, under the hypotheses of Lemma 2.2.2, for each* $a \in \mathbb{I}$ *and* $\xi \in \mathbb{R}^n$, $\mathcal{CP}(D)$ *has at least one global solution.*

Problem 2.3. *Let* $f : \mathbb{R} \times \mathbb{R} \to \mathbb{R}$ *be defined by*

$$f(t,x) = \begin{cases} 0 & \text{if } t \in \mathbb{R} \text{ and } x = 0 \\ \dfrac{t}{x} & \text{if } t \in \mathbb{R} \text{ and } x \in \mathbb{R} \setminus \{0\}. \end{cases}$$

Prove that, for each $(a,\xi) \in \mathbb{R} \times \mathbb{R}$, $\mathcal{CP}(\mathbb{R}, \mathbb{R}, f, a, \xi)$ *has at least one right global solution but, nevertheless,* f *is not continuous on* $\mathbb{R} \times \mathbb{R}$. *Consequently, the continuity of the function* f *is not a necessary condition for the existence of solutions.*

Problem 2.4. *Let* $f : \mathbb{R} \times \mathbb{R} \to \mathbb{R}$ *be defined by*

$$f(t,x) = \begin{cases} -1 & \text{if } t \in \mathbb{R} \text{ and } x \geq 0 \\ 1 & \text{if } t \in \mathbb{R} \text{ and } x < 0. \end{cases}$$

As we have seen in Example 2.2.1, $\mathcal{CP}(\mathbb{R},\mathbb{R},f,0,0)$ *has no local right solution. Show that* $\mathcal{CP}(\mathbb{R},\mathbb{R},f,0,0)$ *has a unique left saturated solution and find it.*

Problem 2.5. *Let* $f : \mathbb{I}\times\Omega \to \mathbb{R}^n$ *be a function with the property that, for each* $(a,\xi)\in\mathbb{I}\times\Omega$ *there exist a neighborhood* $\mathcal{V}$ *of* (a,ξ), $\mathcal{V}\subset\mathbb{I}\times\Omega$ *and* $L = L(\mathcal{V}) > 0$ *such that for each* $(t,x),(t,y)\in\mathcal{V}$, *we have* $\|f(t,x)-f(t,y)\|\leq L\|x-y\|$. *Prove that* f *is locally Lipschitz on* Ω *in the sense of Definition* 2.3.2.

Problem 2.6. *Let* $f : \mathbb{I}\times\Omega \to \mathbb{R}^n$ *be a function which has partial derivatives with respect to the last n-arguments with the property that* $\partial f_i/\partial x_j$ *are continuous on* $\mathbb{I}\times\Omega$ *for each* $i,j = 1,2,\ldots,n$. *Prove that* f *is locally Lipschitz on* Ω.

Problem 2.7. *Let* $f,g : \mathbb{R}\times\mathbb{R}\to\mathbb{R}$, $f(t,x) = \sqrt[3]{(x-t)^2}+1$ *and* $g(t,x) = 2f(t,x)$ *for each* $(t,x)\in\mathbb{R}\times\mathbb{R}$. *Prove that for each* $(a,\xi)\in\mathbb{R}\times\mathbb{R}$, $\mathcal{CP}(\mathbb{R},\mathbb{R},g,a,\xi)$ *has the uniqueness property but, nevertheless, for each* $a\in\mathbb{R}$, $\mathcal{CP}(\mathbb{R},\mathbb{R},f,a,a)$ *has at least two solutions.* ([Halanay (1972)], pp. 79–80)

Problem 2.8. *Let* $f : \mathbb{R}\times\mathbb{R}\to\mathbb{R}$ *be a continuous function, let* $(a,\xi)\in\mathbb{R}\times\mathbb{R}$ *and let* $x,y : \mathbb{J}\to\mathbb{R}$ *be two solutions of* $\mathcal{CP}(\mathbb{R},\mathbb{R},f,a,\xi)$. *Show that both* $x\vee y$ *and* $x\wedge y$ *defined by* $(x\vee y)(t) = \max\{x(t),y(t)\}$, $(x\wedge y)(t) = \min\{x(t),y(t)\}$ *for each* $t\in\mathbb{J}$, *are solutions of* $\mathcal{CP}(\mathbb{R},\mathbb{R},f,a,\xi)$.

Problem 2.9. *Let* $f : \mathbb{R}\times\mathbb{R}\to\mathbb{R}$ *be a continuous function such that* $\mathcal{CP}(\mathcal{D})$ *has the uniqueness property. Let* $a\in\mathbb{R}$ *be fixed,* $\xi\in\mathbb{R}$ *and let* $x(\cdot,\xi) : [\,a,b_\xi)\to\mathbb{R}$ *be the unique saturated solution of* $\mathcal{CP}(\mathbb{R},\mathbb{R},f,a,\xi)$. *Show that, whenever* $\xi\leq\eta$, *we have* $x(t,\xi)\leq x(t,\eta)$ *for each* t *in* $[\,a,b_\xi)\cap[\,a,b_\eta)$.

Problem 2.10. *Let* $\mathbb{I}$ *and* Ω *be two nonempty open intervals in* $\mathbb{R}$, *and let* $f : \mathbb{I}\times\Omega\to\mathbb{R}$ *be a continuous function such that* $\mathcal{CP}(\mathcal{D})$ *has the uniqueness property. Let* $(a,\xi)\in\mathbb{I}\times\Omega$ *and let* $x : [a,b)\to\Omega$ *the unique saturated solution of* $\mathcal{CP}(\mathbb{I},\Omega,f,a,\xi)$. *Let* $y : [\,a,b)\to\Omega$ *be a function of class* C^1 *satisfying*

$$\begin{cases} y'(t)\leq f(t,y(t)) \\ y(a)\leq\xi \end{cases}$$

for each $t\in[\,a,b)$. *Show that, for each* $t\in[\,a,b)$, $y(t)\leq x(t)$.

Problem 2.11. *Let* $\mathbb{I}$ *be a nonempty and open interval in* $\mathbb{R}$, Ω *a nonempty and open subset of* $\mathbb{R}^n$ *and* $f : \mathbb{I}\times\Omega\to\mathbb{R}^n$ *a continuous function for which there exists a continuous function* $\omega : \mathbb{I}\times\mathbb{R}_+\to\mathbb{R}_+$ *such that, for each* $(t,x),(t,y)\in\mathbb{I}\times\Omega$, *we have* $\langle f(t,x)-f(t,y),x-y\rangle\leq\omega(t,\|x-y\|)\|x-y\|$. *Show that if, for a certain* $a\in\mathbb{I}$, *the unique saturated solution of* $\mathcal{CP}(\mathbb{I},\mathbb{R}_+,\omega,a,0)$ *is identically* 0, *then for each* $\xi\in\Omega$, $\mathcal{CP}(\mathbb{I},\Omega,f,a,\xi)$ *has at most one solution on a given interval. Prove Theorems* 2.3.1 *and* 2.3.3 *by using this result.*

Problem 2.12. *Let* $\omega : \mathbb{R}_+\to\mathbb{R}_+$ *be a continuous function with* $\omega(r) > 0$ *for each* $r > 0$ *and* $\omega(0) = 0$. *If* $\int_0^1\frac{d\eta}{\omega(\eta)} = +\infty$, *then the only saturated solution of the Cauchy problem* $x' = \omega(x)$, $x(0) = 0$ *is* $x\equiv 0$.

Problem 2.13. *Let $f, g : \mathbb{I} \times \Omega \to \mathbb{R}^n$ be two functions, continuous on $\mathbb{I} \times \Omega$ with f Lipschitz and g dissipative on Ω. Then $\mathcal{CP}(\mathbb{I}, \Omega, f + g, a, \xi)$ has the uniqueness property.*

Problem 2.14. *If $f : \mathbb{I} \times \Omega \to \mathbb{R}^n$ is continuous and there exists a continuous function $\omega : \mathbb{R}_+ \to \mathbb{R}_+$, with $\omega(r) > 0$ for each $r > 0$, $\omega(0) = 0$, $\int_0^1 \frac{d\eta}{\omega(\eta)} = +\infty$ and $\|f(t,x) - f(t,y)\| \leq \omega(\|x - y\|)$ for each $t \in \mathbb{I}$ and $x, y \in \Omega$, then $\mathcal{CP}(\mathcal{D})$ has the uniqueness property. This is the Osgood's uniqueness theorem.*

Problem 2.15. *Prove Theorem* 2.3.1 *by using Theorem* 2.3.3 *and the method of integrating factor.*

Problem 2.16. *Let $f : \mathbb{R} \times \mathbb{R} \to \mathbb{R}$ be defined by $f(t,x) = \sqrt[3]{x^2}$ for each (t,x) in $\mathbb{R} \times \mathbb{R}$. Show that the solution $x : [-1, 0] \to \mathbb{R}$, $x(t) = 0$ for each $t \in [-1, 0]$, has at least two saturated solutions of $\mathcal{CP}(\mathbb{R}, \mathbb{R}, f, -1, 0)$ extending it.*

Problem 2.17. *Find two nontrivial intervals $\mathbb{I}$ and Ω in $\mathbb{R}$ and a continuous function $f : \mathbb{I} \times \Omega \to \mathbb{R}$ which does not map bounded subsets in $\mathbb{I} \times \Omega$ into bounded subsets in $\mathbb{R}$.*

Problem 2.18. *Prove a result analogous to Theorem* 2.4.5 *under the hypothesis that the function $f : \mathbb{I} \times \Omega \to \mathbb{R}^n$ is continuous and for each $\mathbb{J} \times B \subset \mathbb{I} \times \Omega$ with $\mathbb{J}$ compact and B bounded, $f(\mathbb{J} \times B)$ is bounded in $\mathbb{R}^n$. Is the class of functions f satisfying the condition above strictly broader than that of functions f which carry bounded subsets into $\mathbb{I} \times \Omega$ in bounded subsets in $\mathbb{R}^n$?*

Problem 2.19. *Let $\mathcal{K}$ be compact in $\mathbb{R}^n$ and $\mathcal{F}$ closed $\mathbb{R}^n$, with $\mathcal{K} \cap \mathcal{F} = \emptyset$. As we already know by Lemma* 2.5.1, $\text{dist}(\mathcal{K}, \mathcal{F}) > 0$. *Could this result be extended to the more general case in which both subsets $\mathcal{K}$ and $\mathcal{F}$ are only closed?*

Problem 2.20. *Let $f, g : \mathbb{R} \to \mathbb{R}$ be two continuous functions, and let $G : \mathbb{R} \to \mathbb{R}$ be defined by $G(x) = \int_0^x g(s)ds$ for each $x \in \mathbb{R}$. Assume that there exists $a > 0$ such that, for each $x, y \in \mathbb{R}$, we have $G(x) \geq ax^2$ and $yf(y) \geq 0$. Show that, for each $\xi_1, \xi_2 \in \mathbb{R}$, each saturated solution of the Cauchy problem*

$$\begin{cases} x'' + f(x') + g(x) = 0 \\ x(0) = \xi_1, \quad x'(0) = \xi_2, \end{cases}$$

is defined on $\mathbb{R}_+$.

Problem 2.21. *Let $f, g : \mathbb{R}_+ \times \mathbb{R}^n \to \mathbb{R}^n$ be continuous, with f Lipschitz and g dissipative on $\mathbb{R}^n$. Then, for each $(a, \xi) \in \mathbb{R}_+ \times \mathbb{R}^n$, $\mathcal{CP}(\mathbb{R}_+, \mathbb{R}^n, f + g, a, \xi)$ has a unique global solution.*

Problem 2.22. *Let $f : (t_1, t_2) \times (\omega_1, \omega_2) \to \mathbb{R}$ be continuous, let $a \in (t_1, t_2)$, $\xi \in (\omega_1, \omega_2)$ and let $x : [a, b) \to \mathbb{R}$ be a saturated solution of the Cauchy problem*

$$\begin{cases} x' = f(t,x) \\ x(a) = \xi. \end{cases}$$

Show that, if $b < t_2$ and x is bounded, then there exists $\lim_{t\uparrow b} x(t) = x^*$. Prove a generalization to the case in which $f : (t_1, t_2) \times \Omega \to \mathbb{R}^n$, where Ω is a nonempty and open subset of $\mathbb{R}^n$ whose boundary contains only isolated points.

Problem 2.23. *Let $f : \mathbb{R} \to \mathbb{R}$ be a continuous function and let $x : [\,a, b\,] \to \mathbb{R}$ be a function of class C^1 satisfying*

$$\begin{cases} x' = f(x) \\ x(a) = x(b). \end{cases}$$

Show that x is constant. Extend this result to the case in which $f : \mathbb{R}^n \to \mathbb{R}^n$ with $n > 1$. Is the continuity of f sufficient in this case?

Problem 2.24. *Let $\mathbb{I}$ be a nonempty and open interval and $f : \mathbb{I} \times \mathbb{R}^n \to \mathbb{R}^n$ a continuous and bounded function such that $\mathcal{CP}(\mathcal{D})$ has the uniqueness property. Show that, for each $[\,a, b\,] \subset \mathbb{I}$, the mapping $\xi \mapsto x(\cdot, \xi)$ is continuous from $\mathbb{R}^n$ in $C([\,a, b\,]; \mathbb{R}^n)$, the latter being endowed with the uniform convergence topology.*

Problem 2.25. *Let $f : \mathbb{R} \times \mathbb{R} \to \mathbb{R}$ be as in Problem 2.3. For $\xi \in (0, +\infty)$, denote by $x(\cdot, \xi)$ the unique global solution of $\mathcal{CP}(\mathbb{R}, \mathbb{R}, f, 0, \xi)$. Show that, although $\lim\limits_{\xi\downarrow 0} x(t, \xi) = t$ uniformly for $t \in [0, +\infty)$, the function $y(t) = t$, for $t \in [0, +\infty)$, is not a solution of $\mathcal{CP}(\mathbb{R}, \mathbb{R}, f, 0, 0)$. Explain the result.*

Problem 2.26. *Let $f : \mathbb{R} \times \mathbb{R} \times \mathbb{R} \to \mathbb{R}$ be defined by*

$$f(t, x, p) = \begin{cases} 0 & \text{if } t \in \mathbb{R} \text{ and } x + p = 0 \\ \dfrac{t}{x + p} & \text{if } t \in \mathbb{R} \text{ and } x + p \neq 0. \end{cases}$$

For $p \in (0, +\infty)$ we denote by $x(\cdot, p) : \mathbb{R}_+ \to \mathbb{R}$ the unique global solution of $\mathcal{CP}(\mathbb{R}, \mathbb{R}, f, 0, 0)_p$. Show that, although $\lim\limits_{p\downarrow 0} x(t, p) = t$ uniformly for $t \in [\,0, 1\,]$, the function $y(t) = t$, for $t \in [\,0, 1\,]$, is not a solution of $\mathcal{CP}(\mathbb{R}, \mathbb{R}, f, 0, 0)_0$. Explain the result.

Problem 2.27. *Let $f : \mathbb{R} \times \mathbb{R} \times \mathbb{R} \to \mathbb{R}$ be defined by $f(t, x, p) = 3\sqrt[3]{x^2 + p^2}$ for each $(t, x, p) \in \mathbb{R} \times \mathbb{R} \times \mathbb{R}$. Show that, for each $p \neq 0$, $\mathcal{CP}(\mathcal{D})_p$ has the uniqueness property. In addition, $\lim_{p\to 0} x(t, x, p) = f(t, x, 0)$ uniformly for $(t, x) \in \mathbb{R} \times \mathbb{R}$, but, nevertheless, $\mathcal{CP}(\mathcal{D})_0$ lacks the uniqueness property. Thus, the uniqueness property does not depend "continuously" on parameters.*

Problem 2.28. *Let $D \subseteq \mathbb{R}^{n+1}$, let $F : D \to \mathbb{R}^n$ a C^1-function, and let $(a, \xi) \in D$, where $a \in \mathbb{R}$ and $\xi \in \mathbb{R}^n$. Show that, if*

$$\det\left(\frac{\partial F_i}{\partial x_j}(a, \xi)\right)_{n\times n} \neq 0,$$

then, there exist $b > a$ and a unique C^1-function, $x : [\,a, b\,] \to \mathbb{R}^n$, implicitly defined by $F(t, x) = F(a, \xi)$.

Chapter 3

Approximation Methods

This chapter is entirely dedicated to the presentation of several approximation methods for the solution of a given Cauchy problem. Although these methods are no longer used in their original form, they are still of interest in many effective numerical algorithms. In the first section we prove that a Cauchy problem has only analytic solutions whenever the right-hand side of the corresponding differential equation is an analytic function. This theorem is, on one hand, an approximation result (ensuring the possibility to develop any solution in power series), and on the other hand a regularity result. In the next three sections we discuss: *the method of successive approximations*, *the method of polygonal lines*, known also as *the Euler explicit method*, and *the Euler implicit method*. The chapter ends with a set of exercises and problems.

3.1 The Power Series Method

In this section, by using the so-called majorant series method proposed by Cauchy and improved by Lindelöff, we shall prove that, whenever f is analytic on $\mathbb{I}\times\Omega$, the unique solution x of the Cauchy problem $\mathcal{CP}(\mathcal{D})$ is also analytic on its domain and, moreover, the coefficients in the powers series expansion of the solution can be determined explicitly. Therefore, one can always find approximations for the solution in the form of the partial sums of its power series expansion, and, in some cases, the solution itself may be determined explicitly.

3.1.1 *An Example*

This method of solving a Cauchy problem by means of power series is one of the oldest and most effective. In order to illustrate it, let us analyze the following example.

Example 3.1.1. Solve the Cauchy problem

$$\begin{cases} (1-t^2)x'' - 4tx' - 2x = 0 \\ x(0) = 1 \\ x'(0) = 0 \end{cases}$$

by looking for the solution in the form of a power series:

$$x(t) = \sum_{n=0}^{\infty} c_n t^n \qquad (\mathcal{S})$$

for $|t| < r$, with r suitably chosen.

We solve this problem in two steps. First, we find the coefficients c_n for $n = 0, 1, \ldots$, and then we estimate the radius of convergence of the power series thus obtained. A last step, which solves the problem, consists in the continuation of the solution to a maximal domain of existence.

In order to find the coefficients, we impose the condition that x, given by $(\mathcal{S})$, satisfies the differential equation and the initial conditions. The initial conditions are equivalent to

$$c_0 = 1, \; c_1 = 0,$$

while the fact that x satisfies $(1-t^2)x'' - 4tx' - 2x = 0$ is expressed as

$$\sum_{n=2}^{\infty} n(n-1)c_n t^{n-2} - \sum_{n=2}^{\infty} n(n-1)c_n t^n - \sum_{n=1}^{\infty} 4nc_n t^n - \sum_{n=0}^{\infty} 2c_n t^n = 0$$

for every t with $|t| < r$. The previous equality may be rewritten as

$$\sum_{n=0}^{\infty} (n+2)(n+1)(c_{n+2} - c_n)t^n = 0$$

for every t with $|t| < r$. However, a power series with radius of convergence $r > 0$ is identically zero on $|t| < r$ if and only if all the coefficients are zero. So we have $c_{n+2} - c_n = 0$ for $n = 0, 1, \ldots$. From here and the initial conditions, we deduce $c_{2k} = 1$ and $c_{2k+1} = 0$ for $k = 0, 1, \ldots$. Consequently, $x(t) = \sum_{k=0}^{\infty} t^{2k}$ for every t with $|t| < r$. However, in this case $r = 1$ and $x(t) = 1/(1-t^2)$ for every t with $|t| < 1$.

3.1.2 *The Existence of Analytic Solutions*

In what follows we will identify some circumstances under which any solution of a certain Cauchy problem can be developed in a power series.

More precisely, let $\mathbb{I}$ be a nontrivial interval in $\mathbb{R}$, Ω a nonempty and open subset of $\mathbb{R}^n$, $f : \mathbb{I} \times \Omega \to \mathbb{R}^n$ an analytic function on $\mathbb{I} \times \Omega$, $a \in \mathbb{I}$, $\xi \in \Omega$ and let us consider the Cauchy problem with data $\mathcal{D} = (f, \mathbb{I}, \Omega, a, \xi)$

$$\begin{cases} x' = f(t,x) \\ x(a) = \xi. \end{cases} \qquad \mathcal{CP}(\mathcal{D})$$

The main result in this section is

Theorem 3.1.1. (Cauchy–Lindelöff) *If $f : \mathbb{I} \times \Omega \to \mathbb{R}^n$ is analytic on $\mathbb{I} \times \Omega$ then, for each $(a, \xi) \in \mathbb{I} \times \Omega$, the unique saturated solution of $\mathcal{CP}(\mathcal{D})$ is analytic on its interval of existence.*

Proof. Since f is analytic on $\mathbb{I} \times \Omega$, it satisfies the conditions of both Theorem 2.3.2 and Corollary 2.4.1 and therefore, for every $(a, \xi) \in \mathbb{I} \times \Omega$, $\mathcal{CP}(\mathcal{D})$ has a unique saturated solution. Hence, in order to complete the proof, it suffices to show that, for every $(a, \xi) \in \mathbb{I} \times \Omega$, the unique saturated solution of $\mathcal{CP}(\mathcal{D})$ can be developed into a power series in a neighborhood of a. From here one can easily deduce that the solution can be developed into a power series about any point b in its domain of definition.

First, let us observe that we may assume with no loss of generality that $a = 0$ and $\xi = 0$. Indeed, if this is not the case, then after two translations, i.e. $t - a = s$ and $x - \xi = y$, we are in the specific situation described above.

Next, we will show that, for every $i = 1, 2, \ldots, n$, we have

$$x_i(t) = \sum_{s=0}^{\infty} b_s^{(i)} t^s, \ \textit{with} \ b_s^{(i)} = \frac{1}{s!} \frac{d^s x_i}{dt^s}(0) \tag{3.1.1}$$

for $t \in \mathbb{R}$, $|t| < r_0$, with $r_0 > 0$ suitably chosen.

We recall that the unique saturated solution x of the Cauchy problem $\mathcal{CP}(\mathcal{D})$ is of class C^∞ because f is of class C^∞. See Remark 2.1.3. So all the coefficients in the series (3.1.1) are well defined. It remains then to prove that the series (3.1.1) are convergent on a neighborhood of $a = 0$. To this aim let us remark that all the coefficients $b_s^{(i)}$ in (3.1.1) can be expressed by means of $f(0,0)$ and of the partial derivatives of the function f calculated at $(0,0)$. Indeed,

$$b_0^{(i)} = x_i(0) = 0, \ b_1^{(i)} = x_i'(0) = f_i(0,0),$$

$$b_2^{(i)} = x_i''(0) = \frac{1}{2!}\left(\frac{\partial f_i}{\partial t}(0,0) + \sum_{j=1}^{n} \frac{\partial f_i}{\partial x_j}(0,0) f_j(0,0)\right) \quad \text{and so on.}$$

The idea of proof consists in finding a function $g : (-r,r) \times (-\rho,\rho)^n \to \mathbb{R}$ with $r > 0$ and $\rho > 0$, suitably defined, such that both g and all its partial derivatives are positive at $(0,0)$ and satisfy

$$\left|\frac{\partial^{p_0+p_1+\cdots+p_n} f_i}{\partial t^{p_0}\partial x_1^{p_1}\ldots\partial x_n^{p_n}}(0,0)\right| \le \frac{\partial^{p_0+p_1+\cdots+p_n} g}{\partial t^{p_0}\partial x_1^{p_1}\ldots\partial x_n^{p_n}}(0,0), \quad i = 1,2,\ldots,n. \tag{3.1.2}$$

Let g be such a function, let us define $F : (-r,r) \times (-\rho,\rho)^n \to \mathbb{R}^n$ by

$$F(t,x) = \begin{pmatrix} g(t,x) \\ g(t,x) \\ \vdots \\ g(t,x) \end{pmatrix}$$

and let us consider the auxiliary Cauchy problem

$$\begin{cases} y' = F(t,y) \\ y(0) = 0. \end{cases} \tag{3.1.3}$$

If we were able to show that the problem (3.1.3) admits a local analytic solution

$$y_i(t) = \sum_{s=0}^{\infty} c_s^{(i)} t^s, \ \textit{with} \ c_s^{(i)} = \frac{1}{s!}\frac{d^s y_i}{dt^s}(0), \quad i = 1,2\ldots,n,$$

then, since all the coefficients $c_s^{(i)}$ can be expressed by means of the values of both F and its derivatives at $(0,0)$ in the same manner as the coefficients $b_s^{(i)}$ can be expressed by means of both f and its derivatives at $(0,0)$, from (3.1.2) it would follow $\left|b_s^{(i)}\right| \le c_s^{(i)}$ for $i = 1,2,\ldots,n$ and $s \in \mathbb{N}^*$. But these inequalities, along with the convergence of the series which define the partial functions y_i, would prove the convergence of the series in (3.1.1).

In order to define the function g, we recall that, from the analyticity of f, there exist $r > 0$ and $\rho > 0$ such that

$$f_i(t,x) = \sum_{p_0,p_1,\ldots,p_n \in \mathbb{N}} a_{p_0,p_1,\ldots,p_n}^{(i)} t^{p_0} x_1^{p_1} \ldots x_n^{p_n}, \quad i = 1,2,\ldots,n, \tag{3.1.4}$$

for every $(t,x) \in \mathbb{I} \times \Omega$ with $|t| < r$ and $|x_i| < \rho$, $i = 1,2,\ldots,n$, where the coefficients $a_{p_0,p_1,\ldots,p_n}^{(i)}$ are given by

$$a_{p_0,p_1,\ldots,p_n}^{(i)} = \frac{1}{p_0!p_1!\ldots p_n!} \cdot \frac{\partial^{p_0+p_1+\cdots+p_n} f_i}{\partial t^{p_0}\partial x_1^{p_1}\ldots\partial x_n^{p_n}}(0,0). \tag{3.1.5}$$

Moreover, as the series (3.1.4) are convergent, it follows that there exists $M > 0$ such that $\left|a^{(i)}_{p_0,p_1,\ldots,p_n} r^{p_0}\rho^{p_1+p_2+\cdots+p_n}\right| \leq M$ for every $i = 1, 2, \ldots, n$ and every $p_0, p_1, \ldots, p_n \in \mathbb{N}$. From this inequality and from (3.1.5), we get

$$\left|\frac{\partial^{p_0+p_1+\cdots+p_n} f_i}{\partial t^{p_0}\partial x_1^{p_1}\ldots\partial x_n^{p_n}}(0,0)\right| \leq M \cdot \frac{p_0!p_1!\ldots p_n!}{r^{p_0}\rho^{p_1+p_2+\cdots+p_n}}$$

for every $i = 1, 2, \ldots, n$ and every $p_0, p_1, \ldots, p_n \in \mathbb{N}$. This inequality suggests that g ought to be such that

$$\frac{\partial^{p_0+p_1+\cdots+p_n} g}{\partial t^{p_0}\partial x_1^{p_1}\ldots\partial x_n^{p_n}}(0,0) = M \cdot \frac{p_0!p_1!\ldots p_n!}{r^{p_0}\rho^{p_1+p_2+\cdots+p_n}}$$

for every $p_0, p_1, \ldots, p_n \in \mathbb{N}$. Such a function is

$$\begin{aligned} g(t,x) &= M \sum_{p_0,p_1,\ldots,p_n\in\mathbb{N}} \left(\frac{t}{r}\right)^{p_0}\left(\frac{x_1}{\rho}\right)^{p_1}\left(\frac{x_2}{\rho}\right)^{p_2}\ldots\left(\frac{x_n}{\rho}\right)^{p_n} \\ &= \frac{M}{\left(1-\frac{t}{r}\right)\left(1-\frac{x_1}{\rho}\right)\left(1-\frac{x_2}{\rho}\right)\ldots\left(1-\frac{x_n}{\rho}\right)} \end{aligned}$$

for every $t \in (-r, r)$ and every $x \in \mathbb{R}^n$ with $x_i \in (\rho, \rho)$ for $i = 1, 2, \ldots, n$.

Now, let us observe that the problem (3.1.3) may be rewritten in the form

$$\begin{cases} y_i' = g(t, y_1, y_2, \ldots, y_n) \\ y_i(0) = 0 \end{cases}$$

for $i = 1, 2, \ldots, n$ and $t \in (-r, r)$. From $y_1'(t) = y_2'(t) = \cdots = y_n'(t)$ for every $t \in (-r, r)$ and $y_1(0) = y_2(0) = \cdots = y_n(0)$, we deduce that $y_1(t) = y_2(t) = \cdots = y_n(t) = y(t)$ for every $t \in (-r, r)$, where y is the unique saturated solution of the problem

$$\begin{cases} y' = g(t, y, y, \ldots, y) \\ y(0) = 0 \end{cases}$$

for $t \in (-r, r)$. Taking into account the definition of g, the problem is equivalent to

$$\begin{cases} y' = M\left(1-\dfrac{t}{r}\right)^{-1}\left(1-\dfrac{y}{\rho}\right)^{-n} \\ y(0) = 0 \end{cases}$$

for $t \in (-r, r)$. This equation is with separable variables and can be solved explicitly. More precisely, its solution is given by

$$y(t) = \rho\left(1 - \sqrt[n+1]{1 + \frac{(n+1)Mr}{\rho}\ln\left(1-\frac{t}{r}\right)}\right)$$

for every $t \in (-r_0, r_0)$ with $r_0 \in (0, r\,]$ suitably chosen. Since this function is analytic on $(-r_0, r_0)$, this completes the proof. □

Remark 3.1.1. We mention that the power series method is applicable in many situations, even for some Cauchy problems for which the differential equation, due to the presence of some singularities in the coefficients, cannot be put into the normal form. Indeed, in order to better understand this observation, let us analyze the following example.

Example 3.1.2. Solve the Cauchy problem

$$\begin{cases} tx'' - tx' - x = 0 \\ x(0) = 0 \\ x'(0) = 1 \end{cases}$$

looking for the solution in the form of a power series:

$$x(t) = \sum_{n=0}^{\infty} c_n t^n \tag{$\mathcal{S}$}$$

for $t \in \mathbb{R}$ with $|t| < r$.

One may easily see that the differential equation $tx'' - tx' - x = 0$ is neither in the normal form, nor can it be put into the normal form in any neighborhood of 0 (with an analytic right-hand side, of course). Nevertheless, one may solve the problem in the same manner as in Example 3.1.1. In order to find the coefficients of the power series above, we impose the condition that x, given by ($\mathcal{S}$), satisfies both the differential equation and the initial conditions. The initial conditions lead to $c_0 = 0$ and $c_1 = 1$, while from the differential equation $tx'' - tx' - x = 0$ we get

$$\sum_{n=2}^{\infty} n(n-1)c_n t^{n-1} - \sum_{n=1}^{\infty} nc_n t^n - \sum_{n=0}^{\infty} c_n t^n = 0$$

for every $t \in \mathbb{R}$.

The previous equality may be rewritten in the form

$$\sum_{n=0}^{\infty} [(n+1)nc_{n+1} - (n+1)c_n]t^n = 0$$

for every $t \in \mathbb{R}$ with $|t| < r$. However, a power series with radius of convergence $r > 0$ is identically zero on $|t| < r$ if and only if all the coefficients are zero. So we have $(n+1)nc_{n+1} - (n+1)c_n = 0$ for $n = 0, 1, \ldots$. From here we deduce $c_n = 1/(n-1)!$ for $n = 1, 2, \ldots$ and $c_0 = 0$. Consequently $r = +\infty$ and

$$x(t) = t\sum_{n=1}^{\infty} \frac{t^{n-1}}{(n-1)!} = te^t$$

for every $t \in \mathbb{R}$.

3.2 The Successive Approximations Method

In many situations it is of great importance to know, not only that a Cauchy problem has exactly one solution on a given interval, but also how to find this solution. Unfortunately, the class of functions f for which we can obtain an explicit representation of the solution is extremely narrow. This explains why it would be very useful to have some effective methods to find some "approximate solutions", i.e. to get explicit representation of some functions which, although not solutions of the problem, are in some sense "as close as we wish" to the "exact solutions". We have already presented three such methods: the first one in Section 2.2, where the functions $\{x_m;\ m \in \mathbb{N}^*\}$ defined by $(\mathcal{IE}_m)$ "approximate" the solution of the Cauchy problem $\mathcal{CP}(\mathcal{D})$ on $[a, b]$, under the assumption that the solution is unique, the second one in Remark 2.6.4 combined with Example 2.6.1, and the third one in the preceding section, where each term of the sequence of partial sums of the series which defines the solution approximates the latter.

The aim of this section, as well as of the next two, is to present three more such methods which are at the core of several very efficient algorithms in the numerical analysis of differential systems.

We begin with the so-called *method of successive approximations* due to [Picard (1890)]. We note that this method is completely unspecific, in the sense that it is applicable not only for differential equations and systems, but also for *Volterra integral equations, Fredholm integral equations, integro-differential equations, neutral equations, partial differential equations, etc.* See Problems 3.7, 3.8 and 3.11.

To begin with, let $a \in \mathbb{R}$, $\xi \in \mathbb{R}^n$, $h > 0$ and $r > 0$, and let us consider the cylinder $\Delta = [a, a+h] \times B(\xi, r)$. Let $f : \Delta \to \mathbb{R}^n$ be a continuous function and let us consider the Cauchy problem

$$\begin{cases} x' = f(t,x) \\ x(a) = \xi, \end{cases} \tag{$\mathcal{CP}$}$$

which, as shown in Proposition 2.1.1, is equivalent to the integral equation

$$x(t) = \xi + \int_a^t f(\tau, x(\tau))\, d\tau \tag{$\mathcal{IE}$}$$

for every $t \in \mathbb{J}$.

Remark 3.2.1. We emphasize that, the framework here considered is quite different from that used in the preceding sections, simply because both

$\mathbb{I} = [a, a+h]$ and $\Omega = B(\xi, r)$ are closed sets. Nevertheless, we can easily see that all the concepts and results there extend in a natural manner to this framework, as long as we refer to the right solutions, since *a is the left end-point of the interval $\mathbb{I}$ and ξ is an interior point of the set Ω*. The sole exception which should be noticed is that every saturated solution of the problem $(\mathcal{CP})$ (whose existence is ensured by Corollary 2.4.1) is defined on a closed interval. Indeed, as Δ is a compact set and f is continuous, there exists $M > 0$ such that, for every $(t, u) \in \Delta$, we have

$$\|f(t,u)\| \leq M. \tag{3.2.1}$$

Now, if we assume that $x : [a, b) \to B(\xi, r)$ is a saturated solution of $(\mathcal{CP})$, from (3.2.1) and Proposition 2.4.1, it follows that x can be extended to $[a, b]$ as a solution of $(\mathcal{CP})$. This contradiction can be eliminated only if every saturated solution of $(\mathcal{CP})$ is defined on a closed interval.

Let $x : [a, b] \to B(\xi, r)$ be a saturated solution of $(\mathcal{CP})$. The next lemma gives a lower bound for the length of the interval of existence of saturated solutions.

Lemma 3.2.1. *Let $f : \Delta \to \mathbb{R}^n$ be a continuous function on Δ and let*

$$\delta = \min\left\{h, \frac{r}{M}\right\}, \tag{3.2.2}$$

where $M > 0$ satisfies (3.2.1). *Then each saturated solution of* $(\mathcal{CP})$ *is defined at least on $[a, a+\delta]$.*

Proof. Let $x : [a, b] \to B(\xi, r)$ be a saturated solution of $(\mathcal{CP})$. Since x is also a solution of $(\mathcal{IE})$, for every $t \in [a, b]$, we have

$$\|x(t) - \xi\| \leq \int_a^t \|f(\tau, x(\tau))\|\, d\tau \leq (t-a)M.$$

So, if $b < a + \delta$, then it necessarily follows that

$$\|x(b) - \xi\| \leq (b-a)M < \delta M \leq r,$$

and therefore $x(b)$ is an interior point of $B(\xi, r)$. Consequently, x is not saturated. This contradiction is due to the assumption that $b < a + \delta$. Hence $a + \delta \leq b$ and this completes the proof. □

Lemma 3.2.2. *Let $f : \Delta \to \mathbb{R}^n$ be continuous on Δ and let δ be defined by* (3.2.2). *Then, for each continuous function $y : [a, a+\delta] \to B(\xi, r)$, the function $x : [a, a+\delta] \to \mathbb{R}^n$, defined by*

$$x(t) = \xi + \int_a^t f(\tau, y(\tau))\, d\tau$$

for every $t \in [a, a+\delta]$, maps $[a, a+\delta]$ into $B(\xi, r)$.

Proof. In view of (3.2.1), (3.2.2) and the definition of x, we have

$$\|x(t)-\xi\| \leq \int_a^t \|f(\tau, y(\tau))\|\, d\tau \leq \delta M \leq r$$

for every $t \in [a, a+\delta]$ and this completes the proof. □

Now we proceed to the definition of the sequence of successive approximations $(x_k)_{k\in\mathbb{N}}$ corresponding to the problem $(\mathcal{CP})$. Let us consider $x_0 : [a, a+\delta] \to B(\xi, r)$ defined by $x_0(t) = \xi$ for every $t \in [a, a+\delta]$ and let us define $x_k : [a, a+\delta] \to B(\xi, r)$, for $k \geq 1$, by

$$x_k(t) = \xi + \int_a^t f(\tau, x_{k-1}(\tau))\, d\tau, \quad \text{for each } \ t \in [a, a+\delta]. \tag{3.2.3}$$

A simple inductive argument, combined with Lemma 3.2.2, shows that x_k is well-defined for every $k \in \mathbb{N}$. For obvious reasons, $(x_k)_{k\in\mathbb{N}}$ is called *the sequence of successive approximations corresponding to* $\mathcal{CP}(\mathcal{D})$.

The main result in this section is Picard's Theorem below referring to the uniform convergence of $(x_k)_{k\in\mathbb{N}}$.

Theorem 3.2.1. (Picard) *Let us assume that* $f : \Delta \to \mathbb{R}^n$ *is continuous on* Δ *and Lipschitz on* $B(\xi, r)$, *i.e. there exists* $L > 0$ *such that for every* $(t, u), (t, v) \in \Delta$, *we have*

$$\|f(t, u) - f(t, v)\| \leq L\|u - v\|. \tag{3.2.4}$$

Then the sequence of successive approximations corresponding to $\mathcal{CP}(\mathcal{D})$ *is uniformly convergent on* $[a, a+\delta]$ *to the unique solution* x *of* $(\mathcal{CP})$ *defined on that interval. In addition, we have the following error estimate formula*

$$\|x_k(t) - x(t)\| \leq M\frac{L^k\delta^{k+1}}{(k+1)!}, \tag{3.2.5}$$

for every $k \in \mathbb{N}$ *and* $t \in [a, a+\delta]$.

Proof. From Theorem 2.3.2 and both Theorem 2.4.2 and Remark 3.2.1, it follows that $(\mathcal{CP})$ has a unique solution defined at least on $[a, a+\delta]$. Let $x : [a, a+\delta] \to B(\xi, r)$ be this solution and let us observe that, from $(\mathcal{IE})$ and (3.2.1), for every $t \in [a, a+\delta]$, we have

$$\|x_0(t) - x(t)\| = \|\xi - x(t)\| \leq \int_a^t \|f(\tau, x(\tau))\|\, d\tau \leq M(t-a).$$

From $(\mathcal{IE})$, (3.2.3), (3.2.4) and from the inequality above, we deduce

$$\|x_1(t) - x(t)\| \leq \int_a^t \|f(\tau, \xi) - f(\tau, x(\tau))\|\, d\tau$$

$$\leq L\int_a^t \|x_0(\tau)-x(\tau)\|\,d\tau \leq M\frac{L(t-a)^2}{2!}\,,$$

for every $t\in[\,a,a+\delta\,]$. This inequality suggests that, for every $k\in\mathbb{N}$ and $t\in[\,a,a+\delta\,]$, we ought to have

$$\|x_k(t)-x(t)\|\leq M\frac{L^k(t-a)^{k+1}}{(k+1)!}\,. \tag{3.2.6}$$

We shall prove (3.2.6) by induction. Since for $k=0$, or $k=1$ this inequality is obviously satisfied, let us assume that it holds for some $k\in\mathbb{N}$ and for every $t\in[\,a,a+\delta\,]$. Then, from (3.2.3), (3.2.4) and from the inductive assumption, we deduce

$$\|x_{k+1}(t)-x(t)\|\leq \int_a^t \|f(\tau,x_k(\tau))-f(\tau,x(\tau))\|\,d\tau$$

$$\leq L\int_a^t \|x_k(\tau)-x(\tau)\|\,d\tau \leq L\int_a^t M\frac{L^k(\tau-a)^{k+1}}{(k+1)!}\,d\tau$$

$$=M\frac{L^{k+1}(t-a)^{k+2}}{(k+2)!}\,,$$

for every $t\in[\,a,a+\delta\,]$, and thus (3.2.6) holds for $k+1$ as well. Hence (3.2.6) holds true for each $k\in\mathbb{N}$. Obviously (3.2.6) implies (3.2.5), while from (3.2.5) we get

$$\sup_{t\in[\,a,a+\delta\,]}\|x_k(t)-x(t)\|\leq M\frac{L^k\delta^{k+1}}{(k+1)!}$$

for every $k\in\mathbb{N}$ and $t\in[\,a,a+\delta\,]$. Since

$$\lim_{k\to\infty} M\frac{L^k\delta^{k+1}}{(k+1)!}=0,$$

the last inequality shows that $(x_k)_{k\in\mathbb{N}}$ is uniformly convergent on $[\,a,a+\delta\,]$ to x and this completes the proof. □

3.3 The Method of Polygonal Lines

In this section we present another method to approximate the solution of a Cauchy problem, i.e. *the method of polygonal lines* due to Euler.

As in the preceding section, let $a\in\mathbb{R}$, $\xi\in\mathbb{R}^n$, $h>0$ and $r>0$ and let us denote by $\Delta=[\,a,a+h\,]\times B(\xi,r)$. Let $f:\Delta\to\mathbb{R}^n$ be a continuous function and let us consider the Cauchy problem

$$\begin{cases} x'=f(t,x)\\ x(a)=\xi. \end{cases} \tag{$\mathcal{CP}$}$$

Since f is continuous on the compact set Δ, there exists $M > 0$ such that, for every $(t, u) \in \Delta$, we have

$$\|f(t,u)\| \leq M. \tag{3.3.1}$$

We recall that, by virtue of Lemma 3.2.1, the Cauchy problem $\mathcal{CP}(\mathcal{D})$ has at least one solution $x : [\,a, a+\delta\,] \to B(\xi, r)$, where $\delta > 0$ is given by

$$\delta = \min\left\{h, \frac{r}{M}\right\}. \tag{3.3.2}$$

In what follows we assume that x is the unique solution of $\mathcal{CP}(\mathcal{D})$ defined on $[\,a, a+\delta\,]$. Let $k \in \mathbb{N}^*$ and let us consider the partition

$$\mathcal{P}_k : a = t_0 < t_1 < \cdots < t_k = a + \delta.$$

For simplicity, we assume that $t_{i+1} - t_i = \delta/k$ for $i = 0, 1, \ldots, k-1$. The main idea of the method of polygonal lines consists in replacing the integral equation

$$x(t) = \xi + \int_a^t f(\tau, x(\tau))\, d\tau, \tag{$\mathcal{IE}$}$$

which is equivalent to $\mathcal{CP}(\mathcal{D})$, by the system

$$\begin{cases} \xi_0 = \xi \\ \xi_{i+1} = \xi_i + (t_{i+1} - t_i) f(t_i, \xi_i), \text{ for } i = 0, 1, \ldots, k-1, \end{cases} \tag{$\mathcal{S}_k$}$$

and in showing that the function $y_k : [\,a, a+\delta\,] \to \mathbb{R}^n$, defined by

$$y_k(t) = \xi_i + (t - t_i) f(t_i, \xi_i) \tag{3.3.3}$$

for $t \in [t_i, t_{i+1})$ if $i = 0, 1, \ldots, k-2$, or $t \in [\,t_i, t_{i+1}\,]$ if $i = k-1$, is "sufficiently close" to the unique solution $x : [\,a, a+\delta\,] \to B(\xi, r)$ of $(\mathcal{IE})$, for k "large enough".

We notice that the name of the method is suggested by the remark that the graph of the function y_k is the polygonal line passing through $(t_0, \xi_0), (t_1, \xi_1), \ldots, (t_k, \xi_k)$. See Figure 3.3.1.

Lemma 3.3.1. *If $f : \Delta \to \mathbb{R}^n$ is continuous and δ is defined by* (3.3.2) *then, for every $k \in \mathbb{N}^*$,* $(\mathcal{S}_k)$ *has a unique solution* $(\xi_0, \xi_1, \ldots, \xi_k)$.

Proof. The uniqueness is obvious. As concerns the existence, it suffices to show that, whenever $\xi_i \in B(\xi, r)$ for some $i = 0, 1, \ldots, k-2$, then $\xi_{i+1} \in B(\xi, r)$. To this aim, we shall prove that for every $i = 0, 1, \ldots, k-1$, we have $\|\xi_i - \xi\| \leq \frac{ir}{k}$.

For $i = 0$ the inequality above is obviously satisfied because $\xi_0 = \xi$. Next, let us assume that the inequality holds for some $i \in \{1, 2, \ldots, k-2\}$. From (3.3.1), (3.3.2) and $(\mathcal{S}_k)$, we have

$$\|\xi_{i+1} - \xi\| \leq \|\xi_{i+1} - \xi_i\| + \|\xi_i - \xi\| \leq \frac{\delta}{k} M + \frac{ir}{k} \leq \frac{(i+1)r}{k},$$

and this completes the proof. □

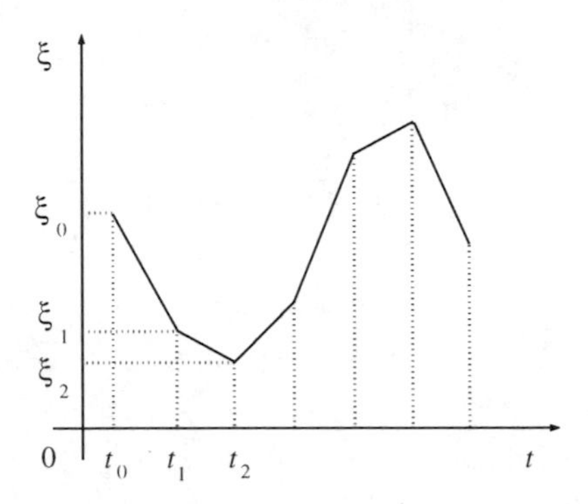

Fig. 3.3.1

Now, let us define the step function $\sigma_k : [\,a, a+\delta\,] \to [\,a, a+\delta\,]$ by $\sigma_k(t) = t_i$, for $t \in [\,t_i, t_{i+1})$ if $i = 0, 1, \ldots, k-2$, or $t \in [\,t_i, t_{i+1}\,]$ if $i = k-1$. Obviously

$$|\sigma_k(t) - t| \leq \frac{\delta}{k}$$

for every $t \in [\,a, a+\delta\,]$, and therefore

$$\lim_{k\to\infty} \sigma_k(t) = t$$

uniformly for $t \in [\,a, a+\delta\,]$.

Lemma 3.3.2. *Let $f : \Delta \to \mathbb{R}^n$ be a continuous function, let $\delta > 0$ be defined by* (3.3.2) *and let $y_k : [\,a, a+\delta\,] \to \mathbb{R}^n$ be defined by* (3.3.3), *where $(\xi_0, \xi_1, \ldots, \xi_k)$ is the unique solution of the system $(\mathcal{S}_k)$. Then y_k satisfies*

$$y_k(t) = \xi + \int_a^t f(\sigma_k(\tau), y_k(\sigma_k(\tau)))\,d\tau, \tag{3.3.4}$$

for every $t \in [\,a, a+\delta\,]$, where σ_k is defined as above.

Proof. For $t \in [\,a, t_1\,]$, we have

$$y_k(t) = \xi + \int_a^t f(a, \xi)\,d\tau = \xi + \int_a^t f(\sigma_k(\tau), y_k(\sigma(\tau)))\,d\tau.$$

Furthermore, let us assume that y_k satisfies (3.3.4) for $t \in [\,a, t_i\,]$, where $i \leq k-1$. Then, for $t \in [\,t_i, t_{i+1}\,]$, we have

$$y_k(t) = y_k(t_i) + \int_{t_i}^t f(t_i, y_k(t_i))\,d\tau$$

$$= \xi + \int_a^{t_i} (\sigma_k(\tau), y_k(\sigma_k(\tau)))\, d\tau + \int_{t_i}^{t} f(\sigma_k(\tau), y_k(\sigma_k(\tau)))\, d\tau$$

$$= \xi + \int_a^{t} (\sigma_k(\tau), y_k(\sigma_k(\tau)))\, d\tau.$$

So y_k satisfies (3.3.4) and this completes the proof. □

We can now proceed to the statement of the main result of this section.

Theorem 3.3.1. *If $f : \Delta \to \mathbb{R}^n$ is continuous on Δ and $\mathcal{CP}(\mathcal{D})$ has the uniqueness property, then $(y_k)_{k\in\mathbb{N}^*}$, defined by (3.3.3), is uniformly convergent on $[\,a, a+\delta\,]$ to the unique solution x of $\mathcal{CP}(\mathcal{D})$ on that interval. If, in addition, f is Lipschitz on Δ, i.e. there exists $L > 0$ such that, for every $(t,u), (s,v) \in \Delta$*

$$\|f(t,u) - f(s,v)\| \leq L(|t-s| + \|u-v\|),$$

then the following error estimate formula holds

$$\|y_k(t) - x(t)\| \leq \frac{L\delta^2(M+1)}{k} e^{L\delta}, \tag{3.3.5}$$

for every $k \in \mathbb{N}^*$ and every $t \in [\,a, a+\delta\,]$.

Proof. From Lemma 3.3.2 we know that, for every $k \in \mathbb{N}^*$, y_k satisfies (3.3.4). For this reason, for every $k \in \mathbb{N}^*$ and $t \in [\,a, a+\delta\,]$

$$\|y_k(t)\| \leq \|\xi\| + \int_a^t \|(\sigma_k(\tau), y_k(\sigma_k(\tau)))\| d\tau \leq \|\xi\| + \delta M,$$

and so $\{y_k;\ k \in \mathbb{N}^*\}$ is uniformly bounded on $[\,a, a+\delta\,]$. Furthermore, for every $k \in \mathbb{N}^*$ and $t, \tilde{t} \in [\,a, a+\delta\,]$, we have

$$\|y_k(t) - y_k(\tilde{t})\| \leq \left| \int_{\tilde{t}}^t \|(\sigma_k(\tau), y_k(\sigma_k(\tau)))\|\, d\tau \right| \leq M|t - \tilde{t}|$$

and therefore the family $\{y_k;\ k \in \mathbb{N}^*\}$ is equicontinuous on $[\,a, a+\delta\,]$. By Arzelà-Ascoli's Theorem 12.2.1, it follows that $\{y_k;\ k \in \mathbb{N}^*\}$ is relatively compact in $C([\,a, a+\delta\,]; \mathbb{R}^n)$ endowed with the uniform convergence topology. Since, from (3.3.4) and Corollary 12.2.2, every limit point of the sequence $(y_k)_{k\in\mathbb{N}^*}$ is a solution of $\mathcal{CP}(\mathcal{D})$ and the latter has the uniqueness property, we conclude that the sequence $(y_k)_{k\in\mathbb{N}^*}$ is uniformly convergent on $[\,a, a+\delta\,]$ to the unique solution of $\mathcal{CP}(\mathcal{D})$ defined on that interval.

Finally, let us assume that f is Lipschitz on Δ. We then have

$$\|y_k(t) - x(t)\| \leq \int_a^t \|(\sigma_k(\tau), y_k(\sigma_k(\tau))) - f(\tau, x(\tau))\|\, d\tau$$

$$\leq \int_a^t \|(\sigma_k(\tau), y_k(\sigma_k(\tau))) - f(\tau, y_k(\tau))\| d\tau + \int_a^t \|f(\tau, y_k(\tau)) - f(\tau, x(\tau))\| d\tau$$

$$\leq L \int_a^t |\sigma_k(\tau) - \tau| \, d\tau + L \int_a^t \|y_k(\sigma_k(\tau)) - y_k(\tau)\| \, d\tau + L \int_a^t \|y_k(\tau) - x(\tau)\| \, d\tau,$$

for every $k \in \mathbb{N}^*$ and $t \in [\, a, a + \delta \,]$. Since by (3.3.3) and (3.3.1), we have

$$\|y_k(\sigma_k(\tau)) - y_k(\tau)\| \leq \frac{M\delta}{k} \text{ and } |\sigma_k(\tau) - \tau| \leq \frac{\delta}{k},$$

for every $k \in \mathbb{N}^*$ and $\tau \in [\, a, a + \delta \,]$, from the last three inequalities we deduce that

$$\|y_k(t) - x(t)\| \leq \frac{L\delta^2(M+1)}{k} + L \int_a^t \|y_k(\tau) - x(\tau)\| \, d\tau,$$

for every $k \in \mathbb{N}^*$ and $t \in [\, a, a+\delta \,]$. This inequality together with Gronwall's Lemma 1.5.2 shows that (3.3.5) holds. The proof is complete. □

Remark 3.3.1. In many books of differential equations, the method of polygonal lines is the main tool in the proof of Peano's local existence theorem. As an exercise, we suggest the interested reader to prove the latter theorem by using this method.

3.4 The Euler Implicit Method. The Exponential Formula

In this section we will present another method to approximate the solution of a Cauchy problem, a method closely related to the Euler method of polygonal lines. This method proved extremely useful for constructing effective algorithms for the numerical treatment of both ordinary differential equations, but especially of partial differential equations.

3.4.1 *The Semigroup Generated by $\mathcal{A}$*

Let $\mathcal{A} : \mathbb{R}^n \to \mathbb{R}^n$ be a continuous function. We recall that $\mathcal{A}$ is called dissipative if

$$\langle \mathcal{A}(x) - \mathcal{A}(y), x - y \rangle \leq 0$$

for every $x, y \in \mathbb{R}^n$. Let us consider the autonomous equation

$$x' = \mathcal{A}(x). \tag{3.4.1}$$

For each $a \in \mathbb{R}_+$ and $\xi \in \mathbb{R}^n$, we denote $x(\cdot, a, \xi)$ the unique saturated solution of equation (3.4.1) with $x(a, a, \xi) = \xi$. From Theorem 2.4.7 it follows that this solution is global, i.e. defined on $[a, +\infty)$.

For any fixed $t \geq 0$, let us define the operator, possibly non-linear, $\mathcal{S}(t) : \mathbb{R}^n \to \mathbb{R}^n$, by

$$\mathcal{S}(t)\xi = x(t, 0, \xi). \tag{3.4.2}$$

Here and thereafter $\mathcal{S}(t)\mathcal{S}(s)$ denotes the superposition operator defined by $[\mathcal{S}(t)\mathcal{S}(s)]\xi = \mathcal{S}(t)[\mathcal{S}(s)\xi]$ for each $\xi \in \mathbb{R}^n$.

Theorem 3.4.1. *The family of operators* $\{\mathcal{S}(t);\ t \geq 0\}$ *defined by means of the relation* (3.4.2) *satisfies*

(S_1) $\mathcal{S}(t+s) = \mathcal{S}(t)\mathcal{S}(s)$, *for every* $t, s \geq 0$;
(S_2) $\mathcal{S}(0) = \mathcal{I}$;
(S_3) $\lim_{t \downarrow 0} \mathcal{S}(t)\xi = \xi$ *for every* $\xi \in \mathbb{R}^n$;
(S_4) $\|\mathcal{S}(t)\xi - \mathcal{S}(t)\eta\| \leq \|\xi - \eta\|$ *for every* $t \geq 0$ *and every* $\xi, \eta \in \mathbb{R}^n$;
(S_5) $\lim_{t \downarrow 0} \frac{1}{t}\left(\mathcal{S}(t)\xi - \xi\right) = \mathcal{A}(\xi)$ *for every* $\xi \in \mathbb{R}^n$.

Proof. Let us remark that, for every $\xi \in \mathbb{R}^n$ and every $t, s \geq 0$, we have

$$x(t+s, 0, \xi) = x(t, 0, x(s, 0, \xi))$$

since both $t \mapsto x(t+s, 0, \xi)$ and $t \mapsto x(t, 0, x(s, 0, \xi))$ are solutions of the same Cauchy problem

$$\begin{cases} y' = \mathcal{A}(y) \\ y(0) = x(s, 0, \xi), \end{cases}$$

problem which, by virtue of the fact that $\mathcal{A}$ is dissipative, has the uniqueness property. But the equality above is equivalent to (S_1).

Obviously, (S_2) expresses nothing else but the initial condition satisfied by $x(\cdot, 0, \xi)$, i.e. $x(0, 0, \xi) = \xi$, while (S_3) follows from the continuity of the solution at $t = 0$.

In order to prove (S_4), let us observe that

$$x'(t, 0, \xi) - x'(t, 0, \eta) = \mathcal{A}(x(t, 0, \xi)) - \mathcal{A}(x(t, 0, \eta)),$$

from which, taking the inner product on both sides with $x(t, 0, \xi) - x(t, 0, \eta)$ and using Lemma 12.1.2, we deduce

$$\frac{1}{2}\frac{d}{dt}\left(\|x(t, 0, \xi) - x(t, 0, \eta)\|^2\right) \leq 0$$

for every $t \geq 0$. But this inequality obviously implies (S_4). Since (S_5) follows directly from the definition of the solution for the Cauchy problem, this completes the proof. □

Definition 3.4.1. The family of operators $\{\mathcal{S}(t);\ t \geq 0\}$ defined by means of (3.4.2) is *the semigroup of non-expansive operators generated by* $\mathcal{A}$.

Remark 3.4.1. The term semigroup is justified by the property (S_1), called *semigroup property*, which shows that the family $\{\mathcal{S}(t);\ t \geq 0\}$ is a semigroup with respect to the usual composition of mappings. The property (S_4) shows that each of the operators $\mathcal{S}(t)$ in the semigroup is non-expansive in the sense of Definition 2.5.2, while (S_5) expresses the fact that the semigroup $\{\mathcal{S}(t);\ t \geq 0\}$ is "generated" by $\mathcal{A}$.

3.4.2 *Two Auxiliary Lemmas*

In what follows we will present an approximation method for $\mathcal{S}(t)\xi$, called *the Euler implicit method.* In order to define the suitable approximating sequence, we need the following two lemmas, which are also of interest in themselves. In order to simplify the exposition, we will assume that $\mathcal{A}(0) = 0$, although all the results which follow remain valid without this technical restriction.

Lemma 3.4.1. *Let* $\mathcal{A} : \mathbb{R}^n \to \mathbb{R}^n$ *be a continuous and dissipative function with* $\mathcal{A}(0) = 0$. *Then, for every* $\lambda > 0$, *every* $y \in \mathbb{R}^n$ *and every* $\xi \in \mathbb{R}^n$, *the unique saturated solution* $x(\cdot, 0, \xi, y)$ *of the Cauchy problem*

$$\begin{cases} x' = \lambda\mathcal{A}(x) - x + y \\ x(0) = \xi \end{cases} \tag{3.4.3}$$

is defined on $[\,0, +\infty)$ *and there exists* $\lim_{t\uparrow+\infty} x(t, 0, \xi, y) = \tilde{y}$. *In addition,* $\tilde{y}$ *is independent of* ξ *and satisfies* $\tilde{y} - \lambda\mathcal{A}(\tilde{y}) = y$.

Proof. Let $\lambda > 0$, $y \in \mathbb{R}^n$ and $\xi \in \mathbb{R}^n$. Since $\lambda > 0$ and $\mathcal{A}$ is dissipative, the function $x \mapsto \lambda\mathcal{A}(x) - x + y$ is also dissipative. Therefore, in view of Theorem 2.4.7, for every $\xi \in \mathbb{R}^n$, the problem (3.4.3) has a unique global solution. Since in the first part of the proof, λ, ξ and y are fixed, for the sake of simplicity, we denote this solution by $x : \mathbb{R}_+ \to \mathbb{R}^n$.

We will show first that x is bounded on $\mathbb{R}_+$. To this aim, let us take the inner product on both sides in (3.4.3) with $x(t)$ and let us observe that, from Lemma 12.1.2[1] and from the dissipativity condition, recalling that $\mathcal{A}(0) = 0$, we have

$$\frac{1}{2}\frac{d}{dt}\left(\|x(t)\|^2\right) \leq \langle y, x(t)\rangle - \|x(t)\|^2$$

[1]See Chapter 10 Auxiliary Results.

for every $t \geq 0$. Using the Cauchy-Schwarz inequality in order to evaluate the first term on the right-hand side and then the mean's inequality, we deduce $\frac{d}{dt}\left(\|x(t)\|^2\right) \leq \|y\|^2 - \|x(t)\|^2$ for every $t \in \mathbb{R}_+$. Multiplying both sides of this inequality by e^t, rearranging the terms and integrating from 0 to t, we get successively

$$e^t\|x(t)\|^2 \leq \|\xi\|^2 + \int_0^t e^s\|y\|^2 ds$$

and $\|x(t)\|^2 \leq e^{-t}\|\xi\|^2 + \left(1 - e^{-t}\right)\|y\|^2$ for every $t \geq 0$. From here, it readily follows that x is bounded on $\mathbb{R}_+$.

We can now proceed to the proof of the fact that there exists

$$\lim_{t\uparrow+\infty} x(t) = \tilde{y}. \tag{3.4.4}$$

To this aim, let $t \geq 0$ and $h \geq 0$. Multiplying both sides of the equality

$$x'(t+h) - x'(t) = \lambda(\mathcal{A}(x(t+h)) - \mathcal{A}(x(t))) - (x(t+h) - x(t))$$

by $(x(t+h) - x(t))$ and taking into account the dissipativity of the function $\lambda\mathcal{A}$, we get

$$\frac{1}{2}\frac{d}{dt}\left(\|x(t+h) - x(t)\|^2\right) \leq -\|x(t+h) - x(t)\|^2.$$

Multiplying both sides of this inequality by e^{2t} and integrating from 0 to t, we deduce that

$$\|x(t+h) - x(t)\| \leq e^{-t}\|x(h) - \xi\|.$$

But the inequality above, combined with the boundedness of the function x, shows that x satisfies Cauchy's condition for the existence of the limit as t tends to $+\infty$, and this proves (3.4.4).

Finally, let us remark that, from the mean value's theorem, for every $m \in \mathbb{N}$ and every $i = 1, 2, \ldots, n$, there exists $\theta_m^i \in [\,m, m+1\,]$ such that

$$x_i(m+1) - x_i(m) = x_i'(\theta_m^i).$$

Passing to the limit for m tending to $+\infty$ in these equalities, we deduce that $\lim_{m\uparrow+\infty} x_i'(\theta_m^i) = 0$ for $i = 1, 2, \ldots, n$. Since $\lambda\mathcal{A} - \mathcal{I}$ is continuous, from (3.4.4), we conclude that there exists $\lim_{t\uparrow+\infty} x'(t) = 0$. Passing to the limit in equation (3.4.3) we deduce that $\tilde{y}$ is the solution of the equation $\tilde{y} - \lambda\mathcal{A}(\tilde{y}) = y$. Finally, in order to prove that $\tilde{y}$ does not depend on ξ, it suffices to show that the equation above has a unique solution. So, let $\tilde{z}$ be a solution of the equation. Then $\tilde{y} - \tilde{z} - \lambda(\mathcal{A}(\tilde{y}) - \mathcal{A}(\tilde{z})) = 0$. Taking the inner product on both sides with $\tilde{y} - \tilde{z}$, recalling that $\lambda > 0$ and $\mathcal{A}$ is dissipative, we deduce that $\|\tilde{y} - \tilde{z}\|^2 \leq 0$, which proves the uniqueness of the solution of the problem above. The proof is complete. □

Lemma 3.4.2. *Let $\mathcal{A} : \mathbb{R}^n \to \mathbb{R}^n$ be a continuous and dissipative function with $\mathcal{A}(0) = 0$. Then, for every $\lambda > 0$, $\mathcal{I} - \lambda\mathcal{A}$ is bijective, and its inverse in the sense of relations, $(\mathcal{I} - \lambda\mathcal{A})^{-1}$, is a non-expansive function.*

Proof. From Lemma 3.4.1 we know that, for every $\lambda > 0$, every $y \in \mathbb{R}^n$ and every $\xi \in \mathbb{R}^n$, there exists the limit as t approaches $+\infty$ of the unique saturated solution $x(\cdot, 0, \xi, y)$ of the problem (3.4.3), and this limit $\tilde{y}$ is the unique solution of the equation $\tilde{y} - \lambda\mathcal{A}(\tilde{y}) = y$. Since y is arbitrary in $\mathbb{R}^n$, it follows that $\mathcal{I} - \lambda\mathcal{A}$ is surjective. On the other hand, from the uniqueness of the solution of the equation above, it follows that $\mathcal{I} - \lambda\mathcal{A}$ is injective.

Finally, let $y_1, y_2 \in \mathbb{R}^n$ and let $\tilde{y}_1, \tilde{y}_2 \in \mathbb{R}^n$ such that $\tilde{y}_1 - \lambda\mathcal{A}(\tilde{y}_1) = y_1$ and $\tilde{y}_2 - \lambda\mathcal{A}(\tilde{y}_2) = y_2$. Subtracting the two equalities and multiplying the equality thus obtained by $\tilde{y}_1 - \tilde{y}_2$, we deduce

$$\|\tilde{y}_1 - \tilde{y}_2\|^2 \leq \langle y_1 - y_2, \tilde{y}_1 - \tilde{y}_2\rangle \leq \|y_1 - y_2\|\|\tilde{y}_1 - \tilde{y}_2\|.$$

Taking into account that $\tilde{y}_i = (\mathcal{I} - \lambda\mathcal{A})^{-1}(y_i)$ for $i = 1, 2$, from the preceding inequality, we get

$$\|(\mathcal{I} - \lambda\mathcal{A})^{-1}(y_1) - (\mathcal{I} - \lambda\mathcal{A})^{-1}(y_2)\| \leq \|y_1 - y_2\|,$$

for every $y_1, y_2 \in \mathbb{R}^n$. So $(\mathcal{I} - \lambda\mathcal{A})^{-1}$ is non-expansive. The proof of Lemma 3.4.2 is complete. □

3.4.3 *The Exponential Formula*

Let $T > 0$, $k \in \mathbb{N}^*$ and let $\mathcal{P}_k : 0 = t_0 < t_1 < \cdots < t_k = T$ be a partition with equidistant points, i.e. $t_{i+1} - t_i = T/k$ for every $i = 0, 1, \ldots, k-1$.

The main idea of the Euler implicit method is to approximate the solution of the integral equation

$$x(t) = \xi + \int_0^t \mathcal{A}(x(\tau))\, d\tau, \tag{$\mathcal{IE}$}$$

which is equivalent to the Cauchy problem

$$\begin{cases} x' = \mathcal{A}(x) \\ x(0) = \xi, \end{cases} \tag{$\mathcal{CP}$}$$

by means of the solution $\xi_0, \xi_1, \ldots, \xi_k$ of the system

$$\begin{cases} \xi_0 = \xi \\ \xi_{i+1} = \xi_i + (t_{i+1} - t_i)\mathcal{A}(\xi_{i+1}), \text{ for } i = 0, 1, \ldots, k-1. \end{cases} \tag{$\mathcal{S}_k$}$$

Namely, we shall show that the function $y_k : [0,T] \to \mathbb{R}^n$, defined by

$$y_k(t) = \xi_i + (t - t_i)\mathcal{A}(\xi_{i+1}) \tag{3.4.5}$$

for $t \in [t_i, t_{i+1})$ if $i = 0, 1, \ldots, k-2$, or $t \in [t_i, t_{i+1}]$ if $i = k-1$, is "sufficiently close" to the unique solution $x : [0,T] \to \mathbb{R}^n$ of $(\mathcal{IE})$, for k "big enough".

We notice that the name of the method is suggested by the remark that, this time, ξ_{i+1} is implicitly defined as a function of ξ_i, and not explicitly, as in the case of the method of polygonal lines described in the preceding section. Although, at first glance, more complicated than the latter, the Euler implicit method is extremely useful, especially in the case (which we will not touch upon here) when $\mathcal{A}$ is defined on a "rather small" subset D of $\mathbb{R}^n$, the case in which the method of polygonal lines might not be applicable, simply because for a certain i, ξ_{i+1} could leave D. We emphasize that this weak point of the method of polygonal lines is much more evident in the case of partial differential equations, where $\mathcal{A}$ is discontinuous and defined on a proper subset D of a function space X, while $(\mathcal{I} - \lambda\mathcal{A})^{-1}$ is non-expansive and defined on the whole space X.

Lemma 3.4.3. *If $\mathcal{A} : \mathbb{R}^n \to \mathbb{R}^n$ is continuous, dissipative and, in addition, $\mathcal{A}(0) = 0$ then, for every $k \in \mathbb{N}^*$, $(\mathcal{S}_k)$ has a unique solution $(\xi_0, \xi_1, \ldots, \xi_k)$. Moreover, for every $k \in \mathbb{N}$ and every $i = 1, 2, \ldots k$, we have*

$$\|\xi_i\| \leq \|\xi\|. \tag{3.4.6}$$

Proof. Let us observe that $(\mathcal{S}_k)$ may be equivalently rewritten in the form

$$\begin{cases} \xi_0 = \xi \\ \xi_{i+1} = (\mathcal{I} - (t_{i+1} - t_i)\mathcal{A})^{-1}(\xi_i), \text{ for } i = 0, 1, \ldots, k-1. \end{cases} \tag{3.4.7}$$

According to Lemma 3.4.2, the operator $(\mathcal{I} - (t_{i+1} - t_i)\mathcal{A})^{-1}$ is defined on $\mathbb{R}^n$, is non-expansive and satisfies $(\mathcal{I} - (t_{i+1} - t_i)\mathcal{A})^{-1}(0) = 0$, for $i = 1, 2, \ldots, k$. Therefore, it follows that (3.4.7) has a unique solution. In addition, as $(\mathcal{I} - (t_{i+1} - t_i)\mathcal{A})^{-1}$ is non-expansive and vanishes at 0, we have

$$\begin{aligned} \|(\mathcal{I} - (t_{i+1} - t_i)\mathcal{A})^{-1}(\xi_i)\| \leq \|\xi_i\| &= \|(\mathcal{I} - (t_i - t_{i-1})\mathcal{A})^{-1}(\xi_{i-1})\| \\ &\leq \|\xi_{i-1}\| \leq \cdots \leq \|\xi\| \end{aligned}$$

and this completes the proof. □

Theorem 3.4.2. *If $\mathcal{A} : \mathbb{R}^n \to \mathbb{R}^n$ is continuous, dissipative and $\mathcal{A}(0) = 0$, then the sequence of functions defined by* (3.4.5) *is uniformly convergent on $[0,T]$ to the unique solution of $\mathcal{CP}(\mathcal{D})$.*

Proof. Let $k \in \mathbb{R}^n$ and let us define the function $\sigma_k : [\,0,T\,] \to [\,0,T\,]$ by

$$\sigma_k(t) = t_{i+1}$$

for $t \in [\,t_i, t_{i+1})$ if $i = 1, 2, \ldots, k-2$ or $t \in [\,t_i, t_{i+1}\,]$ for $i = k-1$. Since $t_{i+1} - t_i = T/k$, we have

$$\lim_{k\to\infty} |\sigma_k(t) - t| = 0$$

uniformly on $[\,0,T\,]$. Now, let us observe that equality (3.4.5) may be equivalently rewritten in the integral form

$$y_k(t) = \xi + \int_0^t \mathcal{A}(y_k(\sigma_k(\tau)))\,d\tau. \tag{3.4.8}$$

Since $\mathcal{A}$ is continuous, its restriction to any compact set is bounded. Hence, there exists $M > 0$ such that

$$\|\mathcal{A}(y)\| \le M$$

for every $y \in \mathbb{R}^n$ with $\|y\| \le \|\xi\|$. Then, by virtue of the inequality (3.4.6) and the definition of the function σ_k, we have

$$\|\mathcal{A}(y_k(\sigma_k(t)))\| = \|\mathcal{A}(y_k(t_{i+1}))\| = \|\mathcal{A}(\xi_{i+1})\| \le M$$

for every $t \in [\,t_i, t_{i+1}\,]$, $i = 0, 1, \ldots k-1$. From this inequality and from (3.4.8), we deduce

$$\|y_k(t)\| \le \|\xi\| + TM$$

for every $k \in \mathbb{N}$ and every $t \in [\,0,T\,]$. Consequently, the family of functions $\{y_k;\ k \in \mathbb{N}\}$ is uniformly bounded on $[\,0,T\,]$. Again from (3.4.8), we deduce

$$\|y_k(t) - y_k(s)\| \le M|t - s|$$

for every $k \in \mathbb{N}$ and every $t, s \in [\,0,T\,]$. From here it follows that the family $\{y_k;\ k \in \mathbb{N}\}$ is equi-continuous on $[\,0,T\,]$. According to Arzelà-Ascoli's Theorem 12.2.1, it follows that, at least on a subsequence, we have

$$\lim_{k\to\infty} y_k(t) = y(t)$$

uniformly on $[\,0,T\,]$, where y is the unique solution of $\mathcal{CP}(\mathcal{D})$. Since the limit of any convergent subsequence of the sequence $(y_k)_{k\in\mathbb{N}}$ is a solution of $\mathcal{CP}(\mathcal{D})$ which has the uniqueness property, using once again the fact that the family $\{y_k;\ k \in \mathbb{N}\}$ is relatively compact in the uniform convergence topology, we conclude that the sequence $(y_k)_{k\in\mathbb{N}}$ itself converges uniformly on $[\,0,T\,]$ to y. The proof is complete. □

Theorem 3.4.3. (The Exponential Formula) *If* $\mathcal{A} : \mathbb{R}^n \to \mathbb{R}^n$ *is continuous, dissipative and* $\mathcal{A}(0) = 0$ *then, for every* $\xi \in \mathbb{R}^n$, *we have*

$$\lim_{k\to\infty} \left(\mathcal{I} - \frac{t}{k}\mathcal{A}\right)^{-k} (\xi) = \mathcal{S}(t)\xi$$

uniformly for t *in bounded subsets of* $\mathbb{R}_+$.

Proof. Let $t > 0$, $k \in \mathbb{N}^*$ and let

$$\xi_{i+1} = \left(\mathcal{I} - \frac{t}{k}\mathcal{A}\right)^{-1} (\xi_i), \quad \xi_0 = \xi$$

for $i = 0, 1, \ldots k-1$. Obviously ξ_i represents the solution of the system $(\mathcal{S}_k)$ in the case when T is replaced by t. Taking $i = k-1$ in the preceding relation, we successively deduce

$$\xi_k = \left(\mathcal{I} - \frac{t}{k}\mathcal{A}\right)^{-1} (\xi_{k-1}) = \left(\mathcal{I} - \frac{t}{k}\mathcal{A}\right)^{-2} (\xi_{k-2}) = \cdots = \left(\mathcal{I} - \frac{t}{k}\mathcal{A}\right)^{-k} (\xi).$$

The conclusion follows from Theorem 3.4.2 by observing that, for any fixed $\xi \in \mathbb{R}^n$, and $T > 0$, the sequence $(y_k)_{k\in\mathbb{N}^*}$ is uniformly bounded (with respect to $k \in \mathbb{N}^*$) on $[0, T]$ by a constant $C(\xi, T) = \|\xi\| + TM$ which is nondecreasing as a function of T. The proof is complete. □

We notice that the name "Exponential Formula" comes from the simple observation that, in the one-dimensional linear case, i.e. when the function $\mathcal{A} : \mathbb{R} \to \mathbb{R}$ is defined by $\mathcal{A}(y) = ay$, with $a < 0$, this formula reduces to the well-known equality

$$\lim_{k\to\infty} \left(1 - \frac{ta}{k}\right)^{-k} \xi = e^{ta}\xi.$$

We conclude this section with the mention that, if $\mathcal{A}(0) \neq 0$, then, by defining the operator $\mathcal{B} : \mathbb{R}^n \to \mathbb{R}^n$ by $\mathcal{B}(y) = \mathcal{A}(y) - \mathcal{A}(0)$, for $y \in \mathbb{R}^n$, we can easily see that this is continuous, dissipative and satisfies $\mathcal{B}(0) = 0$. In addition $\mathcal{CP}(\mathcal{D})$ may be rewritten in the form

$$\begin{cases} y' = \mathcal{B}(y) + h \\ y(0) = \xi, \end{cases}$$

with $h = \mathcal{A}(0)$. We leave it to the reader to prove that, with some obvious modifications, all previous considerations also remain valid in this case. To all readers interested in extensions of the results in this section we recommend the monograph [Barbu (1976)].

3.5 Exercises and Problems

Exercise 3.1. *Solve the following Cauchy problems by expanding the solution in power series and by identifying the coefficients:*

$$(1)\begin{cases}(1-t)x' = 1+t-x\\ x(0)=0.\end{cases} \qquad (2)\begin{cases}tx''+x=0\\ x(0)=0,\ x'(0)=1.\end{cases}$$

$$(3)\begin{cases}tx''+x'+tx=0\\ x(0)=1,\ x'(0)=0.\end{cases} \qquad (4)\begin{cases}tx''+2x'+tx=0\\ x(0)=1,\ x'(0)=0.\end{cases}$$

$$(5)\begin{cases}(1-t)x''-x'+tx=0\\ x(0)=1,\ x'(0)=1.\end{cases} \qquad (6)\begin{cases}x''-2tx'-2x=0\\ x(0)=1,\ x'(0)=0.\end{cases}$$

Problem 3.1. *Integrate the Hermite equation*

$$x'' - 2tx' + 2\lambda x = 0,$$

where $\lambda \in \mathbb{R}$. *Show that the equation has as solution a non-identically zero polynomial if and only if* $\lambda \in \mathbb{N}$.

Problem 3.2. *Show that every solution of the equation* [2]

$$-x'' + t^2x = (2\lambda+1)x,$$

where $\lambda \in \mathbb{R}$, *is of the form* $x(t) = y(t)e^{-t^2/2}$, *where* y *satisfies the Hermite equation*

$$y'' - 2ty' + 2\lambda y = 0.$$

Prove that, if $\lambda \in \mathbb{N}$, *at least one non-identically zero solution* x *of the equation is bounded on* $\mathbb{R}_+$.

Exercise 3.2. *Find the general solution of the Airy equation*

$$x'' - tx = 0.$$

Problem 3.3. *Find the solutions of the form* $x(t) = t^\alpha \sum_{k=0}^\infty c_k t^k$ *with* $c_0 \neq 0$ *of the Bessel equation*

$$t^2x'' + tx' + (t^2 - n^2)x = 0,$$

where $n \in \mathbb{N}$. *Show that*

$$x(t) = \sum_{k=0}^{\infty} \frac{(-1)^k (t/2)^{n+2k}}{k!(n+k)!}$$

is such a solution.

[2]Equation used in the oscillator theory in Quantum Mechanics.

Exercise 3.3. *Find the solution of the Cauchy problem for the Gauss equation*

$$\begin{cases} t(1-t)x'' + [c-(a+b+1)t]x' - abx = 0 \\ x(0) = 1, \ x'(0) = (ab)/c, \end{cases}$$

where $c > 0$.

Problem 3.4. *Show that Legendre equation*

$$(1-t^2)x'' - 2tx' + \lambda(\lambda+1)x = 0,$$

where $\lambda \in \mathbb{R}$, *has polynomials as local solutions about* $t = 0$ *if and only if* $\lambda \in \mathbb{N}$.

Problem 3.5. *Show that the first term in the sequence of successive approximations may be any continuous function* $x_0 : [\,a, a+\delta\,] \to B(\xi, r)$, *without affecting its uniform convergence to the unique solution of* $\mathcal{PC}(\mathcal{D})$ *on* $[\,a, a+\delta\,]$. *Find an error estimate of the type* (3.4.6) *in this general case.*

Problem 3.6. *Prove Theorem* 2.3.2 *by the method of successive approximations.*

Problem 3.7. *Let* $f : [\,a,b\,] \to \mathbb{R}^n$ *and* $g : [\,a,b\,] \times [\,a,b\,] \times \mathbb{R}^n \to \mathbb{R}^n$ *be continuous and let us assume in addition that* g *is Lipschitz on* $\mathbb{R}^n$, *i.e. there exists* $L > 0$ *such that for every* $(t,s,x), (t,s,y) \in [\,a,b\,] \times [\,a,b\,] \times \mathbb{R}^n$, *we have*

$$\|g(t,s,x) - g(t,s,y)\| \le L\|x-y\|.$$

Using the sequence of successive approximations, show that the Volterra integral equation

$$x(t) = f(t) + \int_a^t g(t,\tau,x(\tau))d\tau$$

has a unique solution $x : [\,a,b\,] \to \mathbb{R}^n$.

Problem 3.8. *Let* $f : [\,a,b\,] \times \mathbb{R}^n \to \mathbb{R}^n$ *and* $g : [\,a,b\,] \times [\,a,b\,] \times \mathbb{R}^n \to \mathbb{R}^n$ *be two continuous functions on* $[\,a,b\,] \times \mathbb{R}^n$ *and on* $[\,a,b\,] \times [\,a,b\,] \times \mathbb{R}^n$ *respectively, and Lipschitz on* $\mathbb{R}^n$. *Using the sequence of successive approximations, show that, for every* $\xi \in \mathbb{R}^n$, *the Cauchy problem for the integro-differential equation*

$$\begin{cases} x'(t) = f(t,x(t)) + \displaystyle\int_a^t g(t,\tau,x(\tau))d\tau \\ x(a) = \xi \end{cases}$$

has a unique solution $x : [\,a,b\,] \to \mathbb{R}^n$.

Problem 3.9. *Let* $\mathcal{A} : \mathbb{R}^n \to \mathbb{R}^n$ *be continuous and dissipative. Prove that for every continuous function* $h : [\,0,T\,] \to \mathbb{R}^n$ *and every* $\xi \in \mathbb{R}^n$, *the Cauchy problem*

$$\begin{cases} x' = \mathcal{A}(x) + h(t) \\ x(0) = \xi \end{cases} \qquad \mathcal{CP}(h,\xi)$$

has a unique solution $x(\cdot,h,\xi) : [0,T] \to \mathbb{R}^n$. *Show that for every continuous functions* $h_i : [0,T] \to \mathbb{R}^n$ *and every* $\xi_i \in \mathbb{R}^n$, *the functions* $x_i = x(\cdot,h_i,\xi_i)$, $i = 1,2$, *satisfy*

$$\|x_1(t) - x_2(t)\|^2 \leq \|\xi_1 - \xi_2\|^2 + 2\int_0^t \langle h_1(s) - h_2(s), x_1(s) - x_2(s)\rangle ds$$

and

$$\|x_1(t) - x_2(t)\| \leq \|\xi_1 - \xi_2\| + \int_0^t \|h_1(s) - h_2(s)\| ds$$

for every $t \in [0,T]$.

Problem 3.10. *Let* $\mathcal{A} : \mathbb{R}^n \to \mathbb{R}^n$ *be a continuous and dissipative function on* $\mathbb{R}^n$ *and let* $f : [0,T] \times \mathbb{R}^n \to \mathbb{R}^n$ *be continuous on* $[0,T] \times \mathbb{R}^n$ *and Lipschitz on* $\mathbb{R}^n$. *Let* $\xi \in \mathbb{R}^n$, *and let us define the sequence of successive approximations:* $x_0(t) = \xi$, $x_k(t) = x(t, f(t, x_{k-1}(t)), \xi)$ *for* $k = 1,2,\ldots$ *and* $t \in [0,T]$, *where* $x(\cdot, f(\cdot, x_{k-1}(\cdot)), \xi) : [0,T] \to \mathbb{R}^n$ *is the unique solution of the Cauchy problem*

$$\begin{cases} x_k' = \mathcal{A}(x_k) + f(t, x_{k-1}(t)) \\ x_k(0) = \xi \end{cases} \qquad \mathcal{CP}(k)$$

for $k = 1,2,\ldots$. *Using the second inequality established in Problem* 3.9, *prove that the sequence* $(x_k)_{k\in\mathbb{N}}$ *is uniformly convergent on* $[0,T]$ *to the unique solution of the Cauchy problem*

$$\begin{cases} x' = \mathcal{A}(x) + f(t,x) \\ x(0) = \xi. \end{cases} \qquad \mathcal{CP}(f,\xi)$$

Problem 3.11. *Let* $\mathcal{A} : \mathbb{R}^n \to \mathbb{R}^n$ *and* $f : \mathbb{R} \to \mathbb{R}^n$ *be two continuous functions. Let us assume that there exists* $\omega > 0$ *such that* $\langle \mathcal{A}(x) - \mathcal{A}(y), x - y\rangle \leq -\omega^2\|x-y\|^2$ *for every* $x, y \in \mathbb{R}^n$ *and let* $T > 0$ *be fixed. Let us define the function*[3] $\mathcal{P} : \mathbb{R}^n \to \mathbb{R}^n$ *by* $\mathcal{P}(\xi) = x(T,0,\xi)$, *where* $x(\cdot,0,\xi)$ *is the unique global solution of the Cauchy problem*

$$\begin{cases} x' = \mathcal{A}(x) + f(t) \\ x(0) = \xi. \end{cases} \qquad \mathcal{CP}(\xi)$$

We define the sequence of successive approximations $\xi_0 = \xi$ *and* $\xi_k = \mathcal{P}(\xi_{k-1})$ *for* $k \in \mathbb{N}^*$.

(1) *Prove that the sequence* $(\xi_k)_{k\in\mathbb{N}}$ *is convergent to an element* $\eta \in \mathbb{R}^n$.

(2) *Prove that the unique global solution of* $\mathcal{CP}(\eta)$ *with* $\eta = \lim_{k\to\infty} \xi_k$ *satisfies* $x(T,0,\eta) = x(0,0,\eta) = \eta$.

(3) *If, in addition,* f *is periodic of period* $T > 0$, *then the unique global solution* $x(\cdot,0,\eta)$ *of* $\mathcal{CP}(\eta)$ *is periodic of period* T.

(4) *The equation* $x' = \mathcal{A}(x) + f(t)$ *has at most one* T*-periodic solution.*

[3]This function is known as *the Poincaré's* mapping.

Chapter 4

Systems of Linear Differential Equations

This chapter contains the most important results referring to the Cauchy problem governed by a system of n first-order linear differential equations with n unknown functions. In the first section we show that the set of all saturated solutions of such a homogeneous system is an n-dimensional vector space over $\mathbb{R}$. The second section is dedicated to the study of non-homogeneous systems. It includes the celebrated *variation of constants formula.* In the third and fourth sections we present two methods of finding an algebraic basis in the space of all saturated solutions of a homogeneous system with constant coefficients. The aim of the fifth section is to rephrase the previously proved results in order to handle as particular case the nth-order linear differential equation, while the sixth section is dedicated to a simple method of solving explicitly such equations with constant coefficients. The chapter ends with a section of exercises and problems.

4.1 Homogeneous Systems. The Space of Solutions

Let $a_{ij} : \mathbb{I} \to \mathbb{R}$ and $b_i : \mathbb{I} \to \mathbb{R}$ be continuous for $i, j = 1, 2, \ldots, n$, and let us consider the system of first-order linear differential equations

$$\begin{cases} x_1' = a_{11}(t)x_1 + a_{12}(t)x_2 + \cdots + a_{1n}(t)x_n + b_1(t) \\ x_2' = a_{21}(t)x_1 + a_{22}(t)x_2 + \cdots + a_{2n}(t)x_n + b_2(t) \\ \vdots \\ x_n' = a_{n1}(t)x_1 + a_{n2}(t)x_2 + \cdots + a_{nn}(t)x_n + b_n(t), \end{cases} \tag{4.1.1}$$

which, with the notations

$$x = \begin{pmatrix} x_1 \\ x_2 \\ \vdots \\ x_n \end{pmatrix}, \quad b(t) = \begin{pmatrix} b_1(t) \\ b_2(t) \\ \vdots \\ b_n(t) \end{pmatrix}$$

and

$$\mathcal{A}(t) = \begin{pmatrix} a_{11}(t) & a_{12}(t) \dots & a_{1n}(t) \\ a_{21}(t) & a_{22}(t) \dots & a_{2n}(t) \\ \vdots & & \\ a_{n1}(t) & a_{n2}(t) \dots & a_{nn}(t) \end{pmatrix},$$

for $t \in \mathbb{I}$, can be rewritten as a first-order vector differential equation

$$x' = \mathcal{A}(t)x + b(t). \tag{4.1.2}$$

For the sake of simplicity, in all what follows, we will write the system (4.1.1) only in the form of a vector differential equation (4.1.2), and we will call it by extension *first-order system of linear differential equations.* We also point out that, throughout this section, all the vectors considered are column vectors.

Definition 4.1.1. The system (4.1.2) is said to be *homogeneous* if b is identically 0 on $\mathbb{I}$ and *non-homogeneous* if b is not identically 0 on $\mathbb{I}$.

From Corollary 2.4.4, we deduce:

Theorem 4.1.1. *For every $a \in \mathbb{I}$ and every $\xi \in \mathbb{R}^n$ the Cauchy problem*

$$\begin{cases} x' = \mathcal{A}(t)x + b(t) \\ x(a) = \xi \end{cases}$$

has a unique global solution.

In turn, Theorem 4.1.1 implies:

Theorem 4.1.2. *Every saturated solution of the system* (4.1.2) *is defined on* $\mathbb{I}$.

Let us consider also the homogeneous system, i.e.

$$x' = \mathcal{A}(t)x, \tag{4.1.3}$$

called *the homogeneous system attached to* (4.1.2).

Theorem 4.1.3. *The set of all saturated solutions of the homogeneous system* (4.1.3) *is an n-dimensional vector space over $\mathbb{R}$.*

Proof. In view of Theorem 4.1.2, each saturated solution of (4.1.3) is global, and thus defined on $\mathbb{I}$. We will show that the set of all these solutions, which is included in $C^1(\mathbb{I};\mathbb{R}^n)$ is a vector subspace isomorphic to $\mathbb{R}^n$. Let $\mathcal{S}$ be the set of all saturated solutions of (4.1.3), let $x, y \in \mathcal{S}$ and $\alpha, \beta \in \mathbb{R}$. We will check that $\alpha x + \beta y \in \mathcal{S}$, from where it will follow that $\mathcal{S}$ is a vector subspace of $C^1(\mathbb{I};\mathbb{R}^n)$. Indeed, let us observe that

$$(\alpha x + \beta y)'(t) = \alpha x'(t) + \beta y'(t) = \alpha \mathcal{A}(t)x(t) + \beta \mathcal{A}(t)y(t)$$

$$= \mathcal{A}(t)[\alpha x(t) + \beta y(t)] = \mathcal{A}(t)[(\alpha x + \beta y)(t)]$$

for every $t \in \mathbb{I}$, relation which proves that $\alpha x + \beta y \in \mathcal{S}$.

We may now proceed to the definition of an isomorphism between $\mathcal{S}$ and $\mathbb{R}^n$. More precisely, let us fix $a \in \mathbb{I}$, and let us define $\mathcal{T} : \mathcal{S} \to \mathbb{R}^n$ by

$$\mathcal{T}(x) = x(a)$$

for every $x \in \mathcal{S}$. Obviously $\mathcal{T}$ is linear. In addition, from the uniqueness part of Theorem 4.1.1, we conclude that $\mathcal{T}$ is injective, while from the existence part of the same Theorem 4.1.1 it follows that $\mathcal{T}$ is surjective. So $\mathcal{T}$ is an isomorphism of vector spaces. As each two isomorphic vector spaces have the same dimension, the dimension of $\mathcal{S}$ is n. The proof is complete. □

Remark 4.1.1. Theorem 4.1.3 is of crucial importance in the theory of first-order homogeneous systems of linear differential equations because it shows that, within this framework, in order to find the general solution of the system, it suffices to find only n linearly independent saturated solutions.[1] Indeed, Theorem 4.1.3 asserts that, in $\mathcal{S}$, every algebraic basis has exactly n elements. On the other hand, if $x^1, x^2, \ldots, x^n \in \mathcal{S}$ is an algebraic basis, every element $x \in \mathcal{S}$ can be written in a unique way as a linear combination of elements in the basis, i.e. there exist $c_1, c_2, \ldots, c_n \in \mathbb{R}$, uniquely determined, such that

$$x(t) = \sum_{i=1}^{n} c_i x^i(t) \tag{4.1.4}$$

for every $t \in \mathbb{I}$. In other words, if we know a family of n linear independent saturated solutions of the system (4.1.3), we know any other solution, and thus the general solution.

[1] We recall that $x^1, x^2, \ldots, x^n \in \mathcal{S}$ are linearly independent if from $\sum_{i=1}^n c_i x^i(t) = 0$ for every $t \in \mathbb{I}$ it follows $c_1 = c_2 = \cdots = c_n = 0$.

Remark 4.1.2. In accordance with Remark 4.1.1, a fundamental problem in the study of the system (4.1.3) consists in finding at least one algebraic basis in the space of all saturated solutions. We emphasize that no general method of finding such basis is known. A remarkable and, at the same time, very important exception is that when the matrix $\mathcal{A}$ is constant. This case will be thoroughly analyzed in a forthcoming section.

We will present next a simple method of checking whether or not n saturated solutions of the system (4.1.3) are linearly independent. Let $x^1, x^2, \dots, x^n$ be n saturated solutions of the system (4.1.3), and let us define the matrix $\mathcal{X} : \mathbb{I} \to \mathcal{M}_{n\times n}(\mathbb{R})$ by $\mathcal{X}(t) = \mathrm{col}(x^1(t), x^2(t), \dots, x^n(t))$ for every $t \in \mathbb{I}$, i.e. the matrix whose columns at $t \in \mathbb{I}$ are the vectors $x^1(t), x^2(t), \dots, x^n(t)$. More precisely

$$\mathcal{X}(t) = \begin{pmatrix} x_1^1(t) & x_1^2(t) \dots & x_1^n(t) \\ x_2^1(t) & x_2^2(t) \dots & x_2^n(t) \\ \vdots & & \\ x_n^1(t) & x_n^2(t) \dots & x_n^n(t) \end{pmatrix} \tag{4.1.5}$$

for every $t \in \mathbb{I}$.

Definition 4.1.2. The matrix $\mathcal{X}$ defined by (4.1.5) is called the *associated matrix* of the system of solutions $x^1, x^2, \dots, x^n \in \mathcal{S}$.

Remark 4.1.3. Since each column of the associated matrix $\mathcal{X}$ of (4.1.3) is a solution of that system, it follows that $\mathcal{X} : \mathbb{I} \to \mathcal{M}_{n\times n}(\mathbb{R})$ is a solution of the matrix system

$$\mathcal{X}' = \mathcal{A}(t)\mathcal{X}.$$

Definition 4.1.3. The system $x^1, x^2, \dots, x^n \in \mathcal{S}$ is called a *fundamental system of solutions* of equation (4.1.3) if it is an algebraic basis in $\mathcal{S}$.

Definition 4.1.4. The matrix associated to a fundamental system of solutions of equation (4.1.3) is called a *fundamental matrix* of the system (4.1.3).

Remark 4.1.4. We notice that (4.1.3) has infinitely many fundamental matrices. This follows from the observation that the space of saturated solutions of the system (4.1.3) has infinitely many algebraic bases.

Remark 4.1.5. If $\mathcal{X}$ is a fundamental matrix for the system (4.1.3), then the general solution of (4.1.3) is given by

$$x(t, c) = \mathcal{X}(t)c \tag{4.1.6}$$

for $t \in \mathbb{I}$ and $c \in \mathbb{R}^n$. Indeed, (4.1.6) represents nothing but the matrix version of the relation (4.1.4), because

$$\mathcal{X}(t)c = \begin{pmatrix} x_1^1(t) \; x_1^2(t) \ldots \; x_1^n(t) \\ x_2^1(t) \; x_2^2(t) \ldots \; x_2^n(t) \\ \vdots \\ x_n^1(t) \; x_n^2(t) \ldots \; x_n^n(t) \end{pmatrix} \begin{pmatrix} c_1 \\ c_2 \\ \vdots \\ c_n \end{pmatrix} = \sum_{i=1}^n c_i \begin{pmatrix} x_1^i(t) \\ x_2^i(t) \\ \vdots \\ x_n^i(t) \end{pmatrix}$$

$$= \sum_{i=1}^n c_i x^i(t).$$

Definition 4.1.5. If $\mathcal{X}$ is the matrix associated to a system of solutions $x^1, x^2, \ldots, x^n$ in $\mathcal{S}$, its determinant, denoted by $\mathcal{W} : \mathbb{I} \to \mathbb{R}$, i.e.

$$\mathcal{W}(t) = \det \mathcal{X}(t)$$

for every $t \in \mathbb{I}$, is called *the Wronskian associated to the system of solutions*[2] $x^1, x^2, \ldots, x^n$.

Theorem 4.1.4. *Let $x^1, x^2, \ldots, x^n$ be a system of saturated solutions of equation* (4.1.3), *let $\mathcal{X}$ be the associated matrix and let $\mathcal{W}$ be the associated Wronskian. Then the conditions below are equivalent:*

(*i*) *the matrix $\mathcal{X}$ is fundamental;*
(*ii*) *for every $t \in \mathbb{I}$, $\mathcal{W}(t) \neq 0$;*
(*iii*) *there exists $a \in \mathbb{I}$ such that $\mathcal{W}(a) \neq 0$.*

Proof. We begin by showing that (i) implies (ii). So, let us assume that the matrix $\mathcal{X}$ is fundamental, which amounts to saying that the system $x^1, x^2, \ldots, x^n$ is linearly independent. Let us assume by contradiction that there exists $a \in \mathbb{I}$ with $\mathcal{W}(a) = 0$. Therefore the linear and homogeneous system of algebraic equations

$$\mathcal{X}(a)c = 0$$

with the unknowns $c_1, c_2, \ldots, c_n$ has at least one nontrivial solution $\xi_1, \xi_2, \ldots, \xi_n$. On the other hand, the function $x : \mathbb{I} \to \mathbb{R}^n$, defined by

$$x(t) = \mathcal{X}(t)\xi$$

[2]The name of this determinant comes from the name of the polish mathematician Höene Joseph Maria Wronski (1776–1853) who was the first to define and study it.

for every $t \in \mathbb{I}$ is, in view of Remark 4.1.5, a solution of the system (4.1.3) which satisfies $x(a) = 0$. From the uniqueness part of Theorem 4.1.1 it follows that $x(t) = 0$ for every $t \in \mathbb{I}$, relation which is equivalent to

$$\mathcal{X}(t)\xi = \sum_{i=1}^{n} \xi_i x^i(t) = 0$$

for every $t \in \mathbb{I}$, where at least one of $\xi_1, \xi_2, \ldots, \xi_n$ is not zero. Hence the system $x^1, x^2, \ldots, x^n$ is not linearly independent, assertion which contradicts (i). This contradiction can be eliminated only if (ii) holds.

Obviously, (ii) implies (iii).

Finally, we shall prove that (iii) implies (i). Let $c_1, c_2, \ldots, c_n$ be some constants such that

$$\sum_{i=1}^{n} c_i x^i(t) = \mathcal{X}(t)c = 0$$

for every $t \in \mathbb{I}$. In particular, we have

$$\sum_{i=1}^{n} c_i x^i(a) = \mathcal{X}(a)c = 0.$$

But this linear homogeneous system has only the trivial solution $c_1 = c_2 = \cdots = c_n = 0$, because its determinant, $W(a)$, is nonzero. So, $x^1, x^2, \ldots, x^n$ are linearly independent and thus $\mathcal{X}(t)$ is fundamental, as claimed. The proof is complete. □

Remark 4.1.6. Let $a \in \mathbb{I}$, $\xi \in \mathbb{R}^n$ and $\mathcal{X}$ be a fundamental matrix for the homogeneous system (4.1.3). Then, the unique solution of the Cauchy problem

$$\begin{cases} x' = \mathcal{A}(t)x \\ x(a) = \xi \end{cases}$$

is given by

$$x(t, a, \xi) = \mathcal{X}(t)\mathcal{X}^{-1}(a)\xi \tag{4.1.7}$$

for every $t \in \mathbb{I}$. Indeed, from Remark 4.1.5, we know that $x(\cdot, a, \xi)$ is given by (4.1.6), i.e.

$$x(t, a, \xi) = \mathcal{X}(t)c$$

for every $t \in \mathbb{I}$, where $c \in \mathbb{R}^n$. Imposing the condition $x(a, a, \xi) = \xi$, we deduce $\mathcal{X}(a)c = \xi$. In view of Theorem 4.1.4, $\mathcal{X}(a)$ is invertible and consequently $c = \mathcal{X}^{-1}(a)\xi$, which proves (4.1.7).

Lemma 4.1.1. *Let $\mathcal{X}$ be a fundamental matrix of the system* (4.1.3). *Then, the matrix-valued function $\mathcal{U} : \mathbb{I} \times \mathbb{I} \to \mathcal{M}_{n\times n}(\mathbb{R})$, defined by*

$$\mathcal{U}(t, s) = \mathcal{X}(t)\mathcal{X}^{-1}(s)$$

for every $t, s \in \mathbb{I}$, is independent of the choice of the fundamental matrix $\mathcal{X}$. In addition, for every $s \in \mathbb{I}$, $\mathcal{U}(\cdot, s)$ satisfies

$$\begin{cases} \dfrac{\partial \mathcal{U}}{\partial t}(t, s) = \mathcal{A}(t)\mathcal{U}(t, s) \\ \\ \mathcal{U}(s, s) = \mathcal{I} \end{cases} \tag{4.1.8}$$

for every $t \in \mathbb{I}$, where $\mathcal{I}$ is the unit $n \times n$ matrix.

Proof. The fact that $\mathcal{U}(\cdot, s)$ satisfies (4.1.8) follows from Remark 4.1.6 with $a = s$ by taking successively $\xi = e^1, \xi = e^2, \dots, \xi = e^n$, with $e^1, e^2, \dots, e^n$ the canonical basis in $\mathbb{R}^n$. Since the Cauchy problem (4.1.8) has the uniqueness property, we deduce that $\mathcal{U}$ does not depend on the choice of the fundamental matrix $\mathcal{X}$. The proof is complete. □

Definition 4.1.6. The family of matrices $\{\mathcal{U}(t, s)\,;\ t, s \in \mathbb{I}\}$, defined in Lemma 4.1.1, is called the *evolutor*, or the *evolution operator* generated by $\mathcal{A}$.

Remark 4.1.7. The evolutor has the following properties:

($\mathcal{E}_1$) $\mathcal{U}(s, s) = \mathcal{I}$ for every $s \in \mathbb{I}$;
($\mathcal{E}_2$) $\mathcal{U}(t, s)\mathcal{U}(s, \tau) = \mathcal{U}(t, \tau)$ for every $\tau, s, t \in \mathbb{I}$;
($\mathcal{E}_3$) $\lim\limits_{t \to s} \|\mathcal{U}(t, s) - \mathcal{I}\|_{\mathcal{M}} = 0$.

Indeed, ($\mathcal{E}_1$) and ($\mathcal{E}_3$) follow from Lemma 4.1.1, while ($\mathcal{E}_2$) is a direct consequence of the definition of the operators $\mathcal{U}(t, s)$.

Remark 4.1.8. Remark 4.1.6 can be restated in terms of the evolution operator generated by $\mathcal{A}$. More precisely, for every $a \in \mathbb{I}$ and every $\xi \in \mathbb{R}^n$, the unique saturated solution of the Cauchy problem for the system (4.1.3), which satisfies $x(a, a, \xi) = \xi$, is given by

$$x(t, a, \xi) = \mathcal{U}(t, a)\xi$$

for every $t \in \mathbb{I}$.

In what follows, we shall prove a result which shows explicitly how the Wronskian of a system of solutions depends on the elements of the associated matrix $\mathcal{A}$. We begin with

Lemma 4.1.2. *Let* $d_{ij} : \mathbb{I} \to \mathbb{R}$ *be differentiable on* $\mathbb{I}$, $i, j = 1, 2, \ldots, n$. *Then the function* $\mathcal{D} : \mathbb{I} \to \mathbb{R}$ *defined by:*

$$\mathcal{D}(t) = \begin{vmatrix} d_{11}(t) \; d_{12}(t) \ldots \; d_{1n}(t) \\ d_{21}(t) \; d_{22}(t) \ldots \; d_{2n}(t) \\ \vdots \\ d_{n1}(t) \; d_{n2}(t) \ldots \; d_{nn}(t) \end{vmatrix}$$

for every $t \in \mathbb{I}$ *is differentiable on* $\mathbb{I}$ *and*

$$\mathcal{D}'(t) = \sum_{k=1}^{n} \mathcal{D}_k(t)$$

for every $t \in \mathbb{I}$, *where* $\mathcal{D}_k$ *is the determinant obtained from* $\mathcal{D}$ *by replacing the elements* $d_{k1}(t), d_{k2}(t), \ldots, d_{kn}(t)$ *of the* k^{th} *row by the corresponding derivatives* $d'_{k1}(t), d'_{k2}(t), \ldots d'_{kn}(t)$, $k = 1, 2, \ldots, n$.

Proof. We denote by $\mathcal{S}(n)$ the set of substitutions of $\{1, 2, \ldots, n\}$ and by $\varepsilon(\sigma)$ the signature of the substitution $\sigma \in \mathcal{S}(n)$, $\sigma = \begin{pmatrix} 1 \; 2 \; \ldots \; n \\ i_1 \; i_2 \; \ldots \; i_n \end{pmatrix}$. Since, in view of the definition of the determinant, we have

$$\mathcal{D}(t) = \sum_{\sigma \in \mathcal{S}(n)} \varepsilon(\sigma) d_{1i_1}(t) d_{2i_2}(t) \ldots d_{ni_n}(t)$$

for every $t \in \mathbb{I}$, it follows that $\mathcal{D}$ is differentiable on $\mathbb{I}$. In addition, we have

$$\mathcal{D}'(t) = \sum_{k=1}^{n} \sum_{\sigma \in \mathcal{S}(n)} \varepsilon(\sigma) d_{1i_1}(t) d_{2i_2}(t) \ldots d'_{ki_k}(t) \ldots d_{ni_n}(t) = \sum_{k=1}^{n} \mathcal{D}_k(t),$$

which completes the proof. □

Theorem 4.1.5. (Liouville[3]) *If* $\mathcal{W}$ *is the Wronskian of a system of* n *solutions of* (4.1.3), *then*

$$\mathcal{W}(t) = \mathcal{W}(t_0) \exp\left(\int_{t_0}^{t} \operatorname{tr} \mathcal{A}(s) \, ds \right) \tag{4.1.9}$$

for every $t \in \mathbb{I}$, *where* $t_0 \in \mathbb{I}$ *is fixed, while* $\operatorname{tr}\mathcal{A}$ *is the trace of* $\mathcal{A}$, *i.e.* $\operatorname{tr}\mathcal{A}(s) = \sum_{i=1}^{n} a_{ii}(s)$ *for every* $s \in \mathbb{I}$.

[3] Joseph Liouville (1809–1882) French mathematician known for his contributions to the study of the transcendental functions and that of double-periodical functions.

Proof. From Lemma 4.1.2, it follows that $\mathcal{W}$ is differentiable on $\mathbb{I}$, and in addition

$$\mathcal{W}'(t) = \sum_{k=1}^{n} \begin{vmatrix} x_1^1(t) & x_1^2(t) & \dots & x_1^n(t) \\ \vdots & \vdots & \vdots & \vdots \\ (x_k^1)'(t) & (x_k^2)'(t) & \dots & (x_k^n)'(t) \\ \vdots & \vdots & \vdots & \vdots \\ x_n^1(t) & x_n^2(t) & \dots & x_n^n(t) \end{vmatrix}.$$

Taking into account that $x^1, x^2, \dots, x^n$ are solutions of the system (4.1.3), we get

$$\mathcal{W}'(t) = \sum_{k=1}^{n} \begin{vmatrix} x_1^1(t) & x_1^2(t) & \dots & x_1^n(t) \\ \vdots & \vdots & \vdots & \vdots \\ \sum_{j=1}^n a_{kj}(t)x_j^1(t) & \sum_{j=1}^n a_{kj}(t)x_j^2(t) & \dots & \sum_{j=1}^n a_{kj}(t)x_j^n(t) \\ \vdots & \vdots & \vdots & \vdots \\ x_n^1(t) & x_n^2(t) & \dots & x_n^n(t) \end{vmatrix}$$

$$= \sum_{k=1}^{n}\sum_{j=1}^{n} a_{kj}(t) \begin{vmatrix} x_1^1(t) & x_1^2(t) & \dots & x_1^n(t) \\ \vdots & \vdots & \vdots & \vdots \\ x_j^1(t) & x_j^2(t) & \dots & x_j^n(t) \\ \vdots & \vdots & \vdots & \vdots \\ x_n^1(t) & x_n^2(t) & \dots & x_n^n(t) \end{vmatrix}.$$

Since, for each $k \neq j$, the corresponding determinants in the last sum has two equal rows, i.e. the kth and the jth row, it follows that all these determinants are zero, and therefore

$$\mathcal{W}'(t) = \sum_{j=1}^{n} a_{jj}(t) \begin{vmatrix} x_1^1(t) & x_1^2(t) & \dots & x_1^n(t) \\ \vdots & \vdots & \vdots & \vdots \\ x_j^1(t) & x_j^2(t) & \dots & x_j^n(t) \\ \vdots & \vdots & \vdots & \vdots \\ x_n^1(t) & x_n^2(t) & \dots & x_n^n(t) \end{vmatrix},$$

or equivalently $\mathcal{W}$ satisfies $\mathcal{W}'(t) = \text{tr}\mathcal{A}(t)\mathcal{W}(t)$ for every $t \in \mathbb{I}$. But the equation above is linear and homogeneous, and therefore $\mathcal{W}$ is given by (4.1.9). □

Remark 4.1.9. We notice that Theorem 4.1.4 may also be proved with the help of Theorem 4.1.5. We leave it as an exercise to prove Theorem 4.1.4 by using Theorem 4.1.5.

4.2 Non-homogeneous Systems. The Variation of Constants Formula

Let us consider the first-order linear non-homogeneous system

$$x' = \mathcal{A}(t)x + b(t), \tag{4.2.1}$$

where $\mathcal{A} : \mathbb{I} \to \mathcal{M}_{n\times n}(\mathbb{R})$ and $b : \mathbb{I} \to \mathbb{R}^n$ are continuous functions. At the same time, let us consider the homogeneous system

$$x' = \mathcal{A}(t)x. \tag{4.2.2}$$

In this section we will present a method of determining the general solution of the system (4.2.1) with the help of the general solution of the attached homogeneous system.

We begin with the following simple, but very useful theorem in applications.

Theorem 4.2.1. *Let $\mathcal{X}$ be a fundamental matrix of the system* (4.2.2) *and let $y : \mathbb{I} \to \mathbb{R}^n$ be a solution of the system* (4.2.1). *A function $x : \mathbb{I} \to \mathbb{R}^n$ is a solution of the system* (4.2.1) *if and only if x is of the form*

$$x(t) = \mathcal{X}(t)c + y(t) \tag{4.2.3}$$

for every $t \in \mathbb{I}$, where $c \in \mathbb{R}^n$.

Proof. Necessity. Let $x : \mathbb{I} \to \mathbb{R}^n$ be a solution of the system (4.2.1) and let us define the function $z : \mathbb{I} \to \mathbb{R}^n$ by

$$z(t) = x(t) - y(t)$$

for every $t \in \mathbb{I}$. Obviously z is differentiable on $\mathbb{I}$ and we have

$$z'(t) = x'(t) - y'(t) = \mathcal{A}(t)x(t) + b(t) - \mathcal{A}(t)y(t) - b(t)$$

$$= \mathcal{A}(t)(x(t) - y(t)) = \mathcal{A}(t)z(t)$$

for every $t \in \mathbb{I}$. Hence z is a solution of the homogeneous system (4.2.2) and, in view of Remark 4.1.5, it is of the form

$$z(t) = \mathcal{X}(t)c$$

for every $t \in \mathbb{I}$, where $c \in \mathbb{R}^n$. From this relation and from the definition of the function z, we deduce (4.2.3), which completes the proof of the necessity.

Sufficiency. Let x be the function defined by (4.2.3). Since $t \mapsto \mathcal{X}(t)c$ is a solution of the homogeneous system (4.2.2), it follows that x is differentiable on $\mathbb{I}$. In addition

$$x'(t) = \mathcal{X}'(t)c + y'(t) = \mathcal{A}(t)\mathcal{X}(t)c + \mathcal{A}(t)y(t) + b(t)$$

$$= \mathcal{A}(t)(\mathcal{X}(t)c + y(t)) + b(t) = \mathcal{A}(t)x(t) + b(t)$$

for every $t \in \mathbb{I}$, and therefore x is a solution of the system (4.2.1). The proof is complete. □

Remark 4.2.1. Theorem 4.2.1 asserts that the general solution of the system (4.2.1) is of the form (4.2.3), with y a particular solution of the system (4.2.1) and $c \in \mathbb{R}^n$.

Now let $a \in \mathbb{I}$, $\xi \in \mathbb{R}^n$ and let us consider the Cauchy problem

$$\begin{cases} x' = \mathcal{A}(t)x + b(t) \\ x(a) = \xi. \end{cases} \tag{4.2.4}$$

Theorem 4.2.2. *Let $\mathcal{X}$ be a fundamental matrix of the homogeneous system (4.2.2). Then the unique saturated solution of the Cauchy problem (4.2.4) is given by*

$$x(t, a, \xi) = \mathcal{X}(t)\mathcal{X}^{-1}(a)\xi + \int_a^t \mathcal{X}(t)\mathcal{X}^{-1}(s)b(s)\,ds \tag{4.2.5}$$

for every $t \in \mathbb{I}$.

Proof. Remark 4.1.5 suggests to look for the unique solution of the Cauchy problem of the form

$$x(t, a, \xi) = \mathcal{X}(t)c(t) \tag{4.2.6}$$

for $t \in \mathbb{I}$, where $c : \mathbb{I} \to \mathbb{R}^n$ is a function of class C^1 to be determined. We will find c by requiring x defined by (4.2.6) is a solution of the Cauchy problem (4.2.4). We have

$$x'(t, a, \xi) = \mathcal{X}'(t)c(t) + \mathcal{X}(t)c'(t)$$

for every $t \in \mathbb{I}$. Hence x, given by (4.2.6), is a solution of the system (4.2.1) if and only if

$$\mathcal{X}'(t)c(t) + \mathcal{X}(t)c'(t) = \mathcal{A}(t)\mathcal{X}(t)c(t) + b(t)$$

for every $t \in \mathbb{I}$. Recalling that $\mathcal{X}$ satisfies $\mathcal{X}'(t) = \mathcal{A}(t)\mathcal{X}(t)$ for every $t \in \mathbb{I}$, the last equality is equivalent to

$$\mathcal{A}(t)\mathcal{X}(t)c(t) + \mathcal{X}(t)c'(t) = \mathcal{A}(t)\mathcal{X}(t)c(t) + b(t)$$

for every $t \in \mathbb{I}$ which, in its turn, can be rewritten in the form

$$\mathcal{X}(t)c'(t) = b(t) \tag{4.2.7}$$

for every $t \in \mathbb{I}$. Since $\mathcal{X}(t)$ is non-singular, we deduce that $c'(t) = \mathcal{X}^{-1}(t)b(t)$ for $t \in \mathbb{I}$. Integrating this relation from a to t, we get

$$c(t) = c(a) + \int_a^t \mathcal{X}^{-1}(s)b(s)\, ds,$$

relation which, in view of (4.2.6), leads to

$$x(t, a, \xi) = \mathcal{X}(t)c(a) + \mathcal{X}(t) \int_a^t \mathcal{X}^{-1}(s)b(s)\, ds$$

for every $t \in \mathbb{I}$. According to (i) in Lemma 12.1.3,[4] $\mathcal{X}(t)$ commutes with the integral. So

$$x(t, a, \xi) = \mathcal{X}(t)c(a) + \int_a^t \mathcal{X}(t)\mathcal{X}^{-1}(s)b(s)\, ds$$

for every $t \in \mathbb{I}$. By requiring that $x(a, a, \xi) = \xi$, we deduce that $c(a) = \mathcal{X}^{-1}(a)\xi$ relation which, along with the preceding one, implies (4.2.5). $\square$

Remark 4.2.2. Formula (4.2.5), called *the variation of constants formula*, can be equivalently rewritten in the form

$$x(t, a, \xi) = \mathcal{U}(t, a)\xi + \int_a^t \mathcal{U}(t, s)b(s)\, ds$$

for every $t \in \mathbb{I}$, where $\{\mathcal{U}(t, s)\,;\ t, s \in \mathbb{I}\}$ is the evolutor generated by $\mathcal{A}$, i.e.

$$\mathcal{U}(t, s) = \mathcal{X}(t)\mathcal{X}^{-1}(s)$$

for every $t, s \in \mathbb{I}$. See Definition 4.1.6.

4.3 The Exponential of a Matrix

Let us consider the first-order linear homogeneous system with constant coefficients

$$x' = \mathcal{A}x, \tag{4.3.1}$$

where $\mathcal{A} \in \mathcal{M}_{n\times n}(\mathbb{R})$. Since the right-hand side is an analytic function on $\mathbb{R}^n$, according to Theorem 3.1.1, it follows that all the solutions of the system (4.3.1) are analytic on $\mathbb{R}$. On the other hand, in the case of the

[4] See Chapter 10 Auxiliary Results.

scalar equation $x' = ax$, the general solution is given by $x(t) = \xi e^{ta}$ for $t \in \mathbb{R}$, where

$$e^{ta} = \sum_{k=0}^{\infty} \frac{t^k a^k}{k!},$$

the convergence being uniform on every bounded subset of $\mathbb{R}$. Concerning the n-dimensional case, i.e. the case of the system (4.3.1), these two remarks suggest to define (formally for the moment) a candidate to the title of a fundamental matrix by

$$e^{t\mathcal{A}} = \sum_{k=0}^{\infty} \frac{t^k}{k!} \mathcal{A}^k.$$

We notice that $\mathcal{A}^k$ is the k-times product of the matrix $\mathcal{A}$ by itself, while $\mathcal{A}^0 = \mathcal{I}$. By analogy with the scalar case, we shall prove that the series on the right-hand side is uniformly convergent, for t in bounded sets in $\mathbb{R}$, in the sense of the norm $\|\cdot\|_{\mathcal{M}}$ defined in Section 12.1. Finally, we will show that the sum of this series is the unique fundamental matrix $\mathcal{X}(t)$ of the system (4.3.1) which satisfies $\mathcal{X}(0) = \mathcal{I}$.

To fix the ideas, we begin with

Definition 4.3.1. The series $\sum_{k=0}^{\infty} \mathcal{C}_k$, whose terms belong to $\mathcal{M}_{n\times n}(\mathbb{R})$, is said to be *convergent* to $\mathcal{C}$ if

$$\lim_{m\to+\infty} \left\| \sum_{k=0}^{m} \mathcal{C}_k - \mathcal{C} \right\|_{\mathcal{M}} = 0$$

where $\|\cdot\|_{\mathcal{M}}$ is the norm defined in Section 12.1. The series $\sum_{k=0}^{\infty} \mathcal{C}_k$, whose terms belong to $\mathcal{M}_{n\times n}(\mathbb{R})$, is *absolutely convergent* if the series $\sum_{k=0}^{\infty} \|\mathcal{C}_k\|_{\mathcal{M}}$ is convergent.

Remark 4.3.1. One may easily see that, for every absolutely convergent series of matrices, $\sum_{k=0}^{\infty} \mathcal{C}_k$, there exists one matrix $\mathcal{C}$ such that the series converges to $\mathcal{C}$. This follows from the simple observation that the sequence of partial sums of any absolutely convergent series of matrices is fundamental in the norm of the space $\mathcal{M}_{n\times n}(\mathbb{R})$, which is complete (it can be identified with $\mathbb{R}^{n\times n}$ endowed with the Euclidean norm). See Remark 12.1.1.

Definition 4.3.2. Let $\mathcal{C}_k : \mathbb{I} \to \mathcal{M}_{n\times n}(\mathbb{R})$, $k \in \mathbb{N}$. We say that the series of matrix-valued functions $\sum_{k=0}^{\infty} \mathcal{C}_k(t)$ is *uniformly convergent* on $\mathbb{I}$ to $\mathcal{C}$:

$\mathbb{I} \to \mathcal{M}_{n\times n}(\mathbb{R})$ if for every $\varepsilon > 0$ there exists $m(\varepsilon) \in \mathbb{N}$ such that, for every $m \in \mathbb{N}$, $m \geq m(\varepsilon)$, we have

$$\left\| \sum_{k=0}^{m} \mathcal{C}_k(t) - \mathcal{C}(t) \right\|_{\mathcal{M}} \leq \varepsilon$$

for every $t \in \mathbb{I}$.

Theorem 4.3.1. *For every $\mathcal{A} \in \mathcal{M}_{n\times n}(\mathbb{R})$, the series*

$$\sum_{k=0}^{\infty} \frac{t^k}{k!} \mathcal{A}^k$$

is uniformly convergent on every bounded interval $\mathbb{I}$ in $\mathbb{R}$. In addition, its sum $e^{t\mathcal{A}}$ is differentiable on $\mathbb{R}$ and

$$\frac{d}{dt}\left(e^{t\mathcal{A}}\right) = \mathcal{A}e^{t\mathcal{A}} = e^{t\mathcal{A}}\mathcal{A} \tag{4.3.2}$$

for every $t \in \mathbb{R}$.

Proof. According to Corollary 12.1.1,[5] we have

$$\left\| \sum_{k=m}^{m+p} \frac{t^k}{k!} \mathcal{A}^k \right\|_{\mathcal{M}} \leq \sum_{k=m}^{m+p} \frac{(|t|\|\mathcal{A}\|_{\mathcal{M}})^k}{k!}$$

for every $m,\ p \in \mathbb{N}$ and every $t \in \mathbb{R}$. This inequality shows that the series in question satisfies Cauchy's condition uniformly for t in every bounded subset, simply because the numerical comparison series has this property. Hence the sequence of partial sums is a uniformly Cauchy sequence on every bounded interval $\mathbb{I}$, and therefore the series considered is uniformly convergent on $\mathbb{I}$.

In order to prove the second part of the theorem, we begin by observing that the series is termwise differentiable and that the series of derivatives is, in turn, uniformly convergent on every bounded interval in $\mathbb{R}$. Indeed, it is easy to see that

$$\frac{d}{dt}(\mathcal{I}) = 0 \quad \text{and} \quad \frac{d}{dt}\left(\frac{t^k}{k!}\mathcal{A}^k\right) = \mathcal{A}\frac{t^{k-1}}{(k-1)!}\mathcal{A}^{k-1} = \frac{t^{k-1}}{(k-1)!}\mathcal{A}^{k-1}\mathcal{A}$$

for every $k \in \mathbb{N}^*$ and every $t \in \mathbb{R}$. From here it follows that

$$\left\| \sum_{k=m}^{m+p} \frac{d}{dt}\left(\frac{t^k}{k!}\mathcal{A}^k\right) \right\|_{\mathcal{M}} \leq \|\mathcal{A}\|_{\mathcal{M}} \sum_{k=m}^{m+p} \frac{(|t|\|\mathcal{A}\|_{\mathcal{M}})^{k-1}}{(k-1)!}.$$

[5] See Chapter 10 Auxiliary Results.

Therefore the series of derivatives satisfies Cauchy's condition uniformly for t in every bounded interval. So, the sum of the initial series is differentiable and its derivative satisfies

$$\frac{d}{dt}\left(e^{t\mathcal{A}}\right) = \mathcal{A}\left(\sum_{k=1}^{\infty}\frac{t^{k-1}}{(k-1)!}\mathcal{A}^{k-1}\right) = \left(\sum_{k=1}^{\infty}\frac{t^{k-1}}{(k-1)!}\mathcal{A}^{k-1}\right)\mathcal{A},$$

relations which obviously are equivalent to (4.3.2), which completes the proof. □

Remark 4.3.2. The first equality in (4.3.2) proves that every column of the matrix $e^{t\mathcal{A}}$, regarded as a function from $\mathbb{R}$ to $\mathbb{R}^n$, is a solution of the homogeneous system (4.3.1). Since $e^{0\mathcal{A}} = \mathcal{I}$ and $\mathcal{I}$ is non-singular, it follows that $e^{t\mathcal{A}}$ is a fundamental matrix for the system (4.3.1).

Some useful consequences of Theorem 4.3.1 are stated below.

Proposition 4.3.1. *For every $\mathcal{A} \in \mathcal{M}_{n\times n}(\mathbb{R})$ the series*

$$\sum_{k=0}^{\infty}\frac{1}{k!}\mathcal{A}^k$$

is convergent. In addition, the function $\mathcal{A} \mapsto e^{\mathcal{A}}$ defined on $\mathcal{M}_{n\times n}(\mathbb{R})$ and taking values in $\mathcal{M}_{n\times n}(\mathbb{R})$, where $e^{\mathcal{A}}$ is the sum of the series above, has the following properties:

(i) $e^{\mathcal{I}} = e\mathcal{I}$, $e^{\mathcal{O}} = \mathcal{I}$, *where $\mathcal{O}$ is the null $n \times n$-matrix;*
(ii) *if $\mathcal{A}\mathcal{B} = \mathcal{B}\mathcal{A}$ then $e^{\mathcal{A}+\mathcal{B}} = e^{\mathcal{A}}e^{\mathcal{B}}$;*
(iii) *if $\mathcal{A} = \mathcal{Q}^{-1}\mathcal{B}\mathcal{Q}$ then $e^{\mathcal{A}} = \mathcal{Q}^{-1}e^{\mathcal{B}}\mathcal{Q}$;*
(iv) $e^{-\mathcal{A}} = \left(e^{\mathcal{A}}\right)^{-1}$.

Proof. Property (i) is an immediate consequence of the definition of the matrix $e^{\mathcal{A}}$. In order to prove (ii), let us observe that, if $\mathcal{A}\mathcal{B} = \mathcal{B}\mathcal{A}$, then

$$e^{t\mathcal{A}}\mathcal{B} = \mathcal{B}e^{t\mathcal{A}} \tag{4.3.3}$$

for every $t \in \mathbb{R}$. Indeed, if $\mathcal{A}\mathcal{B} = \mathcal{B}\mathcal{A}$, then $\mathcal{A}^k\mathcal{B} = \mathcal{B}\mathcal{A}^k$ for every $k \in \mathbb{N}$, relation which, together with the definition of the matrix $e^{t\mathcal{A}}$, implies (4.3.3). From (4.3.3) and (4.3.2), it follows that

$$\frac{d}{dt}\left(e^{t\mathcal{A}}e^{t\mathcal{B}}\right) = \frac{d}{dt}\left(e^{t\mathcal{A}}\right)e^{t\mathcal{B}} + e^{t\mathcal{A}}\frac{d}{dt}\left(e^{t\mathcal{B}}\right)$$

$$= \mathcal{A}e^{t\mathcal{A}}e^{t\mathcal{B}} + e^{t\mathcal{A}}\mathcal{B}e^{t\mathcal{B}} = (\mathcal{A}+\mathcal{B})e^{t\mathcal{A}}e^{t\mathcal{B}}$$

for every $t \in \mathbb{R}$. Consequently $\mathcal{X}(t) = e^{t\mathcal{A}}e^{t\mathcal{B}}$ is a fundamental matrix for the system

$$x' = (\mathcal{A} + \mathcal{B})x$$

which satisfies $\mathcal{X}(0) = \mathcal{I}$. From the uniqueness part of Theorem 4.1.1 and from Remark 4.3.2, it follows that $e^{t\mathcal{A}}e^{t\mathcal{B}} = e^{t(\mathcal{A}+\mathcal{B})}$ for every $t \in \mathbb{R}$, which obviously implies (ii).

If $\mathcal{A} = \mathcal{Q}^{-1}\mathcal{B}\mathcal{Q}$ then $\mathcal{A}^k = \mathcal{Q}^{-1}\mathcal{B}^k\mathcal{Q}$ for every $k \in \mathbb{N}$, and so

$$\sum_{k=0}^{\infty} \frac{t^k}{k!}\mathcal{A}^k = \sum_{k=0}^{\infty} \frac{t^k}{k!}\mathcal{Q}^{-1}\mathcal{B}^k\mathcal{Q} = \mathcal{Q}^{-1}\left(\sum_{k=0}^{\infty} \frac{t^k}{k!}\mathcal{B}^k\right)\mathcal{Q},$$

which proves (iii).

Finally, since $\mathcal{A}$ and $-\mathcal{A}$ commute, from (ii), we deduce

$$e^{\mathcal{A}}e^{-\mathcal{A}} = e^{-\mathcal{A}}e^{\mathcal{A}} = e^{\mathcal{O}} = \mathcal{I}.$$

Consequently, $e^{\mathcal{A}}$ is invertible and its inverse is $e^{-\mathcal{A}}$, which completes the proof. □

Remark 4.3.3. Let us consider the Cauchy problem

$$\begin{cases} x' = \mathcal{A}x + b(t) \\ x(a) = \xi \end{cases}$$

where $\mathcal{A} \in \mathcal{M}_{n\times n}(\mathbb{R})$, $b : \mathbb{I} \to \mathbb{R}^n$ is a continuous function, $a \in \mathbb{I}$ and $\xi \in \mathbb{R}^n$. Then the unique solution of this problem is given by

$$x(t, a, \xi) = e^{(t-a)\mathcal{A}}\xi + \int_a^t e^{(t-s)\mathcal{A}}b(s)\,ds \tag{4.3.4}$$

for every $t \in \mathbb{I}$.

We notice that (4.3.4) is a consequence of the variation of constants formula (4.2.5). Indeed, taking $\mathcal{X}(t) = e^{t\mathcal{A}}$ and making use of (ii) and (iv) in Proposition 4.3.1, we deduce that $\mathcal{X}(t)\mathcal{X}^{-1}(s) = e^{(t-s)\mathcal{A}}$ for every $t, s \in \mathbb{R}$. From here and (4.2.5), we deduce (4.3.4).

Remark 4.3.4. All the considerations in this section can be extended without difficulty to the case of first-order differential systems of linear equations with constant complex coefficients. More precisely, let us consider the first-order linear differential homogeneous system

$$w' = \mathcal{A}w,$$

where $\mathcal{A} \in \mathcal{M}_{n\times n}(\mathbb{C})$. By a solution of this system we mean a function $w : D \to \mathbb{C}^n$, holomorphic on $D \subseteq \mathbb{C}$, and which satisfies $w'(z) = \mathcal{A}w(z)$ for every $z \in D$.

We endow $\mathbb{C}^n$ with the standard inner product, i.e. $\langle\cdot,\cdot\rangle$, defined by

$$\langle v, w\rangle = \sum_{i=1}^{n} v_i \overline{w_i}$$

for every $v, w \in \mathbb{C}^n$, and we define the induced norm $\|\cdot\|_e : \mathbb{C}^n \to \mathbb{R}_+$ by $\|v\|_e = \sqrt{\langle v, v\rangle}$, and the norm $\|\cdot\|_{\mathcal{M}} : \mathcal{M}_{n\times n}(\mathbb{C}) \to \mathbb{R}_+$ by

$$\|\mathcal{A}\|_{\mathcal{M}} = \sup\{\|\mathcal{A}v\|_e;\ \|v\|_e \leq 1\}$$

for every $\mathcal{A} \in \mathcal{M}_{n\times n}(\mathbb{C})$. Now, let us observe that the series $\sum_{k=0}^{\infty} \frac{z^k}{k!}\mathcal{A}^k$ is uniformly convergent on every bounded subset of $\mathbb{C}$ and its sum is a matrix whose elements are entire functions (holomorphic on $\mathbb{C}$). We denote this matrix by $e^{z\mathcal{A}}$. From a classical theorem concerning the differentiability of complex power series, we deduce that the matrix above is a solution (in $\mathbb{C}$) of the Cauchy problem

$$\begin{cases} \mathcal{W}' = \mathcal{A}\mathcal{W} \\ \mathcal{W}(0) = \mathcal{I}. \end{cases}$$

4.4 A Method of Finding $e^{t\mathcal{A}}$

In this section we will present a method of finding the matrix $e^{t\mathcal{A}}$ using an algorithm due to [Putzer (1966)]. This method is much simpler than the one presented in most textbooks, which involves the Jordan normal form of the matrix A.

We begin by recalling a fundamental result in Linear Algebra.

Theorem 4.4.1. (Cayley–Hamilton) *Every matrix* $\mathcal{A} \in \mathcal{M}_{n\times n}(\mathbb{R})$ *satisfies its characteristic equation* $\det(\mathcal{A} - \lambda\mathcal{I}) = 0$.

For a very simple and elegant proof of Theorem 4.4.1, see [Braun (1983)], Theorem 1.3, pp. 349–350.

Remark 4.4.1. During the computation of $e^{t\mathcal{A}}$ in concrete situations, when some eigenvalues of $\mathcal{A}$ are complex, one needs to know what is the meaning of e^{λ}, where $\lambda = \alpha + i\beta$ is a complex number. Since

$$e^{\alpha+i\beta} = \sum_{k=0}^{\infty} \frac{(\alpha + i\beta)^k}{k!},$$

by observing that the real part of the above series is $e^{\alpha}\cos\beta$, while the imaginary part is $e^{\alpha}\sin\beta$, we conclude that

$$e^{\alpha+i\beta} = e^{\alpha}(\cos\beta + i\sin\beta). \tag{4.4.1}$$

Theorem 4.4.2. (Putzer's Algorithm) *Let $\mu_1, \mu_2, \ldots, \mu_n$ be the roots, not necessarily distinct, of the characteristic equation* $\det(\mathcal{A} - \mu\mathcal{I}) = 0$. *Then*

$$e^{t\mathcal{A}} = \sum_{k=0}^{n-1} p_{k+1}(t)\mathcal{M}_k, \tag{4.4.2}$$

where

$$\begin{cases} \mathcal{M}_0 = \mathcal{I}, \\ \mathcal{M}_k = \prod_{j=1}^{k}(\mathcal{A} - \mu_j\mathcal{I}), \quad k = 1, 2, \ldots, n \end{cases} \tag{4.4.3}$$

and $p : \mathbb{R} \to \mathbb{C}^n$,

$$p(t) = \begin{pmatrix} p_1(t) \\ p_2(t) \\ \vdots \\ p_n(t) \end{pmatrix} \tag{4.4.4}$$

for $t \in \mathbb{R}$, is the unique solution of the Cauchy problem

$$\begin{cases} p'(t) = \mathcal{J}p(t), \ t \in \mathbb{R}, \\ p(0) = e_1, \end{cases} \tag{4.4.5}$$

$$\mathcal{J} = \begin{pmatrix} \mu_1 & 0 & 0 & \ldots & 0 \\ 1 & \mu_2 & 0 & \ldots & 0 \\ \vdots & & & & \\ 0 & 0 & 0 & \ldots 1 & \mu_n \end{pmatrix} \quad \text{and} \quad e_1 = \begin{pmatrix} 1 \\ 0 \\ \vdots \\ 0 \end{pmatrix}. \tag{4.4.6}$$

Proof. First, let us observe that from (4.4.5) and (4.4.6), we have

$$\begin{cases} p_1'(t) = \mu_1 p_1(t), t \in \mathbb{R}, \\ p_k'(t) = p_{k-1}(t) + \mu_k p_k(t), \quad k = 2, 3, \ldots, n, \quad t \in \mathbb{R}. \end{cases} \tag{4.4.7}$$

Let us consider the function $\mathcal{X} : \mathbb{R} \to \mathcal{M}_{n\times n}(\mathbb{R})$, defined by

$$\mathcal{X}(t) = \sum_{k=0}^{n-1} p_{k+1}(t)\mathcal{M}_k$$

for each $t \in \mathbb{R}$, where $p_{k+1}(t)$ is the $(k+1)^{\text{th}}$ component of $p(t)$ given (4.4.4) and satisfying (4.4.5) and $\mathcal{M}_k$ is defined as in (4.4.3), $k = 0, 1, \ldots, n-1$. At this point, let us observe that, in order to prove (4.4.2), it suffices to show that $\mathcal{X}$ is a solution of the Cauchy problem

$$\begin{cases} \mathcal{X}'(t) = \mathcal{A}\mathcal{X}(t), \ t \in \mathbb{R}, \\ \mathcal{X}(0) = \mathcal{I}. \end{cases}$$

Indeed, recalling that $t \mapsto e^{t\mathcal{A}}$ is a solution of the problem above and that this solution is unique, this would imply (4.4.2). After some calculation involving (4.4.7), we get

$$\mathcal{X}'(t) - \mathcal{A}\mathcal{X}(t) = \sum_{k=0}^{n-1} p'_{k+1}(t)\mathcal{M}_k - \mathcal{A}\sum_{k=0}^{n-1} p_{k+1}(t)\mathcal{M}_k$$

$$= \mu_1 p_1(t)\mathcal{M}_0 + \sum_{k=1}^{n-1}[\mu_{k+1}p_{k+1}(t) + p_k(t)]\mathcal{M}_k - \sum_{k=0}^{n-1} p_{k+1}(t)\mathcal{A}\mathcal{M}_k$$

$$= \mu_1 p_1(t)\mathcal{M}_0 + \sum_{k=1}^{n-1}[\mu_{k+1}p_{k+1}(t) + p_k(t)]\mathcal{M}_k - \sum_{k=0}^{n-1} p_{k+1}(t)[\mathcal{M}_{k+1} + \mu_{k+1}\mathcal{M}_k]$$

$$= \sum_{k=1}^{n-1} p_k(t)\mathcal{M}_k - \sum_{k=0}^{n-1} p_{k+1}(t)\mathcal{M}_{k+1} = -p_n(t)\mathcal{M}_n.$$

Since, from the Cayley–Hamilton Theorem 4.4.1, we have $\mathcal{M}_n = \mathcal{O}$, this completes the proof. □

Example 4.4.1. The next example is from [Kelley and Peterson (2010)], Example 2.3.7, p. 46. Find $e^{t\mathcal{A}}$ for

$$\mathcal{A} = \begin{pmatrix} 1 & -1 \\ 5 & -1 \end{pmatrix}.$$

The characteristic equation

$$\begin{vmatrix} \mu - 1 & 1 \\ -5 & \mu + 1 \end{vmatrix} = \mu^2 + 4 = 0$$

has the complex roots, $\mu_1 = 2i$ and $\mu_2 = -2i$. In this case, formula (4.4.2) has the specific form

$$e^{t\mathcal{A}} = \sum_{k=0}^{1} p_{k+1}(t)\mathcal{M}_k = p_1(t)\mathcal{M}_0 + p_2(t)\mathcal{M}_1,$$

where

$$\mathcal{M}_0 = \begin{pmatrix} 1 & 0 \\ 0 & 1 \end{pmatrix} \quad \text{and} \quad \mathcal{M}_1 = \mathcal{A} - \mu_1 \mathcal{I} = \begin{pmatrix} 1-2i & -1 \\ 5 & -1-2i \end{pmatrix}.$$

According to (4.4.5) – see also (4.4.7) – the function $p : \mathbb{R} \to \mathbb{C}^2$, $p = \begin{pmatrix} p_1 \\ p_2 \end{pmatrix}$, satisfies:

$$\begin{cases} p_1'(t) = 2ip_1(t) \\ p_1(0) = 1, \end{cases} \qquad \begin{cases} p_2'(t) = p_1(t) - 2ip_2(t) \\ p_2(0) = 0. \end{cases}$$

Solving these equations as in the real case, and using (4.4.1), we obtain

$$\begin{cases} p_1(t) = e^{2it} = \cos 2t + i \sin 2t, & t \in \mathbb{R} \\ p_2(t) = \dfrac{1}{4i} e^{2it} - \dfrac{1}{4i} e^{-2it} = \dfrac{1}{2} \sin 2t, & t \in \mathbb{R}. \end{cases}$$

Consequently

$$e^{t\mathcal{A}} = p_1(t)\mathcal{M}_0 + p_2(t)\mathcal{M}_1 = e^{2it} \begin{pmatrix} 1 & 0 \\ 0 & 1 \end{pmatrix} + \frac{1}{2} \sin 2t \begin{pmatrix} 1-2i & -1 \\ 5 & -1-2i \end{pmatrix}.$$

After a simple computation, we get

$$e^{t\mathcal{A}} = \begin{pmatrix} \cos 2t + \dfrac{1}{2} \sin 2t & -\dfrac{1}{2} \sin 2t \\ \dfrac{5}{2} \sin 2t & \cos 2t - \dfrac{1}{2} \sin 2t \end{pmatrix}.$$

We conclude with a fundamental result in the theory of systems of linear differential equations with constant coefficients, result known as *the structure theorem of the matrix* $e^{t\mathcal{A}}$.

For simplicity reasons, in the next result, it is convenient to relabel the roots $\mu_1, \mu_2, \ldots, \mu_n$ of the characteristic equation, not necessarily distinct, as $\lambda_1, \lambda_2, \ldots, \lambda_s$, with $\lambda_j \neq \lambda_p$ for each $j \neq p$, by specifying that the order of multiplicity of λ_j is m_j, for $j = 1, 2, \ldots, s$, with $\sum_{j=1}^{s} m_j = n$. More precisely, the root λ_j, which in Theorem 4.4.2 is labeled m_j-times as $\mu_{p+1} = \mu_{p+2} = \cdots = \mu_{p+m_j} = \lambda_j$, here is considered only once, with the mention that its order of multiplicity is m_j.

Theorem 4.4.3. *All the elements of the matrix* $e^{t\mathcal{A}}$ *are of the form*

$$\sum_{j=1}^{s} e^{\alpha_j t} \left[P_j(t) \cos(\beta_j t) + Q_j(t) \sin(\beta_j t) \right],$$

where $\lambda_j = \alpha_j + i\beta_j$, $j = 1, 2, \ldots, s$, *are the roots of the characteristic equation* $\det(\mathcal{A} - \lambda \mathcal{I}) = 0$, *while* P_j *and* Q_j *are polynomials with real coefficients, of degree not exceeding* $m_j - 1$, m_j *being the order of multiplicity of the root* $\alpha_j + i\beta_j$, $j = 1, 2, \ldots, s$.

Proof. Let $\lambda_j = \alpha_j + i\beta_j$, $j = 1, 2, \ldots, s$, be any root of the equation $\det(\mathcal{A} - \lambda\mathcal{I}) = 0$, whose order of multiplicity is m_j. Taking into account that, by (4.4.1), we have $e^{t\lambda_j} = e^{\alpha_j t}\left[\cos(\beta_j t) + i\sin(\beta_j t)\right]$, we perform step by step the calculation in (4.4.7) as follows. We use the first equation to get p_1, then we use a complex variant of (1.3.4) to get p_2, and so on. Then use (4.4.3) in order to get $\mathcal{M}_k$, $k = 1, 2, \ldots, n$, and (4.4.2) to obtain $e^{t\mathcal{A}}$.

Clearly, the functions $t \mapsto p_k(t)$, $k = 1, 2, \ldots, n$, are linear combinations of $e^{\alpha_1 t}\cos(\beta_1 t), \ldots, e^{\alpha_s t}\cos(\beta_s t)$ and $e^{\alpha_1 t}\sin(\beta_1 t), \ldots, e^{\alpha_s t}\sin(\beta_s t)$, whose coefficients, are complex polynomials with degrees not exceeding $m_1 - 1, m_2 - 1, \ldots, m_s - 1$. Then, for each pair of complex conjugate roots, if any, adding their two corresponding "contributions" to $e^{t\mathcal{A}}$, we get only linear combinations with real polynomial coefficients, P_j and Q_j, for $j = 1, 2, \ldots, s$, of $e^{\alpha_1 t}\cos(\beta_1 t), \ldots, e^{\alpha_s t}\cos(\beta_s t)$ and $e^{\alpha_1 t}\sin(\beta_1 t), \ldots, e^{\alpha_s t}\sin(\beta_s t)$. So, the conclusion follows, and this completes the proof. □

The functions of the form specified in Theorem 4.4.3 are known in the literature under the name of *quasi-polynomials*.

4.5 The nth-Order Linear Differential Equation

Let us consider the nth-order linear differential equation

$$y^{(n)} + a_1(t)y^{(n-1)} + \cdots + a_n(t)y = f(t), \tag{4.5.1}$$

where $a_1, a_2, \ldots a_n, f$ are continuous functions from a nontrivial interval $\mathbb{I}$ in $\mathbb{R}$. As we have seen in Section 1.2, equation (4.5.1) can be rewritten as a system of first-order differential equations. Indeed, by means of the transformations

$$x = (x_1, x_2, \ldots, x_n) = (y, y', \ldots, y^{(n-1)}) \tag{$\mathcal{T}$}$$

(4.5.1) may be rewritten as a system of n linear differential equations:

$$\begin{cases} x_1' = x_2 \\ x_2' = x_3 \\ \vdots \\ x_{n-1}' = x_n \\ x_n' = -a_n(t)x_1 - a_{n-1}(t)x_2 - \cdots - a_1(t)x_n + f(t). \end{cases} \tag{4.5.2}$$

With the notations

$$x = \begin{pmatrix} x_1 \\ x_2 \\ \vdots \\ x_n \end{pmatrix}, \quad b(t) = \begin{pmatrix} 0 \\ 0 \\ \vdots \\ f(t) \end{pmatrix}$$

and

$$\mathcal{A}(t) = \begin{pmatrix} 0 & 1 & 0 & \dots & 0 \\ 0 & 0 & 1 & \dots & 0 \\ \vdots & & & & \\ -a_n(t) & -a_{n-1}(t) & -a_{n-2}(t) & \dots & -a_1(t) \end{pmatrix},$$

for $t \in \mathbb{I}$, the system (4.5.2) can be written as a first-order vector differential equation

$$x' = \mathcal{A}(t)x + b(t). \tag{4.5.3}$$

At this point, it is completely clear that all the considerations in the preceding sections of this chapter may be rephrased in order to be applicable to equation (4.5.1). We begin by introducing some concepts and by establishing some variants of the previously proved results.

Definition 4.5.1. If in (4.5.1) $f(t) = 0$ on $\mathbb{I}$, then equation (4.5.1) is called *homogeneous.* Otherwise, it is called *non-homogeneous.*

Theorem 4.5.1. *For every $a \in \mathbb{I}$ and every $\xi \in \mathbb{R}^n$, the Cauchy problem*

$$\begin{cases} y^{(n)} + a_1(t)y^{(n-1)} + \cdots + a_n(t)y = f(t) \\ y(a) = \xi_1, y'(a) = \xi_2, \dots, y^{(n-1)}(a) = \xi_n \end{cases}$$

has a unique global solution.

Theorem 4.5.2. *Every saturated solution of* (4.5.1) *is defined on* $\mathbb{I}$.

Let us consider now the homogeneous equation attached to equation (4.5.1), i.e.

$$y^{(n)} + a_1(t)y^{(n-1)} + \cdots + a_n(t)y = 0. \tag{4.5.4}$$

We denote by $\mathcal{S}_n$ the set of all saturated solutions of the homogeneous equation (4.5.4) and with $\mathcal{S}$ the set of all saturated solutions of the linear homogeneous system attached to (4.5.3).

Lemma 4.5.1. *The map $\mathcal{T} : \mathcal{S}_n \to \mathcal{S}$, defined by*

$$\mathcal{T}(y) = (y, y', \dots, y^{(n-1)}) = (x_1, x_2, \dots, x_n)$$

for every $y \in \mathcal{S}_n$, is an isomorphism of vector spaces.

Proof. We may easily see that $\mathcal{S}_n$ is a vector subspace of $C^n(\mathbb{I};\mathbb{R})$ and that the map $\mathcal{T}$ is linear. In addition, $\mathcal{T}$ is surjective because given a solution $(x_1, x_2, \ldots, x_n)$ of the homogeneous system attached to (4.5.3) it is obvious that the function $y = x_1$ is of class C^n from $\mathbb{I}$ to $\mathbb{R}$ and $\mathcal{T}(y) = (x_1, x_2, \ldots, x_n)$. Finally, $\mathcal{T}$ is injective because $\mathcal{T}(y) = \mathcal{T}(z)$ is equivalent to $(y, y', \ldots, y^{(n-1)}) = (z, z', \ldots, z^{(n-1)})$, equality which clearly implies $y = z$. Hence $\mathcal{T}$ is an isomorphism, and this completes the proof. □

An immediate consequence of Lemma 4.5.1 is:

Theorem 4.5.3. *The set of all saturated solutions of the homogeneous equation* (4.5.4) *is an n-dimensional vector space over $\mathbb{R}$.*

Remark 4.5.1. By virtue of Theorem 4.5.3, the determination of the general solution of equation (4.5.4) is equivalent to the determination of n saturated linearly independent solutions.

From Theorem 4.2.1, we deduce

Theorem 4.5.4. *The general solution $y(\cdot, c)$, $c \in \mathbb{R}^n$, of* (4.5.1) *has the form*

$$y(t, c) = y_{\mathcal{M}}(t, c) + y_p(t),$$

where $y_{\mathcal{M}}(\cdot, c)$ is the general solution of the homogeneous equation (4.5.4), *while y_p is a particular saturated solution of equation* (4.5.1).

Now let $y_1, y_2, \ldots, y_n$ be a system of saturated solutions of equation (4.5.4), and let us define the matrix $\mathcal{Y} : \mathbb{I} \to \mathcal{M}_{n\times n}(\mathbb{R})$ by

$$\mathcal{Y}(t) = \begin{pmatrix} y_1(t) & y_2(t) \ldots & y_n(t) \\ y_1'(t) & y_2'(t) \ldots & y_n'(t) \\ \vdots & & \\ y_1^{(n-1)}(t) & y_2^{(n-1)}(t) \ldots & y_n^{(n-1)}(t) \end{pmatrix} \tag{4.5.5}$$

for every $t \in \mathbb{I}$.

Definition 4.5.2. The matrix $\mathcal{Y}$ defined by (4.5.5) is called *the associated matrix* of the system of solutions $y_1, y_2, \ldots, y_n \in \mathcal{S}_n$.

Definition 4.5.3. The system $y_1, y_2, \ldots, y_n \in \mathcal{S}_n$ is said to be *a fundamental system of solutions* of equation (4.5.4) if it is an algebraic basis for the vector space $\mathcal{S}_n$ of all saturated solutions of equation (4.5.4).

Definition 4.5.4. The associated matrix corresponding to a fundamental system of solutions of (4.5.4) is *a fundamental matrix* of (4.5.4).

Remark 4.5.2. Recalling that the map $\mathcal{T}$, defined in Lemma 4.5.1, is an isomorphism between $\mathcal{S}_n$ and $\mathcal{S}$, it follows that a system of saturated solutions $y_1, y_2, \ldots, y_n$ of equation (4.5.4) is fundamental if and only if $x^1, x^2, \ldots, x^n$, with $x^i = \mathcal{T}(y_i)$ for $i = 1, 2, \ldots, n$, is a fundamental system of solutions for the homogeneous system associated to the system (4.5.3). This simple observation allows us to reformulate several results, established for homogeneous linear systems, in this new framework of the nth-order linear differential equation.

More precisely, let $y_1, y_2, \ldots, y_n$ be a system of saturated solutions of equation (4.5.4), let $\mathcal{Y}$ be the associated matrix of this system, and let $\mathcal{W}(t) = \det \mathcal{Y}(t)$, for $t \in \mathbb{I}$, be the determinant which, by analogy with the case previously studied, is called *the Wronskian* of the system of solutions.

Theorem 4.5.5. (Liouville) *Let $\mathcal{W}$ be the Wronskian of a given system of n saturated solutions of* (4.5.4). *Then*

$$\mathcal{W}(t) = \mathcal{W}(t_0)\exp\left(-\int_{t_0}^{t} a_1(s)\,ds\right)$$

for every $t \in \mathbb{I}$, where $t_0 \in \mathbb{I}$ is fixed.

Proof. The conclusion follows from Theorem 4.1.5, by observing that, in the case of the homogeneous system attached to (4.5.3), the trace of the matrix $\mathcal{A}$ equals $-a_1$. □

Theorem 4.5.6. *Let $y_1, y_2, \ldots, y_n$ be a system of saturated solutions of* (4.5.4), *let $\mathcal{Y}$ and $\mathcal{W}$ be the matrix, and respectively the Wronskian, associated to the system of solutions. The following conditions are equivalent:*

(i) *the matrix $\mathcal{Y}$ is fundamental;*
(ii) *for every $t \in \mathbb{I}$, $\mathcal{W}(t) \neq 0$;*
(iii) *there exists $a \in \mathbb{I}$ such that $\mathcal{W}(a) \neq 0$.*

Proof. The conclusion follows from Theorem 4.1.4. □

Let $y_1, y_2, \ldots, y_n$ be a fundamental system of solutions of equation (4.5.4). Then, from Theorem 4.5.3, it follows that the general solution of the homogeneous equation (4.5.4) is given by

$$y(t) = \sum_{i=1}^{n} c_i y_i(t)$$

with $c_i \in \mathbb{R}$ for $i = 1, 2, \ldots, n$. Concerning the non-homogeneous equation (4.5.1), we have:

Theorem 4.5.7. *Let $y_1, y_2, \ldots, y_n$ be a fundamental system of solutions of equation* (4.5.4). *Then, the general solution of the non-homogeneous equation* (4.5.1) *is given by*

$$y(t) = \sum_{i=1}^{n} c_i(t) y_i(t),$$

where $c_i : \mathbb{I} \to \mathbb{R}$ for $i = 1, 2, \ldots, n$ are arbitrary functions of class C^1 which satisfy the system

$$\begin{cases} c_1'(t)y_1(t) + c_2'(t)y_2(t) + \cdots + c_n'(t)y_n(t) = 0 \\ c_1'(t)y_1'(t) + c_2'(t)y_2'(t) + \cdots + c_n'(t)y_n'(t) = 0 \\ \vdots \\ c_1'(t)y_1^{(n-2)}(t) + c_2'(t)y_2^{(n-2)}(t) + \cdots + c_n'(t)y_n^{(n-2)}(t) = 0 \\ c_1'(t)y_1^{(n-1)}(t) + c_2'(t)y_2^{(n-1)}(t) + \cdots + c_n'(t)y_n^{(n-1)}(t) = f(t) \end{cases} \tag{4.5.6}$$

for every $t \in \mathbb{I}$.

Proof. Let us observe that $y(t) = \sum_{i=1}^{n} c_i(t)y_i(t)$ is a solution of equation (4.5.1) if and only if $x(t) = \mathcal{Y}(t)c(t)$ is a solution of the system (4.5.3), where $c(t)$ is the column vector whose components are $c_1(t), c_2(t), \ldots, c_n(t)$. Reasoning as in the proof of Theorem 4.2.2, we deduce that c must satisfy (4.2.7). But the system (4.5.6) is nothing but the specific form taken by (4.2.7) in this case. The proof is complete. □

The method of finding the general solution of the non-homogeneous equation (4.5.1) as specified in Theorem 4.5.7 is called *the variation of constants method.*

4.6 The nth-Order Linear Differential Equation with Constant Coefficients

Next, we describe a method of finding a fundamental system of solutions in the case of the nth-order linear homogeneous differential equation with constant coefficients. We emphasize that, for the general case of an equation with variable coefficients, no such methods are known.

Let us consider the nth-order linear homogeneous differential equation with constant coefficients

$$y^{(n)} + a_1 y^{(n-1)} + \cdots + a_n y = 0, \tag{4.6.1}$$

where $a_1, a_2, \ldots, a_n \in \mathbb{R}$. By means of the transformations

$$x = (x_1, x_2, \ldots, x_n) = (y, y', \ldots, y^{(n-1)}), \tag{$\mathcal{T}$}$$

(4.6.1) may be rewritten as a first-order linear homogeneous vector differential equation

$$x' = \mathcal{A}x, \tag{4.6.2}$$

where

$$x = \begin{pmatrix} x_1 \\ x_2 \\ \vdots \\ x_n \end{pmatrix} \text{ and } \mathcal{A} = \begin{pmatrix} 0 & 1 & 0 & \ldots & 0 \\ 0 & 0 & 1 & \ldots & 0 \\ \vdots & & & & \\ -a_n & -a_{n-1} & -a_{n-2} & \ldots & -a_1 \end{pmatrix}.$$

Remark 4.6.1. It is easy to see, by direct computation, that, in this case, the equation $\det(\mathcal{A} - \lambda\mathcal{I}) = 0$ has the form

$$\lambda^n + a_1\lambda^{n-1} + \cdots + a_n = 0. \tag{4.6.3}$$

This is called *the characteristic equation* attached to equation (4.6.1), while the corresponding polynomial on the left-hand side is known as *the characteristic polynomial* attached to equation (4.6.1).

The main result referring to the determination of a fundamental system of solutions for equation (4.6.1) is:

Theorem 4.6.1. *Let $\lambda_1, \lambda_2, \ldots, \lambda_s$ be the roots of equation* (4.6.3) *with orders of multiplicity $m_1, m_2, \ldots, m_s$. Then, a fundamental system of solutions for equation* (4.6.1) *is $\mathcal{F} = \bigcup_{j=1}^{s} \mathcal{F}_j$, where, if λ_j is real, then*

$$\mathcal{F}_j = \left\{ e^{\lambda_j t}, te^{\lambda_j t}, t^2 e^{\lambda_j t}, \ldots, t^{(m_j - 1)} e^{\lambda_j t} \right\},$$

while if λ_j is not real, then $\mathcal{F}_j = \mathcal{G}_j \cup \mathcal{H}_j$, with

$$\mathcal{G}_j = \left\{ e^{\alpha_j t}\cos(\beta_j t), te^{\alpha_j t}\cos(\beta_j t), t^2 e^{\alpha_j t}\cos(\beta_j t), \ldots, t^{m_j - 1} e^{\alpha_j t}\cos(\beta_j t) \right\}$$

and

$$\mathcal{H}_j = \left\{ e^{\alpha_j t}\sin(\beta_j t), te^{\alpha_j t}\sin(\beta_j t), t^2 e^{\alpha_j t}\sin(\beta_j t), \ldots, t^{m_j - 1} e^{\alpha_j t}\sin(\beta_j t) \right\}.$$

In the latter case, α_j is the real part of λ_j, while β_j is the modulus of the imaginary part of the same root.

Proof. One may easily see that the family $\mathcal{F}$ contains at most n elements. So, in order to prove the theorem, it suffices to show that every solution of equation (4.6.1) is a linear combination of elements in $\mathcal{F}$. Indeed, if we assume that this is the case, then $\mathcal{F}$ is a spanning set for the space of all saturated solutions $\mathcal{S}_n$ of equation (4.6.1), which, according to Theorem 4.5.3, is an n-dimensional vector space over $\mathbb{R}$. Then, $\mathcal{F}$ must have exactly n elements, and so it is a basis for $\mathcal{S}_n$, as required.

So, let $y \in \mathcal{S}_n$. Then the function $\mathcal{T}(y) = x$ defined in Lemma 4.5.1 is a solution of the homogeneous equation (4.6.2). According to Remark 4.1.5, there exists $c \in \mathbb{R}^n$ such that $x(t) = e^{t\mathcal{A}}c$ for every $t \in \mathbb{R}$. On the other hand, from Theorem 4.4.3 and Remark 4.6.1, it follows that all components of x are linear combinations of elements in $\mathcal{F}$. In particular, $y = x_1$ enjoys the same property, which completes the proof. □

We can now analyze an example which gives a mathematical explanation of the resonance phenomenon in the case of *forced harmonic oscillations*.

Example 4.6.1. Let us consider the second-order linear differential equation

$$x'' + \omega^2 x = f(t) \tag{4.6.4}$$

which describes the oscillations of a material point P of mass m which moves on the Ox axis under the action of two forces: the first, an elastic one $F(x) = -kx$ for $x \in \mathbb{R}$, and the second, a periodic one of the form $G(t) = mf(t)$ for $t \in \mathbb{R}$. We recall that $\omega^2 = k/m$. We emphasize that here we have two systems: the first one, characterized by the elastic force called *receptor*, and the second one, called *excitatory* and characterized by the perturbing force G, exterior to the receptor system. Equation (4.6.4) describes the action of the excitatory system on the receptor system. The equation above is called *the equation of forced oscillations* of the material point P. We recall that here $x(t)$ represents the elongation of the point P at the moment t. Let us remark that, by Theorem 4.6.1, the general solution of the corresponding homogeneous equation is given by $x(t) = c_1 \sin \omega t + c_2 \cos \omega t$ for every $t \in \mathbb{R}$, where $c_1, c_2 \in \mathbb{R}$. We analyze next the case in which the excitatory force G is itself a solution of the homogeneous equation, situation in which G amplifies the oscillations of the material point. A proper understanding of the mechanism of this phenomenon, known under the name of *resonance*, represents a first step towards the explanation of many other phenomena, much more complex, but essentially of the same nature. More precisely, let us assume that $f(t) = k_1 \sin \omega t + k_2 \cos \omega t$ for $t \in \mathbb{R}$, where at least one of the numbers $k_1, k_2 \in \mathbb{R}$ is non-zero. In view of the variation of constants method, presented at the end of the preceding section, we conclude that the general solution of (4.6.4) is of the form $x(t) = c_1(t)\sin\omega t + c_2(t)\cos\omega t$, where c_1, c_2 are functions of class C^1 which satisfy

$$\begin{cases} c_1'(t)\sin\omega t + c_2'(t)\cos\omega t = 0 \\ \omega c_1'(t)\cos\omega t - \omega c_2'(t)\sin\omega t = k_1\sin\omega t + k_2\cos\omega t. \end{cases}$$

Solving this system, after a simple integration, we get

$$\begin{cases} c_1(t) = \frac{1}{4\omega^2}\left(-k_1\cos 2\omega t + k_2\sin 2\omega t\right) + \frac{k_2}{2\omega}t + k_3 \\ c_2(t) = \frac{1}{4\omega^2}\left(k_1\sin 2\omega t + k_2\cos 2\omega t\right) - \frac{k_1}{2\omega}t + k_4 \end{cases}$$

for $t \in \mathbb{R}$, where $k_3, k_4 \in \mathbb{R}$. Accordingly, the solution of equation (4.6.4) is $x(t) = \left(\frac{k_1}{4\omega^2} + k_3\right)\sin\omega t + \frac{k_2}{2\omega}t\sin\omega t + \left(\frac{k_2}{4\omega^2} + k_3\right)\cos\omega t - \frac{k_1}{2\omega}t\cos\omega t$. One can easily see that, unlike the solution of the homogeneous equation, which is bounded on $\mathbb{R}$, this solution is unbounded. This observation is very important in practice. Namely, it shows that the components of any structure subjected to vibrations have to be chosen such that their own frequencies are different from any rational multiple of the frequency of the excitatory force.

We conclude this section with the presentation of a class of nth-order linear differential equations with variable coefficients which, by a simple substitution, reduce to n^{th}-order linear differential equations with constant coefficients. More precisely, let us consider the equation

$$t^n y^{(n)} + t^{n-1}a_1 y^{(n-1)} + \cdots + a_n y = f(t), \tag{4.6.5}$$

with $a_1, a_2, \ldots, a_n \in \mathbb{R}$ and $f : \mathbb{R}_+^* \to \mathbb{R}$, equation known as *the Euler equation.*

Theorem 4.6.2. *By means of the substitutions*

$$\begin{cases} t = e^s \\ y(t) = z(s) \end{cases}$$

for $s \in \mathbb{R}$, (4.6.5) reduces to an nth-order linear differential equation with constant coefficients, with the new unknown function z depending on the new argument s.

Proof. Let us remark that, for $k = 1, 2, \ldots, n$, the kth-order derivative of y has the form

$$\frac{d^k y}{dt^k} = e^{-ks}\left(c_1\frac{dz}{ds} + c_2\frac{d^2 z}{ds^2} + \cdots + c_k\frac{d^k z}{ds^k}\right) \tag{4.6.6}$$

with $c_1, c_2, \ldots, c_n$ constants. Indeed, for $k = 1$, we have

$$\frac{dy}{dt} = e^{-s}\frac{dz}{ds}.$$

Assuming that (4.6.6) holds true for some $k \leq n-1$ and differentiating, we deduce that

$$\frac{d^{k+1}y}{dt^{k+1}} = e^{-(k+1)s}\left(c_1\frac{d^2z}{ds^2} + c_2\frac{d^3z}{ds^3} + \cdots + c_k\frac{d^{k+1}z}{ds^{k+1}}\right)$$

$$-ke^{-(k+1)s}\left(c_1\frac{dz}{ds} + c_2\frac{d^2z}{ds^2} + \cdots + c_k\frac{d^kz}{ds^k}\right)$$

$$= e^{-(k+1)s}\left(d_1\frac{d^z}{ds} + d_2\frac{d^2z}{ds^2} + \cdots + d_{k+1}\frac{d^{k+1}z}{ds^{k+1}}\right)$$

with $d_1, d_2, \ldots, d_{k+1}$ real constants. It follows that (4.6.6) holds true for every $k = 1, 2, \ldots, n$. Computing the derivatives of y, substituting these in (4.6.5) and taking into account that, for every $k = 1, 2, \ldots, n$, we have $t^k e^{-ks} = 1$, we easily conclude that z is the solution of an nth-order linear differential equation with constant coefficients. The proof is complete. □

Remark 4.6.2. Analogously, by means of the transformations

$$\begin{cases} \alpha t + \beta = e^s \\ y(t) = z(s) \end{cases}$$

for $\alpha t + \beta > 0$ and $s \in \mathbb{R}$, the equation

$$(\alpha t + \beta)^n y^{(n)}(t) + (\alpha t + \beta)^{n-1} a_1 y^{(n-1)}(t) + \cdots + a_n y(t) = f(t),$$

with $\alpha > 0$ and $\beta \in \mathbb{R}$, reduces to an nth-order linear differential equation with constant coefficients.

4.7 Exercises and Problems

Problem 4.1. *Let $a, b : \mathbb{R}_+ \to \mathbb{R}$ be two continuous functions with $\lim_{t\to+\infty} a(t) = 1$ and b absolutely integrable on $\mathbb{R}_+$. Let us consider the system*

$$\begin{cases} x' = a(t)y \\ y' = b(t)x. \end{cases} \tag{$\mathcal{S}$}$$

Prove that

(i) *if* (x, y) *is a solution of the system* $(\mathcal{S})$ *with* x *bounded on* $\mathbb{R}_+$, *then*

$$\lim_{t\to+\infty} y(t) = 0\,;$$

(ii) *there exists at least one solution of the system* $(\mathcal{S})$ *which is unbounded on* $\mathbb{R}_+$.

Problem 4.2. *Let* $f : \mathbb{I}\times\mathbb{R}^n \to \mathbb{R}^n$ *be a function of class* C^1 *with the property that*

$$\mathrm{div}_x f(t,x) = \sum_{i=1}^{n} \frac{\partial f_i}{\partial x_i}(t,x) = 0$$

on $\mathbb{I}\times\mathbb{R}^n$. *For* $a \in \mathbb{I}$ *and* $\xi \in \mathbb{R}^n$, *we denote by* $S(\cdot)\xi : [\,a,b) \to \mathbb{R}^n$ *the unique saturated solution of the problem* $\mathcal{CP}(\mathbb{I},\mathbb{R}^n, f, a, \xi)$. *Let* D *be a domain of finite volume in* $\mathbb{R}^n$ *and let* $D(t) = S(t)D$ *for* $t \in [\,a,b)$. *Prove that the volume of* $D(t)$ *i.e.*

$$\mathrm{Vol}\,(D(t)) = \iint \cdots \int_D \det\left(\frac{\partial S_i(t)x}{\partial x_j}\right) dx_1\, dx_2 \ldots dx_n$$

is constant on $[\,a,b)$. *This result is known as Liouville's theorem and is especially useful in Statistical Physics.*

Problem 4.3. *Let* $H : \mathbb{R}^n\times\mathbb{R}^n \to \mathbb{R}$ *be a function of class* C^2 *and let us consider the Hamiltonian system*

$$\begin{cases} \dfrac{dp_i}{dt} = -\dfrac{\partial H}{\partial q_i}(p,q) \\[2ex] \dfrac{dq_i}{dt} = \dfrac{\partial H}{\partial p_i}(p,q) \end{cases} \qquad i = 1,2,\ldots,n.$$

Let $\xi,\eta \in \mathbb{R}^n$ *and let* $S(\cdot)(\xi,\eta) = (p(\cdot), q(\cdot))$, *where* $(p,q) : [\,a,b) \to \mathbb{R}^n\times\mathbb{R}^n$ *is the unique saturated solution of the system which satisfies* $p(a) = \xi$ *and* $q(a) = \eta$. *Prove that, for each domain* D *of finite volume* $\mathrm{Vol}(D)$ *in* $\mathbb{R}^n\times\mathbb{R}^n$, *we have* $\mathrm{Vol}(S(t)D) = \mathrm{Vol}(D)$ *for every* $t \in [\,a,b)$.

Problem 4.4. *Let* $\mathcal{A} \in \mathcal{M}_{n\times n}(\mathbb{R})$ *be a matrix whose transpose* $\mathcal{A}^\tau$ *satisfies* $\mathcal{A}^\tau = -\mathcal{A}$. *Show that, for every* $t \in \mathbb{R}$, *the matrix* $e^{t\mathcal{A}}$ *is orthogonal. We recall that a matrix* $\mathcal{B}$ *is orthogonal if it is non-singular and* $\mathcal{B}^\tau = \mathcal{B}^{-1}$.

Problem 4.5. *Let* $\mathcal{A} \in \mathcal{M}_{n\times n}(\mathbb{R})$ *be a matrix whose transpose* $\mathcal{A}^\tau$ *satisfies* $\mathcal{A}^\tau = -\mathcal{A}$. *Show that every fundamental matrix* $\mathcal{X}$ *of the system*

$$x' = \mathcal{A}x,$$

which is orthogonal at $t = 0$, *is orthogonal at every* $t \in \mathbb{R}$.

Problem 4.6. *Let $\mathcal{A} : \mathbb{R} \to \mathcal{M}_{n\times n}(\mathbb{R})$ be a continuous function with the property that, for every $t \in \mathbb{R}$, $\mathcal{A}^{\tau}(t) = -\mathcal{A}(t)$. Prove that every fundamental matrix $\mathcal{X}$ of the system*

$$x' = \mathcal{A}(t)x,$$

which is orthogonal at $t = 0$, is orthogonal at every $t \in \mathbb{R}$.

Problem 4.7. *Let $\mathcal{A} \in \mathcal{M}_{n\times n}(\mathbb{R})$. Show that, if $\lambda \in \mathbb{C}$ is a root of the equation* $\det(\mathcal{A} - \lambda\mathcal{I}) = 0$, *then, for every $t \in \mathbb{R}$, $e^{t\lambda}$ is a root of* $\det(e^{t\mathcal{A}} - \mu\mathcal{I}) = 0$.

Problem 4.8. *If $\mathcal{A} \in \mathcal{M}_{n\times n}(\mathbb{R})$ is symmetric, i.e. $\mathcal{A}^{\tau} = \mathcal{A}$ then $e^{t\mathcal{A}}$ is symmetric for every $t \in \mathbb{R}$.*

Problem 4.9. *Let $\mathcal{A} : \mathbb{R} \to \mathcal{M}_{n\times n}(\mathbb{R})$ be a continuous function with the property that $\mathcal{A}^*(t) = \mathcal{A}(t)$ for every $t \in \mathbb{R}$. Prove that every fundamental matrix $\mathcal{X}$ of the system*

$$x' = \mathcal{A}(t)x,$$

which is symmetric at $t = 0$, is symmetric at every $t \in \mathbb{R}$.

Problem 4.10. *Let $\mathcal{A} \in \mathcal{M}_{n\times n}(\mathbb{R})$. A necessary and sufficient condition for all the elements of the matrix $e^{t\mathcal{A}}$ to be positive for every $t \geq 0$ is that all the non-diagonal elements of the matrix $\mathcal{A}$ be positive.* ([Halanay (1972)], p. 190)

Problem 4.11. *Let $\mathcal{A}, \mathcal{B}, \mathcal{C} \in \mathcal{M}_{n\times n}(\mathbb{R})$. Prove that the solution of the Cauchy problem*

$$\begin{cases} \mathcal{X}' = \mathcal{A}\mathcal{X} + \mathcal{X}\mathcal{B} \\ \mathcal{X}(0) = \mathcal{C} \end{cases}$$

is given by $\mathcal{X}(t) = e^{t\mathcal{A}}\mathcal{C}e^{t\mathcal{B}}$. ([Halanay (1972)], p. 191)

Problem 4.12. *Let $\mathcal{A}, \mathcal{B}, \mathcal{C} \in \mathcal{M}_{n\times n}(\mathbb{R})$. Prove that if the integral*

$$\mathcal{X} = -\int_0^{+\infty} e^{s\mathcal{A}}\mathcal{C}e^{s\mathcal{B}}\,ds$$

is convergent, then it satisfies $\mathcal{A}\mathcal{X} + \mathcal{X}\mathcal{B} = \mathcal{C}$. ([Halanay (1972)], p. 191)

Problem 4.13. *Let $\mathcal{A} \in \mathcal{M}_{n\times n}(\mathbb{R})$ and let*

$$\begin{cases} \cos\mathcal{A} = \displaystyle\sum_{k=0}^{\infty}(-1)^k\frac{\mathcal{A}^{2k}}{(2k)!} \\ \sin\mathcal{A} = \displaystyle\sum_{k=0}^{\infty}(-1)^k\frac{\mathcal{A}^{2k+1}}{(2k+1)!}. \end{cases}$$

(1) *Compute* $\dfrac{d}{dt}(\cos t\mathcal{A})$ *and* $\dfrac{d}{dt}(\sin t\mathcal{A})$;

(2) *Show that the* $2n \times 2n$ *matrix*

$$\mathcal{Z}(t) = \begin{pmatrix} \cos t\mathcal{A} & \sin t\mathcal{A} \\ -\mathcal{A}\sin t\mathcal{A} & \mathcal{A}\cos t\mathcal{A} \end{pmatrix}$$

is an associated matrix of a certain system of solutions for the first-order system of linear differential equations with $2n$ *unknown functions:* $x_1, x_2, \ldots, x_n, y_1, y_2, \ldots, y_n$

$$\begin{cases} x' = y \\ y' = -\mathcal{A}^2 x. \end{cases}$$

Under what circumstances is this a fundamental matrix?

([Halanay (1972)], p. 191)

Problem 4.14. *Let* $f : \mathbb{I} \times \mathbb{R}^n \to \mathbb{R}^n$ *be continuous on* $\mathbb{I} \times \mathbb{R}^n$ *and Lipschitz on* $\mathbb{R}^n$*, let* $\xi \in \mathbb{R}^n$*,* $a \in \mathbb{I}$ *and* $\mathcal{A} \in \mathcal{M}_{n\times n}(\mathbb{R})$*. We define the following sequence of successive approximations:* x_0 *is the unique global solution of the system*

$$\begin{cases} x' = \mathcal{A}x \\ x(a) = \xi, \end{cases}$$

while x_m *is the unique global solution of the system*

$$\begin{cases} x_m' = \mathcal{A}x_m + f(t, x_{m-1}(t)) - \mathcal{A}x_{m-1}(t) \\ x_m(a) = \xi. \end{cases}$$

Prove that, for every $b > a$ *with* $[a, b] \subset \mathbb{I}$*,* $(x_m)_{m\in\mathbb{N}}$ *converges uniformly on* $[a, b]$ *to the unique solution* $x : [a, b] \to \mathbb{R}^n$ *of the Cauchy problem*

$$\begin{cases} x' = f(t, x) \\ x(a) = \xi. \end{cases}$$

([Halanay (1972)], p. 196)

Exercise 4.1. *Solve the following systems of linear differential equations:*

$$(1)\begin{cases} x_1' = x_1 + 2x_2 \\ x_2' = 4x_1 + 3x_2. \end{cases} \qquad (2)\begin{cases} x_1' = x_2 \\ x_2' = -x_1. \end{cases}$$

$$(3)\begin{cases} x_1' = x_1 + 5x_2 \\ x_2' = -x_1 - 3x_2. \end{cases} \qquad (4)\begin{cases} x_1' = x_1 + x_2 \\ x_2' = x_1 + x_2 + t. \end{cases}$$

$$(5)\begin{cases} x_1' + 2x_1 + x_2 = \sin t \\ x_2' - 4x_1 - 2x_2 = \cos t. \end{cases} \qquad (6)\begin{cases} x_1' + 2x_1 + 4x_2 = 1 + 4t \\ x_2' + x_1 - x_2 = \frac{3}{2}t^2. \end{cases}$$

$$(7)\begin{cases} x_1' = x_2 \\ x_2' = x_3 \\ x_3' = x_1. \end{cases} \qquad (8)\begin{cases} x_1' = x_2 + x_3 \\ x_2' = x_3 + x_1 \\ x_3' = x_1 + x_2. \end{cases}$$

Exercise 4.2. *Solve the following second-order linear differential equations*:

(1) $x'' - 5x' + 4x = 0$. (2) $x'' + 2x' + x = 0$. (3) $x'' + 4x = 0$.
(4) $x'' - 4x = t^2e^{2t}$. (5) $x'' + 9x = \cos 2t$. (6) $x'' + x = \dfrac{1}{\sin t}$.
(7) $x'' + x = 2t\cos t\cos 2t$. (8) $x'' - 4x' + 4x = te^{2t}$. (9) $x'' - 2x = 4t^2e^{t^2}$.

Exercise 4.3. *Solve the following higher-order linear differential equations*:

(1) $x''' - 13x'' + 12x' = 0$. (2) $x''' - x' = 0$. (3) $x''' + x = 0$.
(4) $x^{IV} + 4x = 0$. (5) $x''' - 3x'' + 3x' - x = t$. (6) $x^{IV} + 2x'' + x = 0$.
(7) $x^{IV} - 2x''' + x'' = e^t$. (8) $x''' + x'' + x' + x = te^t$. (9) $x''' + 6x'' + 9x' = t$.

Exercise 4.4. *Solve the following Euler, or reducible to Euler equations*:

(1) $t^2x'' + 3tx' + x = 0$. (2) $t^2x'' - tx' - 3x = 0$.
(3) $t^2x'' + tx' + 4x = 0$. (4) $t^3x''' - 3t^2x'' + 6tx' - 6x = 0$.
(5) $(3t + 2)x'' + 7x' = 0$. (6) $x'' = \dfrac{2x}{t^2}$.
(7) $x'' + \dfrac{x'}{t} + \dfrac{x}{t^2} - 0$. (8) $t^2x'' - 4tx' + 6x = t$.

(9) $(1 + t)^2x'' - 3(1 + t)x' + 4x = (1 + t)^3$. (10) $t^2x'' - tx' + x = 2t$.

Chapter 5

Elements of Stability

This chapter is entirely dedicated to the study of the stability of solutions to certain systems of differential equations. In the first section we introduce and illustrate the main concepts referring to stability. The second one is concerned with several necessary and sufficient conditions for various types of stability in the particular case of first-order systems of linear differential equations. In the third section we present some sufficient conditions under which the asymptotic stability of the null solution of a first-order differential system is inherited by the null solution of a certain perturbed system, provided the perturbation is small enough. In the fourth section we prove several sufficient conditions for stability expressed by means of some functions decreasing along the trajectories, while in the fifth section we include several results regarding the stability of solutions of dissipative systems. In the sixth section we analyze the stability problem referring to automatic control systems, while the seventh section is dedicated to some considerations concerning instability and chaos. As each chapter of this book, this one also ends with an Exercises and Problems section.

5.1 Types of Stability

In its usual meaning, *stability* is that property of a particular state of a given system of preserving the features of its evolution, as long as the perturbations of the initial data are sufficiently small. This meaning comes from Mechanics, where it describes that property of the equilibrium state of a conservative system of being insensitive "*à la longue*" to any kind of perturbations of "small intensity". Mathematically speaking, this notion has many other senses, all coming from the preceding one, and describing various kinds of continuity of a given global solution of a system as function of the initial data, senses which are more or less different from one another. The rigorous study of stability has its origins in the works of Celestial Mechanics of both Poincaré and Maxwell, and has culminated in 1892 with the doctoral thesis of the founder of this modern branch of differential equations, Lyapunov.

As we have already shown in Theorem 2.5.2, under certain regularity conditions on the function f, the map $\eta \mapsto x(\cdot, a, \eta)$ — the unique saturated solution of the Cauchy problem

$$\begin{cases} x' = f(t, x) \\ x(a) = \eta \end{cases} \qquad \mathcal{CP}(\mathcal{D})$$

— is locally Lipschitz from Ω to $C([\,a, b\,]; \mathbb{R}^n)$, for each $b \in (a, b_\xi)$, where $[\,a, b_\xi)$ is the domain of definition of the saturated solution $x(\cdot, a, \xi)$. A much more delicate problem, and of great practical interest, is that of finding sufficient conditions on the function f such that, on one hand, $x(\cdot, a, \xi)$ be defined on $[\,a, +\infty)$ and, on the other hand, the map $\eta \mapsto x(\cdot, a, \eta)$ be continuous from a neighborhood of ξ to the space of continuous functions from $[\,a, +\infty)$ to $\mathbb{R}^n$, endowed with the uniform convergence topology.

Let Ω be a nonempty and open subset of $\mathbb{R}^n$, let $f : \mathbb{R}_+ \times \Omega \to \mathbb{R}^n$ be continuous on $\mathbb{R}_+ \times \Omega$ and either locally Lipschitz, or dissipative on Ω, and let us consider the differential system

$$x' = f(t, x). \tag{5.1.1}$$

Let us assume that (5.1.1) has a global solution $\phi : \mathbb{R}_+ \to \Omega$.

Definition 5.1.1. The solution $\phi : \mathbb{R}_+ \to \Omega$ of (5.1.1) is *stable* if:

(i) for every $a \geq 0$ there exists $\mu(a) > 0$ such that for every $\xi \in \Omega$ with $\|\xi - \phi(a)\| \leq \mu(a)$, the unique saturated solution $x(\cdot, a, \xi)$, of the system (5.1.1), satisfying $x(a, a, \xi) = \xi$, is defined on $[\,a, +\infty)$ and

(ii) for every $a \geq 0$ and every $\varepsilon > 0$, there exists $\delta(\varepsilon, a) \in (0, \mu(a)\,]$ such that, for each $\xi \in \Omega$ with $\|\xi - \phi(a)\| \leq \delta(\varepsilon, a)$, the unique saturated solution $x(\cdot, a, \xi)$, of the system (5.1.1), satisfying $x(a, a, \xi) = \xi$, also satisfies $\|x(t, a, \xi) - \phi(t)\| \leq \varepsilon$ for every $t \in [\,a, +\infty)$.

For a suggestive illustration of the situation described in Definition 5.1.1, in the case $n = 2$, see Figure 5.1.1.

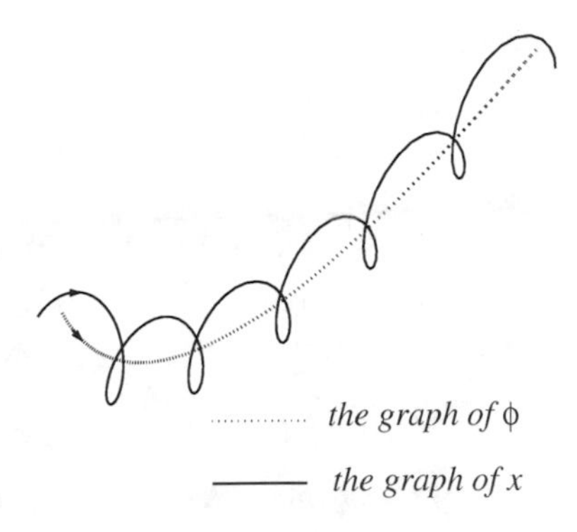

Fig. 5.1.1

Definition 5.1.2. The solution $\phi : \mathbb{R}_+ \to \Omega$ of (5.1.1) is *uniformly stable* if it is stable and both $\mu(a) > 0$ and $\delta(\varepsilon, a) > 0$ in Definition 5.1.1 can be chosen independent of $a \geq 0$, i.e. $\mu(a) = \mu$ and $\delta(\varepsilon, a) = \delta(\varepsilon)$.

Definition 5.1.3. The solution $\phi : \mathbb{R}_+ \to \Omega$ of (5.1.1) is *asymptotically stable* if it is stable and, for every $a \geq 0$, $\mu(a) > 0$ in Definition 5.1.1 can be chosen such that, for each $\xi \in \Omega$ with $\|\xi - \phi(a)\| \leq \mu(a)$, the unique saturated solution $x(\cdot, a, \xi)$ of the system (5.1.1), satisfying $x(a, a, \xi) = \xi$, also satisfies $\lim_{t \to +\infty} \|x(t, a, \xi) - \phi(t)\| = 0$.

The situation described in Definition 5.1.2 is illustrated in Figure 5.1.2.

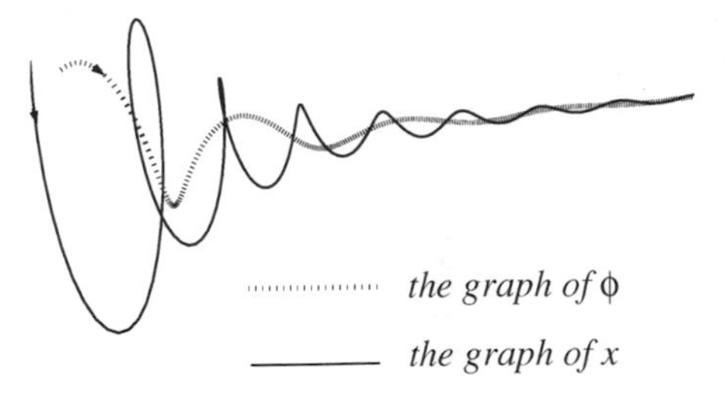

Fig. 5.1.2

Definition 5.1.4. The solution $\phi : \mathbb{R}_+ \to \Omega$ of (5.1.1) is *uniformly asymptotically stable* if it is uniformly stable and, for every $\varepsilon > 0$, there exists $r(\varepsilon) > 0$ such that, for every $a \geq 0$, every $\xi \in \Omega$ with $\|\xi - \phi(a)\| \leq \mu$ (where

$\mu > 0$ is given by Definition 5.1.2), and every $t \geq a + r(\varepsilon)$, we have

$$\|x(t, a, \xi) - \phi(t)\| \leq \varepsilon.$$

Remark 5.1.1. These four concepts of stability refer to a property of a certain solution of (5.1.1), and not to a property of the system itself. More precisely, there exist systems which have both stable and unstable solutions. Indeed, let us consider the differential equation

$$x' = ax(p - x),$$

where $a > 0$ and $p > 0$ are constants. As we have seen in Section 1.4, this equation describes the spread of a disease within a population p, $x(t)$ representing the number of the infected individuals at the moment t. We recall that for every $\tau \geq 0$ and every $\xi \in \mathbb{R}$ the unique global solution $x(\cdot, \tau, \xi) : [\,\tau, +\infty) \to \mathbb{R}$ of this equation, which satisfies $x(\tau, \tau, \xi) = \xi$, is

$$x(t, \tau, \xi) = \frac{p\xi e^{ap(t-\tau)}}{p + \xi(e^{ap(t-\tau)} - 1)}$$

for $t \in [\,\tau, +\infty)$. Among the two stationary solutions $x \equiv 0$ and $x \equiv p$ of the equation, the first one is unstable, while the second one is uniformly stable. Rephrasing this observation in the terms of the modelled phenomenon, we can say that: *in an isolated biological system, the state of health* ($x \equiv 0$) *is fragile to small perturbations, i.e. unstable, while the state of illness* ($x \equiv p$) *is uniformly stable.*

Remark 5.1.2. Every solution ϕ of the system (5.1.1) which is uniformly asymptotically stable is both uniformly stable and asymptotically stable. Moreover, every uniformly, or asymptotically stable solution is stable. We emphasize that: (1) stability does not imply uniform stability; (2) the concepts of uniform stability and asymptotic stability are independent; (3) uniform stability does not imply uniform asymptotic stability. See the example below.

Example 5.1.1. In order to prove the item (1) in Remark 5.1.2, let us consider the equation $x' = a(t)x$, where

$$a(t) = \frac{d}{dt}\left[t(1 - t\cos t)\cos t\right].$$

It is not difficult to see that a satisfies the condition (1) in Problem 5.1 with $M : \mathbb{R}_+ \to \mathbb{R}$ defined by $M(t_0) = (t_0 \cos t_0 - \frac{1}{2})^2$ for every $t_0 \in \mathbb{R}_+$, but it satisfies none of the other three conditions. Hence the null solution of the equation above is stable but is neither uniformly, nor asymptotically stable, which proves (1).

The identically zero solution of the equation $x' = 0$ is uniformly stable but it is neither asymptotically stable, nor uniformly asymptotically stable. This observation proves (3) and the fact that uniformly stability does not imply asymptotic stability. In order to complete the proof of item (2) in Remark 5.1.2, we will show that the asymptotic stability does not imply uniformly stability. To this aim let us consider the equation $x' = a(t)x$, where

$$a(t) = \frac{d}{dt}\left[t\left(\sin t - \alpha t\right)\right]$$

for every $t \geq 0$, where $\alpha \in (0, 1/\pi)$. We leave it to the reader to prove the fact that a satisfies condition (3) in Problem 5.1 but does not satisfy condition (2) in the same problem. Hence the null solution of the equation is asymptotically stable, but it is not uniformly stable.

Remark 5.1.3. By means of the transformation $y = x - \phi$ the study of any type of stability, referring to the solution ϕ of the system (5.1.1), reduces to the study of the same type of stability referring to the identically zero solution of the system $y' = f(t, y + \phi(t)) - \phi'(t)$. Therefore, in the following, we will assume that $0 \in \Omega$, $f(t, 0) = 0$, and we will confine ourselves only to the study of the stability of the identically zero solution of the system (5.1.1).

A *stationary point* or *equilibrium point* of (5.1.1) is an element x^* in Ω with the property that $f(t, x^*) = 0$ for every $t \in \mathbb{R}_+$. Obviously, if x^* is a stationary point for the system (5.1.1), the function $x \equiv x^*$ is a constant solution of (5.1.1), called *stationary solution*. Let us observe that the identically zero solution of the system (5.1.1) is in fact a stationary solution or an equilibrium point for that system in the just mentioned sense. In the case of autonomous systems, i.e. of systems (5.1.1) for which f does not depend explicitly on the variable $t \in \mathbb{R}_+$, we have the following result on the behavior of the solutions as t approaches $+\infty$.

Theorem 5.1.1. *Let $f : \Omega \to \mathbb{R}^n$ be continuous and let $x : [a, +\infty) \to \Omega$ be a solution of*

$$x' = f(x). \tag{5.1.2}$$

If there exists $\lim_{t \to +\infty} x(t) = x^*$ and $x^* \in \Omega$, then x^* is an equilibrium point for the system (5.1.2).

We mention that a rather similar result has already been established in Section 3.4. See Lemma 3.4.1.

Proof. From the mean-value theorem applied to the component x_i of the solution on the interval $[m, m+1]$, with $m \in \mathbb{N} \cap [a, +\infty)$ and $i = 1, 2, \ldots, n$, it follows that there exists θ_{im} in $(m, m+1)$ such that

$$x_i(m+1) - x_i(m) = x_i'(\theta_{im}) = f_i(x(\theta_{im}))$$

for $i = 1, 2, \ldots, n$ and $m \in \mathbb{N} \cap [a, +\infty)$. As $\lim_m (x_i(m+1) - x_i(m)) = 0$ and $\lim_m f_i(x(\theta_{im})) = f_i(x^*)$, it follows that

$$\lim_{t \to +\infty} f_i(x(t)) = f_i(x^*) = 0$$

for $i = 1, 2, \ldots, n$ and therefore $f(x^*) = 0$. The proof is complete. □

For the sake of simplicity we will restate the preceding definitions in the particular case $\phi \equiv 0$.

Definition 5.1.5. The null solution of (5.1.1) is *stable* if:

(i) for every $a \geq 0$ there exists $\mu(a) > 0$ such that, for every $\xi \in \Omega$ with $\|\xi\| \leq \mu(a)$, the unique saturated solution $x(\cdot, a, \xi)$, of the system (5.1.1), satisfying $x(a, a, \xi) = \xi$, is defined on $[a, +\infty)$ and
(ii) for every $a \geq 0$ and every $\varepsilon > 0$, there exists $\delta(\varepsilon, a) \in (0, \mu(a)]$ such that, for each $\xi \in \Omega$ with $\|\xi\| \leq \delta(\varepsilon, a)$, the unique saturated solution $x(\cdot, a, \xi)$, of the system (5.1.1), satisfying $x(a, a, \xi) = \xi$, also satisfies $\|x(t, a, \xi)\| \leq \varepsilon$ for every $t \in [a, +\infty)$.

Definition 5.1.6. The null solution of (5.1.1) is *uniformly stable* if it is stable and both $\mu(a) > 0$ and $\delta(\varepsilon, a) > 0$ in Definition 5.1.5 can be chosen independent of $a \geq 0$, i.e. $\mu(a) = \mu$ and $\delta(\varepsilon, a) = \delta(\varepsilon)$.

Definition 5.1.7. The null solution of (5.1.1) is *asymptotically stable* if it is stable and, for every $a \geq 0$, $\mu(a) > 0$ in Definition 5.1.1 can be chosen such that, for each $\xi \in \Omega$ with $\|\xi\| \leq \mu(a)$, the unique saturated solution $x(\cdot, a, \xi)$ of the system (5.1.1), which satisfies $x(a, a, \xi) = \xi$, also satisfies $\lim_{t \to +\infty} \|x(t, a, \xi)\| = 0$.

Definition 5.1.8. The null solution of (5.1.1) is *uniformly asymptotically stable* if it is uniformly stable and, for every $\varepsilon > 0$, there exists $r(\varepsilon) > 0$ such that, for every $a \geq 0$, every $\xi \in \Omega$ with $\|\xi\| \leq \mu$ (where $\mu > 0$ is given by Definition 5.1.6), and every $t \geq a + r(\varepsilon)$, we have

$$\|x(t, a, \xi)\| \leq \varepsilon.$$

We conclude this section with the definition of a stability concept which, this time, describes a property of the system (5.1.1) and not one of a certain solution.

Definition 5.1.9. The system (5.1.1) is *globally asymptotically stable* if, for every $a \geq 0$ and every $\xi \in \Omega$, its unique saturated solution, $x(\cdot, a, \xi)$ which satisfies $x(a, a, \xi) = \xi$, is defined on $[a, +\infty)$ and

$$\lim_{t \to +\infty} x(t, a, \xi) = 0.$$

5.2 Stability of Linear Systems

The aim of this section is to present several results referring to various types of stability in the particular case of first-order systems of linear differential equations. More precisely, let us consider the system

$$x' = \mathcal{A}(t)x, \tag{5.2.1}$$

where $\mathcal{A} = (a_{ij})_{n \times n}$ is a matrix whose elements a_{ij} are continuous functions from $\mathbb{R}_+$ in $\mathbb{R}$.

Theorem 5.2.1. *The null solution of* (5.2.1) *is stable* (*asymptotically stable*), (*uniformly stable*), (*uniformly asymptotically stable*) *if and only if each of its saturated solutions is stable* (*asymptotically stable*), (*uniformly stable*), (*uniformly asymptotically stable*).

Proof. If $x = \phi$ is a saturated solution of the system (5.2.1), by the transformation $y = x - \phi$, this solution corresponds to the saturated solution $y \equiv 0$. The conclusion of the theorem follows from the simple observation that ϕ satisfies the conditions in Definition 5.1.1, (5.1.2), (5.1.3), (5.1.4) if and only if $y \equiv 0$ satisfies the corresponding conditions in Definition 5.1.5, (5.1.6), (5.1.7), (5.1.8). □

Remark 5.2.1. According to Theorem 5.2.1, in the case of linear systems, the stability of one saturated solution is equivalent to the stability of any saturated solution whatsoever. Therefore, within this framework, we will speak about the stability, or instability of the system itself, understanding by this the stability, or instability, of the null solution, or of any one of its saturated solutions.

We continue with a fundamental result referring to stability.

Theorem 5.2.2. *The following assertions are equivalent*:

(i) *the system* (5.2.1) *is stable*;
(ii) *the system* (5.2.1) *has a fundamental system of solutions which are bounded on* $\mathbb{R}_+$;
(iii) *all saturated solutions of the system* (5.2.1) *are bounded on* $\mathbb{R}_+$;
(iv) *all fundamental matrices of the system* (5.2.1) *are bounded on* $\mathbb{R}_+$;
(v) *the system* (5.2.1) *has a bounded fundamental matrix on* $\mathbb{R}_+$.

Proof. If (5.2.1) is stable, then for $\varepsilon = 1$ and $a = 0$ there exists $\delta > 0$ such that, for every $\xi \in \mathbb{R}^n$ with $\|\xi\| \le \delta$, the unique saturated solution $x(\cdot, 0, \xi)$ of the system (5.2.1) satisfies

$$\|x(t, 0, \xi)\| \le 1$$

for every $t \in \mathbb{R}_+$. Take n linearly independent vectors in the ball $B(0, \delta)$, and let us observe that the n saturated solutions, which have as initial data at $t = 0$ those n chosen vectors, are bounded on $\mathbb{R}_+$ and constitute a fundamental system of solutions for (5.2.1). Hence (i) implies (ii). If (5.2.1) has a fundamental system of solutions, bounded on $\mathbb{R}_+$, as every solution is a linear combination of elements in the fundamental system, it is bounded and therefore (ii) implies (iii). Obviously (iii) implies (iv) which in its turn implies (v). Finally, let us consider a fundamental matrix $\mathcal{X}(t)$ of the system (5.2.1) and let us recall that, for every $a \ge 0$ and every $\xi \in \mathbb{R}^n$, the unique saturated solution $x(\cdot, a, \xi)$ of the system (5.2.1) is given by

$$x(t, a, \xi) = \mathcal{X}(t)\mathcal{X}^{-1}(a)\xi$$

for every $t \ge a$. Assuming that (v) holds true, we can choose $\mathcal{X}(t)$ such that there exists $M > 0$ with the property

$$\|\mathcal{X}(t)\|_{\mathcal{M}} \le M$$

for every $t \in \mathbb{R}_+$, where $\|\mathcal{X}(t)\|_{\mathcal{M}}$ is the norm defined in Section 12.1, norm which, according to Remark 12.1.1, is equivalent to the Euclidean norm of a matrix $\mathcal{X}(t)$, i.e. with the square root of the sum of its squared elements. From the last two relations, we have

$$\|x(t, a, \xi)\| \le M\|\mathcal{X}^{-1}(a)\|_{\mathcal{M}}\|\xi\|$$

for every $t \ge a$. Consequently, for every $\varepsilon > 0$ and every $a \ge 0$, there exists

$$\delta(\varepsilon, a) = \varepsilon \left(M\|\mathcal{X}^{-1}(a)\|_{\mathcal{M}}\right)^{-1} > 0$$

such that, for every $\xi \in \mathbb{R}^n$ with $\|\xi\| \le \delta(\varepsilon, a)$ we have

$$\|x(t, a, \xi)\| \le \varepsilon$$

for every $t \ge a$. Hence (v) implies (i) and this completes the proof. □

In the linear case, a necessary and sufficient condition of the asymptotic stability is:

Proposition 5.2.1. *The system* (5.2.1) *is asymptotically stable if and only if for every* $a \geq 0$ *there exists* $\mu(a) > 0$ *such that for every* $\xi \in \mathbb{R}^n$ *with* $\|\xi\| \leq \mu(a)$, *we have*

$$\lim_{t \to +\infty} x(t, a, \xi) = 0.$$

Proof. The necessity is obvious. In order to prove the sufficiency let us observe that, for $a = 0$ there exists $\mu > 0$ such that all the saturated solutions of the system (5.2.1), which have as initial data at $t = 0$ vectors in $B(0, \mu)$, tends to 0 as t approaches $+\infty$. Consequently, all these solutions are bounded on $\mathbb{R}_+$. In particular, every fundamental system of solutions of the system (5.2.1), having as initial data at $t = 0$ vectors in $B(0, \mu)$, contains only functions which are bounded on $\mathbb{R}_+$. From the equivalence of the assertions (i) and (ii) in Theorem 5.2.2, it follows that (5.2.1) is stable, which completes the proof. □

Concerning the asymptotic stability of linear systems, we have:

Theorem 5.2.3. *The following assertions are equivalent*:

(i) *the system* (5.2.1) *is asymptotically stable*;
(ii) *the system* (5.2.1) *has a fundamental system of solutions which tend to* 0 *as* t *approaches* $+\infty$;
(iii) *the system* (5.2.1) *is globally asymptotically stable*;
(iv) *the norm of any fundamental matrix of the system* (5.2.1) *tends to* 0 *as* t *approaches* $+\infty$;
(v) *there exists a fundamental matrix of the system* (5.2.1) *whose norm tends to* 0 *as* t *approaches* $+\infty$.

Proof. If (5.2.1) is asymptotically stable, for $a = 0$, there exists $\mu > 0$ such that, for every $\xi \in \mathbb{R}^n$ with $\|\xi\| \leq \mu$, the unique saturated solution $x(\cdot, 0, \xi)$ of the system (5.2.1), which satisfies $x(0, 0, \xi) = \xi$, tends to 0 as t approaches $+\infty$. Let us consider a fundamental system of solutions of (5.2.1) consisting of functions whose values at $t = 0$ belong to $B(0, \mu)$, and let us observe that, from the manner of choice of $\mu > 0$, this fundamental system contains only functions which tend to 0 as t approaches $+\infty$. So (i) implies (ii). If (5.2.1) has a fundamental system of solutions which tend to 0 as t approaches $+\infty$, then every solution of (5.2.1) enjoys the same property, being a linear combination of elements in the fundamental system considered. Hence (ii) implies (iii). Obviously (iii) implies (iv) which, in its turn, implies (v). Finally, if $\mathcal{X}(t)$ is a fundamental matrix of the system (5.2.1) with

$$\lim_{t\to+\infty} \|\mathcal{X}(t)\|_{\mathcal{M}} = 0,$$

from the representation formula of the solution: $x(t,a,\xi) = \mathcal{X}(t)\mathcal{X}^{-1}(a)\xi$ for every $t \geq a$, we deduce that every saturated solution of the system (5.2.1) tends to 0 as t approaches $+\infty$. According to Proposition 5.2.1, the system (5.2.1) is stable, and therefore (v) implies (i) which completes the proof. □

Theorem 5.2.4. *The system* (5.2.1) *is uniformly stable if and only if it has a fundamental matrix* $\mathcal{X}(t)$ *for which there exists* $M > 0$ *such that*

$$\|\mathcal{X}(t)\mathcal{X}^{-1}(s)\|_{\mathcal{M}} \leq M \tag{5.2.2}$$

for every $t, s \in \mathbb{R}_+$, $s \leq t$.

Proof. In order to prove the sufficiency let us assume that the system (5.2.1) has a fundamental matrix which satisfies (5.2.2). Let $\xi \in \mathbb{R}^n$ and $a \in \mathbb{R}_+$. Since the unique saturated solution of (5.2.1), $x(\cdot,a,\xi)$, is given by

$$x(t,a,\xi) = \mathcal{X}(t)\mathcal{X}^{-1}(a)\xi,$$

from (5.2.2), it follows that

$$\|x(t,a,\xi)\| \leq \|\mathcal{X}(t)\mathcal{X}^{-1}(a)\|_{\mathcal{M}}\|\xi\| \leq M\|\xi\|$$

for every $t \geq a$. Let $\varepsilon > 0$. From the preceding inequality we deduce that, for every $\xi \in \Omega$, with $\|\xi\| \leq \varepsilon M^{-1}$, we have

$$\|x(t,a,\xi)\| \leq \varepsilon$$

for every $t \geq a$. So (5.2.1) is uniformly stable.

In order to prove the necessity, let us assume that (5.2.1) is uniformly stable. Then, for $\varepsilon = 1$ there exists $\delta > 0$ such that, for every $a \in \mathbb{R}_+$ and every $\xi \in \Omega$ with $\|\xi\| \leq \delta$, the unique saturated solution $x(\cdot,a,\xi)$ of (5.2.1), which satisfies $x(a,a,\xi) = \xi$, satisfies the inequality

$$\|x(t,a,\xi)\| \leq 1$$

for every $t \geq a$. Let $\mathcal{X}(t)$ be that fundamental matrix of the system (5.2.1) which satisfies $\mathcal{X}(0) = \mathcal{I}_n$. Let $\lambda \in (0,\delta)$ and $t, s \in \mathbb{R}_+$ with $s \leq t$. Let us remark that the matrix $\lambda\mathcal{X}(t)\mathcal{X}^{-1}(s)$ has as column of rank $i \in \{1,2,\ldots,n\}$

that saturated solution x^i of the system (5.2.1) which for $t = s$ takes the value ξ^i, where ξ^i is the vector with all components zero, except that one on the row i which equals λ. Then $\|\xi^i\| = \lambda < \delta$ and therefore $\|x^i(t)\| \leq 1$ for every $i \in \{1, 2, \ldots, n\}$ and every $t \geq s$. Since, from Remark 12.1.1, we have

$$\lambda\|\mathcal{X}(t)\mathcal{X}^{-1}(s)\|_{\mathcal{M}} \leq \lambda\|\mathcal{X}(t)\mathcal{X}^{-1}(s)\|_e = \left(\sum_{i=1}^{n} \|x^i(t)\|^2\right)^{1/2}$$

for every $t \geq s$, from the preceding inequalities, we get

$$\|\mathcal{X}(t)\mathcal{X}^{-1}(s)\|_{\mathcal{M}} \leq \lambda^{-1}n^{1/2}$$

for every $t, s \in \mathbb{R}_+$, $s \leq t$. The proof is complete. □

With regard to the uniform asymptotic stability we prove:

Theorem 5.2.5. *The system* (5.2.1) *is uniformly asymptotically stable if and only if it has a fundamental matrix* $\mathcal{X}(t)$ *which satisfies*

$$\lim_{t-s\to+\infty} \|\mathcal{X}(t)\mathcal{X}^{-1}(s)\|_{\mathcal{M}} = 0.$$

Proof. Let us remark that (5.2.1) is uniformly asymptotically stable if and only if it is at the same time uniformly stable, and asymptotically stable. The conclusion follows from Theorems 5.2.3 and 5.2.4. □

Now, let us consider the system

$$x' = \mathcal{A}x \tag{5.2.3}$$

where $\mathcal{A} \in \mathcal{M}_{n\times n}(\mathbb{R})$ is a constant matrix.

Definition 5.2.1. The matrix $\mathcal{A}$ is *Hurwitzian*[1] if all the roots of the characteristic equation $\det(\mathcal{A} - \lambda\mathcal{I}) = 0$ have strictly negative real parts.

Lemma 5.2.1. *If* $\mathcal{A}$ *is Hurwitzian then there exist the constants* $M \geq 1$ *and* $\omega > 0$ *such that*

$$\|e^{t\mathcal{A}}\|_{\mathcal{M}} \leq Me^{-t\omega}$$

for every $t \geq 0$.

[1]The name of this property comes from the name of the German mathematician Adolf Hurwitz (1859–1919) which has defined and studied this class of matrices.

Proof. According to Theorem 4.4.3, all the elements of $e^{t\mathcal{A}}$ are of the form $\sum_{j=1}^{s}\left[P_j(t)e^{\alpha_j t}\cos(\beta_j t)+Q_j(t)e^{\alpha_j t}\sin(\beta_j t)\right]$, where $\alpha_j+i\beta_j$ is a root of the characteristic equation $\det(\mathcal{A}-\lambda\mathcal{I})=0$, of order of multiplicity m_j, while P_j and Q_j are polynomials with real coefficients, of degree not exceeding m_j-1. If $\mathcal{A}$ is Hurwitzian then there exists $\omega>0$ such that every root $\alpha+i\beta$ of the characteristic equation satisfies

$$\alpha<-\omega.$$

Indeed, let $\lambda_1,\lambda_2,\ldots,\lambda_s$ be the roots of the characteristic equation, ordered like their real parts: $\alpha_1\leq\alpha_2\leq\cdots\leq\alpha_s<0$. Then $\omega=-\frac{1}{2}\alpha_j$ satisfies the property mentioned above. So we have

$$\|e^{t\mathcal{A}}\|_{\mathcal{M}}=e^{-t\omega}\|\mathcal{B}(t)\|_{\mathcal{M}},$$

where all the elements of the matrix $\mathcal{B}$ are of the form

$$\sum_{j=1}^{s}\left[P_j(t)e^{-t\gamma_j}\cos(\beta_j t)+Q_j(t)e^{-t\gamma_j}\sin(\beta_j t)\right],$$

where $\gamma_j>0$, while P_j and Q_j are polynomials. Accordingly, there exists $M\geq 1$ such that

$$\|\mathcal{B}(t)\|_{\mathcal{M}}\leq M$$

for every $t\geq 0$. This inequality along with the relation above completes the proof. □

Theorem 5.2.6. *If the system* (5.2.3) *is asymptotically stable then the matrix $\mathcal{A}$ is Hurwitzian. If the matrix $\mathcal{A}$ is Hurwitzian then the system* (5.2.3) *is globally and uniformly asymptotically stable.*

Proof. In order to prove the first assertion, let us assume for contradiction that, although the system (5.2.3) is asymptotically stable, the matrix $\mathcal{A}$ is not Hurwitzian. This means that there exists at least one root of the characteristic equation $\det(\mathcal{A}-\lambda\mathcal{I})=0$ with nonnegative real part. Let $\lambda=\alpha+i\beta$ be that root. Then the matrix $\mathcal{A}$, thought as an element in $\mathcal{M}_{n\times n}(\mathbb{C})$, has at least one eigenvector $z\in\mathbb{C}^n$ corresponding to the eigenvalue λ. We denote this vector by $z=\xi+i\eta$, where $\xi,\eta\in\mathbb{R}^n$. If the eigenvalue λ is real, then the corresponding eigenvector has only real components ($z=\xi$) and in this case the function $x(t,0,c\xi)=ce^{\lambda t}\xi$, with $c\in\mathbb{R}$, is a solution of the system (5.2.3). Since ξ, as an eigenvector of the matrix $\mathcal{A}$, is nonzero, while $\lambda\geq 0$, it follows that for every $c\in\mathbb{R}^*$ the function $x(\cdot,0,c\xi)$ cannot tend to 0. So, the null solution of the system

(5.2.3) cannot be asymptotically stable. If λ is not real, then its complex conjugate $\overline{\lambda}$ is also an eigenvalue of $\mathcal{A}$, while $\overline{z} = \xi - i\eta$ is a corresponding eigenvector. Moreover, let us observe that $\eta \neq 0$. Indeed if η would be zero, then $z \in \mathbb{R}^n$ and from $\mathcal{A}z = \lambda z$ would follow $\lambda \in \mathbb{R}$, in contradiction with the initial supposition. In these conditions, let us observe that the function $y(\cdot, 0, c\eta) : \mathbb{R} \to \mathbb{R}^n$, defined by

$$y(t, 0, c\eta) = \frac{c}{2i}\left(e^{\lambda t}z - e^{\overline{\lambda}t}\overline{z}\right) = ce^{\alpha t}(\sin \beta t \cdot \xi + \cos \beta t \cdot \eta)$$

for every $t \in \mathbb{R}$, where $c \in \mathbb{R}^*$, is nonzero being the imaginary part of the function $ce^{\lambda t}z$, which at $t = 0$ equals $c\eta$. But this function $y(\cdot, 0, c\eta)$, which is a solution of the system (5.2.3), cannot tend to 0 for any choice of $c \in \mathbb{R}^*$. Hence the null solution of the system is not asymptotically stable. This contradiction can be eliminated only if $\mathcal{A}$ is Hurwitzian. The proof of the first assertion is complete.

On the other hand, if the matrix $\mathcal{A}$ is Hurwitzian, from Lemma 5.2.1, it follows that

$$\lim_{t-s \to +\infty} \|e^{t\mathcal{A}}e^{-s\mathcal{A}}\|_{\mathcal{M}} = \|e^{(t-s)\mathcal{A}}\|_{\mathcal{M}} = 0,$$

relation which, by virtue of Theorem 5.2.5, completes the proof of the second assertion. □

Remark 5.2.2. Theorem 5.2.6 shows that, for systems of first-order linear differential equations with constant coefficients, the asymptotic stability is equivalent to both global, and uniform asymptotic stability. Moreover, all these are equivalent to the property of $\mathcal{A}$ being Hurwitzian.

A useful completion of Theorem 5.2.6 is:

Theorem 5.2.7. *If the characteristic equation* $\det(\mathcal{A} - \lambda \mathcal{I}) = 0$ *has at least one root with strictly positive real part, then the system* (5.2.3) *is unstable. If all the roots of the characteristic equation have non-positive real parts and all the roots having the real part* 0 *are simple, then the system* (5.2.3) *is uniformly stable.*

Proof. Using similar considerations as those in the proof of the first part of Theorem 5.2.6, we deduce that, if $\alpha + i\beta$ is a root of the characteristic equation and $\xi + i\eta \in \mathbb{C}^n$ is a corresponding eigenvector, then, at least one of the functions $x(t, 0, c\xi) = ce^{\lambda t}\xi$ or $y(t, 0, c\eta) = ce^{\alpha t}(\sin \beta t \cdot \xi + \cos \beta t \cdot \eta)$, with $c \in \mathbb{R}^*$, is a nontrivial solution of system. Obviously, if $\alpha > 0$, that solution is unbounded. In accordance with Theorem 5.2.2, the system (5.2.3) is unstable. If all the roots of the characteristic equation have non-positive real parts and all the roots having the real part 0 are simple, it follows that all the elements of $t \mapsto e^{t\mathcal{A}}$ are bounded on $\mathbb{R}_+$. As a consequence, there exists $M > 0$ such that $\|e^{t\mathcal{A}}e^{-s\mathcal{A}}\|_{\mathcal{M}} = \|e^{(t-s)\mathcal{A}}\|_{\mathcal{M}} \leq M$ for every $t, s \in \mathbb{R}_+$, $s \leq t$. By virtue of Theorem 5.4.5, the system (5.2.3) is uniformly stable. □

We conclude this section with a necessary and sufficient condition for a given matrix $\mathcal{A}$ to be Hurwitzian or, equivalently, for a polynomial with real coefficients to have all the roots with strictly negative real parts. Let $p(z) = \alpha_0 z^n + \alpha_1 z^{n-1} + \cdots + \alpha_n$ be a polynomial with real coefficients. To this polynomial we associate the so-called *Hurwitz matrix*

$$H = \begin{pmatrix} \alpha_1 & \alpha_0 & 0 & 0 & \dots & 0 \\ \alpha_3 & \alpha_2 & \alpha_1 & \alpha_0 & \dots & 0 \\ \vdots & & & & & \\ \alpha_{2k-1} & \alpha_{2k-2} & \alpha_{2k-3} & \alpha_{2k-4} & \dots & \alpha_{2k-n} \\ \vdots & & & & & \\ 0 & 0 & 0 & 0 & \dots & \alpha_n \end{pmatrix},$$

where $\alpha_i = 0$ if $i < 0$ or $i > n$.

Theorem 5.2.8. (Routh–Hurwitz) *A polynomial $p(z) = \alpha_0 z^n + \alpha_1 z^{n-1} + \cdots + \alpha_n$ has all the roots with strictly negative real parts if and only if all the principal minors of the associated Hurwitz matrix are positive, i.e.*

$$D_1 = \alpha_1 > 0,\ D_2 = \begin{vmatrix} \alpha_1 & \alpha_0 \\ \alpha_3 & \alpha_2 \end{vmatrix} > 0,\ D_3 = \begin{vmatrix} \alpha_1 & \alpha_0 & 0 \\ \alpha_3 & \alpha_2 & \alpha_1 \\ \alpha_5 & \alpha_4 & \alpha_3 \end{vmatrix} > 0,\ \dots,$$

$$D_n = \det(H) > 0.$$

For the proof of this theorem see [Nistor and Tofan (1997)], pp. 176–177 or [Gantmacher (1959)], pp. 177–180.

5.3 The Case of Perturbed Systems

Let Ω be an open neighborhood of $0 \in \mathbb{R}^n$ and let $F : \mathbb{R}_+ \times \Omega \to \mathbb{R}^n$ be continuous on $\mathbb{R}_+ \times \Omega$ and locally Lipschitz on Ω with $F(t, 0) = 0$ for every $t \in \mathbb{R}_+$. Let $\mathcal{A} \in \mathcal{M}_{n\times n}(\mathbb{R})$ and let us consider the system

$$x' = \mathcal{A}x + F(t, x). \tag{5.3.1}$$

Since, in what follows this system will be thought of as a modified version of the linear and homogeneous system

$$x' = \mathcal{A}x \tag{5.3.2}$$

obtained by adding the so-called *perturbing function* $F(t,x)$, it will be called *perturbed system.*

In this section, we present some sufficient conditions for the stability properties of the system (5.3.2) to be inherited by the system (5.3.1). As we will see later, if (5.3.2) is asymptotically stable and F is "dominated in a certain sense" by $\mathcal{A}$, then the null solution of the system (5.3.1) is asymptotically stable. This is no longer the case for the stability, which is fragile to perturbations. Indeed, (5.3.2) may be stable if all the roots of the characteristic equation have 0 real parts. On the other hand, one can construct linear perturbations "as small as we wish" leading to a linear system governed by a matrix for which at least one of the roots of the characteristic equation has a strictly positive real part, situation generating instability. For instance, the scalar equation $x'(t) = 0$ is stable, while the perturbed equation $x'(t) = \varepsilon x(t)$, with $\varepsilon > 0$, is not, and this regardless of how small $\varepsilon > 0$ is.

We begin with the following fundamental result.

Theorem 5.3.1. (Poincaré–Lyapunov) *Let* $\mathcal{A} \in \mathcal{M}_{n\times n}(\mathbb{R})$, *let* Ω *be a neighborhood of* $0 \in \mathbb{R}^n$, *and let* $F : \mathbb{R}_+ \times \Omega \to \mathbb{R}^n$ *be continuous on* $\mathbb{R}_+ \times \Omega$ *and locally Lipschitz on* Ω. *If there exist* $M \geq 1$, $\omega > 0$ *and* $L > 0$ *such that*

$$\|e^{t\mathcal{A}}\|_{\mathcal{M}} \leq Me^{-\omega t} \tag{5.3.3}$$

for every $t \in \mathbb{R}_+$,

$$\|F(t,x)\| \leq L\|x\| \tag{5.3.4}$$

for every $(t,x) \in \mathbb{R}_+ \times \Omega$ *and*

$$LM - \omega < 0, \tag{5.3.5}$$

then the null solution of the system (5.3.1) *is asymptotically stable.*

Proof. Let $\xi \in \Omega$, $a \in \mathbb{R}_+$ and let $x(\cdot, a, \xi) : [a, T_m) \to \Omega$ be the unique saturated solution of the system (5.3.1) which satisfies the initial condition $x(a, a, \xi) = \xi$. To begin with, we will show that, if $\|\xi\|$ is sufficiently small, then $x(\cdot, a, \xi)$ is defined on $[a, +\infty)$. In order to do this, let us observe that, by virtue of the variation of constants formula (4.3.4), with $b(t) = F(t, x(t))$ for $t \in [a, T_m)$, we have

$$x(t, a, \xi) = e^{(t-a)\mathcal{A}}\xi + \int_a^t e^{(t-s)\mathcal{A}} F(s, x(s, a, \xi))\, ds$$

for every $t \in [a, T_m)$. From this relation we deduce

$$\|x(t, a, \xi)\| \leq \|e^{(t-a)\mathcal{A}}\|_{\mathcal{M}}\|\xi\| + \int_a^t \|e^{(t-s)\mathcal{A}}\|_{\mathcal{M}}\|F(s, x(s, a, \xi))\|\, ds$$

from where, by virtue of conditions (5.3.3) and (5.3.4), it follows that

$$\|x(t, a, \xi)\| \leq Me^{-\omega(t-a)}\|\xi\| + \int_a^t LMe^{-\omega(t-s)}\|x(s, a, \xi)\|\, ds$$

for every $t \in [a, T_m)$. Multiplying both sides of this inequality by $e^{\omega t} > 0$, we get

$$e^{\omega t}\|x(t, a, \xi)\| \leq Me^{\omega a}\|\xi\| + \int_a^t LMe^{\omega s}\|x(s, a, \xi)\|\, ds$$

for every $t \in [a, T_m)$. Denoting by $y : [a, T_m) \to \mathbb{R}_+$ the function defined by

$$y(t) = e^{\omega t}\|x(t, a, \xi)\|$$

for $t \in [a, T_m)$, the preceding inequality rewrites equivalently in the form

$$y(t) \leq Me^{\omega a}\|\xi\| + \int_a^t LMy(s)\, ds$$

for every $t \in [a, T_m)$. From Gronwall's Lemma 1.5.2, it follows

$$y(t) \leq Me^{\omega a}\|\xi\|e^{LM(t-a)}$$

from where, recalling the definition of the function y, we deduce

$$\|x(t, a, \xi)\| \leq M\|\xi\|e^{(LM-\omega)(t-a)} \tag{5.3.6}$$

for every $t \in [a, T_m)$.

Now, let $\rho > 0$ be such that $B(0, \rho) \subset \Omega$ and let $\mu(a) > 0$ be defined by

$$\mu(a) = \frac{\rho}{2M}.$$

Then, according to the inequality (5.3.6), for every $\xi \in \Omega$ with $\|\xi\| \leq \mu(a)$, we have

$$\|x(t, a, \xi)\| \leq \frac{\rho}{2}$$

for every $t \in [a, T_m)$. Assuming that $T_m < +\infty$, from this inequality and Proposition 2.4.1, it follows that there exists

$$\lim_{t \uparrow T_m} x(t, a, \xi) = x^*,$$

$x^* \in B(0, \frac{\rho}{2}) \subset \Omega$ relation which, by virtue of (ii) in Theorem 2.4.3, contradicts the fact that $x(\cdot, a, \xi)$ is saturated. This contradiction can be eliminated only if, for every $\xi \in \Omega$ satisfying $\|\xi\| \leq \mu(a)$, we have $T_m = +\infty$.

Finally let us observe that, from what we have already shown combined with (5.3.6), it follows that, for every $\xi \in \Omega$ with $\|\xi\| \leq \mu(a)$, we have

$$\lim_{t \uparrow +\infty} x(t, a, \xi) = 0,$$

which completes the proof. □

A useful consequence is stated below.

Theorem 5.3.2. *Let $\mathcal{A} \in \mathcal{M}_{n\times n}(\mathbb{R})$ be Hurwitzian, let Ω be a neighborhood of $0 \in \mathbb{R}^n$, and let $F : \mathbb{R}_+ \times \Omega \to \mathbb{R}^n$ be continuous on $\mathbb{R}_+ \times \Omega$ and locally Lipschitz on Ω. If there exists a function $\alpha : \mathbb{R}_+ \to \mathbb{R}_+$ such that*

$$\|F(t, x)\| \leq \alpha(\|x\|)$$

for every $(t, x) \in \mathbb{R}_+ \times \Omega$ and

$$\lim_{r \downarrow 0} \frac{\alpha(r)}{r} = 0,$$

then the null solution of the system (5.3.1) is asymptotically stable.

Proof. Since $\mathcal{A}$ is Hurwitzian, by virtue of Lemma 5.2.1, there exist $M \geq 1$ and $\omega > 0$ such that (5.3.3) holds. Fix $L > 0$ with the property (5.3.5), and choose $\delta > 0$ such that

$$\alpha(r) \leq Lr$$

for every $r \in [0, \delta)$. Clearly both the matrix $\mathcal{A}$ and the restriction of F to $\mathbb{R}_+ \times \{x \in \Omega\,;\, \|x\| < \delta\}$ satisfy the hypotheses of Theorem 5.3.1, which completes the proof. □

Now, we proceed to the study of *the stability by the first approximation method*, which proves very useful in applications. Let $f : \Omega \to \mathbb{R}^n$ be a function of class C^1 with $f(0) = 0$ and let us consider the system

$$x' = f(x) \tag{5.3.7}$$

which obviously has the identically zero solution.

Theorem 5.3.3. *Let Ω be a neighborhood of $0 \in \mathbb{R}^n$. If $f : \Omega \to \mathbb{R}^n$ is a function of class C^1 with $f(0) = 0$ and whose Jacobian matrix $\mathcal{A} = f_x(0)$ is Hurwitzian, then the null solution of the system (5.3.7) is asymptotically stable.*

Proof. Since f is of class C^1, it is differentiable and therefore

$$f(x) = f(0) + f_x(0)x + F(x) = \mathcal{A}x + F(x)$$

for every $x \in \Omega$, where

$$\lim_{x \to 0} \frac{\|F(x)\|}{\|x\|} = 0.$$

So, we are in the hypotheses Theorem 5.3.2 with

$$\alpha(r) = \sup\{\|f_x(\theta) - f_x(0)\|_{\mathcal{M}}\,;\ \theta \in \Omega,\ \|\theta\| \leq r\}$$

for $r \geq 0$. □

The next example shows that, though it is a corollary of Theorem 5.3.1, Theorem 5.3.3 proves effective even in situations in which Theorem 5.3.1 is not directly applicable.

Example 5.3.1. Let us consider the *Liénard equation*

$$z'' + g'(z)z' + z = 0,$$

where $g : \mathbb{R} \to \mathbb{R}$ is a function of class C^1 with $g(0) = 0$. This equation rewrites as a first-order differential system

$$\begin{cases} z' = y - g(z) \\ y' = -z. \end{cases}$$

The system above is of the form (5.3.1) with $n = 2$,

$$x = \begin{pmatrix} z \\ y \end{pmatrix}, \quad \mathcal{A} = \begin{pmatrix} 0 & 1 \\ -1 & 0 \end{pmatrix} \quad \text{and} \quad F(t, x) = \begin{pmatrix} -g(z) \\ 0 \end{pmatrix}.$$

Since the matrix $\mathcal{A}$ is not Hurwitzian, the condition (5.3.3) in Theorem 5.3.1 is not fulfilled and therefore Theorem 5.3.1 is not directly applicable.

Nevertheless, let us observe that the system above may be regarded as a system of the form (5.3.7) in which the function $f : \mathbb{R}^2 \to \mathbb{R}^2$ is defined by

$$f(x) = \begin{pmatrix} y - g(z) \\ -z \end{pmatrix}$$

for every $x \in \mathbb{R}^2$ and whose Jacobian matrix at $(0, 0)$ is

$$\mathcal{A} = \begin{pmatrix} -g'(0) & 1 \\ -1 & 0 \end{pmatrix}.$$

Since this matrix is Hurwitzian if $g'(0) > 0$, in view of Theorem 5.3.3, it follows that, in this case ($g'(0) > 0$), the null solution of the system above is asymptotically stable. As a consequence of this result, we deduce that the null solution of the *Van der Pol equation*, i.e. the equation corresponding to the particular case $g(z) = z - z^3$ for every $z \in \mathbb{R}$, is asymptotically stable.

We state without proof a completion of Theorem 5.3.3. For details see [Malkin (1952)], Chapter 4.

Theorem 5.3.4. *Let $f : \Omega \to \mathbb{R}^n$ be a function of class C^1 with $f(0) = 0$ and let $\mathcal{A} = f_x(0)$. If there exist $\alpha > 1$, $M > 0$ and $r > 0$ such that $\|f(x) - \mathcal{A}x\| \leq M\|x\|^\alpha$ for every $x \in \mathbb{R}^n$ with $\|x\| \leq r$ and at least one root of the characteristic equation* $\det(\mathcal{A} - \lambda\mathcal{I}) = 0$ *has strictly positive real part, then the null solution of the system* (5.3.7) *is unstable.*

The case when at least one of the characteristic roots of the matrix $f_x(0)$ has 0 real part needs a much more delicate analysis involving the signs of the higher order derivatives of the function f at $x = 0$. See also [Malkin (1952)], *loc. cit.*

5.4 The Lyapunov Function Method

A very refined and powerful method of establishing sufficient conditions of stability consists in finding a certain real-valued function, decreasing along the trajectories of a given system, function which may increase only with the norm of the argument. Suggested by the evolution of certain phenomena in Classical Mechanics, where this function represents, in some sense, the potential energy of the system, this method, invented by the Russian mathematician Lyapunov in 1892, proves extremely effective and useful and is still far of being obsolete.

Let Ω be an open neighborhood of $0 \in \mathbb{R}^n$, and let $f : \mathbb{R}_+ \times \Omega \to \mathbb{R}^n$ be continuous on $\mathbb{R}_+ \times \Omega$ and locally Lipschitz on Ω, with $f(t, 0) = 0$ for every $t \in \mathbb{R}_+$. This last condition shows that $\varphi \equiv 0$ is a solution of the system

$$x' = f(t, x). \tag{5.4.1}$$

Definition 5.4.1. A function $V : \mathbb{R}_+ \times \Omega \to \mathbb{R}_+$ is *positive definite* on $\mathbb{R}_+ \times \Omega$ if there exists a function $\omega : \mathbb{R}_+ \to \mathbb{R}_+$ continuous, nondecreasing, with $\omega(r) = 0$ if and only if $r = 0$, and such that

$$V(t, x) \geq \omega(\|x\|) \tag{5.4.2}$$

for every $(t, x) \in \mathbb{R}_+ \times \Omega$. A function $V : \mathbb{R}_+ \times \Omega \to \mathbb{R}_-$ is *negative definite* on $\mathbb{R}_+ \times \Omega$ if $-V$ is positive definite on $\mathbb{R}_+ \times \Omega$.

Definition 5.4.2. A function $V : \mathbb{R}_+ \times \Omega \to \mathbb{R}_+$ is a *Lyapunov function* for the system (5.4.1) if:

(i) V is of class C^1 on $\mathbb{R}_+ \times \Omega$ and $V(t, 0) = 0$ for every $t \in \mathbb{R}_+$;
(ii) V is positive definite on $\mathbb{R}_+ \times \Omega$;
(iii) for every $(t, x) \in \mathbb{R}_+ \times \Omega$ we have

$$\frac{\partial V}{\partial t}(t, x) + \sum_{i=1}^{n} f_i(t, x) \frac{\partial V}{\partial x_i}(t, x) \leq 0. \tag{5.4.3}$$

Theorem 5.4.1. (Lyapunov) *If* (5.4.1) *has a Lyapunov function then its null solution is stable.*

Proof. Let $a \in \mathbb{R}_+$, $\xi \in \Omega$ and let $x(\cdot, 0, \xi) : [a, T_m) \to \Omega$ be the unique saturated solution of the system (5.4.1) which satisfies the initial condition $x(a, a, \xi) = \xi$. First we will show that, if $\|\xi\|$ is sufficiently small, then $T_m = +\infty$. To this aim let us define the function $g : [a, T_m) \to \mathbb{R}_+$ by

$$g(t) = V(t, x(t, a, \xi))$$

for every $t \in [a, T_m)$, where V is a Lyapunov function for the system (5.4.1). Obviously g is of class C^1 on $[a, T_m)$. In addition, from (5.4.3), it follows that

$$g'(t) = \frac{\partial V}{\partial t}(t, x(t, a, \xi)) + \sum_{i=1}^{n} \frac{\partial V}{\partial x_i}(t, x(t, a, \xi)) \frac{dx_i}{dt}(t, a, \xi)$$

$$= \frac{\partial V}{\partial t}(t, x(t, a, \xi)) + \sum_{i=1}^{n} \frac{\partial V}{\partial x_i}(t, x(t, a, \xi)) f_i(t, x(t, a, \xi)) \leq 0$$

for every $t \in [a, T_m)$. So g is non-increasing, and hence $g(t) \leq g(a)$ for every $t \in [a, T_m)$. Recalling the definition of the function g, this inequality rewrites equivalently in the form $V(t, x(t, a, \xi)) \leq V(a, \xi)$ for $t \in [a, T_m)$. From (5.4.2) and the preceding inequality, we deduce

$$\omega(\|x(t, a, \xi)\|) \leq V(a, \xi)$$

for every $t \in [a, T_m)$.

Let $\rho > 0$ with $B(0, \rho) \subset \Omega$. Since $V(a, \cdot)$ is continuous at 0 and $V(a, 0) = 0$, for $\omega(\rho) > 0$, with $\rho > 0$ as above, there exists $r = r(a) \in (0, \rho)$ such that $V(a, \xi) < \omega(\rho)$ for every $\xi \in \Omega$ with $\|\xi\| \leq r$. From this inequality and from the preceding one, we deduce $\omega(\|x(t, a, \xi)\|) < \omega(\rho)$ for every $t \in [a, T_m)$. Recalling that ω is nondecreasing, we deduce

$$\|x(t, a, \xi)\| \leq \rho$$

for every $t \in [a, T_m)$. Since $B(0, \rho) \subset \Omega$ and $x(\cdot, a, \xi)$ is saturated, this inequality proves that, for every $\xi \in \Omega$ with $\|\xi\| \leq r(a)$, $T_m = +\infty$. Finally, a similar argument shows that, for every $a \in \mathbb{R}_+$ and every $\varepsilon > 0$ there exists $\delta(a, \varepsilon) > 0$ such that

$$\|x(t, a, \xi)\| \leq \varepsilon$$

for every $\xi \in \Omega$ with $\|\xi\| \leq \delta(a, \varepsilon)$ and every $t \geq a$. Hence the null solution of the system (5.4.1) is stable, which completes the proof. □

Theorem 5.4.2. (Lyapunov) *If* (5.4.1) *has a Lyapunov function* V *and there exist* $\lambda, \eta : \mathbb{R}_+ \to \mathbb{R}_+$ *continuous, nondecreasing, satisfying* $\lambda(r) = 0$ *if and only if* $r = 0$, $\eta(s) = 0$ *if and only if* $s = 0$,

$$V(t, x) \leq \lambda(\|x\|) \tag{5.4.4}$$

and

$$\frac{\partial V}{\partial t}(t, x) + \sum_{i=1}^{n} f_i(t, x) \frac{\partial V}{\partial x_i}(t, x) \leq -\eta(\|x\|) \tag{5.4.5}$$

for every $(t, x) \in \mathbb{R}_+ \times \Omega$, *then its null solution is asymptotically stable.*

Proof. By virtue of Theorem 5.4.1 the null solution of the system (5.4.1) is stable. So, for every $a \in \mathbb{R}_+$, there exists $\mu(a) > 0$ such that, for every $\xi \in \Omega$ with $\|\xi\| \leq \mu(a)$, the unique saturated solution of the system (5.4.1), with initial data a and ξ, is defined on $[\,a, +\infty)$. Let $x(\cdot, a, \xi) : [\,a, +\infty) \to \Omega$ be such a solution and let us define the function $g : [\,a, +\infty) \to \mathbb{R}_+$ by $g(t) = V(t, x(t, a, \xi))$ for $t \in [\,a, +\infty)$, where V is a Lyapunov function with the properties (5.4.4) and (5.4.5). The function g is of class C^1 and

$$g'(t) = \frac{\partial V}{\partial t}(t, x(t, a, \xi)) + \sum_{i=1}^{n} \frac{\partial V}{\partial x_i}(t, x(t, a, \xi)) \frac{dx_i}{dt}(t, a, \xi)$$

$$= \frac{\partial V}{\partial t}(t, x(t, a, \xi)) + \sum_{i=1}^{n} \frac{\partial V}{\partial x_i}(t, x(t, a, \xi)) f_i(t, x(t, a, \xi)) \leq -\eta(\|x(t, a, \xi)\|)$$

for every $t \in [\,a, +\infty)$. Integrating this relation both sides from a to t, we deduce

$$\int_a^t \eta(\|x(s, a, \xi)\|)\, ds + V(t, x(t, a, \xi)) \leq V(a, \xi),$$

for every $t \in [\,a, +\infty)$. Since both η and V are positive, this inequality ensures, on one hand, that the integral

$$\int_a^{+\infty} \eta(\|x(s, a, \xi)\|)\, ds$$

is convergent and, on the other hand, that there exists the finite limit

$$\lim_{t\uparrow+\infty} V(t, x(t, a, \xi)) = \ell.$$

Now, let us observe that, from the positivity of the function η and the convergence of the integral above, it follows that there exists at least one sequence $(t_n)_{n\in\mathbb{N}}$ with

$$\begin{cases} \lim_{n\to\infty} t_n = +\infty \\ \lim_{n\to\infty} \eta(\|x(t_n, a, \xi)\|) = 0. \end{cases}$$

Since η is nondecreasing and $\eta(r) = 0$ if and only if $r = 0$, it follows that

$$\lim_{n\to\infty} \|x(t_n, a, \xi)\| = 0.$$

Recalling that V satisfies (5.4.4) and that λ is continuous and vanishes at 0, we deduce that $\ell = 0$. From the remark above and from (5.4.2), we have

$$\limsup_{t\uparrow+\infty} \omega(\|x(t, a, \xi)\|) \leq \lim_{t\uparrow+\infty} V(t, x(t, a, \xi)) = 0$$

and therefore

$$\lim_{t\uparrow+\infty} \omega(\|x(t, a, \xi)\|) = 0.$$

Since ω is nondecreasing and $\omega(r) = 0$ if and only if $r = 0$, the relation above holds true only if $\lim_{t\uparrow+\infty} \|x(t, a, \xi)\| = 0$. The proof is complete. □

As concerns the global asymptotic stability of the system (5.4.1), we have:

Theorem 5.4.3. *Let us assume that $\Omega = \mathbb{R}^n$ and that the system* (5.4.1) *has a Lyapunov function $V : \mathbb{R}_+ \times \mathbb{R}^n \to \mathbb{R}_+$ satisfying all the hypotheses of Theorem* 5.4.2. *If, in addition, $\omega : \mathbb{R}_+ \to \mathbb{R}_+$ in Definition* 5.4.1 *satisfies*

$$\lim_{r\uparrow+\infty} \omega(r) = +\infty,$$

then the system (5.4.1) *is globally asymptotically stable.*

Proof. The only difference from the proof of Theorem 5.4.2 consists in that all the evaluations performed are still valid for every $\xi \in \mathbb{R}^n$. We emphasize that the hypothesis imposed on the function ω is needed only in proving that, for every $a \in \mathbb{R}_+$ and every $\xi \in \mathbb{R}^n$, the unique saturated solution $x(\cdot, a, \xi)$ of (5.4.1), satisfying $x(a, a, \xi) = \xi$, is global, i.e. defined on $[a, +\infty)$. □

For autonomous differential systems, i.e. for systems of the type

$$x' = f(x), \tag{5.4.6}$$

where $f : \Omega \to \mathbb{R}^n$, we look for *autonomous Lyapunov functions*, i.e. we look for functions $V : \Omega \to \mathbb{R}$ which do not depend on t. Lemma 5.4.1 below gives a sufficient condition for such a function to be positive definite.

Lemma 5.4.1. *If $V : \Omega \to \mathbb{R}$ is continuous on Ω, $V(0) = 0$ and $V(x) > 0$ for every $x \in \Omega$, $x \neq 0$, then there exists a neighborhood of the origin, $\Omega_0 \subset \Omega$, such that V is positive definite on Ω_0.*

Proof. It suffices to show that V is positive definite on a set of the form $B(0,\rho) \subset \Omega$, with $\rho > 0$ suitably chosen. To this aim, let $\rho > 0$ such that $B(0,\rho) \subset \Omega$, and let $\omega : \mathbb{R}_+ \to \mathbb{R}_+$ be defined by

$$\omega(r) = \begin{cases} \inf\{V(x);\ r \leq \|x\| \leq \rho\} \text{ for } 0 \leq r \leq \rho \\ \omega(\rho) \qquad\qquad\qquad\qquad \text{for } r > \rho. \end{cases}$$

Clearly the function ω is continuous and nondecreasing on $\mathbb{R}_+$. In addition, one can easily see that $\omega(r) = 0$ if and only if $r = 0$. Consequently V is positive definite, which completes the proof of the lemma. □

Corollary 5.4.1. *If $V : \Omega \to \mathbb{R}$ satisfies*

(i) *V is of class C^1 on Ω and $V(0) = 0$;*
(ii) *for every $x \in \Omega$, $x \neq 0$, we have $V(x) > 0$;*
(iii) *for every $x \in \Omega$ we have*

$$\sum_{i=1}^{n} f_i(x) \frac{\partial V}{\partial x_i}(x) \leq 0,$$

then there exists a neighborhood of the origin, $\Omega_0 \subset \Omega$, such that V is a Lyapunov function for the autonomous equation (5.4.6) *on Ω_0.*

The image of the trajectory of an autonomous differential system in $\mathbb{R}^2$ through a Lyapunov function V is illustrated in Figure 5.4.1 below. The sense indicated corresponds to the sense of increase of the argument t. The practical effectiveness of Corollary 5.4.1 is illustrated by the next example.

Example 5.4.1. Check for stability the null solution of the differential system

$$\begin{cases} x_1' = -x_1 x_2 - x_2 \\ x_2' = x_1 + x_1 x_2. \end{cases}$$

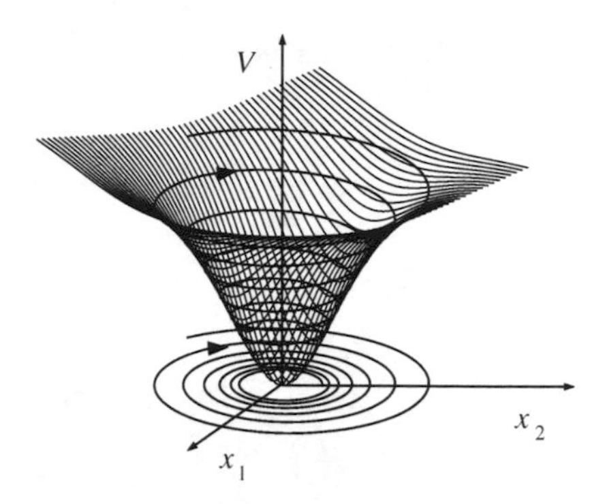

Fig. 5.4.1

Let us observe from the beginning that the system above may be rewritten as a vector differential equation $x' = f(x)$, where $f : \mathbb{R}^2 \to \mathbb{R}^2$ is defined by $f(x) = (f_1(x), f_2(x)) = (-x_1x_2 - x_2, x_1 + x_1x_2)$, for every $x = (x_1, x_2) \in \mathbb{R}^2$. Obviously f is of class C^∞ while its Jacobian matrix at 0 is

$$\mathcal{A} = \begin{pmatrix} 0 & -1 \\ 1 & 0 \end{pmatrix}.$$

The equation $\det(\mathcal{A} - \lambda\mathcal{I}) = 0$ has the roots $\pm i$ and therefore $\mathcal{A}$ is not Hurwitzian. For this reason, none of the stability results proved in the preceding section can be used in this case. Nevertheless, we will show that the system above possesses a Lyapunov function defined on a suitably chosen neighborhood Ω of 0. To this aim, let us observe that the function $V : (-1,1) \times (-1,1) \to \mathbb{R}$ defined by $V(x) = x_1 + x_2 - \ln(1+x_1) - \ln(1+x_2)$ for every $x = (x_1, x_2) \in (-1,1) \times (-1,1)$ is of class C^1, $V(0) = 0$, $V(x) > 0$ for every $x \neq 0$ and

$$f_1(x)\frac{\partial V}{\partial x_1}(x) + f_2(x)\frac{\partial V}{\partial x_2}(x) = 0$$

for every $x \in (-1,1) \times (-1,1)$. From Corollary 5.4.1 it follows that V is a Lyapunov function for the system and therefore, by virtue of Theorem 5.4.1, the null solution is stable.

We conclude with the remark that the manner, somehow obscure, in which we have found the function V will be much more understandable and completely clarified in the next chapter where we will present a necessary and sufficient condition for a function V to remain constant along the trajectories of an autonomous system. For this reason we do not enter into details which could distract us from the essence of the problem: the fact that Theorem 5.4.1 is useful in situations in which Theorem 5.3.3 is not.

In the case of first-order systems of linear differential equations with constant coefficients

$$x' = \mathcal{A}x, \tag{5.4.7}$$

where $\mathcal{A} \in \mathcal{M}_{n\times n}(\mathbb{R})$, we have the following characterization of global and uniform asymptotic stability.

Theorem 5.4.4. (Lyapunov) *The system* (5.4.7) *is globally and uniformly asymptotically stable if and only if there exists a symmetric, positive definite matrix* $\mathcal{P} \in \mathcal{M}_{n\times n}(\mathbb{R})$ *satisfying*

$$\mathcal{A}^*\mathcal{P} + \mathcal{P}\mathcal{A} = -\mathcal{I}, \tag{5.4.8}$$

where $\mathcal{A}^*$ *is the transpose of the matrix* $\mathcal{A}$.

Proof. According to Theorem 5.2.6, (5.4.7) is globally and uniformly asymptotically stable if and only if $\mathcal{A}$ is Hurwitzian. For the necessity let us assume that $\mathcal{A}$ is Hurwitzian. Then, by virtue of the obvious equality $\det(\mathcal{A} - \lambda\mathcal{I}) = \det(\mathcal{A}^* - \lambda\mathcal{I})$, it follows that $\mathcal{A}^*$ is Hurwitzian too. So, from Lemma 5.2.1, there exist $M \geq 1$ and $\omega > 0$ such that

$$\|e^{t\mathcal{A}}\|_{\mathcal{M}} \leq Me^{-t\omega} \text{ and } \|e^{t\mathcal{A}^*}\|_{\mathcal{M}} \leq Me^{-t\omega}$$

for every $t \in \mathbb{R}_+$. We can then define

$$\mathcal{P} = \int_0^{+\infty} e^{t\mathcal{A}^*}e^{t\mathcal{A}}dt$$

since the integral on the right-hand side is convergent. Let us observe that the matrix $\mathcal{P}$ is symmetric. Indeed, as $\left(e^{t\mathcal{A}}\right)^* = e^{t\mathcal{A}^*}$ and $(\mathcal{B}\mathcal{C})^* = \mathcal{C}^*\mathcal{B}^*$, from (ii) in Lemma 12.1.3 we have

$$\mathcal{P}^* = \left(\int_0^{+\infty} e^{t\mathcal{A}^*}e^{t\mathcal{A}}dt\right)^* = \int_0^{+\infty} \left(e^{t\mathcal{A}^*}e^{t\mathcal{A}}\right)^* dt = \int_0^{+\infty} e^{t\mathcal{A}^*}e^{t\mathcal{A}}dt = \mathcal{P}.$$

From (iii) of the same Lemma 12.1.3, we deduce

$$\langle \mathcal{P}x, x\rangle = \int_0^{+\infty} \langle e^{t\mathcal{A}^*}e^{t\mathcal{A}}x, x\rangle\, dt = \int_0^{+\infty} \|e^{t\mathcal{A}}x\|^2 dt.$$

So $\langle \mathcal{P}x, x\rangle > 0$ for $x \neq 0$, and therefore $\mathcal{P}$ is positive definite. To conclude the proof of the necessity let us observe that, from (i) in Lemma 12.1.3, it follows that

$$\mathcal{A}^*\mathcal{P} = \int_0^{+\infty} \mathcal{A}^*e^{t\mathcal{A}^*}e^{t\mathcal{A}}dt = \int_0^{+\infty} \frac{d}{dt}\left(e^{t\mathcal{A}^*}\right)e^{t\mathcal{A}}dt,$$

from where, integrating by parts, we get

$$\mathcal{A}^*\mathcal{P} = e^{t\mathcal{A}^*}e^{t\mathcal{A}}\Big|_0^\infty - \int_0^{+\infty} e^{t\mathcal{A}^*}\frac{d}{dt}\left(e^{t\mathcal{A}}\right)dt = -\mathcal{I} - \mathcal{P}\mathcal{A}.$$

The sufficiency follows by observing that, if $\mathcal{P}$ is a positive definite and symmetric solution of (5.4.8), then $V : \mathbb{R}^n \to \mathbb{R}$ defined by $V(x) = \frac{1}{2}\langle \mathcal{P}x, x\rangle$ is a Lyapunov function for the system (5.4.7). Indeed, we can easily see that $\nabla V(x) = \mathcal{P}x$, while from (5.4.8), it follows that $\langle \mathcal{P}x, \mathcal{A}x\rangle = -\frac{1}{2}\|x\|^2$ for every $x \in \mathbb{R}^n$. Hence V is a Lyapunov function for the system (5.4.7). In addition, because $\mathcal{P}$ is positive definite, there exists $\eta > 0$ such that $V(x) \geq \eta\|x\|^2$ for every $x \in \mathbb{R}^n$. From this observation and from the obvious inequality $V(x) \leq \frac{1}{2}\|\mathcal{P}\|_{\mathcal{M}}\|x\|^2$ for every $x \in \mathbb{R}^n$, we deduce that V satisfies the conditions of Theorem 5.4.3. Hence the system (5.4.7) is global and asymptotically stable. According to Theorem 5.2.6, the matrix $\mathcal{A}$ is Hurwitzian and therefore, again from the same Theorem 5.2.6, it follows that the system (5.4.7) is globally and uniformly asymptotically stable. The proof is complete. □

Remark 5.4.1. The existence of a matrix $\mathcal{P}$, satisfying a more general equation than (5.4.8), has been considered in Problem 4.12.

We conclude this section with a direct application of the previously proved results to the study of stability of the null solution for a class of autonomous dissipative systems. A much more detailed analysis of the stability problems in connection with these systems will be done in the next section.

Let Ω be an open neighborhood of the origin and let $\mathcal{A} : \Omega \to \mathbb{R}^n$. We recall that $\mathcal{A}$ is *dissipative* if $\langle \mathcal{A}(x) - \mathcal{A}(y), x - y\rangle \leq 0$ for every $x, y \in \Omega$. Let us consider the autonomous equation

$$x' = \mathcal{A}(x). \tag{5.4.9}$$

Theorem 5.4.5. *If $\mathcal{A} : \Omega \to \mathbb{R}^n$ is a continuous dissipative function with $\mathcal{A}(0) = 0$, then the null solution of* (5.4.9) *is stable. If, in addition, $\Omega = \mathbb{R}^n$ and for every $x \in \mathbb{R}^n$, $x \neq 0$, $\langle \mathcal{A}(x), x\rangle < 0$, then the system* (5.4.9) *is globally asymptotically stable, i.e. for every $\xi \in \mathbb{R}^n$, we have $\lim_{t\uparrow+\infty} u(t, 0, \xi) = 0$, where $u(\cdot, 0, \xi) : [0, +\infty) \to X$ is the unique global solution of* (5.4.9) *satisfying $u(0, 0, \xi) = \xi$.*

A closely related result has been proved in Section 3.4. See Lemma 3.4.1.

Proof. Since $\mathcal{A}$ is dissipative and $\mathcal{A}(0) = 0$ it follows that $V : \Omega \to \mathbb{R}$, defined by $V(x) = \frac{1}{2}\|x\|^2$ for every $x \in \Omega$, is a Lyapunov function for the system (5.4.9). Indeed, V is of class C^1, $V(0) = 0$ and is positive definite; the function ω being in this case defined by $\omega(r) = \frac{1}{2}r^2$ for $r \in \mathbb{R}$. In addition, as for every $x \in \Omega$

$$\sum_{i=1}^{n} \mathcal{A}_i(x)\frac{\partial V}{\partial x_i}(x) = \sum_{i=1}^{n} \mathcal{A}_i(x)x_i = \langle \mathcal{A}(x), x\rangle,$$

from the dissipativity of the function $\mathcal{A}$ and from the condition $\mathcal{A}(0) = 0$, we deduce that V, defined as above, is a Lyapunov function for (5.4.9). From this observation and Theorem 5.4.1, it follows that the null solution of the system (5.4.9) is stable. If in addition $\Omega = \mathbb{R}^n$ and $\langle \mathcal{A}(x), x\rangle < 0$ for every $x \in \mathbb{R}^n$ with $x \neq 0$, then, according to Lemma 5.4.1, the functions $\lambda, \eta : \mathbb{R}^n \to \mathbb{R}$, defined by $\lambda(x) = \frac{1}{2}\|x\|^2$ and $\eta(x) = -\langle \mathcal{A}(x), x\rangle$ for every $x \in \mathbb{R}^n$, satisfy the conditions of Theorem 5.4.3. The proof is complete. □

5.5 The Case of Dissipative Systems

As we have seen in the preceding section, in the case of dissipative systems for which 0 is an equilibrium point, the function $V : \mathbb{R}^n \to \mathbb{R}$, defined by $V(x) = \frac{1}{2}\|x\|^2$ for every $x \in \mathbb{R}^n$, is a Lyapunov function. Now, we will show that, due to the particularities of these systems, we can reveal new and interesting properties of global solutions with regard to their asymptotic behavior as t approaches $+\infty$.

Let $\mathcal{A} : \mathbb{R}^n \to \mathbb{R}^n$ be a continuous and dissipative function and let us consider the autonomous equation

$$x' = \mathcal{A}(x). \tag{5.5.1}$$

For $a \in \mathbb{R}_+$ and $\xi \in \mathbb{R}^n$, we denote by $x(\cdot, a, \xi)$ the unique saturated solution of equation (5.5.1) which satisfies $x(a, a, \xi) = \xi$. From Theorem 2.4.7, it follows that this solution is global, i.e. defined on $\mathbb{R}_+$.

Let $t \geq 0$ and let $S(t) : \mathbb{R}^n \to \mathbb{R}^n$ be defined by $S(t)\xi = x(t, 0, \xi)$ for every $\xi \in \mathbb{R}^n$. As we have seen in Theorem 3.4.1 in Section 3.4, the family of operators $\{S(t);\ t \geq 0\}$, defined by means of the relation above, satisfies:

- (S_1) $S(t + s) = S(t)S(s)$, for every $t, s \geq 0$;
- (S_2) $S(0) = \mathcal{I}$;
- (S_3) $\lim_{t\downarrow 0} S(t)\xi = \xi$ for every $\xi \in \mathbb{R}^n$;
- (S_4) $\|S(t)\xi - S(t)\eta\| \leq \|\xi - \eta\|$ for every $t \geq 0$ and every $\xi, \eta \in \mathbb{R}^n$;
- (S_5) $\lim_{t\downarrow 0} \frac{1}{t}(S(t)\xi - \xi) = \mathcal{A}(\xi)$ for every $\xi \in \mathbb{R}^n$.

We recall that this family of operators is *the semigroup of non-expansive operators generated by* $\mathcal{A}$.

Definition 5.5.1. Let $\xi \in \mathbb{R}^n$. The set

$$\gamma(\xi) = \{S(t)\xi;\ t \geq 0\}$$

is *the trajectory*, or *the orbit* of the solution of (5.5.1) starting from ξ.

Definition 5.5.2. Let $\xi \in \mathbb{R}^n$. The set

$$\omega(\xi) = \left\{p \in \mathbb{R}^n; \exists t_k \to +\infty \text{ such that } \lim_{k\to\infty} S(t_k)\xi = p\right\}$$

is *the* ω*-limit set* of the trajectory $\gamma(\xi)$. In the specific case $n = 2$, an ω-limit set which is the trajectory of a periodic solution is called *limit cycle.*

One trajectory of the dissipative system

$$\begin{cases} x_1' = -2x_1^3 + x_2 \\ x_2' = -x_1 - x_2^3 \end{cases}$$

approaching its ω-limit set is illustrated in Figure 5.5.1 (a), which in this case is a limit cycle, while Figure 5.5.1 (b) shows the graph of the corresponding solution.

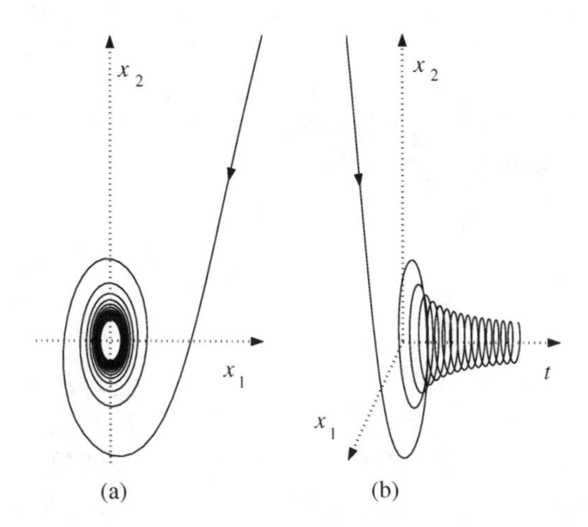

Fig. 5.5.1

We denote by $\mathcal{A}^{-1}(0) = \{y \in \mathbb{R}^n;\ \mathcal{A}(y) = 0\}$ and let us recall that a subset D in $\mathbb{R}^n$ is *convex* if, for every $x, y \in D$ and every $\lambda \in (0, 1)$, we have $\lambda x + (1 - \lambda)y \in D$.

Lemma 5.5.1. *Let $\mathcal{A} : \mathbb{R}^n \to \mathbb{R}^n$ be continuous and dissipative. If the set $\mathcal{A}^{-1}(0)$ is nonempty, then it is closed and convex.*

Proof. Since $\mathcal{A}$ is continuous, it follows that $\mathcal{A}^{-1}(0)$ is closed. In order to prove the convexity, let $x, y \in \mathcal{A}^{-1}(0)$ and $\xi \in \mathbb{R}^n$. From the dissipativity condition, we deduce that

$$\langle \mathcal{A}(\xi), \xi - x\rangle = \langle \mathcal{A}(\xi) - \mathcal{A}(x), \xi - x\rangle \leq 0.$$

Analogously

$$\langle \mathcal{A}(\xi), \xi - y\rangle \leq 0.$$

Multiplying both sides of the two inequalities: the first one by $\lambda \in (0,1)$, the second one by $1 - \lambda \in (0,1)$, and adding them side by side, we get

$$\langle \mathcal{A}(\xi), \xi - \lambda x - (1-\lambda)y\rangle \leq 0.$$

Denoting by $x_\lambda = \lambda x + (1-\lambda)y$, the inequality above can be rewritten in the form

$$\langle \mathcal{A}(\xi), \xi - x_\lambda\rangle \leq 0. \tag{5.5.2}$$

Now let $\varepsilon \in (0,1)$ and $z \in \mathbb{R}^n$. Taking $\xi = \xi_\varepsilon = \varepsilon x_\lambda + (1-\varepsilon)z$ in (5.5.2), we deduce

$$(1-\varepsilon)\langle \mathcal{A}(\xi_\varepsilon), z - x_\lambda\rangle \leq 0.$$

Dividing this inequality on both sides by $1-\varepsilon > 0$, and then passing to the limit for ε tending to 1, we obtain

$$\langle \mathcal{A}(x_\lambda), z - x_\lambda\rangle \leq 0$$

for every $z \in \mathbb{R}^n$. Taking $z = x_\lambda + \mathcal{A}(x_\lambda)$, we get $\langle \mathcal{A}(x_\lambda), \mathcal{A}(x_\lambda)\rangle \leq 0$, which is equivalent to $\mathcal{A}(x_\lambda) = 0$. Hence $\lambda x + (1-\lambda)y \in \mathcal{A}^{-1}(0)$ for every $x, y \in \mathcal{A}^{-1}(0)$ and every $\lambda \in (0,1)$. The proof is complete. □

Lemma 5.5.2. *Let $\mathcal{A} : \mathbb{R}^n \to \mathbb{R}^n$ be continuous and dissipative. Then*

$$\mathcal{A}^{-1}(0) = \{y \in \mathbb{R}^n;\ S(t)y = y \text{ for each } t \geq 0\},$$

where $\{S(t);\ t \geq 0\}$, is the semigroup of non-expansive operators generated by $\mathcal{A}$.

Proof. We begin by showing that $\mathcal{A}^{-1}(0) \subset \{y \in \mathbb{R}^n;\ S(t)y = y,\ t \geq 0\}$. Let $y \in \mathcal{A}^{-1}(0)$ and let us define the function $\phi : \mathbb{R}_+ \to \mathbb{R}^n$ by $\phi(t) = y$ for every $t \geq 0$. Obviously $\phi'(t) = 0 = \mathcal{A}(y) = \mathcal{A}(\phi(t))$ for every $t \geq 0$. On the other hand, $y = \phi(t) = S(t)y$, for every $t \geq 0$, and since y is arbitrary in $\mathcal{A}^{-1}(0)$, we have

$$\mathcal{A}^{-1}(0) \subset \{y \in \mathbb{R}^n;\ S(t)y = y \text{ for every } t \geq 0\}.$$

Now let $y \in \mathbb{R}^n$ with $S(t)y = y$ for every $t \geq 0$. Since $S(t)y = x(t, 0, y)$ for every $t \geq 0$, we deduce $0 = (S(t)y)' = \mathcal{A}(S(t)y)$ for every $t \geq 0$, equality which is equivalent to $y \in \mathcal{A}^{-1}(0)$. It then follows

$$\{y \in \mathbb{R}^n;\ S(t)y = y \text{ for every } t \geq 0\} \subset \mathcal{A}^{-1}(0)$$

and this completes the proof. □

Lemma 5.5.3. *Let $\eta \in \omega(\xi)$ and let $(t_k)_{k\in\mathbb{N}}$ with $\lim_{k\to\infty}(t_{k+1} - t_k) = +\infty$ and $\lim_{k\to\infty} S(t_k)\xi = \eta$. Then*

$$\lim_{k\to\infty} S(t_{k+1} - t_k)\eta = \eta. \tag{5.5.3}$$

Proof. We denote by $s_k = t_{k+1} - t_k$ and let us observe that, from (S_1) and (S_4), it follows

$$\|S(s_k)\eta - \eta\| = \|S(s_k)\eta - S(s_k + t_k)\xi + S(s_k + t_k)\xi - \eta\|$$

$$\leq \|S(s_k)\eta - S(s_k)S(t_k)\xi\| + \|S(t_{k+1})\xi - \eta\| \leq \|\eta - S(t_k)\xi\| + \|S(t_{k+1})\xi - \eta\|$$

for every $k \in \mathbb{N}$. Since $\lim_{k\to\infty} S(t_k)\xi = \eta$, this inequality completes the proof of (5.5.3). □

Remark 5.5.1. We mention that, for every $\eta \in \omega(\xi)$, the sequence $(t_k)_{k\in\mathbb{N}}$ with the property that $\lim_{k\to\infty} t_k = +\infty$ and $\lim_{k\to\infty} S(t_k)\xi = \eta$, can always be chosen in order to satisfy the condition imposed by Lemma 5.5.3, i.e. $\lim_{k\to\infty}(t_{k+1} - t_k) = +\infty$.

Theorem 5.5.1. *Let $\mathcal{A} : \mathbb{R}^n \to \mathbb{R}^n$ be continuous and dissipative. If the set $\mathcal{A}^{-1}(0)$ is nonempty and $\xi \in \mathbb{R}^n$, then:*

(i) *$\omega(\xi)$ is nonempty, bounded and closed;*
(ii) *for every $t \geq 0$, $S(t)\omega(\xi) \subset \omega(\xi)$;*
(iii) *for every $\eta, \nu \in \omega(\xi)$, we have $\|S(t)\eta - S(t)\nu\| = \|\eta - \nu\|$;*
(iv) *for every $\eta \in \mathcal{A}^{-1}(0)$, there exists $r \geq 0$ such that $\omega(\xi) \subset \{x \in \mathbb{R}^n;\ \|x - \eta\| = r\}$;*
(v) *if $\omega(\xi) \subset \mathcal{A}^{-1}(0)$ then $\omega(\xi)$ contains only one point η and*

$$\lim_{t\uparrow+\infty} S(t)\xi = \eta.$$

Proof. In order to prove (i) it suffices to show that the trajectory of equation (5.5.1) issuing from ξ, $\gamma(\xi)$, is bounded. Then, it will follow that the function $t \mapsto S(t)\xi$ has at least one limit point as t approaches $+\infty$. Since $\mathcal{A}^{-1}(0)$ is nonempty, there exists at least one $y \in \mathbb{R}^n$ such that $\mathcal{A}(y) = 0$. Taking into account that $S(t)\xi = x(t, 0, \xi)$, we have

$$(S(t)\xi)' = \mathcal{A}(S(t)\xi) - \mathcal{A}(y)$$

for every $t \geq 0$. Taking the inner product on both sides of this inequality by $S(t)\xi - y$, and using the dissipativity condition and Lemma 12.1.2, we deduce

$$\frac{d}{dt}\left(\|S(t)\xi - y\|^2\right) \leq 0$$

for every $t \geq 0$. From here, it follows that $t \mapsto \|S(t)\xi - y\|$ is non-increasing on $[0, +\infty)$ and by consequence $\|S(t)\xi - y\| \leq \|\xi - y\|$ for every $t \geq 0$. But this inequality shows that the function $t \mapsto S(t)\xi$ is bounded on $[0, +\infty)$, or equivalently that $\gamma(\xi)$ is bounded. From the remark above and Cesaró's lemma, it follows that $\omega(\xi)$ is nonempty, bounded and closed, which shows that (i) holds true.

Let $\eta \in \omega(\xi)$ and $t > 0$. Then there exists $(t_k)_{k\in\mathbb{N}}$ with $\lim_{k\to\infty} t_k = +\infty$ and $\lim_{k\to\infty} S(t_k)\xi = \eta$. In order to check (ii) it suffices to show that

$$\lim_{k\to\infty} S(t + t_k)\xi = S(t)\eta. \tag{5.5.4}$$

To this aim, let us observe that, from (S_1) and (S_4), we have

$$\|S(t + t_k)\xi - S(t)\eta\| = \|S(t)S(t_k)\xi - S(t)\eta\| \leq \|S(t_k)\xi - \eta\|$$

for every $k \in \mathbb{N}$, inequality which proves (5.5.4).

In order to prove (iii), we will show first that, for every $\eta \in \omega(\xi)$, we have

$$\omega(\eta) = \omega(\xi). \tag{5.5.5}$$

From (5.5.3) and Remark 5.5.1, it follows that $\omega(\xi) \subset \omega(\eta)$. In order to prove the converse inclusion, let $\nu \in \omega(\eta)$ and let $(t_k)_{k\in\mathbb{N}}$ with $\lim_{k\to\infty} t_k = +\infty$ and $\lim_{k\to\infty} S(t_k)\eta = \nu$. Since $\eta \in \omega(\xi)$, we conclude that there exists $(\tau_k)_{k\in\mathbb{N}}$ with $\lim_{k\to\infty} \tau_k = +\infty$ and $\lim_{k\to\infty} S(\tau_k)\xi = \eta$. Denoting by $\theta_k = t_k + \tau_k$, we have $\lim_{k\to\infty} S(\theta_k)\xi = \nu$. From this relation, we deduce that $\omega(\eta) \subset \omega(\xi)$, inclusion which, along with the preceding one, proves (5.5.5).

Now let $\eta, \nu \in \omega(\xi)$. From (5.5.5), we know that $\eta \in \omega(\nu)$, and therefore there exists a sequence $(\tau_k)_{k\in\mathbb{N}}$ with $\lim_{k\to\infty} \tau_k = +\infty$ and

$\lim_{k\to\infty} S(\tau_k)\nu = \eta$. On the other hand, as $\nu \in \omega(\xi)$, according to Lemma 5.5.3, there exists $(s_k)_{k\in\mathbb{N}}$ with $\lim_{k\to\infty} s_k = +\infty$ and

$$\lim_{k\to\infty} \|S(s_k)\nu - \nu\| = 0. \tag{5.5.6}$$

Let us remark that

$$\|S(s_k)\eta-\eta\| \leq \|S(s_k)\eta - S(\tau_k+s_k)\nu\| + \|S(\tau_k+s_k)\nu - S(\tau_k)\nu\| + \|S(\tau_k)\nu-\eta\|$$

$$\leq 2\|S(\tau_k)\nu - \eta\| + \|S(s_k)\nu - \nu\|$$

for every $k \in \mathbb{N}$, inequality which, along with (5.5.6), shows that

$$\begin{cases} \lim\limits_{k\to\infty} S(s_k)\nu = \nu \\ \lim\limits_{k\to\infty} S(s_k)\eta = \eta. \end{cases} \tag{5.5.7}$$

From (5.5.7), (S_1) and (S_2), we deduce

$$\|\eta - \nu\| = \lim_{k\to\infty} \|S(s_k)\eta - S(s_k)\nu\|$$

$$= \lim_{k\to\infty} \|S(s_k - t)S(t)\eta - S(s_k - t)S(t)\nu\| \leq \|S(t)\eta - S(t)\nu\| \leq \|\eta - \nu\|$$

for every $t \geq 0$, which proves (iii).

Let $\eta \in \mathcal{A}^{-1}(0)$. Then the function $t \mapsto S(t)\xi$ satisfies

$$\begin{cases} (S(t)\xi)' = \mathcal{A}(S(t)\xi) - \mathcal{A}(\eta) \\ S(0)\xi = \xi. \end{cases}$$

Multiplying both sides of the equation above by $S(t)\xi - \eta$, from Lemma 12.1.2 and the dissipativity condition, we deduce

$$\frac{1}{2}\frac{d}{dt}(\|S(t)\xi - \eta\|^2) \leq 0$$

for every $t \geq 0$. Consequently the function $t \mapsto \|S(t)\xi - \eta\|$ is non-increasing on $\mathbb{R}_+$ and, as it is bounded from below (being positive), there exists

$$\lim_{t\uparrow+\infty} \|S(t)\xi - \eta\| = r.$$

This relation shows that, for every $\nu \in \omega(\xi)$, we have $\|\nu - \eta\| = r$ and therefore $\omega(\xi) \subset \{x \in \mathbb{R}^n;\ \|x - \eta\| = r\}$, which proves (iv).

Let us remark that, in accordance with (iv), for every $\eta \in \mathcal{A}^{-1}(0)$, there exists $r \geq 0$ such that $\omega(\xi) \subset \{x \in \mathbb{R}^n;\ \|x - \eta\| = r\}$. Taking $\eta \in \omega(\xi) \subset \mathcal{A}^{-1}(0)$, it follows $r = \|\eta - \eta\| = 0$, which shows that $\omega(\xi) = \{\eta\}$. Obviously this relation is equivalent to $\lim\limits_{t\uparrow+\infty} S(t)\xi = \eta$ and this completes the proof. □

5.6 The Case of Controlled Systems

All the equations and systems of differential equations considered by now were abstract expressions of some mathematical models describing the *free* or *uninfluenced* evolution of some phenomena from Physics, Demography, Biology, *etc.* In contrast with these models which offer only a contemplative description of the evolution, there are some others trying to catch the possible external interventions done during the evolution with the precise purpose to modify it according to some performance criteria. For instance, as we have already seen in Section 1.4, the free evolution of a species of bacteria is described by the so-called logistic equation

$$x' = cx(b - x),$$

where $x(t)$ stands for the number of bacteria at the moment t, $b > 0$ and $c > 0$. Since the null solution of this equation is unstable (see Remark 5.1.1), it is clear that, once we intend to make this number $x(t)$ tend to 0 as t approaches $+\infty$, we have to intervene in the evolution of the system trying to minimize, as much as possible, the instantaneous rate of growth of these bacteria. In order to fix the ideas, let us assume that the case in question is that of a contagious disease whose complete remission rests heavily on the action of a certain drug, having the role of bringing the number of bacteria involved as close to 0 as possible in a fairly reasonable short time. This external intervention can be expressed by means of the new mathematical model:

$$x' = cx(b - x) + ku. \tag{5.6.1}$$

More precisely, $u : \mathbb{R}_+ \to \mathbb{R}_+$ is a function which gives a quantitative description of the intervention, while $k < 0$ signifies the fact that this intervention takes place in the desired sense, i.e. *in order to minimize* x'. We can imagine, for instance, that $u(t)$ represents the amount of penicillin used at the moment t with the scope of diminishing as much as possible $x(t)$.

At this moment, some observations are needed. Firstly, it is clear that, from medical reasons, the values of the function u cannot exceed a certain admissible maximal level. Second, let us remark that, from rather obvious practical considerations, the function u cannot be continuous. Instead, it may be (and, in this case, it certainly is) a step function. Finally, we have to take into consideration a very important economical criteria, i.e. the total cost of the recovering, amount expressed mathematically as $Pu(+\infty)$, where $P > 0$ is the per/unit price of the penicillin while $u(+\infty)$ is the total quantity of penicillin used, be minimal.

We conclude with a special mention on such kind of problems, having a general character. In practice one uses the fact that the function u^*, which enjoys all the required properties, can be expressed in the form $u^*(t) = Q(x^*(t))$ for every $t \in \mathbb{R}_+$, where x^* is the solution of the problem (5.6.1) corresponding to $u = u^*$, while $Q : \mathbb{R} \to \mathbb{R}$ is the so-called *synthesis operator* or *feedback operator*. Substituting ku by $kQ(x)$ in equation (5.6.1), we get

$$x' = cx(b - x) + kQ(x), \tag{5.6.2}$$

called *the closed loop system*. Recalling once again the fact, of maximal importance in this context, that the null solution of the logistic equation is unstable, it is easy to understand why the function Q which, in general, is neither unique, nor known, has to be chosen such that the null solution of the closed loop system (5.6.2) be asymptotically stable.

This is an extremely simple example of a control system which, we believe, is convincing enough with regard to the importance of the study of this branch of Applied Mathematics which is in a very fast process of development.

We will now proceed to the statement, within a general abstract framework, of a large class of such control problems. For the sake of simplicity we will confine ourselves only to the linear case in which the *control function*, or *controller* u, is real valued. Let us consider the system

$$x' = \mathcal{A}x + u(t)b,$$

where $\mathcal{A} \in \mathcal{M}_{n\times n}(\mathbb{R})$, $b \in \mathbb{R}^n$ and $u : \mathbb{R}_+ \to \mathbb{R}$. Such a system is called *control system*. The function u is called *controller* or *input*, while x *the state function*, or *output*. In many cases the state is known only in an indirect form $y = \eta(x)$, where η is the so-called *observation operator*, while y is *the observed variable* or *the observed output*. For instance, in the case of the example presented before, the total number of bacteria x cannot be effectively counted. Nevertheless, we can make a pretty good idea on its size once we have access to some variable which is "drastically dependent" on it, as for instance the temperature $y = \eta(x)$ of the infected subject. In general, the observation operator η depends on x by means of a differential equation of the type

$$\eta'(t) = \langle c, x(t)\rangle - \alpha\varphi(\eta(t)),$$

where $\varphi : \mathbb{R} \to \mathbb{R}$ is a nonlinear law, usually known, while $\alpha > 0$, $c \in \mathbb{R}^n$ and $b \in \mathbb{R}^n$ are the so-called *regulating parameters*. The problem in this context is to find some simple sufficient conditions for the system

$$\begin{cases} x' = \mathcal{A}x + \varphi(\eta(t))b \\ \eta'(t) = \langle c, x(t) \rangle - \alpha\varphi(\eta(t)) \end{cases} \tag{5.6.3}$$

to be globally asymptotically stable for any choice of the function $\varphi : \mathbb{R} \to \mathbb{R}$ which satisfies the conditions

$$\eta\varphi(\eta) > 0 \text{ for every } \eta \in \mathbb{R},\ \eta \neq 0 \tag{5.6.4}$$

and

$$\lim_{\eta \to +\infty} \int_0^\eta \varphi(r)\, dr = +\infty. \tag{5.6.5}$$

The problem above is the so-called *Lurie-Postnikov problem.*

Theorem 5.6.1. *If $\mathcal{A}$ is Hurwitzian, then there exist $\alpha > 0$, $b \in \mathbb{R}^n$ and $c \in \mathbb{R}^n$ such that, for every $\varphi : \mathbb{R} \to \mathbb{R}$ which satisfies* (5.6.4) *and* (5.6.5), *the system* (5.6.3) *is globally asymptotically stable.*

Proof. The idea is to construct a Lyapunov function which satisfies all conditions of Theorem 5.4.3. More precisely, let $V : \mathbb{R}^n \times \mathbb{R} \to \mathbb{R}$ be defined by

$$V(x, \eta) = \frac{1}{2}\langle \mathcal{P}x, x \rangle + \int_0^\eta \varphi(r)\, dr,$$

where $\mathcal{P}$ is the symmetric and positive definite matrix whose existence is ensured by Theorem 5.4.4. Also from Theorem 5.4.4 combined with (5.6.4) and (5.6.5), it follows that V is positive definite and

$$\lim_{\|x\|+|\eta| \to +\infty} V(x, \eta) = +\infty.$$

On the other hand, we have

$$\nabla V(x, \eta) = ((\mathcal{P}x)_1, (\mathcal{P}x)_2, \ldots, (\mathcal{P}x)_n, \varphi(\eta))$$

where $(\mathcal{P}x)_i$ is the component of rank $i = 1, 2, \ldots, n$ of the vector $\mathcal{P}x$. Let us observe that the system (5.6.3) may be rewritten in the form $z' = f(z)$, where $f : \mathbb{R}^{n+1} \to \mathbb{R}^{n+1}$ is defined by $f(z) = f(x, \eta) = (\mathcal{A}x + \varphi(\eta)b, \langle c, x \rangle - \alpha\varphi(\eta))$ for every $z \in \mathbb{R}^{n+1}$, $z = (x, \eta)$. Then we have

$$\langle \nabla V(x, \eta), f(x, \eta) \rangle = \langle \mathcal{A}x, \mathcal{P}x \rangle + \varphi(\eta)\langle b, \mathcal{P}x \rangle + \varphi(\eta)\langle c, x \rangle - \alpha\varphi^2(\eta).$$

From this relation, taking into account that $\mathcal{P}$ is symmetric and satisfies $\mathcal{A}^*\mathcal{P} + \mathcal{P}\mathcal{A} = -\mathcal{I}$, it follows that $\langle \mathcal{A}x, \mathcal{P}x \rangle = -\frac{1}{2}\|x\|^2$, and therefore

$$\begin{aligned} \langle \nabla V(x, \eta), f(x, \eta) \rangle &= -\frac{1}{2}\|x\|^2 - \alpha\varphi^2(\eta) + \varphi(\eta)\langle c + \mathcal{P}b, x \rangle \\ &\leq -\frac{1}{2}\|x\|^2 - \alpha\varphi^2(\eta) + \varphi(\eta)\|c + \mathcal{P}b\|\|x\|. \end{aligned}$$

Consequently, if α, b and c satisfy the so-called *regulating inequality*

$$\|\mathcal{P}b + c\| \leq \sqrt{2\alpha},$$

then V fulfills all the conditions of Theorem 5.4.3, from where it follows that (5.6.3) is globally asymptotically stable. □

5.7 Unpredictability and Chaos

As we have already noticed occasionally, the study of differential equations and systems of differential equations owes the main part of its tremendous development to the efficiency in predicting with great accuracy the future evolution of many phenomena of practical interest. We emphasize that, in the mathematical treatment of concrete problems, in general, we cannot dispose of the exact initial data of the Cauchy problem which models the phenomenon in question. This inconvenience is due to the technological impossibility to diminish a certain degree of imprecision in the process of measurement, and also to the inevitable errors: of reading, of rounding, random, *etc.*, appearing during both the collection and the processing of data. So, the user of the techniques and previously established abstract results has to confine himself, almost at any time, with approximate data. A fundamental problem raised in this context is to establish the "degree of credibility" of a conclusion obtained on this basis. It is easy to understand that, in order to make such a prediction which could be accepted and took into consideration in getting conclusions with a very low level of ambiguity, we need:

(i) a mathematical model describing the real phenomenon as accurate as possible, and whose solutions be stable;
(ii) numerical data as close as possible to the real values of the initial state of the system;
(iii) efficient numerical methods and suitable computing equipment in order to find as fast as possible the approximate solutions of the system with an error not exceeding a certain preassigned critical level imposed by practice.

In order to understand these requirements, let us analyze the following examples. We begin with one, with no apparent connection with differential equations, example adapted from [Hubbard and West (1995)], Example 522, p. 216.

Example 5.7.1. Let us consider a system in which, to every bit of a clock a certain angle doubles. The state of the system is the angle and it doubles again and again at each second. So, the evolution of the system is completely described by the sequence of angles: $\theta_0, \theta_1, \ldots$, where θ_0 is the initial state, while $\theta_{n+1} = 2\theta_n$, for $n = 0, 1, 2, \ldots$. Let us remark that, if θ_0 is very slightly perturbed, say with 10^{-8}, after only 30 seconds, the state of the system is completely unknown, simply because the degree of uncertainty grew up to $2^{30} \times 10^{-8} > 1$. As a consequence, the evolution of this system, although extremely simple and ordered, is practically unpredictable even on short intervals of time.

Example 5.7.2. Let us consider the differential equation $x' = x$ and we aim to determine the approximate values of the solution $x(\cdot, 0, \pi)$ at $t = 10^{10}$ with an error not exceeding 10^{-2}. We have chosen the initial data π simply because, this one being an irrational number we are forced to replace it during the numerical processing by some of its rational approximates. Let us remark that "the exact solution" is given by

$$x(10^{10}, 0, \pi) = \pi e^{10^{10}},$$

while the approximate solution corresponding to the approximate value π_a of π is

$$x(10^{10}, 0, \pi) = \pi_a e^{10^{10}}.$$

Then, the error is

$$\mathcal{E}(\pi_a) = |\pi - \pi_a| e^{10^{10}}.$$

In order for the error not to exceed 10^{-2}, the approximate value π_a of π must satisfy

$$|\pi - \pi_a| \leq 10^{-2} e^{-10^{10}}.$$

To do this we must choose π_a with more than $3,000,000$ exact digits, a fact which is very hard to achieve practically.

Let us observe that, if we take as approximate initial data a truncation π_a of π satisfying the realistic and feasible condition $|\pi - \pi_a| \leq 10^{-3}$, the absolute error of the corresponding approximate solution is greater than $10^{-3} \times e^{10^{10}}$. Since $e^x > x^4$ for $x > 100$, the error exceeds 10^{37}. So, the information thus obtained is completely inaccurate and of no practical use.

Consequently, we may consider that the evolutions described by the equation $x' = x$, although ordered, are unpredictable on long term. This fact is nothing but a simple consequence of the instability of the linear differential equation above.

These two examples which reveal the great importance, in the study of such kind of problems, of the so-called *entropy*. In this context, the entropy is *the rate at which the information dissipates in time*, which may be identified with *the rate of growth of the degree of uncertainty*, or with *the minimal time needed to double the uncertainty.* For instance, in the case previously analyzed, the rate of growth of the degree of uncertainty on an interval of time T is e^T. This is due to the fact that an initial error (uncertainty) $\mathcal{E}$ multiplies after the time T by e^T. In other words, the remark above asserts that, on every interval of time of length $\ln 2$, the degree of uncertainty doubles.

Besides these very simple situations when the evolution of the system, although "ordered", is unpredictable on long term, there are examples of systems having stationary solutions with very strange behavior, in the sense that, for initial data very close to the stationary solution, the corresponding solutions have a highly disordered even chaotic evolution, on relatively short intervals of time. An example of this type is the celebrated *Lorenz system* in [Lorenz (1963)].

Example 5.7.3. Let us consider the so-called Lorenz nonlinear differential system

$$\begin{cases} x' = -\sigma x + \sigma y \\ y' = rx - y - xz \\ z' = -bz + xy \end{cases} \tag{5.7.1}$$

which is a simplified model, deducted from a system of partial differential equations describing thermal convection within an incompressible fluid moving in a horizontal plane. Here $b > 0$, $\sigma > b + 1$ and $r > 0$ are some parameters which characterize the fluid. For $r \in (0, 1)$, the system above has a unique stationary solution, namely $(0, 0, 0)$, which is stable. For $r > 1$, the null solution is no longer stable and, in addition, the system has two more stationary solutions $(\sqrt{b(r-1)}, \sqrt{b(r-1)}, r-1)$ and $(-\sqrt{b(r-1)}, -\sqrt{b(r-1)}, r-1)$. Actually, these two new stationary solutions mark the beginning of an extremely stirred convection process which starts when r crosses from the left the value 1.

This strange behavior has been observed in 1963, by the meteorologist Edward N. Lorenz from Massachusetts Institute of Technology, who used this system in order to get meteorological predictions on short term. More precisely, during some numerical simulations, he has observed that certain solutions of the nonlinear differential system (5.7.1), for the specific choice $\sigma = 10$, $r = 28$ and $b = 8/3$, seem to become closer and closer, in a very disorderly manner — and therefore highly unpredictable — either to one stationary solution, or to the other. Lorenz has stated that very small perturbations of the initial data produce considerable modifications in the evolution of the system. The projection on the xOz plane of one

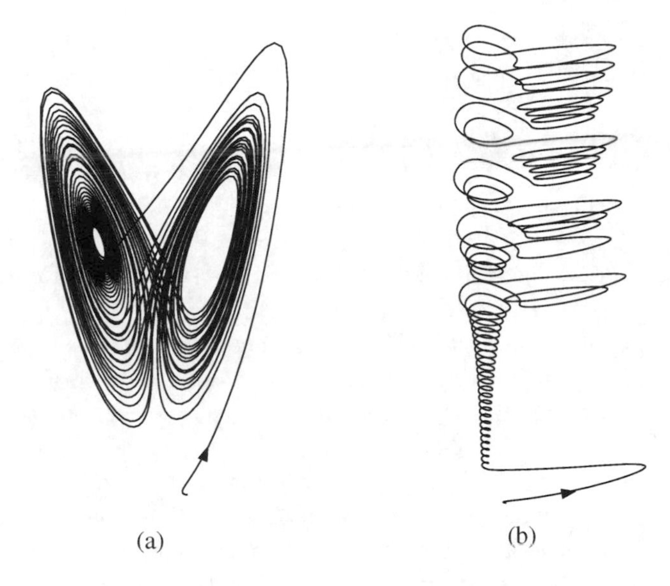

Fig. 5.7.1

trajectory of the Lorenz system is illustrated in Figure 5.7.1 (a), while the graph of $t \mapsto (x(t), z(t))$ (the t-axis is vertical) in Figure 5.7.1 (b). One may easily observe the chaotic evolution of the solution corresponding to this trajectory which "becomes closer and closer" both to the first and the second stationary solution.

This disorderly behavior generating unpredictable evolutions on medium, or even short term, is known in the literature under the name of *chaos*, or *chaotic behavior*.

In order to understand and to explain what is really hidden behind this apparently paradoxical phenomenon, we will define the so-called *modulus of continuity* of the solution of a Cauchy problem as a function of the initial data. To this aim let $f : \mathbb{R}^n \to \mathbb{R}^n$ be a locally Lipschitz function and let us consider the Cauchy problem

$$\begin{cases} x' = f(x) \\ x(0) = \xi. \end{cases} \tag{5.7.2}$$

Let us assume that for every $\xi \in \mathbb{R}^n$ the unique saturated solution of (5.7.2), $x(\cdot, \xi)$, is global and let us observe that, by virtue of Theorem 2.5.2, for every $T > 0$, the function $\xi \mapsto x(\cdot, \xi)$ is continuous from $\mathbb{R}^n$ to $C([0, T]; \mathbb{R}^n)$. This means that, for every $\xi \in \mathbb{R}^n$, every $T > 0$ and every $\varepsilon > 0$ there exists $\delta(\xi, T, \varepsilon) > 0$ such that, for every $\eta \in \mathbb{R}^n$ with $\|\xi - \eta\| \leq \delta(\xi, T, \varepsilon)$ we have $\|x(t, \xi) - x(t, \eta)\| \leq \varepsilon$ uniformly with respect to $t \in [0, T]$. We denote by $m(\xi, T, \varepsilon)$ the larger real number $\delta(\xi, T, \varepsilon)$ from $(0, 1]$ with the properties above. The function $\varepsilon \mapsto m(\xi, T, \varepsilon)$ is called *the modulus of continuity* of the function $\eta \mapsto x(\cdot, \eta)$, from $\mathbb{R}^n$ in $C([0, T]; \mathbb{R}^n)$, at a point ξ. One may easily observe that, in general, for $\xi \in \mathbb{R}^n$ and $\varepsilon > 0$ fixed, the function $T \mapsto m(\xi, T, \varepsilon)$ is non-increasing. One may also see that a stationary solution ξ of the system $x' = f(x)$ is stable if and only if, for every $\varepsilon > 0$, $\inf\{m(\xi, T, \varepsilon);\ T > 0\} = m(\xi, \varepsilon) > 0$.

The last two examples dealt with the manner in which m varies as a function of T. More precisely, the extremely drastic growth of the degree of uncertainty is caused by the very abrupt decrease of the modulus of continuity m with respect to T.

In other situations, which we did not touch upon by now, the unpredictability and chaos are generated by the instability of the solutions with respect to one or several parameters. Many mathematical models describe phenomena whose evolution laws modify as functions of a certain parameter. So, the evolutions of such a phenomenon are described by a differential system of the type

$$x' = f(t, x, p), \tag{5.7.3}$$

where $f : \mathbb{R}_+ \times \mathbb{R}^n \times \mathbb{P} \to \mathbb{R}^n$ is a continuous function, while $\mathbb{P}$ is an open subset of $\mathbb{R}^m$. We denote by $x(\cdot, a, \xi, p) : [a, T) \to \mathbb{R}^n$ the unique saturated solution of the system (5.7.3) which satisfies $x(a, a, \xi, p) = \xi$ and let us assume that $0 \in \mathbb{P}$ and $f(t, 0, 0) = 0$ for every $t \in \mathbb{R}_+$, which means that, for the value 0 of the parameter, (5.7.3) has the null global solution. Let us also assume that f is locally Lipschitz on $\mathbb{R}^n \times \mathbb{P}$. In these conditions, paraphrasing the definition of stability we introduce:

Definition 5.7.1. The null solution of the system (5.7.3) is *persistently stable*, or *robust* if:

(i) for every $a \geq 0$ there exists $\mu(a) > 0$ such, that for every $p \in \mathcal{P}$ with $\|p\| \leq \mu(a)$, the unique saturated solution $x(\cdot, a, 0, p)$, of (5.7.3), satisfying $x(a, a, 0, p) = 0$, is defined on $[a, +\infty)$ and
(ii) for every $a \geq 0$ and every $\varepsilon > 0$, there exists $\delta(\varepsilon, a) \in (0, \mu(a)]$ such that, for each $p \in \mathcal{P}$ with $\|p\| \leq \delta(\varepsilon, a)$, the unique saturated solution $x(\cdot, a, 0, p)$, of (5.7.3), satisfying $x(a, a, 0, p) = \xi$, also satisfies $\|x(t, a, 0, p)\| \leq \varepsilon$ for every $t \in [a, +\infty)$.

We leave it to the reader to define an analogous concept referring to an arbitrary solution of the system (5.7.3) as well as all other concepts which paraphrase those of uniform, asymptotic and uniform asymptotic stability.

As one can easily realize, the lack of robustness of a certain solution could cause unpredictability and even chaos. A very interesting example of this sort is that of the nonlinear oscillator.

Example 5.7.4. [Arecchi and Lisi (1982)] Let us consider the second-order nonlinear differential equation

$$x'' + kx' - x + 4x^3 = A\cos\omega t$$

where $x(t)$ is the abscissa, at the moment t, of a material point of mass 1 moving under the action of a force $F(t,x,x') = -kx' + x - 4x^3 + A\cos\omega t$ centered at the origin. It has been proved that, for every initial datum, the solution evolves towards a periodic one, called by extension *limit cycle*. For a special choice of k, A and ω, the graph of the function $t \mapsto (x(t), x'(t))$, with $x : [\,0, 50\,] \to \mathbb{R}$ the solution of the equation above satisfying $x(0) = 10$ and $x'(0) = 0$, is illustrated in Figure 5.7.2 (a), while the graph of $t \mapsto x(t)$ in Figure 5.7.2 (b).

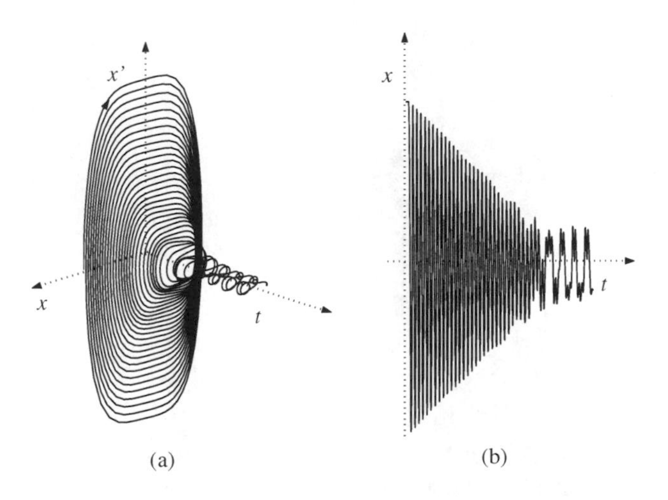

Fig. 5.7.2

Moreover, one has observed that, the decreasing of the parameter[2] k, has as effect the growing up to $+\infty$ of the period of the limit cycle. As a

[2]In this model, the parameter k represents the coefficient of friction, while $-kx'$ the force of friction.

consequence, there exists a sequence $(k_n)_{n\in\mathbb{N}}$ tending to 0, with the property that the corresponding sequence of periods $(T_n)_{n\in\mathbb{N}}$ satisfies $T_{n+1} = 2T_n$ for every $n \in \mathbb{N}$. From this reason, for very small values of k the evolution described by the equation above becomes unpredictable, even chaotic.

In Figure 5.7.3, we reconsider the case in Figure 5.7.2 with the very same data, excepting for k which is now ten times smaller.

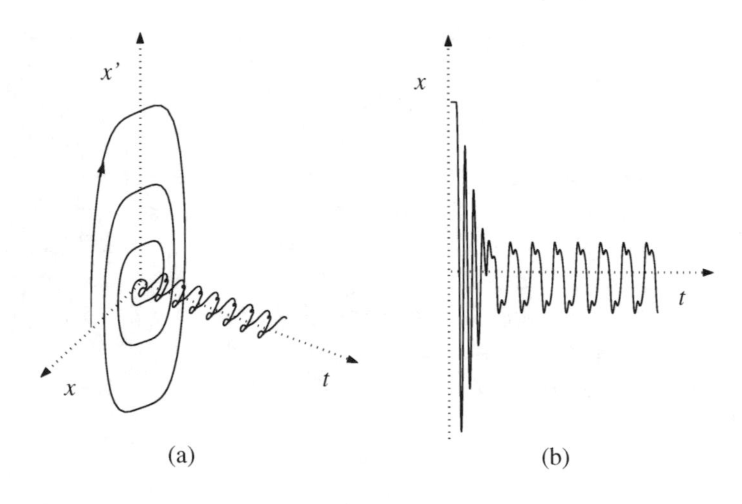

Fig. 5.7.3

For a detailed and systematic presentation of theory of limit cycles the interested reader is referred to [Ye *et al.* (1986)].

All these examples lead to the conclusion that many phenomena which evolve according to deterministic laws can be studied on the basis of these laws only on very short intervals of time, and this because extremely small perturbations of the initial data produce dramatic changes in the evolution of the system. For this reason one may assert that "the determinism has a local character". On the other hand, some of these phenomena which evolve towards chaos admit statistical models which may bring some extra information exactly where the deterministic techniques say nothing, or very few. In many cases of this kind, the evolution, although chaotic from a purely deterministic point of view, is, statistically speaking, very smooth and regular, in the sense of the existence of the *mean ergodic*, i.e. of $\lim_{T\to\infty} \frac{1}{T}\int_0^T x(t)\,dt$. Here, we have in mind first the turbulence in Fluid Mechanics, but also the disorders of the cardiac rhythm, the noises in the electric lines, the evolution of a certain society, *etc.* For details on this subject, the reader is referred to [Arnold and Avez (1967)].

5.8 Exercises and Problems

Problem 5.1. *Let us consider the scalar linear differential equation*

$$x' = a(t)x, \tag{$\mathcal{E}$}$$

$t \geq 0$, *where* $a : [0, +\infty) \to \mathbb{R}$ *is a continuous function. Prove that:*

(1) $(\mathcal{E})$ *is stable if and only if there exists a function* $K : [0, +\infty) \to \mathbb{R}$ *such that:*

$$\int_{t_0}^{t} a(s)\, ds \leq K(t_0)$$

for every $t_0 \geq 0$ *and every* $t \geq t_0$;

(2) $(\mathcal{E})$ *is uniformly stable if and only if there exists* $M \in \mathbb{R}$ *such that:*

$$\int_{t_0}^{t} a(s)\, ds \leq M$$

for every $t_0 \geq 0$ *and every* $t \geq t_0$;

(3) $(\mathcal{E})$ *is asymptotically stable if and only if*

$$\lim_{t \to +\infty} \int_{0}^{t} a(s)\, ds = -\infty;$$

(4) $(\mathcal{E})$ *is uniformly asymptotically stable if and only if there exist* $K \geq 0$ *and* $\alpha > 0$ *such that*

$$\int_{t_0}^{t} a(s)\, ds \leq K - \alpha(t - t_0)$$

for every $t_0 \geq 0$ *and every* $t \geq t_0$.

([Corduneanu (1977)], p. 117)

Exercise 5.1. *Check for stability the null solution of*:

(1) $x' = x$.	(2) $x' = 0$.	(3) $x' = -x$.
(4) $x' = -2x + \sin x$.	(5) $x' = x^2$.	(6) $x' = -x^2$.
(7) $x' = -\tan x$.	(8) $x' = -\sin x$.	(9) $x' = -x + x^2$.

Exercise 5.2. *Check for stability the following systems of first-order linear differential equations :*

$$(1)\begin{cases} x_1' = -x_1 + x_2 \\ x_2' = 2x_1 - x_2. \end{cases} \quad (2)\begin{cases} x_1' = x_2 \\ x_2' = -x_1. \end{cases} \quad (3)\begin{cases} x_1' = x_1 + 5x_2 \\ x_2' = -x_1 - 3x_2. \end{cases}$$

$$(4)\begin{cases} x_1' = -x_1 + x_2 \\ x_2' = x_1 + 2x_2. \end{cases} \quad (5)\begin{cases} x_1' = -3x_1 + x_2 \\ x_2' = 4x_1 - 3x_2. \end{cases} \quad (6)\begin{cases} x_1' = -2x_1 + 4x_2 \\ x_2' = x_1 - 2x_2. \end{cases}$$

$$(7)\begin{cases} x_1' = x_2 \\ x_2' = x_3 \\ x_3' = x_1. \end{cases} \quad (8)\begin{cases} x_1' = x_2 + x_3 \\ x_2' = x_3 + x_1 \\ x_3' = x_1 + x_2. \end{cases} \quad (9)\begin{cases} x_1' = x_2 - x_3 \\ x_2' = x_3 - x_1 \\ x_3' = x_1 - x_2. \end{cases}$$

Problem 5.2. *Let $\omega > 0$ and $f : [0, +\infty) \to \mathbb{R}$ be a continuous function which is absolutely integrable on $[0, +\infty)$. Prove that every global solution of the equation $x'' + \omega^2 x = f(t)$, $t \geq 0$, is bounded on $[0, +\infty)$.* ([Corduneanu (1977)], p. 152)

Problem 5.3. *Let $\omega > 0$ and $f : [0, +\infty) \to \mathbb{R}$ be a continuous function which is absolutely integrable on $[0, +\infty)$. Prove that the null solution of the equation $x'' + [\omega^2 + f(t)]x = 0$, $t \geq 0$, is uniformly stable.* ([Corduneanu (1977)], p. 152)

Problem 5.4. *Let $\mathcal{A} \in \mathcal{M}_{n\times n}(\mathbb{R})$ be Hurwitzian. Let $\mathcal{B} : [0, +\infty) \to \mathcal{M}_{n\times n}(\mathbb{R})$ be continuous with $\lim_{t\to+\infty} \|\mathcal{B}(t)\|_{\mathcal{M}} = 0$. Prove that the null solution of the system $x' = [\mathcal{A} + \mathcal{B}(t)]x$, $t \geq 0$, is asymptotically stable.* ([Corduneanu (1977)], p. 153)

Problem 5.5. *Let $\mathcal{A} \in \mathcal{M}_{n\times n}(\mathbb{R})$ be Hurwitzian and let $\mathcal{B} : [0, +\infty) \to \mathcal{M}_{n\times n}(\mathbb{R})$ be continuous with*

$$\int_0^{+\infty} \|\mathcal{B}(s)\|_{\mathcal{M}} ds < +\infty.$$

Prove that there exist $k > 0$ and $\alpha > 0$ such that every solution $x : [0, +\infty) \to \mathbb{R}^n$ of the system $x' = [\mathcal{A} + \mathcal{B}(t)]x$, $t \geq 0$, satisfies

$$\|x(t)\| \leq ke^{-\alpha t}\|x(0)\|$$

for every $t \geq 0$. In particular the null solution of the system is asymptotically stable. ([Halanay (1972)], p. 194)

Problem 5.6. *Let $\mathcal{A}_k \in \mathcal{M}_{n\times n}(\mathbb{R})$, $k = 0, 1, \ldots, m$. If $\mathcal{A}_0$ is Hurwitzian then, for every $a > 0$ there exists $\delta(a) > 0$ such that, for every $\xi \in B(0, \delta(a))$, the unique global solution $x(\cdot, a, \xi)$ of the Cauchy problem*

$$\begin{cases} x' = (t^m\mathcal{A}_0 + t^{m-1}\mathcal{A}_1 + \cdots + \mathcal{A}_m)x \\ x(a) = \xi \end{cases}$$

satisfies

$$\lim_{t\to+\infty} x(t, a, \xi) = 0.$$

Problem 5.7. *Let $f : \mathbb{R}_+ \times \mathbb{R} \to \mathbb{R}$ be continuous on $\mathbb{R}_+ \times \mathbb{R}$ and locally Lipschitz on $\mathbb{R}$ and let $x(\cdot, \xi_i) : [0, +\infty) \to \mathbb{R}$, $\xi_1 < \xi_2$, $i = 1, 2$, be two solutions of the differential equation*

$$x' = f(t, x)$$

with $x(0, \xi_i) = \xi_i$, $i = 1, 2$ and $\lim_{t\to+\infty} x(t, \xi_1) = \lim_{t\to+\infty} x(t, \xi_2) = x^* \in \mathbb{R}$. Prove that for every $\xi \in (\xi_1, \xi_2)$, the saturated solution of the equation above, $x(\cdot, \xi)$, which satisfies $x(0, \xi) = \xi$, is globally and asymptotically stable. ([Glăvan *et al.* (1993)], p. 178)

Problem 5.8. *Let $f : \Omega \subset \mathbb{R}^n \to \mathbb{R}^n$ be a locally Lipschitz function with $f(0) = 0$. If all saturated solutions of the differential equation*

$$x' = f(x)$$

are global and bounded on $[0, +\infty)$, is the null solution of the equation above stable? ([Glăvan *et al.* (1993)], p. 179)

Problem 5.9. *Let $f : \mathbb{R} \to \mathbb{R}$ be a function of class C^1 with $f(0) = 0$ and having the property that $f'(0) = \lambda > 0$. Then the null solution of the equation $x' = f(x)$ is not asymptotically stable.*

Exercise 5.3. *Check for stability the null solution of the following first-order nonlinear differential systems :*

$$(1)\begin{cases} x_1' = -x_1 + x_2^2 \\ x_2' = -x_1^3 - 2x_2. \end{cases} \quad (2)\begin{cases} x_1' = x_1 + 3x_2^5 \\ x_2' = -x_1^4 - 4x_2. \end{cases} \quad (3)\begin{cases} x_1' = -\sin x_1 + 5x_2 \\ x_2' = -x_1^3 - x_2. \end{cases}$$

$$(4)\begin{cases} x_1' = 2x_1 - x_2^2 \\ x_2' = x_1 x_2 - x_2. \end{cases} \quad (5)\begin{cases} x_1' = -\sin x_1 + x_2^2 \\ x_2' = -4x_1 - 5x_2. \end{cases} \quad (6)\begin{cases} x_1' = 2\,\mathrm{sh}\, x_2 \\ x_2' = -x_1^2 - 3x_2. \end{cases}$$

Exercise 5.4. *Check for stability the null solution of the following first-order nonlinear differential systems:*

$$(1)\begin{cases} x_1' = -x_1^3 + x_2 \\ x_2' = -x_1 - 2x_2^3. \end{cases} \quad (2)\begin{cases} x_1' = -x_1^5 - 3x_2 \\ x_2' = 3x_1 - 4x_2^3. \end{cases} \quad (3)\begin{cases} x_1' = -x_1 + 5x_2^3 \\ x_2' = -x_1^3 - 3x_2. \end{cases}$$

$$(4)\begin{cases} x_1' = x_1 - x_2^2 \\ x_2' = x_1 x_2 - x_2. \end{cases} \quad (5)\begin{cases} x_1' = -\sin x_1 + x_2 \\ x_2' = -4x_1 - 3\tan x_2. \end{cases} \quad (6)\begin{cases} x_1' = -2\,\mathrm{sh}\, x_1 + 4x_2^3 \\ x_2' = -x_1^3 - 2x_2. \end{cases}$$

See also Problems 6.5 and 6.6 in the next chapter.

Chapter 6

Prime Integrals

This chapter is dedicated to the introduction and study of the concept of prime integral for a system of first-order differential equations. In the first two sections we present the main notions and results referring to this problem in the case of both autonomous and non-autonomous systems. The third section is concerned with the study of first-order linear and quasi-linear partial differential equations, while the fourth section contains a fundamental existence and uniqueness theorem with regard to the Cauchy problem for a class of first-order quasi-linear partial differential equations. In the fifth section we collect some specific properties of the so-called conservation law. The chapter ends with an exercises and problems section.

6.1 Prime Integrals for Autonomous Systems

Let Ω be a nonempty and open subset of $\mathbb{R}^n$, let $f : \Omega \to \mathbb{R}^n$ be a continuous function and let us consider the autonomous system

$$x' = f(x). \tag{6.1.1}$$

In many specific situations, considerations of extra-mathematical nature based on the physical meaning of the functions involved in (6.1.1), prove the existence of some functions of class C^1, $U : \Omega \to \mathbb{R}$ which, although non-constant on Ω, are constant along the trajectories[1] of the system (6.1.1). Any family of functionally independent such functions could be of real help in obtaining information on the solutions of (6.1.1) which, in most cases,

[1] Here and thereafter, by *trajectories* of (6.1.1), we mean the trajectories corresponding to the solutions of (6.1.1).

cannot be explicitly solved. Moreover, the larger is such a family, the bigger are the chances to solve (6.1.1) explicitly or, at least, to obtain crucial information on its solutions. This is because, from a set of relations of the form $U_i(x_1, x_2, \dots, x_n) = c_i$, $i = 1, 2, \dots, p$, with U_i functionally independent and c_i constants, one can express (locally at least) p components of x as functions of the other $n - p$. So, (6.1.1) is equivalent (locally) to a system of $n - p$ equations with $n - p$ unknown functions.

In order to be more specific and explicit, let us analyze the following example.

Example 6.1.1. Let us consider the second-order differential equation

$$x'' = g(x),$$

where $g : \mathbb{R} \to \mathbb{R}$ is a continuous function. This equation, obtained from Newton's second law, describes the movement of a material point of mass 1, along the Ox axis, under the action of a force parallel to Ox, and whose intensity at the point of abscissa x is $g(x)$. We mention that $x(t)$ is the position, $x'(t)$ the speed and $x''(t)$ the acceleration of the point at the moment t. We recall that, in accordance with Remark 2.1.2, the preceding equation may be equivalently rewritten as a first-order system of differential equations of the form

$$\begin{cases} x' = y \\ y' = g(x). \end{cases}$$

Multiplying the second equality in this system on both sides by $y = x'$, we deduce

$$\frac{1}{2}\frac{d}{dt}\left(y^2\right) = g(x)x'$$

for every t in the interval $[0, T)$ of existence of the solution. Integrating the equality above on both sides from 0 to t, we get

$$\frac{1}{2}y^2(t) - G(x(t)) = \frac{1}{2}y^2(0) - G(x(0))$$

for every $t \in [0, T)$, where G is a primitive of the function g.

So, the function $U : \mathbb{R}^2 \to \mathbb{R}$, defined by

$$U(x, y) = \frac{1}{2}y^2 - G(x)$$

for $(x, y) \in \mathbb{R}^2$, which is obviously of class C^1 and nonconstant on $\mathbb{R}^2$, remains constant along the trajectories of the system.

Let us observe that the preceding equality may be rewritten in the equivalent form

$$\frac{1}{2}x'^2(t) - G(x(t)) = \frac{1}{2}x'^2(0) - G(x(0))$$

for every $t \in [0, T)$, which asserts that *the total energy of the material point remains constant on the trajectories.*

An advantage of this observation consists in the possibility to reduce the order of the equations by one unit, expressing either x, or x', as a function of the other one by means of the equality $U(x, x') = c$, where c is a real constant.

Another situation quite frequently encountered in applications is the one in which the explicit solving of a system of differential equations is practically impossible, but the determination of one unknown as a function of the other one suffices in order to obtain the information we need. We hope that the next example is convincing enough in this sense.

Example 6.1.2. Let us consider the prey-predator system

$$\begin{cases} x' = (a - ky)x \\ y' = -(b - hx)y \end{cases}$$

and let us assume that we intend to find out the number of individuals from the predator species to a given moment $T > 0$. In order to solve this problem it suffices to know $x(0)$ and $y(0)$, and then to determine explicitly the solution of the corresponding Cauchy problem. Unfortunately, due to the nonlinearity of the system, this way is not easy to go through. Therefore, it is of great importance to find a simpler procedure of getting $y(T)$ avoiding the explicit solving of the Cauchy problem. To this aim, let us assume that we have at our disposal the technical devices to determine the number $x(t)$ of individuals from the prey species at any time t of its evolution. Then, in order to determine $y(T)$, it suffices to express y as a function of x, $x(0)$ and $y(0)$. In the case considered, this is clearly feasible because, considering y as function of class C^1 of x, from the system, we get

$$\frac{dy}{dx} = -\frac{y(b - hx)}{x(a - ky)}.$$

This is an equation with separable variables and its general solution is defined implicitly by

$$hx + ky - b \ln x - a \ln y = c$$

where $x > 0$, $y > 0$ and $c \in \mathbb{R}$. Consequently, in order to find $y(T)$, it suffices to know $x(0) = \xi$, $y(0) = \eta$ and $x(T)$. In these circumstances, we can obtain $y(T)$ solving the equation

$$hx(T) + ky(T) - b \ln x(T) - a \ln y(T) = h\xi + k\eta - b \ln \xi - a \ln \eta.$$

We emphasize the fact of extreme importance that, in order to solve this problem by the method described before, we have to determine only three values $x(0) = \xi$, $y(0) = \eta$ and $x(T)$. Thereafter, *without solving* the corresponding Cauchy problem, we can get $y(T)$ from the equation above.

Definition 6.1.1. Let $\Omega_0 \subset \Omega$ be nonempty and open. A *prime, or first integral* of the system (6.1.1) on Ω_0 is a function $U : \Omega_0 \to \mathbb{R}$ satisfying

(i) U is nonconstant on Ω_0;
(ii) U is of class C^1 on Ω_0;
(iii) for every solution $x : \mathbb{I} \to \Omega_0$ of the system (6.1.1) there exists a constant $c \in \mathbb{R}$ such that $U(x(t)) = c$ for every $t \in \mathbb{I}$.

For $n = 2$, the situation described in Definition 6.1.1 is illustrated in Figure 6.1.1.

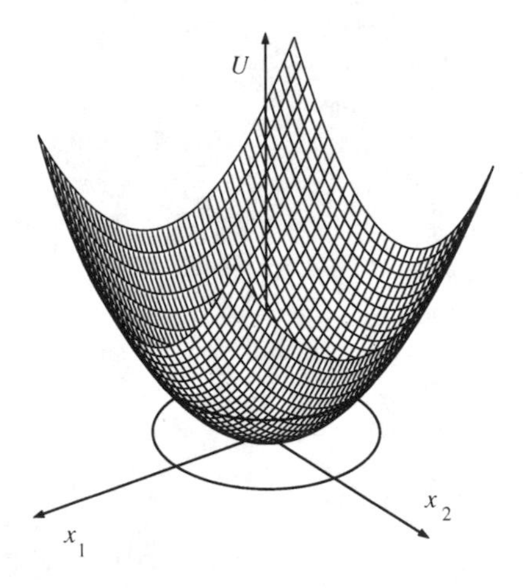

Fig. 6.1.1

Remark 6.1.1. Since (6.1.1) is autonomous, (iii) in Definition 6.1.1 is equivalent to

(iv) for every solution $x : [0, T) \to \Omega_0$ of the system (6.1.1) there exists a constant $c \in \mathbb{R}$ such that $U(x(t)) = c$ for every $t \in [0, T)$.

Theorem 6.1.1. *Let $f : \Omega \to \mathbb{R}^n$ be continuous, let Ω_0 be a nonempty and open subset of Ω and let $U : \Omega_0 \to \mathbb{R}$ be a function of class C^1, nonconstant on Ω_0. The necessary and sufficient condition for U to be a prime integral of* (6.1.1) *on Ω_0 is that*

$$\sum_{i=1}^{n} f_i(\xi)\frac{\partial U}{\partial x_i}(\xi) = 0 \tag{6.1.2}$$

for every $\xi \in \Omega_0$.

Proof. Necessity. Let U be a prime integral of the system (6.1.1) on Ω_0, let $\xi \in \Omega_0$ and let $x(\cdot, 0, \xi) : [0, T_m) \to \Omega_0$ be a saturated solution of the system (6.1.1) which satisfies $x(0, 0, \xi) = \xi$. Since $U(x(t)) = c$ for every $t \in [0, T_m)$, it follows that

$$0 = \frac{d}{dt}(U(x))(t) = \sum_{i=1}^{n} \frac{\partial U}{\partial x_i}(x(t))\frac{dx_i}{dt}(t) = \sum_{i=1}^{n} \frac{\partial U}{\partial x_i}(x(t))f_i(x(t)).$$

Taking $t = 0$ in equality above, we get (6.1.2).

Sufficiency. Let $x : [0, T) \to \Omega_0$ be a solution of the system (6.1.1). Let us define the function $g : [0, T) \to \mathbb{R}$ by $g(t) = U(x(t))$ for every $t \in [0, T)$. Obviously g is of class C^1 and, by virtue of the relation (6.1.2), we have

$$g'(t) = \frac{d}{dt}(U(x))(t) = \sum_{i=1}^{n} \frac{\partial U}{\partial x_i}(x(t))\frac{dx_i}{dt}(t) = \sum_{i=1}^{n} \frac{\partial U}{\partial x_i}(x(t))f_i(x(t)) = 0.$$

Hence $U(x)$ is constant on $[0, T)$, which completes the proof. □

Remark 6.1.2. The condition (6.1.2) has a very suggestive geometrical interpretation. Essentially, it asserts that, for every $\xi \in \Omega_0$ for which $\nabla U(\xi) \neq 0$, the vector $f(\xi)$ is parallel to the tangent plane to the surface of equation $U(x) = U(\xi)$ at ξ. Indeed, the condition (6.1.2) expresses the fact that $f(\xi)$ is orthogonal to $\nabla U(\xi)$ which, in its turn, is orthogonal to the surface $U(x) = U(\xi)$ at the point ξ.

In view of the preceding observation, we have:

Theorem 6.1.2. *Let $f : \Omega \to \mathbb{R}^n$ be continuous, let Ω_0 be a nonempty and open subset of Ω and let $U : \Omega_0 \to \mathbb{R}$ be a function of class C^1 with the property that $\nabla U(x) \neq 0$ on Ω_0. The necessary and sufficient condition for every trajectory of* (6.1.1)*, passing through a point of the surface of constant level, corresponding to any $\xi \in \Omega_0$,*

$$\Sigma_\xi = \{x \in \Omega_0;\ U(x) = U(\xi)\},$$

to remain entirely on this surface is that, for every $\xi \in \Omega_0$ and every $\eta \in \Sigma_\xi$, $f(\eta)$ be parallel to the tangent plane to Σ_ξ at η.

Proof. The condition that, "for every $\xi \in \Omega_0$, all trajectories of equation (6.1.1) starting from the surface Σ_ξ remain entirely in Σ_ξ" is equivalent to the condition that, "the function U be constant on every trajectory of the differential equation (6.1.1) having the initial datum in Ω_0". In accordance with Theorem 6.1.1, the latter condition is equivalent to $\langle f(\xi), \nabla U(\xi)\rangle = 0$ which, in its turn, is equivalent to the condition that, for every $\eta \in \Sigma_\xi$, $f(\eta)$ be parallel to the tangent plane to Σ_ξ at η, and this completes the proof of the theorem. □

We recall that a point $a \in \Omega$ is a *stationary point*, or an *equilibrium point* for the system (6.1.1) if $f(a) = 0$.

Definition 6.1.2. Let $a \in \Omega$ and let Ω_0 be an open neighborhood of a included in Ω. The prime integrals $U_1, U_2, \ldots, U_k : \Omega_0 \to \mathbb{R}$ of the system (6.1.1) are *independent* at a if

$$rank\left(\frac{\partial U_i}{\partial x_j}(a)\right)_{k\times n} = k.$$

The prime integrals $U_1, U_2, \ldots, U_k : \Omega_0 \to \mathbb{R}$ of the system (6.1.1) are *independent* on Ω_0 if they are independent at any $b \in \Omega_0$.

Obviously, (6.1.1) can have at most n prime integrals which are independent at a point $a \in \Omega$. The following theorems bring precise information in this respect, in the case in which $a \in \Omega$ is not a stationary point of the system (6.1.1).

Theorem 6.1.3. *Let $f : \Omega \to \mathbb{R}^n$ be continuous and let $a \in \Omega$ be a non-stationary point of the system* (6.1.1). *Then, on every open neighborhood Ω_0 of a included in Ω, there exist at most $n-1$ prime integrals of the system* (6.1.1) *independent at a.*

Proof. Let us assume, for contradiction, that there exist at least one non-stationary point a of the system (6.1.1) and one open neighborhood Ω_0 of a, included in Ω, such that (6.1.1) has n prime integrals $U_1, U_2, \ldots, U_n$ on Ω_0, which are independent at a. From (6.1.2) in Theorem 6.1.1, it follows that

$$\begin{cases} f_1(a)\dfrac{\partial U_1}{\partial x_1}(a) + f_2(a)\dfrac{\partial U_1}{\partial x_2}(a) + \cdots + f_n(a)\dfrac{\partial U_1}{\partial x_n}(a) = 0 \\ f_1(a)\dfrac{\partial U_2}{\partial x_1}(a) + f_2(a)\dfrac{\partial U_2}{\partial x_2}(a) + \cdots + f_n(a)\dfrac{\partial U_2}{\partial x_n}(a) = 0 \\ \vdots \\ f_1(a)\dfrac{\partial U_n}{\partial x_1}(a) + f_2(a)\dfrac{\partial U_n}{\partial x_2}(a) + \cdots + f_n(a)\dfrac{\partial U_n}{\partial x_n}(a) = 0. \end{cases} \tag{6.1.3}$$

Obviously, (6.1.3) can be interpreted as an algebraic linear homogeneous system with the unknowns $f_1(a), f_2(a), \ldots, f_n(a)$. Taking into account that $U_1, U_2, \ldots, U_n$ are independent at a, it follows that the determinant of this system is nonzero. Consequently the system (6.1.3) admits only the trivial solution $f_1(a) = f_2(a) = \cdots = f_n(a) = 0$, which is in contradiction with the fact that a is non-stationary. This contradiction can be eliminated only if $U_1, U_2, \ldots, U_n$ are not independent at a. The proof is complete. □

If f satisfies some extra-regularity conditions, the preceding result may be considerably improved. More precisely, we have:

Theorem 6.1.4. *Let $f : \Omega \to \mathbb{R}^n$ be a function of class C^1 and let $a \in \Omega$ be a non-stationary point of the system* (6.1.1). *Then, there exists an open neighborhood Ω_0 of a, included in Ω, on which there are defined $n-1$ prime integrals of the system* (6.1.1), *independent on Ω_0.*

Proof. Let $a \in \Omega$ be a non-stationary point of (6.1.1). Relabelling the components of f and those of x if necessary, we may assume with no loss of generality that $f_n(a) \neq 0$. Let

$$\Omega(a) = \{(\lambda_1, \lambda_2, \ldots, \lambda_{n-1}) \in \mathbb{R}^{n-1};\ (\lambda_1, \lambda_2, \ldots, \lambda_{n-1}, a_n) \in \Omega\}.$$

We denote by $\phi : [0, T_m) \times \Omega(a) \to \Omega$ the function defined by

$$\phi(t, \lambda_1, \lambda_2, \ldots, \lambda_{n-1}) = x(t, \lambda_1, \lambda_2, \ldots, \lambda_{n-1}, a_n)$$

for $(t, \lambda_1, \lambda_2, \ldots, \lambda_{n-1}) \in [0, T_m) \times \Omega(a)$, where $x(\cdot, \lambda_1, \lambda_2, \ldots, \lambda_{n-1}, a_n)$ is the unique saturated right solution of (6.1.1) satisfying

$$x(0, \lambda_1, \lambda_2, \ldots, \lambda_{n-1}, a_n) = (\lambda_1, \lambda_2, \ldots, \lambda_{n-1}, a_n).$$

From the fact that $x(\cdot, \lambda_1, \lambda_2, \ldots, \lambda_{n-1}, a_n)$ is a solution of the system (6.1.1), and from (2.6.2) in Theorem 2.6.1, we deduce

$$\frac{D(\phi_1, \phi_2, \ldots, \phi_n)}{D(t, \lambda_1, \ldots, \lambda_{n-1})}(0, a_1, a_2, \ldots, a_{n-1})$$

$$= \begin{vmatrix} \dfrac{\partial x_1}{\partial t} & \dfrac{\partial x_1}{\partial \lambda_1} & \cdots & \dfrac{\partial x_1}{\partial \lambda_{n-1}} \\ \dfrac{\partial x_2}{\partial t} & \dfrac{\partial x_2}{\partial \lambda_1} & \cdots & \dfrac{\partial x_2}{\partial \lambda_{n-1}} \\ \vdots & & & \\ \dfrac{\partial x_n}{\partial t} & \dfrac{\partial x_n}{\partial \lambda_1} & \cdots & \dfrac{\partial x_n}{\partial \lambda_{n-1}} \end{vmatrix} (0, a)$$

$$= \begin{vmatrix} f_1(a) & 1 & 0 & 0 \dots 0 \\ f_2(a) & 0 & 1 & 0 \dots 0 \\ \vdots & & & \\ f_n(a) & 0 & 0 & 0 \dots 0 \end{vmatrix} = (-1)^{n+1} f_n(a) \neq 0. \tag{6.1.4}$$

So, we are in the hypotheses of the local inversion theorem, from where we deduce that there exists an open neighborhood D_0 of $(0, a_1, a_2, \dots, a_{n-1})$, included in $[0, T_m) \times \Omega(a)$, such that ϕ is invertible on D_0, with inverse of class C^1. We denote this inverse by $U : \Omega_0 \to D_0$, $U = (U_0, U_1, \dots, U_{n-1})$, where Ω_0 is an open neighborhood of a included in Ω. From the definition of the function U, it follows

$$\begin{cases} U_0(x(t, \lambda_1, \lambda_2, \dots, \lambda_{n-1}, a_n)) = t \\ U_i(x(t, \lambda_1, \lambda_2, \dots, \lambda_{n-1}, a_n)) = \lambda_i, \ \ i = 1, 2, \dots n-1. \end{cases} \tag{6.1.5}$$

We will prove in what follows that $U_1, U_2, \dots, U_{n-1}$ are prime integrals of the system (6.1.1), defined on Ω_0 and independent at a. To this aim, let us observe that, by virtue of the last $n-1$ relations in (6.1.5), these functions are constants along any solution of (6.1.1) with values in Ω_0, and which at 0 satisfies $x_n(0) = a_n$. Let $\xi \in \Omega_0$. Since U is the inverse of the function ϕ, it follows that $(\tau, \lambda_1, \dots, \lambda_{n-1}) = U(\xi)$ belongs to the set $[0, T_m) \times \Omega(a)$ and $\xi = x(\tau, \lambda_1, \dots, \lambda_{n-1}, a_n)$. So

$$x(t, \xi_1, \xi_2, \dots, \xi_n) = x(t, x(\tau, \lambda_1, \dots, \lambda_{n-1}, a_n)) = x(t + \tau, \lambda_1, \dots, \lambda_{n-1}, a_n).$$

From this equality and from (6.1.5), we deduce that

$$U_i(x(t, \xi_1, \xi_2, \dots, \xi_n)) = U_i(x(t + \tau, \lambda_1, \dots, \lambda_{n-1}, a_n)) = \lambda_i$$

for $i = 1, 2, \dots, n-1$. Consequently, U_i, $i = 1, 2, \dots, n-1$ are of class C^1 and remain constants on all the trajectories of the system (6.1.1) included in Ω_0.

Finally, as U satisfies (6.1.4) on Ω_0, it follows that the functions U_i, $i = 1, 2, \dots, n-1$ are non-constant and independent on Ω_0. See [Nicolescu *et al.* (1971a)], Corollary p. 677 or [Rudin (1976)], Theorem 9.32, p. 229. The proof is complete. □

Theorem 6.1.5. *Let $f : \Omega \to \mathbb{R}^n$ be a continuous function, let $a \in \Omega$ be a non-stationary point of the system (6.1.1) and let $U_1, U_2, \ldots, U_{n-1}$ be prime integrals of (6.1.1), defined on an open neighborhood Ω_0 of a included in Ω, and independent at a. Then, for every prime integral $U : \Omega_0 \to \mathbb{R}$ of the system (6.1.1), there exist an open neighborhood $\Omega_1 \subset \Omega_0$ of the point a, an open subset D in $\mathbb{R}^{n-1}$, with $(U_1(a), U_2(a), \ldots, U_{n-1}(a)) \in D$, and a function of class C^1 $F : D \to \mathbb{R}$, such that*

$$U(x) = F(U_1(x), U_2(x), \ldots, U_{n-1}(x))$$

for every $x \in \Omega_1$.

Proof. Since $U_1, U_2, \ldots, U_{n-1}$ are independent at a, it follows that for every open neighborhood $\Omega_1 \subset \Omega_0$ of a there exists at least one function of class C^1, $U_n : \Omega_1 \to \mathbb{R}$, such that

$$det\left(\frac{\partial U_i}{\partial x_j}(a)\right)_{n\times n} \neq 0.$$

A simple example of such function is $U_n(x) = x_j$ for each $x \in \Omega_1$, where $j \in \{1, 2, \ldots, n\}$ is so that the determinant obtained from the matrix

$$\left(\frac{\partial U_i}{\partial x_j}(a)\right)_{(n-1)\times n}$$

by cancelling the column j is nonzero.

Clearly the open neighborhood Ω_1 of a can be chosen such that, the transformation $G = (U_1, U_2, \ldots, U_n)$ is a diffeomorphism from Ω_1 to an open set Δ in $\mathbb{R}^n$. Let $H : \Delta \to \Omega_1$ be the inverse of this transformation, and let us observe that

$$U(H(U_1(x), U_2(x), \ldots, U_n(x))) = U(x)$$

for every $x \in \Omega_1$. So, denoting by $F = U \circ H$, in order to complete the proof, it suffices to show that F, defined as above, does not depend on the last variable y_n. Let us observe that

$$\frac{\partial F}{\partial y_n}(y) = \sum_{i=1}^{n} \frac{\partial U}{\partial x_i}(H(y))\frac{\partial H_i}{\partial y_n}(y). \tag{6.1.6}$$

Recalling that $U_1, U_2, \ldots, U_{n-1}, U$ are prime integrals of the system (6.1.1) on Ω_0, by virtue of Theorem 6.1.1, it follows that

$$\begin{cases} f_1(x)\dfrac{\partial U_1}{\partial x_1}(x) + f_2(x)\dfrac{\partial U_1}{\partial x_2}(x) + \cdots + f_n(x)\dfrac{\partial U_1}{\partial x_n}(x) = 0 \\ \\ f_1(x)\dfrac{\partial U_2}{\partial x_1}(x) + f_2(x)\dfrac{\partial U_2}{\partial x_2}(x) + \cdots + f_n(x)\dfrac{\partial U_2}{\partial x_n}(x) = 0 \\ \vdots \\ f_1(x)\dfrac{\partial U}{\partial x_1}(x) + f_2(x)\dfrac{\partial U}{\partial x_2}(x) + \cdots + f_n(x)\dfrac{\partial U}{\partial x_n}(x) = 0. \end{cases}$$

Since a is a non-stationary point, we have $f(a) \neq 0$, and therefore we can choose an open neighborhood $\Omega_1 \subset \Omega_0$ of a such that

$$f(x) \neq 0 \quad \textit{and} \quad \textit{rank}\left(\frac{\partial U_i}{\partial x_j}(x)\right)_{(n-1)\times n} = n-1$$

for every $x \in \Omega_1$. In these conditions, interpreting the system above as a linear and homogeneous system with unknowns $f_1(x), f_2(x), \ldots, f_n(x)$, it follows that its determinant is identically zero on Ω_1. Since this determinant $D(x)$ has at least one minor of order $n-1$, whose first $n-1$ rows correspond to the first $n-1$ rows of $D(x)$, which is nonzero, it follows that the last row of $D(x)$ is a linear combination of the others. More precisely, there exist the functions $a_i : \Omega_1 \to \mathbb{R}$ with $i \in \{1, 2, \ldots, n-1\}$ such that

$$\frac{\partial U}{\partial x_j}(x) = \sum_{i=1}^{n-1} a_i(x) \frac{\partial U_i}{\partial x_j}(x)$$

for every $j \in \{1, 2, \ldots, n\}$ and $x \in \Omega_1$. From (6.1.6), using these equalities, we deduce

$$\frac{\partial F}{\partial y_n}(y) = \sum_{j=1}^{n} \sum_{i=1}^{n-1} a_i(H(y)) \frac{\partial U_i}{\partial x_j}(H(y)) \frac{\partial H_j}{\partial y_n}(y).$$

Observing that, from the definition of H, we have $x = H(y)$ if and only if $y = (U_1(x), U_2(x), \ldots, U_n(x))$, we conclude that

$$\frac{\partial U_i}{\partial x_j}(H(y)) \frac{\partial H_j}{\partial y_n}(y) = \frac{\partial y_i}{\partial y_n} = 0$$

for $i = 1, 2, \ldots, n-1$. Consequently

$$\frac{\partial F}{\partial y_n}(y) = 0$$

for every $y \in \Delta$. Since Δ can be chosen convex (diminishing the set Ω_1 if necessary), and $D = \{(y_1, \ldots, y_{n-1});\ (y_1, \ldots, y_{n-1}, y_n) \in \Delta\}$, this relation proves that F does not depend on y_n, which completes the proof. □

Remark 6.1.3. If we know p prime integrals of the differential system (6.1.1) which are independent at a non-stationary point $a \in \Omega$, then there exists one neighborhood of a on which the system (6.1.1) is equivalent to another differential system with $n-p$ unknown functions. In particular, for $p = n-1$, there exists a neighborhood of a on which the system (6.1.1) is equivalent to a scalar differential equation, i.e. with only one unknown function. Indeed, let $U_1, U_2, \ldots, U_p : \Omega_0 \to \mathbb{R}$ be those p prime integrals of (6.1.1), independent at a, and let $x : \mathbb{I} \to \Omega_0$ be a generic solution of the system (6.1.1). Taking into account that there exist the constants $c_1, c_2, \ldots, c_p$ such that

$$U_j(x_1, x_2, \ldots, x_n) = c_j, \quad j = 1, 2, \ldots, p,$$

by virtue of the fact that $U_1, U_2, \ldots, U_p$ are independent at a, and of the implicit functions theorem, it follows that there exists a neighborhood of a, on which, p components of x can be uniquely expressed as functions of class C^1 of the other $n - p$ components. Relabelling if necessary, we may assume that those components which express as functions of the others are the last p. Substituting these components of x in the first $n - p$ equations of (6.1.1), we get a differential system with $n - p$ unknown functions.

6.2 Prime Integrals for Non-Autonomous Systems

In this section we will extend the preceding considerations to the case of non-autonomous systems of the form

$$x' = f(t, x), \tag{6.2.1}$$

where $f : \mathbb{I} \times \Omega \to \mathbb{R}^n$ is a continuous function, by reducing these to the autonomous case. More precisely, let $D = \mathbb{I} \times \Omega \subset \mathbb{R}^{n+1}$, let

$$z = \begin{pmatrix} t \\ x \end{pmatrix},$$

and let $F : D \to \mathbb{R}^{n+1}$ be defined by

$$F(z) = \begin{pmatrix} 1 \\ f(t, x) \end{pmatrix}$$

for every $z \in D$. Obviously, F is of class C^1 and (6.2.1) may be equivalently rewritten in the autonomous form

$$z' = F(z). \tag{6.2.2}$$

Taking into consideration the equivalence between (6.2.1) and (6.2.2), we will define the concept of prime integral for (6.2.1) as follows.

Definition 6.2.1. Let $D_0 \subset \mathbb{I} \times \Omega$ be nonempty and open. A function $U : D_0 \to \mathbb{R}$ is called *prime integral* of the system (6.2.1) on D_0 if

(i) U is nonconstant on D_0;
(ii) U is of class C^1 on D_0;

(iii) for every solution $x : \mathbb{J} \to \Omega$ of (6.2.1) with $(t, x(t)) \in D_0$ for every $t \in \mathbb{J}$, there exists $c \in \mathbb{R}$ such that $U(t, x(t)) = c$ for every $t \in \mathbb{J}$.

We state next some of the most important results referring to prime integrals for systems of type (6.2.1). Since, due to the equivalence between (6.2.1) and (6.2.2), all these results are consequences of the theorems proved in the autonomous case, we do not give proofs.

Theorem 6.2.1. *Let $f : \mathbb{I} \times \Omega \to \mathbb{R}^n$ be continuous, let D_0 be a nonempty and open subset of $\mathbb{I} \times \Omega$ and let $U : D_0 \to \mathbb{R}$ be a function of class C^1, nonconstant on D_0. The necessary and sufficient condition for U to be a prime integral for* (6.2.1) *is that*

$$\frac{\partial U}{\partial t}(s,\xi) + \sum_{i=1}^{n} f_i(s,\xi)\frac{\partial U}{\partial x_i}(s,\xi) = 0$$

for every $(s,\xi) \in D_0$.

Due to the particular form of the function F, it follows that every point in D is non-stationary. So we have:

Theorem 6.2.2. *Let $f : \mathbb{I} \times \Omega \to \mathbb{R}^n$ be continuous. Then on every open neighborhood of any point in $\mathbb{I} \times \Omega$, there exist at most n prime integrals of the system* (6.2.1) *which are independent at a.*

Theorem 6.2.3. *Let $f : \mathbb{I} \times \Omega \to \mathbb{R}^n$ be a function of class C^1. Then for every $(s, a) \in \mathbb{I} \times \Omega$ there exists an open neighborhood D_0 of (s, a), included in $\mathbb{I} \times \Omega$, on which there are defined n prime integrals of the system* (6.2.1) *which are independent on D_0.*

6.3 First Order Partial Differential Equations

Let Ω be a nonempty and open subset of $\mathbb{R}^3$, let $f : \Omega \to \mathbb{R}^3$ be a function of class C^1 (vector field) and let us consider the following problem with geometrical character: *determine all surfaces Σ of class C^1 in $\mathbb{R}^3$ with the property that at any point of coordinates $(x_1, x_2, x_3) \in \Sigma$, $f(x_1, x_2, x_3)$ is parallel to the tangent plane to the surface.* From the formulation of the problem itself, we are led to look for these surfaces either explicitly, i.e.

$$x_3 = x_3(x_1, x_2) \tag{$\mathcal{E}$}$$

with (x_1, x_2) in a nonempty and open subset D in $\mathbb{R}^2$, or implicitly, i.e.

$$\phi(x_1, x_2, x_3) = c, \tag{$\mathcal{I}$}$$

where $\phi : \Omega \to \mathbb{R}$ is of class C^1, while $c \in \mathbb{R}$.

Let us remark that, a necessary and sufficient condition for a surface Σ to have the desired property is that, at every point (x_1, x_2, x_3) in Σ, the normal vector to Σ at that point, $n(x_1, x_2, x_3)$, be orthogonal to $f(x_1, x_2, x_3)$. This condition may be equivalently written as

$$\langle f(x_1, x_2, x_3), N(x_1, x_2, x_3)\rangle = 0$$

where $N(x_1, x_2, x_3)$ is any vector parallel to $n(x_1, x_2, x_3)$. So, if we decide to find those surfaces in the explicit form $(\mathcal{E})$, taking into account that, in that case, $N(x_1, x_2, x_3)$ can be taken as

$$N(x_1, x_2, x_3) = \left(\frac{\partial x_3}{\partial x_1}(x_1, x_2), \frac{\partial x_3}{\partial x_2}(x_1, x_2), -1\right),$$

the necessary and sufficient condition above can be rewritten in the form

$$\sum_{i=1}^{2} f_i(x_1, x_2, x_3(x_1, x_2))\frac{\partial x_3}{\partial x_i}(x_1, x_2) = f_3(x_1, x_2, x_3(x_1, x_2)) \tag{6.3.1}$$

for every $(x_1, x_2) \in D$.

If we choose the implicit variant $(\mathcal{I})$, as in this case a normal vector to the surface is

$$N(x_1, x_2, x_3) = \left(\frac{\partial \phi}{\partial x_1}(x_1, x_2, x_3), \frac{\partial \phi}{\partial x_2}(x_1, x_2, x_3), \frac{\partial \phi}{\partial x_3}(x_1, x_2, x_3)\right),$$

the previous necessary and sufficient condition takes the form

$$\sum_{i=1}^{3} f_i(x_1, x_2, x_3)\frac{\partial \phi}{\partial x_i}(x_1, x_2, x_3) = 0 \tag{6.3.2}$$

for every $(x_1, x_2, x_3) \in \Omega$.

So, the determination of these surfaces reduces to the determination, either of all functions x_3, of class C^1, satisfying (6.3.1), or of all functions ϕ, also of class C^1, satisfying (6.3.2). Hence, in order to solve the problem, we have to solve an equation in which the unknown function is involved together with its first-order partial derivatives. In what follows we will present the most important results referring to such kind of equations.

Let Ω be a nonempty and open subset of $\mathbb{R}^{n+1}$ and let $f_i, f : \Omega \to \mathbb{R}$ with $i = 1, 2, \ldots, n$, be functions of class C^1.

Definition 6.3.1. A *first-order quasi-linear partial differential equation* is an equation of the form

$$\sum_{i=1}^{n} f_i(x, z(x))\frac{\partial z}{\partial x_i}(x) = f(x, z(x)), \tag{6.3.3}$$

where

$$\sum_{i=1}^{n} f_i^2(x,z) \neq 0$$

at least for one $(x,z) \in \Omega$. A *solution* of (6.3.3) is a function $z : D \to \mathbb{R}$ of class C^1, with D nonempty and open in $\mathbb{R}^n$, such that $(x, z(x)) \in \Omega$ and z satisfies (6.3.3) for every $x \in D$. The set of all solutions of equation (6.3.3) is called *the general solution* of (6.3.3).

If $f = 0$ on Ω and f_i, $i = 1, 2, \ldots, n$, do not depend on z, equation (6.3.3) is called *linear*. More precisely, let D be a nonempty and open subset of $\mathbb{R}^n$ and let $f_i : D \to \mathbb{R}$, $i = 1, 2, \ldots, n$, be functions of class C^1 on D.

Definition 6.3.2. A *first-order linear partial differential equation* is an equation of the form

$$\sum_{i=1}^{n} f_i(x) \frac{\partial \phi}{\partial x_i}(x) = 0. \tag{6.3.4}$$

A *solution* of this equation is a function $\phi : D_0 \to \mathbb{R}$, of class C^1, with D_0 nonempty and open in D, such that ϕ satisfies (6.3.4). The set of all solutions of (6.3.4) is called *the general solution* of equation (6.3.4).

Obviously every constant function on D is a solution of equation (6.3.4). Therefore, in all what follows, we will refer only to solutions of equation (6.3.4) which are nonconstant.

We begin with the study of equation (6.3.4). Then, we will show how the study of the problem (6.3.3) reduces to that of a problem of the type (6.3.4). The price we have to pay in order to do that consists in the introducing of a new unknown function of $n+1$ variables, $\phi(x_1, x_2, \ldots, x_n, z)$, which by means of $\phi(x_1, x_2, \ldots, x_n, z) = 0$ implicitly defines z as a function of $x_1, x_2, \ldots, x_n$.

Definition 6.3.3. The differential system

$$x_i' = f_i(x), \quad i = 1, 2, \ldots, n \tag{6.3.5}$$

is called *the characteristic system* attached to the linear equation (6.3.4).

Remark 6.3.1. From traditional reasons, very frequently, this system is formally written under the so-called *symmetric form*

$$\frac{dx_1}{f_1(x)} = \frac{dx_2}{f_2(x)} = \cdots = \frac{dx_n}{f_n(x)}. \tag{6.3.6}$$

We make the convention that whenever, for some $i = 1, 2, \ldots, n$, $f_i \equiv 0$ on a certain open subset Ω_0 in Ω, the "fraction" $dx_i/0$ in (6.3.6) should be interpreted as $dx_i \equiv 0$ on Ω_0.

We begin with the following reformulation of Theorem 6.1.1.

Theorem 6.3.1. *Let D_0 be a nonempty and open subset in D, and let $\phi : D_0 \to \mathbb{R}$ be a nonconstant function of class C^1. The necessary and sufficient condition for ϕ to be a solution of equation* (6.3.4) *is that ϕ be a prime integral, on D_0, of the characteristic system* (6.3.5).

An immediate consequence of Theorem 6.1.5 is:

Theorem 6.3.2. *Let $a \in D$ be a non-stationary point of the characteristic system* (6.3.5), *let D_0 be an open neighborhood of a, included in D, and let $U_1, U_2, \ldots, U_{n-1} : D_0 \to \mathbb{R}$ be prime integrals of the system* (6.3.5), *independent at a. Then, there exists an open neighborhood $D_1 \subset D_0$ of a, such that the general solution of equation* (6.3.4) *on D_1 is given by*

$$\phi(x) = F(U_1(x), U_2(x), \ldots, U_{n-1}(x))$$

for $x \in D_1$, where F belongs to the set of all real-valued functions of class C^1, defined on the range of $U = (U_1, U_2, \ldots, U_{n-1}) : D_1 \to \mathbb{R}^{n-1}$.

Example 6.3.1. Find the general solution of the equation

$$(x_2 - x_3)\frac{\partial \phi}{\partial x_1} + (x_3 - x_1)\frac{\partial \phi}{\partial x_2} + (x_1 - x_2)\frac{\partial \phi}{\partial x_3} = 0$$

on the set of all non-stationary points. The characteristic system in the symmetric form is

$$\frac{dx_1}{x_2 - x_3} = \frac{dx_2}{x_3 - x_1} = \frac{dx_3}{x_1 - x_2}.$$

We have $dx_1 + dx_2 + dx_3 = 0$ and $x_1 dx_1 + x_2 dx_2 + x_3 dx_3 = 0$. So, the functions $U_1, U_2 : \mathbb{R}^3 \to \mathbb{R}$, defined by $U_1(x_1, x_2, x_3) = x_1 + x_2 + x_3$ and by $U_2(x_1, x_2, x_3) = x_1^2 + x_2^2 + x_3^2$ respectively, are prime integrals for this system. The stationary points of the system are of the form (x_1, x_2, x_3) with $x_1 = x_2 = x_3$. One may easily see that the prime integrals above are independent at any of the non-stationary points. So, the general solution of the equation is $\phi(x_1, x_2, x_3) = F(x_1 + x_2 + x_3, x_1^2 + x_2^2 + x_3^2)$, where $F : \mathbb{R}^2 \to \mathbb{R}$ is a function of class C^1.

As we can state from the example at the beginning of this section, a function x_3, of class C^1, implicitly defined by a relation of the form $\phi(x_1, x_2, x_3) = c$ is a solution of the problem (6.3.1) if and only if ϕ is a solution of the problem (6.3.2). This observation suggests to look for the solution of the problem (6.3.3) as a function z, implicitly defined by a relation of the form $\phi(x, z) = c$. From the theorem on the differentiation of implicitly defined functions, we have

$$\frac{\partial z}{\partial x_i}(x) = -\frac{\dfrac{\partial \phi}{\partial x_i}}{\dfrac{\partial \phi}{\partial z}}(x, z(x))$$

for every $i = 1, 2, \ldots, n$. Substituting $\partial z / \partial x_i$ in (6.3.3) and eliminating the denominator, we get

$$\sum_{i=1}^{n} f_i(x, z)\frac{\partial \phi}{\partial x_i}(x, z) + f(x, z)\frac{\partial \phi}{\partial z}(x, z) = 0, \tag{6.3.7}$$

equation which is of the type (6.3.4). From Theorem 6.3.2 we deduce

Theorem 6.3.3. *Let $(a, \zeta) \in \Omega$ be a non-stationary point of the characteristic system*

$$\begin{cases} x_i' = f_i(x, z), & i = 1, 2, \ldots, n \\ z' = f(x, z) \end{cases} \tag{6.3.8}$$

attached to equation (6.3.7), let Ω_0 be an open neighborhood of the point (a, ζ), included in Ω, and let $U_1, U_2, \ldots, U_n : \Omega_0 \to \mathbb{R}$ be prime integrals of the system (6.3.8), independent at the point (a, ζ). Then there exists an open neighborhood $\Omega_1 \subset \Omega_0$ of (a, ζ) such that the general solution of equation (6.3.3) on Ω_1 is defined implicitly by

$$F(U_1(x, z(x)), U_2(x, z(x)), \ldots, U_n(x, z(x))) = c,$$

where F belongs to the set of all real-valued functions of class C^1 defined on the range of the transformation $U = (U_1, U_2, \ldots, U_n) : \Omega_1 \to \mathbb{R}^n$, while c belongs to $\mathbb{R}$.

6.4 The Cauchy Problem for Quasi-Linear Equations

In this section we prove an existence and uniqueness result concerning the solution of a first-order quasi-linear partial differential equation, solution which satisfies a certain generalized Cauchy condition. In order to understand the geometrical significance of this condition, we will consider first the specific case corresponding to the dimension $n = 2$. So, let $\Omega \subset \mathbb{R}^3$ be a nonempty and open subset, let $f_1, f_2, f : \Omega \to \mathbb{R}$ be functions of class C^1 and let us consider the equation

$$\sum_{i=1}^{2} f_i(x_1, x_2, z(x_1, x_2)) \frac{\partial z}{\partial x_i}(x_1, x_2) = f(x_1, x_2, z(x_1, x_2)). \tag{6.4.1}$$

Let Γ be a curve of class C^1, included in Ω. The *Cauchy problem* for equation (6.4.1) on the curve Γ consists in: *the determination of a nonempty and open subset $D \subset \mathbb{R}^2$ and of a surface Σ of equation $z = z(x_1, x_2)$ with $(x_1, x_2) \in D$, where $z : D \to \mathbb{R}$ is a solution of equation* (6.4.1) *on D with the property that Γ is contained in Σ*. See Figure 6.4.1.

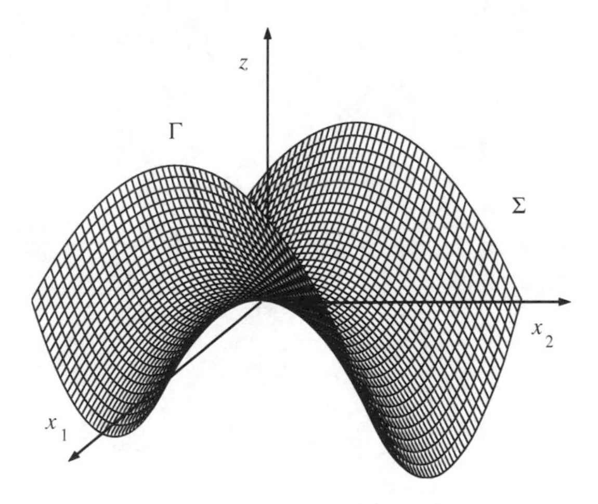

Fig. 6.4.1

Let us observe that this problem has no solution in the case in which Γ is contained in no surface defined by a solution of (6.4.1), and it has at least two solutions when Γ is defined as an intersection of two surfaces defined by two distinct solutions of equation (6.4.1). Finally, it has exactly one solution when Γ is contained in one and only one surface defined by a solution of (6.4.1). So, the problem above might be successfully approached only if Γ lies in none of the first two situations described above. If this is the case, the idea to solve the problem is the following. Let (ξ, ζ) be a current point on Γ, $\xi = (\xi_1, \xi_2)$ and let us consider the Cauchy problem for the characteristic system attached to equation (6.4.1)

$$\begin{cases} x_i' = f_i(x,z), & i = 1,2 \\ z' = f(x,z) \\ x(0) = \xi, & z(0) = \zeta. \end{cases}$$

This problem admits a unique local solution, whose graph is, according to Theorem 6.3.3, a curve $\Gamma(\xi, \zeta)$ defined as the intersection of two surfaces $z_i = z_i(x_1, x_2)$, with z_i, $i = 1, 2$, independent solutions of equation (6.4.1). Such a curve is called *characteristic curve*. So, if (ξ, ζ) moves on Γ, the family of characteristic curves $\Gamma(\xi, \zeta)$ describes a surface Σ, which, as we shall see later, is explicitly defined by an equation $z = z(x_1, x_2)$ with z a solution of (6.4.1), and which, obviously, contains Γ. See also Figure 6.4.1.

We can now proceed to the presentation of the problem in the general case. Let Ω be a nonempty and open subset of $\mathbb{R}^{n+1}$, let $f_i, f : \Omega \to \mathbb{R}$ with $i = 1, 2, \ldots, n$ be functions of class C^1, and let us consider the first-order quasi-linear partial differential equation

$$\sum_{i=1}^{n} f_i(x, z(x)) \frac{\partial z}{\partial x_i}(x) = f(x, z(x)). \tag{6.4.2}$$

Let U be a nonempty and open subset of $\mathbb{R}^{n-1}$, and let us consider the $(n-1)$-dimensional manifold Γ of equations

$$\begin{cases} x = \varphi(u) \\ z = \theta(u), \quad u = (u_1, u_2, \ldots, u_{n-1}) \in U, \end{cases} \tag{6.4.3}$$

where $\varphi : U \to \mathbb{R}^n$ and $\theta : U \to \mathbb{R}$ are of class C^1 and satisfy

$$rank \left(\frac{\partial \varphi_i}{\partial u_j} \right)_{n \times (n-1)} = n - 1.$$

We also assume that Γ is included in Ω, i.e. $\{(\varphi(u), \theta(u))\,;\; u \in U\} \subset \Omega$. *The Cauchy problem* (6.4.3) *for equation* (6.4.2) consists in finding a solution $z : D \to \mathbb{R}$ of (6.4.2), which contains the manifold Γ, i.e.

$$\theta(u) = z(\varphi(u))$$

for every $u \in U$. Such a solution z of equation (6.4.2), defined on $D \subset \mathbb{R}^n$, with the property that the set $\Sigma_0 = \{(x, z(x))\,;\; x \in D\}$ is a neighborhood of the manifold Γ, is called *a local solution* of the Cauchy problem (6.4.3) for equation (6.4.2).

Theorem 6.4.1. *If $\varphi : U \to \mathbb{R}^n$ and $\theta : U \to \mathbb{R}$ satisfy*

$$\Delta(u) = \begin{vmatrix} f_1(\varphi,\theta) & \dfrac{\partial \varphi_1}{\partial u_1} & \cdots & \dfrac{\partial \varphi_1}{\partial u_{n-1}} \\ f_2(\varphi,\theta) & \dfrac{\partial \varphi_2}{\partial u_1} & \cdots & \dfrac{\partial \varphi_2}{\partial u_{n-1}} \\ \vdots & & & \\ f_n(\varphi,\theta) & \dfrac{\partial \varphi_n}{\partial u_1} & \cdots & \dfrac{\partial \varphi_n}{\partial u_{n-1}} \end{vmatrix} (u) \neq 0$$

for every $u \in U$, then there exists $D \subset \mathbb{R}^n$ such that the Cauchy problem (6.4.3) *for equation* (6.4.2) *have a unique local solution defined on D.*

Proof. [2] Let $u \in U$ be arbitrary, and let us consider the Cauchy problem for the characteristic system attached to equation (6.4.2) with data $(\varphi(u), \theta(u)) \in \Omega$, i.e.

$$\begin{cases} x_i' = f_i(x,z), \quad i = 1,2,\ldots,n \\ z' = f(x,z) \\ x(0) = \varphi(u), \quad z(0) = \theta(u). \end{cases} \tag{6.4.4}$$

This problem admits a unique saturated solution

$$\begin{cases} X = X(\cdot, \varphi(u), \theta(u)) \\ Z = Z(\cdot, \varphi(u), \theta(u)) \end{cases}$$

defined on a neighborhood of the origin (t_u, T_u).

Let us consider the map $x : \{(t,u) \in \mathbb{R} \times U\,;\; u \in U,\; t \in (t_u, T_u)\} \to \mathbb{R}^n$ defined by

$$x(t,u) = X(t, \varphi(u), \theta(u)) \tag{6.4.5}$$

and let us observe that, in view of Theorem 2.6.2, we have

$$\frac{D(x_1, x_2, \ldots, x_n)}{D(s, u_1, \ldots, u_{n-1})}(0,u) = \Delta(u) \neq 0$$

for every $u \in U$. The determinant above being jointly continuous, there exists an open set V, of the form

$$\{(t,u) \in \mathbb{R} \times U\,;\; u \in U,\; t \in (t_u^*, T_u^*)\},$$

with $t_u \leq t_u^* < 0 < T_u^* \leq T_u$ for every $u \in U$, such that

$$\frac{D(x_1, x_2, \ldots, x_n)}{D(s, u_1, \ldots, u_{n-1})}(t,u) \neq 0$$

[2] The method of proof, due to Cauchy, relies on the generation of the surface, i.e. solution, by the family of characteristic curves supported by the manifold Γ and, from this reason, is called *the characteristic method* or *the Cauchy's method.*

for every $(t,u) \in V$. From the local inversion theorem, it follows that, the map (6.4.5) is a local diffeomorphism from V into an open subset D in $\mathbb{R}^n$. We denote by $\Phi : D \to V$ the local inverse of this application. Then we have

$$\begin{cases} u_i = \Phi_i(x), & i = 1, 2, \ldots, n-1 \\ t = \Phi_n(x), \end{cases} \tag{6.4.6}$$

for $x \in D$. Substituting u_i and t given by (6.4.6) in

$$z(t,u) = Z(t, \varphi(u), \theta(u)),$$

we obtain the explicit equation

$$z = z(x_1, x_2, \ldots, x_n)$$

of a surface Σ, which obviously contains the manifold Γ.

We will prove in what follows that z is a solution of equation (6.4.2), and that it is the only one defined on D. To this aim, let us observe that, for every $x \in D$, $x = x(t,u)$, we have

$$f(x, z(x)) = \frac{dz(x)}{dt} = \sum_{i=1}^{n} \frac{\partial z}{\partial x_i}(x) \frac{\partial x_i}{\partial t} = \sum_{i=1}^{n} \frac{\partial z}{\partial x_i}(x) f_i(x, z(x)),$$

which shows that z is a solution of equation (6.4.2) on D.

In order to prove the uniqueness on D of the solution z of the Cauchy problem (6.4.3) for equation (6.4.2), let $w : D \to \mathbb{R}^n$ be another solution of this problem. Let us consider the Cauchy problem

$$\begin{cases} x_i'(t) = f_i(x(t), w(x(t))) \quad i = 1, 2, \ldots, n \\ x(0) = \varphi(u), \end{cases}$$

which has a unique saturated solution x defined on $[0, T)$. Since w is a solution of equation (6.4.2), we have

$$\frac{dw(x)}{dt} = \sum_{i=1}^{n} \frac{\partial w}{\partial x_i}(x) \frac{\partial x_i}{\partial t} = \sum_{i=1}^{n} \frac{\partial w}{\partial x_i}(x) f_i(x, w(x)) = f(x, w(x))$$

for every $t \in [0, T)$. Inasmuch as $w(x(0, \varphi(u))) = \theta(u)$, it follows that the pair $(x, w) : [0, T) \to \Omega$ is a solution of the Cauchy problem (6.4.4). But this problem has the unique solution (x, w) given by

$$x(t,u) = X(t, \varphi(u), \theta(u)) \quad \textit{and} \quad w(t,u) = Z(t, \varphi(u), \theta(u))$$

for every $t \in [0, T_u)$. Hence $w(x) = z(x)$ for every $x \in D$ and the proof is complete. □

6.5 Conservation Laws

In this section we present a first-order partial differential equation which describes the evolution of several phenomena whose characteristic feature is the conservation of a certain physical property as for instance: the mass, the energy, the kinetic momentum, *etc.*, on the whole duration of the evolution.

6.5.1 *Some Examples*

To fix the ideas, let us analyze:

Example 6.5.1. Let us consider a highway which, for simplicity, will be assumed infinite and oriented in the direction of the Oy axis, and let us denote by $\rho(t,y)$ the density of vehicles at the point $y \in \mathbb{R}$ and moment $t \in \mathbb{R}_+$. Let us denote also by $v(t,y)$ the speed of the vehicles in traffic at the point y and moment t. Then, the flux of vehicles at the point y and moment t is given by

$$q(t,y) = \rho(t,y)v(t,y). \tag{6.5.1}$$

We may assume that the speed is a function of density, i.e. $v = V(\rho)$. We emphasize also the fact, very important for the point of view of future considerations, that on the highway there is a unique sense. Namely, let us assume that sense allowed is the positive sense of the Oy axis. Then, by observing that, for every $y \in \mathbb{R}$ and every $\delta > 0$, all the vehicles which pass through y reach $y + \delta$ and leave, and that the rate of decrease of the number of vehicles, situated at the moment t on the part $[\,y, y+\delta\,]$ of the highway, equals the variation of the flux on that portion, we deduce

$$\frac{\partial}{\partial t}\left(\int_y^{y+\delta} \rho(t,x)\,dx\right) + q(t,y+\delta) - q(t,y) = 0$$

for every $y \in \mathbb{R}$ and every $t \in \mathbb{R}_+$. Dividing the equality above by $\delta > 0$ and passing to the limit for δ tending to 0, we get

$$\frac{\partial \rho}{\partial t} + \frac{\partial q}{\partial y} = 0.$$

From this equation and from (6.5.1), denoting by $W(\rho) = V(\rho) + \rho V'(\rho)$, it follows

$$\frac{\partial \rho}{\partial t} + W(\rho)\frac{\partial \rho}{\partial y} = 0$$

for every $y \in \mathbb{R}$ and every $t \in \mathbb{R}_+$, equation known under the name of *the traffic equation.*

Example 6.5.2. Let us consider an infinite channel which, for simplicity, is modelled as a straight line having the same direction and sense as the Oy axis. The channel contains water which is flowing in the positive sense of the Oy axis. We assume that the transversal section of this channel is a rectangle whose width equals 1. Let us assume also that, for every $y \in \mathbb{R}$ and every $t \in \mathbb{R}_+$, the flux of water per/unit of surface on the section through the point y and at the moment t is constant on the entire section and, more than this, the height of the water in the channel at the moment t is also constant on the whole section through the point y, i.e. it does not depend on the x-variable. See Figure 6.5.1. We denote by $q(t, y)$ the flux of

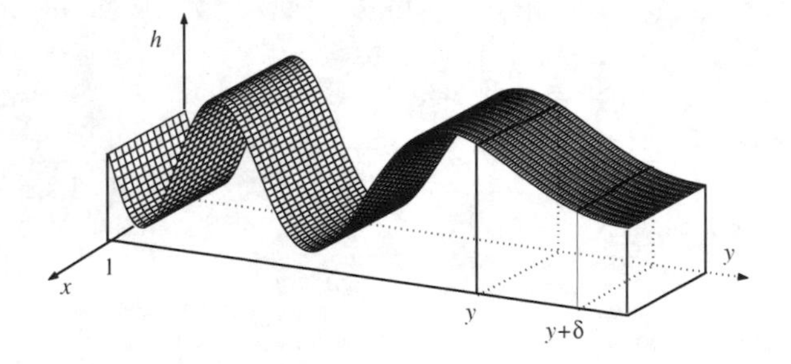

Fig. 6.5.1

the water through the section at the point y and moment t, and by $h(t, y)$ the height of the water at the same section and same moment.

The mathematical expression of the fact that, for every $y \in \mathbb{R}$ and every $t \in \mathbb{R}_+$, the rate of decrease of the mass of water on the portion of the channel $[\,y, y + \delta\,]$ equals the variation of the flux at the two endpoints of the interval, is

$$\frac{\partial}{\partial t}\left(\int_y^{y+\delta} h(t, x)\, dx\right) + q(t, y + \delta) - q(t, y) = 0$$

for every $y \in \mathbb{R}$ and $t \in \mathbb{R}_+$. See also Figure 6.5.1.

Dividing by δ the equality above and making δ tend to 0, we get

$$\frac{\partial h}{\partial t} + \frac{\partial q}{\partial y} = 0$$

for every $y \in \mathbb{R}$ and $t \in \mathbb{R}_+$. Now, taking into account that the flux is a function of the height h, i.e. $q = Q(h)$, where $Q : \mathbb{R}_+ \to \mathbb{R}$, from the last equation, it follows

$$\frac{\partial h}{\partial t} + Q'(h)\frac{\partial h}{\partial y} = 0$$

for every $y \in \mathbb{R}$ and $t \in \mathbb{R}_+$. Experimental considerations show that the function Q is of the form $Q(h) = \alpha h^{3/2}$ for every $h \in \mathbb{R}_+$, where $\alpha > 0$. So the equation above may be rewritten in the form

$$\frac{\partial h}{\partial t} + \frac{3}{2}\alpha h^{1/2}\frac{\partial h}{\partial y} = 0 \tag{6.5.2}$$

for every $y \in \mathbb{R}$ and $t \in \mathbb{R}_+$. This is the so-called *equation of big waves in long rivers.*

6.5.2 *A Local Existence and Uniqueness Result*

These examples justify the importance given in the last decades to the study of the following Cauchy problem for the first-order quasi-linear partial differential equation, called *the conservation law*

$$\begin{cases} \dfrac{\partial z}{\partial t} + a(z)\dfrac{\partial z}{\partial y} = 0 \\ \\ z(0,y) = \psi(y), \end{cases} \tag{6.5.3}$$

for $(t, y) \in \mathbb{R}_+ \times \mathbb{R}$, where $a : \mathbb{R} \to \mathbb{R}$ and $\psi : \mathbb{R} \to \mathbb{R}$ are of class C^1.

In order to solve this problem, we shall use the methods and the results established in the preceding sections. More precisely, we have:

Theorem 6.5.1. *If $a : \mathbb{R} \to \mathbb{R}$ and $\psi : \mathbb{R} \to \mathbb{R}$ are functions of class C^1 then there exists a function $b : \mathbb{R} \to \mathbb{R}_+^*$ such that the problem* (6.5.3) *has a unique solution $z : D \to \mathbb{R}$, where $D = \{(t, y) \in \mathbb{R}_+ \times \mathbb{R};\ t \in [0, b(y))\}$. In addition, this solution is implicitly defined by*

$$z = \psi(y - ta(z)). \tag{6.5.4}$$

Proof. With the notations in Section 6.4, we have: $n = 2$, $x_1 = t$, $x_2 = y$, $\Omega = \mathbb{R}^2$, $f_1(x_1, x_2, z) = 1$, $f_2(x_1, x_2, z) = a(z)$, $f(x_1, x_2, z) = 0$, $U = \mathbb{R}$, $\varphi_1(u) = 0$, $\varphi_2(u) = u$ and $\theta(u) = \psi(u)$ for every $u \in \mathbb{R}$. Let us observe that $\Delta(u)$ in Theorem 6.4.1 is given by

$$\Delta(u) = \begin{vmatrix} 1 & 0 \\ a(\psi(u)) & 1 \end{vmatrix} = 1 \neq 0$$

for every $u \in \mathbb{R}$. So, the existence and uniqueness part of Theorem 6.5.1 follows directly from Theorem 6.4.1. In order to prove (6.5.3), let us observe that the characteristic system attached to the conservation law is

$$\frac{dt}{1} = \frac{dy}{a(z)} = \frac{dz}{0}.$$

Solving this system, we get the general equation of the characteristic curves

$$\begin{cases} t = \tau + \xi \\ y = a(\zeta)\tau + \eta, & \tau \in \mathbb{R}. \\ z = \zeta \end{cases}$$

From here, we deduce that the parametric equations of the surface z are

$$\begin{cases} y = a(\psi(\tau))x + \tau \\ z = \psi(\tau) \end{cases}, \quad (\tau, x) \in \mathbb{R} \times \mathbb{R}.$$

Eliminating the parameter τ in the system above, we get (6.5.4), which is the implicit equation of the surface. The proof is complete. □

6.5.3 *Weak Solutions*

We will present next some qualitative properties specific to the solutions of the conservation law.

From (6.5.3), it follows that if ψ is bounded on $\mathbb{R}$, say by $M > 0$, then, at the point (t, y), the solution z depends only on the initial-values $z(0, w) = \psi(w)$ for w in the interval $\{w \in \mathbb{R} : |w - y| \leq \|a\|_M\}$, where $\|a\|_M = \sup\{|a(z)| : |z| \leq M\}$. From here, it follows that, if the initial datum ψ vanishes outside the set $\{y \in \mathbb{R};\ |y| \leq r\}$, then z vanishes outside the set $\{(t, y) \in \mathbb{R}_+ \times \mathbb{R};\ |y| \leq r + \|a\|_M t\}$. This property justifies the assertion that *the solution has finite speed of propagation.*

At this point, let us observe that, in order that equation (6.5.4) fulfil the hypotheses of the implicit functions theorem, we must have

$$t\psi'(y - ta(z))a'(z) \neq -1.$$

Whenever this condition fails to be satisfied, a certain loss of regularity of the solution may occur, as a consequence of the loss of its character of being a classical single-valued function. For instance, for equation (6.5.2), of big waves in rivers, with the initial condition $\psi(y) = \sin y + 1$, the solution, which at $t = 0$ has the form in Figure 6.5.2 (a), for those values of t for which $t\cos(y - \frac{3}{2}\alpha\sqrt{z}) = -\frac{4}{3}\sqrt{z}$, takes the form illustrated in Figure 6.5.2 (b), form which corresponds to *the overturn of the crest of the wave.*

These situations which completely agree with the evolution of the real phenomenon have imposed the relaxation of the concept of solution for the conservation law, having as main goal the possibility of handling the singular cases just mentioned as well. In what follows we will present such an extension of the class of functions which are candidates to the title of solution.

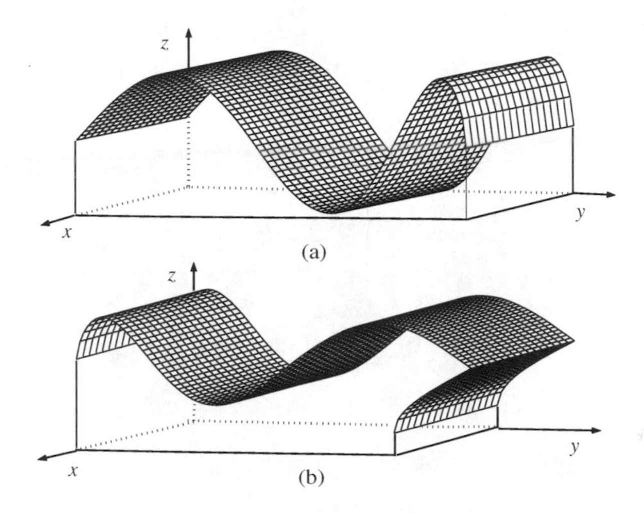

Fig. 6.5.2

Let $D = \mathbb{R}_+ \times \mathbb{R}$. We denote by $C_0^1(D)$ the set of all functions $\varphi : D \to \mathbb{R}$ of class C^1 on D for which there exists a compact set $K \subset D$ with the property that $\varphi(t, y) = 0$ for every $(t, y) \in D \setminus K$.

Definition 6.5.1. Let $\psi : \mathbb{R} \to \mathbb{R}$ be locally Lebesgue integrable on $\mathbb{R}$.[3] A locally Lebesgue integrable function $z : D \to \mathbb{R}$ is a *weak solution* of equation (6.5.3) if, for every $\varphi \in C_0^1(D)$, we have

$$\iint_D \left(z(t,y)\frac{\partial \varphi}{\partial t}(t,y) + A(z(t,y))\frac{\partial \varphi}{\partial y}(t,y) \right) dt\, dy + \int_{-\infty}^{+\infty} \psi(y)\varphi(0,y)\, dy = 0, \tag{6.5.5}$$

where $A : \mathbb{R} \to \mathbb{R}$ is defined by

$$A(r) = \int_0^r a(\eta)\, d\eta$$

for every $r \in \mathbb{R}$.

The precise meaning of this apparently strange condition (6.5.5), will be completely clarified during the proof of Theorem 6.5.2.

In what follows, by a *classical solution* of equation (6.5.3) we mean any solution in the sense of Definition 6.3.2. In order to understand the relationship between the two types of solution of the problem (6.5.3), the next auxiliary result is needed.

[3]We recall that a function $u : \Omega \subset \mathbb{R}^n \to \mathbb{R}$ is called *locally Lebesgue integrable* on Ω if its restriction to any closed ball included in Ω is Lebesgue integrable on that ball.

Lemma 6.5.1. *Let $\Omega \subset \mathbb{R}^n$ be a nonempty and open subset and $u : \Omega \to \mathbb{R}$ a locally Lebesgue integrable function. If for every function $\varphi : \Omega \to \mathbb{R}$ of class C^1 vanishing outside of a compact subset $K \subset \Omega$*

$$\iint \cdots \int_\Omega u(x)\varphi(x)\, dx_1 dx_2 \ldots dx_n = 0$$

then, at every point $\xi \in \Omega$ of continuity of u, we have $u(\xi) = 0$.

Proof. Let $\xi \in \Omega$ be a point of continuity of the function u, let $r > 0$ be such that $B(\xi, r) \subset \Omega$ and let us consider a function $\varphi : \Omega \to \mathbb{R}$, of class C^1 on Ω, which vanishes outside the set $B(\xi, r)$ and which satisfies

$$\iint \cdots \int_\Omega \varphi(x)\, dx_1 dx_2 \ldots dx_n = 1.$$

An example of such a function $\varphi : \Omega \to \mathbb{R}$ is:

$$\varphi(x) = \begin{cases} ce^{-\frac{\|x-\xi\|^2}{\|x-\xi\|^2 - r^2}} & \text{for } \|x - \xi\| < r \\ 0 & \textit{for } \|x - \xi\| \geq r, \end{cases}$$

where

$$c = \left(\iint \cdots \int_{B(\xi,r)} e^{-\frac{\|x-\xi\|^2}{\|x-\xi\|^2 - r^2}} dx_1 dx_2 \ldots dx_n \right)^{-1}.$$

For every $\varepsilon \in (0, 1)$ let us define $\varphi_\varepsilon : \Omega \to \mathbb{R}$ by

$$\varphi_\varepsilon(x) = \frac{1}{\varepsilon^n} \varphi\left(\frac{x - \xi}{\varepsilon} + \xi \right)$$

for every $x \in \Omega$. One may easily see that φ_ε vanishes outside the ball $B(\xi, \varepsilon r)$ and that

$$\iint \cdots \int_\Omega \varphi_\varepsilon(x)\, dx_1 dx_2 \ldots dx_n = 1.$$

From the last relation, the hypothesis and from the fact that u is continuous at ξ, we have

$$u(\xi) = \lim_{\varepsilon \downarrow 0} \iint \cdots \int_\Omega \varphi_\varepsilon(x) u(\xi)\, dx_1 dx_2 \ldots dx_n$$

$$= \lim_{\varepsilon \downarrow 0} \iint \cdots \int_\Omega \varphi_\varepsilon(x) [\, u(\xi) - u(x) \,]\, dx_1 dx_2 \ldots dx_n$$

$$= \lim_{\varepsilon \downarrow 0} \iint \cdots \int_{B(\xi, \varepsilon r)} \varphi_\varepsilon(x) [\, u(\xi) - u(x) \,]\, dx_1 dx_2 \ldots dx_n = 0.$$

The proof is complete. □

Theorem 6.5.2. *A function $z : D \to \mathbb{R}$ is a classical solution of equation (6.5.3) if and only if z is of class C^1 and is a weak solution of the same equation.*

Proof. Necessity. Let $z : D \to \mathbb{R}$ be a classical solution of equation (6.5.3). Let $\varphi \in C_0^1(D)$ and let us consider a rectangle $[0, b] \times [c, d]$ with $K \subset [0, b) \times (c, d)$, where K is the compact set outside of which φ vanishes. Multiplying the first equality in (6.5.3) by $\varphi(t, y)$ and integrating over D, we get

$$0 = \iint_D \left(\frac{\partial z}{\partial t}(t, y) + a(z(t, y)) \frac{\partial z}{\partial y}(t, y) \right) \varphi(t, y)\, dt\, dy$$

$$= \int_c^d \int_0^b \left(\frac{\partial z}{\partial t}(t, y)\varphi(t, y) + a(z(t, y)) \frac{\partial z}{\partial y}(t, y)\varphi(t, y) \right) dt\, dy$$

$$= \int_c^d z(t, y)\varphi(t, y) \Big|_0^b dy - \int_c^d \int_0^b z(t, y) \frac{\partial \varphi}{\partial t}(t, y)\, dt\, dy$$

$$+ \int_c^d \int_0^b \frac{\partial}{\partial y} \left(A(z(t, y)) \right) \varphi(t, y)\, dt\, dy.$$

Since φ vanishes for $t = b$, $y = c$, or $y = d$, from the equality above, Fubini's theorem on the interchanging the order of integration (see [Dunford and Schwartz (1958)], Theorem 9, p. 190), taking into account that z satisfies $z(0, y) = \psi(y)$ for $y \in \mathbb{R}$, we deduce

$$\int_c^d \int_0^b \left(z(t, y) \frac{\partial \varphi}{\partial t}(t, y) + A(z(t, y)) \frac{\partial \varphi}{\partial y}(t, y) \right) dt\, dy + \int_c^d \psi(y)\varphi(0, y)\, dy = 0.$$

Since φ vanishes outside the rectangle $[0, b] \times [c, d]$, the equality above is equivalent to (6.5.5).

Sufficiency. Let $z : D \to \mathbb{R}$ be a weak solution of equation (6.5.3) which is of class C^1. Starting from (6.5.5), and repeating backward the calculations done in the necessity part, we deduce that z satisfies

$$\iint_D \left(\frac{\partial z}{\partial t}(t, y) + a(z(t, y)) \frac{\partial z}{\partial y}(t, y) \right) \varphi(t, y)\, dt\, dy$$

$$+ \int_{-\infty}^{+\infty} (z(0, y) - \psi(y))\varphi(0, y)\, dy = 0 \tag{6.5.6}$$

for every function $\varphi \in C_0^1(D)$. Now let φ be a function of class C^1 which vanishes outside of a compact subset included in the interior of the set D. Then, for every $y \in \mathbb{R}$, $\varphi(0,y) = 0$, and from the last relation, it follows

$$\iint_D \left(\frac{\partial z}{\partial t}(t,y) + a(z(t,y)) \frac{\partial z}{\partial y}(t,y) \right) \varphi(t,y)\, dt\, dy = 0. \tag{6.5.7}$$

Hence the function $u : D \to \mathbb{R}$, defined by

$$u(t,y) = \frac{\partial z}{\partial t}(t,y) + a(z(t,y)) \frac{\partial z}{\partial y}(t,y)$$

for every $(t,y) \in D$, satisfies the hypotheses of Lemma 6.5.1 on $\Omega = \overset{\circ}{D}$. So, z satisfies the first equality in (6.5.3) for every $(t,y) \in \overset{\circ}{D}$. Now let $\eta : \mathbb{R} \to \mathbb{R}$ be a function of class C^1 which vanishes outside of a compact interval $[\,c,d\,]$. Let us define $\varphi : D \to \mathbb{R}$ by

$$\varphi(t,y) = \begin{cases} t^2(1-t)^2\eta(y) & \text{for } (t,y) \in [\,0,1\,] \times \mathbb{R} \\ 0 & \text{otherwise.} \end{cases}$$

We can easily state that φ is of class C^1 and vanishes outside the compact set $[\,0,1\,] \times [\,c,d\,]$. From (6.5.6) and (6.5.7), it follows

$$\int_{-\infty}^{+\infty} (z(0,y) - \psi(y))\eta(y)\, dy = 0.$$

From Lemma 6.5.1, we deduce that $z(0,y) = \psi(y)$ for $y \in \mathbb{R}$, which completes the proof. □

Remark 6.5.1. Actually, analyzing the proof of the sufficiency, we can see that if a function z is a weak solution of the problem (6.5.3) and is of class C^1 on an open subset D_0 in D, then z is a classical solution on D_0 of the partial differential equation in (6.5.3) and satisfies the initial condition relative to D_0, i.e. $z(0,y) = \psi(y)$ for every $y \in \mathbb{R}$ for which $(0,y) \in D_0$.

The example below illustrates the fact that the conservation law can have weak solutions which are discontinuous.

Example 6.5.3. ([Barbu (1985)], p. 184) Let us consider the Cauchy problem

$$\begin{cases} \dfrac{\partial z}{\partial t}(t,y) + z^2(t,y) \dfrac{\partial z}{\partial y}(t,y) = 0 \\ \\ z(0,y) = \psi(y), \end{cases}$$

where $\psi : \mathbb{R} \to \mathbb{R}$ is defined by

$$\psi(y) = \begin{cases} 0 \text{ if } y \leq 0 \\ 1 \text{ if } y > 0. \end{cases}$$

Then, the function $z : \mathbb{R}_+ \times \mathbb{R} \to \mathbb{R}$, defined by

$$z(t, y) = \begin{cases} 0 \text{ if } y \leq t/3 \\ 1 \text{ if } y > t/3, \end{cases}$$

is a discontinuous weak solution of the Cauchy problem.

Now let $z : D \to \mathbb{R}$ be a weak solution of the Cauchy problem (6.5.3) with the property that there exists a simple curve Γ of equation $y = f(t)$, separating D into two sub-domains $D_- = \{(t, y) \in D;\ y < f(t)\}$ and $D_+ = \{(t, y) \in D;\ y > f(t)\}$ on which z is of class C^1. For the sake of simplicity, we will assume that the transformation f is a non-increasing bijection from $\mathbb{R}_+^*$ to $\mathbb{R}$. See Figure 6.5.3.

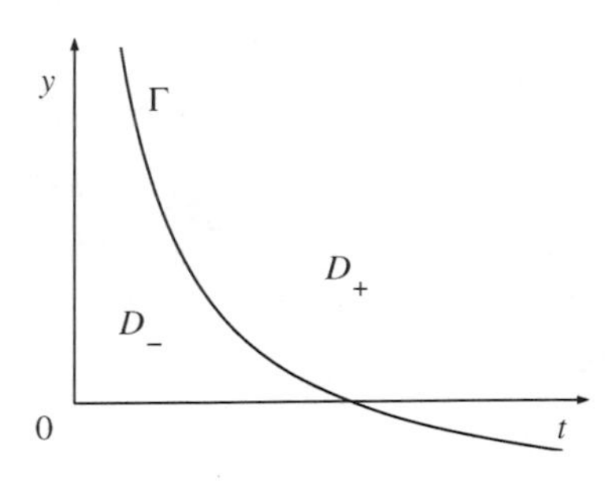

Fig. 6.5.3

The next theorem describes the behavior of weak solutions along the curve Γ.

Theorem 6.5.3. *Let $z : D \to \mathbb{R}$ be a weak solution of the Cauchy problem (6.5.3) which is of class C^1 on both domains D_- and D_+ and for which there exist finite limits*

$$\begin{cases} \lim\limits_{\substack{(\tau,\eta)\to(t,f(t)) \\ (\tau,\eta)\in D_+}} z(\tau, \eta) = z_+(t, f(t)) \\ \\ \lim\limits_{\substack{(\tau,\eta)\to(t,f(t)) \\ (\tau,\eta)\in D_-}} z(\tau, \eta) = z_-(t, f(t)) \end{cases}$$

uniformly for t in bounded subsets in $\mathbb{R}_+^*$. Then, at every point $t \in \mathbb{R}_+^*$, z satisfies the jump condition on Γ[4]

$$A(z_+(t, f(t))) - A(z_-(t, f(t))) = f'(t)[\, z_+(t, f(t)) - z_-(t, f(t))\,], \quad (6.5.8)$$

where A is the function from Definition 6.5.1.

Proof. From Remark 6.5.1, we deduce that z satisfies the first equality in (6.5.3) both on D_- and D_+. Let $\varphi \in C_0^1(D)$. Multiplying the first equality in (6.5.3) by φ and integrating over D_-, we get

$$\begin{aligned}
0 &= \iint_{D_-} \left(\frac{\partial z}{\partial t}(t, y) + a(z(t, y)) \frac{\partial z}{\partial y}(t, y) \right) \varphi(t, y)\, dt\, dy \\
&= \int_{-\infty}^{+\infty} dy \int_0^{f^{-1}(y)} \frac{\partial z}{\partial t}(t, y) \varphi(t, y)\, dt \\
&+ \int_0^{+\infty} dt \int_{-\infty}^{f(t)} \frac{\partial}{\partial y} \left(A(z(t, y)) \right) \varphi(t, y)\, dy \\
&= \int_{-\infty}^{+\infty} z(t, y) \varphi(t, y) \Big|_0^{f^{-1}(y)} dy - \int_{-\infty}^{+\infty} dy \int_0^{f^{-1}(y)} z(t, y) \frac{\partial \varphi}{\partial t}(t, y)\, dt \\
&+ \int_0^{+\infty} A(z(t, y)) \varphi(t, y) \Big|_{-\infty}^{f(t)} dt - \int_0^{+\infty} dt \int_{-\infty}^{f(t)} A(z(t, y)) \frac{\partial \varphi}{\partial y}(t, y)\, dy \\
&= \int_{-\infty}^{+\infty} \left(z_-(f^{-1}(y), y) \varphi(f^{-1}(y), y) - \psi(y) \varphi(0, y) \right) dy
\end{aligned}$$

[4]This is known as *Hugoniot–Rankine's condition.*

$$-\iint_{D_-} z(t,y)\frac{\partial\varphi}{\partial t}(t,y)\,dt\,dy$$

$$+\int_0^{+\infty} A(z_-(t,f(t))\varphi(t,f(t))\,dt - \iint_{D_-} A(z(t,y))\frac{\partial\varphi}{\partial y}(t,y)\,dt\,dy$$

$$= -\int_{-\infty}^{+\infty} \psi(y)\varphi(0,y)\,dy$$

$$-\iint_{D_-}\left(z(t,y)\frac{\partial\varphi}{\partial t}(t,y) + A(z(t,y))\frac{\partial\varphi}{\partial y}(t,y)\right)dt\,dy$$

$$-\int_0^{+\infty} (f'(t)z_-(t,f(t)) - A(z_-(t,f(t)))\,\varphi(t,f(t))\,dt.$$

A similar calculus leads to

$$0 = \iint_{D_+}\left(z(t,y)\frac{\partial\varphi}{\partial t}(t,y) + A(z(t,y))\frac{\partial\varphi}{\partial y}(t,y)\right)dt\,dy$$

$$-\int_0^{+\infty} (f'(t)z_+(t,f(t)) - A(z_+(t,f(t)))\,\varphi(t,f(t)))\,dt.$$

These two relations and (6.5.5) show that $\int_0^{+\infty} g(t)\varphi(t,f(t))\,dt = 0$ for every $\varphi \in C_0^1$, where g is defined by

$$g(t) = f'(t)[\,z_-(t,f(t)) - z_+(t,f(t))\,] - A(z_-(t,f(t))) + A(z_+(t,f(t)))$$

for $t > 0$.

Finally, we shall prove that the equality above implies (6.5.8). To this aim let us observe that the functions $t \mapsto z_-(t,f(t))$ and $t \mapsto z_+(t,f(t))$ are continuous on $\mathbb{R}_+^*$ because the two sequences of functions of class C^1 $(v_n)_{n\in\mathbb{N}^*}$ and $(w_n)_{n\in\mathbb{N}^*}$, defined by $v_n(t) = z(t,f((1-1/n)t))$ and respectively by $w_n(t) = z(t,f((1+1/n)t))$ for $n \in \mathbb{N}^*$ and $t \in \mathbb{R}$, are uniformly convergent on every bounded interval in $\mathbb{R}_+^*$ to z_- and respectively to z_+. So, the function $g : \mathbb{R}_+^* \to \mathbb{R}$, defined as above, is continuous on $\mathbb{R}_+^*$. From Lemma 6.5.1, we deduce that $g(t) = 0$ for every $t \in \mathbb{R}_+^*$, which achieves the proof. □

We conclude this section with the remark that the uniqueness property of the classical solution for the problem (6.5.3) is no longer true in the case of weak solutions, as we can see from the example below.

Example 6.5.4. ([Barbu (1985)], p. 185) Let us consider the Cauchy problem in Example 6.5.3. Then, besides the weak solution there mentioned, the problem admits a second weak solution $w : \mathbb{R}_+ \times \mathbb{R} \to \mathbb{R}$, defined by

$$w(t,y) = \begin{cases} 0 \quad for\ \dfrac{y}{t} < 0 & or\ t = 0 \text{ and } y \le 0 \\ \sqrt{\dfrac{y}{t}}\ for\ 0 \le \dfrac{y}{t} \le 1 & \\ 1 \quad for\ \dfrac{y}{t} > 1 & or\ t = 0 \text{ and } \ y > 0. \end{cases}$$

In order to individualize one weak solution of (6.5.3), customarily one imposes an extra-qualifying criterium expressing an essential property of the solutions of the system which, although very important, was not taken into consideration in the process of the mathematical modelling. Such a criterium, suggested by the second principle of thermodynamics, is that of the *entropy's growth*. More precisely, it can be proved that, among the weak solutions of the problem (6.5.3), one and only one evolves in the sense of the entropy's growth.

For more details on conservation laws see [Courant and Hilbert (1962)]. A more advanced approach can be found in [Lu (2003)].

6.6 Exercises and Problems

Exercise 6.1. *For each of the following autonomous differential systems, find two prime integrals which are independent at any non-stationary point*:

$$(1)\begin{cases} x_1' = x_2 - x_3 \\ x_2' = x_3 - x_1 \\ x_3' = x_1 - x_2. \end{cases} \qquad (2)\begin{cases} x_1' = x_2 \\ x_2' = x_1 \\ x_3' = x_1 - x_2. \end{cases}$$

$$(3)\begin{cases} x_1' = x_1 x_2 \\ x_2' = -x_1^2 \\ x_3' = x_2 x_3. \end{cases} \qquad (4)\begin{cases} x_1' = x_2 + x_1 x_3 \\ x_2' = x_1 + x_2 x_3 \\ x_3' = x_3^2 - 1. \end{cases}$$

$$(5)\begin{cases} x_1' = x_1 \\ x_2' = x_2 \\ x_3' = -2x_1 x_2. \end{cases} \qquad (6)\begin{cases} x_1' = x_1 x_2 \\ x_2' = -x_2^2 \\ x_3' = -x_1(1 + x_1^2). \end{cases}$$

$$(7)\begin{cases} x_1' = x_1 x_2^2 \\ x_2' = x_1^2 x_2 \\ x_3' = x_3(x_1^2 + x_2^2). \end{cases} \qquad (8)\begin{cases} x_1' = 2x_2(2 - x_1) \\ x_2' = x_1^2 - x_2^2 + x_3^2 - 4x_1 \\ x_3' = -x_2 x_3. \end{cases}$$

Problem 6.1. *Prove that the function* $U : \mathbb{R}_+^* \times \mathbb{R}_+^* \to \mathbb{R}$, *defined by*

$$U(x, y) = x^{-b} y^{-a} e^{hx+ky}$$

for every $(x, y) \in \mathbb{R}^2$, *is a prime integral for the prey-predator system, known also as the Lotka–Volterra system*:

$$\begin{cases} x' = (a - ky)x \\ y' = -(b - hx)y, \end{cases}$$

where a, b, k, h *are positive constants.* ([Arrowsmith and Place (1982)], p. 145)

Problem 6.2. *Prove that all trajectories of the differential system*

$$\begin{cases} x_1' = x_3 - x_2 \\ x_2' = x_1 - x_3 \\ x_3' = x_2 - x_1 \end{cases}$$

are circles.

Problem 6.3. *Prove that all trajectories of the prey-predator system which start in the first quadrant, except for the two semi-axes, remain in the first quadrant and are closed curves.* ([Glăvan *et al.* (1993)], p. 134)

Problem 6.4. *Under another formulation, Problem* 6.3 *says that each solution of the prey-predator system which starts in the first quadrant, except for the two semi-axes, is periodic with period* T *depending on the initial data. Prove that the medium populations of both species on an interval whose length equals the period* T, *i.e.*

$$x_m = \frac{1}{T} \int_t^{t+T} x(s)\, ds \quad \text{and} \quad y_m = \frac{1}{T} \int_t^{t+T} y(s)\, ds$$

are independent of the initial data. ([Arrowsmith and Place (1982)], p. 172)

Problem 6.5. *Let* $f : \Omega \subset \mathbb{R}^n \to \mathbb{R}^n$ *be a locally Lipschitz function. Prove that all the points of strict local minimum of a prime integral of the differential system*

$$x' = f(x),$$

which are stationary solutions of the system above, are stable. ([Glăvan *et al.* (1993)], p. 180)

Problem 6.6. *Prove that the stationary solution* $(b/h, a/k)$ *of the prey-predator system is stable.*

Problem 6.7. *Find the first-order autonomous differential systems which have injective prime integrals. Are there non-autonomous systems having injective prime integrals?*

Problem 6.8. *Let $f : \mathbb{R}^n \to \mathbb{R}^n$ be a continuous function. If there exists a prime integral $U : \mathbb{R}^n \to \mathbb{R}$ of the autonomous system*

$$x' = f(x)$$

which is coercive, i.e.

$$\lim_{\|x\| \to +\infty} U(x) = +\infty,$$

then all saturated solutions of the system are global. Does this conclusion remain valid if the limit is $-\infty$?

Problem 6.9. *The evolution of many phenomena in physics is described by the so-called Hamiltonian systems*

$$\begin{cases} \dfrac{dp_i}{dt} = -\dfrac{\partial H}{\partial q_i}(p, q) \\ & i = 1, 2, \ldots, n \\ \dfrac{dq_i}{dt} = \dfrac{\partial H}{\partial p_i}(p, q), \end{cases}$$

where $H : \Omega \subset \mathbb{R}^{2n} \to \mathbb{R}$ is a function of class C^1, nonconstant, known as the Hamilton function, depending on $p_1, p_2, \ldots, p_n$, called generalized momenta and on $q_1, q_2, \ldots, q_n$, called configuration coordinates. Prove that the Hamilton function is a prime integral for the Hamiltonian system. ([Glăvan *et al.* (1993)] p. 134)[5]

Problem 6.10. *Let Ω be a nonempty and open subset of $\mathbb{R}^n$ and let $f : \Omega \to \mathbb{R}^n$ be a continuous function. Prove that, for every $\lambda \neq 0$, the set of prime integrals for the equation $x' = f(x)$ coincides with the set of prime integrals for the equation $x' = \lambda f(x)$.*

Problem 6.11. *Show that, for the decoupled autonomous system below, there is no prime integral defined on $\mathbb{R}^2$*

$$\begin{cases} x_1' = 2x_1 \\ x_2' = x_2. \end{cases}$$

Prove that there exist prime integrals defined on $\{(x_1, x_2) \in \mathbb{R}^2;\ x_1 > 0\}$. ([Glăvan *et al.* (1993)], p. 135)

Problem 6.12. *Let $\mathcal{A} : \mathbb{I} \to \mathcal{M}_{n\times n}(\mathbb{R})$ be continuous, with $a_{ij}(t) = -a_{ji}(t)$ for every $i, j = 1, 2, \ldots, n$ and every $t \in \mathbb{I}$. Prove that every global solution of the system*

$$x'(t) = \mathcal{A}(t)x(t)$$

[5] The fact that H is constant along the trajectories of the system represents a special instance of *the conservation law of the energy* and this because, in all concrete cases, $H(p, q)$ is nothing but the energy of the system corresponding to the values of the parameters $(p, q) = (p_1, \ldots, p_n, q_1, \ldots, q_n)$.

is bounded on $\mathbb{I}$. *In the case in which* $\mathbb{I} = [0, +\infty)$, *is the system above stable?* ([Glăvan *et al.* (1993)], p. 136)

Exercise 6.2. *Find the general solutions of the first-order partial differential equations below*:

$$(1)\ (x_2^2 - x_3^2)\frac{\partial z}{\partial x_1} + (x_3^2 - x_1^2)\frac{\partial z}{\partial x_2} + (x_1^2 - x_2^2)\frac{\partial z}{\partial x_3} = 0.$$

$$(2)\ -x_1 e^{x_2}\frac{\partial z}{\partial x_1} + \frac{\partial z}{\partial x_2} + x_3 e^{x_2}\frac{\partial z}{\partial x_3} = 0.$$

$$(3)\ x_1(x_2 - x_3)\frac{\partial z}{\partial x_1} + x_2(x_3 - x_1)\frac{\partial z}{\partial x_2} + x_3(x_1 - x_2)\frac{\partial z}{\partial x_3} = 0.$$

$$(4)\ (x_1 - x_3)\frac{\partial x_3}{\partial x_1} + (x_2 - x_3)\frac{\partial x_3}{\partial x_2} = 2x_3.$$

$$(5)\ x_3\frac{\partial x_3}{\partial x_1} - x_3\frac{\partial x_3}{\partial x_2} = x_2 - x_1.$$

$$(6)\ x_1\frac{\partial x_3}{\partial x_1} + x_2\frac{\partial x_3}{\partial x_2} = x_3 + \frac{x_1 x_2}{x_3}.$$

$$(7)\ x_2 x_3\frac{\partial x_3}{\partial x_1} + x_1 x_3\frac{\partial x_3}{\partial x_2} = 2x_1 x_2.$$

$$(8)\ (1 + \sqrt{z - a_1 x_1 - a_2 x_2 - a_3 x_3})\frac{\partial z}{\partial x_1} + \frac{\partial z}{\partial x_2} + \frac{\partial z}{\partial x_3} = a_1 + a_2 + a_3.$$

([Craiu and Roşculeţ (1971)], pp. 48–60)

Exercise 6.3. *Solve the following Cauchy problems*:

$$(1)\ \begin{cases} x\dfrac{\partial z}{\partial y} - y\dfrac{\partial z}{\partial x} = 0 \\ z(x, 0) = \cos x. \end{cases} \qquad (2)\ \begin{cases} x\dfrac{\partial z}{\partial x} - z\dfrac{\partial z}{\partial y} = 0 \\ z(x, x^2) = x^3. \end{cases}$$

$$(3)\ \begin{cases} x\dfrac{\partial u}{\partial x} - y\dfrac{\partial u}{\partial y} - 2z\dfrac{\partial u}{\partial z} = 0 \\ u(1, y, z) = \sin(y + z). \end{cases} \qquad (4)\ \begin{cases} x\dfrac{\partial z}{\partial x} + (y + x^2)\dfrac{\partial z}{\partial y} = z \\ z(2, y) = (y - 4)^3. \end{cases}$$

([Glăvan *et al.* (1993)], pp. 188–192)

Problem 6.13. *Let* $f : \mathbb{R} \times \mathbb{R} \to \mathbb{R}$ *and* $\varphi : \mathbb{R} \to \mathbb{R}$ *be two functions of class* C^1

and $a \in \mathbb{R}$. Find the solution of the Cauchy problem

$$\begin{cases} \dfrac{\partial z}{\partial t} + a\dfrac{\partial z}{\partial x} = f(t,x) \\ \\ z(0,x) = \varphi(x). \end{cases}$$

Problem 6.14. *Let $f : \mathbb{R} \times \mathbb{R} \to \mathbb{R}$ and $\varphi : \mathbb{R} \to \mathbb{R}$ be two functions of class C^1 and $a : \mathbb{R} \to \mathbb{R}$ a continuous function. Find the solution of the Cauchy problem*

$$\begin{cases} \dfrac{\partial z}{\partial t} + a(t)\dfrac{\partial z}{\partial x} = f(t,x) \\ \\ z(0,x) = \varphi(x). \end{cases}$$

Problem 6.15. *Let $f : \mathbb{R} \times \mathbb{R}^n \to \mathbb{R}$ and $\varphi : \mathbb{R}^n \to \mathbb{R}$ be two functions of class C^1 and $a \in \mathbb{R}^n$. Find the solution of the Cauchy problem*

$$\begin{cases} \dfrac{\partial z}{\partial t} + \displaystyle\sum_{i=1}^{n} a_i \dfrac{\partial z}{\partial x_i} = f(t,x) \\ \\ z(0,x) = \varphi(x). \end{cases}$$

([Barbu (1985)], p. 200)

Problem 6.16. *Let $f : \mathbb{R} \times \mathbb{R}^n \to \mathbb{R}$ and $\varphi : \mathbb{R}^n \to \mathbb{R}$ be two functions of class C^1 and $a : \mathbb{R} \to \mathbb{R}^n$ a continuous function. Find the solution of the Cauchy problem*

$$\begin{cases} \dfrac{\partial z}{\partial t} + \displaystyle\sum_{i=1}^{n} a_i(t) \dfrac{\partial z}{\partial x_i} = f(t,x) \\ \\ z(0,x) = \varphi(x). \end{cases}$$

Problem 6.17. *Let $f : \mathbb{R} \times \mathbb{R} \to \mathbb{R}$ and $\varphi : \mathbb{R} \to \mathbb{R}$ be two functions of class C^1 and $a \in \mathbb{R}$. Find the solution of the Cauchy problem*

$$\begin{cases} \dfrac{\partial z}{\partial t} + ax\dfrac{\partial z}{\partial x} = f(t,x) \\ \\ z(0,x) = \varphi(x). \end{cases}$$

Problem 6.18. *Let $f : \mathbb{R} \times \mathbb{R} \to \mathbb{R}$ and $\varphi : \mathbb{R} \to \mathbb{R}$ be two functions of class C^1 and $a : \mathbb{R} \to \mathbb{R}$ a continuous function. Find the solution of the Cauchy problem*

$$\begin{cases} \dfrac{\partial z}{\partial t} + a(t)x\dfrac{\partial z}{\partial x} = f(t,x) \\ \\ z(0,x) = \varphi(x). \end{cases}$$

Problem 6.19. *Let $f : \mathbb{R} \times \mathbb{R}^n \to \mathbb{R}$ and $\varphi : \mathbb{R}^n \to \mathbb{R}$ be two functions of class C^1 and let $\mathcal{A} \in \mathcal{M}_{n\times n}(\mathbb{R})$. Find the solution of the Cauchy problem*

$$\begin{cases} \dfrac{\partial z}{\partial t} + \langle \mathcal{A}x, \nabla_x z\rangle = f(t,x) \\ \\ z(0,x) = \varphi(x). \end{cases}$$

([Barbu (1985)], p. 200)

Problem 6.20. *Let $f : \mathbb{R} \times \mathbb{R}^n \to \mathbb{R}$ and $\varphi : \mathbb{R}^n \to \mathbb{R}$ be two functions of class C^1 and let $\mathcal{A} : \mathbb{R} \to \mathcal{M}_{n\times n}(\mathbb{R})$ a continuous function. Find the solution of the Cauchy problem*

$$\begin{cases} \dfrac{\partial z}{\partial t} + \langle \mathcal{A}(t)x, \nabla_x z\rangle = f(t,x) \\ \\ z(0,x) = \varphi(x). \end{cases}$$

Problem 6.21. *Find the surface containing the circle $x_1^2 + x_3^2 = 1$, $x_2 = 2$, orthogonal to the family of cones $x_1x_2 = \alpha x_3^2$, $\alpha \in \mathbb{R}^*$.* ([Craiu and Roşculeţ (1971)], p. 58)

Chapter 7

Extensions and Generalizations

The main goal of this chapter is to present several methods to approach some Cauchy problems which, from various reasons, do not find their place in the preceding theoretical framework. In order to extend the concept of solution in the case of linear differential equations and systems with discontinuous right-hand sides, in the first three sections we introduce and study the notion of distribution as a generalization of an infinitely many differentiable function. In the same spirit, in the fourth section, we present another type of solution suitable for the nonlinear case when the function f on the right-hand side is discontinuous with respect to the t variable. In the next two sections, we discuss two variants of approaching some Cauchy problems for which f is discontinuous as a function of the state variable x, situation involving much more difficulties than the preceding one. In both cases, the manner of approach consists in replacing the differential equation with a so-called differential inclusion. The sixth section is concerned with the study of a class of variational inequalities, while in the next four sections we deal with a Cauchy problem in which the function on the right-hand side of the equation is defined on a set which is not open. In the eleventh section, we present an existence and uniqueness result referring to the Cauchy problem for a class of systems of first-order nonlinear partial differential equations of type gradient. The chapter ends with a section of Exercises and Problems.

7.1 Distributions of One Variable

In many cases, the right-hand side of a differential equation does not satisfy the minimal requirements ensuring the existence of at least one *classical solution*, i.e. of a function of class C^n, where n is order of the equation considered and which satisfy the desired equality for every t in its domain of definition. In all these situations, in order to recuperate as many results already established as possible, one has tried to introduce a new type of solution by enlarging the class of candidates to this title. We emphasize that all these extensions have been done in such a way that, whenever we are in the "classical" hypotheses of regularity, the only "generalized" solutions are the classical ones. As expected, in the process of definition of the concept of any kind of generalized solution, we have postulated as characteristic some minimal properties satisfied by all classical solutions. In this way, a necessary, but not sufficient, condition for a certain mathematical object to be a classical solution, has been promoted to the rank of definition.

The aim of this section is to present one of the deepest extensions of this kind, having a great impact in the development of the theory of both ordinary, but especially of partial differential equations. The main idea of this extension is very simple and extremely efficient. More precisely, it is based on the simple remark that, if $x : \mathbb{R} \to \mathbb{R}$ is a solution of class C^1 of the first-order differential equation

$$x' = f(t, x), \tag{$\mathcal{E}$}$$

with $f : \mathbb{R} \times \mathbb{R} \to \mathbb{R}$ continuous, then, for each function, $\phi : \mathbb{R} \to \mathbb{R}$, of class C^1, for which there exists $[\,a, b\,]$ such that $\phi(t) = 0$ for every $t \in \mathbb{R} \setminus [\,a, b\,]$, we have

$$\int_{\mathbb{R}} \phi'(t)x(t)\,dt + \int_{\mathbb{R}} \phi(t)f(t, x(t))\,dt = 0. \tag{7.1.1}$$

Indeed, in order to deduce (7.1.1), let us multiply ($\mathcal{E}$) on both sides by $\phi(t)$ and then, let us integrate the equality thus obtained over $\mathbb{R}$. This is always possible because, due to the conditions imposed on the function ϕ, the improper integral on $\mathbb{R}$ reduces to a proper one on $[\,a, b\,]$. So, integrating by parts, and taking into account that $\phi(a) = \phi(b) = 0$ we get (7.1.1).

At this point, let us observe that, through these simple manipulations, the action of the differential operator has moved from the solution x to the function ϕ. Obviously, (7.1.1) can take place for any function x which is only Lebesgue integrable on every compact interval, but might fail to be of class C^1. At this moment, (7.1.1) suggests that we may extend the notion of classical solution by defining, for example, as *generalized solution* of the equation ($\mathcal{E}$) any function $x : \mathbb{R} \to \mathbb{R}$, Lebesgue integrable on every compact interval and which satisfies (7.1.1) for every function of class C^1 ϕ which vanishes outside an interval of the form $[\,a_\phi, b_\phi\,]$. This is a very brief presentation of one of the main ideas which partially motivates the birth of the distribution theory.

In order to introduce the concept of distribution, some preliminaries are needed.

Definition 7.1.1. *The support* of the function $\phi : \mathbb{R} \to \mathbb{R}$ is the set

$$\operatorname{supp} \phi = \overline{\{t;\ t \in \mathbb{R},\ \phi(t) \neq 0\}}.$$

We denote by

$$\mathcal{D}(\mathbb{R}) = \{\phi;\ \phi \in C^\infty(\mathbb{R}; \mathbb{R}),\ \phi \text{ with compact support}\}$$

and we call $\mathcal{D}(\mathbb{R})$ *the space of testing, or test functions on* $\mathbb{R}$.

Example 7.1.1. The function $\phi : \mathbb{R} \to \mathbb{R}$, defined by

$$\phi(t) = \begin{cases} e^{\frac{1}{t^2-1}} & \textit{for } t \in (-1, 1) \\ 0 & \textit{for } t \in \mathbb{R} \setminus (-1, 1), \end{cases}$$

is of class C^∞ and has the compact support $[-1, 1]$. Hence the set $\mathcal{D}(\mathbb{R})$ is nonempty.

Proposition 7.1.1. *The set $\mathcal{D}(\mathbb{R})$ is a vector space over $\mathbb{R}$ with respect to the usual operations. In addition, for each $\phi \in \mathcal{D}(\mathbb{R})$ and $\psi \in C^\infty(\mathbb{R}; \mathbb{R})$, $\phi\psi \in \mathcal{D}(\mathbb{R})$.*

Proof. Obviously every linear combination of functions of class C^∞ from $\mathbb{R}$ to $\mathbb{R}$ is a function enjoying the same property. In addition, the product of a function with compact support and an arbitrary function from $\mathbb{R}$ to $\mathbb{R}$ is a function with compact support. Hence $\mathcal{D}(\mathbb{R})$ is a vector subspace of $C^\infty(\mathbb{R}; \mathbb{R})$. Since the last part of the conclusion also follows from the considerations above, this completes the proof. □

In what follows, we introduce a convergence structure on $\mathcal{D}(\mathbb{R})$, allowing us to define the class of functions which are sequentially continuous from $\mathcal{D}(\mathbb{R})$ in $\mathbb{R}$. We emphasize that this convergence structure is essential in the process of definition of the concept of distribution.

Definition 7.1.2. A sequence $(\phi_k)_{k \in \mathbb{N}}$ in $\mathcal{D}(\mathbb{R})$ is *convergent in* $\mathcal{D}(\mathbb{R})$ to $\phi \in \mathcal{D}(\mathbb{R})$ if there exists a compact interval $[a, b]$ such that:

(i) $\operatorname{supp} \phi_k \subset [a, b]$ for every $k \in \mathbb{N}$;
(ii) for every $p \in \mathbb{N}$, $\lim_{k \to \infty} \phi_k^{(p)}(t) = \phi^{(p)}(t)$ uniformly for $t \in [a, b]$, or equivalently, uniformly for $t \in \mathbb{R}$.

We denote this situation by $\phi_k \xrightarrow{\mathcal{D}(\mathbb{R})} \phi$.

Definition 7.1.3. A *distribution* on $\mathcal{D}(\mathbb{R})$ is a linear continuous functional $x : \mathcal{D}(\mathbb{R}) \to \mathbb{R}$, i.e. a function x which satisfies:

(i) $x(\alpha\phi + \beta\psi) = \alpha x(\phi) + \beta x(\psi)$ for every $\phi, \psi \in \mathcal{D}(\mathbb{R})$ and every $\alpha, \beta \in \mathbb{R}$;

(ii) if $\phi_k \xrightarrow{\mathcal{D}(\mathbb{R})} \phi$ then $\lim_{k\to\infty} x(\phi_k) = x(\phi)$.

Remark 7.1.1. A linear mapping $x : \mathcal{D}(\mathbb{R}) \to \mathbb{R}$ is a distribution if and only if it is continuous at $\phi = 0$, i.e. if and only if from $\phi_k \xrightarrow{\mathcal{D}(\mathbb{R})} 0$ it follows $\lim_{k\to\infty} x(\phi_k) = 0$.

In all what follows we denote by $\mathcal{D}'(\mathbb{R})$ the set of all distributions defined on $\mathcal{D}(\mathbb{R})$.

Example 7.1.2. Distributions of type function. Let $x : \mathbb{R} \to \mathbb{R}$ be a *locally Lebesgue integrable function*, i.e. a function whose restriction to each compact interval is Lebesgue integrable on that interval. We define the map $\tilde{x} : \mathcal{D}(\mathbb{R}) \to \mathbb{R}$ by

$$\tilde{x}(\phi) = \int_{\mathbb{R}} x(t)\phi(t)\, dt$$

for every $\phi \in \mathcal{D}(\mathbb{R})$. Since $\phi \in \mathcal{D}(\mathbb{R})$, there exists $[\,a, b\,]$ with $\operatorname{supp}\phi \subset [\,a, b\,]$ and it follows that $\tilde{x}$ is well-defined, in the sense that the integral on the right-hand side, apparently on a set of infinite measure, is in fact a Lebesgue integral defined on a compact interval. Since $\tilde{x}$ is obviously linear and continuous, it follows that it is a distribution on $\mathcal{D}(\mathbb{R})$. We call such a distribution a *distribution of type function*. Let us remark that, if $x, y : \mathbb{R} \to \mathbb{R}$ are two locally Lebesgue integrable functions which are almost everywhere equal on $\mathbb{R}$, then $\tilde{x} = \tilde{y}$. Indeed,

$$\int_{\mathbb{R}} x(t)\phi(t)\, dt = \int_{\mathbb{R}} y(t)\phi(t)\, dt$$

for every $\phi \in \mathcal{D}(\mathbb{R})$. Let us denote by $\mathcal{L}^1_{\text{loc}}(\mathbb{R})$ the space of all locally Lebesgue integrable functions $x : \mathbb{R} \to \mathbb{R}$, and let us define the relation $\rho \subset \mathcal{L}^1_{\text{loc}}(\mathbb{R}) \times \mathcal{L}^1_{\text{loc}}(\mathbb{R})$ by $x\rho y$ if and only if $x = y$ almost everywhere on $\mathbb{R}$. Then, ρ is an equivalence relation on $\mathcal{L}^1_{\text{loc}}(\mathbb{R})$. One may prove that any two equal distributions of type function are defined by two functions which are almost everywhere equal. According to this observations, the set of distributions of type function can be identified with the quotient space

$\mathcal{L}^1_{\text{loc}}(\mathbb{R})/\rho$, which is nothing but the well-known space $L^1_{\text{loc}}(\mathbb{R})$. For this reason, we agree to denote a distribution of type function by $\tilde{x}$, $\tilde{x}$ being the ρ-equivalence class of the element x, or even by x.

Example 7.1.3. The Dirac Delta. Let $\delta : \mathcal{D}(\mathbb{R}) \to \mathbb{R}$ be defined by

$$\delta(\phi) = \phi(0)$$

for every $\phi \in \mathcal{D}(\mathbb{R})$. One may easily see that δ is linear and continuous. So, it is a distribution on $\mathcal{D}(\mathbb{R})$ called *the Dirac delta*. We notice that this distribution has been introduced in 1926 by Paul Dirac[1] with the aim of explaining certain phenomena with impulsive character. For the sake of simplicity, in the example that follows, we will confine ourselves to the presentation of an extremely simple such physical situation whose description from a rigorous mathematical viewpoint cannot avoid the use of Dirac delta.

Example 7.1.4. Density of a point of mass m. Let us consider a material point of mass m whose density we intend to determine. Since the point has zero measure, at first glance we could be tempted to assert that the density is $+\infty$. On the other hand, if we think of the point as to a geometrical object obtained as a "limit of a sequence of objects", all of mass m and all having well-defined densities, for which we can give a sense to the limit of the corresponding sequence of densities, that limit could be a good candidate for the density of the point, with the condition to be independent of the sequence of approximates in question. Approximating, for example, the point by an interval of length 2ℓ, centered at the origin, of mass m and of density $d : \mathbb{R} \to \mathbb{R}$ uniformly distributed on the interval $[-\ell, \ell]$, i.e.

$$d_\ell(t) = \begin{cases} \dfrac{m}{2\ell} & \text{if } t \in [-\ell, \ell] \\ 0 & \textit{if } t \in \mathbb{R} \setminus [-\ell, \ell] \end{cases}$$

and expressing the mass as function of density, we deduce

$$m = \int_{\mathbb{R}} d_\ell(t)\, dt.$$

At this point, let us observe that there exists

$$\lim_{\ell \downarrow 0} d_\ell(t) = d_0(t)$$

[1] British physicist (1902–1984). He was one of the founders of Quantum Mechanics. Using the mathematical formalism introduced by himself, he has succeeded to predict the existence of the positive electron, or positron. He received the Nobel Prize in 1933.

point-wise on $\mathbb{R}$, where

$$d_0(t) = \begin{cases} +\infty & \text{if } t = 0 \\ \\ 0 & \textit{if } t \in \mathbb{R} \setminus \{0\}. \end{cases}$$

On the other hand

$$\int_{\mathbb{R}} d_\ell(t)dt = m$$

for each $\ell > 0$, and therefore we could be tempted to conclude that "d_0 is a function from $\mathbb{R}$ to $\overline{\mathbb{R}}$, taking the value $+\infty$ at $t = 0$, and zero otherwise, but whose integral over $\mathbb{R}$ equals $m > 0$". Obviously such a conclusion is unacceptable because there exists no function with the properties above.

However, to give a sense to the limit above (other than the usual one), let us observe that, by virtue of the mean-value theorem, for every $\phi \in \mathcal{D}(\mathbb{R})$ and every $\ell > 0$ there exists $\theta_\ell \in [-\ell, \ell]$ such that we have

$$\lim_{\ell \downarrow 0} \int_{\mathbb{R}} d_\ell(t)\phi(t)\, dt = \lim_{\ell \downarrow 0} \int_{-\ell}^{\ell} \frac{m}{2\ell}\phi(t)\, dt = \lim_{\ell \downarrow 0} m\phi(\theta_\ell) = m\phi(0) = m\delta(\phi).$$

Thus, the density of a material point of mass m can be identified with $m\delta$ where δ is the Dirac delta.

Proposition 7.1.2. *The set $\mathcal{D}'(\mathbb{R})$ is a vector space over $\mathbb{R}$ with respect to the usual operations with functions, i.e. addition, and multiplication by a real number.*

If $x \in \mathcal{D}'(\mathbb{R})$ and $\phi \in \mathcal{D}(\mathbb{R})$, we agree to denote by

$$x(\phi) = (x, \phi) = (x(t), \phi(t)).$$

We emphasize that *the notation $x(t)$ is somehow improper because x is not a function of a real variable t*, but is useful whenever we want to specify which one of the arguments of a function ϕ of several variables is that one with respect to which ϕ is considered as a test function. See Example 7.1.2. We will face such situations in some of the next definitions in which, in order to avoid possible ambiguities, we will be led to use this notation.

Operations with distributions. Besides the usual operations of addition of two distributions and of multiplying a distribution by a real number, we can define some other new ones, which endow the space $\mathcal{D}'(\mathbb{R})$ with a particularly rich and, at the same time, useful structure.

Let $x \in \mathcal{D}'(\mathbb{R})$ and $\eta \in C^\infty(\mathbb{R}; \mathbb{R})$. One can easily see that the function $\eta x : \mathcal{D}(\mathbb{R}) \to \mathbb{R}$, defined by

$$(\eta x)(\phi) = (x(t), \eta(t)\phi(t)) \tag{7.1.2}$$

for every $\phi \in \mathcal{D}(\mathbb{R})$, is a distribution.

Remark 7.1.2. The function ηx is well-defined because, if $\eta \in C^{\infty}(\mathbb{R};\mathbb{R})$ and $\phi \in \mathcal{D}(\mathbb{R})$, then $\eta\phi \in \mathcal{D}(\mathbb{R})$.

Definition 7.1.4. By definition, *the product of the distribution x by the function of class C^{∞} η* is the distribution ηx defined by means of the relation (7.1.2).

If $x \in \mathcal{D}'(\mathbb{R})$ and $\alpha > 0$, then the function $x(\alpha t) : \mathcal{D}(\mathbb{R}) \to \mathbb{R}$, defined by

$$x(\alpha t)(\phi(t)) = \frac{1}{\alpha}\left(x(t), \phi\left(\frac{t}{\alpha}\right)\right) \tag{7.1.3}$$

for every $\phi \in \mathcal{D}(\mathbb{R})$, is a distribution.

Definition 7.1.5. By definition, *the homothety of coefficient $\alpha > 0$ of the distribution x* is the distribution $x(\alpha t)$ defined by means of the relation (7.1.3).

If $x \in \mathcal{D}'(\mathbb{R})$ and $a \in \mathbb{R}$, then the function $x(t-a) : \mathcal{D}(\mathbb{R}) \to \mathbb{R}$, defined by

$$x(t-a)(\phi(t)) = (x(t), \phi(t+a)) \tag{7.1.4}$$

for every $\phi \in \mathcal{D}(\mathbb{R})$, is a distribution.

Definition 7.1.6. By definition, *the translation by a of the distribution x* is the distribution $x(t-a)$, defined by means of the relation (7.1.4).

Example 7.1.5. For instance, the translation by a of the Dirac delta, is the so-called *Dirac delta concentrated at a*, i.e. $\delta(t-a)$, defined by

$$(\delta(t-a), \phi(t)) = \phi(a)$$

for every $\phi \in \mathcal{D}(\mathbb{R})$.

Remark 7.1.3. The idea of defining the operations above was suggested by the particular case of distributions of type function, case in which we have

$$(\eta(t)x(t), \phi(t)) = \int_{\mathbb{R}} \eta(t)x(t)\phi(t)\,dt = \int_{\mathbb{R}} x(t)\eta(t)\phi(t)\,dt = (x(t), \eta(t)\phi(t)),$$

$$(x(\alpha t), \phi(t)) = \int_{\mathbb{R}} x(\alpha t)\phi(t)\,dt = \frac{1}{\alpha}\int_{\mathbb{R}} x(t)\phi\left(\frac{t}{\alpha}\right)dt = \frac{1}{\alpha}\left(x(t), \phi\left(\frac{t}{\alpha}\right)\right),$$

and respectively

$$(x(t-a),\phi(t)) = \int_{\mathbb{R}} x(t-a)\phi(t)\,dt = \int_{\mathbb{R}} x(t)\phi(t+a)\,dt = (x(t),\phi(t+a))$$

for the product by a function of class C^∞, for the homothety of coefficient $\alpha > 0$ and for the translation by a, respectively.

Let $x \in \mathcal{D}'(\mathbb{R})$ and $k \in \mathbb{N}$. Taking into account the definition of the convergence on $\mathcal{D}(\mathbb{R})$, we can easily conclude that the function $x^{(k)} : \mathcal{D}(\mathbb{R}) \to \mathbb{R}$, defined by

$$(x^{(k)}(t),\phi(t)) = (-1)^k(x(t),\phi^{(k)}(t)) \tag{7.1.5}$$

for every $\phi \in \mathcal{D}(\mathbb{R})$, is a distribution.

Definition 7.1.7. *The derivative of order* k of the distribution x is the distribution $x^{(k)}$ defined by means of the relation (7.1.5).

Remark 7.1.4. The definition of the derivative of a distribution was also suggested by the case of distributions of type function which are of class C^k and for which the equality (7.1.5) follows by a successive application of k-times by parts integrations. At this point, we are in the position to notice the advantage of these new mathematical objects, i.e. distributions versus classical functions. More precisely this advantage consists in that, unlike the usual functions, the distributions are always infinitely-many differentiable. For this reason, the distributional framework is very suitable to the construction of a general theory of linear differential equations and systems.

Next, we prove a simple but useful result which completely clarifies the relationship between the distributional derivative of a function, i.e. the derivative of a distribution of type function in the sense of Definition 7.1.7, and the classical derivative of that function. In order to avoid possible confusion, we denote by $\dot{x}$ the *distributional derivative of the function* x, i.e. the derivative given by Definition 7.1.7, and by x' the *classical derivative of the function* x, i.e. in the usual sense of Real Analysis.

Proposition 7.1.3. *Let $x : \mathbb{R} \to \mathbb{R}$ be a function of class C^1 on $\mathbb{R} \setminus \{a\}$ with the property that x' is locally Lebesgue integrable. If a is a discontinuity point of the first kind*[2] *of x then*

$$\dot{x}(t) = x'(t) + \omega(x,a)\delta(t-a), \tag{7.1.6}$$

[2]We recall that a discontinuity point of the first kind of a function $x : \mathbb{R} \to \mathbb{R}$ is a discontinuity point $a \in \mathbb{R}$ at which there exist both one-sided limits $x(a+0)$ and $x(a-0)$.

where $\omega(x,a) = x(a+0) - x(a-0)$ *is the jump of the function* x *at the point* a. *In particular, if* $\omega(x,a) = 0$, *then*

$$\dot{x} = x'.$$

Proof. Since x' is locally Lebesgue integrable, it generates a distribution of type function. Let $\phi \in \mathcal{D}(\mathbb{R})$. According to Definition 7.1.7, we have

$$(\dot{x}(t), \phi(t)) = -(x(t), \phi'(t)) = -\int_{\mathbb{R}} x(t)\phi'(t)\,dt$$

$$= -\int_{-\infty}^{a} x(t)\phi'(t)\,dt - \int_{a}^{+\infty} x(t)\phi'(t)\,dt$$

$$= -x(t)\phi(t)|_{-\infty}^{a} + \int_{-\infty}^{a} x'(t)\phi(t)\,dt - x(t)\phi(t)|_{a}^{+\infty} + \int_{a}^{+\infty} x'(t)\phi(t)\,dt.$$

Since ϕ has compact support, we deduce that

$$\lim_{t\downarrow -\infty} x(t)\phi(t) = \lim_{t\uparrow +\infty} x(t)\phi(t) = 0$$

and therefore

$$(\dot{x}(t), \phi(t)) = \int_{\mathbb{R}} x'(t)\phi(t)\,dt + \omega(x,a)\phi(a).$$

Since this equality is obviously equivalent to (7.1.6), this completes the proof. □

Corollary 7.1.1. *Let* $\eta \in C^\infty(\mathbb{R};\mathbb{R})$ *and* $x \in \mathcal{D}'(\mathbb{R})$. *Then, the derivative of the distribution* ηx *is given by the Leibniz rule*

$$(\dot{\eta x}) = \eta' x + \eta \dot{x}.$$

Proof. Let $\phi \in \mathcal{D}(\mathbb{R})$. We have

$$((\dot{\eta x}), \phi) = -(\eta x, \phi') = -(x, \eta\phi')$$

$$= -(x, (\eta\phi)' - \eta'\phi) = -(x, (\eta\phi)') + (x, \eta'\phi)$$

$$= (\dot{x}, \eta\phi) + (\eta' x, \phi) = (\eta' x, \phi) + (\eta\dot{x}, \phi)$$

$$= (\eta' x + \eta\dot{x}, \phi),$$

relation which completes the proof. □

7.2 The Convolution Product

We denote by $\mathcal{D}'_+(\mathbb{R})$ the set of all distributions x with the property that for every $\phi \in \mathcal{D}(\mathbb{R})$ with $\operatorname{supp}\phi \subset (-\infty, 0)$, we have $(x(t), \phi(t)) = 0$. In other words, $\mathcal{D}'_+(\mathbb{R})$ is the set of all distributions which "depend" only on the values at $t \in [0, +\infty)$ of the test functions. Indeed, $x \in \mathcal{D}'_+(\mathbb{R})$ if and only if for every $\phi, \psi \in \mathcal{D}(\mathbb{R})$ with $\phi(t) = \psi(t)$ for every $t \in [0, +\infty)$, we have $(x(t), \phi(t)) = (x(t), \psi(t))$.

Lemma 7.2.1. *Let $x \in \mathcal{D}'_+(\mathbb{R})$, $b \in \mathbb{R}$ and $\eta, \mu, \psi \in C^\infty(\mathbb{R}; \mathbb{R})$ be such that $\eta(t) = \mu(t) = 1$ for every $t \in [-1, +\infty)$, $\eta(t) = \mu(t) = 0$ for every $t \in (-\infty, -2]$ and $\operatorname{supp}\psi \subset (-\infty, b]$. Then $\eta\psi, \mu\psi \in \mathcal{D}(\mathbb{R})$ and*

$$(x(t), \eta(t)\psi(t)) = (x(t), \mu(t)\psi(t)).$$

The graph of such function η, with the properties in Lemma 7.2.1, is illustrated in Figure 7.2.1.

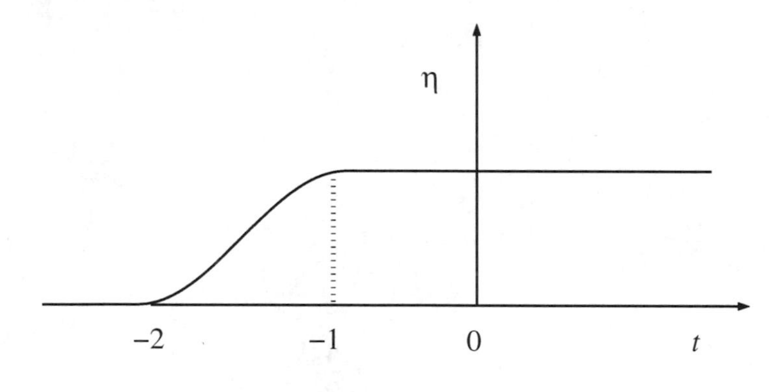

Fig. 7.2.1

Proof. Obviously $\eta\psi, \mu\psi \in C^\infty(\mathbb{R}; \mathbb{R})$. If $b \leq -2$, then $\eta\psi = \mu\psi = 0$. If $b > -2$, $\operatorname{supp}(\eta\psi) \subset [-2, b]$ and $\operatorname{supp}(\mu\psi) \subset [-2, b]$, which shows that $\eta\psi, \mu\psi \in \mathcal{D}(\mathbb{R})$. On the other hand, $\operatorname{supp}((\eta - \mu)\psi) \subset (-\infty, -1] \subset (-\infty, 0)$ and, as $x \in \mathcal{D}'_+(\mathbb{R})$, it follows that

$$(x(t), (\eta(t) - \mu(t))\psi(t)) = 0.$$

The proof is complete. □

Lemma 7.2.2. *If $y \in \mathcal{D}'_+(\mathbb{R})$ and $\phi \in \mathcal{D}(\mathbb{R})$ satisfies $\operatorname{supp}\phi \subset [a, b]$, then the function $t \mapsto (y(s), \phi(t+s))$ is of class C^∞ and its support is included in $(-\infty, b]$.*

Proof. Since y is linear and continuous and ϕ is of class C^∞, by a simple inductive argument, we deduce that the function $t \mapsto (y(s), \phi(t+s))$ is of class C^∞ because

$$\frac{d^k}{dt^k}\,(y(s), \phi(t+s)) = \left(y(s), \frac{d^k}{dt^k}(\phi(t+s))\right)$$

for every $k \in \mathbb{N}$.

In order to prove that the support of this function is included in $(-\infty, b\,]$, let us observe that, for each $t \in \mathbb{R}$, $\operatorname{supp} \phi(t+\cdot) \subset [a-t, b-t\,]$. So, if $b-t < 0$, we have $\operatorname{supp} \phi(t+\cdot) \subset (-\infty, 0)$ and, by consequence $(y(s), \phi(t+s)) = 0$. But this relation shows that the support of the function $t \mapsto (y(s), \phi(t+s))$ is included in $(-\infty, b\,]$ and this completes the proof. □

Corollary 7.2.1. *Let $x, y \in \mathcal{D}'_+(\mathbb{R})$ and $\eta, \mu \in C^\infty(\mathbb{R}; \mathbb{R})$ be such that $\eta(t) = \mu(t) = 1$ for every $t \in [-1, +\infty)$ and $\eta(t) = \mu(t) = 0$ for every $t \in (-\infty, -2\,]$. Then, for every $\phi \in \mathcal{D}(\mathbb{R})$*

$$(x(t), \eta(t)(y(s), \phi(t+s))) = (x(t), \mu(t)(y(s), \phi(t+s))).$$

Proof. In view of Lemma 7.2.2, it follows that the function $\psi : \mathbb{R} \to \mathbb{R}$, defined by $\psi(t) = (y(s), \phi(t+s))$ for every $t \in \mathbb{R}$ is of class C^∞ and its support is included in $(-\infty, b\,]$. So we are in the hypotheses of Lemma 7.2.1, from where the conclusion follows. The proof is complete. □

Corollary 7.2.1 allows us to introduce:

Definition 7.2.1. Let $x, y \in \mathcal{D}'_+(\mathbb{R})$. *The convolution product* of the distributions x and y is the functional $x * y : \mathcal{D}(\mathbb{R}) \to \mathbb{R}$, defined by

$$((x * y)(t), \phi(t)) = (x(t), \eta(t)(y(s), \phi(t+s)))$$

for every $\phi \in \mathcal{D}(\mathbb{R})$, where $\eta \in C^\infty$ is a function with $\eta(t) = 1$ for $t \geq -1$ and $\eta(t) = 0$ for $t \leq -2$.

Remark 7.2.1. For every $x, y \in \mathcal{D}'_+(\mathbb{R})$ the convolution product $x * y$ is well-defined because, according to Corollary 7.2.1, $(x(t), \eta(t)(y(s), \phi(t+s)))$ is independent of the choice of the function η. Moreover, we can show that $x * y$ is a distribution.

Example 7.2.1. Let $x, y : \mathbb{R} \to \mathbb{R}$ be two locally Lebesgue integrable functions with the property that $x(t) = y(t) = 0$ a.e. for $t < 0$. Then, the distributions of type function x and y belong to $\mathcal{D}_+(\mathbb{R})$ and their convolution product $x * y$ is a distribution of type function, $x * y : \mathbb{R} \to \mathbb{R}$, defined by

$$(x * y)(t) = \int_0^t x(\tau) y(t-\tau)\, d\tau \tag{7.2.1}$$

for every $t \in \mathbb{R}$. In order to prove the equality above, let us observe that, for every $\phi \in \mathcal{D}(\mathbb{R})$, we have

$$((x * y)(t), \phi(t)) = (x(t), \eta(t)\,(y(s), \phi(s + t)))$$

$$= \left(x(t), \eta(t) \int_{\mathbb{R}} y(s)\phi(s+t)\,ds\right) = \int_{\mathbb{R}} x(t)\eta(t) \left(\int_{\mathbb{R}} y(s)\phi(s+t)\,ds\right) dt$$

$$= \int_{\mathbb{R}} \int_{\mathbb{R}} x(t)\eta(t)y(s)\phi(t+s)\,ds\,dt,$$

where η is a function of class C^∞ with $\eta(t) = 1$ for $t \in [-1, +\infty)$ and $\eta(t) = 0$ for $t \in (-\infty, -2\,]$. Making the substitution $t + s = \tau$ and using Fubini's theorem (see [Dunford and Schwartz (1958)], Theorem 9, p. 190) in order to interchange the order of integration, we get

$$((x * y)(t), \phi(t)) = \int_{\mathbb{R}} \int_{\mathbb{R}} x(t)\eta(t)y(\tau - t)\phi(\tau)\,d\tau\,dt$$

$$= \int_{\mathbb{R}} \int_{\mathbb{R}} x(t)\eta(t)y(\tau - t)\phi(\tau)\,dt\,d\tau$$

$$= \int_{\mathbb{R}} \phi(\tau) \left(\int_{-\infty}^{0} x(t)\eta(t)y(\tau - t)\,dt\right.$$

$$\left. + \int_0^\tau x(t)\eta(t)y(\tau - t)\,dt + \int_\tau^{+\infty} x(t)\eta(t)y(\tau - t)\,dt\right) d\tau$$

$$= \int_{\mathbb{R}} \phi(\tau) \int_0^\tau x(t)y(\tau - t)\,dt\,d\tau,$$

equality which proves that $x * y$ is of the form (7.2.1).

We notice that we have already used the convolution product of two functions, as expressed by formula (7.2.1), without naming it explicitly. Indeed, let us consider the Cauchy problem for the linear non-homogeneous differential equation

$$\begin{cases} x'(t) = ax(t) + b(t) \\ x(0) = \xi, \end{cases}$$

where $a \in \mathbb{R}$ and $b \in C(\mathbb{R}; \mathbb{R})$. We know that the solution of this problem is given by the variation of constants formula

$$x(t) = e^{at}\xi + \int_0^t e^{a(t-s)}b(s)\,ds$$

for every $t \in \mathbb{R}$. Let us observe that this formula can be rewritten under the form

$$x(t) = e^{at}\xi + b(t) * e^{at},$$

where $b(t) * e^{at}$ is given by (7.2.1). This observation suggests one to define the solution of the equation above by the very same formula, even in the case when b is no longer a function but a distribution in $\mathcal{D}'_+(\mathbb{R})$. We shall develop this idea in the next section.

Proposition 7.2.1. *The convolution product has the following properties:*

(i) $x * (y + z) = x * y + x * z$ *for every* $x, y, z \in \mathcal{D}'_+(\mathbb{R})$;
(ii) $x * \delta = \delta * x = x$ *for every* $x \in \mathcal{D}'_+(\mathbb{R})$;
(iii) $(x \dot{*} y) = x * \dot{y}$ *for every* $x, y \in \mathcal{D}'_+(\mathbb{R})$.

Proof. The property (i) follows from the definition of the convolution product combined with the linearity of the distribution x.

Let $\phi \in \mathcal{D}(\mathbb{R})$ and η be a function as in Definition 7.2.1. We have

$$((x*\delta)(t), \phi(t)) = (x(t), \eta(t)(\delta(s), \phi(t+s))) = (x(t), \eta(t)\phi(t)) = (x(t), \phi(t))$$

because $\eta(t)\phi(t) = \phi(t)$ for every $t \geq 0$ and

$$((\delta * x)(t), \phi(t)) = (\delta(t), \eta(t)(x(s), \phi(s+t))) = (x(s), \phi(s)),$$

equalities which prove (ii).

Finally, let us observe that

$$((x \dot{*} y)(t), \phi(t)) = -((x * y)(t), \phi'(t)) = -(x(t), \eta(t)(y(s), \phi'(s+t)))$$

$$= (x(t), \eta(t)(\dot{y}(s), \phi(s+t))) = ((x * \dot{y})(t), \phi(t))$$

for every $\phi \in \mathcal{D}(\mathbb{R})$, relation which completes the proof. □

Remark 7.2.2. One can show that, for every $x, y \in \mathcal{D}'_+(\mathbb{R})$, we have $x * y = y * x$ (the commutativity of the convolution product). The proof of this property is not elementary and therefore we will not include it here, especially because we will not make use of it in this book.

7.3 Generalized Solutions

In this section, we will present a distributional approach to those equations, or systems of linear non-homogeneous differential equations which, due to the lack of regularity of the right-hand side(s), cannot be handled by classical methods. We begin with the first-order system of linear non-homogeneous differential equations with constant coefficients:

$$\dot{x} = \mathcal{A}x + f \tag{7.3.1}$$

where $\mathcal{A} \in \mathcal{M}_{n\times n}(\mathbb{R})$ and $f \in [\mathcal{D}'(\mathbb{R})]^n$. Since the right-hand side of the equation is an n-tuple whose components are distributions, we have to clarify from the very beginning what do we mean by a solution of equation (7.3.1).

Definition 7.3.1. *A generalized solution*, or *solution distribution* of the equation (7.3.1) is an element $x = (x_1, x_2, \ldots, x_n) \in [\mathcal{D}'(\mathbb{R})]^n$ which satisfies the relation

$$\dot{x} = \mathcal{A}x + f,$$

where $\dot{x} = (\dot{x_1}, \dot{x_2}, \ldots, \dot{x_n})$, while $\dot{x_i}$ stands for the derivative of x_i, $i = 1, 2, \ldots, n$ in the sense of Definition 7.1.7, i.e. the distributional derivative.

Example 7.3.1. Let us consider the first-order linear differential scalar equation

$$\dot{x} = 2x + \delta,$$

where δ is the Dirac delta. Here and in all that follows, we denote by $\theta : \mathbb{R} \to \mathbb{R}$ the function defined by

$$\theta(t) = \begin{cases} 0 & if\, t < 0 \\ 1 & if\, t \geq 0, \end{cases}$$

called *the unit function*, or *the Heaviside function*, and let us observe that $\dot{\theta}(t) = \delta(t)$. Then, a generalized solution for this equation is a distribution x, of type function,

$$x(t) = e^{2t}\theta(t),$$

where $e^{2t}\theta(t)$ represents the product of the function of class C^∞ e^{2t} by the distribution of type function $\theta(t)$.

Indeed, using Leibniz rule, established in the preceding section, one can easily state that

$$\dot{x} = \left(e^{2t}\right)' \theta(t) + e^{2t}\dot{\theta} = 2e^{2t}\theta(t) + e^{2t}\delta(t),$$

where $e^{2t}\delta(t)$ denotes the product of the function of class C^∞ e^{2t} by the distribution δ. See Definition 7.1.4. In order to calculate this product, let us observe that

$$(e^{2t}\delta(t), \phi(t)) = (\delta(t), e^{2t}\phi(t)) = e^0\phi(0) = \phi(0) = (\delta(t), \phi(t))$$

for every $\phi \in \mathcal{D}(\mathbb{R})$. Hence $e^{2t}\delta(t) = \delta(t)$ in the sense of distributions.

We mention that, in general, for every function of class C^∞, η, and for every $a \in \mathbb{R}$, we have

$$\eta(t)\delta(t-a) = \eta(a)\delta(t-a).$$

It then follows that

$$\dot{x} = 2x + \delta,$$

which means that x is a generalized solution of the equation considered.

The next two lemmas will prove useful in the sequel.

Lemma 7.3.1. *A function $\eta \in \mathcal{D}(\mathbb{R})$ is of the form*

$$\eta = \phi', \tag{7.3.2}$$

with $\phi \in \mathcal{D}(\mathbb{R})$, if and only if

$$\int_{\mathbb{R}} \eta(s)\, ds = 0. \tag{7.3.3}$$

Proof. Let us assume that η is of the form (7.3.2). As ϕ is with compact support, there exists $[a,b]$ such that $\phi(t) = 0$ for every $t \in \mathbb{R} \setminus [a,b]$. So

$$\int_{\mathbb{R}} \eta(s)\, ds = \int_{\mathbb{R}} \phi'(s)\, ds = \phi(s)|_{-\infty}^{+\infty} = 0,$$

which proves (7.3.3).

Conversely, if (7.3.3) holds true, then taking $\phi : \mathbb{R} \to \mathbb{R}$, defined by

$$\phi(t) = \int_{-\infty}^{t} \eta(s)\, ds$$

and taking into account the fact that there exists $[a,b]$ such that $\eta(t) = 0$ for every $t \in \mathbb{R} \setminus [a,b]$, we deduce that ϕ is with compact support. More precisely

$$\phi(t) = 0$$

for every $t \in \mathbb{R} \setminus [a,b]$. Since ϕ is obviously of class C^∞, it follows that $\phi \in \mathcal{D}(\mathbb{R})$, and it satisfies (7.3.2). The proof is complete. □

Lemma 7.3.2. *The only generalized solutions of the first-order linear scalar differential equation*

$$\dot{x} = g, \tag{7.3.4}$$

where $g : \mathbb{R} \to \mathbb{R}$ is a continuous function, are the primitives of g.

Proof. We begin by showing that the only generalized solutions of the differential scalar equation

$$\dot{x} = 0 \tag{7.3.5}$$

are the constant functions. To this aim, let $\psi \in \mathcal{D}(\mathbb{R})$, with $\int_{\mathbb{R}} \psi(s)\,ds = 1$. Then

$$\phi(t) = \psi(t)\int_{\mathbb{R}} \phi(s)\,ds + \phi(t) - \psi(t)\int_{\mathbb{R}} \phi(s)\,ds$$

and, by consequence

$$(x(t), \phi(t)) = \left(x(t), \psi(t)\int_{\mathbb{R}} \phi(s)\,ds\right) + \left(x(t), \phi(t) - \psi(t)\int_{\mathbb{R}} \phi(s)\,ds\right).$$

Since

$$\int_{\mathbb{R}} \left(\phi(t) - \psi(t)\int_{\mathbb{R}} \phi(s)\,ds\right) dt = 0,$$

from Lemma 7.3.1, it follows that there exists $\mu \in \mathcal{D}(\mathbb{R})$ such that

$$\phi(t) - \psi(t)\int_{\mathbb{R}} \phi(s)\,ds = \mu'(t)$$

for every $t \in \mathbb{R}$. Then, from (7.3.5), we deduce that

$$\left(x(t), \phi(t) - \psi(t)\int_{\mathbb{R}} \phi(s)\,ds\right) = (x(t), \mu'(t)) = -(\dot{x}(t), \mu(t)) = 0.$$

Hence

$$(x(t), \phi(t)) = \left(x(t), \psi(t)\int_{\mathbb{R}} \phi(s)\,ds\right) = (x(t), \psi(t))\int_{\mathbb{R}} \phi(s)\,ds.$$

Denoting by $c = (x(t), \psi(t))$, the last equality rewrites

$$(x(t), \phi(t)) = \int_{\mathbb{R}} c\phi(s)\,ds,$$

which proves that x is the distribution generated by the function $x(t) = c$ for $t \in \mathbb{R}$. Consequently, the only generalized solutions of equation (7.3.5) are the constant functions.

Now, let us consider equation (7.3.4). Let $G : \mathbb{R} \to \mathbb{R}$ be a primitive of the function g and let us observe that (7.3.4) may be rewritten in the form $\dot{x} = G'$ equality which, by virtue of Proposition 7.1.3, is equivalent to $\dot{x} = \dot{G}$. Since the last relation may be written in the form $\overset{\bullet}{\overbrace{x - G}} = 0$, from what we have previously proved, it follows that $x - G = c$ with $c \in \mathbb{R}$, which completes the proof of the lemma. □

Theorem 7.3.1. *If $f = (f_1, f_2, \dots, f_n)$, where f_i are distributions of type function generated by continuous functions from $\mathbb{R}$ to $\mathbb{R}$, $i = 1, 2, \dots, n$, then the only generalized solutions of equation* (7.3.1) *are the classical ones.*

Proof. From Lemma 7.3.2 we deduce that the only generalized solutions of the system

$$\dot{x}_i = g_i$$

with $g_i : \mathbb{R} \to \mathbb{R}$ continuous, $i = 1, 2, \dots, n$, are of the form $x_i = G_i$, with G_i a primitive of the function g_i, $i = 1, 2, \dots, n$.

Let $y \in [\mathcal{D}'(\mathbb{R})]^n$ be a generalized solution of the system (7.3.1). We emphasize that, by $e^{-t\mathcal{A}}y(t)$ we mean the n-tuple of distributions formally obtained by multiplying the matrix $e^{-t\mathcal{A}}$ by the vector $y(t)$, with the specification that the product of one element $p_{ij}(t)$ of the matrix by a component $y_j(t)$ of the solution $y(t)$ should be understood as the product of the function of class C^∞, $p_{ij}(t)$, by the distribution $y_j(t)$. See Definition 7.1.4.

Then, taking into account that $\mathcal{A}$ commutes with $e^{-t\mathcal{A}}$, we deduce

$$\overbrace{e^{-t\mathcal{A}}y(t)}^{\cdot} = -\mathcal{A}e^{-t\mathcal{A}}y(t) + e^{-t\mathcal{A}}\dot{y}(t)$$
$$= -\mathcal{A}e^{-t\mathcal{A}}y(t) + e^{-t\mathcal{A}}(\mathcal{A}y(t) + f(t)) = e^{-t\mathcal{A}}f(t).$$

So, y satisfies

$$\overbrace{e^{-t\mathcal{A}}y(t)}^{\cdot} = e^{-t\mathcal{A}}f(t).$$

Since the right-hand side of this equation is a continuous function, according to what we have already proved, it follows that

$$e^{-t\mathcal{A}}y(t) = c + \int_0^t e^{-s\mathcal{A}}f(s)\,ds$$

or equivalently

$$y(t) = e^{t\mathcal{A}}c + \int_0^t e^{(t-s)\mathcal{A}}f(s)\,ds.$$

The proof is complete. □

Let us consider now the nth-order linear differential equation

$$y^{(n)}(t) + a_1(t)y^{(n-1)}(t) + \cdots + a_n(t)y(t) = f(t), \tag{7.3.6}$$

where $a_1, a_2, \dots, a_n$ are functions of class C^∞ from $\mathbb{R}$ to $\mathbb{R}$ and $f \in \mathcal{D}'(\mathbb{R})$. For simplicity, we denote by $\mathcal{L} : \mathcal{D}'(\mathbb{R}) \to \mathcal{D}'(\mathbb{R})$ *the nth-order differential operator associated to* (7.3.6), i.e. the operator defined by

$$\mathcal{L}[\,y\,](t) = y^{(n)}(t) + a_1(t)y^{(n-1)}(t) + \cdots + a_n(t)y(t)$$

for every $y \in \mathcal{D}'(\mathbb{R})$, where $a_k(t)y^{(n-k)}(t)$ is the product of the function of class C^∞ a_k by the distributional derivative of order $n-k$, $y^{(n-k)}$, of the distribution y. See Definitions 7.1.4 and 7.1.7.

Definition 7.3.2. By a *generalized solution* of equation (7.3.6), we mean a distribution y which satisfies

$$\mathcal{L}[\,y\,] = f,$$

where $\mathcal{L}$ is the nth-order differential operator associated to equation (7.3.6).

Remark 7.3.1. By Proposition 7.1.3, we have $\mathcal{L}\left(C^n(\mathbb{R};\mathbb{R})\right) \subset C(\mathbb{R};\mathbb{R})$. Then, if $f:\mathbb{R}\to\mathbb{R}$ is a continuous function, every classical solution of equation (7.3.6) is a generalized solution. We recall that (7.3.6) can be rewritten as a first-order linear differential system. So, if a_i are constants, $i = 1, 2, \ldots, n$, and f is continuous, by virtue of Theorem 7.3.1, it follows that the only generalized solutions of equation (7.3.6) are the classical ones.

Definition 7.3.3. By an *elementary solution* of the nth-order differential operator $\mathcal{L}$ we mean a distribution $\mathcal{E}$ which satisfies

$$\mathcal{L}[\,\mathcal{E}\,] = \delta,$$

where δ is the Dirac delta (concentrated at 0).

The next theorem is fundamental in order to understand the mechanism of solving equation (7.3.6).

Theorem 7.3.2. *If $\mathcal{E} \in \mathcal{D}'_+(\mathbb{R})$ is an elementary solution of the nth-order differential operator $\mathcal{L}$, then, for every $f \in \mathcal{D}'_+(\mathbb{R})$, the distribution*

$$y = f * \mathcal{E}$$

is a solution of equation (7.3.6).

Proof. In accordance with Proposition 7.2.1, we have

$$\mathcal{L}[\,y\,] = \mathcal{L}[\,f * \mathcal{E}\,] = f * \mathcal{L}[\,\mathcal{E}\,] = f * \delta = f,$$

which achieves the proof. □

Remark 7.3.2. Essentially, Theorem 7.3.2 asserts that if one knows the response of the system to a "unitary impulse" concentrated at $t = 0$, then one can find out the response of the system to any perturbation f.

In Example 7.3.1 we have determined an elementary solution for a first-order differential operator. In what follows we will present a method to determine an elementary solution to the nth-order differential operator $\mathcal{L}$, associated to equation (7.3.6), whenever the latter has constant coefficients. More precisely, let $\mathcal{L} : \mathcal{D}'(\mathbb{R}) \to \mathcal{D}'(\mathbb{R})$ be defined by

$$\mathcal{L}[\,y\,] = y^{(n)} + a_1 y^{(n-1)} + \cdots + a_n y,$$

where $a_i \in \mathbb{R}$, for $i = 1, 2, \ldots, n$. In order to find out an elementary solution $\mathcal{E} \in \mathcal{D}'_+(\mathbb{R})$ of this operator, we shall use an extension of the variation of constants formula to the framework of generalized solutions. More precisely, let $y_1, y_2, \ldots, y_n$ be a fundamental system of solutions for the homogeneous equation $\mathcal{L}[\,y\,] = 0$. By analogy with the method presented in Section 4.5, we will look for an elementary solution of the form

$$\mathcal{E} = \sum_{i=1}^{n} y_i c_i,$$

where $c_i \in \mathcal{D}'(\mathbb{R})$, for $i = 1, 2, \ldots, n$, satisfy the system

$$\begin{cases} y_1 \dot{c_1} + y_2 \dot{c_2} + \cdots + y_n \dot{c_n} = 0 \\ y_1' \dot{c_1} + y_2' \dot{c_2} + \cdots + y_n' \dot{c_n} = 0 \\ \vdots \\ y_1^{(n-2)} \dot{c_1} + y_2^{(n-2)} \dot{c_2} + \cdots + y_n^{(n-2)} \dot{c_n} = 0 \\ y_1^{(n-1)} \dot{c_1} + y_2^{(n-1)} \dot{c_2} + \cdots + y_n^{(n-1)} \dot{c_n} = \delta \end{cases} \tag{7.3.7}$$

in a distributional sense. See Theorem 4.5.7. We emphasize that in the system (7.3.7) $y^{(n-k)} \dot{c_i}$ is the product of the function of class C^∞ $y^{(n-k)}$ by the distribution $\dot{c_i}$, for $k = 1, 2, \ldots, n$ and $i = 1, 2, \ldots, n$.

This system with unknowns $\dot{c_i}$, $i = 1, 2, \ldots, n$ has the solution

$$\dot{c_i}(t) = \frac{\mathcal{W}_i(t)}{\mathcal{W}(t)} \delta(t) \tag{7.3.8}$$

for $i = 1, 2, \ldots, n$, where $\mathcal{W}$ is the Wronskian of the fundamental system of solutions $y_1, y_2, \ldots, y_n$, while $\mathcal{W}_i$ is a determinant obtained from $\mathcal{W}$ by substituting the column of rank i by a column containing $n-1$ zeros on the first $n-1$ rows and 1 on the last row. As we have already observed in Example 7.3.1, the product of the function of class C^∞ η by the distribution δ is $\eta(t)\delta(t) = \eta(0)\delta(t)$. So

$$\frac{\mathcal{W}_i(t)}{\mathcal{W}(t)} \delta(t) = \frac{\mathcal{W}_i(0)}{\mathcal{W}(0)} \delta(t)$$

and the relation (7.3.8) may be rewritten in the form

$$\dot{c}_i(t) = \frac{\mathcal{W}_i(0)}{\mathcal{W}(0)}\delta(t).$$

At this moment, let us observe that if we choose $y_1, y_2, \dots, y_n$ so that $y_i(0) = e_i$ for $i = 1, 2, \dots, n$, where $e_1, e_2, \dots, e_n$ are the vectors of the canonical basis in $\mathbb{R}^n$, then $\mathcal{W}(0) = 1$ and

$$\mathcal{W}_i(0) = \begin{cases} 0 \textit{ for } i = 1, 2, \dots, n-1 \\ 1 \textit{ for } i = n. \end{cases}$$

Hence, a possible solution of the system above is $c_1 = c_2 = \dots = c_{n-1} = 0$ and $c_n(t) = \theta(t)$, where θ is the distribution generated by the Heaviside function. So, we have

$$\mathcal{E}(t) = y_n(t)\theta(t), \tag{7.3.9}$$

where y_n is the unique global solution, of the homogeneous Cauchy problem

$$\begin{cases} \mathcal{L}[\,y\,] = 0 \\ y(0) = y'(0) = \dots = y^{(n-2)}(0) = 0, \;\; y^{(n-1)}(0) = 1. \end{cases}$$

Obviously $\mathcal{E} \in \mathcal{D}'_+(\mathbb{R})$.

Remark 7.3.3. From Theorem 7.3.2 and from the preceding arguments, it follows that the general solution of the non-homogeneous linear equation $\mathcal{L}[\,y\,] = f$ with $f \in \mathcal{D}'_+(\mathbb{R})$ is given by

$$y = \sum_{i=1}^{n} k_i y_i + f * \mathcal{E},$$

where $k_1, k_2, \dots, k_n \in \mathbb{R}$, $y_1, y_2, \dots, y_n$ is the fundamental system of solutions of equation $\mathcal{L}[\,y\,] = 0$ satisfying $(y_i(0), y_i'(0), \dots y_i^{(n-1)}(0)) = e_i$ for $i = 1, 2, \dots, n$, where $e_1, e_2, \dots, e_n$ are the vectors of the canonical basis in $\mathbb{R}^n$, while $\mathcal{E}$ is given by (7.3.9).

7.4 Carathéodory Solutions

As we have already seen in Section 2.2, there are Cauchy problems which do not admit solutions of class C^1. We recall that, if $f : \mathbb{R} \to \mathbb{R}$ is defined by

$$f(x) = \begin{cases} -1 & \textit{if } x \geq 0 \\ \;\;\,1 & \textit{if } x < 0, \end{cases}$$

then the Cauchy problem

$$\begin{cases} x' = f(x) \\ x(0) = 0 \end{cases}$$

has no C^1 local solution at the right. See Example 2.2.1. This phenomenon is a consequence of the fact that the right-hand side f of the equation is discontinuous with respect to "the state variable" x. One may easily see that not only the discontinuity of f with respect to x might be incompatible with the existence of C^1 solutions. Actually, the discontinuity of f with respect to t is equally responsible for such nonexistence phenomena. For instance, for $f : \mathbb{R} \to \mathbb{R}$ defined by

$$f(t) = \begin{cases} -1 & \textit{if } t \geq 0 \\ 1 & \textit{if } t < 0, \end{cases}$$

the Cauchy problem

$$\begin{cases} x' = f(t) \\ x(0) = 0 \end{cases}$$

has no C^1 local right solutions. On the other hand, instead of C^1 functions, we may allow as candidates for solutions continuous functions $x : \mathbb{J} \to \mathbb{R}$ which are almost everywhere differentiable. Furthermore, we may impose as a qualification criterion for solutions to satisfy the differential equation for every $t \in \mathbb{J} \setminus \mathbb{E}$, where $\mathbb{E}$ is a set of Lebesgue null measure. If this is the case, the second equation above has as solution the function $x : \mathbb{R} \to \mathbb{R}$ defined by $x(t) = -|t|$ for every $t \in \mathbb{R}$. One may easily see that the function x, which is obviously continuous, satisfies the differential equation for every $t \in \mathbb{R} \setminus \{0\}$. So, in the second example, in which the function f is discontinuous only with respect to the t variable, by suitably redefining the concept of solution and by paying the price of some slight modifications, we may completely rebuild the whole theory referring to C^1 solutions developed previously for the continuous right-hand side case.[3] We recall that we have already done such a construction for systems of linear differential equations, when we have introduced the generalized solutions, i.e. solutions distributions. See Section 7.3. Unfortunately, due to its "starting linear philosophy", the

[3]The situation is completely different in the case of the first example in which the discontinuity of the function f with respect to the state variable x leads to a law of evolution which is contradictory by itself. In this case, only the redefinition of the concept of solution in the class of continuous functions cannot ensure the existence. As we shall see in the next sections, in such situations, besides the introduction of a new concept of solution, a "minimal correction" of the evolution law f is necessary, and this, in order to eliminate the existential self-contradiction.

theory there developed can be adapted only to very few nonlinear cases and therefore, in the general nonlinear setting, the construction of a completely different theory is needed. The aim of this section is to present briefly such a theory, initiated at the beginning of the XX century by Constantin Carathéodory by using the framework offered by the very new (at that time) Lebesgue integral.

More precisely, we begin with the definition of the class of functions allowed as right-hand sides in the differential equation corresponding to the Cauchy problem

$$\begin{cases} x' = f(t,x) \\ x(a) = \xi. \end{cases} \tag{$\mathcal{CP}(\mathcal{D})$}$$

Definition 7.4.1. The function $f : \mathbb{I}\times\Omega \to \mathbb{R}^n$ is a *Carathéodory function* if:

(i) for almost all $t \in \mathbb{I}$, the function $x \mapsto f(t,x)$ is continuous from Ω to $\mathbb{R}^n$;

(ii) for every $x \in \Omega$, the function $t \mapsto f(t,x)$ is Lebesgue measurable on $\mathbb{I}$;

(iii) for every $(a,\xi) \in \mathbb{I}\times\Omega$ there exist $r > 0$, $\delta > 0$ and a Lebesgue integrable function $h : [\,a, a+\delta\,] \to \mathbb{R}_+$ such that $B(\xi,r) \subset \Omega$, $[\,a,a+\delta\,] \subset \mathbb{I}$ and

$$\|f(t,x)\| \le h(t) \tag{7.4.1}$$

for every $x \in B(\xi,r)$ and for almost all $t \in [\,a,a+\delta\,]$.

Let $a \in \mathbb{I}$ and let $\delta > 0$ be such that $[\,a,a+\delta\,] \subset \mathbb{I}$.

Definition 7.4.2. A function $x : [\,a,a+\delta\,] \to \Omega$ is *a Carathéodory solution* of the problem $\mathcal{CP}(\mathcal{D})$ if x is absolutely continuous[4] on $[\,a,a+\delta\,]$, $x(a) = \xi$ and x satisfies $x'(t) = f(t,x(t))$ for almost all $t \in [\,a,a+\delta\,]$.

In what follows we will show that, if f is a Carathéodory function, then, for every $(a,\xi) \in \mathbb{I}\times\Omega$, there exists $\delta > 0$ such that $\mathcal{CP}(\mathcal{D})$ has at least one Carathéodory solution defined on $[\,a,a+\delta\,]$.

[4] We recall that a function $x : [\,a,a+\delta\,] \to \mathbb{R}^n$ is *absolutely continuous* if for each $\varepsilon > 0$ there exists $\eta(\varepsilon) > 0$ such that, for all points $t_i, s_i \in [\,a,a+\delta\,]$, $i = 1,2,\ldots,m$, with $\sum_{i=1}^m |t_i - s_i| \le \eta(\varepsilon)$, we have $\sum_{i=1}^m \|x(t_i) - x(s_i)\| \le \varepsilon$. It is known that each absolutely continuous function is almost everywhere differentiable, its derivative is Lebesgue integrable on $[\,a,a+\delta\,]$ and, for each $t,s \in [\,a,a+\delta\,]$, $x(t) - x(s) = \int_s^t x'(\tau)\,d\tau$.

We begin with the particular case in which $\Omega = \mathbb{R}^n$ and f is a Carathéodory function which satisfies (7.4.1) for every $x \in \mathbb{R}^n$, and then we will show how the general case reduces to the latter.

So, let $f : \mathbb{I} \times \mathbb{R}^n \to \mathbb{R}^n$, let $a \in \mathbb{I}$, $\xi \in \Omega$ and let us consider the integral equation with the delay $\lambda > 0$

$$x_\lambda(t) = \begin{cases} \xi, & \text{for } t \in [\,a - \lambda, a\,] \\ \xi + \displaystyle\int_a^t f(\tau, x_\lambda(\tau - \lambda))\, d\tau, & \text{for } t \in (a, a + \delta\,]. \end{cases} \qquad (\mathcal{E}\mathcal{I})_\lambda$$

We begin with the following copy of Lemma 2.2.1. First we recall that a function $h : \mathbb{I} \to \mathbb{R}^n$ is *locally Lebesgue integrable* if its restriction to any compact interval included in $\mathbb{I}$ is Lebesgue integrable on that interval.

Lemma 7.4.1. *Let $f : \mathbb{I} \times \mathbb{R}^n \to \mathbb{R}^n$ be a Carathéodory function for which there exists a locally Lebesgue integrable function $h : \mathbb{I} \to \mathbb{R}_+$, such that*

$$\|f(t, x)\| \le h(t) \qquad (7.4.2)$$

for every $x \in \mathbb{R}^n$ and for almost all $t \in \mathbb{I}$. Then for every $(a, \xi) \in \mathbb{I} \times \mathbb{R}^n$ and every $\delta > 0$ such that $[\,a, a + \delta\,] \subset \mathbb{I}$, $(\mathcal{E}\mathcal{I})_\lambda$ has one and only one absolutely continuous solution defined on $[\,a - \lambda, a + \delta\,]$.

Proof. Let us remark that, if $y : \mathbb{I} \to \mathbb{R}^n$ is a continuous function, then the function $t \mapsto f(t, y(t))$ is Lebesgue measurable on $\mathbb{I}$ and, by virtue of the inequality (7.4.2), it satisfies

$$\|f(t, y(t))\| \le h(t)$$

for almost all $t \in \mathbb{I}$. It follows then that $t \mapsto f(t, y(t))$ is locally Lebesgue integrable. From here, it follows that if x_λ is defined and continuous on an interval of the form $[\,a-\lambda, a+i\lambda\,]$ with $a+i\lambda < a+\delta$, then $s \mapsto f(s, x_\lambda(s-\lambda))$ is Lebesgue integrable on $[\,a, a + (i + 1)\lambda\,]$ and therefore x_λ can be uniquely extended to $[\,a - \lambda, a + (i + 1)\lambda\,]$.

Clearly x_λ is uniquely determined on $[\,a - \lambda, a\,]$ from equation $(\mathcal{E}\mathcal{I})_\lambda$ itself. Let then $t \in [\,a, a + \lambda\,]$. Let us remark that, for every $\tau \in [\,a, t\,]$, we have $\tau - \lambda \in [\,a - \lambda, a\,]$ and therefore $x_\lambda(\tau - \lambda) = \xi$. Therefore

$$x_\lambda(t) = \xi + \int_a^t f(\tau, \xi)\, d\tau$$

and x_λ is uniquely determined on $[\,a, a + \lambda\,]$. Similarly, we can uniquely determine x_λ on $[\,a + \lambda, a + 2\lambda\,]$, $[\,a + 2\lambda, a + 3\lambda\,]$, a.s.o. After m steps, with $m\lambda \ge a+\delta$, we can define x_λ on the whole interval $[\,a, a+\delta\,]$. Observing that x_λ is absolutely continuous, being the primitive of a Lebesgue integrable function, this completes the proof of the lemma. $\square$

As we have already mentioned, we shall prove first the following auxiliary existence result which is interesting by itself.

Lemma 7.4.2. *Let $f : \mathbb{I} \times \mathbb{R}^n \to \mathbb{R}^n$ be a Carathéodory function with the property that there exists a locally Lebesgue integrable function $h : \mathbb{I} \to \mathbb{R}_+$, such that (7.4.2) is satisfied. Then, for every $(a, \xi) \in \mathbb{I} \times \mathbb{R}^n$ and every $\delta > 0$ such that $[\,a, a+\delta\,] \subset \mathbb{I}$, $(\mathcal{CP})$ has at least one Carathéodory solution defined on $[\,a, a+\delta\,]$.*

Proof. Let $(a, \xi) \in \mathbb{I} \times \mathbb{R}^n$ and $\delta > 0$ such that $[\,a, a+\delta\,] \subset \mathbb{I}$, let $m \in \mathbb{N}^*$ and let us consider the integral equation "with the delay $\delta_m = \delta/m$"

$$x_m(t) = \begin{cases} \xi, & \text{for } t \in [\,a - \delta_m, a\,] \\ \xi + \displaystyle\int_a^t f(\tau, x_m(\tau - \delta_m))\,d\tau, & \text{for } t \in (a, a+\delta\,]. \end{cases} \qquad (\mathcal{EI})_m$$

Let us remark that, by virtue of Lemma 7.4.1, for every $m \in \mathbb{N}^*$, $(\mathcal{EI})_m$ has a unique absolutely continuous solution $x_m : [\,a - \delta_m, a+\delta\,] \to \mathbb{R}^n$.

We will show in what follows that the family of functions $\{x_m;\ m \in \mathbb{N}^*\}$ is uniformly bounded and equicontinuous on $[\,a, a+\delta\,]$. For the beginning let us observe that, by virtue of the inequality (7.4.2), we have

$$\|x_m(t)\| \le \|\xi\| + \int_a^t h(s)\,ds \le \|\xi\| + \int_a^{a+\delta} h(s)\,ds$$

for every $m \in \mathbb{N}^*$ and $t \in [\,a, a+\delta\,]$. So $\{x_m;\ m \in \mathbb{N}^*\}$ is uniformly bounded on $[\,a, a+\delta\,]$.

Next, let us observe that, also from (7.4.2), we have

$$\|x_m(t) - x_m(s)\| \le \left| \int_s^t h(\tau)\,d\tau \right|$$

for every $m \in \mathbb{N}^*$ and $t, s \in [\,a, a+\delta\,]$. Since h is Lebesgue integrable on $[\,a, a+\delta\,]$, the function $t \mapsto \int_a^t h(\tau)\,d\tau$ is absolutely continuous on $[\,a, a+\delta\,]$ and, by consequence, the preceding inequality shows that $\{x_m;\ m \in \mathbb{N}^*\}$ is equicontinuous on $[\,a, a+\delta\,]$. By virtue of Arzelá-Ascoli's Theorem 12.2.1, it follows that $(x_m)_{m\in\mathbb{N}^*}$ has at least one subsequence, denoted for simplicity again by $(x_m)_{m\in\mathbb{N}^*}$, uniformly convergent on $[\,a, a+\delta\,]$ to a continuous function x. Obviously, we have

$$\lim_{m\to\infty} x_m(\tau - \delta_m) = x(\tau),$$

uniformly for $\tau \in [\,a, a+\delta\,]$. Since f is a Carathéodory function on $\mathbb{I} \times \mathbb{R}^n$, we have

$$\lim_{m\to\infty} f(\tau, x_m(\tau - \delta_m)) = f(\tau, x(\tau)),$$

for almost all $\tau \in [a, a+\delta]$. From this relation, from (7.4.2) and the Lebesgue's dominated convergence theorem, see [Dunford and Schwartz (1958)], Theorem 7, p. 124, it follows that we can pass to the limit in $(\mathcal{EI})_m$ for $m \to \infty$. We deduce that x satisfies

$$x(t) = \xi + \int_a^t f(\tau, x(\tau))\, d\tau$$

for every $t \in [a, a+\delta]$, and therefore x is absolutely continuous on $[a, a+\delta]$ and $x(a) = \xi$. Thus, $x'(t) = f(t, x(t))$ for almost all $t \in [a, a+\delta]$, and this completes the proof of the lemma. □

Remark 7.4.1. In the hypotheses of Lemma 7.4.2, we can prove that for every initial data $(a, \xi) \in \mathbb{I} \times \mathbb{R}^n$, $\mathcal{CP}(\mathcal{D})$ has at least one global solution.

We can now proceed to the formulation of the main result in this section. To this aim, let $\mathbb{I}$ be a nonempty and open interval in $\mathbb{R}$, let Ω be a nonempty and open subset of $\mathbb{R}^n$ and $f : \mathbb{I} \times \Omega \to \mathbb{R}^n$ a given function.

Theorem 7.4.1. (Carathéodory). *If $f : \mathbb{I} \times \Omega \to \mathbb{R}^n$ is a Carathéodory function then, for every $(a, \xi) \in \mathbb{I} \times \Omega$, there exists $\delta > 0$ with $[a, a+\delta] \subset \mathbb{I}$ and such that $\mathcal{CP}(\mathbb{I}, \Omega, f, a, \xi)$ has at least one Carathéodory solution defined on $[a, a+\delta]$.*

Proof. Let $(a, \xi) \in \mathbb{I} \times \Omega$. Since both $\mathbb{I}$ and Ω are open, there exist $d > 0$ and $r > 0$ such that $[a-d, a+d] \subset \mathbb{I}$ and

$$B(\xi, r) = \{\eta \in \mathbb{R}^n;\ \|\eta - \xi\| \le r\} \subset \Omega.$$

Taking into account that f is a Carathéodory function, diminishing r if necessary, we may assume that there exists a Lebesgue integrable function $h : [a, a+d] \to \mathbb{R}_+$, such that (7.4.1) be satisfied. Let us define $\rho : \mathbb{R}^n \to \mathbb{R}^n$ by

$$\rho(y) = \begin{cases} y & \text{for } y \in B(\xi, r) \\ \dfrac{r}{\|y - \xi\|}(y - \xi) + \xi & \text{for } y \in \mathbb{R}^n \setminus B(\xi, r). \end{cases}$$

One can easily see that ρ maps $\mathbb{R}^n$ in $B(\xi, r)$ and is continuous on $\mathbb{R}^n$.

Now, let us define $g : (a-d, a+d) \times \mathbb{R}^n \to \mathbb{R}^n$, by

$$g(t, y) = f(t, \rho(y))$$

for every $(t, y) \in (a-d, a+d) \times \mathbb{R}^n$.

Since f is a Carathéodory function , while r, d and h are chosen such that (7.4.1) holds, it follows that g satisfies all the hypotheses of Lemma 7.4.2. So, for every $k \in (0, d)$, the Cauchy problem

$$\begin{cases} x' = g(t, x) \\ x(a) = \xi \end{cases}$$

has at least one Carathéodory solution $x : [\,a, a + k\,] \to \mathbb{R}^n$. Since $x(a) = \xi$ and x is continuous at $t = a$, for $r > 0$, there exists $\delta \in (0, k]$ with the property that for every $t \in [\,a, a + \delta\,]$, $\|x(t) - \xi\| \leq r$. But, in this case, $g(t, x(t)) = f(t, x(t))$ and therefore $x : [\,a, a + \delta\,] \to \Omega$ is a Carathéodory solution of $\mathcal{CP}(\mathbb{I}, \Omega, f, a, \xi)$. The proof is complete. □

Remark 7.4.2. We mention that, whenever f is continuous, then the only Carathéodory solutions of $\mathcal{CP}(\mathcal{D})$ are the classical ones, i.e. of class C^1. Indeed, in this case, if $x : [\,a, b\,] \to \Omega$ is a Carathéodory solution then it is continuous. Since f is continuous, it follows that x' is continuous too, which shows that x is of class C^1. If f is discontinuous as function of the t variable, even at a single point, then $\mathcal{CP}(\mathcal{D})$ might have Carathéodory solution but no classical solution, as we can see from the example below.

Example 7.4.1. The Cauchy problem

$$\begin{cases} x' = x \operatorname{sgn}(t - 1) \\ x(0) = 1 \end{cases}$$

has as unique saturated Carathéodory solution the function $x : \mathbb{R}_+ \to \mathbb{R}$, defined by $x(t) = e^{|t-1|-1}$ for every $t \in \mathbb{R}_+$, function which is not of class C^1.

7.5 Differential Inclusions

The evolutions of certain phenomena which present one or more states of ambiguity cannot have in general as satisfactory mathematical models differential equations, or systems of differential systems. Some of these phenomena can be fairly well described by the so-called *differential inclusions*. Roughly speaking, these are generalizations of differential equations in that, instead of a single-valued function, they have, on the right-hand side, a set-valued function (called for this reason *multi-valued function*). The next simple but very instructive model of *pursuit-evasion* due to [Brezis (1975)] is illuminating in this respect.

Example 7.5.1. A policeman P chases a gangster G. The policeman's strategy is to run as fast as he can towards the gangster. So, if $g(t)$ and $p(t)$ are the position vectors, at the moment t, of the gangster and of the policeman respectively, then the policeman's speed $p'(t)$ is given by

$$p'(t) = V\frac{g(t) - p(t)}{\|g(t) - p(t)\|}$$

if $p(t) \neq g(t)$ and $p'(t) \in B(0, V)$ if $p(t) = g(t)$, where V represents the maximal speed which P can reach. See Figure 7.5.1 below.

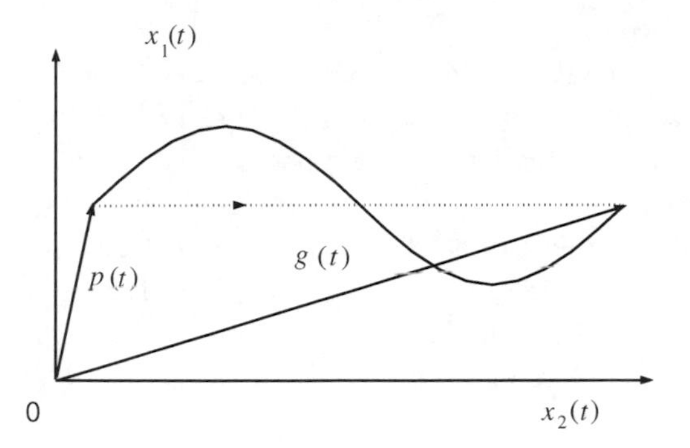

Fig. 7.5.1

Introducing the set-valued function $\mathcal{A} : \mathbb{R}^2 \to 2^{\mathbb{R}^2}$, by

$$\mathcal{A}(x) = \begin{cases} \left\{-V\dfrac{x}{\|x\|}\right\} & \textit{if } x \neq 0 \\ B(0, V) & \textit{if } x = 0 \end{cases}$$

and the new unknown functions $u = p - g$, the relations above may be rewritten in the form

$$u' \in \mathcal{A}(u) + h(t),$$

where $h = g'$. One may easily see that, on $\mathbb{R}^2 \setminus \{0\}$, $\mathcal{A}$ can be identified with a continuous and dissipative function. Let us observe that in fact $\mathcal{A}$ constitutes, up to a multiplicative constant, a generalization of the signum function. We emphasize that, in this example the state of ambiguity of the system is that one at which $p(t) = g(t)$, where the sense of the displacement has no relevance. It should also be noticed that, exactly at this state, one has to consider as value of the right-hand side a set and not a single point.

Another situation, which may suggest the consideration of some differential inclusions as alternative models, is that one in which, from various reasons, the corresponding "single-valued differential models" has no "classical solutions". For instance, let us consider the Cauchy problem

$$\begin{cases} x' = f(x) \\ x(0) = 0, \end{cases} \tag{7.5.1}$$

where $f : \mathbb{R} \to \mathbb{R}$ is defined by

$$f(x) = \begin{cases} 1 & \text{if } x < 0 \\ -1 & \text{if } x \geq 0. \end{cases}$$

As we have already seen in Example 2.2.1, this problem has no local classical right solution. More precisely, there exists no function of class C^1 x, defined on a right neighborhood of 0, to satisfy (7.5.1). For this reason, in order to give a "reasonable sense" to this problem, we must first enlarge the class of all possible candidates to the title of solution of (7.5.1), and second to replace the right-hand side of the equation in (7.5.1) by a "less discontinuous function", but "as close as possible" to the initial one. In this respect, we might replace (7.5.1) by

$$\begin{cases} x' \in F(x) \\ x(0) = 0, \end{cases} \tag{7.5.2}$$

where $F : \mathbb{R} \to 2^{\mathbb{R}}$ is defined by

$$F(x) = \begin{cases} \{1\} & \text{if } x < 0 \\ [-1, 1] & \text{if } x = 0 \\ \{-1\} & \text{if } x > 0. \end{cases} \tag{7.5.3}$$

We leave it to the reader to find out the similarities as well as the differences between the graph of the function f and that one of the set-valued function F as illustrated in Figure 7.5.2 (a) and (b), respectively.

In this context, we might accept as solution of equation (7.5.1) any almost everywhere differentiable function $x : [0, \delta) \to \mathbb{R}$ which satisfies both the differential inclusion in (7.5.2) for almost all $t \in [0, \delta)$, and the initial condition.

In view of a rigorous development of these ideas, some concepts and results concerning multi-valued functions are needed first. In order not to complicate the notations, we will confine ourselves to the autonomous case, leaving to the interested reader the extension to the general case. Let Ω be a nonempty subset in $\mathbb{R}^n$. A *multi-valued*, or *set-valued function*, or *multifunction*, defined on Ω with values in $\mathbb{R}^n$, is a function $F : \Omega \to 2^{\mathbb{R}^n}$.

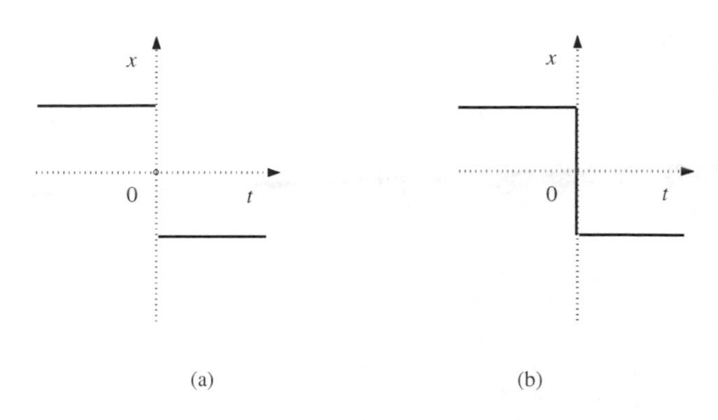

Fig. 7.5.2

Definition 7.5.1. The multi-valued function $F : \Omega \to 2^{\mathbb{R}^n}$ is *upper semi-continuous (u.s.c.)* at $x \in \Omega$ if for every open subset D with $F(x) \subset D$ there exists an open subset $V \subset \mathbb{R}^n$ such that for every $y \in V \cap \Omega$, we have $F(y) \subset D$. A multi-valued function which is upper semi-continuous at every $x \in \Omega$ is called *upper semi-continuous* on Ω.

Remark 7.5.1. In the case in which a multi-valued function F has as values only singletons it can be identified in a natural way with a single-valued function from Ω in $\mathbb{R}^n$ denoted, for simplicity, also by F. In this case F is u.s.c. on Ω if and only if F is continuous on Ω in the usual sense.

Remark 7.5.2. Let us observe that the multi-valued function defined by means of the relation (7.5.3) is u.s.c. on $\mathbb{R}^n$.

In what follows, if A, B are two subsets in $\mathbb{R}^n$, we denote by $A + B$ their sum, i.e. the set of all elements z in $\mathbb{R}^n$ of the form $z = x + y$ with $x \in A$ and $y \in B$. By $S(0, \varepsilon)$ we denote the open ball centered at 0 and of radius ε in $\mathbb{R}^n$.

Proposition 7.5.1. *Let $F : \Omega \to 2^{\mathbb{R}^n}$ and $x \in \Omega$. If $F(x)$ is compact then F is u.s.c. at x if and only if for every $\varepsilon > 0$ there exists $\delta > 0$ such that $F(y) \subset F(x) + S(0, \varepsilon)$ for every $y \in \Omega$ with $\|y - x\| < \delta$.*

Proof. The necessity follows from the simple observation that for every $\varepsilon > 0$, $F(x) + S(0, \varepsilon)$ is an open set which includes $F(x)$. For the sufficiency, let us remark that, if D is open with $F(x) \subset D$, then $F(x) \cap \partial D = \emptyset$. Since $F(x)$ is compact and ∂D is closed, in view of Lemma 2.5.1, it follows that the distance d from $F(x)$ to ∂D is strictly positive. Taking $\varepsilon = d/2$, from the hypothesis and the preceding remark, it follows that there exists $\delta > 0$ such that $F(y) \subset F(x) + S(0, \varepsilon) \subset D$ for each $y \in \Omega$ with $\|y - x\| < \delta$, which completes the proof of the sufficiency. □

Definition 7.5.2. The multi-valued function $F : \Omega \to 2^{\mathbb{R}^n}$ is *bounded* on Ω if

$$\{y;\ y \in F(x),\ x \in \Omega\}$$

is bounded.

The next lemma is an extension of Weierstrass theorem in Real Analysis.

Lemma 7.5.1. *If $\Omega \subset \mathbb{R}^n$ is nonempty and compact and $F : \Omega \to 2^{\mathbb{R}^n}$ is u.s.c. on Ω and, for every $x \in \Omega$, $F(x)$ is a bounded set, then F is bounded on Ω.*

Proof. Let us assume the contrary. Then there would exist a nonempty and compact subset Ω in $\mathbb{R}^n$ and a u.s.c. multi-valued function, F, on Ω, with bounded values, which is not bounded on Ω. This means that for every $n \in \mathbb{N}^*$ there exist $x_n \in \Omega$ and $y_n \in F(x_n)$ such that

$$\|y_n\| \geq n. \tag{7.5.4}$$

Since Ω is compact, we may assume without loss of generality that there exists $x \in \Omega$ such that, on a subsequence at least, we have

$$\lim_{n\to\infty} x_n = x.$$

By virtue of Proposition 7.5.1 combined with the relation above, it follows that for $\varepsilon = 1$ there exists $n_0 \in \mathbb{N}^*$ such that, for every $n \geq n_0$, we have $F(x_n) \subset F(x) + S(0,1)$. Since $F(x) + S(0,1)$ is bounded and $y_n \in F(x_n)$ for every $n \in \mathbb{N}^*$, the preceding inclusion is in contradiction with the inequality (7.5.4). This contradiction is a consequence of the initial supposition that there would exist a nonempty and compact subset Ω in $\mathbb{R}^n$ and a u.s.c. multi-valued function, F, on Ω, with bounded values, which is unbounded on Ω. The proof is complete. □

Let $F : \Omega \to 2^{\mathbb{R}^n}$ be a multi-valued function, $\xi \in \Omega$ and let us consider the Cauchy problem with data $\mathcal{D} = (F, \Omega, 0, \xi)$

$$\begin{cases} x' \in F(x) \\ x(0) = \xi. \end{cases} \tag{$\mathcal{CP}(\mathcal{D})$}$$

Definition 7.5.3. *An a.e. solution* of $\mathcal{CP}(\mathcal{D})$ is a function $x : \mathbb{I} \to \Omega$, with $\mathbb{I}$ a right neighborhood of 0, which is absolutely continuous on $\mathbb{I}$ and which satisfies both $x(0) = \xi$ and $x'(t) \in F(x(t))$ a.e. for $t \in \mathbb{I}$.

Let $\lambda > 0$ and $a_\lambda : [0, +\infty) \to \mathbb{R}$ defined by

$$a_\lambda(t) = (i-1)\lambda$$

if $t \in [(i-1)\lambda, i\lambda)$, $i \in \mathbb{N}$.

Let us consider the differential inclusion with the delay $\lambda > 0$

$$\begin{cases} x'_\lambda(t) \in F(x_\lambda(a_\lambda(t))) & \textit{for } t \in [0, \delta) \\ x_\lambda(t) = \xi & \textit{for } t \in [-\lambda, 0]. \end{cases} \qquad \mathcal{CP}_\lambda(\mathcal{D})$$

By an *a.e. solution* of $\mathcal{CP}_\lambda(\mathcal{D})$ on $[0, \delta)$ we mean an absolutely continuous function $x_\lambda : [-\lambda, \delta) \to \mathbb{R}^n$ which satisfies $x_\lambda(t) = \xi$ for every $t \in [-\lambda, 0]$ and

$$x'_\lambda(t) \in F(x_\lambda(a_\lambda(t)))$$

a.e. $t \in [0, \delta)$. We begin with the following simple, but useful lemma.

Lemma 7.5.2. *If $F : \mathbb{R}^n \to 2^{\mathbb{R}^n}$ is nonempty-valued and $\lambda > 0$ then, for every $\xi \in \mathbb{R}^n$ and every $\delta > 0$, $\mathcal{CP}_\lambda(\mathcal{D})$ has at least one a.e. solution defined on $[0, \delta]$.*

Proof. Obviously x_λ is uniquely determined on $[-\lambda, 0]$ from "the initial condition" itself. Let then $t \in [0, \lambda)$. Let us remark that $a_\lambda(t) = 0$ and therefore $x_\lambda(a_\lambda(t)) = \xi$. Fix y_1 in $F(\xi)$ and let us define $x_\lambda : [0, \lambda] \to \mathbb{R}^n$ by

$$x_\lambda(t) = \xi + ty_1$$

for $t \in [0, \lambda]$. One may easily see that x is a.e. solution of $\mathcal{CP}_\lambda(\mathcal{D})$ on $[0, \lambda)$. Taking $t \in [\lambda, 2\lambda)$, let us observe that, $a_\lambda(t) = \lambda$ and, by consequence, we have

$$x_\lambda(a_\lambda(t)) = x_\lambda(\lambda) = \xi + \lambda y_1.$$

Let us fix y_2 in $F(\xi + \lambda y_1)$ and let us define $x_\lambda : [\lambda, 2\lambda] \to \mathbb{R}^n$ by

$$x_\lambda(t) = \xi + \lambda y_1 + (t - \lambda)y_2$$

for $t \in [\lambda, 2\lambda]$. Similarly, we can uniquely determine x_λ successively on $[2\lambda, 3\lambda]$, $[3\lambda, 4\lambda]$, a.s.o. After m steps, with $m\lambda \geq \delta$, we can define x_λ on the whole interval $[0, \delta]$, and this completes the proof of the lemma. □

By analogy with the single-valued case, we shall prove first the following existence result which, although auxiliary, is interesting by itself.

Lemma 7.5.3. *If $F : \mathbb{R}^n \to 2^{\mathbb{R}^n}$ is a nonempty compact convex valued multi-valued function which is u.s.c. and bounded on $\mathbb{R}^n$, then for every $\xi \in \mathbb{R}^n$ and every $\delta > 0$, $\mathcal{CP}(F, \mathbb{R}^n, 0, \xi)$ has at least one a.e. solution defined on $[0, \delta]$.*

Proof. Let $\xi \in \mathbb{R}^n$ and $\delta > 0$, let $m \in \mathbb{N}^*$, and let us consider the differential inclusion with the delay $\delta_m = \delta/m$

$$\begin{cases} x_m'(t) \in F(x_m(a_m(t))) \\ x_m(t) = \xi \text{ for } t \in [-\delta_m, 0], \end{cases} \qquad \mathcal{CP}_m(\mathcal{D})$$

where $a_m : [0, +\infty) \to \mathbb{R}$ is given by

$$a_m(t) = (i-1)\delta_m$$

if $t \in [(i-1)\delta_m, i\delta_m)$, $i \in \mathbb{N}$. By virtue of Lemma 7.5.2, for every $m \in \mathbb{N}^*$, $\mathcal{CP}_m(\mathcal{D})$ has at least one a.e. solution $x_m : [-\delta_m, \delta] \to \mathbb{R}^n$. For every $m \in \mathbb{N}^*$ let us fix such an a.e. solution x_m with the property that its a.e. derivative is a step function. We can do that, thanks to the construction described in Lemma 7.5.2.

We will show in what follows that $\{x_m;\ m \in \mathbb{N}^*\}$ is uniformly bounded and equicontinuous on $[0, \delta]$. Since F is bounded on $\mathbb{R}^n$, there exists $M > 0$ such that

$$\|x_m'(t)\| \leq M \tag{7.5.5}$$

for every $m \in \mathbb{N}^*$ and a.e. $t \in [0, \delta]$. Since

$$x_m(t) = \xi + \int_0^t x_m'(s)\, ds$$

for every $t \in [0, \delta]$, we conclude that $\|x_m(t)\| \leq \|\xi\| + \delta M$, for every $m \in \mathbb{N}^*$ and $t \in [0, \delta]$. Consequently, $\{x_m;\ m \in \mathbb{N}^*\}$ is uniformly bounded on $[0, \delta]$. Next, let us observe that, from (7.5.5), we have

$$\|x_m(t) - x_m(s)\| \leq \left| \int_s^t \|x_m'(\tau)\|\, d\tau \right| \leq M|t-s| \tag{7.5.6}$$

for every $m \in \mathbb{N}^*$ and $t, s \in [0, \delta]$. So, $\{x_m;\ m \in \mathbb{N}^*\}$ is equicontinuous on $[0, \delta]$. By virtue of Theorem 12.2.1, it follows that $(x_m)_{m\in\mathbb{N}^*}$ has at least one subsequence, denoted for simplicity again by $(x_m)_{m\in\mathbb{N}^*}$, which is uniformly convergent on $[0, \delta]$ to a continuous function x. From (7.5.6), we deduce that

$$\|x(t) - x(s)\| \leq M|t-s|$$

for every $t, s \in [0, \delta]$, and therefore x is absolutely continuous on $[0, \delta]$. In order to complete the proof it suffices to show that

$$x'(t) \in F(x(t))$$

a.e. $t \in [0, \delta]$. To this aim let D be the set of all points $t \in [0, \delta]$ at which, both x, and x_m are differentiable for every $m \in \mathbb{N}^*$. One may easily see that the Lebesgue measure of the set D equals δ. Let $t \in D$, $m \in \mathbb{N}^*$ and $h > 0$. Let us observe that

$$\frac{1}{h}\left(x_m(t+h) - x_m(t)\right) = \frac{1}{h}\int_t^{t+h} x_m'(s)\,ds.$$

Since x_m' is a step function, it follows that the right-hand side of the equality above is a convex combination of elements in

$$\bigcup_{s\in[\,t,t+h\,]} F(x_m(a_m(s))).$$

In other words,

$$\frac{1}{h}\left(x_m(t+h) - x_m(t)\right) \in conv \bigcup_{s\in[\,t,t+h\,]} F(x_m(a_m(s))). \tag{7.5.7}$$

We recall that $\operatorname{conv}(F)$ is the set of all convex combinations of elements of the set F. Let $\varepsilon > 0$. Since F is u.s.c. at $x(t)$ and $F(x(t))$ is compact, by virtue of Proposition 7.5.1 it follows that there exists $\eta = \eta(\varepsilon) > 0$ such that, for every $y \in \mathbb{R}^n$ with $\|y - x(t)\| < \eta$, we have $F(y) \subset F(x(t)) + S(0,\varepsilon)$. Since

$$\|x_m(a_m(s)) - x(t)\| \le \|x_m(a_m(s)) - x_m(s)\| + \|x_m(s) - x(s)\| + \|x(s) - x(t)\|$$

for every $m \in \mathbb{N}^*$ and $t, s \in [\,0,\delta\,]$, recalling that

$$\lim_{m\to\infty} x_m(s) = x(s) \quad and \quad \lim_{m\to\infty} a_m(s) = s$$

uniformly on $[\,0,\delta\,]$ and that the family $\{x_m;\ m \in \mathbb{N}^*\}$ is equicontinuous, we deduce that, for $\eta = \eta(\varepsilon) > 0$, there exist $h_\varepsilon > 0$ and $m_\varepsilon \in \mathbb{N}^*$ such that

$$\|x_m(a_m(s)) - x(t)\| \le \eta \tag{7.5.8}$$

for every $m \in \mathbb{N}^*$, $m \ge m_\varepsilon$ and every $s \in [\,t, t+h_\varepsilon]$. From the definition of η and from (7.5.8), it follows

$$\bigcup_{s\in[\,t,t+h\,]} F(x_m(a_m(s))) \subset F(x(t)) + S(0,\varepsilon)$$

for every $m \in \mathbb{N}^*$, $m \ge m_\varepsilon$ and $h \in (0, h_\varepsilon\,]$. Since $F(x(t)) + S(0,\varepsilon)$ is convex, we deduce

$$conv \bigcup_{s\in[\,t,t+h\,]} F(x_m(a_m(s))) \subset F(x(t)) + S(0,\varepsilon)$$

for every $m \in \mathbb{N}^*$, $m \ge m_\varepsilon$ and $h \in (0, h_\varepsilon\,]$. Using this inclusion and passing to the limit in (7.5.7) first for $m \to \infty$ and second for $h \downarrow 0$, we deduce

$$x'(t) \in F(x(t)) + S(0,\varepsilon)$$

for every $\varepsilon > 0$. Since $F(x(t))$ is closed, this relation implies

$$x'(t) \in F(x(t)).$$

Recalling that the relation above is satisfied for every $t \in D$ and that $[\,0,\delta\,] \setminus D$ has null measure, this completes the proof. □

The main result in this section is the following generalization of Peano's Theorem 2.2.1.

Theorem 7.5.1. *Let $\Omega \subset \mathbb{R}^n$ be nonempty and open. If $F : \Omega \to 2^{\mathbb{R}^n}$ is u.s.c. on Ω and has nonempty compact and convex values then, for every ξ from Ω, $\mathcal{CP}(\mathcal{D})$ has at least one local a.e. solution.*

Proof. Let $\xi \in \Omega$. Since Ω is open, there exists $r > 0$ such that

$$B(\xi, r) = \{\eta \in \mathbb{R}^n;\ \|\eta - \xi\| \leq r\} \subset \Omega.$$

We define $\rho : \mathbb{R}^n \to \mathbb{R}^n$ by

$$\rho(y) = \begin{cases} y & \text{for } y \in B(\xi, r) \\ \dfrac{r}{\|y - \xi\|}(y - \xi) + \xi & \text{for } y \in \mathbb{R}^n \setminus B(\xi, r). \end{cases}$$

One can easily see that ρ maps $\mathbb{R}^n$ in $B(\xi, r)$ and is continuous on $\mathbb{R}^n$.

Let us define $G : \mathbb{R}^n \to 2^{\mathbb{R}^n}$ by

$$G(y) = F(\rho(y)), \text{ for every } y \in \mathbb{R}^n.$$

Since F is u.s.c. and has bounded values, from Lemma 7.5.1, it follows that its restriction to $B(\xi, r)$ is bounded. So, G is bounded on $\mathbb{R}^n$. In addition, G is u.s.c. on $\mathbb{R}^n$ being the superposition of two functions with this property. From Lemma 7.5.3 we know that, for every $d > 0$, the Cauchy problem

$$\begin{cases} x' \in G(x) \\ x(0) = \xi \end{cases}$$

has at least one a.e. solution $x : [\,0, d\,] \to \mathbb{R}^n$. Since $x(0) = \xi$ and x is continuous at $t = 0$, for $r > 0$, there exists $\delta \in (0, d\,]$ such that, for every $t \in [\,0, \delta\,]$, $\|x(t) - \xi\| \leq r$. But in this case $G(x(t)) = F(x(t))$ and therefore $x : [\,0, \delta\,] \to \Omega$ is a solution of $\mathcal{CP}(\mathcal{D})$. The proof is complete. □

For more details on such kind of problems, the reader is referred to [Aubin and Cellina (1984)], [Cârjă (2003)] and [Vrabie (1995)].

7.6 Variational Inequalities

The evolution of certain systems from chemistry, physics, biology, sociology *etc.* is described by mathematical models expressed by differential inequalities with state constraints of the type

$$\begin{cases} x(t) \in K \\ \langle x'(t) - f(x(t)) - g(t), x(t) - u \rangle \leq 0 \\ x(0) = \xi \end{cases} \tag{7.6.1}$$

for every $u \in K$ and a.e. $t \in [0,T]$, where K is a nonempty convex and closed subset of $\mathbb{R}^n$, $f : K \to \mathbb{R}^n$ is Lipschitz continuous, $g : [0,T] \to \mathbb{R}^n$ is continuous and $\xi \in K$. Problems of this type have been considered and studied for the first time by Jaques-Louis Lions and Guido Stampacchia, under the name of *evolution variational inequalities.*

In order to illustrate the importance of such mathematical models, let us analyze an example in population dynamics.

Example 7.6.1. Let us consider a species of fish living in a lake and whose free evolution, i.e. non-influenced by external factors, is described by means of the Cauchy problem associated to the logistic equation

$$\begin{cases} x' = cx(b-x) \\ x(0) = \xi, \end{cases}$$

where $x(t)$ represents the fish population at the time t, c and b are two positive constants, and ξ is the initial fish population. See Subsection 1.4.6. Now, let us assume that, at each moment $t \geq 0$, one harvests a constant number of fishes in such a way that the remaining number of fishes in the lake belongs to a given interval $K = [c_1, c_2]$, with $0 < c_1 < c_2 < b$. In these conditions, it is easy to see that x has to satisfy

$$\begin{cases} x(t) \in K & \text{for every } t \geq 0 \\ x'(t) - cx(t)(b - x(t)) = 0 & \text{if } x(t) \in (c_1, c_2) \\ x'(t) - cx(t)(b - x(t)) > 0 & x(t) = c_1 \\ x'(t) - cx(t)(b - x(t)) < 0 & x(t) = c_2 \\ x(0) = \xi. \end{cases}$$

Let us observe that the system above may be equivalently rewritten in the form (7.6.1), with $f : [c_1, c_2] \to \mathbb{R}$ defined by

$$f(x) = cx(b-x)$$

for every $x \in [c_1, c_2]$ and $g \equiv 0$. So, the problems of the form (7.6.1) are completely justified by practice, in the sense that they furnish a better description of the evolution of those phenomena when the state x is subjected to a point-wise restriction of the form $x(t) \in K$ for every $t \geq 0$.

Let $x \in K$ and let us denote by

$$N(x) = \{w \in \mathbb{R}^n;\ \langle w, x - u\rangle \geq 0 \text{ for every } u \in K\},$$

set called *the normal cone to the set K at the point x*. Let us observe that the problem (7.6.1) may be equivalently rewritten as a Cauchy problem for a differential inclusion of the form

$$\begin{cases} x' \in f(x) + g(t) - N(x) \\ x(0) = \xi. \end{cases} \tag{7.6.2}$$

One may easily see that, for every x in the interior of the set K, we have $N(x) = \{0\}$. So, in the case $K = \mathbb{R}^n$, we have $N(x) = \{0\}$ for every $x \in \mathbb{R}^n$ and the differential inclusion above reduces to the differential equation $x' = f(x) + g$.

The main result referring to (7.6.2) is:

Theorem 7.6.1. *Let $K \subset \mathbb{R}^n$ be a nonempty closed and convex subset, let $f : K \to \mathbb{R}^n$ be a Lipschitz continuous function and $g : [\,0, T\,] \to \mathbb{R}^n$ a function of class C^1. Then, for every $\xi \in K$ there exists a unique absolutely continuous function $x : [\,0, T\,] \to K$ which satisfies* (7.6.2). *In addition x satisfies*

$$x'(t) = f(x(t)) + g(t) - \mathcal{P}_{N(x(t))}(f(x(t)) + g(t)) \tag{7.6.3}$$

a.e. $t \in [\,0, T\,]$[5].

Proof. First let us observe that we may assume with no loss of generality that f is defined and Lipschitz continuous on $\mathbb{R}^n$. Indeed, if this is not the case, let us replace f by $f_K = f \circ \mathcal{P}_K$, where $\mathcal{P}_K : \mathbb{R}^n \to K$ is the projection operator on the set K (see Definition 12.4.1). According to Lemma 12.4.2, f_K is well-defined, Lipschitz continuous on $\mathbb{R}^n$ and coincides with f on K. If we succeed to prove the theorem for f replaced by f_K, from the fact that $x(t) \in K$ for every $t \geq 0$, we can conclude that $f_K(x(t)) = f(x(t))$ for every $t \geq 0$, and this will achieve the proof in the general case, i.e. $f : K \to \mathbb{R}^n$.

Hence, let $f : \mathbb{R}^n \to \mathbb{R}^n$, let $\varepsilon > 0$ and let us consider the ε-approximate problem

$$\begin{cases} x_\varepsilon' = f_\varepsilon(x_\varepsilon) + g(t) \\ x_\varepsilon(0) = \xi, \end{cases} \tag{7.6.4}$$

where $f_\varepsilon : \mathbb{R}^n \to \mathbb{R}^n$ is defined by

$$f_\varepsilon(x) = f(x) + \frac{1}{\varepsilon}(\mathcal{P}_K(x) - x)$$

for every $x \in \mathbb{R}^n$. From the hypothesis and Lemma 12.4.2, it follows that the function f_ε is Lipschitz continuous on $\mathbb{R}^n$ with Lipschitz constant L_ε and satisfies

$$\|f_\varepsilon(x) + g(t)\| \leq \|f_\varepsilon(x) - f_\varepsilon(0)\| + \|f_\varepsilon(0)\| + \|g(t)\| \leq L_\varepsilon\|x\| + \|f_\varepsilon(0)\| + \|g(t)\|$$

for every $x \in \mathbb{R}^n$ and every $t \in [\,0, T\,]$. Accordingly, the problem (7.6.4) has a unique global solution $x_\varepsilon : [\,0, T\,] \to \mathbb{R}^n$. See Theorem 2.4.6.

[5]We recall that $\mathcal{P}_{N(x(t))}(f(x(t)) + g(t))$ is the projection of the vector $f(x(t)) + g(t)$ on the set $N(x(t))$. See Definition 12.4.1.

The idea of proof consists in showing that there exists one sequence $(\varepsilon_k)_{k\in\mathbb{N}}$, of positive numbers, tending to 0, such that the corresponding sequence of solutions of the ε_k-approximate problems (7.6.4) is uniformly convergent on $[0,T]$ to an absolutely continuous function x, which is a solution of the differential inclusion (7.6.2). We have to show next (7.6.3) and to prove the uniqueness of the solution. For the sake of simplicity we divide the proof into four steps.

First step. The family $\{x_\varepsilon;\ \varepsilon > 0\}$ satisfies the hypotheses of Arzelà-Ascoli's Theorem 12.2.1. Indeed, taking the inner product on both sides in (7.6.4) by $x_\varepsilon(t) - \xi$ and taking into account the inequality (12.4.2) which characterizes the projection, we get

$$\frac{1}{2}\frac{d}{dt}\|x_\varepsilon(t) - \xi\|^2 \leq \langle f(x_\varepsilon(t)) + g(t), x_\varepsilon(t) - \xi\rangle$$

$$\leq \|f(x_\varepsilon(t)) - f(\xi)\|\|x_\varepsilon(t) - \xi\| + (\|f(\xi)\| + \|g(t)\|)\|x_\varepsilon(t) - \xi\|$$

for every $t \in [0,T]$. Let $M = \|f(\xi)\| + \sup\{\|g(t)\|;\ t \in [0,T]\}$ and let L be the Lipschitz constant of the function f. From the preceding inequality, we get

$$\|x_\varepsilon(t) - \xi\| \leq Mt + L\int_0^t \|x_\varepsilon(s) - \xi\|\,ds$$

for every $t \in [0,T]$. Using Gronwall's Lemma 1.5.2, we conclude that

$$\|x_\varepsilon(s) - \xi\| \leq Me^{Lt}t \leq Me^{LT}t.$$

Taking $s = t$ in this inequality, we deduce

$$\|x_\varepsilon(t) - \xi\| \leq Me^{LT}t \tag{7.6.5}$$

for every $t \in [0,T]$. From (7.6.5) it follows that $\|x_\varepsilon(t)\| \leq \|\xi\| + Me^{LT}T$ for every $\varepsilon > 0$ and every $t \in [0,T]$, and therefore the family $\{x_\varepsilon;\ \varepsilon > 0\}$ is uniformly bounded. Now let $h > 0$ and $t \in [0,T-h]$. Taking the inner product on both sides in the equality

$$\frac{d}{dt}(x_\varepsilon(t+h) - x_\varepsilon(t)) = f_\varepsilon(x_\varepsilon(t+h)) - f_\varepsilon(x_\varepsilon(t)) + g(t+h) - g(t)$$

by $x_\varepsilon(t+h) - x_\varepsilon(t)$ and recalling that $\frac{1}{\varepsilon}(\mathcal{P} - \mathcal{I})$ is dissipative (see Lemma 12.4.2), we deduce

$$\frac{1}{2}\frac{d}{dt}\|x_\varepsilon(t+h) - x_\varepsilon(t)\|^2$$

$$\leq (\|f(x_\varepsilon(t+h)) - f(x_\varepsilon(t))\| + \|g(t+h) - g(t)\|)\|x_\varepsilon(t+h) - x_\varepsilon(t)\|$$

for every $h > 0$ and every $t \in [0, T-h]$. From here, from the fact that f is Lipschitz continuous of constant L and from Lemma 1.5.3, it follows

$$\|x_\varepsilon(t+h) - x_\varepsilon(t)\| \leq \|x_\varepsilon(h) - \xi\| + L\int_0^t \|x_\varepsilon(s+h) - x_\varepsilon(s)\|\,ds$$

$$+\int_0^t \|g(s+h) - g(s)\|\,ds.$$

In view of Gronwall's Lemma 1.5.2, we get

$$\|x_\varepsilon(t+h) - x_\varepsilon(t)\| \leq \left(\|x_\varepsilon(h) - \xi\| + \int_0^T \|g(s+h) - g(s)\|\,ds\right)e^{LT},$$

for every $h > 0$ and every $t \in [0, T-h]$. This inequality, along with (7.6.5) and with the fact that g is of class C^1, implies

$$\|x_\varepsilon(t+h) - x_\varepsilon(t)\| \leq Ch \tag{7.6.6}$$

for every $h > 0$ and every $t \in [0, T-h]$, where $C > 0$ is independent of $\varepsilon > 0$. But (7.6.6) implies

$$\|x'_\varepsilon(t)\| \leq C \tag{7.6.7}$$

for every $\varepsilon > 0$ and every $t \in [0, T]$. Since

$$x_\varepsilon(t) - x_\varepsilon(s) = \int_s^t x'_\varepsilon(\tau)\,d\tau,$$

for every $t, s \in [0, T]$, from the last inequality, we deduce

$$\|x_\varepsilon(t) - x_\varepsilon(s)\| \leq C|t-s| \tag{7.6.8}$$

for every $t, s \in [0, T]$. Obviously, (7.6.8) shows that the family $\{x_\varepsilon;\ \varepsilon > 0\}$ is equicontinuous on $[0, T]$. According to Arzelà-Ascoli's Theorem 12.2.1, it follows that there exists at least one sequence $(\varepsilon_k)_{k\in\mathbb{N}}$, convergent to 0, such that the corresponding sequence of ε_k-approximate solutions, denoted for simplicity by $(x_k)_{k\in\mathbb{N}}$, converges uniformly on $[0, T]$ to a continuous function $x : [0, T] \to \mathbb{R}^n$. Passing to the limit for k tending to $+\infty$ in (7.6.7) with $\varepsilon = \varepsilon_k$, we deduce that x is Lipschitz continuous on $[0, T]$. So, it is absolutely continuous on $[0, T]$.

Second step. Now, we will prove that the function x is a solution of the evolution variational inequality (7.6.1). To begin with, let us observe that, from (7.6.4), from the definition of f_ε and from (7.6.7), we have

$$\|x_k(t) - \mathcal{P}_K(x_k(t))\| \leq C_1\varepsilon_k$$

for every $k \in \mathbb{N}$ and every $t \in [0,T]$, where $C_1 > 0$ depends neither on k nor on t. Passing to the limit for $k \to \infty$ in this inequality, we deduce $\|x(t) - \mathcal{P}_K(x(t))\| = 0$ for every $t \in [0,T]$, which shows that $x(t) \in K$ for every $t \in [0,T]$. From (7.6.4) and from the characterization of the projection of the point $x_k(s)$ — see (12.4.2) — it follows

$$\frac{1}{2}\frac{d}{ds}\left(\|x_k(s) - u\|^2\right) \leq \langle f(x_k(s)) + g(s), x_k(s) - u\rangle$$

for every $s \in [0,T]$ and every $u \in K$. Now, let us consider a point $t \in [0,T)$ at which x is differentiable and let $h > 0$ with $t + h \in [0,T]$. Integrating the inequality above from t to $t+h$, we get

$$\frac{1}{2}\left(\|x_k(t+h) - x_k(t)\|^2 - \|x_k(t) - u\|^2\right)$$
$$\leq \int_t^{t+h} \langle f(x_k(s)) + g(s), x_k(s) - u\rangle\, ds$$

for every $u \in K$. At this point, let us observe that, in view of Schwarz inequality, it follows

$$\langle w - v, v - u\rangle \leq \frac{1}{2}\left(\|w - u\|^2 - \|v - u\|^2\right)$$

for every $u, v, w \in \mathbb{R}^n$. Taking $w = x_k(t+h)$ and $v = x_k(t)$ in this inequality, using the preceding one, and then dividing by $h > 0$, we deduce

$$\frac{1}{h}\langle x_k(t+h) - x_k(t), x_k(t) - u\rangle \leq \frac{1}{h}\int_t^{t+h} \langle f(x_k(s)) + g(s), x_k(s) - u\rangle\, ds$$

for every $u \in K$. Passing to the limit for k tending to ∞ we get

$$\frac{1}{h}\langle x(t+h) - x(t), x(t) - u\rangle \leq \frac{1}{h}\int_t^{t+h} \langle f(x(s)) + g(s), x(s) - u\rangle\, ds$$

for every $u \in K$. Finally, passing to the limit for h tending to 0, we conclude that x satisfies

$$\langle x'(t) - f(x(t)) - g(t), x(t) - u\rangle \leq 0$$

for every $u \in K$ and a.e. $t \in [0,T]$. So x is a solution of the problem (7.6.1).

Third step. In order to prove the uniqueness, let x and y be two solutions of the problem (7.6.1), and let $s \in [0,T]$ be a point of differentiability of both x and y. Taking successively $u = y(s) \in K$ in the inequality (7.6.1) satisfied by x, and $u = x(s) \in K$ in the inequality (7.6.1) satisfied by y, we get

$$\frac{d}{ds}\left(\|x(s) - y(s)\|^2\right) \leq \langle f(x(s)) - f(y(s)), x(s) - y(s)\rangle \leq L\|x(s) - y(s)\|^2.$$

Integrating this relation from 0 to t and using Gronwall's Lemma 1.5.2, we deduce that $x(t) = y(t)$ for every $t \in [0, T]$, which completes the proof of the uniqueness part.

Fourth step. In order to prove (7.6.3), let us observe that, from (7.6.2), we have

$$\frac{d}{ds}(x(t+s)) - f(x(t+s)) - g(t+s) \in -N(x(t+s))$$

for every $t \in [0, T)$ and a.e. $s \in \mathbb{R}_+^*$ with $t+s \in [0, T]$. From the definition of the set $N(x(t+s))$, it follows that

$$\langle v - u, x(t+s) - x(t) \rangle \geq 0$$

for every $v \in N(x(t+s))$ and every $u \in N(x(t))$. So

$$\left\langle \frac{d}{ds}(x(t+s)), x(t+s) - x(t) \right\rangle$$

$$\leq \langle f(x(t+s)) + g(t+s) - u, x(t+s) - x(t) \rangle$$

for every $u \in N(x(t))$. Integrating with respect to s from 0 to $h > 0$, we get

$$\frac{1}{2}\|x(t+h) - x(t)\|^2 \leq \int_0^h \langle f(x(t+s)) + g(t+s) - u, x(t+s) - x(t) \rangle \, ds$$

for every $t \in [0, T]$, $h > 0$ with $t+h \in [0, T]$ and every $u \in N(x(t))$. From Schwarz inequality and Lemma 1.5.3, it follows

$$\|x(t+h) - x(t)\| \leq \int_0^h \|f(x(t+s)) + g(t+s) - u\| \, ds$$

for every $t \in [0, T]$, $h > 0$ with $t + h \in [0, T]$, and every $u \in N(x(t))$. Taking a point t of differentiability of the function x, dividing by h, and passing to the limit for h tending to 0, we deduce

$$\|x'(t)\| \leq \|f(x(t)) + g(t) - u\|$$

a.e. for $t \in [0, T]$ and for every $u \in N(x(t))$, relation equivalent to (7.6.3). The proof is complete. □

For some extensions and generalizations of Theorem 7.6.1, we refer to [Barbu (1976)] and [Barbu (2010)].

7.7 Problems of Viability

In accordance with Theorem 2.2.1, if Ω is a nonempty and open subset of $\mathbb{R}^n$ and $f : \Omega \to \mathbb{R}^n$ is a continuous function, then for every $\xi \in \Omega$, the Cauchy problem with data $\mathcal{D} = (\mathbb{R}, \Omega, f, 0, \xi)$

$$\begin{cases} x' = f(x) \\ x(0) = \xi \end{cases} \qquad \mathcal{CP}(\mathcal{D})$$

has at least a local solution $x : [0, T] \to \Omega$. The condition which assumes that Ω is open is essential and cannot be removed, unless some other compensating extra-condition is added, as we can see from the simple example below.

Example 7.7.1. Let us consider the plane $\Sigma = \{(x_1, x_2, x_3);\ x_3 = 1\}$ and the function $f : \Sigma \to \mathbb{R}^3$, defined by $f(x_1, x_2, x_3) = (x_2 + x_3, -x_1, -x_1)$ for every $(x_1, x_2, x_3) \in \Sigma$. Then, if ξ is the projection of the origin on this plan (i.e. $\xi = (0, 0, 1)$), $\mathcal{CP}(\mathcal{D})$ has no local solution. Indeed, assuming for contradiction that there exists such a solution $x : [0, T] \to \Sigma$, we have $\langle x'(t), x(t)\rangle = \langle f(x(t)), x(t)\rangle = 0$ and therefore $\|x(t)\| = \|\xi\| = 1$ for every $t \in [0, T]$. Hence $x(t)$ lies on the sphere of center 0 and radius 1 which has only one point in common with Σ, namely ξ. Then, necessarily $x(t) = \xi$ for every $t \in [0, T]$, which is impossible, because, in this case, one would have $x_1(t) = 0$ and $x_1'(t) = x_2(t) + x_3(t) = 1$ for every $t \in [0, T]$. This contradiction can be eliminated only if $\mathcal{CP}(\mathcal{D})$ has no local solution.

We have already seen in Theorem 6.1.2, that if $U : \Omega \to \mathbb{R}$ is a function of class C^1 with $\nabla U(x) \neq 0$ on Ω and if $f : \Omega \to \mathbb{R}^n$ is continuous and parallel to the tangent plane to every level surface $\Sigma_\eta = \{x \in \Omega;\ U(x) = U(\eta)\}$ at every point of this surface, then, the restriction of the function f to any surface Σ_η has the property that, for every $\xi \in \Sigma_\eta$, $\mathcal{CP}(\mathcal{D})$ has at least one local solution $x : [0, T] \to \Sigma_\eta$. This condition constitutes a first step through a partial answer to the question: *what extra-conditions must satisfy the set $\Sigma \subset \mathbb{R}^n$ and the continuous function $f : \Sigma \to \mathbb{R}^n$, in order that, for every $\xi \in \Sigma$ to exist at least one function of class C^1, $x : [0, T] \to \Sigma$, such that $x(0) = \xi$ and $x'(t) = f(x(t))$ for each $t \in [0, T]$*. However, the conditions offered by Theorem 6.1.2 have three weak points. First, they ask f to be defined on the union of all surfaces Σ_η and not on a single one. Second, f must satisfy the mentioned "tangency condition" on each of the surfaces of the family. Finally, the set Σ is in this case of a very specific type, namely it is a surface of constant level for a function $U : \Omega \to \mathbb{R}$, of class C^1 and satisfying $\nabla U(x) \neq 0$ for every $x \in \Omega$.

The possibility of removing the already mentioned three weak points is suggested even by Theorem 6.1.2 which we reformulate below.

Theorem 7.7.1. *Let Σ be a regular surface in $\mathbb{R}^n$ and $f : \Sigma \to \mathbb{R}^n$ a continuous function. Then for every $\xi \in \Sigma$ there exist $T > 0$ and a function of class C^1, $x : [\,0, T\,] \to \Sigma$, such that $x(0) = \xi$ and $x'(t) = f(x(t))$ for every $t \in [\,0, T\,]$, if and only if for every $\eta \in \Sigma$, $f(\eta)$ be tangent to Σ at η.*

We will obtain this theorem as a consequence of a more general result which we will present in what follows.

We begin with some background material we will need subsequently. For the sake of simplicity, we will confine our considerations to the autonomous case, although all the results we shall prove can be extended to the non-autonomous case as well.

Definition 7.7.1. Let $\Sigma \subset \mathbb{R}^n$ be nonempty and $f : \Sigma \to \mathbb{R}^n$. The set Σ is *viable* with respect to the differential equation $x' = f(x)$ if for every $\xi \in \Sigma$ there exist $T > 0$ and at least one function of class C^1, $x : [\,0, T\,] \to \Sigma$, such that $x(0) = \xi$ and $x'(t) = f(x(t))$ for every $t \in [\,0, T\,]$.

Definition 7.7.2. The set $\Sigma \subset \mathbb{R}^n$ is *locally closed* if for every $\xi \in \Sigma$ there exists $r > 0$ such that $\Sigma \cap B(\xi, r)$ be closed.

Remark 7.7.1. Obviously every closed set is locally closed. Furthermore, every open set Σ is locally closed too. Indeed, if Σ is open, for every $\xi \in \Sigma$ there exists $r > 0$ such that $B(\xi, r) \subset \Sigma$ which proves the assertion. There exist however locally closed sets which are neither open, nor closed, as for example $\Sigma \subset \mathbb{R}^3$ defined by $\Sigma = \{(x_1, x_2, x_3) \in \mathbb{R}^3;\ x_3 = 0,\ x_1^2 + x_2^2 < 1\}$. This set, which is in fact the set of points in the interior of the disk of center O and radius 1 in the plane x_1Ox_2 is a locally closed set which is neither open, nor closed.

The next concept has been introduced independently by [Bouligand (1930)] and [Severi (1930)].

Definition 7.7.3. Let $\Sigma \in \mathbb{R}^n$ and $\xi \in \Sigma$. The vector $\eta \in \mathbb{R}^n$ is *tangent in the sense of Bouligand–Severi to the set Σ at the point ξ* if

$$\liminf_{t \downarrow 0} \frac{1}{t} \operatorname{dist}(\xi + t\eta, \Sigma) = 0. \tag{7.7.1}$$

The set of all vectors which are tangent in the sense of Bouligand–Severi to the set Σ at the point ξ is a closed cone[6] (see Problem 7.9) and is called *Bouligand–Severi tangent cone* to the set Σ at the point ξ. We denote this cone by $\mathcal{T}_\Sigma(\xi)$.

Proposition 7.7.1. *A vector $\eta \in \mathbb{R}^n$ belongs to the cone $\mathcal{T}_\Sigma(\xi)$ if and only if for every $\varepsilon > 0$ there exist $h \in (0, \varepsilon)$ and $p_h \in B(0, \varepsilon)$ with the property*

$$\xi + h(\eta + p_h) \in \Sigma.$$

Proof. Obviously $\eta \in \mathcal{T}_\Sigma(\xi)$ if and only if, for every $\varepsilon > 0$ there exists $h \in (0, \varepsilon)$ and $z_h \in \Sigma$ such that $\frac{1}{h}\|\xi + h\eta - z_h\| \leq \varepsilon$. Now, let us define $p_h = \frac{1}{h}(z_h - \xi - h\eta)$, and let us observe that we have both $\|p_h\| \leq \varepsilon$, and $\xi + h(\eta + p_h) = z_h \in \Sigma$, thereby completing the proof. □

Remark 7.7.2. We notice that, if ξ is an interior point of the set Σ, then $\mathcal{T}_\Sigma(\xi) = \mathbb{R}^n$. Indeed, if there exists $r > 0$ such that $B(\xi, r) \subset \Sigma$, it follows that, for each $\eta \in \mathbb{R}^n$ and $t \in (0, r\|\eta\|^{-1})$, $\xi + t\eta \in B(\xi, r) \subset \Sigma$. In these circumstances, we have $\operatorname{dist}(\xi + t\eta, \Sigma) = 0$. So, the condition (7.7.1) in Definition 7.7.3 is satisfied, and therefore $\eta \in \mathcal{T}_\Sigma(\xi)$, as claimed.

We can now proceed to the main result in this section.

Theorem 7.7.2. (Nagumo) *Let $\Sigma \subset \mathbb{R}^n$ be a nonempty and locally closed set and $f : \Sigma \to \mathbb{R}^n$ a continuous function. The necessary and sufficient condition for Σ to be viable with respect to $x' = f(x)$ is that, for every $\xi \in \Sigma$, $f(\xi) \in \mathcal{T}_\Sigma(\xi)$.*

The necessary and sufficient condition for viability in Theorem 7.7.2 can be expressed equivalently as the so-called *Nagumo tangency condition* below

$$\liminf_{h \downarrow 0} \frac{1}{h} \operatorname{dist}(\xi + hf(\xi), \Sigma) = 0 \qquad (7.7.2)$$

for each $\xi \in \Sigma$.

7.8 Proof of the Nagumo's Viability Theorem

Proof. In order to prove the necessity let $\xi \in \Sigma$. Then, there exist $T > 0$ and a function of class C^1, $x : [0, T] \to \Sigma$, with $x(0) = \xi$ and $x'(t) = f(x(t))$

[6]We recall that a *cone* is a set $\mathcal{C} \subset \mathbb{R}^n$, such that, for each $\eta \in \mathcal{C}$ and $\lambda > 0$, we have $\lambda\eta \in \mathcal{C}$.

for every $t \in [0, T]$. Since $x(t) \in \Sigma$ we deduce that

$$\operatorname{dist}(\xi + tf(\xi), \Sigma) = \operatorname{dist}(x(0) + tf(x(0)), \Sigma)$$

$$\leq \|x(0) + tf(x(0)) - x(t)\| = t \left\| f(x(0)) - \frac{x(t) - x(0)}{t} \right\|.$$

Since x is differentiable at $t = 0$ and $x'(0) = f(x(0))$, we have

$$\lim_{t \downarrow 0} \left\| f(x(0)) - \frac{x(t) - x(0)}{t} \right\| = 0.$$

From the last relation and the preceding inequality, we deduce

$$\lim_{t \downarrow 0} \frac{1}{t} \operatorname{dist}(\xi + tf(\xi), \Sigma) = 0.$$

But this relation shows that, for every $\xi \in \Sigma$, $f(\xi) \in \mathcal{T}_\Sigma(\xi)$. The proof of the necessity is complete.

Remark 7.8.1. We have shown that the set $\mathcal{T}_\Sigma(\xi)$ in Theorem 7.7.2 can be replaced by the set $\mathcal{F}_\Sigma(\xi)$ of all vectors $\eta \in \mathbb{R}^n$ which are tangent to Σ at ξ in the sense of Federer, i.e. of all vectors η satisfying

$$\lim_{t \downarrow 0} \frac{1}{t} \operatorname{dist}(\xi + t\eta, \Sigma) = 0.$$

It should be noticed that $\mathcal{F}_\Sigma(\xi)$ is, in general, strictly smaller than $\mathcal{T}_\Sigma(\xi)$.

For the sake of simplicity, we will divide the proof of the sufficiency into three steps. In the first one, we shall prove the existence of a family of approximate solutions for the Cauchy problem

$$\begin{cases} x' = f(x) \\ x(0) = \xi, \end{cases} \tag{7.8.1}$$

defined on intervals of the form $[0, a]$, with $a > 0$. In the second step we will show that the problem (7.8.1) admits such approximate solutions, all defined on an interval $[0, T]$ independent of the "approximation order". Finally, in the last step, we shall prove the uniform convergence on $[0, T]$ of a sequence of such approximate solutions to a solution of the problem (7.8.1).

Let $\xi \in \Sigma$ be arbitrary and let us choose $r > 0$, $M > 0$ and $T > 0$, such that $B(\xi, r) \cap \Sigma$ be closed,

$$\|f(x)\| \leq M \tag{7.8.2}$$

for every $x \in B(\xi, r) \cap \Sigma$ and

$$T(M + 1) \leq r. \tag{7.8.3}$$

The existence of these three numbers is ensured: by the fact that Σ is locally closed (from where it follows the existence of $r > 0$), by the continuity of f which implies its boundedness on $B(\xi, r)$ (and so the existence of $M > 0$) and by the fact that $T > 0$ may be chosen as small as we wish. In the first step, we will show that, once fixed an $\varepsilon \in (0,1)$ and r, M and T as above, there exist three functions: $\sigma : [0,T] \to [0,T]$ — nondecreasing, $g : [0,T] \to \mathbb{R}^n$ — Lebesgue integrable, and $x : [0,T] \to \mathbb{R}^n$ — continuous, satisfying

(i) $\sigma(t) \leq t$ and $t - \sigma(t) \leq \varepsilon$ for every $t \in [0,T]$;
(ii) $\|g(t)\| \leq \varepsilon$ for every $t \in [0,T]$;
(iii) $x(\sigma(t)) \in B(\xi, r) \cap \Sigma$ for every $t \in [0,T]$ and $x(T) \in B(\xi, r) \cap \Sigma$;
(iv) $x(t) = \xi + \int_0^t f(x(\sigma(s)))\, ds + \int_0^t g(s)\, ds$ for every $t \in [0,T]$.

For the sake of simplicity, in all that follows, we will call such a triple (σ, g, x) an *ε-approximate solution* of the Cauchy problem (7.8.1) on the interval $[0,T]$.

The first step. Let $\xi \in \Sigma$ and let $r > 0$, $M > 0$ and $T > 0$ be fixed as above. We begin by showing that, for each $\varepsilon \in (0,1)$, there exists at least one ε-approximate solution on an interval $[0,a]$, with $a \leq T$. Since for every $\xi \in \Sigma$ we have $f(\xi) \in \mathcal{T}_\Sigma(\xi)$, from Proposition 7.7.1, it follows that there exist $a \in (0,T]$, $a \leq \varepsilon$ and $p \in \mathbb{R}^n$ with $\|p\| \leq \varepsilon$ such that $\xi + af(\xi) + ap \in \Sigma$. At this point, we can define the functions $\sigma : [0,a] \to [0,a]$, $g : [0,a] \to \mathbb{R}^n$ and $x : [0,a] \to \mathbb{R}^n$ by

$$\begin{cases} \sigma(t) = 0 & \text{for } t \in [0,a] \\ g(t) = p & \text{for } t \in [0,a] \\ x(t) = \xi + tf(\xi) + tp & \text{for } t \in [0,a]. \end{cases}$$

One can readily see that triple (σ, g, x) is an ε-approximate solution of the Cauchy problem (7.8.1) on the interval $[0,a]$. Indeed the conditions (i), (ii) and (iv) are obviously fulfilled, while (iii) follows from (7.8.2), (7.8.3) and (i), by observing that $x(\sigma(t)) = \xi \in B(\xi, r) \cap \Sigma$ for every $t \in [0,a]$, $x(a) = \xi + af(\xi) + ap \in \Sigma$, and

$$\|x(a) - \xi\| = \|af(\xi) + ap\| \leq a\|f(\xi)\| + a\varepsilon \leq a(M+1) \leq r.$$

The second step. Now, we will prove the existence of an ε-approximate solution defined on the whole interval $[0,T]$. To this aim we shall make use of Zorn's lemma, as follows. Let $\mathcal{S}$ be the set of all ε-approximate solutions of the problem (7.8.1) having the domains of definition of the form $[0,a]$ with $a \leq T$. On $\mathcal{S}$ we define the relation "$\preceq$" by $(\sigma_1, g_1, x_1) \preceq (\sigma_2, g_2, x_2)$

if the domain of definition of the first triple, $[0, a_1]$, is included in the domain of definition, $[0, a_2]$, of the second triple and the two ε-approximate solutions coincide on the common part of the domains. Clearly "$\preceq$" is a partial order relation on $\mathcal{S}$. Let us observe that the set $\mathcal{S}$ endowed with "$\preceq$" is inductively ordered, i.e. every totally ordered subset of $\mathcal{S}$ has a majorant. Indeed, let $\mathcal{L} = \{(\sigma_a, g_a, x_a);\ a \in \Gamma\}$ be such a totally ordered subset. Since $\mathcal{L}$ is totally ordered, we may assume with no loss of generality that Γ is the set of those elements $a \in (0, T\,]$ with the property that (σ_a, g_a, x_a) is defined on $[\,0, a\,]$. If Γ has one last element a^*, then the corresponding ε-approximate solution is a majorant for $\mathcal{L}$. If $\sup \Gamma = a^*$, which is clearly in $[\,0, T\,]$, does not belong to Γ, we will define a majorant for $\mathcal{L}$ as follows. First, let us observe that, because all the functions in the set $\{\sigma_a;\ a \in \Gamma\}$ are nondecreasing, with values in $[\,0, T\,]$ and satisfy $\sigma_a(a) \leq \sigma_b(b)$ for every $a, b \in \Gamma$ with $a \leq b$, there exists $\lim_{a \uparrow a^*} \sigma_a(a)$ and this limit belongs to $[\,0, T\,]$. From the fact that $\mathcal{L}$ is totally ordered, we deduce that, if $a, b \in \Gamma$ and $a \leq b$, then $x_a(a) = x_b(a)$. Taking into account (iii), (iv) and (7.8.2), we deduce

$$\|x_a(a) - x_b(b)\| \leq \int_a^b \|f(x_b(\sigma_b(s))\|\, ds + \int_a^b \|g_b(s)\|\, ds \leq (M + \varepsilon)|b - a|$$

for every $a, b \in \Gamma$, and thus there exists $\lim_{a \uparrow a^*} x_a(a)$. As for every $a \in \Gamma$, $x_a(a) \in B(\xi, r) \cap \Sigma$, and the latter is closed, we have $\lim_{a \uparrow a^*} x_a(a) \in B(\xi, r) \cap \Sigma$. Then, we can define $(\sigma^*, g^*, x^*) : [\,0, a^*] \to [\,0, a^*] \times \mathbb{R}^n \times \mathbb{R}^n$ by

$$\sigma^*(t) = \begin{cases} \sigma_a(t) & \textit{for } t \in [\,0, a\,],\ a \in \Gamma \\ \lim_{a \uparrow a^*} \sigma_a(a) & \textit{for } t = a^* \end{cases}$$

$$g^*(t) = \begin{cases} g_a(t) & \text{for } t \in [\,0, a\,],\ a \in \Gamma \\ 0 & \text{for } t = a^* \end{cases}$$

$$x^*(t) = \begin{cases} x_a(t) & \text{for } t \in [\,0, a\,],\ a \in \Gamma \\ \lim_{a \uparrow a^*} x_a(a) & \text{for } t = a^*. \end{cases}$$

Obviously (σ^*, g^*, x^*) is an ε-approximate solution which majorizes every element in $\mathcal{L}$. According to Zorn's lemma, it follows that $\mathcal{S}$ has maximal elements. Let (σ, g, x) be such an element having the domain of definition $[\,0, a\,]$. We will show that $a = T$. Indeed, let us assume for contradiction that $a < T$. Then, taking into account the fact that $x(a) \in B(\xi, r) \cap \Sigma$, we deduce that

$$\|x(a) - \xi\| \leq \int_0^a \|f(x(\sigma(s))\|\, ds + \int_0^a \|g(s)\|\, ds \leq a(M + \varepsilon) \leq a(M + 1) < r.$$

As $x(a) \in \Sigma$ and $f(x(a)) \in \mathcal{T}_\Sigma(x(a))$, there exist $\delta \in (0, T-a)$, $\delta \le \varepsilon$ and $p \in \mathbb{R}^n$ such that $\|p\| \le \varepsilon$ and $x(a) + \delta f(x(a)) + \delta p \in \Sigma$. Then, from the inequality above, it follows that we can diminish δ if necessary, in order to have $\|x(a) + \delta(f(x(a)) + p) - \xi\| \le r$. Let us define the functions $\widetilde{\sigma} : [\,0, a+\delta\,] \to [\,0, a+\delta\,]$ and $\widetilde{g} : [\,0, a+\delta] \to \mathbb{R}^n$ by

$$\widetilde{\sigma}(t) = \begin{cases} \sigma(t) & \textit{for}\, t \in [\,0, a\,] \\ a & \textit{for}\, t \in (a, a+\delta\,] \end{cases}$$

$$\widetilde{g}(t) = \begin{cases} g(t) & \text{for } t \in [\,0, a\,] \\ p & \text{for } t \in (a, a+\delta\,]. \end{cases}$$

It is no difficult to see that $\widetilde{\sigma}$ is nondecreasing, $\widetilde{g}$ is Lebesgue integrable on $[\,0, a+\delta\,]$ and $\|g(t)\| \le \varepsilon$ for every $t \in [\,0, a+\delta\,]$. In addition, for every $t \in [\,0, a+\delta\,]$, $\widetilde{\sigma}(t) \in [\,0, a\,]$ and therefore $x(\widetilde{\sigma}(t))$ is well-defined and belongs to the set $B(\xi, r) \cap \Sigma$. Accordingly, we can define $\widetilde{x} : [\,0, a+\delta\,] \to \mathbb{R}^n$ by

$$\widetilde{x}(t) = \xi + \int_0^t f(x(\widetilde{\sigma}(s)))\, ds + \int_0^t \widetilde{g}(s)\, ds$$

for every $t \in [\,0, a+\delta\,]$. Clearly $\widetilde{x}$ coincides with x on $[\,0, a\,]$ and then it readily follows that $\widetilde{\sigma}$, $\widetilde{g}$ and $\widetilde{x}$ satisfy all the conditions in (i) and (ii). In order to prove (iii) and (iv), let us observe that

$$\widetilde{x}(t) = \begin{cases} x(t) & \text{for } t \in [\,0, a\,] \\ x(a) + (t-a)f(x(a)) + (t-a)p & \text{for } t \in (a, a+\delta\,]. \end{cases}$$

Then $\widetilde{x}$ satisfies (iv). Since

$$\widetilde{x}(\widetilde{\sigma}(t)) = \begin{cases} x(\sigma(t)) & \textit{for}\;\; t \in [\,0, a\,] \\ x(a) & \textit{for}\;\; t \in (a, a+\delta\,], \end{cases}$$

it follows that $\widetilde{x}(\widetilde{\sigma}(t)) \in B(\xi, r) \cap \Sigma$ for each $t \in [\,a, a+\delta\,]$. Furthermore, from the choice of δ and p, we have both $\widetilde{x}(a+\delta) = x(a) + \delta f(x(a)) + \delta p \in \Sigma$, and

$$\|\widetilde{x}(a+\delta) - \xi\| = \|x(a) + \delta(f(x(a)) + p) - \xi\| \le r.$$

Consequently $\widetilde{x}$ satisfies (iii). It follows that $(\widetilde{\sigma}, \widetilde{g}, \widetilde{x}) \in \mathcal{S}$ and is a strict upper bound for the maximal element (σ, g, x). But this is absurd. This contradiction can be eliminated only if each maximal element in the set $\mathcal{S}$ is defined on $[\,0, T\,]$.

The third step. Let $(\varepsilon_k)_{k\in\mathbb{N}}$ be a sequence in $(0,1)$ decreasing to 0 and let $((\sigma_k, g_k, x_k))_{k\in\mathbb{N}}$ be a sequence of ε_k-approximate solutions defined on $[\,0, T\,]$. From (i) and (ii), it follows that

$$\lim_{k\to\infty} \sigma_k(t) = t \;\text{ and }\; \lim_{k\to\infty} g_k(t) = 0 \tag{7.8.4}$$

uniformly on $[0,T]$. On the other hand, from (iii), (iv) and (7.8.3) we have

$$\|x_k(t)\| \leq \|x_k(t)-\xi\| + \|\xi\| \leq \int_0^T \|f(x_k(\sigma_k(s)))\|\,ds + \int_0^T \|g_k(s)\|\,ds + \|\xi\|$$

$$\leq T(M+1) + \|\xi\| \leq r + \|\xi\|$$

for every $k \in \mathbb{N}$ and every $t \in [0,T]$. Hence the sequence $(x_k)_{k\in\mathbb{N}}$ is uniformly bounded on $[0,T]$. Again from (iii), (iv) and (7.8.3), we have

$$\|x_k(t)-x_k(s)\| \leq \left|\int_s^t \|f(x_k(\sigma_k(\tau)))\|\,d\tau\right| + \left|\int_s^t \|g_k(\tau)\|\,d\tau\right| \leq (M+1)|t-s|$$

for every $t,s \in [0,T]$. Consequently, the set $\{x_k\,;\ k \in \mathbb{N}\}$ is equicontinuous on $[0,T]$. From Theorem 12.2.1, it follows that, at least on a subsequence, $(x_k)_{k\in\mathbb{N}}$ is uniformly convergent on $[0,T]$ to a function $x : [0,T] \to \mathbb{R}^n$. Taking into account (iii), (7.8.4) and the fact that $B(\xi,r)\cap\Sigma$ is closed, we deduce that $x(t) \in B(\xi,r)\cap\Sigma$ for every $t \in [0,T]$. Passing to the limit in the equation

$$x_k(t) = \xi + \int_0^t f(x_k(\sigma_k(s)))\,ds + \int_0^t g_k(s)\,ds$$

and using (7.8.4), we deduce that

$$x(t) = \xi + \int_0^t f(x(s))\,ds$$

for every $t \in [0,T]$, which completes the proof of the theorem. □

From Remark 7.7.2 combined with Theorem 7.7.2, we deduce a variant of Peano's local existence Theorem 2.2.1 referring to the case of autonomous systems. More precisely, we have

Corollary 7.8.1. *Let $\Sigma \subset \mathbb{R}^n$ be nonempty and open and let $f : \Sigma \to \mathbb{R}^n$ be continuous. Then Σ is viable with respect to $x' = f(x)$.*

We have to mention that Theorem 7.7.1 is also a direct consequence of Theorem 7.7.2, combined with the remark that the classical notion of tangency there used is equivalent to the tangency notion introduced in Definition 7.7.3. See also Problem 7.10.

The readers who are interested in the study of viability problems are referred to [Aubin (1991)], [Cârjă (2003)], [Cârjă and Vrabie (2005)], [Pavel (1984)] and [Cârjă *et al.* (2007)].

7.9 Sufficient Conditions for Invariance

Let $\Omega \subset \mathbb{R}^n$ be given, let $\Sigma \subset \Omega$ be nonempty, and let us consider the differential equation

$$x' = f(x), \tag{7.9.1}$$

where $f : \Omega \to \mathbb{R}^n$ is a continuous function. Since, throughout this section, we are dealing with autonomous equations, we may assume that, for all initial conditions considered, the initial time $\tau = 0$. See Proposition 2.1.3.

Definition 7.9.1. Let $\Omega \subset \mathbb{R}^n$ be given, and let $f : \Omega \to \mathbb{R}^n$. The nonempty subset $\Sigma \subset \Omega$ is *invariant* with respect to $x' = f(x)$ if, for every $\xi \in \Sigma$, and every solution $x : [\,0, c\,] \to \Omega$ of (7.9.1) satisfying $x(0) = \xi$, there exists $T \in (0, c\,]$ such that $x(t) \in \Sigma$ for every $t \in [\,0, T\,]$. It is *globally invariant* if it is invariant and $T = c$.

The relationship between viability and invariance is clarified in:

Remark 7.9.1. If f is continuous on Ω which is open, and Σ is invariant with respect to the differential equation $x' = f(x)$, then Σ is viable with respect to $x' = f_{|\Sigma}(x)$. The converse of this assertion is no longer true as we can see from Example 7.9.1 below. Nevertheless, if Σ is viable with respect to $x' = f_{|\Sigma}(x)$ and $x' = f(x)$, $x(0) = \xi$ has the uniqueness property, then Σ is invariant with respect to $x' = f(x)$.

Example 7.9.1. Let $\Omega = \mathbb{R}$, $\Sigma = \{0\}$ and let $f : \Sigma \to \mathbb{R}$ be defined by $f(x) = 3\sqrt[3]{x^2}$ for every $x \in \mathbb{R}$. Then Σ is viable with respect to $x' = f_{|\Sigma}(x)$ but Σ is not invariant with respect to $x' = f(x)$, because the differential equation $x' = f(x)$ has at least two solutions which satisfy both $x(0) = 0$ and $x(t) \neq y(t)$ for each $t > 0$, i.e. $x \equiv 0$ and $y(t) = t^3$. See also Problem 7.8.

A simple necessary and sufficient condition of invariance is stated below.

Theorem 7.9.1. *Let $\Omega \subset \mathbb{R}^n$ be open, $\Sigma \subset \Omega$ a nonempty and locally closed subset and $f : \Omega \to \mathbb{R}^n$ a continuous function with the property that the associated Cauchy problem has the uniqueness property. Then, a necessary and sufficient condition for the set Σ to be invariant with respect to $x' = f(x)$ is that, for every $\xi \in \Sigma$, $f(\xi) \in \mathcal{T}_\Sigma(\xi)$.*

Proof. The conclusion follows from Theorem 7.7.2 and Remark 7.9.1. □

Theorem 7.9.1 says that, if Σ is viable with respect to $x' = f_{|\Sigma}(x)$, and $x' = f(x)$, $x(0) = \xi$ has the uniqueness property, then Σ is invariant with respect to f. The preceding example shows that this is no longer true if we assume that Σ is viable with respect to $x' = f_{|\Sigma}(x)$ and merely $x' = f_{|\Sigma}(x)$, $x(0) = \xi$ has the uniqueness property.

Remark 7.9.2. Moreover, if $f : \Omega \to \mathbb{R}^n$ is continuous and there exists one point $\xi \in \Omega$ such that the differential equation $x' = f(x)$ has at least two solutions, x and y, satisfying both $x(0) = y(0) = \xi$, and $x(t) \neq y(t)$ for each $t \in (0, T\,]$, then the set $\Sigma = \{x(t)\,;\;\; t \in [\,0, T\,]\}$ is viable, but not invariant, with respect to $x' = f(x)$.

The next example reveals another interesting fact about local invariance. It shows that Σ could be invariant with respect to f even though $x' = f_{|\Sigma}(x)$ does not have the uniqueness property.

Example 7.9.2. Let us consider $\Sigma = \{(x_1, x_2) \in \mathbb{R}^2\,;\;\; x_2 \geq 0\}$ and let $f : \mathbb{R}^2 \to \mathbb{R}^2$ be defined by

$$f((x_1, x_2)) = \begin{cases} (1, 0) & \textit{if } (x_1, x_2) \in \mathbb{R}^2 \setminus \Sigma \\ (1, 3\sqrt[3]{x_2^2}) & \textit{if } (x_1, x_2) \in \Sigma. \end{cases}$$

Clearly Σ is invariant with respect to $x' = f_{|\Sigma}(x)$, but $x' = f_{|\Sigma}(x)$, $x(0) = \xi$ does not have the uniqueness property. The latter assertion follows from the remark that, from each point $(\xi, 0)$ (on the boundary of Σ), we have at least two solutions of $x' = f(x)$, $x(t) = (t + \xi, 0)$ and $y(t) = (t + \xi, t^3)$ satisfying $x(0) = y(0) = (\xi, 0)$.

In order to formulate the main sufficient condition for local invariance, we need some preliminaries.

We denote by $[D_+x](t)$ the right lower Dini derivative of the function x at t, i.e.

$$[D_+x](t) = \liminf_{h \downarrow 0} \frac{x(t+h) - x(t)}{h}.$$

If $x, y \in \mathbb{R}^n$, we denote by $[x, y]_+$ *the right directional derivative* of the norm $\|\cdot\|$ calculated at x in the direction y. Similarly, $(x, y)_+$ denotes *the right directional derivative* of $\frac{1}{2}\|\cdot\|^2$ calculated at x in the direction y. More precisely

$$\begin{aligned} [x, y]_+ &= \lim_{h \downarrow 0} \frac{1}{h}(\|x + hy\| - \|x\|) \\ (x, y)_+ &= \lim_{h \downarrow 0} \frac{1}{2h}(\|x + hy\|^2 - \|x\|^2). \end{aligned}$$

One may easily see that

$$(x,y)_+ = \|x\|[x,y]_+$$

for each $x, y \in \mathbb{R}^n$ and, if $\|\cdot\| = \sqrt{\langle\cdot,\cdot\rangle}$, where $\langle\cdot,\cdot\rangle$ is the inner product on $\mathbb{R}^n$, we have

$$(x,y)_+ = \langle x,y\rangle.$$

Definition 7.9.2. A function $\varphi : [0,a) \to \mathbb{R}$ is a *comparison function* if $\varphi(0) = 0$, and the only continuous function $x : [0,T) \to [0,a)$, satisfying

$$\begin{cases} [D_+x](t) \le \varphi(x(t)) \text{ for all } t \in [0,T) \\ x(0) = 0, \end{cases}$$

is the null function.

Now, we can introduce the following *exterior tangency condition*: there exists an open subset $\mathbb{V}$ of Ω with $\Sigma \subset \mathbb{V}$ such that

$$\liminf_{h\downarrow 0} \frac{1}{h}\left[\operatorname{dist}(\xi + hf(\xi);\Sigma) - \operatorname{dist}(\xi;\Sigma)\right] \le \varphi(\operatorname{dist}(\xi;\Sigma)) \tag{7.9.2}$$

for each $\xi \in \mathbb{V}$, where φ is a certain comparison function. Clearly, this condition reduces to the classical Nagumo's tangency condition (7.7.2) when applied to $\xi \in \Sigma$, and this simply because, at each such point $\xi \in \Sigma$, $\operatorname{dist}(\xi;\Sigma) = 0$. The main result in this section is:

Theorem 7.9.2. (Cârjă-Necula-Vrabie) *Let $\Sigma \subset \Omega \subset \mathbb{R}^n$, with Σ locally closed and Ω open, and let $f : \Omega \to \mathbb{R}^n$. If (7.9.2) is satisfied, then Σ is invariant with respect to $x' = f(x)$.*

Proof. Let $\mathbb{V} \subset \Omega$ be the open neighborhood of Σ whose existence is ensured by (7.9.2) and let $\varphi : [0,a) \to \mathbb{R}$ be the corresponding comparison function. Let $\xi \in \Sigma$ and let $x : [0,c] \to \mathbb{V}$ be any local solution of (7.9.1) satisfying $x(0) = \xi$. Diminishing c if necessary, we may assume that there exists $\rho > 0$ such that $B(\xi,\rho) \cap \Sigma$ is closed, $x(t) \in B(\xi,\rho/2)$ and, in addition, $\operatorname{dist}(x(t);\Sigma) < a$ for each $t \in [0,c)$. Let $g : [0,c] \to \mathbb{R}_+$ be defined by $g(t) = \operatorname{dist}(x(t);\Sigma)$ for each $t \in [0,c]$. Let $t \in [0,c)$ and $h > 0$ with $t+h \in [0,c]$. We have

$$g(t+h) = \operatorname{dist}(x(t+h);\Sigma) = \operatorname{dist}\left(x(t) + \int_t^{t+h} f(x(s))\,ds;\Sigma\right)$$

$$\le h\left\|\frac{1}{h}\int_t^{t+h} x'(s)\,ds - f(x(t))\right\| + \operatorname{dist}(x(t) + hf(x(t));\Sigma).$$

Therefore

$$\frac{g(t+h)-g(t)}{h} \le \alpha(h) + \frac{\operatorname{dist}(x(t)+hf(x(t));\Sigma) - \operatorname{dist}(x(t);\Sigma)}{h},$$

where

$$\alpha(h) = \left\| \frac{1}{h}\int_t^{t+h} x'(s)\,ds - f(x(t)) \right\|.$$

Since $\lim_{h\downarrow 0}\alpha(h) = 0$, passing to the lim inf for $h \downarrow 0$, and taking into account that $\mathbb{V}$, Σ and f satisfy (7.9.2), we get

$$[D_+ g](t) \le \varphi(g(t))$$

for each $t \in [\,0, T)$. So, $g(t) \equiv 0$ which means that $x(t) \in \overline{\Sigma} \cap B(\xi, \rho/2)$. But $\overline{\Sigma} \cap B(\xi, \rho/2) \subset \Sigma \cap B(\xi, \rho)$ for each $t \in [\,0, T)$, which completes the proof. □

Remark 7.9.3. Let $\mathbb{V} \subset \Omega$ be an open neighborhood of Σ. It is easy to see that (7.9.2) is satisfied with $\varphi = \varphi_f$, the function $\varphi_f : [\,0, a) \to \mathbb{R}$, $a = \sup_{\xi\in\mathbb{V}} \operatorname{dist}(\xi;\Sigma)$, being defined by

$$\varphi_f(x) = \sup_{\substack{\xi\in\mathbb{V}\\ \operatorname{dist}(\xi;\Sigma)=x}} \liminf_{h\downarrow 0} \frac{1}{h}[\operatorname{dist}(\xi + hf(\xi);\Sigma) - \operatorname{dist}(\xi;\Sigma)] \tag{7.9.3}$$

for each $x \in [\,0, a)$.

So, Theorem 7.9.2 can be reformulated as:

Theorem 7.9.3. *Let $\Sigma \subset \Omega$, with Σ locally closed, and let $f : \Omega \to \mathbb{R}^n$ be continuous. If there exists an open neighborhood $\mathbb{V}$ of Σ with $\mathbb{V} \subset \Omega$ such that φ_f, defined by* (7.9.3), *is a comparison function, then Σ is invariant with respect to $x' = f(x)$.*

7.10 Necessary Conditions for Invariance

We say that $\xi \in \mathbb{R}^n$ *has projection on* Σ if there exists $\eta \in \Sigma$ such that $\|\xi - \eta\| = \operatorname{dist}(\xi;\Sigma)$. Any $\eta \in \Sigma$ enjoying the above property is called *a projection* of ξ on Σ, and the set of all projections of ξ on Σ is denoted by $\Pi_\Sigma(\xi)$.

Definition 7.10.1. An open neighborhood $\mathbb{V}$ of Σ, with $\Pi_\Sigma(\xi) \neq \emptyset$ for each $\xi \in \mathbb{V}$, is called a *proximal neighborhood* of Σ. If $\mathbb{V}$ is a proximal neighborhood of Σ, then every single-valued selection, $\pi_\Sigma : \mathbb{V} \to \Sigma$, of Π_Σ, i.e. $\pi_\Sigma(\xi) \in \Pi_\Sigma(\xi)$ for each $\xi \in \mathbb{V}$, is a *projection subordinated to* $\mathbb{V}$.

The next lemma, proved in [Cârjă and Ursescu (1993)], essentially shows that each locally closed set Σ has one proximal neighborhood.

Lemma 7.10.1. (Cârjă and Ursescu) *Let Σ be locally closed. Then the set of all $\xi \in \mathbb{R}^n$ such that $\Pi_\Sigma(\xi)$ is nonempty is a neighborhood of Σ.*

Proof. Let $\xi \in \Sigma$. Since Σ is locally closed, there exists $\rho > 0$ such that $\Sigma \cap B(\xi, \rho)$ is closed. We will show that, for every $\eta \in \Sigma$ satisfying $\|\xi - \eta\| < \rho/2$, $\Pi_\Sigma(\eta)$ is nonempty, which justifies our conclusion. Indeed, given η as above, there exists a sequence $(\zeta_k)_k$ in Σ such that the sequence $(\|\zeta_k - \eta\|)_k$ converges to $\text{dist}\,(\eta; \Sigma)$. We can suppose, taking a subsequence if necessary, that the sequence $(\zeta_k)_k$ converges to a point $\zeta \in \mathbb{R}^n$. So we have $\text{dist}\,(\eta; \Sigma) = \|\zeta - \eta\|$. Further $\|\zeta_k - \xi\| \leq \|\zeta_k - \eta\| + \|\eta - \xi\|$ for all $k \in \mathbb{N}$, and consequently $\|\zeta - \xi\| \leq \text{dist}\,(\eta; \Sigma) + \|\eta - \xi\| \leq 2\|\eta - \xi\| < \rho$. Finally, $\|\zeta_k - \eta\| < \rho$ for all $k \in \mathbb{N}$ sufficiently large. Hence $\zeta_k \in \Sigma \cap B(\xi, \rho)$, and since the latter is closed, it follows that $\zeta \in \Sigma$. Thus $\Pi_\Sigma(\eta)$ is nonempty, and this achieves the proof. □

Definition 7.10.2. Let $\Sigma \subset \Omega \subset \mathbb{R}^n$. We say that a function $f : \Omega \to \mathbb{R}^n$ has *the comparison property with respect to* (Ω, Σ) if there exist a proximal neighborhood $\mathbb{V} \subset \Omega$ of Σ, one projection $\pi_\Sigma : \mathbb{V} \to \Sigma$ subordinated to $\mathbb{V}$, and one comparison function $\varphi : [0, a) \to \mathbb{R}$, with $a = \sup_{\xi \in \mathbb{V}} \text{dist}\,(\xi; \Sigma)$, such that

$$[\xi - \pi_\Sigma(\xi), f(\xi) - f(\pi_\Sigma(\xi))]_+ \leq \varphi(\|\xi - \pi_\Sigma(\xi)\|) \tag{7.10.1}$$

for each $\xi \in \mathbb{V}$.

Let us observe that (7.10.1) is automatically satisfied for each $\xi \in \Sigma$, and therefore, in Definition 7.10.2, we only have to assume that (7.10.1) holds for each $\xi \in \mathbb{V} \setminus \Sigma$.

Definition 7.10.3. The function $f : \Omega \to \mathbb{R}^n$ is called:

(i) (Ω, Σ)-*Lipschitz* if there exist a proximal neighborhood $\mathbb{V} \subset \Omega$ of Σ, a subordinated projection $\pi_\Sigma : \mathbb{V} \to \Sigma$, and $L > 0$, such that

$$\|f(\xi) - f(\pi_\Sigma(\xi))\| \leq L\|\xi - \pi_\Sigma(\xi)\|$$

for each $\xi \in \mathbb{V} \setminus \Sigma$;

(ii) (Ω, Σ)-*dissipative* if there exist a proximal neighborhood $\mathbb{V} \subset \Omega$ of Σ, and a projection, $\pi_\Sigma : \mathbb{V} \to \Sigma$, subordinated to $\mathbb{V}$, such that

$$[\xi - \pi_\Sigma(\xi), f(\xi) - f(\pi_\Sigma(\xi))]_+ \leq 0$$

for each $\xi \in \mathbb{V} \setminus \Sigma$.

Let $\mathbb{V}$ be a proximal neighborhood of Σ, and let $\pi_\Sigma : \mathbb{V} \to \Sigma$ be a projection subordinated to $\mathbb{V}$. If $f : \mathbb{V} \to \Sigma$ is a continuous function with the property that, for each $\eta \in \Sigma$, its restriction to the "segment"

$$\mathbb{V}_\eta = \{\xi \in \mathbb{V} \setminus \Sigma;\ \pi_\Sigma(\xi) = \eta\}$$

is dissipative, then f is (Ω, Σ)-dissipative.

It is easy to see that if f is either (Ω, Σ)-Lipschitz, or (Ω, Σ)-dissipative, then it has the comparison property with respect to (Ω, Σ). We notice that there are examples showing that there exist functions f which, although neither (Ω, Σ)-Lipschitz, nor (Ω, Σ)-dissipative, do have the comparison property. Moreover, there exist functions which, although (Ω, Σ)-Lipschitz, are not Lipschitz on Ω, as well as, functions which although (Ω, Σ)-dissipative, are not dissipative on Ω. In fact, these two properties describe merely the local behavior of f at the interface between Σ and $\Omega \setminus \Sigma$. We include below two examples: the first one of an (Ω, Σ)-Lipschitz function which is not locally Lipschitz, and the second one of a function which, although non-dissipative, is (Ω, Σ)-dissipative.

Example 7.10.1. The graph of an (Ω, Σ)-Lipschitz function $f : \mathbb{R} \to \mathbb{R}$ which is not Lipschitz is illustrated in Figure 7.10.1. Here $\Sigma = [-a, a]$ and

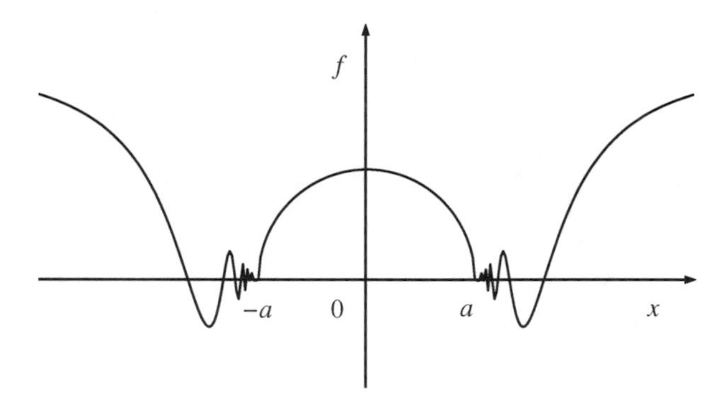

Fig. 7.10.1

Ω is any open subset of $\mathbb{R}$ including Σ.

Example 7.10.2. The graph of a function $f : \mathbb{R} \to \mathbb{R}$ which is (Ω, Σ)-dissipative but not dissipative is illustrated in Figure 7.10.2. This time, Σ is either $(-\infty, \beta]$, or $[\alpha, +\infty)$, or $[\alpha, \beta]$ with $\alpha \le -a \le a \le \beta$, and Ω is any open subset of $\mathbb{R}$ including Σ.

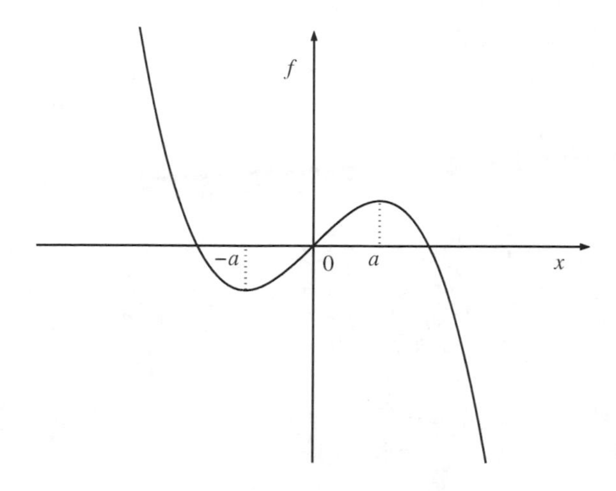

Fig. 7.10.2

Theorem 7.10.1. (Cârjă-Necula-Vrabie) *Let $\Sigma \subset \Omega \subset \mathbb{R}^n$, with Σ locally closed and Ω open, and let $f : \Omega \to \mathbb{R}^n$. If f has the comparison property with respect to (Ω, Σ), and (7.7.2) is satisfied, then (7.9.2) is also satisfied.*

Proof. Let $\mathbb{V} \subset \Omega$ be a proximal neighborhood of Σ, let $\xi \in \mathbb{V}$, and let $\pi_\Sigma : \mathbb{V} \to \Sigma$ be the projection subordinated to $\mathbb{V}$ as in Definition 7.10.2. Let $h > 0$. Since $\|\xi - \pi_\Sigma(\xi)\| = \operatorname{dist}(\xi; \Sigma)$, we have

$$\operatorname{dist}(\xi + hf(\xi); \Sigma) - \operatorname{dist}(\xi; \Sigma) \leq \|\xi - \pi_\Sigma(\xi) + h[f(\xi) - f(\pi_\Sigma(\xi))]\|$$

$$-\|\xi - \pi_\Sigma(\xi)\| + \operatorname{dist}(\pi_\Sigma(\xi) + hf(\pi_\Sigma(\xi)); \Sigma).$$

Dividing by h, passing to the lim inf for $h \downarrow 0$, and using (7.7.2), we get

$$\liminf_{h \downarrow 0} \frac{1}{h}[\operatorname{dist}(\xi + hf(\xi); \Sigma) - \operatorname{dist}(\xi; \Sigma)] \leq [\xi - \pi_\Sigma(\xi), f(\xi) - f(\pi_\Sigma(\xi))]_+$$

$$\leq \varphi(\|\xi - \pi_\Sigma(\xi)\|).$$

This inequality shows that (7.9.2) holds and this completes the proof. □

From Theorems 7.7.2 and 7.10.1, we deduce the following necessary condition for invariance.

Theorem 7.10.2. (Cârjă-Necula-Vrabie) *Let $\Sigma \subset \Omega \subset \mathbb{R}^n$, with Σ locally closed and Ω open, and let $f : \Omega \to \mathbb{R}^n$ be continuous. If f has the comparison property with respect to (Ω, Σ) and Σ is invariant with respect to $x' = f(x)$, then (7.9.2) holds true.*

Proof. As Σ is invariant with respect to $x' = f(x)$ and f is continuous, it follows that Σ is viable with respect to $x' = f(x)$. In view of Theorem 7.7.2 we conclude that (7.7.2) is satisfied, and thus we are in the hypotheses of Theorem 7.10.1. The proof is complete. □

Combining Theorems 7.7.2 and 7.10.1, we deduce:

Theorem 7.10.3. (Cârjă-Necula-Vrabie) *Let $\Sigma \subset \Omega \subset \mathbb{R}^n$, with Σ locally closed and Ω open, and let $f : \Omega \to \mathbb{R}^n$ be continuous. If f has the comparison property with respect to (Ω, Σ), and* (7.7.2) *is satisfied, then Σ is invariant with respect to* (7.9.1).

For details on invariance problems see [Aubin (1991)], [Cârjă *et al.* (2004)] and [Cârjă and Vrabie (2005)].

7.11 Gradient Systems. Frobenius Theorem

Let $\mathcal{U} \subset \mathbb{R}^m$ and $\mathcal{G} \subset \mathbb{R}^n$ be nonempty and open, let $X^i : \mathcal{U} \times \mathcal{G} \to \mathbb{R}^n$, $i = 1, 2, \ldots, m$, be functions of class C^1 and let us consider *the gradient system*

$$dy = \sum_{i=1}^{m} X^i(x, y)\, dx_i. \tag{7.11.1}$$

Definition 7.11.1. Let $\tau \in \mathcal{U}$ and $\xi \in \mathcal{G}$. A *solution* of the system (7.11.1) subjected to the Cauchy condition $y(\tau) = \xi$ is a function $y : \mathcal{V} \subset \mathcal{U} \to \mathcal{G}$, where $\mathcal{V}$ is a neighborhood of the point τ, which satisfies

$$\begin{cases} \dfrac{\partial y}{\partial x_i}(x) = X^i(x, y(x)) \ \textit{for} \ \ x \in \mathcal{V},\, i = 1, 2, \ldots, m \\ y(\tau) = \xi. \end{cases}$$

If $\mathcal{V} = \mathcal{U}$, the solution y is *global*.

The existence problem for such equations has been encountered in some particular cases, i.e. $(n = 1, m = 1)$, $(n = 1, m = 2)$ and $(n = 1, m = 3)$. For $n = 1, m = 1$ equation (7.11.1) reduces to $dy = X(x, y)dx$, or equivalently to $y' = X(x, y)$, situation completely clarified by Peano's local existence Theorem 2.2.1. In the case $n = 1, m = 2$ the equation has the form $dy = X^1(x, y)\, dx_1 + X^2(x, y)\, dx_2$, while the local existence problem equivalently rephrases as the problem of finding sufficient conditions, or even necessary and sufficient, for the right-hand side of the equation above to be an exact differential. In this case, the continuity of the functions X^1, X^2 is no longer sufficient for existence. Moreover, now, even under the very restrictive hypothesis that X^1, X^2 do not depend on y and are of class C^∞ on $\mathcal{G}$, the problem considered may have no solution. We recall that, if X^1, X^2 are independent of y and are of class C^1 on $\mathcal{G}$ which is simply connected, then a necessary and sufficient condition for the existence of a function y, of class C^1 on $\mathcal{G}$ such that $dy = X^1(x_1, x_2)\, dx_1 + X^2(x_1, x_2)\, dx_2$, is that

$$\frac{\partial X^1}{\partial x_2} = \frac{\partial X^2}{\partial x_1}$$

on the set $\mathcal{G}$. See 10.13 (a) in [Wrede and Spiegel (2002)], p. 244.

In the case $n = 1, m = 3$, again in the hypothesis that the functions X^1, X^2, X^3 do not depend on y, are of class C^1 and that $\mathcal{G}$ is a parallelepiped, a necessary and sufficient condition in order for

$$X^1(x_1, x_2, x_3)\, dx_1 + X^2(x_1, x_2, x_3)\, dx_2 + X^3(x_1, x_2, x_3)\, dx_3$$

to be an exact differential is that

$$\frac{\partial X^1}{\partial x_2} = \frac{\partial X^2}{\partial x_1}, \quad \frac{\partial X^2}{\partial x_3} = \frac{\partial X^3}{\partial x_2}, \quad \frac{\partial X^3}{\partial x_1} = \frac{\partial X^1}{\partial x_3} \tag{7.11.2}$$

on the set $\mathcal{G}$. See 10.30 (a) in [Wrede and Spiegel (2002)], p. 255.

The results we have just recalled, from which the last two refer to the case in which X^i for $i = 1, 2, \ldots, m$ are independent of y, reveal the difficulty of finding sufficient conditions of existence for the general case of equation (7.11.1), when at least one of the functions X^i, $i = 1, 2, \ldots, m$ depends on y.

We will consider first the case in which X^i is independent of the variable $x \in \mathcal{U}$, for $i = 1, \ldots, m$, called homogeneous, and we shall prove a necessary and sufficient condition in order that (7.11.1) have *the local existence and uniqueness property.* We will then extend this result to the fully general case.

So, let $X^i : \mathcal{G} \to \mathbb{R}^n$, $i = 1, 2, \ldots, m$, be functions of class C^1 and let us consider the *homogeneous system with exact differentials*

$$\begin{cases} \dfrac{\partial y}{\partial x_i}(x) = X^i(y(x)) \textit{ for } x \in \mathcal{V}, i = 1, 2, \ldots, m \\ y(\tau) = \xi. \end{cases} \tag{7.11.3}$$

One may easily see that, if $y : \mathcal{V} \subset \mathcal{U} \to \mathcal{G}$ is a solution of the system (7.11.3), then y is of class C^2 and, by virtue of Schwarz theorem, one has

$$\frac{\partial^2 y}{\partial x_i \partial x_j}(x) = \frac{\partial^2 y}{\partial x_j \partial x_i}(x)$$

for every $x \in \mathcal{V}$. It then follows that, a necessary condition in order that the system (7.11.1) has the local existence property is that

$$\frac{\partial X^i}{\partial y}(y)X^j(y) - \frac{\partial X^j}{\partial y}(y)X^i(y) = 0 \ \text{ for } \ y \in \mathcal{G}. \tag{7.11.4}$$

Definition 7.11.2. *The Lie–Jacobi bracket* associated to the vector fields $X^i, X^j : \mathcal{G} \to \mathbb{R}^n$, of class C^1 on $\mathcal{G}$, is the function $\left[X^i, X^j\right] : \mathcal{G} \to \mathbb{R}$, defined by

$$\left[X^i, X^j\right](y) = \frac{\partial X^i}{\partial y}(y)X^j(y) - \frac{\partial X^j}{\partial y}(y)X^i(y) \ \text{ for } \ y \in \mathcal{G}.$$

Remark 7.11.1. One may easily state that, for all vector fields X^i, X^j, of class C^1, we have

$$\left[X^i, X^j\right] = -\left[X^j, X^i\right].$$

This property shows that $\left[X^i, X^j\right] = \left[X^j, X^i\right]$ if and only if $\left[X^i, X^j\right] = 0$, which justifies the definition that follows.

Definition 7.11.3. The family $\{X^i : \mathcal{G} \to \mathbb{R}^n\,;\ i = 1, 2, \ldots, m\}$ *commutes* on $\mathcal{G}$ if

$$\left[X^i, X^j\right](y) = 0$$

for every $i, j \in \{1, 2, \ldots, m\}$ and $y \in \mathcal{G}$.

The main local existence and uniqueness result referring to the homogeneous system (7.11.3) is:

Theorem 7.11.1. *Let $X^i : \mathcal{G} \to \mathbb{R}^n$, $i = 1, 2, \ldots, m$ be of class C^1 on $\mathcal{G}$. The necessary and sufficient condition in order that the system* (7.11.3) *has the local existence and uniqueness property is that the family of vector fields $\{X^1, X^2, \ldots, X^m\}$ commute on $\mathcal{G}$.*

Proof. The necessity was already proved when we have shown the relation (7.11.4).

Sufficiency. The system being homogeneous, we may assume without loss of generality that $\tau = 0$. Let $\xi \in \mathcal{G}$ be arbitrary, and let us denote by $\varphi_i(\cdot, \xi) : (-\alpha_{i,\xi}, \beta_{i,\xi}) \to \mathbb{R}^n$ the unique saturated bilateral solution of the Cauchy problem

$$\begin{cases} \varphi'(t) = X^i(\varphi(t)) \\ \varphi(0) = \xi. \end{cases}$$

Let $a = \frac{1}{2}\min\{\alpha_{i,\xi}, \beta_{i,\xi}\,;\; i = 1, 2, \ldots, m\}$ and let us observe that, in view of Theorem 2.5.1, it follows that there exists $r > 0$ such that $S(\xi, r) \subset \mathcal{G}$ and, for every $\lambda \in S(\xi, r)$, the unique saturated solution $\varphi_i(\cdot, \lambda)$ of the Cauchy problem

$$\begin{cases} \varphi'(t) = X^i(\varphi(t)) \\ \varphi(0) = \lambda \end{cases}$$

is defined at least on $(-a, a)$. Let us define $y : (-a, a)^m \times S(\xi, r) \to \mathbb{R}^n$ by

$$y(x_1, \ldots, x_m, \lambda) = \varphi_1(x_1, \cdot) \circ \varphi_2(x_2, \cdot) \circ \cdots \circ \varphi_m(x_m, \lambda), \tag{7.11.5}$$

where by "$\circ$" we denoted the superposition of functions. We will prove, by induction over m, that $y(\cdot, \xi)$, defined by (7.11.5), is the desired solution. For $m = 1$ we have one vector field and $y(x_1, \xi) = \varphi_1(x_1, \xi)$ is obviously a solution of the problem (7.11.3). Let us assume then that, for every family of $m - 1$ vector fields of class C^1 which commutes, $y(\cdot, \xi)$, defined by (7.11.5), is a solution of the Cauchy problem (7.11.3). In particular, since $\{X^2, X^3, \ldots, X^m\}$ commutes, the function

$$\tilde{y}(x_2, \ldots, x_m, \xi) = \varphi_2(x_2, \cdot) \circ \varphi_3(x_3, \cdot) \circ \cdots \circ \varphi_m(x_m, \xi)$$

is the solution of the gradient system

$$\frac{\partial \tilde{y}}{\partial x_j} = X^j(\tilde{y}(\tilde{x}; \xi)), \quad j = 2, 3, \ldots, m,$$

where $\tilde{x} = (x_2, x_3, \ldots, x_m)$. We will prove that the function

$$y(x, \xi) = \varphi_1(x_1, \tilde{y}(\tilde{x}, \xi)) = \varphi_1(x_1, \cdot) \circ \varphi_2(x_2, \cdot) \circ \cdots \circ \varphi_m(x_m, \xi)$$

is a solution of the problem (7.11.3). We have

$$\frac{\partial y}{\partial x_1}(x, \lambda) = \frac{\partial \varphi_1}{\partial x_1}(x_1, \mu) = X^1(\varphi_1(x_1, \mu)) = X^1(y(x, \lambda)),$$

where

$$\mu = \varphi_2(x_2, \cdot) \circ \varphi_3(x_3, \cdot) \circ \cdots \circ \varphi_m(x_m, \lambda).$$

From the semigroup property, we have

$$\varphi_1(-x_1, y(x, \xi)) = \varphi_1(-x_1, \varphi_1(x_1, \tilde{y}(\tilde{x}, \xi))) = \varphi_1(-x_1 + x_1, \tilde{y}(\tilde{x}, \xi)) = \tilde{y}(\tilde{x}, \xi).$$

Differentiating both sides of the equality above with respect to x_j, we deduce

$$\frac{\partial \varphi_1}{\partial y}(-x_1, y(x, \xi)) \frac{\partial y}{\partial x_j}(x, \xi) = \frac{\partial \tilde{y}}{\partial x_j}(\tilde{x}, \xi), \quad j = 2, 3, \ldots, m.$$

From this relation, by observing that

$$\frac{\partial \varphi_1}{\partial y}(-x_1, y(x,\xi)) = \frac{\partial \varphi_1}{\partial \lambda}(-x_1, y(x,\xi)),$$

and taking into account Theorem 2.6.1, we get

$$\frac{\partial y}{\partial x_j}(x,\xi) = \left[\frac{\partial \varphi_1}{\partial \lambda}(-x_1, y(x,\xi))\right]^{-1} \frac{\partial \tilde{y}}{\partial x_j}(\tilde{x},\xi)$$

$$= H_1(-x_1, y(x,\xi))X^j(\tilde{y}(\tilde{x},\xi)) = H_1(-x_1,y)X^j(\varphi_1(-x_1,y)), \qquad (7.11.6)$$

for $y = y(x,\xi)$, and $j = 2, 3, \ldots, m$, where

$$H_1(t,z) = \left[\frac{\partial \varphi_1}{\partial \lambda}(t,z)\right]^{-1}$$

satisfies

$$\begin{cases} \dfrac{dH_1}{dt}(t,z) = -\dfrac{\partial X^1}{\partial y}(\varphi_1(t,z))H_1(t,z) \\ H_1(0,z) = I_n. \end{cases}$$

It then follows that $H_1(-x_1, y)$ satisfies

$$\frac{dH_1}{dx_1}(-x_1,y) = -\frac{\partial X^1}{\partial y}(\varphi_1(-x_1,y))H_1(-x_1,y).$$

In view of Leibniz-Newton formula, the right-hand side of (7.11.6), denoted by $Y^j(x_1, y)$, can be rewritten as

$$Y^j(x_1,y) = H_1(-x_1,y)X^j(\varphi_1(-x_1,y)) = Y^j(0,y) + \int_0^{x_1} \frac{d}{d\tau}Y^j(\tau,y)\,d\tau$$

$$= X^j(y) + \int_0^{x_1} H_1(-\tau,y)\left[X^1, X^j\right](\varphi_1(-\tau,y))\,d\tau = X^j(y) \qquad (7.11.7)$$

because, by hypothesis, $\left[X^1, X^j\right](z) = 0$ for every $z \in \mathcal{G}$. In particular, taking $y = y(x,\xi)$ in (7.11.7), we get

$$Y^j(x_1, y(x,\xi)) = X^j(y(x,\xi)), \quad j = 2, 3, \ldots, m,$$

from where, using (7.11.6), we deduce

$$\frac{\partial y}{\partial x_j}(x,\xi) = X^j(y(x,\xi)), \quad j = 2, 3, \ldots, m$$

and the proof of the local existence part is complete.

In order to prove the uniqueness part, let $y, z : (-a,a)^m \to \mathbb{R}^n$ be two solutions of the Cauchy problem (7.11.3). Let $\theta \in (-a,a)^m$ and let us define the functions $u, v : [0,1] \to \mathbb{R}^n$ by

$$u(t) = y(t\theta) \text{ and } v(t) = z(t\theta)$$

for every $t \in [0,1]$. Further, let us define the function $f : \mathcal{G} \to \mathbb{R}^m$ by $f_i(w) = \langle X^i(w), \theta \rangle$ for every $i = 1,2,\ldots,m$ and every $w \in \mathcal{G}$. From the hypothesis, we know that X^i, $i = 1,2,\ldots,m$ are vector fields of class C^1 and therefore f has the same property. On the other hand, one can easily see that u and v are solutions of the Cauchy problem

$$\begin{cases} w' = f(w) \\ w(0) = \xi, \end{cases}$$

which has the uniqueness property. Then we have $u(1) = v(1)$, which is equivalent to $y(\theta) = z(\theta)$. Since θ is arbitrary in $(-a,a)^m$, this completes the proof. □

We will now proceed to the general non-autonomous case.

Theorem 7.11.2. (Frobenius) *Let $X^i : \mathcal{U} \times \mathcal{G} \to \mathbb{R}^n$, $i = 1,2,\ldots,m$ be of class C^1 on $\mathcal{U} \times \mathcal{G}$. The necessary and sufficient condition for the system* (7.11.1) *to have the local existence and uniqueness property is that the family of vector fields $\{X^1, X^2, \ldots, X^m\}$ satisfy the relations*

$$\frac{\partial X^i}{\partial x_j}(x,y) + \frac{\partial X^i}{\partial y}(x,y)X^j(x,y) = \frac{\partial X^j}{\partial x_i}(x,y) + \frac{\partial X^j}{\partial y}(x,y)X^i(x,y)$$

for every $i,j = 1,2\ldots,m$ and every $(x,y) \in \mathcal{U} \times \mathcal{G}$, called the Frobenius integrability conditions.

Proof. The idea of proof is the same as that used to rewrite a non-autonomous system as an autonomous one by introducing an extra unknown function and an extra equation. To begin with, let us denote by $z = (x,y) \in \mathbb{R}^{n+m}$, and let us define $Z^i : \mathcal{U} \times \mathcal{G} \to \mathbb{R}^{n+m}$ by

$$Z^i(z) = \begin{pmatrix} e_i \\ X^i(x,y) \end{pmatrix}$$

for $i = 1,2,\ldots,m$ and $z = (x,y) \in \mathcal{U} \times \mathcal{G}$, where $e_1, e_2, \ldots, e_m$ is the canonical basis in $\mathbb{R}^m$. Let us observe that the system (7.11.1) admits a solution y satisfying the Cauchy condition $y(\tau) = \xi$, if and only if the system

$$\frac{\partial z}{\partial x_i} = Z^i(z)$$

$i = 1, 2, \ldots, m$ admits the solution $z(x) = \begin{pmatrix} x \\ y(x) \end{pmatrix}$ satisfying the condition $z(\tau) = \eta$, where $\eta = \begin{pmatrix} \tau \\ \xi \end{pmatrix}$. Finally, because

$$\frac{\partial Z^i}{\partial z} = \begin{pmatrix} \mathcal{O}_{m\times m} & \mathcal{O}_{m\times n} \\ \dfrac{\partial X^i}{\partial x} & \dfrac{\partial X^i}{\partial y} \end{pmatrix},$$

we have

$$\frac{\partial Z^i}{\partial z}(z)Z^j(z) = \begin{pmatrix} \mathcal{O}_{m\times 1} \\ \dfrac{\partial X^i}{\partial x_j}(x,y) + \dfrac{\partial X^i}{\partial y}(x,y)X^j(x,y) \end{pmatrix}$$

for every $z = (x, y) \in \mathcal{U} \times \mathcal{G}$. Accordingly, $\{X^1, X^2, \ldots, X^m\}$ fulfils the Frobenius integrability conditions if and only if $\{Z^1, Z^2, \ldots, Z^m\}$ commutes on $\mathcal{U} \times \mathcal{G}$ which, by virtue of Theorem 7.11.1, completes the proof. □

Remark 7.11.2. If $n = 1$, $m = 3$ and the vector fields X^1, X^2, X^3 do not depend on y, the Frobenius integrability conditions reduce to the conditions (7.11.2).

7.12 Exercises and Problems

Exercise 7.1. *Let $\phi \in \mathcal{D}(\mathbb{R})$. Which of the sequences below are convergent in $\mathcal{D}(\mathbb{R})$?*

$$(1)\ \frac{1}{k}\phi(t) \quad (2)\ \frac{1}{k}\phi(kt) \quad (3)\ \frac{1}{k}\phi\left(\frac{t}{k}\right),\ k = 1, 2, \ldots.$$

([Vladimirov *et al.* (1981)], p. 101)

Exercise 7.2. *Find the derivatives in the sense of Definition* 7.1.7 *for the following distributions of type function:*

$$\begin{array}{ll} (1)\ x(t) = \operatorname{sgn}(t) & (2)\ x(t) = \cos t \operatorname{sgn}(t) \\ (3)\ x(t) = t \operatorname{sgn}(t) & (4)\ x(t) = t \operatorname{sgn}(t-1) \\ (5)\ x(t) = \sin t\, \theta(t) & (6)\ x(t) = e^t\theta(t). \end{array}$$

Problem 7.1. *Prove that the product of the function of class C^∞ η by $\dot{\delta}(t)$ is given by*

$$\eta(t)\dot{\delta}(t) = -\eta'(0)\delta(t) + \eta(0)\dot{\delta}(t).$$

Prove that:

$$(1)\ t\delta^{(m)}(t) = -m\delta^{(m-1)}(t),\ m = 1, 2, \ldots$$
$$(2)\ t^m\delta^{(m)}(t) = (-1)^m m!\delta(t),\ m = 0, 1, \ldots$$
$$(3)\ t^k\delta^{(m)}(t) = 0,\ m = 1, 2, \ldots, k-1.$$

([Vladimirov *et al.* (1981)], p. 110).

Problem 7.2. *Let $x : \mathbb{R} \to \mathbb{R}$ be a locally Lebesgue integrable function, let $n \in \mathbb{N}$ and let us define $x_n : \mathbb{R} \to \mathbb{R}$ by $x_n(t) = nx(nt)$ for every $t \in \mathbb{R}$. Prove that the sequence of distributions of type function $(x_n)_{n\in\mathbb{N}}$ is point-wise convergent in $\mathcal{D}(\mathbb{R})$ and find its limit.*

Exercise 7.3. *Find an elementary solution for the differential operator $\mathcal{L}$ if:*

(1) $\mathcal{L}[y] = y'' + 3y' + 2y$. (2) $\mathcal{L}[y] = y'' + 2y' + y$. (3) $\mathcal{L}[y] = y'' + 4y$.
(4) $\mathcal{L}[y] = y''' + 4y'' + 4y'$. (5) $\mathcal{L}[y] = y''' - y'$. (6) $\mathcal{L}[y] = y''' - 3y' + 2y$.

Exercise 7.4. *Using the elementary solutions found in Exercise 7.3, solve the following differential equations:*

(1) $y'' + 3y' + 2y = e^t\,\theta(t)$. (4) $y''' + 4y'' + 4y' = \sin t\,\theta(t)$.
(2) $y'' + 2y' + y = t\,\theta(t)$. (5) $y''' - y' = t\,\theta(t)$.
(3) $y'' + 4y = \cos 2t\,\theta(t)$. (6) $y''' - 3y' + 2y = te^t\,\theta(t)$.

Exercise 7.5. *Find the Carathéodory solutions of the Cauchy problems below:*

$$(1)\begin{cases} x' = -x + \theta(t-1) \\ x(0) = 1. \end{cases} \qquad (4)\begin{cases} x' = (1+x^2)\theta(t-\frac{\pi}{2})\sin t \\ x(\frac{\pi}{2}) = 0. \end{cases}$$
$$(2)\begin{cases} x' = 2x + t\theta(t-3) \\ x(1) = 0. \end{cases} \qquad (5)\begin{cases} x' = \operatorname{sgn}(t-\frac{\pi}{4})\cos^2 x \\ x(0) = 1. \end{cases}$$
$$(3)\begin{cases} tx' = x + \theta(t-2) \\ x(1) = -1. \end{cases} \qquad (6)\begin{cases} 2xx' = \operatorname{sgn}(t-1) \\ x(0) = 1. \end{cases}$$

Problem 7.3. *Indicate the points at which the following multi-valued function $F:\mathbb{R}\to 2^{\mathbb{R}}$ is upper semi-continuous if:*[7]

$$(1)\ F(x)=\begin{cases}0 & \text{if } x<0\\ [0,1] & \text{if } x=0\\ 1 & \text{if } x>0.\end{cases}\quad (2)\ F(x)=\begin{cases}0 & \text{if } x<0\\ [1,2] & \text{if } x=0\\ 1 & \text{if } x>0.\end{cases}$$

$$(3)\ F(x)=\begin{cases}0 & \text{if } x<0\\ (0,1) & \text{if } x=0\\ 1 & \text{if } x>0.\end{cases}\quad (4)\ F(x)=\begin{cases}0 & \text{if } x<0\\ [0,1] & \text{if } x=0\\ x^2+1 & \text{if } x>0.\end{cases}$$

$$(5)\ F(x)=\begin{cases}0 & \text{if } x<0\\ [0,1] & \text{if } x=0\\ x^2 & \text{if } x>0.\end{cases}\quad (6)\ F(x)=\begin{cases}0 & \text{if } x<0\\ [0,1] & \text{if } x=0\\ x+2 & \text{if } x>0.\end{cases}$$

Problem 7.4. *Let K be a nonempty and closed subset in $\mathbb{R}^n$ and let $F:K\to 2^{\mathbb{R}^n}$ be a multi-valued function with closed values, which is upper semi-continuous on K. Then its graph is a closed subset of $\mathbb{R}^n\times\mathbb{R}^n$.*

Problem 7.5. *Let $K\subset\mathbb{R}^n$ be nonempty and let $F:K\to 2^{\mathbb{R}^n}$ be a multi-valued function whose values are included in a compact subset H in $\mathbb{R}^n$. If the graph of F is closed in $\mathbb{R}^n\times\mathbb{R}^n$ then F is upper semi-continuous on K.*

Problem 7.6. *Let $K\subset\mathbb{R}^n$ be nonempty and let $F:K\to 2^{\mathbb{R}^n}$ be a multi-valued function having compact values. If F is upper semi-continuous on K then $\overline{\operatorname{conv} F}$*[8] *is upper semi-continuous on K.*

Problem 7.7. *Let $K\subset\mathbb{R}^n$ be nonempty and let $f:K\to\mathbb{R}^n$ be a bounded function. We define $F:K\to 2^{\mathbb{R}^n}$ by $F(x)=\operatorname{Lim} f(x)$, where $\operatorname{Lim} f(x)$ is the set of all limit points of the function f at $x\in K$. Then F is upper semi-continuous on K.*

Problem 7.8. *Find two nonempty sets $\Omega\in\mathbb{R}$, $\Sigma\subset\Omega$, and a continuous function $f:\Omega\to\mathbb{R}$, such that Σ is viable with respect to $x'=f_{|\Sigma}(x)$, but Σ is not invariant with respect to $x'=f(x)$.*

Problem 7.9. *A subset $\mathcal{C}$ in $\mathbb{R}^n$ is a cone if for every $\eta\in\mathcal{C}$ and every $s>0$, we have $s\eta\in\mathcal{C}$. Let Σ be a nonempty subset of $\mathbb{R}^n$ and $\xi\in\Sigma$. Prove that $\mathcal{T}_\Sigma(\xi)$ is a closed cone.*

Problem 7.10. *Let Ω be a nonempty and open subset of $\mathbb{R}^n$ and let $U:\Omega\to\mathbb{R}$ be a function of class C^1 with $\nabla U(x)\neq 0$ on Ω. Let $c\in\mathbb{R}$ and let us assume*

[7] To simplify the writing we agree that, whenever the value of one multi-valued function F at a point x is a set containing a single element y, to use the notation $F(x)=y$, instead of $F(x)=\{y\}$.

[8] We recall that $\overline{\operatorname{conv} F(x)}$ is the closed convex hull of the set $F(x)$, i.e. the intersection of all closed convex subsets in $\mathbb{R}^n$ which include it.

that $\Sigma = \{x \in \Omega;\ U(x) = c\}$ *is nonempty. Prove that* $\eta \in \mathbb{R}^n$ *is tangent to* Σ *at the point* $\xi \in \Sigma$ *if and only if* $\langle \eta, \nabla U(\xi)\rangle = 0$. *In other words, in this case,* $\mathcal{T}_\Sigma(\xi)$ *coincides with the set of vectors parallel to the tangent plane to* Σ *at* ξ.

Problem 7.11. *Prove that the set* $\Sigma = \{(x_1, x_2, x_3) \in \mathbb{R}^3;\ x_1^2 + x_2^2 + x_3^2 = 1\}$ *is invariant for the differential system*

$$\begin{cases} x_1' = -x_2 + x_3^2 \\ x_2' = x_1 \\ x_3' = -x_1 x_3. \end{cases}$$

Problem 7.12. *Prove that each of the two coordinate axes is an invariant set for the Lotka–Volterra system*

$$\begin{cases} x' = (a - ky)x \\ y' = -(b - hx)y. \end{cases}$$

Using this, prove that every solution of the system issued from the first quadrant remains there on the whole interval of existence.

Problem 7.13. *Prove the following interesting consequence of Theorem* 7.7.2 :
Theorem A *Let* $\Sigma \subset \mathbb{R}^n$ *be nonempty and locally closed and let* $f : \Sigma \to \mathbb{R}^n$ *be continuous. A sufficient condition for* Σ *to be viable with respect to* $x' = f(x)$ *is that, for every* $\xi \in \partial\Sigma$, $f(\xi) \in \mathcal{T}_{\partial\Sigma}(\xi)$, *where* $\partial\Sigma$ *denotes the boundary of the set* Σ.

Is this condition necessary?

Chapter 8

Volterra Equations

The main goal of this chapter is to present a class of integral equations which can be viewed as generalizations of differential equations, i.e. Volterra linear equations. We start with a second-order Cauchy problem leading to a Volterra linear equation of the second kind and we prove a general existence and uniqueness result based on the successive approximation method. Then, we show how to construct the resolvent kernel which is, in some sense, the analogue of the fundamental matrix for first-order linear differential systems. We establish a sufficient condition for a Volterra linear equation of the first kind to be reducible to a Volterra equation of the second kind, and we prove an existence result for the nonlinear case. Finally, we include several exercises and problems.

8.1 Volterra Equations of the Second Kind

In this section we consider the so-called Volterra[1] equation of the second kind

$$x(t) = h(t) + \lambda \int_a^t k(t,s)x(s)\,ds \tag{8.1.1}$$

with the unknown $x : [\,a,b\,] \to \mathbb{C}^n$, where the given functions $h : [\,a,b\,] \to \mathbb{C}^n$ and $k : \{(t,s);\ a \leq s \leq t \leq b\} \to \mathcal{M}_{n\times n}(\mathbb{C})$ are assumed to be continuous, while the parameter $\lambda \in \mathbb{C}$.

A simple nontrivial problem leading to a Volterra equation of the second kind is presented below. Let us consider the second-order differential

[1] This equation was studied for the first time by the Italian mathematician Vito Volterra (1860-1940) in 1886.

equation

$$\begin{cases} y''(t) - \lambda y(t) = h(t) \\ y(a) = y'(a) = 0, \end{cases} \tag{8.1.2}$$

where $\lambda \in \mathbb{C}$ and $h : [a,b] \to \mathbb{C}$ is a given continuous function. Let us denote by $x : [a,b] \to \mathbb{C}$ the function defined by $x(t) = y''(t)$ for $t \in [a,b]$. We will show that x is the solution of a Volterra equation of the second kind. Indeed, integrating from a to t the equality defining the function x, we get

$$\int_a^t x(s)\,ds = y'(t) - y'(a) = y'(t).$$

Integrating again, from a to t, we get

$$\int_a^t \int_a^s x(\tau)\,d\tau\,ds = y(t) - y(a) = y(t).$$

But a simple integration by parts yields

$$\int_a^t \int_a^s x(\tau)\,d\tau\,ds = \int_a^t \frac{ds}{ds} \int_a^s x(\tau)\,d\tau\,ds$$

$$= s \int_a^s x(\tau)\,d\tau \Big|_a^t - \int_a^t s x(s)\,ds = \int_a^t (t-s)x(s)\,ds = y(t),$$

for $t \in [a,b]$. Now, taking into account that y satisfies the differential equation in (8.1.2), we get

$$x(t) = h(t) + \lambda \int_a^t (t-s)x(s)\,ds$$

for $t \in [a,b]$. So, the problem (8.1.2) can be reduced to a Volterra equation of the second kind, with $k(t,s) = t - s$.

Definition 8.1.1. By definition, a *solution of* (8.1.1) is a continuous function $x : [a,b] \to \mathbb{C}^n$ which satisfies the equality in (8.1.1) for all $t \in [a,b]$.

Throughout in this chapter, we denote by

$$\Delta = \{(t,\theta);\ a \le \theta \le t \le b\}.$$

Theorem 8.1.1. *Let $\lambda \in \mathbb{C}$ and let $k : \Delta \to \mathcal{M}_{n\times n}(\mathbb{C})$ and $h : [a,b] \to \mathbb{C}^n$ be continuous. Then,* (8.1.1) *has a unique solution, $x : [a,b] \to \mathbb{C}^n$, which is the uniform limit on $[a,b]$ of the sequence of successive approximations, $(x_p)_p$, defined by*

$$\begin{cases} x_0(t) = h(t) \\ x_{p+1}(t) = h(t) + \lambda \displaystyle\int_a^t k(t,s)x_p(s)\,ds, \quad \text{for } p = 0,1,2,\ldots. \end{cases} \tag{8.1.3}$$

Proof. We will show that the sequence $(x_p)_p$, defined as above, is uniformly convergent on $[a,b]$ to a continuous function, $x : [a,b] \to \mathbb{C}^n$, which is a solution of (8.1.1). Throughout, if $x = \alpha + i\beta \in \mathbb{C}$, we denote by $\overline{x}$ its conjugate, i.e. $\overline{x} = \alpha - i\beta$ and by $|x|$ its absolute value, i.e. $|x| = \sqrt{x \cdot \overline{x}} = \sqrt{\alpha^2 + \beta^2}$. Moreover, if $A \in \mathcal{M}_{n\times n}(\mathbb{C})$, we denote by $\|A\|_{\mathcal{M}}$ the norm of the matrix[2] A, i.e.

$$\|A\|_{\mathcal{M}} = \sup\{\|Ax\|;\ x \in \mathbb{C}^n,\ \|x\| \leq 1\},$$

where, for $x = (x_1, x_2, \ldots, x_n) \in \mathbb{C}^n$,

$$\|x\| = \left(\sum_{k=1}^{n} |x_k|^2\right)^{1/2} = \left(\sum_{k=1}^{n} x_k \cdot \overline{x}_k\right)^{1/2}.$$

To this aim, let us observe that the sequence $(x_p)_p$ coincides with the sequence of partial sums of the series

$$x_0(t) + \sum_{p=0}^{\infty}[x_{p+1}(t) - x_p(t)]. \tag{8.1.4}$$

So, it would be enough to show that the series above is uniformly convergent on $[a,b]$. In order to prove this, we will make use of the Weierstrass Comparison Principle, i.e. we will show that there exists a convergent series, $\sum_{p=0}^{\infty} \alpha_p$, with positive terms, such that

$$\|x_{p+1}(t) - x_p(t)\| \leq \alpha_p,$$

for each $p \in \mathbb{N}$ and each $t \in [a,b]$.

Let us denote by $M = \sup\{\|k(t,s)\|_{\mathcal{M}}; a \leq s \leq t \leq b\}$ which is finite, because $\|k(\cdot,\cdot)\|_{\mathcal{M}}$ is continuous and Δ is compact, and by $H = \sup\{\|h(t)\|; t \in [a,b]\}$ which is also finite because h is continuous and $[a,b]$ is compact.

We have

$$\|x_1(t) - x_0(t)\| \leq \left\|\lambda \int_a^t k(t,s)h(s)\,ds\right\| \leq |\lambda| MH(t-a)$$

for each $t \in [a,b]$. Further, one may easily see that, by (8.1.3), we have

$$\|x_2(t) - x_1(t)\| \leq |\lambda M|^2 H \frac{(t-a)^2}{2!}$$

for each $t \in [a,b]$. The last two inequalities suggest that, for each $k \in \mathbb{N}$ and each $t \in [a,b]$, we have

$$\|x_{p+1}(t) - x_p(t)\| \leq |\lambda M|^{p+1} H \frac{(t-a)^{p+1}}{(p+1)!}. \tag{8.1.5}$$

[2]We note that all the results in Section 12.1, referring to square matrices with real entries, extend in an obvious way to square matrices with complex entries.

We prove (8.1.5) by mathematical induction. Clearly, for $p = 0$, (8.1.5) holds true. Let us now assume that it holds true for an arbitrary $p \in \mathbb{N}$. Then, we have

$$\|x_{p+2}(t) - x_{p+1}(t)\| \leq \left\| \lambda \int_a^t k(t,s)[x_{p+1}(s) - x_p(s)]\, ds \right\|$$

$$\leq |\lambda M| \int_a^t \|x_{p+1}(s) - x_p(s)\|\, ds \leq |\lambda M|^{p+2} H \int_a^t \frac{(s-a)^{p+1}}{(p+1)!}\, ds$$

$$= |\lambda M|^{p+2} H \frac{(t-a)^{p+2}}{(p+2)!}$$

for each $t \in [\,a, b\,]$. Hence

$$\|x_{p+1}(t) - x_p(t)\| \leq H \frac{[|\lambda M|(b-a)]^{p+1}}{(p+1)!}$$

for each $p \in \mathbb{N}$ and each $t \in [\,a, b\,]$. So, we can take as a comparison series

$$\sum_{p=0}^{\infty} H \frac{[|\lambda M|(b-a)]^{p+1}}{(p+1)!} = H \left(e^{|\lambda M|(b-a)} - 1 \right).$$

Consequently, the series (8.1.4) is absolutely and uniformly convergent on $[\,a, b\,]$. This shows that the sequence $(x_p)_p$ is uniformly convergent on $[\,a, b\,]$ to a continuous function $x : [\,a, b\,] \to \mathbb{C}^n$. Passing to the limit for both sides in (8.1.3), we deduce that x is a solution of the problem (8.1.1).

To prove the uniqueness, let $y : [\,a, b\,] \to \mathbb{C}^n$ be another solution of (8.1.1). We have

$$\|x(t) - y(t)\| \leq |\lambda| M \int_a^t \|x(s) - y(s)\|\, ds$$

for each $t \in [\,a, b\,]$. By Gronwall's Lemma 1.5.2, we get that $\|x(t) - y(t)\| = 0$ for each $t \in [\,a, b\,]$. Thus x and y coincide and this completes the proof. □

8.2 The Resolvent Kernel

We will show that, likewise in the case of first-order linear differential systems where, once known a fundamental matrix, each solution can be obtained by the variational of constants formula, for the linear Volterra equation of the second kind, the solution corresponding to the function h and parameter λ can be obtained by means of a general formula involving a matrix-valued function, independent of h, called the resolvent kernel of the equation. More precisely, we have:

Theorem 8.2.1. *Let $k : \Delta \to \mathcal{M}_{n\times n}(\mathbb{C})$ be continuous. Then, there exists a unique continuous function $R(\cdot,\cdot,\cdot) : \Delta \times \mathbb{C} \to \mathcal{M}_{n\times n}(\mathbb{C})$, such that for each continuous $h : [\,a, b\,] \to \mathbb{C}^n$ and each $\lambda \in \mathbb{C}$, the unique solution of the Volterra equation (8.1.1) is given by*

$$x(t) = h(t) + \lambda \int_a^t R(t, \theta, \lambda) h(\theta)\, d\theta. \tag{8.2.1}$$

Proof. Let us first consider the second term of the sequence of successive approximations defined by (8.1.3), i.e.

$$x_2(t) = h(t) + \lambda \int_a^t k(t,s) \left[h(s) + \lambda \int_a^s k(s,\theta) h(\theta)\, d\theta \right] ds.$$

By changing the order of integration in the double integral on the right-hand side, we get

$$x_2(t) = h(t) + \lambda \int_a^t k(t,s)h(s)\, ds + \lambda^2 \int_a^t \left[\int_\theta^t k(t,s)k(s,\theta)\, ds \right] h(\theta)\, d\theta.$$

Denoting by

$$k_1(t,s) = k(t,s) \quad \text{and} \quad k_2(t,\theta) = \int_\theta^t k(t,s)k_1(s,\theta)\, ds,$$

we obtain

$$x_2(t) = h(t) + \lambda \int_a^t k_1(t,s)h(s)\, ds + \lambda^2 \int_a^t k_2(t,s)h(s)\, ds.$$

Inductively, we get that

$$x_p(t) = h(t) + \int_a^t \left[\sum_{m=1}^{p} \lambda^m k_m(t,\theta) \right] h(\theta)\, d\theta,$$

where the *iterated kernels*, $(k_m)_{m\in\mathbb{N}^*}$, are defined by

$$k_1(t,\theta) = k(t,\theta) \quad \text{and} \quad k_m(t,\theta) = \int_\theta^t k(t,s)k_{m-1}(s,\theta)\, ds$$

for $m = 2, 3, \ldots$. We know from the preceding section that the sequence $(x_p)_p$ is uniformly convergent on $[\,a, b\,]$ to the unique solution x of (8.1.1). So, to complete the proof, it would be sufficient to show that, for each $r > 0$, the series

$$\sum_{m=1}^{\infty} \lambda^m k_m(t,\theta)$$

is absolutely and uniformly convergent, on $\Delta \times B(0,r)$, to a function $R(t,\theta,\lambda)$. We recall that $B(0,r) = \{\lambda \in \mathbb{C};\ |\lambda| \leq r\}$.

Let

$$M = \sup\{\|k(t,\theta)\|_{\mathcal{M}};\ (t,\theta) \in \Delta\}.$$

To show the convergence of the series, we will prove by mathematical induction that

$$\|k_m(t,\theta)\|_{\mathcal{M}} \leq M^m \frac{(t-a)^{(m-1)}}{(m-1)!}, \tag{8.2.2}$$

for each $m \in \mathbb{N}$ and each $(t,\theta) \in \Delta$. Indeed, for $m = 1$, the inequality (8.2.2) is obviously satisfied. Next, let us assume that it is satisfied by an arbitrary $m \in \mathbb{N}$. We have

$$\|k_{m+1}(t,\theta)\|_{\mathcal{M}} \leq \left\| \int_\theta^t k(t,s) k_m(s,\theta)\, ds \right\|_{\mathcal{M}} \leq \int_\theta^t \|k(t,s)\|_{\mathcal{M}} \|k_m(s,\theta)\|_{\mathcal{M}}\, ds$$

$$\leq M \int_\theta^t M^m \frac{(s-a)^{(m-1)}}{(m-1)!}\, ds = M^{m+1} \frac{(t-a)^m}{m!}.$$

So, (8.2.2) holds true.

Now, let us observe that

$$\|\lambda^m k_m(t,\theta)\|_{\mathcal{M}} \leq rM \frac{[rM(b-a)]^{(m-1)}}{(m-1)!},$$

for each $\lambda \in B(0,r)$, each $m \in \mathbb{N}$ and each $(t,\theta) \in \mathbb{R}^2$, $a \leq \theta \leq t \leq b$. By the Weierstrass Comparison Principle, the series $\sum_{m=1}^{\infty} \lambda^m k_m(t,\theta)$ is absolutely and uniformly convergent on $\Delta \times B(0,r)$. The proof is complete. □

Definition 8.2.1. The function $R(\cdot,\cdot,\cdot) : \Delta \times \mathbb{C} \to \mathcal{M}_{n \times n}(\mathbb{C})$, defined by

$$R(t,s,\lambda) = \sum_{m=1}^{\infty} \lambda^{m-1} k_m(t,s),$$

is called *the resolvent kernel* of the Volterra equation (8.1.1).

8.3 Volterra Equations of the First Kind

Here we consider the Volterra equation of the first kind, i.e.

$$\int_a^t k(t,s) x(s)\, ds = h(t) \tag{8.3.1}$$

with the unknown $x : [a, b] \to \mathbb{C}^n$, where the functions $k : \Delta \to \mathcal{M}_{n\times n}(\mathbb{C})$ and $h : [a, b] \to \mathbb{C}^n$ are assumed to be continuous. The next result is a sufficient condition for (8.3.1) to be reducible to a Volterra equation of the second kind.

Theorem 8.3.1. *Let us assume that:*

(i) $h : [a, b] \to \mathbb{C}$ *is of class* C^1 *on* $[a, b]$ *and* $h(a) = 0$;
(ii) for each $s \in [a, b]$, $t \mapsto k(t, s)$ *is of class* C^1 *on* $[s, b]$ *and both* k *and* $\dfrac{\partial k}{\partial t}$ *are continuous on* Δ;
(iii) $k(s, s) \neq 0$ *for each* $s \in [a, b]$.

Then, (8.3.1) *can be reduced to a Volterra equation of the second kind.*

Proof. First, let us observe that $h(a) = 0$ is a compatibility condition because, otherwise, (8.3.1) would have no solution. Since h and $t \mapsto k(t, s)$ are of class C^1, we can differentiate (8.3.1) on both sides. We get

$$k(t,t)x(t) + \int_a^t \frac{\partial k}{\partial t}(t,s)x(s)\,ds = h'(t),$$

for each $t \in [a, b]$. Recalling that $k(t, t) \neq 0$ for each $t \in [a, b]$, we obtain

$$x(t) = h'(t)[k(t,t)]^{-1} + \int_a^t \left\{(-\frac{\partial k}{\partial t}(t,s)[k(t,t)]^{-1}\right\} x(s)\,ds,$$

which is a Volterra equation of the second kind, as claimed. □

Remark 8.3.1. In the case in which $h : (a, b] \to \mathbb{C}$, we may relax the condition $h(a) = 0$, of course, enlarging the concept of solution to functions which are continuous on $(a, b]$ and may have a singularity at $t = a$. See for instance (5) in Exercise 8.6. Also, in the case in which the condition *(iii)* in Theorem 8.3.1 is not satisfied, we may still have solutions but we lose the uniqueness. See for instance (3) and (4) in Exercise 8.6. Alternatively, if *(iii)* in Theorem 8.3.1 does not hold, it may happen that (8.3.1) has no solution. See for instance Exercise 8.7.

8.4 The Nonlinear Case

In this section we will prove a local existence result for a nonlinear version of the second kind Volterra equation (8.1.1), i.e.

$$x(t) = h(t) + \lambda \int_a^t k(t,s,x(s))\,ds, \tag{8.4.1}$$

where $h : [a, b] \to \mathbb{C}^n$ and $k : \Delta \times \mathbb{C}^n \to \mathbb{C}^n$ are given continuous functions. The main local existence result concerning (8.4.1) is:

Theorem 8.4.1. *Let us assume that $h : [a, b] \to \mathbb{C}^n$ and $k : \Delta \times \mathbb{C}^n \to \mathbb{C}^n$ are continuous and $\lambda \in \mathbb{C}$. Then, there exists $c \in (a, b]$ such that* (8.4.1) *has at least one solution $x : [a, c] \to \mathbb{C}^n$.*

Proof. Let $r > 0$ be such that

$$\|h(s) - h(t)\| \le r \tag{8.4.2}$$

for each $(t, s) \in \Delta$. Next, let us define the set

$$\mathcal{S} = \{(t, s, x) \in \Delta \times \mathbb{C}^n;\ \|x - h(t)\| \le r\}$$

which is compact because h is continuous and $[a, b]$ is compact. So, the restriction of the continuous function k to $\mathcal{S}$ is bounded, i.e. there exists $M > 0$ such that

$$\|k(t, s, x)\| \le M \tag{8.4.3}$$

for each $(t, s, x) \in \mathcal{S}$. Let c be a sufficiently small number in $(a, b]$, satisfying

$$M|\lambda|(c - a) \le r,$$

and let us define the set

$$\mathcal{C} = \{x \in C([a, c]; \mathbb{C}^n);\ \|x(t) - h(t)\| \le r, \text{ for all } t \in [a, c]\}.$$

Let $\omega > 0$ be arbitrary and let us consider the delayed integral equation

$$\begin{cases} x_\omega(t) = h(a) & \text{for } t \in [a - \omega, a] \\ x_\omega(t) = h(t) + \lambda \displaystyle\int_a^t k(t, s, x_\omega(s - \omega))\, ds & \text{for } t \in (a, c]. \end{cases} \tag{8.4.4}$$

It is not difficult to show that (8.4.4) has a unique solution $x_\omega \in \mathcal{C}$. Indeed, x_ω is well defined on $[a - \omega, a]$. If $t \in [a, a + \omega]$, then $x_\omega(t - \omega) = h(a)$ and thus, by virtue of (8.4.2), we get

$$\|x_\omega(t) - h(t)\| = \|h(a) - h(t)\| \le r$$

for each $t \in [a - \omega, a]$. Next, let us assume that for some $p \in \mathbb{N}$, for which $a + p\omega < c$, we have $\|x_\omega(t) - h(t)\| \le r$ for each $t \in [a - \omega, a + p\omega]$. Let $t \in [a + p\omega, a + (p + 1)\omega] \cap [a, c]$. In view of (8.4.4) and (8.4.3) and of the choice of c, we have

$$\|x_\omega(t) - h(t)\| \le |\lambda| M(c - a) \le r.$$

So, after a finite number of steps, m (m being the least positive integer satisfying $a + m\omega \geq c$), we deduce that

$$\|x_\omega(t) - h(t)\| \leq r$$

for each $t \in [\,a-\omega, c\,]$, which means that $x_\omega \in \mathcal{C}$. Next, we will show that the family of functions $\{x_\omega;\ \omega \in (0,1)\}$ is relatively compact in $C([\,a,c\,];\mathbb{C}^n)$. To this aim, we will prove that $\{x_\omega;\ \omega \in (0,1)\}$ is uniformly bounded and equicontinuous on $[\,a,c\,]$ and then, we will use Ascoli-Arzelà Theorem 12.2.1. We notice that here, instead of $C([\,a,c\,];\mathbb{R}^n)$, as required by Theorem 12.2.1, we have $C([\,a,c\,];\mathbb{C}^n)$. Since $\mathbb{C}^n$ can be identified with $\mathbb{R}^{2n}$, clearly we are allowed to use Theorem 12.2.1. Since h is continuous, it is bounded on $[\,a,c\,]$ by some constant $m > 0$. So, we have

$$\|x_\omega(t)\| \leq \|h(t)\| + |\lambda| \left\| \int_a^t k(t,s,x_\omega(s-\omega)\,ds \right\| \leq m + |\lambda|(c-a)M$$

for each $\omega \in (0,1)$ and each $t \in [\,a,c\,]$. Thus, $\{x_\omega;\ \omega \in (0,1)\}$ is uniformly bounded on $[\,a,c\,]$. To check out the equicontinuity, let $t, \tilde{t} \in [\,a,c\,]$ be arbitrary, and let us observe that

$$\|x_\omega(t) - x_\omega(\tilde{t})\| \leq \|h(t) - h(\tilde{t})\|$$

$$+|\lambda| \left\| \int_a^t k(t,s,x_\omega(s-\omega))\,ds - \int_a^{\tilde{t}} k(\tilde{t},s,x_\omega(s-\omega))\,ds \right\|.$$

To fix the ideas, let us assume that $\tilde{t} \leq t$. We have

$$\left\| \int_a^t k(t,s,x_\omega(s-\omega))\,ds - \int_a^{\tilde{t}} k(\tilde{t},s,x_\omega(s-\omega))\,ds \right\|$$

$$\leq \left\| \int_a^{\tilde{t}} k(t,s,x_\omega(s-\omega))\,ds - \int_a^{\tilde{t}} k(\tilde{t},s,x_\omega(s-\omega))\,ds \right\|$$

$$+ \left\| \int_{\tilde{t}}^t k(t,s,x_\omega(s-\omega))\,ds \right\|$$

$$\leq \int_a^{\tilde{t}} \|k(t,s,x(s-\omega)) - k(\tilde{t},s,x(s-\omega))\|\,ds + \int_{\tilde{t}}^t \|k(\tilde{t},s,x(s-\omega))\|\,ds.$$

Since k is continuous on $\mathcal{S}$ and the latter is compact, it follows that k is uniformly continuous. In particular, this implies that, for each $\varepsilon > 0$ there exists $\delta(\varepsilon) > 0$ such that for each $(t,s,x), (\tilde{t},s,x) \in \mathcal{S}$ with $|t - \tilde{t}| \leq \delta(\varepsilon)$, we have

$$\|k(t,s,x) - k(\tilde{t},s,x)\| \leq \frac{\varepsilon}{3|\lambda|(b-a)}.$$

We may assume with no loss of generality that $\delta(\varepsilon) \leq \frac{\varepsilon}{3M|\lambda|}$, where M is the constant satisfying (8.4.3) and, in addition, that

$$\|h(t) - h(\tilde{t})\| \leq \frac{\varepsilon}{3},$$

for each $t, \tilde{t} \in [\,a,b\,]$ with $|t - \tilde{t}| \leq \delta(\varepsilon)$. So, from the inequality above, we deduce that, for each $t, \tilde{t} \in [\,a,b\,]$ with $|t - \tilde{t}| \leq \delta(\varepsilon)$, we have

$$\left\|x_\omega(t) - x_\omega(\tilde{t})\right\| \leq \|h(t) - h(\tilde{t})\| + \frac{\varepsilon|\lambda|(b-a)}{3|\lambda|(b-a)} + \frac{\varepsilon M|\lambda|}{3M|\lambda|} = \varepsilon,$$

uniformly for $\omega \in (0,1)$. Thus, the family $\{x_\omega;\ \omega \in (0,1)\}$ is equicontinuous on $[\,a,c\,]$. By virtue of Ascoli-Arzelà Theorem 12.2.1, we conclude that there exist a sequence $(\omega_k)_k$, with $\lim_{k\to\infty} \omega_k = 0$, and a continuous function $x : [\,a,c\,] \to \mathbb{C}^n$, such that

$$\lim_{k\to\infty} x_{\omega_k(t)} = \lim_{k\to\infty} x_{\omega_k}(t - \omega_k) = x(t),$$

uniformly for $t \in [\,a,c\,]$. Clearly $x \in \mathcal{C}$. Finally, taking into account the uniform continuity of k on $\mathcal{S}$, and passing to the limit for $k \to \infty$ in the delayed equation

$$x_{\omega_k}(t) = h(t) + \lambda \int_a^t k(t,s,x_{\omega_k}(s - \omega_k))\,ds,$$

we deduce that x is a solution of the nonlinear Volterra equation (8.4.1). This completes the proof. □

8.5 Exercises and Problems

Exercise 8.1. *Find the Volterra second kind equation which leads to the following Cauchy problem:*

$$\begin{cases} y''(t) + y(t) = 0, & t > 0 \\ y(0) = 0, \quad y'(0) = 1. \end{cases}$$

([Krasnov *et al.* (1977)], p. 18)

Exercise 8.2. *Find the Volterra second kind equation which leads to the following Cauchy problem:*

$$\begin{cases} y''(t) + ty'(t) + y(t) = 0, & t > 0 \\ y(0) = 1, \quad y'(0) = 0. \end{cases}$$

([Krasnov *et al.* (1977)], p. 17)

Exercise 8.3. *Find the Volterra second kind equation which leads to the following Cauchy problem:*

$$\begin{cases} y''(t) - (\sin t)y'(t) + e^t y(t) = t, & t > 0 \\ y(0) = 1, \quad y'(0) = -1. \end{cases}$$

([Krasnov *et al.* (1977)], p. 18)

Exercise 8.4. *Find the Volterra second kind equation which leads to the following Cauchy problem:*

$$\begin{cases} y''(t) + y(t) = \cos t, & t > 0 \\ y(0) = 0, \quad y'(0) = 0. \end{cases}$$

([Krasnov *et al.* (1977)], p. 18)

Problem 8.1. *Prove that for every Cauchy problem for a second-order differential equations with constant coefficients, the kernel of the corresponding Volterra the second kind equation is of the form* $k(t, s) = \tilde{k}(t - s)$.
([Krasnov *et al.* (1977)], p. 18)

Problem 8.2. *Solve the Volterra equation of the second kind,*

$$x(t) = h(t) + \int_a^t k_1(t)k_2(s)x(s)\,ds,$$

by reducing it to a second-order differential equation.
([Havârneanu (2007)], p. 9)

Exercise 8.5. *Solve the following Volterra equations of the second kind:*

(1) $x(t) = t - \int_0^t e^{t-s}x(s)\,ds$ ([Krasnov *et al.* (1977)], p. 19);

(2) $x(t) = e^t + \int_0^t x(s)\,ds$ ([Krasnov *et al.* (1977)], p. 18);

(3) $x(t) = e^t + \int_0^t e^{t-s}x(s)\,ds$ ([Krasnov *et al.* (1977)], p. 24);

(4) $x(t) = \sin t + 2\int_0^t \cos(t-s)x(s)\,ds$ ([Havârneanu (2007)], p. 12);

(5) $x(t) = \cos t + \int_0^t x(s)\,ds$ ([Havârneanu (2007)], p. 13);

(6) $x(t) = \sin t + 2\int_0^t e^{t-s}x(s)\,ds$ ([Krasnov *et al.* (1977)], p. 24).

Exercise 8.6. *Solve the following Volterra equations of the first kind:*

(1) $\int_0^t e^{t+s}x(s)\,ds = 2t$ ([Havârneanu (2007)], p. 10);

(2) $\displaystyle\int_0^t e^{t-s}x(s)\,ds = t$ ([Krasnov *et al.* (1977)], p. 24);

(3) $\displaystyle\int_0^t (t-2s)x(s)\,ds = -\frac{t^3}{6}$ ([Havârneanu (2007)], p. 14);

(4) $\displaystyle\int_0^t (t^2-4ts+3s^2)x(s)\,ds = \frac{t^4}{4}$ ([Havârneanu (2007)], p. 14);

(5) $\displaystyle\int_0^t \frac{x(s)}{(t-s)^\alpha}\,ds = h(t)$, *where* $h \in C^1((0,+\infty);\mathbb{R})$. *This is the Generalized Abel Equation. We note that the Abel Equation corresponds to the specific case* $\alpha = 1/2$. ([Krasnov *et al.* (1977)], p. 29)

Exercise 8.7. *Show that the Volterra integral equation of the first kind*

$$\int_0^t (t-s)x(s)\,ds = t$$

has no solution.

Problem 8.3. *Let* $k:\Delta\to\mathbb{R}$, $x:[a,b]\to\mathbb{R}$ *and* $h:[a,b]\to\mathbb{R}$ *be continuous and nonnegative and let* $\lambda > 0$. *Prove that if*

$$x(t) \le h(t) + \lambda\int_a^t k(t,s)x(s)\,ds,$$

for each $t\in[a,b]$, *then*

$$x(t) \le h(t) + \lambda\int_a^t R(t,s,\lambda)h(s)\,ds,$$

for each $t\in[a,b]$, *where* $R(t,s,\lambda)$ *is the resolvent kernel of the associated equation*

$$x(t) = h(t) + \lambda\int_a^t k(t,s)x(s)\,ds.$$

([Corduneanu (1977)], p. 130)

Problem 8.4. *Use the result above to deduce Gronwall's Lemma* 1.5.2. ([Corduneanu (1977)], p. 130)

Problem 8.5. *Let us consider the Volterra equation of the second kind with convolution kernel* $k:\mathbb{R}\to\mathbb{C}$, *i.e.*

$$x(t) = h(t) + \lambda\int_a^t k(t-s)x(s)\,ds.$$

Show that the resolvent kernel is a convolution kernel too, i.e. is of the form $R(t-s,\lambda)$. ([Corduneanu (1977)], p. 130)

Exercise 8.8. *Find the resolvent kernel corresponding to the following kernels:*

(1) $k(t,s) = e^{t-s}$ ([Krasnov *et al.* (1977)], p. 22);
(2) $k(t,s) = t - s$ ([Krasnov *et al.* (1977)], p. 22);
(3) $k(t,s) = e^{t^2-s^2}$ ([Krasnov *et al.* (1977)], p. 22);
(4) $k(t,s) = \frac{\cosh t}{\cosh s}$ ([Krasnov *et al.* (1977)], p. 22).

Exercise 8.9. *By using the resolvent kernels found in Exercise* 8.8, *solve the following integral equations:*

(1) $x(t) = t + \int_0^t e^{t-s} x(s)\, ds$ ([Havârneanu (2007)], p. 12);
(2) $x(t) = 1 + \int_0^t (t-s) x(s)\, ds$ ([Havârneanu (2007)], p. 12);
(3) $x(t) = e^{t^2} + \int_0^t e^{t^2-s^2} x(s)\, ds$ ([Havârneanu (2007)], p. 12).

Exercise 8.10. *Solve the following Volterra equations of the second kind by using the Laplace Transform:*

(1) $x(t) = e^t - \int_0^t e^{t-s} x(s)\, ds$ ([Havârneanu (2007)], p. 13);
(2) $x(t) = \sin t + \int_0^t \cos(t-s) x(s)\, ds$ ([Havârneanu (2007)], p. 12);
(3) $x(t) = \sinh t - \int_0^t \sinh(t-s) x(s)\, ds$;
(3) $x(t) = \sin t + \int_0^t (t-s) x(s)\, ds$ ([Havârneanu (2007)], p. 13).

Exercise 8.11. *Using the Laplace Transform, solve the following Volterra equations of the first kind:*

(1) $\int_0^t \sinh(t-s) x(s)\, ds = t^3 e^{-t}$ ([Havârneanu (2007)], p. 13);
(2) $\int_0^t e^{t-s} x(s)\, ds = t^2$ ([Havârneanu (2007)], p. 13);
(3) $\int_0^t \cos(t-s) x(s)\, ds = t \sin t$ ([Havârneanu (2007)], p. 13).

Chapter 9

Calculus of Variations

The aim of this chapter is to present some fundamental results in the calculus of variations. We start with several classical examples leading to calculus of variations problems, i.e. the problems of finding: the minimal area of revolution, the celebrated brachistochrone problem, and a specific example in Lagrangian mechanics. We prove the main necessary condition for extremum, i.e. the Euler–Lagrange Equation – in the case of scalar-valued functions – and the Euler–Lagrange System – in the case of vector-valued functions. We briefly present two specific cases in which some prime integrals for the Euler–Lagrange Equations can be easily obtained. We deduce the First Compatibility Conditions of Erdmann-Weierstrass and then, by using these conditions, we obtain a regularity result for extremals, in the case of a regular Lagrangian. We prove two higher-order necessary conditions for extremum, i.e. Jacobi's necessary condition and Legendre's necessary condition. We also include two sufficient conditions for local weak minimum and one sufficient condition for global minimum, we derive the canonical form of the Euler–Lagrange System and we prove a necessary and sufficient condition for a function to be a prime integral for the Euler–Lagrange System. Finally, we collect several exercises and problems.

9.1 Some Examples

We start by introducing one of the simplest, but very important, problems in the calculus of variations. To this aim, let us denote by $\mathcal{D}^n$ the set of all functions $q : [\,a, b\,] \to \mathbb{R}^n$ with q continuous and q' piecewise continuous on $[\,a, b\,]$. Throughout, by a *piecewise continuous function* we mean a function

$p : [a, b] \to \mathbb{R}^n$ for which there exists an exceptional finite set $E \subseteq (a, b)$ such that p is continuous on $[a, b] \setminus E$ and, at each discontinuity point $t \in E$, there exist $p(t-0) = \lim_{s\uparrow t} p(s)$ and $p(t+0) = \lim_{s\downarrow t} p(t)$ and both are finite.

On $\mathcal{D}^n$, we introduce the norm

$$\|q\|_{\mathcal{D}^n} = \sup_{t\in[a,b]\setminus E} \max\{\|q(t)\|, \|q'(t)\|\}.$$

Let $q_a, q_b \in \mathbb{R}^n$ be fixed and let us denote by

$$\mathcal{D}^n(q_a, q_b) = \{q \in \mathcal{D}^n;\ q(a) = q_a,\quad q(b) = q_b\}.$$

Let $L : [a, b] \times \mathbb{R}^n \times \mathbb{R}^n \to \mathbb{R}$ be a C^1-function, called in what follows *Lagrangian*, and let us define the functional $J : \mathcal{D}^n(q_a, q_b) \to \mathbb{R}$, by

$$J[q] = \int_a^b L(t, q(t), q'(t))\, dt$$

for each $q \in \mathcal{D}^n(q_a, q_b)$. We notice that we use the square brackets to designate $J[q]$ just in order to emphasize that the argument of the functional J is itself a function.

Now we can formulate the main problem to be studied in this chapter.

($\mathcal{P}$) *Find a function* $q \in \mathcal{D}^n(q_a, q_b)$ *such that*

$$J[q] = \inf_{p\in\mathcal{D}^n(q_a,q_b)} J[p].$$

We notice that many specific problems of practical interest can be modeled under the abstract form ($\mathcal{P}$). We give below some illustrating examples.

Since in the study of the problem ($\mathcal{P}$) a very important tool is the first variation of the functional J – see Definition 9.2.4 –, ($\mathcal{P}$) is known under the name of *variational problem.*

Example 9.1.1. (The minimal area of revolution) Find $q \in \mathcal{D}^1(q_a, q_b)$, such that the area of the surface obtained by rotating the graph of q around the $0t$-axis be minimal. It is well-known that the area in question is given by[1]

$$\mathcal{A}[q] = 2\pi \int_a^b q(t)\sqrt{1 + q'^2(t)}\, dt.$$

Thus, this problem is of the form ($\mathcal{P}$), with the Lagrangian, $L : [a, b] \times \mathbb{R} \times \mathbb{R} \to \mathbb{R}$, defined by $L(t, q, q') = 2\pi q\sqrt{1 + q'^2}$.

---

[1]Here, for the sake of simplicity, we assume that $q_a \geq 0$ and $q_b \geq 0$, and thus we may consider that $q(t) \geq 0$ for each $t \in [a, b]$. Otherwise, we have to substitute, under the integral sign, $q(t)$ by $|q(t)|$.

Example 9.1.2. (The brachistochrone problem) Find a C^2-curve (C) lying in the vertical plane $x0y$, joining the points (x_0, y_0) and (x_1, y_1), such that the time in which a material point of mass m, which starts from (x_0, y_0) and moves down without friction along the curve (C) under the gravitational force, reaches the end point (x_1, y_1) in a minimal time T. Let $y = y(x)$ be the equation of the plane curve (C). Without loss of generality, we may assume that $x_0 = 0$, $y_0 = 0$. We denote by s the length of arc and by v the speed, i.e. the absolute value of the velocity. From the conservation of energy law, we have

$$\frac{1}{2}mv^2 = mgy.$$

On the other hand, we have

$$v = \frac{ds}{dt} = \sqrt{1 + \left(\frac{dy}{dx}\right)^2}\frac{dx}{dt}.$$

We get

$$dt = \frac{\sqrt{1 + \left(\frac{dy}{dx}\right)^2}}{\sqrt{2gy}}dx$$

and therefore

$$T = \int_0^{x_1} \frac{\sqrt{1 + y'^2(x)}}{\sqrt{2gy(x)}}dx,$$

where

$$y' = \frac{dy}{dx}.$$

So, this problem is of the form $(\mathcal{P})$ but, in some sense, is singular because the corresponding Lagrangian $L : [0, x_1] \times (0, +\infty) \times \mathbb{R} \to \mathbb{R}$,

$$L(x, y, y') = \frac{\sqrt{1 + y'^2}}{\sqrt{2gy}},$$

is not defined on the whole $\mathbb{R}$ with respect to its second variable, y.

Example 9.1.3. (An example in Lagrangian Mechanics) Let us consider a system of n material points whose generalized coordinates[2] are denoted by $q = (q^1, q^2, \ldots, q^n)$, with $q^i = (q_1^i, q_2^i, q_3^i)$, $i = 1, 2, \ldots, n$. Then, the

[2] In fact, in many situations, these are local coordinates on an appropriate manifold M.

trajectories, $q : [a, b] \to \mathbb{R}^{n\times 3}$, of the system are points of minima for the functional

$$J[q] = \int_a^b L(t, q(t), \dot{q}(t))\, dt,$$

where L can be conceived as a "generalized energy" of the system. A particular case is that of those systems for which the Lagrangian is of the form

$$L(t, q, \dot{q}) = T(\dot{q}) - U(t, q),$$

where T is the *kinetic energy* of the system and U is the *potential energy.* In this context, an even more specific case is of great interest. Namely, the case of a material particle (point) of mass m which moves under the action of a central force F. We describe below this simple situation.

Let $\overline{r}(t) = x(t)\overline{i} + y(t)\overline{j} + z(t)\overline{k}$, be the position vector of a material point, P, of mass m, let $r = |\overline{r}|$ and let $U = U(r)$ be the potential of the force F acting on P, $F = -\nabla U$, where ∇U denotes the gradient of U, i.e.

$$\nabla U = \frac{\partial U}{\partial x}\overline{i} + \frac{\partial U}{\partial y}\overline{j} + \frac{\partial U}{\partial z}\overline{k}.$$

The Newton equation is, in this case,

$$m\ddot{\overline{r}}(t) = -\frac{U'(r(t))}{r(t)}\overline{r}(t).$$

We notice that $\dot{\overline{r}}(t)$ and $\ddot{\overline{r}}(t)$ denote the first and respectively the second derivative with respect to time of the function $t \mapsto \overline{r}(t)$. Taking the vector product on both sides of the equation by $\overline{r}(t)$, we deduce that

$$\frac{d}{dt}\left[\overline{r}(t) \times \dot{\overline{r}}(t)\right] = 0$$

at each moment t. But this shows that the trajectory lies in a plane π, which is, in our case, the configuration space. Taking the polar coordinates (r, θ) in the plane π including the trajectory, we get

$$|\dot{\overline{r}}|^2 = \dot{r}^2 + r^2\dot{\theta}^2.$$

Hence, for this system, the kinetic energy is

$$T = \frac{m}{2}|\dot{\overline{r}}|^2 = \frac{m}{2}\left(\dot{r}^2 + r\dot{\theta}^2\right),$$

the potential energy is

$$U = U(r),$$

while the Lagrangian is

$$L((r,\theta),(\dot{r},\dot{\theta})) = \frac{m}{2}\left(\dot{r}^2 + r\dot{\theta}^2\right) - U(r).$$

So, according to the Least Action Hamilton's Principle, the trajectories of the system are those curves $(r,\theta) : [\,a,b\,] \to \mathbb{R}^2$ which are points of minimum for the functional

$$J[(r,\theta)] = \int_a^b \left[\frac{m}{2}\left(\dot{r}^2(t) + r(t)\dot{\theta}^2(t)\right) - U(r(t))\right] dt.$$

In other words, the system always chooses those trajectories which keep a certain balance between the kinetic energy and the potential energy.

9.2 Necessary Conditions for Extremum

Definition 9.2.1. Let K be a subset of a real vector space V, $q \in K$ and $\eta \in V$. We say that q is an *interior point of* K *in the direction* η if there exists $\varepsilon_0 > 0$ such that $q + \varepsilon\eta \in K$ for each $\varepsilon \in (-\varepsilon_0, \varepsilon_0)$.

Remark 9.2.1. Let $V = \mathcal{D}^n$ and $K = \mathcal{D}^n(q_a, q_b)$, and let $\eta \in \mathcal{D}^n(0,0)$. Then, every $q \in K$ is an interior point of K in the direction η.

Definition 9.2.2. The function $q \in \mathcal{D}^n(q_a, q_b)$ is called a *local weak minimum* for the problem $(\mathcal{P})$ if there exists $r > 0$ such that for each $p \in \mathcal{D}^n(q_a, q_b)$ with $\|p - q\|_{\mathcal{D}^n} \le r$, we have $J[q] \le J[p]$.

Definition 9.2.3. The function $q \in \mathcal{D}^n(q_a, q_b)$ is called a *local* $\mathcal{D}^n(0,0)$-*minimum* for the problem $(\mathcal{P})$ if for every $\eta \in \mathcal{D}^n(0,0)$, $\varepsilon = 0$ is a local minimum for the function $g_\eta : (-1,1) \to \mathbb{R}$, defined by

$$g_\eta(\varepsilon) = J[q + \varepsilon\eta] = \int_a^b L(t, q(t) + \varepsilon\eta(t), q'(t) + \varepsilon\eta'(t))\, dt, \tag{9.2.1}$$

for every $\varepsilon \in (-1,1)$.

Remark 9.2.2. Each local weak minimum is a local $\mathcal{D}(0,0)^n$-minimum, but not conversely. We might be tempted to introduce the concept of global $\mathcal{D}(0,0)^n$-minimum as being a function $q \in \mathcal{D}^n(q_a, q_b)$, with the property that, for every $\eta \in \mathcal{D}^n(0,0)$, $\varepsilon = 0$ is a global minimum for the function $g_\eta : (-1,1) \to \mathbb{R}$, defined by (9.2.1). If this is the case, q is in fact a global minimum, simply because for each $\tilde{q} \in \mathcal{D}^n(q_a, q_b)$, and each $\varepsilon \in (-1, 1)\setminus\{0\}$, we have

$$\int_a^b L(t, \tilde{q}(t), \tilde{q}'(t))\, dt = \int_a^b L(t, q(t) + \varepsilon\eta(t), q'(t) + \varepsilon\eta'(t))\, dt,$$

where $\eta \in \mathcal{D}^n(0, 0)$ is given by $\eta = \dfrac{1}{\varepsilon}[\tilde{q} - q]$.

Definition 9.2.4. By definition, $g'_\eta(0)$ (if there exists) is called the *first variation* of the functional J calculated at q, in the direction η, and is denoted by

$$g'_\eta(0) = \delta J[q][\eta].$$

Further, $g''_\eta(0)$, if it exists, is called the *second variation* of the functional J at q, in the direction η, and is denoted by

$$g''_\eta(0) = \delta^2 J[q][\eta].$$

The next result is an immediate consequence of a well-known necessary condition for extremum in Real Analysis.

Theorem 9.2.1. *If $q \in \mathcal{D}^n(q_a, q_b)$ is a local $\mathcal{D}^n(0, 0)$-minimum for $\mathcal{P}$, then, for each $\eta \in \mathcal{D}^n(0, 0)$ for which there exists $\delta J[q][\eta]$, we have $\delta J[q][\eta] = 0$. If, in addition, there exists $\delta^2 J[q][\eta]$, then necessarily $\delta^2 J[q](\eta) \geq 0$.*

Definition 9.2.5. A function $q \in \mathcal{D}^n$ satisfying

$$\delta J[q][\eta] = 0$$

for each $\eta \in \mathcal{D}^n(0, 0)$ is called an *extremal for the problem* ($\mathcal{P}$).

Remark 9.2.3. We notice that, in the definition of an extremal q, we do not ask that $q(a) = q_a$ and $q(b) = q_b$. Of course, if we are looking for a solution of the problem ($\mathcal{P}$), we have to analyze only those extremals which satisfy the boundary conditions above.

Remark 9.2.4. If a certain curve $q \in \mathcal{D}^n$ is an extremal for the problem ($\mathcal{P}$) in some coordinate system, then it is an extremal for the same problem in any other coordinate system whatsoever. This is the invariance property of extremals with respect to the change of the coordinate system. To be more precise, for a variational problem in which $q \in \mathcal{D}^1$, we will get the very same extremals – identified with curves in a given plane – in both Cartesian and polar coordinates.

The main goal of this section is to obtain some specific necessary conditions in order that a certain function, $q \in \mathcal{D}^n(q_a, q_b)$, be a local $\mathcal{D}^n(0,0)$-minimum for the functional

$$J[p] = \int_a^b L(t, p(t), p'(t))\, dt.$$

To explain the main idea, let $q \in \mathcal{D}^n(q_a, q_b)$ be a local $\mathcal{D}^n(0,0)$-minimum for J, and let $\eta \in \mathcal{D}^n(0,0)$ be arbitrary. Let us observe that, in view of Definition 9.2.3, 0 is a local minimum point of the function g_η defined by (9.2.1).

So, under some appropriate regularity hypotheses on L, we can use Fermat Theorem to get the family of conditions $\{\delta J[q][\eta] = 0;\ \eta \in \mathcal{D}^n(0,0)\}$, conditions which take a specific form expressed by a second-order nonlinear differential equation.

9.2.1 *The Scalar-Valued Case*

For the sake of simplicity, we begin with the one-dimensional case, i.e. with the case in which $L : [\,a, b\,] \times \mathbb{R} \times \mathbb{R} \to \mathbb{R}$, the function $p : [\,a, b\,] \to \mathbb{R}$ $q_a \in \mathbb{R}$ and $q_b \in \mathbb{R}$.

The first necessary condition in the scalar-valued case is:

Theorem 9.2.2. *Let us assume that $L : [\,a, b\,] \times \mathbb{R} \times \mathbb{R} \to \mathbb{R}$ is of class C^1. Then, a necessary condition for a function, $q \in \mathcal{D}^1(q_a, q_b)$, to be a local $\mathcal{D}^1(0,0)$-minimum for $(\mathcal{P})$ is that q satisfy*

$$\int_a^b \left[\frac{\partial L}{\partial q}(t, q(t), q'(t))\eta(t) + \frac{\partial L}{\partial q'}(t, q(t), q'(t))\eta'(t)\right] dt = 0, \tag{9.2.2}$$

for each $\eta \in \mathcal{D}^1(0,0)$.

Proof. As L is of class C^1, $q \in \mathcal{D}^1(q_a, q_b)$ and $\eta \in \mathcal{D}^1(0,0)$, it follows that the function g, defined by (9.2.1), is differentiable at $\varepsilon = 0$. But $\varepsilon = 0$ is a local minimum for g and thus, by virtue of Fermat's Theorem, we necessarily have $g_\eta'(0) = \delta L[q][\eta] = 0$. In our specific case, the last condition takes the specific form (9.2.2). The proof is complete. □

In order to obtain additional information in the case in which the minimum point is more regular, a simple but useful lemma is needed.

Lemma 9.2.1. (Du Bois Raymond) *Let $h : [\,a, b\,] \to \mathbb{R}$ be piecewise continuous and let us assume that for every $\eta \in \mathcal{D}^1(0,0)$, we have*

$$\int_a^b h(t)\eta(t)\, dt = 0.$$

Then, $h(t) = 0$ at every $t \in [a, b]$ at which h is continuous.

Proof. We will proceed by contraposition. Namely, we will assume that h is not identically 0 on the set of its continuity points and we will construct a function $\eta \in \mathcal{D}^1(0,0)$ such that

$$\int_a^b h(t)\eta(t)\,dt \neq 0.$$

So, let us assume that there exists $t_0 \in [a, b]$ such that $h(t_0) \neq 0$ and h is continuous at t_0. By changing the sign of h if necessary, we may assume without loss of generality that $h(t_0) > 0$. We will analyze only the case in which $t_0 \in (a, b)$. If either $t_0 = a$ or $t_0 = b$, the proof is very similar and it is left to the reader. Next, let us observe that, by the continuity of h, it follows that there exist α and β with $a \leq \alpha < t_0 < \beta \leq b$ such that

$$h(t) \geq \frac{h(t_0)}{2} > 0$$

for each $t \in [\alpha, \beta]$. Let us define the function $\tilde{\eta} : [a, b] \to \mathbb{R}$ by

$$\tilde{\eta}(t) = \begin{cases} (t-\alpha)(\beta-t) & \text{if } t \in [\alpha, \beta] \\ 0 & \text{if } t \in [a, b] \setminus [\alpha, \beta]. \end{cases}$$

Clearly, $\tilde{\eta} \in \mathcal{D}^1(0,0)$. In addition, we have

$$\int_a^b h(t)\tilde{\eta}(t)\,dt = \int_\alpha^\beta h(t)\tilde{\eta}(t)\,dt \geq \frac{h(t_0)}{2}\int_\alpha^\beta (t-\alpha)(\beta-t)\,dt > 0.$$

This shows that

$$\int_a^b h(t)\eta(t)\,dt = 0$$

does not hold true for every $\eta \in \mathcal{D}^1(0,0)$. The proof is complete. □

Definition 9.2.6. Let $F : [a, b] \times \mathbb{R} \times \mathbb{R} \times \mathbb{R} \to \mathbb{R}$. By a *piecewise solution* of the second-order differential equation

$$F(t, q(t), q'(t), q''(t)) = 0,$$

we mean a function $q : [a, b] \to \mathbb{R}$, with q and q' continuous, q'' piecewise continuous, and which satisfies the equation at every point at which q'' is continuous.

The main necessary condition in order that $q \in \mathcal{D}^1(q_a, q_b)$, with q, q' continuous and q'' piecewise continuous, be a local $\mathcal{D}^1(0,0)$-minimum for $(\mathcal{P})$ is stated below.

Theorem 9.2.3. *Let us assume that $L : [a,b] \times \mathbb{R} \times \mathbb{R} \to \mathbb{R}$ is of class C^2. Then, a necessary condition for a function, $q \in \mathcal{D}^1(q_a, q_b)$, with q and q' continuous and q'' piecewise continuous, to be a local $\mathcal{D}^1(0,0)$-minimum for $(\mathcal{P})$ is that q be a piecewise solution to the second-order nonlinear two-point boundary value problem*

$$\begin{cases} \dfrac{\partial L}{\partial q}(t, q(t), q'(t)) - \dfrac{d}{dt}\left[\dfrac{\partial L}{\partial q'}(t, q(t), q'(t))\right] = 0 \\ q(a) = q_a, \quad q(b) = q_b. \end{cases} \tag{9.2.3}$$

We notice that the second-order nonlinear differential equation in (9.2.3) is known as the *Euler–Lagrange Equation associated to the problem* $(\mathcal{P})$.

Proof. By Theorem 9.2.2, we know that q satisfies (9.2.2). Since L is of class C^2, q, q' are continuous and q'' is piecewise continuous, we are allowed to integrate by parts the second term under the integral in (9.2.2). Taking into account that $\eta(a) = \eta(b) = 0$, we get

$$\int_a^b \left\{\frac{\partial L}{\partial q}(t, q(t), q'(t)) - \frac{d}{dt}\left[\frac{\partial L}{\partial q'}(t, q(t), q'(t))\right]\right\} \eta(t)\, dt = 0.$$

Finally, by observing that this equality should take place for every $\eta \in \mathcal{D}^1(0,0)$, from Lemma 9.2.1, it follows that q is a piecewise solution of the Euler–Lagrange Equation (9.2.3). This completes the proof. □

Remark 9.2.5. As in the one-dimensional calculus, where there exist functions $f : I \to \mathbb{R}$ such that the derivative f' has a zero at an interior point, t_0, of the interval $\mathbb{I}$, but nevertheless t_0 is not a point of extremum for f, in this case too, there are examples of problems $(\mathcal{P})$, when a function q, although an extremal which satisfy the boundary conditions $q(a) = q_a$ and $q(b) = q_b$, is not a minimum point for $(\mathcal{P})$.

We give below such an example.

Example 9.2.1. Let us consider the Lagrangian $L : [0,1] \times \mathbb{R} \times \mathbb{R} \to \mathbb{R}$ defined by $L(t, q, q') = 3tq^2q'$, for each $(t, q, q') \in [0,1] \times \mathbb{R} \times \mathbb{R}$. Let us consider the problem:

$(\mathcal{Q})$ Find a piecewise C^1-function $q : [0,1] \to \mathbb{R}$, with $q(0) = q(1) = 0$ which realizes the minimum of the functional

$$J[q] = \int_0^1 3tq^2(t)q'(t)\, dt$$

over all piecewise C^1-functions vanishing at both 0 and 1. In this case, the Euler–Lagrange Equation is degenerate, i.e. $-3q^2(t) = 0$ for $t \in [0,1]$.

Hence, $q \equiv 0$ is the unique extremal satisfying the boundary conditions $q(0) = q(1) = 0$. On the other hand, for each $q \in \mathcal{D}^1(0,0)$, we have

$$J[q] = \int_0^1 t\frac{d}{dt}[q^3(t)]\,dt = tq^3(t)|_{t=0}^{t=1} - \int_0^1 q^3(t)\,dt = -\int_0^1 q^3(t)\,dt.$$

So, for each $\varepsilon > 0$, we can find a function $q_\varepsilon \in \mathcal{D}^1(0,0)$ such that $\|q_\varepsilon\|_{\mathcal{D}^1} < \varepsilon$ and $J[q] < 0$. Therefore the extremal $q \equiv 0$ is not a local weak minimum point for $(\mathcal{Q})$. So, it is not a local $\mathcal{D}^1(0,0)$-minimum for $(\mathcal{Q})$. Similarly, we can show that $q \equiv 0$ is not a local weak maximum[3].

9.2.2 *The Vector-Valued Case*

We give below an n-dimensional counterpart of Theorem 9.2.1. So, we consider the problem $(\mathcal{P})$ with the Lagrangian $L : [\,a,b\,] \times \mathbb{R}^n \times \mathbb{R}^n \to \mathbb{R}$.

The n-dimensional counterpart of Theorem 9.2.2 is:

Theorem 9.2.4. *Let us assume that $L : [\,a,b\,] \times \mathbb{R}^n \times \mathbb{R}^n \to \mathbb{R}$ is of class C^1. Then, a necessary condition for a function, $q \in \mathcal{D}^n(q_a, q_b)$, to be a local $\mathcal{D}^n(0,0)$-minimum for $(\mathcal{P})$ is that*

$$\int_a^b \left[\sum_{i=1}^n \frac{\partial L}{\partial q_i}(t,q(t),q'(t))\eta_i(t) + \sum_{i=1}^n \frac{\partial L}{\partial q_i'}(t,q(t),q'(t))\eta_i'(t)\right] dt = 0, \tag{9.2.4}$$

for each $\eta \in \mathcal{D}^n(0,0)$.

Proof. As in the proof of Theorem 9.2.2, it follows that the function g_η, defined by (9.2.1) is differentiable at $\varepsilon = 0$ which is a local minimum for g_η. Thus, we necessarily have $g_\eta'(0) = \delta L[q](\eta) = 0$, condition which, in this case, takes the particular form (9.2.4), as claimed. □

In order to extend Theorem 9.2.3 to the n-dimensional case, we need a generalization of Lemma 9.2.1 to vector-valued functions.

Lemma 9.2.2. (Du Bois Raymond) *Let $h : [\,a,b\,] \to \mathbb{R}^n$ be piecewise continuous and let us assume that for every $\eta \in \mathcal{D}^n(0,0)$, we have*

$$\int_a^b \langle h(t)\eta(t)\rangle\,dt = 0.$$

Then, $h(t) = 0$ at every $t \in [\,a,b\,]$ at which h is continuous.

[3] The definition of a local weak maximum for J is left to the reader.

Proof. Let $i \in \{1, 2, \ldots, n\}$ be arbitrary, $\eta_i \in \mathcal{D}^1(0,0)$, and let us consider the function $\eta \in \mathcal{D}^n(0,0)$ of the form $\eta = (0, 0, \ldots, \eta_i, \ldots, 0, 0)$, where η_i is the ith component of η. By hypothesis we have

$$\int_a^b \langle h(t)\eta(t)\rangle\, dt = \int_a^b h_i(t)\eta_i(t)\, dt = 0,$$

and thus, from Lemma 9.2.1, we deduce that $h_i(t) = 0$ at every $t \in [\,a, b\,]$ at which h_i is continuous. Since this happens for each $i \in \{1, 2, \ldots, n\}$, the proof is complete. □

As in the one-dimensional case, we introduce the concept of piecewise solution.

Definition 9.2.7. Let $F : [\,a, b\,] \times \mathbb{R}^n \times \mathbb{R}^n \times \mathbb{R}^n \to \mathbb{R}$. By a *piecewise solution* of the second-order differential equation

$$F(t, q(t), q'(t), q''(t)) = 0,$$

we mean a function $q : [\,a, b\,] \to \mathbb{R}^n$, with q and q' continuous, q'' piecewise continuous, and which satisfies the equation at every point at which q'' is continuous.

Theorem 9.2.5. *Let us assume that $L : [\,a, b\,] \times \mathbb{R}^n \times \mathbb{R}^n \to \mathbb{R}$ is of class C^2. Then, a necessary condition for a function, $q \in \mathcal{D}^n(q_a, q_b)$, with q, q' continuous and q'' piecewise continuous, to be a local $\mathcal{D}^n(0,0)$-minimum for $(\mathcal{P})$, is that q be a piecewise solution to the second-order nonlinear two-point boundary value problem*

$$\begin{cases} \dfrac{\partial L}{\partial q_i}(t, q(t), q'(t)) - \dfrac{d}{dt}\left[\dfrac{\partial L}{\partial q_i'}(t, q(t), q'(t))\right] = 0, & \text{for } i = 1, 2, \ldots, n \\ q(a) = q_a, \quad q(b) = q_b. \end{cases} \tag{9.2.5}$$

Proof. Since L is of class C^2, q, q' are continuous and q'' is piecewise continuous, we can integrate by parts the second term under the integral in (9.2.4). Recalling that $\eta(a) = \eta(b) = 0$, we get

$$\int_a^b \sum_{i=1}^n \left\{\frac{\partial L}{\partial q_i}(t, q(t), q'(t)) - \frac{d}{dt}\left[\frac{\partial L}{\partial q_i'}(t, q(t), q'(t))\right]\right\} \eta_i(t)\, dt = 0. \tag{9.2.6}$$

Since (9.2.6) should take place for every function $\eta \in \mathcal{D}^n(0,0)$, from Lemma 9.2.2, it follows that q is a piecewise solution of the Euler–Lagrange Equation (9.2.5). This completes the proof. □

In this case, i.e. $n \geq 2$, we will refer to the system of second-order nonlinear differential equations in (9.2.5) as to the *Euler–Lagrange System associated to* ($\mathcal{P}$).

Remark 9.2.6. The proof of Theorem 9.2.5 leads to a stronger result. Namely, if $L : [\,a, b\,] \times \mathbb{R}^n \times \mathbb{R}^n \to \mathbb{R}$ is of class C^2 and q is an extremal of the problem ($\mathcal{P}$) with q, q' continuous and q'' piecewise continuous, then q is a piecewise solution of the Euler–Lagrange System in (9.2.5).

9.3 Some Particular Cases

As we can easily see, the Euler–Lagrange Equation is a second-order nonlinear differential equation for which there are no general methods of integration. However, in certain specific cases, we can find a prime integral which allows us to reduce the order of the equation with one unit. Here we confine ourselves to the one-dimensional case, i.e. to the case of scalar-valued extremals.

Theorem 9.3.1. *Let us assume that the Lagrangian L is of class C^2 and does not depend explicitly on t, i.e. $L : \mathbb{R} \times \mathbb{R} \to \mathbb{R}$. Then, a prime integral for the Euler–Lagrange equation is $H : \mathbb{R} \times \mathbb{R} \to \mathbb{R}$, defined by*

$$H(q, q') = q' \frac{\partial L}{\partial q'}(q, q') - L(q, q'),$$

i.e. for each extremal q there exists a constant c such that

$$q'(t) \frac{\partial L}{\partial q'}(q(t), q'(t)) - L(q(t), q'(t)) = c,$$

for each $t \in [\,a, b\,]$.

Proof. For the sake of simplicity, in what follows, instead of $L(q(t), q'(t))$, $\frac{\partial L}{\partial q}(q(t), q'(t))$ and so on, we shall write L, $\frac{\partial L}{\partial q}$ and so on. Since L does not depend explicitly on t, we have

$$\frac{d}{dt}\left[\frac{\partial L}{\partial q'}\right] = \frac{\partial^2 L}{\partial q \partial q'} q' + \frac{\partial^2 L}{\partial q'^2} q''.$$

So, the Euler–Lagrange equation is

$$\frac{\partial L}{\partial q} - \frac{\partial^2 L}{\partial q \partial q'} q' - \frac{\partial^2 L}{\partial q'^2} q'' = 0.$$

Multiplying both sides by q' and then adding and subtracting $\frac{\partial L}{\partial q'} q''$, we get

$$\frac{\partial L}{\partial q} q' + \frac{\partial L}{\partial q'} q'' - q' \left(\frac{\partial^2 L}{\partial q \partial q'} q' + \frac{\partial^2 L}{\partial q'^2} q'' \right) - \frac{\partial L}{\partial q'} q'' = 0,$$

for each $t \in [a, b]$. This identity can be equivalently rewritten as

$$\frac{d}{dt}\left(L - q'\frac{\partial L}{\partial q'}\right) = 0,$$

which shows that there exists a constant c such that

$$q'(t)\frac{\partial L}{\partial q'}(q(t), q'(t)) - L(q(t), q'(t)) = c$$

for each $t \in [a, b]$, as claimed. □

The prime integral H defined in Theorem 9.6.1 is called the *energy integral* of the functional J.

Another particular case occurring in applications is that when L does not depend on q. In this case, the Euler–Lagrange Equation has the specific form

$$\frac{d}{dt}\left[\frac{\partial L}{\partial q'}(t, q'(t))\right] = 0,$$

which implies that, along any extremal, we have

$$\frac{\partial L}{\partial q'}(t, q'(t)) = c.$$

So, in this case, a prime integral for the Euler–Lagrange Equation is

$$W(t, q') = \frac{\partial L}{\partial q'}(t, q').$$

9.4 Regularity of the Extremals

We begin with a simple but very important result.

Theorem 9.4.1. *If the Lagrangian L is of class $C^2 a$ and $q : [a, b] \to \mathbb{R}^n$ is an extremal for the problem $(\mathcal{P})$, then, for each $i = 1, 2, \ldots, n$ and each $t \in [a, b]$, we have*

$$\frac{\partial L}{\partial q_i'}(t, q(t), q'(t-0)) = \frac{\partial L}{\partial q_i'}(t, q(t), q'(t+0)). \tag{9.4.1}$$

Proof. We begin by noticing that each extremal q is a solution in the sense of distributions over (a, b) of the Euler–Lagrange System (9.2.3). Indeed, this follows from Definition 7.1.7 and (9.2.6), by observing that each C^∞-function, $\eta : [a, b] \to \mathbb{R}^n$, with compact support in (a, b), belongs to $\mathcal{D}^n(0, 0)$. Clearly, (9.4.1) holds true at each continuity point of q'. So, it remains to prove (9.4.1) only for those $t \in [a,b]$ at which q' is discontinuous. To this aim, let $i = 1, 2, \ldots, n$, let us denote by $p_i : [a,b] \to \mathbb{R}$, the function defined by

$$p_i(t) = \int_a^t \frac{\partial L}{\partial q_i}(s, q(s), q'(s))\, ds,$$

and let us observe that (9.2.4) can be equivalently written as

$$\int_a^b \left[\sum_{i=1}^n \frac{dp_i}{dt}(t)\eta_i(t) + \sum_{i=1}^n \frac{\partial L}{\partial q_i'}(t, q(t), q'(t))\eta_i'(t) \right] dt = 0.$$

Integrating by parts, and taking into account that $\eta(a) = \eta(b) = 0$, we get

$$\int_a^b \sum_{i=1}^n \left[p_i(t) - \frac{\partial L}{\partial q_i'}(t, q(t), q'(t)) \right] \eta_i'(t)\, dt = 0.$$

In particular, taking any C^∞-vector-valued function, η, with compact support in (a,b), we conclude that the distributional derivative of

$$t \mapsto \left[p_i(t) - \frac{\partial L}{\partial q_i'}(t, q(t), q'(t)) \right], \quad i = 1, 2, \ldots, n,$$

– see Definition 7.1.7 –, is zero. So, in view of Lemma 7.3.2, the functions

$$t \mapsto \left[p_i(t) - \frac{\partial L}{\partial q_i'}(t, q(t), q'(t)) \right], \quad i = 1, 2, \ldots, n,$$

must be constant on $[a,b]$. Since p_i, $i = 1, 2, \ldots, n$, are continuous, it follows that

$$t \mapsto \left[\frac{\partial L}{\partial q_i'}(t, q(t), q'(t)) \right], \quad i = 1, 2, \ldots, n$$

are continuous too. Thus, we conclude that (9.4.1) is also satisfied at each discontinuity point of q'. This completes the proof. □

The relations (9.4.1) are called *The First Compatibility Conditions* of Erdmann-Weierstrass.

Definition 9.4.1. The Lagrangian L is called *regular* if it is of class C^2 and, for each $(t, q, q') \in [a,b] \times \mathbb{R}^n \times \mathbb{R}^n$, the Hessian matrix

$$M(t, q, q') = \left(\frac{\partial^2 L}{\partial q_i' \partial q_j'}(t, q, q') \right) \tag{9.4.2}$$

is nonsingular.

Remark 9.4.1. If the Lagrangian L is regular, i.e. if L is of class C^2 and the Hessian matrix is nonsingular, it follows that the Euler–Lagrange System in (9.2.5) is a second-order nondegenerate system, i.e. it can be equivalently rewritten as a second-order differential system in normal form whose right-hand side is continuous. Indeed, in this case, from the Euler–Lagrange System, we get

$$q_i'' = [M(t,q,q')]^{-1}\left[\frac{\partial L}{\partial q_i}(t,q,q') - \frac{\partial^2 L}{\partial t \partial q_i'}(t,q,q') - \sum_{j=1}^{n} \frac{\partial^2 L}{\partial q_j \partial q_i'}(t,q,q')q_j'\right],$$

for $i = 1, 2, \ldots, n$, where $M(t,q,q')$ is given by (9.4.2).

If, in addition, L is of class C^3 with respect to the variables q and q', then the right-hand side of the second-order nonlinear differential system above is locally Lipschitz with respect to q and q'. Thus the Euler–Lagrange System in (9.2.5) has the uniqueness property. So, if a solution of this system is 0 on an open subinterval of $[a,b]$, it is necessarily 0 on the whole interval.

Theorem 9.4.2. *Let $q \in \mathcal{D}^n(q_a, q_b)$ be an extremal for the problem $(\mathcal{P})$. Let us assume that the Lagrangian is regular and for each $t \in [a,b]$, the gradient vector field $\nabla_{q'} L(t, q(t), \cdot) : \mathbb{R}^n \to \mathbb{R}^n$, defined by*

$$\nabla_{q'} L(t,q(t),\cdot) = \left(\frac{\partial L}{\partial q_1'}(t,q(t),\cdot), \frac{\partial L}{\partial q_2'}(t,q(t),\cdot) \ldots, \frac{\partial L}{\partial q_n'}(t,q(t),\cdot)\right),$$

is injective. Under these assumptions if, for some $r \geq 2$, L is of class C^r, then q is of class C^r.

Proof. First, let us observe that by the Erdmann-Weierstrass Compatibility Conditions (9.4.1) and the fact that $\nabla_{q'} L(t,q(t),\cdot)$ is injective, it follows that $q'(t-0) = q'(t+0)$ at each point $t \in [a,b]$. Hence q is of class C^1. Further, a simple inspection of the proof of Theorem 9.4.1, shows that, for each $i = 1, 2, \ldots, n$, there exists a constant c_i such that

$$p_i(t) - \frac{\partial L}{\partial q_i'}(t, q(t), q'(t)) = c_i,$$

for each $t \in [a,b]$. Now, let $r = 2$, and let us define

$$F_i(t,x) = p_i(t) - \frac{\partial L}{\partial x_i}(t, q(t), x), \quad i = 1, 2, \ldots, n.$$

Since for each $t \in [a,b]$ and each $x \in \mathbb{R}$, the Hessian matrix is nonsingular, by the implicit function Theorem – see Problem 2.28 –, it follows that the function $x(t) = q'(t)$, for $t \in [a,b]$, which is continuous, is implicitly defined by $F_i(t, x(t)) = c_i$, $i = 1, 2, \ldots, n$. So q' is of class C^1 and thus q is of class C^2.

Next, let $r \geq 3$. By Remark 9.2.6, it follows that q is a piecewise solution of the Euler–Lagrange system in (9.2.5). Since, by the above proof, q is of class C^2, then it is a solution of the Euler–Lagrange System in (9.2.5) in the classical sense, i.e. in the sense of Definition 1.2.1. Furthermore, inasmuch as the Lagrangian is regular and $r \geq 3$, by Remark 9.4.1, it follows that the Euler–Lagrange system can be rewritten as a second-order differential system in the normal form whose right-hand side is of class C^{r-2}. A simple induction argument shows that the solution is of class C^r and this completes the proof. □

Remark 9.4.2. Unlike the one-dimensional case, in the n-dimensional one, if the Hessian matrix (9.4.2) is nonsingular, it follows that the gradient vector field $\nabla_{q'} L(t, q(t), \cdot) : \mathbb{R}^n \to \mathbb{R}^n$ is only locally injective, but it may fail to be injective. An example of such a function L is given below.

Example 9.4.1. Let $L : \mathbb{R}^2 \to \mathbb{R}$ be defined by

$$L(q') = L(q'_1, q'_2) = e^{q'_1} \cos q'_2.$$

Clearly L, which does not depend on t or on q, is of class C^∞, its gradient $\nabla_{q'} L : \mathbb{R}^2 \to \mathbb{R}^2$ is given by

$$\nabla_{q'} L(q'_1, q'_2) = \left(e^{q'_1} \cos q'_2, -e^{q'_1} \sin q'_2\right)$$

and the corresponding Hessian matrix is

$$H(q'_1, q'_2) = \begin{pmatrix} e^{q'_1} \cos q'_2 & -e^{q'_1} \sin q'_2 \\ -e^{q'_1} \sin q'_2 & -e^{q'_1} \cos q'_2 \end{pmatrix}.$$

Since, for each $(q'_1, q'_2) \in \mathbb{R}^2$, $\det H(q'_1, q'_2) = -e^{2q'_1} < 0$, it follows that L is regular. However, $\nabla_{q'} L$, although locally injective, it is not injective simply because, for each $q'_1 \in \mathbb{R}$, the function $q'_2 \mapsto \nabla_{q'} L(q'_1, q'_2)$ is 2π-periodic.

This example is also instructive due to the fact that it shows that if the Jacobian matrix of a C^1-vector field, $F : \mathbb{R}^n \to \mathbb{R}^n$, is nonsingular then F is only locally injective but may not be injective, even though F is a gradient.

Corollary 9.4.1. *Let $q \in \mathcal{D}^n(q_a, q_b)$ be an extremal for the problem $(\mathcal{P})$. If, for each $(t, q) \in [a, b] \times \mathbb{R}^n$, the function $q' \mapsto L(t, q, q')$ is strictly convex and, for some $r \geq 2$, L is of class C^r and, for each $(t, q, q') \in [a, b] \times \mathbb{R}^n \times \mathbb{R}^n$, the Hessian matrix (9.4.2) is nonsingular, then q is of class C^r.*

Proof. Since $p \mapsto L(t,q,p)$ is strictly convex, it follows that $\nabla_{q'} L(t,q(t),\cdot)$ is strictly accretive, i.e.

$$\langle \nabla_{q'} L(t,q(t),u) - \nabla_{q'} L(t,q(t),v), u-v\rangle > 0,$$

for each $u, v \in \mathbb{R}^n$ with $u \neq v$. Thus $\nabla_{q'} L(t,q(t),\cdot)$ is injective. Since the Hessian matrix (9.4.2) is nonsingular, an appeal to Theorem 9.4.2 completes the proof of corollary. □

Corollary 9.4.2. *Let $q \in \mathcal{D}^1(q_a, q_b)$ be an extremal for the problem $(\mathcal{P})$ and let us assume that for each $t \in [a,b]$, and each $x \in \mathbb{R}^n$, we have*

$$\frac{\partial^2 L}{\partial q'^2}(t,q(t),x) \neq 0. \tag{9.4.3}$$

Under these circumstances, if for some $r \geq 2$, L is of class C^r, then q is of class C^r.

Proof. From (9.4.3), we deduce that, for each $t \in [a,b]$, the partial function $\dfrac{\partial L}{\partial q'}(t,q(t),\cdot) : \mathbb{R} \to \mathbb{R}$ is strictly monotone and thus injective. The conclusion follows from Theorem 9.4.2. □

9.5 Higher-Order Necessary Conditions

Throughout this section we will assume that the Lagrangian L is of class C^3. Let $q^* \in \mathcal{D}^n(q_a, q_b)$, let $\eta \in \mathcal{D}^n(0,0)$ and let us consider the function

$$g_\eta(\varepsilon) = J[q^* + \varepsilon\eta] = \int_a^b L(t, q^*(t) + \varepsilon\eta(t), q^{*\prime}(t) + \varepsilon\eta'(t))\,dt,$$

for every $\varepsilon \in (-1,1)$. Under the regularity assumptions on L and q^*, we have seen that if q^* is a $\mathcal{D}^n(0,0)$-minimum point of $(\mathcal{P})$, then 0 is a minimum point of g_η and thus

$$g'_\eta(0) = \delta L[q^*][\eta] = 0$$

for each $\eta \in \mathcal{D}^n(0,0)$, and

$$g''_\eta(0) = \delta^2 L[q^*][\eta]$$

is positive definite as a quadratic form of argument η. This means that q^* is a solution of the Euler–Lagrange System (9.2.5) and, in addition, that the quadratic form – with argument η – satisfies

$$\delta^2 L[q^*][\eta] \geq 0$$

for each $\eta \in \mathcal{D}^n(0,0)$, the equality being possible only if $\eta \equiv 0$.

But

$$\delta^2 L[q^*](\eta) = \int_a^b \sum_{i,j=1}^n \left[\frac{\partial^2 L}{\partial p_i \partial p_j} \eta_i \eta_j + 2 \frac{\partial^2 L}{\partial p_i \partial p_j'} \eta_i \eta_j' + \frac{\partial^2 L}{\partial p_i' \partial p_j'} \eta_i' \eta_j' \right] dt,$$

where all $3n^2$ second partial derivatives of L are calculated at $(t, q^*(t), {q^*}'(t))$, argument which we have dropped for the simplicity of writing.

Theorem 9.5.1. *Let L be of class C^3 and let $\{q(\cdot, s);\ s \in [0,1]\}$ be a family of extremals of $(\mathcal{P})$ such that $q(\cdot, 0) = q^*$, $q(a,s) = q_a$ and*

$$\begin{cases} \dfrac{\partial q'}{\partial s}(a,0) \neq 0, \\ \\ \dfrac{\partial q}{\partial s}(\tilde{t},0) = 0 \quad \textit{for some} \quad \tilde{t} \in (a,b\,]. \end{cases}$$

Then, the function

$$h(t) = \frac{\partial q}{\partial s}(t,0),$$

for $t \in [\,a, \tilde{t}\,]$, is not identically 0 on $[\,a, \tilde{t}\,]$ and is an extremal for the problem

$$\inf_{\eta \in \mathcal{D}^n(0,0)} \delta^2 L[q^*][\eta].$$

Proof. Since $\{q(\cdot, s);\ s \in (0,1)\}$ is a family of extremals for $(\mathcal{P})$, each $q(\cdot, s)$ satisfies the Euler–Lagrange System in (9.2.5). Differentiating both sides of the equations in (9.2.5) with respect to s and taking $s = 0$, we get

$$\sum_{j=1}^n \frac{\partial^2 L}{\partial q_i \partial q_j} h_j(t) + \sum_{j=1}^n \frac{\partial^2 L}{\partial q_i \partial q_j'} h_j'(t)$$

$$- \frac{d}{dt} \left[\sum_{j=1}^n \frac{\partial^2 L}{\partial q_i' \partial q_j} h_j(t) + \sum_{j=1}^n \frac{\partial^2 L}{\partial q_i' \partial q_j'} h_j'(t) \right]$$

$$= \frac{\partial \mathcal{L}}{\partial \eta_i}(t, h(t), h'(t)) - \frac{d}{dt} \left[\frac{\partial \mathcal{L}}{\partial \eta_i'}(t, h(t), h'(t)) \right] = 0, \text{ for } i = 1, 2, \ldots, n.$$

Here, $\mathcal{L}$ is defined by

$$\mathcal{L}(t, \eta, \eta') = \frac{1}{2} \sum_{i,j=1}^n \left[\frac{\partial^2 L}{\partial p_i \partial p_j} \eta_i \eta_j + 2 \frac{\partial^2 L}{\partial p_i \partial p_j'} \eta_i \eta_j' + \frac{\partial^2 L}{\partial p_i' \partial p_j'} \eta_i' \eta_j' \right], \quad (9.5.1)$$

where, as before, for the simplicity of writing, in all $3n^2$ second-order partial derivatives of L, we have dropped the argument $(t, q^*(t), q^{*\prime}(t))$. But $2\mathcal{L}$ – as a function of (t, η, η') – is the Lagrangian of the problem $\inf_{\eta \in \mathcal{D}^n(0,0)} \delta^2 L[q^*][\eta]$. So, h is a solution of the differential equations in the Euler–Lagrange System associated with the problem $\inf_{\eta \in \mathcal{D}^n(0,0)} \delta^2 L[q^*][\eta]$. Further, by hypothesis, we have that $h'(a) \neq 0$ and accordingly, h is not identically 0. Finally,

$$h(a) = \lim_{r \downarrow 0} \frac{1}{s} [q(a, s) - q(a, 0)] = \lim_{s \downarrow 0} \frac{1}{s} [q_a - q_a] = 0.$$

Since $h(\tilde{t}) = 0$ by hypothesis, the proof is complete. □

Definition 9.5.1. The system of nonlinear second-order differential equation in the Cauchy Problem

$$\begin{cases} \dfrac{\partial \mathcal{L}}{\partial \eta_i}(t, \eta(t), \eta'(t)) - \dfrac{d}{dt}\left[\dfrac{\partial \mathcal{L}}{\partial \eta_i'}(t, \eta(t), \eta'(t))\right] = 0, \quad i = 1, 2, \ldots, n, \\ \eta(a) = 0, \quad \eta(b) = 0, \end{cases} \tag{9.5.2}$$

where $\mathcal{L}$ is defined by (9.5.1), is called the *Euler Second System*. In the simplest case $n = 1$, it is called the *Euler Second Equation.*

Definition 9.5.2. Let q^* be an extremal for L. The pair $(\tilde{t}, q^*(\tilde{t}))$, with $\tilde{t} \in (a, b)$, is called *conjugate* with (a, q_a) along (or with respect to) the extremal q^* if there exists a non-identically zero extremal, h, of Euler Second System (9.5.2), such that $h(a) = h(\tilde{t}) = 0$. In this case, we will say that $\tilde{t}$ is *conjugate* with a with respect to q^*.

Remark 9.5.1. The set of conjugate points on the extremals issuing from (a, q), $q(\cdot, s)$, is the *envelope* of the family of extremals, whose parametric equation is

$$s \mapsto (t(s), q(t(s), s)),$$

where $t = t(s)$ is a solution of the equation

$$\frac{\partial q}{\partial s}(t, s) = 0.$$

Theorem 9.5.2. (The Jacobi Necessary Condition) *If L is regular and of class C^3, and $q^* \in \mathcal{D}^n(q_a, q_b)$ is a $\mathcal{D}^n(0, 0)$-local minimum for $(\mathcal{P})$, then there are no $(\tilde{t}, q^*(\tilde{t}))$, with $\tilde{t} \in (a, b)$, which are conjugate with (a, q) along the extremal q^*.*

Proof. Let h be an extremal for $\inf\limits_{\eta \in \mathcal{D}^n(0,0)} \delta^2 L[q^*][\eta]$. We show first that

$$\delta^2 L[q^*][h] = 0.$$

Since the Lagrangian of the problem above is $2\mathcal{L}$, where $\mathcal{L}$ is given by (9.5.1), from (9.5.2), we readily deduce that

$$2\mathcal{L} = \sum_{i=1}^{n} \left[\eta_i \frac{\partial \mathcal{L}}{\partial \eta_i} + \eta_i' \frac{\partial \mathcal{L}}{\partial \eta_i'} \right] = \sum_{i=1}^{n} \left[\eta_i \frac{d}{dt} \left(\frac{\partial \mathcal{L}}{\partial \eta_i'} \right) + \eta_i' \frac{\partial \mathcal{L}}{\partial \eta_i'} \right] = \sum_{i=1}^{n} \frac{d}{dt} \left[\eta_i \frac{\partial \mathcal{L}}{\partial \eta_i'} \right].$$

Hence

$$\delta^2 L[q^*][h] = \int_a^b 2\mathcal{L}(t, h(t), h'(t))\, dt = \int_a^b \frac{d}{dt} \left[\sum_{i=1}^{n} h_i(t) \frac{\partial \mathcal{L}}{\partial h_i'}(t, h(t), h'(t)) \right] dt$$

$$= \sum_{i=1}^{n} h_i(t) \frac{\partial \mathcal{L}}{\partial h_i'}(t, h(t), h'(t)) \Bigg|_a^b = 0.$$

But q^* is a minimum for $(\mathcal{P})$, and hence it follows that

$$\delta^2 L[q^*][h] = \inf_{\eta \in \mathcal{D}^n(0,0)} \delta^2 L[q^*][\eta] = 0.$$

Next, let us assume that there exists a point $(\tilde{t}, q^*(\tilde{t}))$, with $\tilde{t} \in (a, b)$, which is conjugate with (a, q) along the extremal q^*. This means that on the interval $[\,a, \tilde{t}\,]$ there exists a non-identically 0 extremal h of the problem

$$\inf_{\eta \in \mathcal{D}^n(0,0)} \delta^2 L[q^*][\eta].$$

From what we have already proved, it follows that

$$\delta^2 L[q^*][h] = 0.$$

So, denoting by $\tilde{h}$ the function

$$\tilde{h}(t) = \begin{cases} h(t) & if\, t \in [\,a, \tilde{t}\,] \\ 0 & if\, t \in [\tilde{t}, b\,], \end{cases}$$

we easily see that

$$\delta^2 L[q^*][\tilde{h}] = 0.$$

So, $\tilde{h}$ is a minimum point for the functional $\delta^2 L[q^*][\,\cdot\,]$. But L is regular, and since, for each $(t, q, q') \in [\,a, b\,] \times \mathbb{R}^n \times \mathbb{R}^n$, we have

$$\left(\frac{\partial^2 \mathcal{L}}{\partial h_i' \partial h_j'}(t, q, q') \right) = \frac{1}{2} \left(\frac{\partial^2 L}{\partial q_i' \partial q_j'}(t, q, q') \right), \tag{9.5.3}$$

it follows that $\mathcal{L}$ is regular too. Thus, from Theorem 9.4.2, we deduce that $\tilde{h}$ is C^2. Therefore, it satisfies the second Euler System in (9.5.2). Using once again the fact that L is regular, and observing that $\mathcal{L}$ is regular and of class C^3 with respect to h and h', from Remark 9.4.1 combined with the fact that $\tilde{h}(t) = 0$, for each $t \in [a, \tilde{t}]$, we conclude that $\tilde{h} = 0$ on $[a, b]$, which contradicts the fact that $h \neq 0$ on $[a, \tilde{t}]$. This contradiction can be eliminated only if then there are no points $(\tilde{t}, q^*(\tilde{t}))$, with $\tilde{t} \in (a, b)$, which are conjugate with (a, q) along the extremal q^*. The proof is complete. □

In order to derive the next necessary condition for extremum, some preliminaries are needed. Let $P, R, S : [a, b] \to \mathcal{M}_{n\times n}(\mathbb{R})$ be continuous and let us consider the quadratic form $Q : \mathcal{D}^n(0,0) \to \mathbb{R}$, defined by

$$Q[\eta] = \int_a^b [\langle P(t)\eta'(t), \eta'(t)\rangle + 2\langle S(t)\eta'(t), \eta(t)\rangle + \langle R(t)\eta(t), \eta(t)\rangle]\, dt, \tag{9.5.4}$$

for each $\eta \in \mathcal{D}^n(0,0)$.

Proposition 9.5.1. *If $Q[\eta] \geq 0$ for each $\eta \in \mathcal{D}^n(0,0)$, then*

$$\sum_{i,j=1}^{n} P_{ij}(t)\xi_i\xi_j \geq 0$$

for each $t \in [a, b]$.

Proof. We proceed by contraposition, i.e. we prove that whenever there exist $\tilde{t} \in [a, b]$ and $\xi \in \mathbb{R}^n$ such that $\sum_{i,j=1}^{n} P_{ij}(\tilde{t})\xi_i\xi_j < 0$, then there exists $\tilde{\eta} \in \mathcal{D}^n(0,0)$ such that $Q[\tilde{\eta}] < 0$. So, let $\tilde{t} \in [a, b]$ and $\xi \in \mathbb{R}^n$ be such that

$$\sum_{i,j=1}^{n} P_{ij}(\tilde{t})\xi_i\xi_j < 0.$$

This means that there exist $i, j \in \{1, 2, \dots, n\}$ such that

$$P_{ij}(\tilde{t})\xi_i\xi_j < 0.$$

Let $\varepsilon > 0$ be arbitrary and let us define $\theta_\varepsilon : [a, b] \to \mathbb{R}^n$ by

$$\theta_\varepsilon(t) = \begin{cases} \dfrac{1}{\varepsilon}\left(\varepsilon - |t - \tilde{t}|\right) & \text{if } |t - \tilde{t}| < \varepsilon \\ 0 & \text{if } |t - \tilde{t}| \geq \varepsilon. \end{cases}$$

At this point, let us observe that, since P is continuous, we may always assume that $\tilde{t} \in (a, b)$. So, for sufficiently small $\varepsilon > 0$, we have $\theta_\varepsilon(a) = \theta_\varepsilon(b) = 0$, and thus $\theta_\varepsilon \in \mathcal{D}^n(0,0)$.

Further, let us define $\tilde{\eta}_\varepsilon(t) = \theta_\varepsilon(t)(0,0,\ldots,0,\xi_i,0,\ldots,0,\xi_j,\ldots,0)$, where $\theta_\varepsilon(t)\xi_i$ is the ith-component of $\tilde{\eta}_\varepsilon(t)$ and $\theta_\varepsilon(t)\xi_j$ is the jth component of $\tilde{\eta}_\varepsilon(t)$, all other components being 0. One may easily verify that

$$Q[\tilde{\eta}_\varepsilon] = \int_a^b [\langle P(t)\tilde{\eta}'_\varepsilon(t), \tilde{\eta}'_\varepsilon(t)\rangle + 2\langle S(t)\tilde{\eta}'_\varepsilon(t), S(t)\tilde{\eta}_\varepsilon(t)\rangle\, dt + \langle R(t)\tilde{\eta}_\varepsilon(t), \tilde{\eta}_\varepsilon(t)\rangle]\, dt$$

$$= \int_{\tilde{t}-\varepsilon}^{\tilde{t}+\varepsilon} \sum_{k,l=1}^{n} [P_{kl}(t)\tilde{\eta}'_{\varepsilon k}(t)\tilde{\eta}'_{\varepsilon l}(t) + 2S_{kl}(t)\tilde{\eta}'_{\varepsilon k}(t)\tilde{\eta}_{\varepsilon l}(t) + R_{kl}(t)\tilde{\eta}_{\varepsilon k}(t)\tilde{\eta}_{\varepsilon l}(t)]\, dt$$

$$= \int_{\tilde{t}-\varepsilon}^{\tilde{t}+\varepsilon} \left[P_{ij}(t)\theta'^2_\varepsilon(t)\xi_i\xi_j + 2S_{ij}(t)\theta'_\varepsilon(t)\theta_\varepsilon(t)\xi_i\xi_j + R_{ij}(t)\theta^2_\varepsilon(t)\xi_i\xi_j\right] dt$$

$$= \int_{\tilde{t}-\varepsilon}^{\tilde{t}+\varepsilon} P_{ij}(t)\xi_i\xi_j \left[-\frac{1}{\varepsilon}\mathrm{sgn}(t-\tilde{t})\right]^2 dt$$

$$+2\int_{\tilde{t}-\varepsilon}^{\tilde{t}+\varepsilon} S_{ij}(t)\xi_i\xi_j \left[-\frac{1}{\varepsilon}\mathrm{sgn}(t-\tilde{t})\right]\left[\frac{1}{\varepsilon}\left(\varepsilon - |t-\tilde{t}|\right)\right] dt$$

$$+\int_{\tilde{t}-\varepsilon}^{\tilde{t}+\varepsilon} R_{ij}(t)\xi_i\xi_j \left[\frac{1}{\varepsilon}\left(\varepsilon - |t-\tilde{t}|\right)\right]^2 dt.$$

Now, since $P_{i,j}$, S_{ij} and R_{ij} are bounded on $[a,b]$ and $|t-\tilde{t}| < \varepsilon$, it readily follows that

$$\lim_{\varepsilon\downarrow 0} Q[\tilde{\eta}_\varepsilon] = -\infty.$$

But this shows that, for a sufficiently small $\varepsilon > 0$, we have $Q[\tilde{\eta}_\varepsilon] < 0$. The proof is complete. □

Remark 9.5.2. The previous result essentially asserts that the dominant term of the quadratic form (9.5.4) is

$$\int_a^b \langle P(t)\eta'(t), \eta'(t)\rangle\, dt.$$

In fact, this is, in some sense, a variant of the well-known Poincaré's Inequality which says that there exists $\omega > 0$ such that

$$\int_a^b \|\eta'\|^2(t)\, dt \geq \omega \int_a^b \|\eta^2(t)\|\, dt,$$

for each $\eta \in H_0^1(a,b;\mathbb{R}^n)$. We notice that $H_0^1(a,b;\mathbb{R}^n)$ is the space of all functions $\eta : [a,b] \to \mathbb{R}^n$, with η as well as its derivative in the sense of distributions, η', square Lebesgue integrable over $[a, b]$ and satisfying $\eta(a) = \eta(b) = 0$. The natural norm on $H_0^1(a, b; \mathbb{R}^n)$ is defined by

$$\|\eta\|_1 = \left(\int_a^b \|\eta^2(t)\| \, dt + \int_a^b \|\eta'\|^2(t) \, dt \right)^{1/2},$$

and the Poincaré's Inequality amounts to saying that an equivalent norm on $H_0^1(a, b; \mathbb{R}^n)$ is

$$\|\eta\|_0 = \left(\int_a^b \|\eta'\|^2(t) \, dt \right)^{1/2},$$

for each $\eta \in H_0^1(a, b; \mathbb{R}^n)$.

Theorem 9.5.3. (The Legendre Necessary Condition) *If L is of class C^2 and q^* is a local $\mathcal{D}^n(0,0)$-minimum for J, then, for each $t \in [a, b]$ at which ${q^*}'$ is continuous, the Hessian matrix*

$$M^*(t) = \left(\frac{\partial^2 L}{\partial q_i' \partial q_j'}(t, q^*(t), {q^*}'(t)) \right) \tag{9.5.5}$$

is positive definite.

Proof. Just use Theorem 9.2.1 to get $\delta^2 L[q^*][\eta] \geq 0$ for each $\eta \in \mathcal{D}^n(0, 0)$. Then apply Proposition 9.5.1 to the functional $\delta^2 L[q^*][\,\cdot\,]$ to conclude that, for each $t \in [a, b]$, the matrix

$$\left(\frac{\partial^2 \mathcal{L}}{\partial q_i' \partial q_j'}(t, q^*(t), {q^*}'(t)) \right)$$

is positive definite, and finally use (9.5.3) to get the conclusion. □

9.6 Sufficient Conditions for Extremum

The problem is to find sufficient conditions for an extremal, i.e. a solution q^* of (9.2.3), to be a local weak minimum for $(\mathcal{P})$. Namely, we are looking for conditions on L in order that a function $q^* \in \mathcal{D}^n(q_a, q_b)$ for which $\delta L[q^*][\eta] = 0$ for each $\eta \in \mathcal{D}^n(0, 0)$, is a local weak minimum for $(\mathcal{P})$.

As we have already seen in Theorem 9.5.3, if q^* is a local weak minimum for $(\mathcal{P})$, then for each $t \in [a, b]$, the matrix $M^*(t)$, defined in (9.5.5), is positive definite. So, we may ask whether or not the stronger condition that $M^*(t)$ be strictly positive definite for each $t \in [a, b]$, would be enough for q^* to be a local weak minimum. The answer to this question is in the negative because the condition above has a local character and not a global one. More precisely, it might happen that two extremals q_1 and q_2 defined on $[a, c]$ and on $[c, b]$ respectively, satisfy the above condition, but even though q_1 is a local weak minimum for J "on $[a, c]$" and q_2 is a local weak minimum for J "on $[c, b]$", with $q_1(c) = q_2(c)$, the concatenate function, although an extremal on $[a, b]$, is not a local weak minimum for J "on $[a, b]$". Although the next example, taken from [Gelfand and Fomin (1963)], p. 104, refers to a variational problem with constraints, it is very simple to follow in order to understand the mechanism behind the above situation. Namely, the shortest curve joining two points on a sphere is a great circle arc, if the length of this arc is less than one half of the length of the whole great circle. So, each such arc is an extremal, but if we join two arcs situated on the very same great circle and the length of the resulting arc is greater than one half of the length of the whole great circle, this arc is not a local weak minimum.

Similarly, the condition that $\delta^2 J[q^*][\eta]$ is positive definite is also necessary but not sufficient for q^* to be a local weak minimum for J. So, in order to state the first sufficient condition for local weak minimum, we introduce the following concept.

Definition 9.6.1. We say that $\delta^2 J[q^*][\cdot]$, is *strongly positive* if there exists $\omega > 0$ such that, for each $h \in \mathcal{D}(0,0)$, we have

$$\delta^2 J[q^*][h] \geq \omega \|h\|^2_{\mathcal{D}^n}.$$

Theorem 9.6.1. *Let us assume that the Lagrangian L is of class C^3. Let $q^* \in \mathcal{D}(q_a, q_b)$ be an extremal of the problem $(\mathcal{P})$ such that $\delta^2 J[q^*][\cdot]$ is strongly positive. Then, q^* is a local weak minimum for J on $\mathcal{D}(q_a, q_b)$.*

Proof. Since $J[q^*]$ is twice differentiable along each $h \in \mathcal{D}^n(0,0)$, we have that

$$\Delta J[q^*][h] = \delta J[q^*][h] + \delta^2 J[q^*][h] + \varphi(\|h\|_{\mathcal{D}^n})\|h\|^2_{\mathcal{D}^n},$$

where $\lim_{r\downarrow 0} \varphi(r) = 0$. But q^* is an extremal, and thus $\delta J[q^*][h] = 0$ for each $h \in \mathcal{D}^n(0,0)$. Hence, it follows that

$$\Delta J[q^*][h] = \delta^2 J[q^*][h] + \varphi(\|h\|_{\mathcal{D}^n})\|h\|^2_{\mathcal{D}^n}.$$

Since $\lim_{r\downarrow 0} \varphi(r) = 0$, there exists $\delta > 0$ such that, for each $h \in \mathcal{D}^n(0,0)$ with $\|h\|_{\mathcal{D}^n} < \delta$, we have

$$\varphi(\|h\|_{\mathcal{D}^n}) < \frac{\omega}{2},$$

where $\omega > 0$ is the constant appearing in the strong positivity condition on $\delta^2 J[q^*][\cdot]$. Consequently, if $h \in \mathcal{D}^n(0,0)$ is such that $\|h\|_{\mathcal{D}^n} < \delta$, it follows that

$$\Delta J[q^*][h] = \delta^2 J[q^*][h] + \varphi(\|h\|_{\mathcal{D}^n})\|h\|^2_{\mathcal{D}^n} > \frac{\omega}{2}\|h\|^2_{\mathcal{D}^n} > 0.$$

But $\Delta J[q^*][h] = J[q^* + h] - J[q^*]$, which shows that q^* is the only minimum of J in the set $\{q \in \mathcal{D}^n(q_a, q_b);\ \|q^* - q\|_{\mathcal{D}^n} < \delta\}$. In other words, q^* is a local weak minimum for $(\mathcal{P})$, as claimed. □

Theorem 9.6.2. *Let us assume that the Lagrangian L is regular and of class C^3. If q^* is an extremal of the problem $(\mathcal{P})$, i.e. a solution of the Euler–Lagrange System (9.2.5), and the interval $[a, b]$ does not contain points which are conjugate with a along q^*, then q^* is a local weak minimum for $(\mathcal{P})$.*

For the proof of this result, the interested reader is referred to [Gelfand and Fomin (1963)], p. 116 and p. 125.

We conclude this section with a simple but useful sufficient condition for an extremal to be a global minimum.

Theorem 9.6.3. *Let us assume that the Lagrangian L is of class C^1 and strictly convex with respect to the variables q and q'. If $q^* \in \mathcal{D}^n(q_a, q_b)$ is an extremal of the problem $(\mathcal{P})$ then q^* is a global minimum for $(\mathcal{P})$.*

Proof. Since L is strictly convex with respect to the variables q and q', it follows that, for each $\eta \in \mathcal{D}^n(0,0)$, the function $g_\eta : (-1,1) \to \mathbb{R}$, defined by

$$g_\eta(\varepsilon) = \int_a^b L(t, q^*(t) + \varepsilon\eta(t), {q^*}'(t) + \varepsilon\eta'(t))\, dt$$

for $\varepsilon \in (-1, 1)$, is strictly convex too. As $q^* \in \mathcal{D}^n(q_a, q_b)$ is an extremal for $(\mathcal{P})$, we have $g'_\eta(0) = 0$. But g_η is strictly convex and thus 0 is a global minimum for g_η. In view of Remark 9.2.2, q^* is a global minimum for $(\mathcal{P})$. The proof is complete. □

9.7 The Canonical Euler–Lagrange System

Let us consider the Euler–Lagrange System

$$\frac{\partial L}{\partial q_i}(t, q(t), q'(t)) - \frac{d}{dt}\left[\frac{\partial L}{\partial q'_i}(t, q(t), q'(t))\right] = 0, \text{ for } i = 1, 2, \ldots, n. \quad (9.7.1)$$

We will write this system of n second-order nonlinear equations as a system of $2n$ first-order nonlinear differential equations in normal form. First, we introduce the new variables

$$p_i = \frac{\partial L}{\partial q_i'}(t, q_1, q_2, \dots, q_n, q_1', q_2', \dots, q_n'), \quad i = 1, 2, \dots, n, \tag{9.7.2}$$

and, under some regularity assumptions on L, we obtain $q_1', q_2', \dots, q_n'$ as functions of the new variables $t, q_1, q_2, \dots, q_n, p_1, p_2, \dots, p_n$, implicitly defined by the equations above. Next, let us define the function

$$H(t, q_1, q_2, \dots, q_n, p_1, p_2, \dots, p_n) = -L(t, q_1, q_2, \dots, q_1', q_2', \dots, q_n') + \sum_{i=1}^{n} q_i' p_i,$$

where, as we have already mentioned, q_i' are regarded as functions of $t, q_1, q_2, \dots, q_n, p_1, p_2, \dots, p_n$. Now, in order to obtain the partial derivatives of H expressed in terms of the partial derivatives of L, we have to compute the differential, dH, of H. We have

$$dH = -dL + \sum_{i=1}^{n} p_i dq_i' + \sum_{i=1}^{n} q_i' dp_i$$

$$= -\frac{\partial L}{\partial t} dt - \sum_{i=1}^{n} \frac{\partial L}{\partial q_i} dq_i - \sum_{i=1}^{n} \frac{\partial L}{\partial q_i'} dq_i' + \sum_{i=1}^{n} p_i dq_i' + \sum_{i=1}^{n} q_i' dp_i.$$

Taking into account of (9.7.2) we have

$$-\sum_{i=1}^{n} \frac{\partial L}{\partial q_i'} dq_i' + \sum_{i=1}^{n} p_i dq_i' = 0,$$

and thus

$$dH = -\frac{\partial L}{\partial t} dt - \sum_{i=1}^{n} \frac{\partial L}{\partial q_i} dq_i + \sum_{i=1}^{n} q_i' dp_i.$$

Accordingly,

$$\frac{\partial H}{\partial t} = -\frac{\partial L}{\partial t}, \quad \frac{\partial H}{\partial q_i} = -\frac{\partial L}{\partial q_i}, \quad \frac{\partial H}{\partial p_i} = q_i'.$$

From these equalities, we get

$$\begin{cases} q_i' = \dfrac{\partial H}{\partial p_i}(t, p, q), \\ \\ p_i' = -\dfrac{\partial H}{\partial q_i}(t, p, q), \end{cases} \quad i = 1, 2, \dots, n. \tag{9.7.3}$$

This system of $2n$ first-order differential equations in normal form is equivalent to the Euler–Lagrange system (9.7.1). This system is called the *Hamiltonian System* associated to ($\mathcal{P}$), or the *canonical form of the Euler–Lagrange System* (9.7.1).

9.8 Prime Integrals of the Euler–Lagrange System

The aim of this section is to analyze some cases in which one can obtain explicit representations of certain prime integrals for the canonical form of the Euler–Lagrange System (9.7.3). We recall that a prime integral for a differential system is a C^1-nonconstant function which is constant along the trajectories of the system. See Definition 6.1.1. A first result in this direction is:

Theorem 9.8.1. *Let us assume that the Lagrangian L does not depend explicitly on t and it is not constant. Then, H is a prime integral of the canonical form of the Euler–Lagrange System.*

Proof. Since the Lagrangian L does not depend explicitly on t, it follows that H, which is defined by

$$H = -L + \sum_{i=1}^{n} q_i' p_i,$$

enjoys the very same property, i.e. it does not depend explicitly on t. Consequently, we have

$$\frac{dH}{dt} = \sum_{i=1}^{n} \left(\frac{\partial H}{\partial q_i} \frac{dq_i}{dt} + \frac{\partial H}{\partial p_i} \frac{dp_i}{dt} \right).$$

From the canonical form of the Euler–Lagrange System (9.7.3), we deduce that, along each extremal (p, q), we have

$$\frac{dH}{dt} = \sum_{i=1}^{n} \left(\frac{\partial H}{\partial q_i} \frac{\partial H}{\partial p_i} - \frac{\partial H}{\partial p_i} \frac{\partial H}{\partial q_i} \right) = 0.$$

Thus $t \mapsto H(p(t), q(t))$ is constant which means that H is a prime integral for the canonical form of the Euler–Lagrange System (9.7.3). The proof is complete. □

Theorem 9.8.2. *If H does not depend explicitly on t, a necessary and sufficient condition for a nonconstant C^1-function $W : \mathbb{R}^n \times \mathbb{R}^n \to \mathbb{R}$ to be a prime integral for the canonical of the Euler–Lagrange System* (9.7.3) *is that the Poisson bracket $\{ W, H \}$ be identically* 0, *i.e.*

$$\{ W, H \} = \sum_{i=1}^{n} \left(\frac{\partial W}{\partial q_i} \frac{\partial H}{\partial p_i} - \frac{\partial W}{\partial p_i} \frac{\partial H}{\partial q_i} \right) = 0.$$

Proof. Let $W : \mathbb{R}^n \times \mathbb{R}^n \to \mathbb{R}$ be a prime integral for the canonical form of the Euler–Lagrange System (9.7.3). Since H is C^1, by Peano's Theorem 2.2.1, it follows that for each $(\xi, \eta) \in \mathbb{R}^n \times \mathbb{R}^n$, there exists at least one local solution $(p, q) : [0, T) \to \mathbb{R}^n \times \mathbb{R}^n$ of (9.7.3), satisfying $(p(0), q(0)) = (\xi, \eta)$. Since the function $g(t) = W(p(t), q(t))$ is constant, we have $\frac{dg}{dt}(t) = 0$ for each $t \in [0, T)$. From (9.7.3), we deduce that

$$0 = \frac{dg}{dt} = \sum_{i=1}^{n} \left(\frac{\partial W}{\partial p_i} \frac{dp_i}{dt} + \frac{\partial W}{\partial q_i} \frac{dq_i}{dt} \right) = \sum_{i=1}^{n} \left(\frac{\partial W}{\partial q_i} \frac{\partial H}{\partial p_i} - \frac{\partial W}{\partial p_i} \frac{\partial H}{\partial q_i} \right)$$

for each $t \in [0, T)$. Taking $t = 0$, we get that $\{ W, H \}(\xi, \eta) = 0$. Since (ξ, η) is arbitrary this completes the proof of the necessity. To prove the sufficiency, we have to proceed backwards. □

Remark 9.8.1. In the general case, when H and W depend explicitly on t, the necessary and sufficient condition for W to be a prime integral for (9.7.3) is that

$$\frac{\partial W}{\partial t} + \{ W, H \} = 0$$

for each $(t, p, q) \in [a, b] \times \mathbb{R}^n \times \mathbb{R}^n$.

For details on this topic see [Arnold (1978)], [Bliss (1946)], [Elsgolts (1980)], [Gelfand and Fomin (1963)], [Giusti (2003)] and [Lefter (2006a)].

9.9 Exercises and Problems

Exercise 9.1. *Analyze the variational problems corresponding to the following functionals. Here, in each case $q(0) = 0$ and $q(1) = 1$.*

(1) $\displaystyle\int_0^1 q'(t)\, dt$ ([Gelfand and Fomin (1963)], p. 32) ;

(2) $\displaystyle\int_0^1 q(t)q'(t)\, dt$ ([Gelfand and Fomin (1963)], p. 32) ;

(3) $\displaystyle\int_0^1 tq(t)q'(t)\, dt$ ([Gelfand and Fomin (1963)], p. 32).

Exercise 9.2. *Find the extremals of the following functionals:*

(1) $\displaystyle\int_a^b \left[q(t)^2 - {q'}^2(t) - 2q(t) \sin t \right] dt$ ([Gelfand and Fomin (1963)], p. 32) ;

(2) $\displaystyle\int_a^b \left[q(t)^2 + q'^2(t) - 2q(t)\cosh t\right] dt$;

(3) $\displaystyle\int_a^b \left[q(t)^2 + q'^2(t) - 2q(t)\sin t\right] dt$ ([Gelfand and Fomin (1963)], p. 32).

Exercise 9.3. *Analyze the following minimum problems:*

(1) $\displaystyle\min_{q\in\mathcal{D}^1(0,1)} \int_0^1 [t - q(t)]^2\, dt$; ([Gelfand and Fomin (1963)], p. 21);

(2) $\displaystyle\min_{q\in\mathcal{D}^1(0,2)} \int_0^1 [t - q(t)]^2\, dt$ ([Gelfand and Fomin (1963)], p. 21);

(3) $\displaystyle\min_{q\in\mathcal{D}^1(0,1)} \int_1^2 \frac{\sqrt{1+q'^2(t)}}{t}\, dt$ ([Gelfand and Fomin (1963)], p. 19).

Problem 9.1. *Find a plane curve $q : [a,b] \to \mathbb{R}$, with q continuous and q' piecewise continuous on $[a,b]$, joining the points (a, q_a) and (b, q_b), and having minimal length.*

Problem 9.2. *Find the extremals for the minimal area of revolution problem, in the case in which $q_a > 0$ and $q_b > 0$. Discuss the result obtained.*

Problem 9.3. *Let $a, b \in \mathbb{R}$ be such that $a < b$ and let $\ell_{t=a}$ and $\ell_{t=b}$ two vertical lines in the plane $t0q$. Find the necessary condition for minimum for the free endpoints variational problem:*

$$(\mathcal{P}_0)\ \min\left\{\int_a^b L(t, q(t), q'(t))\, dt;\ q \in \mathcal{D}^n,\ q(a) \in \ell_{t=a},\ q(b) \in \ell_{t=b}\right\}.$$

Problem 9.4. *With the notations in Problem 9.3, find the necessary condition for minimum in the case of the final free endpoint variational problem:*

$$(\mathcal{P}_1)\ \min\left\{\int_a^b L(t, q(t), q'(t))\, dt;\ q \in \mathcal{D}^n,\ q(a) = q_a,\ q(b) \in \ell_{t=b}\right\}.$$

Problem 9.5. *As an application of Problem 9.4, solve the following variant of the brachistochrone problem: find a vertical plane curve with the property that a point which slides down the curve from $(0,0)$, reaches the line $\ell_{t=1}$ in the shortest time.* [Gelfand and Fomin (1963)], p. 26.

Problem 9.6. *Let us consider a system of n particles, M_i, having mass m_i, and coordinates (x_i, y_i, z_i), for $i = 1, 2, \ldots, n$. Assume that no constraints whatsoever are imposed on the system. The kinetic energy of the system is*

$$T = \frac{1}{2}\sum_{i=1}^n m_i(x_i'^2 + y_i'^2 + z_i'^2).$$

Further, let us assume that the system has potential energy, i.e. there exists a C^1-function $U : \mathbb{R} \times \mathbb{R}^3 \times \mathbb{R}^3 \times \mathbb{R}^3 \to \mathbb{R}$ such that the force acting on the ith

particle is $F_i = -\left(\frac{\partial U}{\partial x_i}, \frac{\partial U}{\partial y_i}, \frac{\partial U}{\partial z_i}\right)$. *Let us define the Lagrangian of the system by* $L = T - U$. *The result below, due to Hamilton, is known as the Principle of Least Action.*

Theorem 9.9.1. *The motion of a system of n particles with the initial position* $(x_i(t_0), y_i(t_0), z_i(t_0)) = (x_{i0}, y_{i0}, z_{i0})$, *during the time interval* $[t_0, t_1]$, *is described by those functions* $t \mapsto (x_i(t), y_i(t), z_i(t))$, $i = 1, 2, \ldots, n$, *at which the integral*[4]

$$\int_{t_0}^{t_1} L(t, x(t), y(t), z(t))\, dt$$

attaints its minimum[5].

Prove that the Principle of Least Action implies the Newton's Equations of Motion. [Gelfand and Fomin (1963)], p. 84.

Problem 9.7. *Let us assume that the system above is conservative, i.e. the function U does not depend explicitly on t. Show that, in this case, along each trajectory the total energy of the system is constant, i.e. there is a conservation of energy.* [Gelfand and Fomin (1963)], p. 85.

[4]This integral is called the *action of the system.*

[5]A more general form of this principle says that: *the motion of a system of n particles, during the time interval* $[t_0, t_1]$, *is described by those functions* $t \mapsto (x_i(t), y_i(t), z_i(t))$, $i = 1, 2, \ldots, n$, *which are extremals of the action.* See [Arnold (1978)], p. 59.

Chapter 10

Nonlocal Problems

In this chapter we present some basic results referring to a class of differential equations with general nonlocal initial conditions including T-periodic, T-anti-periodic, discrete mean and integral mean conditions. Finally, we include several exercises and problems.

10.1 Nonlocal Initial Conditions

In this section we consider a class of non-standard problems of the form

$$\begin{cases} x'(t) = \mathcal{A}x(t) + f(t, x(t)), & t \in \mathbb{R}_+ \\ x(0) = g(x), \end{cases} \tag{10.1.1}$$

where $\mathcal{A} \in \mathcal{M}_{n\times n}(\mathbb{R})$ and $f : \mathbb{R}_+ \times \mathbb{R}^n \to \mathbb{R}^n$ and $g : C_b(\mathbb{R}_+; \mathbb{R}^n) \to \mathbb{R}^n$ are given continuous functions. For any $a \in \mathbb{R}_+$, $C_b([\,a, +\infty); \mathbb{R}^n)$ denotes the space of all bounded continuous functions from $[\,a, +\infty)$ to $\mathbb{R}^n$. Due to the presence in the initial condition of the term g, which depends on x as a function, we call this type of problems *nonlocal*. The interesting feature of the initial condition in (10.1.1) is that $x(0)$ is assumed to depend not only on a single datum $\xi \in \mathbb{R}^n$, as in the case of a standard Cauchy problem, but on the whole future behavior of the state x. This requirement, which may seem a little bit unusual at first glance, was suggested by simple examples of economical models for which the initial state of the system ought to be chosen according to some future constraints. Of course, this is not the only motivation for the study of nonlocal problems; simple statistical facts have shown that in some cases (as for instance in the study of forecasting problems) a standard initial condition of the form $x(0) = \xi$ is less reliable than the mean condition

$$x(0) = \xi + \sum_{i=1}^{n} \alpha_i x(t_i),$$

where $\alpha_i \in \mathbb{R}_+$ satisfy $\sum_{i=1}^{n} \alpha_i \leq 1$ and $0 < t_1 < t_2 < \cdots < t_n$ are given points at which one has easy access to perform accurate measurements of the values of the real states, $x(t_i)$, $i = 1, 2, \ldots, n$. Moreover, T-periodic problems can be formulated as particular cases of (10.1.1) by taking $g(x) = x(T)$. The same remark holds true for the case of T-anti-periodic problems, i.e. when the function g is defined by $g(x) = -x(T)$.

10.2 An Existence and Uniqueness Result

If $x \in C_b([a, +\infty); \mathbb{R}^n)$, we denote by

$$\|x\|_{C_b([a,+\infty);\mathbb{R}^n)} := \sup\{\|x(t)\|;\ t \in [a, +\infty)\}.$$

We recall that the mapping $\|\cdot\|_{C_b([a,+\infty);\mathbb{R}^n)} : C_b([a, +\infty); \mathbb{R}^n) \to \mathbb{R}_+$ is a norm and $(C_b([a, +\infty); \mathbb{R}^n), \|\cdot\|)$ is a real Banach space. See Definitions 12.3.1, 12.3.4 and (v) in Remark 12.3.1.

Theorem 10.2.1. *Let $\mathcal{A} \in \mathcal{M}_{n\times n}(\mathbb{R})$ be given and let $f : \mathbb{R}_+ \times \mathbb{R}^n \to \mathbb{R}^n$ and $g : C_b(\mathbb{R}_+; \mathbb{R}^n) \to \mathbb{R}^n$ be continuous functions. If there exist $\omega > 0$, $\ell > 0$, $m \geq 0$ and $a > 0$ such that*

- (i) $\|e^{t\mathcal{A}}\|_{\mathcal{M}} \leq e^{-\omega t}$ *for each* $t \in \mathbb{R}_+$;
- (ii) $\|f(t, x) - f(t, y)\| \leq \ell\|x - y\|$ *for each* $t \in \mathbb{R}_+$ *and each* $x, y \in \mathbb{R}^n$;
- (iii) $\|f(t, 0)\| \leq m$ *for each* $t \in \mathbb{R}_+$;
- (iv) $\|g(x) - g(y)\| \leq \|x - y\|_{C_b([a,+\infty);\mathbb{R}^n)}$ *for each* $x, y \in C_b(\mathbb{R}_+; \mathbb{R}^n)$;
- (v) $\omega - \ell > 0$,

then the problem (10.1.1) *has a unique solution* $x \in C_b(\mathbb{R}_+; \mathbb{R}^n)$.

10.3 A Simple Existence Result

Proposition 10.3.1. *Let us assume that the hypotheses* (i) *and* (iv) *in Theorem* 10.2.1 *are satisfied. Then, for each* $h \in C_b(\mathbb{R}_+; \mathbb{R}^n)$, *the problem*

$$\begin{cases} x'(t) = \mathcal{A}x(t) + h(t), & t \in \mathbb{R}_+, \\ x(0) = g(x) \end{cases}$$

has a unique solution, $x \in C_b(\mathbb{R}_+; \mathbb{R}^n)$.

We begin with the following Weierstrass type lemma which is needed in several places within this chapter.

Lemma 10.3.1. *Let $\mathbb{I}$ be a nonempty interval in $\mathbb{R}$, let $(x_k)_{k\in\mathbb{N}}$ be a sequence in $C_b(\mathbb{I};\mathbb{R}^n)$ and let $(\alpha_k)_{k\in\mathbb{N}}$ be a sequence of real positive numbers such that*

$$\|x_{k+1}-x_k\|_{C_b(\mathbb{I};\mathbb{R}^n)}\leq\alpha_k \tag{10.3.1}$$

for each $k\in\mathbb{N}$. Assume that the series with positive terms $\sum_{k=0}^{\infty}\alpha_k$ is convergent. Then $(x_k)_{k\in\mathbb{N}}$ is uniformly convergent on $\mathbb{I}$ to some function $x\in C_b(\mathbb{I};\mathbb{R}^n)$.

Proof. Let us denote by $S_k=\sum_{p=0}^{k-1}\alpha_p$ for each $k\geq 1$. From (10.3.1), we have:

$$\|x_{k+p}(t)-x_k(t)\|\leq\|x_{k+p}(t)-x_{k+p-1}(t)\|+\cdots+\|x_{k+1}(t)-x_k(t)\|$$

$$\leq\alpha_{k+p-1}+\cdots+\alpha_k\leq S_{k+p}-S_k$$

for each $k,p\in\mathbb{N}$ and each $t\in\mathbb{I}$. The sequence $(S_k)_{k\in\mathbb{N}}$ is convergent since the series $\sum_{k=0}^{\infty}\alpha_k$ is convergent. So $(S_k)_{k\in\mathbb{N}}$ is a Cauchy sequence and from the preceding relations we deduce that: for each $\varepsilon>0$ there exists $k(\varepsilon)\in\mathbb{N}$ such that, for each $k\geq k(\varepsilon)$ and each $p\in\mathbb{N}$, we have

$$\|x_{k+p}(t)-x_k(t)\|\leq\varepsilon \tag{10.3.2}$$

for each $t\in\mathbb{I}$. We conclude that, for each arbitrary but fixed $t\in\mathbb{I}$, the sequence $(x_k(t))_{k\in\mathbb{N}}$ is Cauchy, and thus convergent to some $x(t)\in\mathbb{R}^n$. Passing to the limit for $p\to\infty$ in (10.3.2), we deduce that the sequence $(x_k)_{k\in\mathbb{N}}$ is uniformly convergent to the function x. Obviously x is continuous and bounded, i.e. $x\in C_b(\mathbb{I};\mathbb{R}^n)$. □

We can now proceed to the proof of Proposition 10.3.1.

Proof. Let $h\in C_b(\mathbb{R}_+;\mathbb{R}^n)$ be arbitrary but fixed. Let $x\in C_b(\mathbb{R}_+;\mathbb{R}^n)$ and let us consider the auxiliary problem

$$\begin{cases} y'(t)=\mathcal{A}y(t)+h(t), & t\in\mathbb{R}_+,\\ y(0)=g(x). \end{cases} \tag{10.3.3}$$

From Theorem 4.1.1, it follows that the problem (10.3.3) has a unique solution $y : \mathbb{R}_+ \to \mathbb{R}^n$ given by the variation of constants formula (4.3.4), namely

$$y(t) = e^{tA}g(x) + \int_0^t e^{(t-s)A}h(s)\,ds.$$

Then

$$\|y(t)\| \le e^{-\omega t}\|g(x)\| + \int_0^t e^{-\omega(t-s)}\|h(s)\|\,ds$$

and thus

$$\|y(t)\| \le e^{-\omega t}\left[\|x\|_{C_b([a,+\infty);\mathbb{R}^n)} + \|g(0)\|\right] + \left(1 - e^{-\omega t}\right)\frac{1}{\omega}\|h\|_{C_b(\mathbb{R}_+;\mathbb{R}^n)}$$

for each $t \in \mathbb{R}_+$. So, $y \in C_b(\mathbb{R}_+;\mathbb{R}^n)$.

Next, fix an arbitrary $x_0 \in C_b(\mathbb{R}_+;\mathbb{R}^n)$ and consider the sequence of successive approximations $(x_k)_{k\in\mathbb{N}}$, given by

$$x_{k+1}(t) = e^{tA}g(x_k) + \int_0^t e^{(t-s)A}h(s)\,ds \tag{10.3.4}$$

for each $k \in \mathbb{N}$ and $t \in [\,a,+\infty)$. We have

$$\|x_{k+1}(t) - x_k(t)\| \le e^{-\omega a}\|x_k - x_{k-1}\|_{C_b([\,a,+\infty);\mathbb{R}^n)} \tag{10.3.5}$$

and so

$$\begin{aligned}\|x_{k+1} - x_k\|_{C_b([\,a,+\infty);\mathbb{R}^n)} &\le e^{-\omega a}\|x_k - x_{k-1}\|_{C_b([\,a,+\infty);\mathbb{R}^n)}\\ &\le \cdots \le \left(e^{-\omega a}\right)^k\|x_1 - x_0\|_{C_b([\,a,+\infty);\mathbb{R}^n)}\end{aligned}$$

for $k = 1, 2, \ldots$ and $t \in [\,a,+\infty)$. From Lemma 10.3.1, it follows that there exists a function $x \in C_b([\,a,+\infty);\mathbb{R}^n)$ such that

$$\lim_{k\to\infty} x_k(t) = x(t)$$

uniformly for $t \in [\,a,+\infty)$. As a consequence,

$$\lim_{k\to\infty} g(x_k) = g(x).$$

Passing to the limit for $k \to \infty$ in (10.3.4), from (10.3.5), we conclude that $(x_k)_{k\in\mathbb{N}}$ converges to a function extending x from $[\,a,+\infty)$ to $\mathbb{R}_+$, denoted also by x, and

$$x(t) = e^{tA}g(x) + \int_0^t e^{(t-s)A}h(s)\,ds$$

for each $t \in \mathbb{R}_+$. This completes the proof. □

10.4 Proof of Theorem 10.2.1

Proof. We begin with the proof of existence. Let $x_0 \in C_b(\mathbb{R}_+;\mathbb{R}^n)$ be arbitrary and let us consider the sequence of successive approximations $(x_k)_{k\in\mathbb{N}}$, defined by

$$x_{k+1}(t) = e^{tA}g(x_{k+1}) + \int_0^t e^{(t-s)A}f(s,x_k(s))\,ds \tag{10.4.1}$$

for each $k \in \mathbb{N}$ and $t \in \mathbb{R}_+$. Clearly $(x_k)_{k\in\mathbb{N}}$ is well-defined since, for each $k \in \mathbb{N}$, from Proposition 10.3.1, we deduce that the problem (10.4.1) has a unique solution $x_{k+1} \in C_b(\mathbb{R}_+;\mathbb{R}^n)$.

We will prove next that the sequence $(x_k)_{k\in\mathbb{N}}$ is uniformly convergent on $\mathbb{R}_+$. To this goal, we will show that the telescopic series

$$x_0(t) + \sum_{k=0}^{\infty}[x_{k+1}(t) - x_k(t)]$$

is uniformly convergent on $\mathbb{R}_+$.

First, let us observe that, for each $k \in \mathbb{N}^*$ and each $t \in \mathbb{R}_+$, we have

$$\|x_{k+1}(t)-x_k(t)\| \leq e^{-\omega t}\|x_{k+1}(0)-x_k(0)\|+\ell\int_0^t e^{-(t-s)\omega}\|x_k(s)-x_{k-1}(s)\|\,ds.$$

But

$$\|x_{k+1}(0) - x_k(0)\| \leq \|x_{k+1} - x_k\|_{C_b(\mathbb{R}_+;\mathbb{R}^n)}$$

and thus

$$\|x_{k+1}(t) - x_k(t)\| \leq e^{-\omega t}\|x_{k+1} - x_k\|_{C_b(\mathbb{R}_+;\mathbb{R}^n)} + \left(1 - e^{-\omega t}\right)\frac{\ell}{\omega}\|x_k - x_{k-1}\|_{C_b(\mathbb{R}_+;\mathbb{R}^n)}$$

for each $k = 1,2,\ldots$ and $t \in \mathbb{R}_+$. Hence

$$\|x_{k+1} - x_k\|_{C_b(\mathbb{R}_+;\mathbb{R}^n)} \leq \frac{\ell}{\omega}\|x_k - x_{k-1}\|_{C_b(\mathbb{R}_+;\mathbb{R}^n)} \leq \left(\frac{\ell}{\omega}\right)^k\|x_1 - x_0\|_{C_b(\mathbb{R}_+;\mathbb{R}^n)}$$

for each $k = 1,2,\ldots$. Since $\frac{\ell}{\omega} \in (0,1)$, the last inequalities show that we are in the hypotheses of Lemma 10.3.1 with $\alpha_k = \left(\frac{\ell}{\omega}\right)^k\|x_1 - x_0\|_{C_b(\mathbb{R}_+;\mathbb{R}^n)}$ for $k \in \mathbb{N}$ and thus the sequence $(x_k)_{k\in\mathbb{N}}$ is uniformly convergent to some function $x \in C_b(\mathbb{R}_+;\mathbb{R}^n)$. Passing to the limit for $k \to \infty$ in (10.4.1), we conclude that x is a solution of the problem (10.1.1) and this completes the proof of the existence.

To prove the uniqueness, consider x and y two solutions of (10.1.1). Reasoning as above, we get

$$\|x(t)-y(t)\| \leq e^{-\omega t}\|x-y\|_{C_b(\mathbb{R}_+;\mathbb{R}^n)} + \frac{\ell}{\omega}\left(1-e^{-\omega t}\right)\|x-y\|_{C_b(\mathbb{R}_+;\mathbb{R}^n)}$$

for each $t \in \mathbb{R}_+$. It follows that

$$\|x-y\|_{C_b(\mathbb{R}_+;\mathbb{R}^n)} \leq \frac{\ell}{\omega}\|x-y\|_{C_b(\mathbb{R}_+;\mathbb{R}^n)}.$$

Since by (v), $\ell < \omega$, we conclude that $\|x-y\|_{C_b(\mathbb{R}_+;\mathbb{R}^n)} = 0$, which is equivalent to $x = y$. The proof is complete. □

10.5 The Dissipative Case

We analyze next the case in which f is dissipative on $\mathbb{R}^n$. See Definition 2.3.3.

Theorem 10.5.1. *Let $\mathcal{A} \in \mathcal{M}_{n\times n}(\mathbb{R})$ be a given matrix and let $f : \mathbb{R}_+ \times \mathbb{R}^n \to \mathbb{R}^n$ and $g : C_b(\mathbb{R}_+;\mathbb{R}^n) \to \mathbb{R}^n$ be continuous functions. If there exist $\omega > 0$, $\ell > 0$, $m \geq 0$ and $a > 0$ such that*

- (i') *$\langle \mathcal{A}x, x\rangle \leq -\omega\|x\|^2$ for each $x \in \mathbb{R}^n$;*
- (ii') *f is dissipative on $\mathbb{R}^n$;*
- (iii') *$\|f(t,x)\| \leq \ell\|x\| + m$ for each $t \in \mathbb{R}_+$ and each $x \in \mathbb{R}^n$;*
- (iv) *$\|g(x)-g(y)\| \leq \|x-y\|_{C_b([\,a,+\infty);\mathbb{R}^n)}$ for each $x, y \in C_b(\mathbb{R}_+;\mathbb{R}^n)$;*
- (v) *$\omega - \ell > 0$,*

then the problem (10.1.1) has a unique solution $x \in C_b(\mathbb{R}_+;\mathbb{R}^n)$.

Remark 10.5.1. Condition (i') implies (i). Indeed, let us assume that (i') is satisfied, let $\xi \in \mathbb{R}^n$ be arbitrary but fixed and let $x : \mathbb{R}_+ \to \mathbb{R}^n$, $x(t) = e^{t\mathcal{A}}\xi$ for $t \in \mathbb{R}_+$, be the unique solution of the Cauchy problem

$$\begin{cases} x'(t) = \mathcal{A}x(t), \ t \in \mathbb{R}_+, \\ x(0) = \xi. \end{cases}$$

Multiplying both sides of the equation above by $x(t)$, we get

$$\frac{1}{2}\frac{d}{dt}(\|x(t)\|^2) = \langle \mathcal{A}x(t), x(t)\rangle \leq -\omega\|x(t)\|^2$$

for each $t \in \mathbb{R}_+$. Integrating this inequality from 0 to t, we conclude that

$$\|x(t)\|^2 \leq e^{-2\omega t}\|x(0)\|^2$$

for each $t \in \mathbb{R}_+$. Consequently, we obtain

$$\|e^{t\mathcal{A}}\xi\| \leq e^{-\omega t}\|\xi\|$$

for each $\xi \in \mathbb{R}^n$ and each $t \in \mathbb{R}_+$. Recalling the definition of $\|\cdot\|_{\mathcal{M}}$, we deduce

$$\|e^{t\mathcal{A}}\|_{\mathcal{M}} = \sup_{\|\xi\|\leq 1} \|e^{t\mathcal{A}}\xi\| \leq e^{-\omega t}$$

for each $t \in \mathbb{R}_+$ which proves (i).

We may now proceed to the proof of Theorem 10.5.1.

Proof. We will show that the sequence of successive approximations $(x_k)_{k\in\mathbb{N}}$, given by

$$x_{k+1}(t) = e^{t\mathcal{A}}g(x_k) + \int_0^t e^{(t-s)\mathcal{A}}f(s, x_{k+1}(s))\,ds, \tag{10.5.1}$$

for each $k \in \mathbb{N}$ and $t \in \mathbb{R}_+$, is uniformly convergent on $\mathbb{R}_+$ to some function $x \in C_b(\mathbb{R}_+;\mathbb{R}^n)$ which is the unique solution of (10.1.1). Clearly, for each $k \in \mathbb{N}$, x_{k+1} is the unique solution of the problem

$$\begin{cases} x'_{k+1}(t) = \mathcal{A}x_{k+1}(t) + f(t, x_{k+1}(t)), \ t \in \mathbb{R}_+, \\ x_{k+1}(0) = g(x_k), \end{cases}$$

where the first term $x_0 \in C_b(\mathbb{R}_+;\mathbb{R}^n)$ is arbitrary but fixed. To see that $(x_k)_{k\in\mathbb{N}}$ is well-defined, we have to check that whenever, for some $k \in \mathbb{N}$, $x_k \in C_b(\mathbb{R}_+;\mathbb{R}^n)$, it follows that $x_{k+1} \in C_b(\mathbb{R}_+;\mathbb{R}^n)$. From (10.5.1), we successively get

$$\begin{aligned} \|x_{k+1}(t)\| &\leq e^{-\omega t}\left[\|x_k\|_{C_b([a,+\infty);\mathbb{R}^n)} + \|g(0)\|\right] \\ &\quad + \frac{\ell}{\omega}\left(1 - e^{-\omega t}\right)\left[\|x_{k+1}\|_{C([0,T];\mathbb{R}^n)} + \frac{m}{\ell}\right], \\ \|x_{k+1}\|_{C([0,T];\mathbb{R}^n)} &\leq \left[\|x_k\|_{C_b([a,+\infty);\mathbb{R}^n)} + \|g(0)\|\right] \\ &\quad + \frac{\ell}{\omega}\left[\|x_{k+1}\|_{C([0,T];\mathbb{R}^n)} + \frac{m}{\ell}\right] \end{aligned}$$

and

$$\left(1 - \frac{\ell}{\omega}\right)\|x_{k+1}\|_{C([0,T];\mathbb{R}^n)} \leq \|x_k\|_{C_b([a,+\infty);\mathbb{R}^n)} + \|g(0)\| + \frac{m}{\omega},$$

for each $k \in \mathbb{N}^*$, each $T > 0$ and each $t \in [0,T]$. The above inequality shows that if $x_k \in C_b(\mathbb{R}_+;\mathbb{R}^n)$, then $x_{k+1} \in C_b(\mathbb{R}_+;\mathbb{R}^n)$ and hence $(x_k)_{k\in\mathbb{N}}$ is well-defined.

In order to prove that $(x_k)_{k\in\mathbb{N}}$ is uniformly convergent on $\mathbb{R}_+$, we will show that the telescopic series

$$x_0(t) + \sum_{k=0}^{\infty} [x_{k+1}(t) - x_k(t)],\ t \in \mathbb{R}_+,$$

satisfies the hypotheses of Lemma 10.3.1. Let us observe first that

$$x'_{k+1}(t) - x'_k(t) = \mathcal{A}x_{k+1}(t) - \mathcal{A}x_k(t) + f(t, x_{k+1}(t)) - f(t, x_k(t))$$

for each $k \in \mathbb{N}$ and $t \in \mathbb{R}_+$. Hence

$$\langle x'_{k+1}(t) - x'_k(t), x_{k+1}(t) - x_k(t)\rangle \leq -\omega\|x_{k+1}(t) - x_k(t)\|^2$$

for each $k \in \mathbb{N}^*$ and $t \in \mathbb{R}_+$. From (12.1.3) in Lemma 12.1.2, we conclude that

$$\frac{d}{dt}\left[\|x_{k+1}(t) - x_k(t)\|^2\right] \leq -2\omega\|x_{k+1}(t) - x_k(t)\|^2$$

for each $k \in \mathbb{N}$ and $t \in \mathbb{R}_+$. A simple computational argument shows that

$$\|x_{k+1}(t) - x_k(t)\|^2 \leq e^{-2\omega t}\|x_{k+1}(0) - x_k(0)\|^2$$

for each $k \in \mathbb{N}$ and $t \in \mathbb{R}_+$. In turn, this inequality implies

$$\|x_{k+1}(t) - x_k(t)\| \leq e^{-\omega t}\|x_{k+1}(0) - x_k(0)\|$$

for each $k \in \mathbb{N}$ and $t \in \mathbb{R}_+$. Consequently

$$\|x_{k+1}(t) - x_k(t)\| \leq e^{-\omega t}\|x_k - x_{k-1}\|_{C_b([a,+\infty);\mathbb{R}^n)},\ k = 1, 2, \ldots.$$

for each $k \in \mathbb{N}^*$ and $t \in \mathbb{R}_+$. Now, let us observe that

$$\|x_{k+1} - x_k\|_{C_b([a,+\infty);\mathbb{R}^n)} \leq e^{-\omega a}\|x_k - x_{k-1}\|_{C_b([a,+\infty);\mathbb{R}^n)}$$

$$\leq \cdots \leq \left(e^{-\omega a}\right)^k \|x_1 - x_0\|_{C_b([a,+\infty);\mathbb{R}^n)}.$$

Now, recalling that $a > 0$ and $\omega > 0$, the preceding inequalities show that we are in the hypotheses of Lemma 10.3.1 with

$$\alpha_k = \left(e^{-\omega a}\right)^k \|x_1 - x_0\|_{C_b([a,+\infty);\mathbb{R}^n)}$$

for $k \in \mathbb{N}$. Consequently, the sequence $(x_k)_{k\in\mathbb{N}}$ is uniformly convergent on $[a + \infty)$ to some $\widetilde{x} \in C_b(\mathbb{R}_+; \mathbb{R}^n)$. On the other hand, from (iv), it follows that

$$\lim_{k\to\infty} x_{k+1}(0) = \lim_{k\to\infty} g(x_k) = g(\widetilde{x}).$$

Since

$$\|x_{k+p}(t) - x_k(t)\| \leq e^{-\omega t}\|x_{k+p}(0) - x_k(0)\|$$

we readily deduce that $(x_k)_{k\in\mathbb{N}}$ is uniformly convergent on $\mathbb{R}_+$ to a function $x \in C_b(\mathbb{R}_+;\mathbb{R}^n)$, satisfying $\widetilde{x}(t) = x(t)$ for each $t \in [\,a,+\infty)$. Also from (iv), we get $g(\widetilde{x}) = g(x)$. Passing to the limit for $k \to \infty$ in (10.5.1), we conclude that x is a solution of (10.1.1).

As concerns the uniqueness, we have merely to note that, if x and y are solutions of (10.1.1), then

$$\|x(t) - y(t)\| \le e^{-\omega t}\|x - y\|_{C_b([\,a,+\infty);\mathbb{R}^n)}$$

for each $t \in \mathbb{R}_+$. We deduce

$$\|x - y\|_{C_b([\,a,+\infty);\mathbb{R}^n)} \le e^{-\omega a}\|x - y\|_{C_b([\,a,+\infty);\mathbb{R}^n)}.$$

Since $0 < e^{-\omega a} < 1$, it follows that $x(t) = y(t)$ for each $t \in [\,a,+\infty)$. Due to (iv), we get $x(0) = y(0)$ and so x coincides with y on $\mathbb{R}_+$. The proof is complete. □

10.6 A Stability Result

We begin with

Definition 10.6.1. A solution $x \in C_b(\mathbb{R}_+;\mathbb{R}^n)$ of the problem (10.1.1) is called *globally uniformly stable* if for each solution $y \in C(\mathbb{R}_+;\mathbb{R}^n)$ of the problem

$$y'(t) = \mathcal{A}y(t) + f(t, y(t)),\ t \in \mathbb{R}_+, \tag{10.6.1}$$

we have

$$\lim_{t\to+\infty} \|x(t) - y(t)\| = 0.$$

As concerns the global asymptotic stability of solutions of the problem (10.1.1), we have the following two results whose proofs are inspired by the one of Poincaré–Lyapunov's Theorem 5.3.1.

Theorem 10.6.1. *If the hypotheses of Theorem* 10.2.1 *are satisfied, then the unique solution* $x \in C_b(\mathbb{R}_+;\mathbb{R}^n)$ *of the problem* (10.1.1) *is globally uniformly stable in the sense of Definition* 10.6.1.

Theorem 10.6.2. *If the hypotheses of Theorem* 10.5.1 *are satisfied, then the unique solution* $x \in C_b(\mathbb{R}_+;\mathbb{R}^n)$ *of the problem* (10.1.1) *is globally uniformly stable in the sense of Definition* 10.6.1.

We begin with the proof of Theorem 10.6.1.

Proof. Let us observe that, for each $\xi \in \mathbb{R}^n$ and each solution $y \in C(\mathbb{R}_+; \mathbb{R}^n)$ of the problem (10.6.1), we have

$$\|x(t) - y(t)\| \le e^{-\omega t}\|x(0) - y(0)\| + \int_0^t e^{-\omega(t-s)}\ell\|x(s) - y(s)\| \, ds$$

for each $t \in \mathbb{R}_+$. Multiplying both sides of this inequality by $e^{\omega t}$, we get

$$e^{\omega t}\|x(t) - y(t)\| \le \|x(0) - y(0)\| + \int_0^t e^{\omega s}\ell\|x(s) - y(s)\| \, ds$$

for each $t \in \mathbb{R}_+$. Then, the function $z : \mathbb{R}_+ \to \mathbb{R}_+$, defined by

$$z(t) := e^{\omega t}\|x(t) - y(t)\|$$

for $t \in \mathbb{R}_+$, satisfies

$$z(t) \le \|x(0) - y(0)\| + \int_0^t \ell z(s) \, ds$$

for each $t \in \mathbb{R}_+$. From Gronwall's Lemma 1.5.2, we deduce

$$z(t) \le \|x(0) - y(0)\|e^{\ell t}$$

for each $t \in \mathbb{R}_+$. Recalling the definition of z, we obtain

$$\|x(t) - y(t)\| \le \|x(0) - y(0)\|e^{(\ell - \omega)t}$$

for each $t \in \mathbb{R}_+$. Since, by (v) $\ell - \omega < 0$, this inequality shows that

$$\lim_{t \to +\infty} \|x(t) - y(t)\| = 0$$

and this completes the proof. □

We may now proceed to the proof of Theorem 10.6.2.

Proof. In this case, by exploiting the dissipativity conditions (i') and (ii'), we get

$$\frac{d}{dt}\left(\|x(t) - y(t)\|^2\right) \le -2\omega\|x(t) - y(t)\|^2$$

for each $t \in \mathbb{R}_+$. From this inequality, we deduce

$$\|x(t) - y(t)\|^2 \le e^{-2\omega t}\|x(0) - y(0)\|^2$$

for each $t \in \mathbb{R}_+$, which clearly shows that

$$\lim_{t \to +\infty} \|x(t) - y(t)\| = 0,$$

as claimed. □

10.7 Particular Cases

We show next how the abstract just developed theory applies to the study of two types of problems of great importance in applications: T-periodic and T-anti-periodic problems. We begin with some results concerning T-periodic problems.

Let $\mathcal{A} \in \mathcal{M}_{n\times n}(\mathbb{R})$ be a given matrix and let $f : \mathbb{R}_+ \times \mathbb{R}^n \to \mathbb{R}^n$ be a continuous function which is T-periodic with respect to $t \in \mathbb{R}_+$, i.e. $f(t,x) = f(t+T,x)$ for each $t \in \mathbb{R}_+$ and each $x \in \mathbb{R}^n$. Here, for each $x \in \mathbb{R}^n$, $T > 0$ is the *principal period* of $f(\cdot,x)$, i.e. the minimal positive number satisfying the condition above.

Consider the T-periodic problem

$$\begin{cases} x'(t) = \mathcal{A}x(t) + f(t,x(t)), \ t \in \mathbb{R}_+, \\ x(t) = x(t+T), \ t \in \mathbb{R}_+. \end{cases} \tag{10.7.1}$$

Remark 10.7.1. If f is T-periodic with respect to $t \in \mathbb{R}_+$, the problem (10.7.1) is equivalent – in a sense we will make precise below – with the nonlocal problem

$$\begin{cases} x'(t) = \mathcal{A}x(t) + f(t,x(t)), \ t \in \mathbb{R}_+, \\ x(0) = g(x), \end{cases} \tag{10.7.2}$$

where $g : C_b(\mathbb{R}_+;\mathbb{R}^n) \to \mathbb{R}^n$ is defined by

$$g(x) = x(T) \tag{10.7.3}$$

for each $x \in C_b(\mathbb{R}_+;\mathbb{R}^n)$. Indeed, if x is a solution of (10.7.1), then it is a solution of (10.7.2), where g is given by (10.7.3). Conversely, if x is a solution of (10.7.2), then x verifies

$$\begin{cases} x'(t) = \mathcal{A}x(t) + f(t,x(t)), \ t \in [\,0,T\,], \\ x(0) = x(T), \end{cases} \tag{10.7.4}$$

and it can be continued by T-periodicity on $\mathbb{R}_+$ obtaining a T-periodic solution of (10.7.1). If (10.7.1) has the uniqueness property, then (10.7.4) has the uniqueness property and each solution of (10.7.1) is a solution of (10.7.2) and conversely.

Theorem 10.7.1. *Let $\mathcal{A} \in \mathcal{M}_{n\times n}(\mathbb{R})$ be a given matrix and let $f : \mathbb{R}_+ \times \mathbb{R}^n \to \mathbb{R}^n$ be a continuous function which is T-periodic with respect to its first argument on $\mathbb{R}_+$. If there exist $\omega > 0$ and $\ell > 0$ such that*

(*i*) $\|e^{t\mathcal{A}}\|_{\mathcal{M}} \leq e^{-\omega t}$ *for each* $t \in \mathbb{R}_+$;

(*ii*) $\|f(t,x)-f(t,y)\| \le \ell\|x-y\|$ *for each* $t \in \mathbb{R}_+$ *and each* $x,y \in \mathbb{R}^n$;
(*v*) $\omega - \ell > 0$,

then the problem (10.7.1) *has a unique solution* $x \in C_b(\mathbb{R}_+;\mathbb{R}^n)$. *Moreover,* x *is globally asymptotically stable.*

Proof. We shall apply Theorems 10.2.1 and 10.6.1 to problem (10.7.2) and then we will use Remark 10.7.1. We begin by showing that (*iii*) and (*iv*) in Theorem 10.2.1 are satisfied. Since $f(\cdot,0)$ is continuous from $\mathbb{R}_+$ to $\mathbb{R}^n$ and $[\,0,T\,]$ is compact, there exists $m > 0$ such that

$$\|f(t,0)\| \le m$$

for each $t \in [\,0,T\,]$. But $f(\cdot,0)$ is T-periodic, and consequently f satisfies (*iii*).

Next, let us observe that the function $g : C_b(\mathbb{R}_+;\mathbb{R}^n) \to \mathbb{R}^n$, defined by (10.7.3), satisfies

$$\|g(x)-g(y)\| = \|x(T)-y(T)\| \le \|x-y\|_{C_b([\,T,+\infty);\mathbb{R}^n)}$$

for each $x,y \in C_b(\mathbb{R}_+;\mathbb{R}^n)$. Hence g satisfies (*iv*) with $a = T > 0$. Consequently the hypotheses of both Theorems 10.2.1 and 10.6.1 are satisfied. If follows that the problem (10.7.2) has a unique solution which is globally asymptotically stable. Finally, from Remark 10.7.1, we conclude that x is the unique solution of the T-periodic problem (10.7.1). In addition, x is globally asymptotically stable and this completes the proof. □

Theorem 10.7.2. *Let* $\mathcal{A} \in \mathcal{M}_{n\times n}(\mathbb{R})$ *be a given matrix and let* $f : \mathbb{R}_+ \times \mathbb{R}^n \to \mathbb{R}^n$ *be a continuous function which is* T*-periodic with respect to its first argument on* $\mathbb{R}_+$. *If there exist* $\omega > 0$, $m > 0$ *and* $\ell > 0$ *such that*

(*i'*) $\langle \mathcal{A}x, x\rangle \le -\omega\|x\|^2$ *for each* $x \in \mathbb{R}^n$;
(*ii'*) f *is dissipative on* $\mathbb{R}^n$;
(*iii'*) $\|f(t,x)\| \le \ell\|x\| + m$ *for each* $t \in \mathbb{R}_+$ *and each* $x \in \mathbb{R}^n$;
(*v*) $\omega - \ell > 0$,

then the problem (10.7.1) *has a unique solution* $x \in C_b(\mathbb{R}_+;\mathbb{R}^n)$. *Moreover,* x *is globally asymptotically stable.*

Proof. Let us define the function g as in the proof of Theorem 10.7.1. Then g satisfies the hypothesis (*iv*) in Theorems 10.5.1 and 10.6.2 from where the conclusion follows. □

Next, we will focus our attention on T-anti-periodic problems, which are important in Physics in the study of certain phenomena whose in-time-evolution obeys some symmetries. First, we say that $x : \mathbb{R}_+ \to \mathbb{R}^n$ is *T-anti-periodic* if for each $t \in \mathbb{R}_+$, we have

$$x(t) = -x(t+T).$$

Now, let us consider the problem

$$\begin{cases} x'(t) = \mathcal{A}x(t) + f(t, x(t)), & t \in \mathbb{R}_+, \\ x(t) = -x(t+T), & t \in \mathbb{R}_+. \end{cases} \tag{10.7.5}$$

Remark 10.7.2. If f satisfies

$$-f(t+T, -y) = f(t, y), \tag{10.7.6}$$

for each $t \in \mathbb{R}_+$ and each $y \in \mathbb{R}^n$, it follows that, for each $y \in \mathbb{R}^n$, the function $t \mapsto f(t, y)$ is $2T$-periodic, i.e.

$$f(t+2T, y) = f(t, y)$$

for each $t \in \mathbb{R}_+$. Indeed, we have

$$f(t+2T, y) = -f(t+T, -y) = f(t, y)$$

for each $t \in \mathbb{R}_+$ and $y \in \mathbb{R}^n$, as claimed.

If (i), (ii) and (v) in Theorem 10.2.1 are satisfied, f is continuous and satisfies (10.7.6), then the problem (10.7.5) is equivalent to

$$\begin{cases} x'(t) = \mathcal{A}x(t) + f(t, x(t)), & t \in [0, T], \\ x(0) = -x(T). \end{cases} \tag{10.7.7}$$

Clearly, each solution of (10.7.5) is a solution of (10.7.7). Now, if x is a solution of (10.7.7), taking into account that f satisfies (10.7.6), we deduce that this solution can be continued by T-anti-periodicity on $\mathbb{R}_+$. So, we obtain a solution for the problem (10.7.5). Finally, due to the uniqueness ensured by the condition (ii), we observe that (10.7.7) is equivalent to the nonlocal problem

$$\begin{cases} x'(t) = \mathcal{A}x(t) + f(t, x(t)), & t \in \mathbb{R}_+, \\ x(0) = g(x), \end{cases} \tag{10.7.8}$$

where $g : C_b(\mathbb{R}_+; \mathbb{R}^n) \to \mathbb{R}^n$ is defined by

$$g(x) = -x(T), \tag{10.7.9}$$

for each $x \in C_b(\mathbb{R}_+; \mathbb{R}^n)$.

The next result deals with T-anti-periodic solutions.

Theorem 10.7.3. *Let $\mathcal{A} \in \mathcal{M}_{n\times n}(\mathbb{R})$ be a given matrix and let $f : \mathbb{R}_+ \times \mathbb{R}^n \to \mathbb{R}^n$ be a continuous function which satisfies* (10.7.6). *If there exist $\omega > 0$ and $\ell > 0$ such that*

(*i*) $\|e^{t\mathcal{A}}\|_{\mathcal{M}} \le e^{-\omega t}$ *for each* $t \in \mathbb{R}_+$;
(*ii*) $\|f(t,x) - f(t,y)\| \le \ell\|x - y\|$ *for each* $t \in \mathbb{R}_+$ *and each* $x, y \in \mathbb{R}^n$;
(*v*) $\omega - \ell > 0$,

then the problem (10.7.5) *has a unique solution $x \in C_b(\mathbb{R}_+; \mathbb{R}^n)$. Moreover, x is globally asymptotically stable and $2T$-periodic.*

Proof. Reasoning as in the proof of Theorem 10.7.1, we deduce that the function g defined by (10.7.9) satisfies condition (*iv*) in Theorem 10.2.1. Hence, all the hypotheses of Theorem 10.2.1 are satisfied and thus, the problem (10.7.8) has a unique solution x, and x is globally asymptotically stable. From Remark 10.7.2, it follows that x is a solution for (10.7.5) and this completes the proof. □

Theorem 10.7.4. *Let $\mathcal{A} \in \mathcal{M}_{n\times n}(\mathbb{R})$ be a given matrix and let $f : \mathbb{R}_+ \times \mathbb{R}^n \to \mathbb{R}^n$ be a continuous function which satisfies* (10.7.6). *If there exist $\omega > 0$ and $m > 0$ such that*

(*i'*) $\langle \mathcal{A}x, x\rangle \le -\omega\|x\|^2$ *for each* $x \in \mathbb{R}^n$;
(*ii'*) f *is dissipative on* $\mathbb{R}^n$;
(*iii'*) $\|f(t,0)\| \le m$ *for each* $t \in \mathbb{R}_+$,

then the problem (10.7.5) *has a unique solution $x \in C_b(\mathbb{R}_+; \mathbb{R}^n)$. Moreover, x is globally asymptotically stable and $2T$-periodic.*

Proof. The proof repeats, with minor modifications, the arguments in the proof of Theorem 10.8.1 – see Problem 10.1 below – and it is left to the reader. □

10.8 Exercises and Problems

Exercise 10.1. *Let $0 < t_1 < t_2 \ldots$ be a given sequence and let $\alpha_i \in (0, 1)$, $i = 1, 2, \ldots$, be such that*

$$\sum_{i=1}^{\infty} \alpha_i \le 1$$

and let $\xi \in \mathbb{R}^n$. *Show that the function* $g : C_b(\mathbb{R}_+; \mathbb{R}^n) \to \mathbb{R}^n$ *defined by*

$$g(x) := \xi + \sum_{i=1}^{\infty} \alpha_i x(t_i)$$

for each $x \in C_b(\mathbb{R}_+; \mathbb{R}^n)$ *satisfies assumption (iv) in Theorem* 10.2.1.

Exercise 10.2. *Let* $a > 0$ *and let* $k : [\,a, +\infty) \to \mathbb{R}_+$ *be a given function with*

$$\int_a^{\infty} k(s)\, ds \le 1$$

and let $\xi \in \mathbb{R}^n$. *Show that the function* $g : C_b(\mathbb{R}_+; \mathbb{R}^n) \to \mathbb{R}^n$, *defined by*

$$g(x) := \xi + \int_a^{\infty} k(s) x(s)\, ds$$

for each $x \in C_b(\mathbb{R}_+; \mathbb{R}^n)$, *satisfies the hypothesis (iv) in Theorem* 10.2.1.

Problem 10.1. *Prove the following version of Theorem* 10.7.2 :

Theorem 10.8.1. *Let* $\mathcal{A} \in \mathcal{M}_{n\times n}(\mathbb{R})$ *be a given matrix and let* $f : \mathbb{R}_+ \times \mathbb{R}^n \to \mathbb{R}^n$ *be a continuous function which is* T*-periodic with respect to its first argument on* $\mathbb{R}_+$. *If there exist* $\omega > 0$ *and* $m > 0$ *such that*

(*i''*) $\langle \mathcal{A}x, x\rangle \le 0$ *for each* $x \in \mathbb{R}^n$;

(*ii''*) $\langle f(t,x) - f(t,y), x - y\rangle \le -\omega \|x - y\|^2$ *each* $t \in \mathbb{R}_+$ *and each* $x, y \in \mathbb{R}^n$;

(*iii''*) $\|f(t,0)\| \le m$ *for each* $t \in \mathbb{R}_+$,

then the problem (10.1.1) *has a unique solution* $x \in C_b(\mathbb{R}_+; \mathbb{R}^n)$ *which is globally asymptotically stable.*

Problem 10.2. *Let us consider the nonlocal problem*

$$\begin{cases} x'(t) = f(t, x(t)), & t \in [\,0, T\,], \\ x(0) = \xi + h\left(\displaystyle\int_0^T x(s)\, ds\right), \end{cases} \tag{10.8.1}$$

where $f : \mathbb{R}_+ \times \mathbb{R}^n \to \mathbb{R}^n$, $\xi \in \mathbb{R}^n$ *and* $h : \mathbb{R}^n \to \mathbb{R}^n$. *Prove the following existence and uniqueness result:*

Theorem 10.8.2. *Let us assume that there exist* $\ell_1 > 0$ *and* $\ell_2 > 0$ *such that*

(*j*) $f : \mathbb{R}_+ \times \mathbb{R}^n \to \mathbb{R}^n$ *is continuous* ;

(*jj*) f *is locally Lipschitz on bounded sets with respect to its second argument, i.e. for each* $T > 0$ *and each bounded subset* K *in* $\mathbb{R}^n$, *there exists* $\ell_1 = \ell_1(T, K)$ *such that*

$$\|f(t,x) - f(t,y)\| \le \ell_1 \|x - y\|$$

for each $t \in [0,T]$ *and* $x, y \in \mathbb{R}^n$;

(*jjj*) $h(0) = 0$;

(*jv*) $\|h(x) - h(y)\| \leq \ell_2 \|x - y\|$ *for each* $x, y \in \mathbb{R}^n$.

Then, for each $\xi \in \mathbb{R}^n$*, there exists* $T > 0$ *such that the nonlocal problem* (10.8.1) *has a unique solution* $x \in C([0,T]; \mathbb{R}^n)$.

Problem 10.3. *Let* a, b, c, d *be positive constants, let* $h_1 : \mathbb{R} \to \mathbb{R}_+$ *be continuous and increasing and* $h_2 : \mathbb{R} \to \mathbb{R}_-$ *be continuous and decreasing. Assume that* $h_1(0) = h_2(0) = 0$ *and there exists* $\ell_2 > 0$ *such that*

$$\|(h_1(y), h_2(x)) - (h_1(\widetilde{y}), h_2(\widetilde{x}))\| \leq \ell_2 \|(x, y) - (\widetilde{x}, \widetilde{y})\|$$

for each $x, \widetilde{x}, y, \widetilde{y} \in \mathbb{R}$. *Let* $\xi_1, \xi_2 \in \mathbb{R}$ *and consider the nonlocal problem for the prey-predator model*

$$\begin{cases} x'(t) = (a - by(t))x(t), & t \in [0,T], \\ y'(t) = -(c - dx(t))y(t), & t \in [0,T], \\ x(0) = \xi_1 + h_1\left(\displaystyle\int_0^T y(s)\,ds\right) \\ y(0) = \xi_2 + h_2\left(\displaystyle\int_0^T x(s)\,ds\right), \end{cases} \tag{10.8.2}$$

where $x(t)$ *and respectively by* $y(t)$ *denote the population of the prey and respectively predator species at the time* $t \in [0,T]$. *Using Theorem* 10.8.2*, show that there exists* $T > 0$*, independent of* ξ_1 *and* ξ_2*, such that the problem* (10.8.2) *has a unique solution on* $[0,T]$. *Explain the contribution of the functions* h_1 *and* h_2 *to the behavior of this solution.*

Chapter 11

Delay Functional Differential Equations

In this chapter we present some local and global existence results for the functional delay differential equation

$$\begin{cases} x'(t) = f(t, x_t), & t \in [\sigma, T], \\ x(t) = \psi(t - \sigma), & t \in [\sigma - \tau, \sigma], \end{cases}$$

where $\tau \geq 0$, $\sigma \in \mathbb{R}$, $f : [\sigma, +\infty) \times C([-\tau, 0]; \mathbb{R}^n) \to \mathbb{R}^n$ is continuous and the initial history ψ is an element of $C([-\tau, 0]; \mathbb{R}^n)$. Some uniqueness and uniform asymptotic stability results are also given.

11.1 Preliminaries

We begin with some useful notation. If $\mathbb{I}$ is an interval and D is a closed subset of $\mathbb{R}^n$, $C_b(\mathbb{I}; \mathbb{R}^n)$ denotes the space of all bounded and continuous functions from $\mathbb{I}$ to $\mathbb{R}^n$, while $C_b(\mathbb{I}; D)$ denotes the closed subset of $C_b(\mathbb{I}; \mathbb{R}^n)$ consisting of all elements $x \in C_b(\mathbb{I}; \mathbb{R}^n)$ satisfying $x(t) \in D$ for each $t \in \mathbb{I}$. In the case when $\mathbb{I}$ is the compact interval $[a, b]$, $C_b(\mathbb{I}; \mathbb{R}^n)$ coincides with $C([a, b]; \mathbb{R}^n)$ – we recall that each continuous function from a compact interval to $\mathbb{R}^n$ is bounded – and $C([a, b]; D)$ is the closed subset of $C([a, b]; \mathbb{R}^n)$ containing all $x \in C([a, b]; \mathbb{R}^n)$ with $x(t) \in D$ for each $t \in [a, b]$. If $\tau \geq 0$, $\sigma \in \mathbb{R}$, $x \in C([\sigma - \tau, +\infty); \mathbb{R}^n)$ and $t \in [\sigma, +\infty)$, then $x_t \in C([-\tau, 0]; \mathbb{R}^n)$ is defined by

$$x_t(s) = x(t + s)$$

for each $s \in [-\tau, 0]$. If $\tau = 0$, $C([-\tau, 0]; \mathbb{R}^n)$ can be identified with $\mathbb{R}^n$ and x_t can be identified with $x(t)$.

Let $\tau \geq 0$, let $\sigma \in \mathbb{R}$, let $f : [\sigma, +\infty) \times C([-\tau, 0]; \mathbb{R}^n) \to \mathbb{R}^n$ be a continuous function and $\psi \in C([-\tau, 0]; \mathbb{R}^n)$.

In all that follows, we use the term *delay functional differential equation* to designate an equation of the form

$$x'(t) = f(t, x_t),$$

emphasizing the fact that the term $f(t, \cdot)$ in the right-hand side depends on a function, x_t, which represents the delayed states on $[t - \tau, t]$, rather than on a single value of the state, $x(t)$.

We say that the function $f : [\sigma, +\infty) \times C([-\tau, 0]; \mathbb{R}^n) \to \mathbb{R}^n$ is *continuous at* $(t, \psi) \in [\sigma, +\infty) \times C([-\tau, 0]; \mathbb{R}^n)$ if for each sequence $((t_k, \psi_k))_{k \in \mathbb{N}}$ whose terms belong to $[\sigma, +\infty) \times C([-\tau, 0]; \mathbb{R}^n)$ and satisfying

$$\begin{cases} \lim_{k \to \infty} t_k = t \\ \lim_{k \to \infty} \psi_k(s) = \psi(s) \text{ uniformly for } s \in [-\tau, 0], \end{cases}$$

we have

$$\lim_{k \to \infty} f(t_k, \psi_k) = f(t, \psi).$$

We say that the function $f : [\sigma, +\infty) \times C([-\tau, 0]; \mathbb{R}^n) \to \mathbb{R}^n$ is *continuous on* $C \subseteq [\sigma, +\infty) \times C([-\tau, 0]; \mathbb{R}^n)$ if it is continuous at each $(t, \psi) \in C$ in the sense specified above.

We consider first the Cauchy problem for the delay functional differential equation

$$\begin{cases} x'(t) = f(t, x_t), & t \in [\sigma, T], \\ x(t) = \psi(t - \sigma), & t \in [\sigma - \tau, \sigma]. \end{cases} \tag{11.1.1}$$

Definition 11.1.1. By a *solution* of the problem (11.1.1) on $[\sigma - \tau, T]$ we mean a function $x : [\sigma - \tau, T] \to \mathbb{R}^n$ which is of class C^1 on $[\sigma, T]$, satisfies the equality $x'(t) = f(t, x_t)$ for each $t \in [\sigma, T]$ and the initial condition $x(t) = \psi(t - \sigma)$ for each $t \in [\sigma - \tau, \sigma]$. By a *solution* of the problem (11.1.1) on $[\sigma - \tau, +\infty)$ we mean a function $x : [\sigma - \tau, +\infty) \to \mathbb{R}^n$ which is a solution of (11.1.1) on $[\sigma - \tau, T]$ for each $T > \sigma$.

11.2 Local Existence Under Lipschitz Condition

Let $\tau \geq 0$, $\sigma \in \mathbb{R}$, let $f : [\sigma, +\infty) \times C([-\tau, 0]; \mathbb{R}^n) \to \mathbb{R}^n$ be a continuous function and $\psi \in C([-\tau, 0]; \mathbb{R}^n)$. In this section we will give a direct and rather elementary approach to the existence and uniqueness issues for (11.1.1).

The concept of solution we will use is given by Definition 11.1.1. Our next local existence and uniqueness result we assume a local Lipschitz condition on f with respect to its second argument. To make this condition precise, we define

$$\|\varphi\|_{C([-\tau,0];\mathbb{R}^n)} := \sup_{t\in[-\tau,0]} \|\varphi(t)\|$$

for each $\varphi \in C([-\tau,0];\mathbb{R}^n)$ and we introduce:

Definition 11.2.1. The function $f : [\sigma,+\infty) \times C([-\tau,0];\mathbb{R}^n) \to \mathbb{R}^n$ is called *Lipschitz on bounded sets with respect to its second argument* if for each $T > \sigma$ and $r > 0$ there exists $\ell = \ell(T,r) > 0$ such that

$$\|f(t,v) - f(t,w)\| \leq \ell\|v - w\|_{C([-\tau,0];\mathbb{R}^n)}$$

for each $t \in [\sigma,T]$ and each $v, w \in C([-\tau,0];\mathbb{R}^n)$ satisfying

$$\begin{cases} \|v\|_{C([-\tau,0];\mathbb{R}^n)} \leq r, \\ \|w\|_{C([-\tau,0];\mathbb{R}^n)} \leq r. \end{cases}$$

The following existence and uniqueness result is of the same type as the one proved earlier for ODEs.

Theorem 11.2.1. *Let $\tau \geq 0$, $\sigma \in \mathbb{R}$ and let $f : [\sigma,+\infty)\times C([-\tau,0];\mathbb{R}^n) \to \mathbb{R}^n$ be a continuous function which is Lipschitz on bounded sets with respect to its second argument. Then, for each $\psi \in C([-\tau,0];\mathbb{R}^n)$, there exist $T = T(\psi) > \sigma$ and a unique solution $x : [\sigma-\tau,T] \to \mathbb{R}^n$ of the problem* (11.1.1).

Proof. Let $\psi \in C([-\tau,0];\mathbb{R}^n)$, let $T > \sigma$ be arbitrary but fixed and let us define

$$r = \|\psi\|_{C([-\tau,0];\mathbb{R}^n)} + 1. \tag{11.2.1}$$

Let $\ell = \ell(T,r)$ be given by Definition 11.2.1, let

$$C_\psi([\sigma-\tau,T];\mathbb{R}^n)$$

$$= \{y \in C([\sigma-\tau,T];\mathbb{R}^n);\ y(s) = \psi(s-\sigma) \text{ for all } s \in [\sigma-\tau,\sigma]\}$$

and let

$$K = \{y \in C_\psi([\sigma-\tau,T];\mathbb{R}^n);\ \|y_t\|_{C([-\tau,0];\mathbb{R}^n)} \leq r \text{ for all } t \in [\sigma,T]\}.$$

Clearly K is nonempty since $\widetilde{\psi} \in K$, where

$$\widetilde{\psi}(s) = \begin{cases} \psi(s-\sigma), \; s \in [\,\sigma - \tau, \sigma\,] \\ \psi(0), \qquad\;\; t \in (\sigma, T\,]. \end{cases}$$

Moreover, one may easily verify that the uniform limit of any uniformly convergent sequence in K belongs to K. This means that K is closed in $C_\psi([\,\sigma - \tau, T\,]; \mathbb{R}^n)$. Since f is Lipschitz on bounded sets with respect to its second argument, it follows that there exists $m > 0$ such that

$$\|f(t, x_t)\| \le \ell r + m \tag{11.2.2}$$

for each $(t, x) \in [\,\sigma, T\,] \times K$. Indeed, we have

$$\|f(t,x_t)\| \le \|f(t,x_t) - f(t,0)\| + \|f(t,0)\| \le \ell \|x_t\|_{C([\,-\tau,0\,];\mathbb{R}^n)} + \|f(t,0)\|$$

for each $(t, x) \in [\,\sigma, T\,] \times K$. But $f(\cdot, 0)$ is continuous and $[\,\sigma, T\,]$ is compact. Hence there exists $m > 0$ such that

$$\|f(t,0)\| \le m$$

for each $t \in [\,\sigma, T\,]$. From the definition of K and the preceding inequality, we deduce (11.2.2).

Now, taking a smaller $T > \sigma$ if necessary, we may assume that

$$\max\{(T-\sigma)\ell, (T-\sigma)(\ell r + m)\} < 1. \tag{11.2.3}$$

Let $x_0 \in K$ be fixed and let us define the sequence of successive approximations $(x_k)_{k \in \mathbb{N}}$ given by

$$x_{k+1}(t) = \begin{cases} \psi(t-\sigma), & t \in [\,\sigma-\tau, \sigma\,] \\ \psi(0) + \displaystyle\int_\sigma^t f(s, x_{k\,s})\, ds, & t \in (\sigma, T\,], \end{cases} \tag{11.2.4}$$

for each $k \in \mathbb{N}$, where, for any fixed $s \in [\,\sigma, T\,]$, $x_{k\,s}$ denotes the function $(x_k)_s : [\,-\tau, 0\,] \to \mathbb{R}^n$, given by $(x_k)_s(\theta) = x_k(s+\theta)$ for each $\theta \in [\,-\tau, 0\,]$.

Clearly, for each $k \in \mathbb{N}$, x_{k+1} is the unique solution of the problem

$$\begin{cases} x'_{k+1}(t) = f(t, x_{k\,t}), & t \in [\,\sigma, T\,], \\ x_{k+1}(t) = \psi(t-\sigma), & t \in [\,\sigma - \tau, \sigma\,], \end{cases}$$

whose existence and uniqueness is ensured by Theorem 2.4.6. To complete the proof, it suffices to show that, if $T > \sigma$ is such that $T - \sigma$ is "small enough", then the sequence of successive approximations is uniformly convergent on $[\,\sigma - \tau, T\,]$ to some function x which is a solution of the problem (11.1.1).

First, we prove that, for each $k \in \mathbb{N}$, $x_k \in K$. Clearly $x_0 \in K$. Now, let us assume that $x_k \in K$ and let us observe that, from (11.2.1), (11.2.2) and (11.2.3), we get

$$\begin{aligned}\|x_{k+1}(t)\| &\leq \|\psi(0)\| + \int_\sigma^t \|f(s, x_{k\,s})\|\,ds \\ &\leq \|\psi\|_{C([-\tau,0];\mathbb{R}^n)} + (T-\sigma)(\ell r + m) \\ &\leq \|\psi\|_{C([-\tau,0];\mathbb{R}^n)} + 1 = r\end{aligned}$$

for each $k \in \mathbb{N}$ and each $t \in [\sigma, T]$. So,

$$\sup_{t\in[\sigma,T]} \|x_{k+1}(t)\| \leq r.$$

Since

$$\sup_{s\in[\sigma-\tau,\sigma]} \|x_{k+1}(s)\| = \sup_{s\in[\sigma-\tau,\sigma]} \|\psi(s-\sigma)\| = \|\psi\|_{C([-\tau,0];\mathbb{R}^n)},$$

we conclude that

$$\sup_{t\in[\sigma-\tau,T]} \|x_{k+1}(t)\| \leq r,$$

which proves that $x_{k+1} \in K$, for each $k \in \mathbb{N}$.

Next we have

$$\begin{aligned}\|x_{k+1}(t) - x_k(t)\| &\leq \int_\sigma^t \ell \sup_{\theta\in[-\tau,0]} \|x_{k\,s}(\theta) - x_{k-1\,s}(\theta)\|\,ds \\ &\leq (T-\sigma)\ell \sup_{s\in[\sigma,T]} \sup_{\theta\in[-\tau,0]} \|x_k(s+\theta) - x_{k-1}(s+\theta)\| \\ &= (T-\sigma)\ell \sup_{t\in[\sigma-\tau,\sigma]} \|x_k(t) - x_{k-1}(t)\| + (T-\sigma)\ell \sup_{t\in[\sigma,T]} \|x_k(t) - x_{k-1}(t)\|\end{aligned}$$

for each $k \in \mathbb{N}^*$ and each $t \in [\sigma, T]$. Hence

$$\begin{aligned}\sup_{t\in[\sigma,T]} \|x_{k+1}(t) - x_k(t)\| &\leq (T-\sigma)\ell \sup_{t\in[\sigma,T]} \|x_k(t) - x_{k-1}(t)\| \leq \dots \\ &\leq [(T-\sigma)\ell]^k \sup_{t\in[\sigma,T]} \|x_1(t) - x_0(t)\|\end{aligned}$$

for each $k \in \mathbb{N}^*$. Since by (11.2.3), $(T-\sigma)\ell < 1$, these inequalities show that the telescopic series

$$x_0(t) + \sum_{k=0}^{\infty} (x_{k+1}(t) - x_k(t))$$

satisfies the hypotheses of Lemma 10.3.1 and thus $(x_k)_{k\in\mathbb{N}}$ is uniformly convergent on $[\sigma, T]$. Since $x_k(t) = \psi(t-\sigma)$ for each $t \in [\sigma - \tau, \sigma]$ and each $k \in \mathbb{N}$, it follows that $(x_k)_{k\in\mathbb{N}}$ is uniformly convergent on $[\sigma - \tau, T]$ to a function $x \in K$. Passing to the limit in (11.2.4), we conclude that x is a solution of the problem (11.1.1). This completes the proof of the existence part of Theorem 11.2.1.

To prove the uniqueness part, we have only to observe that for any solutions x and y of (11.1.1), we have $x(t) = y(t)$ for each $t \in [\sigma - \tau, \sigma]$. Hence

$$\sup_{t\in[\sigma-\tau,T]} \|x(t) - y(t)\| = \sup_{t\in[\sigma,T]} \|x(t) - y(t)\|.$$

Furthermore, we obtain

$$\|x(t) - y(t)\| \leq \int_\sigma^t \|f(s, x_s) - f(s, y_s)\|\, ds \leq \ell \int_\sigma^t \|x_s - y_s\|_{C([-\tau,0];\mathbb{R}^n)}\, ds$$

$$\leq (T-\sigma)\ell \sup_{s\in[\sigma,T]} \sup_{\theta\in[-\tau,0]} \|x(s+\theta) - y(s+\theta)\| = (T-\sigma)\ell\|x-y\|_{C([\sigma,T];\mathbb{R}^n)}$$

for each $t \in [\sigma, T]$. Accordingly

$$\sup_{t\in[\sigma,T]} \|x(t) - y(t)\| \leq \ell(T-\sigma) \sup_{t\in[\sigma,T]} \|x(t) - y(t)\|.$$

But $(T-\sigma)\ell < 1$ – see (11.2.3) – which shows that $x(t) = y(t)$ for all $t \in [\sigma - \tau, T]$. The proof is complete. □

11.3 Some Auxiliary Results

We begin this section with an extension of Bellman's Inequality to integral inequalities with delay. See Problem 1.16. Lemma 11.3.1 is a simplified variant of Lemma 1.13.3, p. 52 in [Burlică *et al.* (2016)].

Lemma 11.3.1. *Let $T \in [0, +\infty]$ and let $\alpha_0 : [0, T) \to \mathbb{R}_+$ be continuous and nondecreasing and let $\beta : [0, T) \to \mathbb{R}_+$ be continuous. Let $y : [-\tau, T) \to \mathbb{R}_+$ be a continuous function satisfying*

$$y(t) \leq \alpha_0(t) + \int_0^t \beta(s)\|y_s\|_{C([-\tau,0];\mathbb{R})}\, ds \tag{11.3.1}$$

for each $t \in [0, T)$. Then

$$y(t) \leq \alpha(t) + \int_0^t \alpha(s)\beta(s)e^{\int_s^t \beta(\theta)\, d\theta} ds, \tag{11.3.2}$$

for each $t \in \mathbb{R}_+$, where

$$\alpha(t) = \|y_0\|_{C([-\tau,0];\mathbb{R})} + \alpha_0(t)$$

for each $t \in [0, T)$.

Proof. Define $z : [-\tau, T) \to \mathbb{R}_+$ by

$$z(t) = \sup\{\|y_s\|_{C([-\tau,0];\mathbb{R})};\ s \in [0,t]\} = \sup\{y(\theta);\ \theta \in [-\tau, t]\}$$

for each $t \in [-\tau, T)$. Clearly

$$y(t) \le z(t)$$

for each $t \in [0, T)$. Since y is continuous, it follows that for each $t \in [0, T)$ there exists $s_t \in [-\tau, t]$ such that

$$z(t) = y(s_t).$$

Let us observe that there are only two possible cases.

Case 1. If $s_t \in [-\tau, 0]$ and $y(s) < z(s)$ for each $s \in (0, t]$, then

$$z(t) = y(s_t) \le \|y_0\|_{C([-\tau,0];\mathbb{R})} \le \alpha(t) + \int_0^t \beta(s)z(s)\,ds. \tag{11.3.3}$$

Case 2. If $s_t \in (0, t]$, by (11.3.1), we have

$$z(t) = y(s_t) \le \alpha_0(s_t) + \int_0^{s_t} \beta(s)\|y_s\|_{C([-\tau,0];\mathbb{R})}\,ds$$

$$\le \alpha_0(t) + \|y_0\|_{C([-\tau,0];\mathbb{R})} + \int_0^t \beta(s)z(s)\,ds = \alpha(t) + \int_0^t \beta(s)z(s)\,ds.$$

So, (11.3.3) holds true for each $t \in \mathbb{R}_+$. From Problem 1.16, we get

$$z(t) \le \alpha(t) + \int_0^t \alpha(s)\beta(s)e^{\int_s^t \beta(\theta)\,d\theta}ds,$$

for each $t \in [0, T)$. Since $y(t) \le z(t)$ for each $t \in [0, T)$, the preceding inequality implies (11.3.2). The proof is complete. □

The next boundedness lemma is a slight extension of the Weierstrass Theorem on continuous real-valued functions defined on compact sets. See Lemma 1.13.3, p. 52 in [Burlică *et al.* (2016)].

Lemma 11.3.2. *Let $f : [\sigma, T] \times C([-\tau, 0]; \mathbb{R}^n) \to \mathbb{R}^n$ be a function which is continuous on $[\sigma, T] \times C$, where $C \subseteq C([-\tau, 0]; \mathbb{R}^n)$ is a compact set.*[1] *Then there exist $r > 0$ and $M > 0$ such that*

$$\|f(t, y)\| \le M$$

for each $t \in [\sigma, T]$ and $y \in \cup_{\eta\in C} D(\eta, r)$, where

$$D(\eta, r) := \{\varphi \in C([-\tau, 0]; \mathbb{R}^n);\ \|\varphi(t) - \eta(t)\| \le r,\ \text{for all } t \in [-\tau, 0]\}.$$

[1] See Definition 12.2.1.

Proof. Let us assume for contradiction that even though C is compact and f is continuous on $[\sigma,T]\times C$, there exist two sequences $(r_k)_{k\in\mathbb{N}}$ and $(M_k)_{k\in\mathbb{N}}$ in $(0,+\infty)$ with $\lim_{k\to\infty} r_k=0$ and $\lim_{k\to\infty} M_k=+\infty$ and three sequences $(t_k)_{k\in\mathbb{N}}$ in $[\sigma,T]$, $(\eta_k)_{k\in\mathbb{N}}$ in C and $(y_k)_{k\in\mathbb{N}}$ in $C([-\tau,0];\mathbb{R}^n)$ satisfying:

$$\|\eta_k-y_k\|\leq r_k$$

and

$$\|f(t_k,y_k)\|>M_k$$

for each $n\in\mathbb{N}$. Since both $[\sigma,T]$ and C are compact, we can assume with no loss of generality that there exist $t\in[\sigma,T]$ and $\eta\in C$ such that $\lim_{k\to\infty} t_k=t$ and $\lim_{k\to\infty}\eta_k(s)=\eta(s)$ uniformly for $s\in[-\tau,0]$. Using the preceding inequalities and the continuity of f at (t,η), we deduce that $\lim_{k\to\infty} y_k(s)=\eta(s)$ uniformly for $s\in[-\tau,0]$ and $\|f(t,\eta)\|\geq+\infty$ the latter inequality being absurd. This contradiction can be eliminated only if the conclusion of lemma holds true. □

11.4 Local Existence. The Continuous Case

Theorem 11.4.1. *Let $\tau\geq 0$ and $\sigma\in\mathbb{R}$. If $f:[\sigma,+\infty)\times C([-\tau,0];\mathbb{R}^n)\to\mathbb{R}^n$ is continuous, then for each $\psi\in C([-\tau,0];\mathbb{R}^n)$ there exists $T=T(\psi)>\sigma$ such that the problem* (11.1.1) *has at least one solution $x:[\sigma-\tau,T]\to\mathbb{R}^n$ in the sense of Definition* 11.1.1.

Proof. Let $\psi\in C([-\tau,0];\mathbb{R}^n)$, let $T>\sigma$ and $r>0$ be arbitrary but fixed. Let

$$C_\psi([\sigma-\tau,T];\mathbb{R}^n)=\{y\in C([\sigma-\tau,T];\mathbb{R}^n);\ y(s)=\psi(s-\sigma),\ \text{for all } s\in[\sigma-\tau,\sigma]\}.$$

Let us define the function $\widetilde{\psi}\in C([\sigma-\tau,T];\mathbb{R}^n)$ by

$$\widetilde{\psi}(t)=\begin{cases}\psi(t-\sigma), & t\in[\sigma-\tau,\sigma],\\ \psi(0), & t\in(\sigma,T],\end{cases}\tag{11.4.1}$$

which clearly belongs to $C_\psi([\sigma-\tau,T];\mathbb{R}^n)$ and let

$$K=\{y\in C_\psi([\sigma-\tau,T];\mathbb{R}^n);\ \|y(t)-\widetilde{\psi}(t)\|\leq r,\ \forall t\in[\sigma-\tau,T]\}.$$

Clearly $\widetilde{\psi}$ given by (11.4.1) belongs to K. Hence the latter is nonempty. Furthermore, K is closed and convex in $C([\sigma-\tau,T];\mathbb{R}^n)$. In addition, for each $y\in K$, we have $\|y_t-\widetilde{\psi}_t\|_{C([-\tau,0];\mathbb{R}^n)}\leq r$ for all $t\in[\sigma,T]$.

Now, let us remark that $\{\widetilde{\psi}_t;\ t \in [\sigma, T]\}$ is compact in $C([-\tau, 0]; \mathbb{R}^n)$. Indeed, since $\widetilde{\psi}([\sigma - \tau, T])$ is compact, it follows that $\{\widetilde{\psi}_t;\ t \in [\sigma, T]\}$ has relatively compact cross-sections. Moreover $\widetilde{\psi}$ is uniformly continuous, being continuous on a compact interval. We deduce that the family $\{\widetilde{\psi}_t;\ t \in [\sigma, T]\}$ is equicontinuous. From the Ascoli-Arzelà Theorem 12.2.1, we conclude that $\{\widetilde{\psi}_t;\ t \in [\sigma, T]\}$ is relatively compact in $C([-\tau, 0]; \mathbb{R}^n)$. Since it is also closed, we conclude that it is compact in $C([-\tau, 0]; \mathbb{R}^n)$. Thus Lemma 11.3.2 is applicable and so, taking a smaller $r > 0$ if necessary, we infer that there exists $M > 0$ such that

$$\|f(t, y_s)\| \leq M \tag{11.4.2}$$

for each $y \in K$, each $t \in [\sigma, T]$ and each $s \in [\sigma, T]$.

Let $\lambda > 0$, let $x_0 \in K$ be arbitrary but fixed and let us consider the integral equation with the augmented delay $\tau + \lambda$

$$x^\lambda(t) = \begin{cases} \psi(-\tau), & t \in [\sigma - \tau - \lambda, \sigma - \tau] \\ \psi(t - \sigma), & t \in (\sigma - \tau, \sigma], \\ \widetilde{\psi}(\sigma) + \displaystyle\int_\sigma^t f(s, x^\lambda_{s-\lambda})\, ds, & t \in (\sigma, T]. \end{cases} \tag{11.4.3}$$

Clearly, the problem has a unique solution defined step by step on $[\sigma - \tau - \lambda, \sigma]$, $[\sigma, \sigma + \lambda] \dots [\sigma + k\lambda, \sigma + (k+1)\lambda] \cap [\sigma, T]$, where $k+1$ is the first natural number for which $\sigma + (k+1)\lambda \geq T$. Indeed, since x^λ is uniquely determined on $[\sigma - \tau - \lambda, \sigma]$, take $t \in [\sigma, \sigma + \lambda]$ and define $x^\lambda(t)$ as

$$x^\lambda(t) = \widetilde{\psi}(\sigma) + \int_\sigma^t f(s, x^\lambda_{s-\lambda})\, ds.$$

At this point, x^λ is known on $[\sigma, \sigma + \lambda]$ and consequently we can define it at each $t \in [\sigma + \lambda, \sigma + 2\lambda]$, by setting

$$x^\lambda(t) = x^\lambda(\sigma + \lambda) + \int_{\sigma+\lambda}^t f(s, x^\lambda_{s-\lambda})\, ds.$$

Repeating this procedure, after a finite number of steps, we can define x^λ on $[\sigma - \tau - \lambda, T]$.

We will show next that, if $T > \sigma$ is such that $T - \sigma$ is sufficiently small then, for each $\lambda > 0$, the restriction of x^λ to $[\sigma - \tau, T]$ belongs to K. Namely, take $T > \sigma$ such that

$$(T - \sigma)M \leq r.$$

Clearly, the preceding requirements in which the original T is involved, are all satisfied. Moreover

$$x^\lambda(t) = \psi(t)$$

for each $t \in [\sigma - \tau, \sigma]$ and hence

$$\|x^\lambda(t) - \widetilde{\psi}(t)\| = \|\psi(0) - \psi(0)\| \leq r.$$

Let $t \in [\sigma, \sigma + \lambda]$ be arbitrary. Observing that the function

$$v(s) = \begin{cases} x^\lambda(s - \lambda), \ s \in [\sigma, \sigma + \lambda], \\ x^\lambda(\sigma), \qquad s \in (\sigma + \lambda, T] \end{cases}$$

belongs to K, by (11.4.2), it follows that

$$\|f(s, x^\lambda_{s-\lambda})\| \leq M$$

for each $s \in [\sigma, T]$. We then conclude that

$$\begin{aligned} \|x^\lambda(t) - \widetilde{\psi}(t)\| &\leq \|\psi(0) - \psi(0)\| + \int_\sigma^t \|f(s, x^\lambda_{s-\lambda})\| \, ds \\ &\leq (T - \sigma)M \leq r \end{aligned}$$

for each $t \in [\sigma, \sigma + \lambda]$. Repeating this argument on $[\sigma + \lambda, \sigma + 2\lambda]$ and so on, we deduce that $x^\lambda \in K$, as claimed.

Next, we will prove that the set

$$\mathcal{S} := \{x^\lambda; \ x^\lambda \text{ satisfies (11.4.3)}, \ \lambda > 0\}$$

is relatively compact in $C([\sigma, T]; \mathbb{R}^n)$. Since $\mathcal{S}$ is included in K and the latter is uniformly bounded, it follows that $\mathcal{S}$ is uniformly bounded. Accordingly, there exists $m_0 > 0$ such that

$$\|x^\lambda(t)\| \leq m_0$$

for each $\lambda > 0$ and $t \in [\sigma, T]$.

To check the equicontinuity of $\mathcal{S}$, let us observe that

$$x^\lambda(t) = \psi(0) + \int_\sigma^t f(s, x^\lambda_{s-\lambda}) \, ds$$

and thus

$$\|x^\lambda(t) - x^\lambda(\widetilde{t})\| \leq M|t - \widetilde{t}|$$

for each $\lambda > 0$ and each $t, \widetilde{t} \in [\sigma, T]$. So $\mathcal{S}$ is equicontinuous on $[\sigma, T]$ and, in view of Ascoli-Arzelà Theorem 12.2.1, we conclude that it is relatively compact in $C([\sigma, T]; \mathbb{R}^n)$. Consequently, if $(\lambda_k)_{k \in \mathbb{N}}$ is a strictly decreasing sequence with $\lim_{k \to \infty} \lambda_k = 0$, on a subsequence at least, denoted for simplicity again by $(\lambda_k)_{k \in \mathbb{N}}$, the sequence $(x^{\lambda_k})_{k \in \mathbb{N}}$ converges uniformly on $[\sigma, T]$ to some function x. Taking $\lambda = \lambda_k$ in (11.4.3) and passing to the limit for $k \to \infty$, we conclude that x is a solution of (11.1.1). This completes the proof of Theorem 11.4.1. □

11.5 Global Existence

In this section we will provide sufficient conditions which ensure that the solution whose local existence is ensured either by Theorem 11.2.1 or by Theorem 11.4.1 can be continued up to a global one. In fact, all the results in this section are inspired by those in Section 2.4. For the sake of simplicity, we will focus our attention only on the case $\sigma = 0$. Namely, we consider

$$\begin{cases} x'(t) = f(t, x_t), & t \in [0, T], \\ x(t) = \psi(t), & t \in [-\tau, 0], \end{cases} \tag{11.5.1}$$

where $\tau \geq 0$, $f : \mathbb{R}_+ \times C([-\tau, 0]; \mathbb{R}^n) \to X$ is continuous and $\psi \in C([-\tau, 0]; \mathbb{R}^n)$.

Definition 11.5.1. A solution $x : [-\tau, T) \to \mathbb{R}^n$ of the problem (11.5.1) is called *continuable* if there exist $\widetilde{T} > T$ and a solution $\widetilde{x} : [-\tau, \widetilde{T}) \to \mathbb{R}^n$ of (11.5.1) such that $x(t) = \widetilde{x}(t)$ for each $t \in [-\tau, T)$.

A solution $x : [-\tau, T) \to \mathbb{R}^n$ of the problem (11.5.1) is called *non-continuable* or *saturated* if it is not continuable. A saturated solution $x : [-\tau, T) \to \mathbb{R}^n$ of the problem (11.5.1) is called *global* if $T = +\infty$.

We begin with a simple but fundamental lemma.

Lemma 11.5.1. *Let $f : \mathbb{R}_+ \times C([-\tau, 0]; \mathbb{R}^n) \to \mathbb{R}^n$ be continuous and let $\psi \in C([-\tau, 0]; \mathbb{R}^n)$. Then a solution $x : [-\tau, T) \to \mathbb{R}^n$ of (11.5.1) is continuable if and only if there exists*

$$x^* = \lim_{t \uparrow T} x(t). \tag{11.5.2}$$

Proof. Since the necessity is obvious, we will restrict our attention only to the proof of the sufficiency. So, let us assume that (11.5.2) holds true. Clearly, for the initial history

$$\eta(t) = \begin{cases} x(T + t), & t \in [-\tau, 0), \\ x^*, & t = 0, \end{cases}$$

there exists $\delta > 0$ such that the problem

$$\begin{cases} w'(t) = f(t, w_t), & t \in [T, T + \delta], \\ w(t) = \eta(t - T), & t \in [T - \tau, T], \end{cases}$$

has at least one solution $w : [\,T-\tau, T+\delta\,] \to \mathbb{R}^n$. Finally, we observe that $\widetilde{x} : [\,-\tau, T+\delta\,] \to \mathbb{R}^n$, defined by

$$\widetilde{x}(t) = \begin{cases} x(t), \ t \in [\,-\tau, T), \\ w(t), \ t \in [\,T, T+\delta\,], \end{cases}$$

is a solution for the problem (11.5.1) which coincides with x on $[\,-\tau, T)$. We conclude that x is continuable and this completes the proof. □

Remark 11.5.1. From Lemma 11.5.1, it follows that each non-continuable solution of the problem (11.5.1) is necessarily defined on an interval, open at the right, i.e. of the form $[\,-\tau, T)$.

A sufficient condition for the existence of the limit (11.5.2) is stated below.

Proposition 11.5.1. *Let $f : \mathbb{R}_+ \times C([\,-\tau, 0\,]; \mathbb{R}^n) \to \mathbb{R}^n$ be continuous and let $\psi \in C([\,-\tau, 0\,]; \mathbb{R}^n)$. Let $x : [\,-\tau, T) \to \mathbb{R}^n$ be a solution of* (11.5.1) *and let us assume that $T < +\infty$ and $t \mapsto f(t, x_t)$ is bounded, i.e. there exists $M > 0$ such that*

$$\|f(t, x_t)\| \leq M$$

for each $t \in [\,0, T)$. Then there exists

$$x^* = \lim_{t \uparrow T} x(t).$$

Proof. Since $t \mapsto f(t, x_t)$ is continuous and bounded on $[\,0, T)$, it follows that it is Riemann integrable on $[\,0, T\,]$ and therefore the Cauchy problem

$$\begin{cases} v'(t) = f(t, x_t), \ t \in [\,0, T\,], \\ v(0) = x(0) \end{cases}$$

has a unique solution $v : [\,0, T\,] \to \mathbb{R}^n$ which, must coincide with x on $[\,0, T)$. But v is continuous. Hence it follows that there exists

$$\lim_{t \uparrow T} x(t) = \lim_{t \uparrow T} v(t) = v(T)$$

and this completes the proof. □

Definition 11.5.2. We say that the function $f : \mathbb{R}_+ \times C([\,-\tau, 0\,]; \mathbb{R}^n) \to \mathbb{R}^n$ *maps uniformly bounded subsets in $\mathbb{R}_+ \times C([\,-\tau, 0\,]; \mathbb{R}^n)$ into bounded subsets in $\mathbb{R}^n$* if for each $T > 0$ and each uniformly bounded subset U in $C([\,-\tau, 0\,]; \mathbb{R}^n)$ the set $f([\,0, T) \times U)$ is bounded in $\mathbb{R}^n$.

For examples of functions which map uniformly bounded subsets in $\mathbb{R}_+ \times C([-\tau,0];\mathbb{R}^n)$ into bounded subsets in $\mathbb{R}^n$ see Exercises 11.3 and 11.6. A useful characterization of continuable solutions of (11.5.1) is:

Theorem 11.5.1. *Let $f : \mathbb{R}_+ \times C([-\tau,0];\mathbb{R}^n) \to \mathbb{R}^n$ be a continuous function which maps uniformly bounded subsets in $\mathbb{R}_+ \times C([-\tau,0];\mathbb{R}^n)$ into bounded subsets in $\mathbb{R}^n$ and let $\psi \in C([-\tau,0];\mathbb{R}^n)$. A necessary and sufficient condition for a solution $x : [-\tau,T) \to \mathbb{R}^n$ of (11.5.1) to be continuable is that the graph of x, i.e.*

$$\text{graph } x = \{(t,x(t)) \in [-\tau,+\infty) \times \mathbb{R}^n;\ t \in [-\tau,T)\}$$

is included in a compact subset of $\mathbb{R} \times \mathbb{R}^n$.

Proof. Necessity. Let $x : [-\tau,T) \to \mathbb{R}^n$ be a continuable solution of (11.5.1). Then there exists at least one solution $\widetilde{x} : [-\tau,\widetilde{T}) \to \mathbb{R}^n$ of the same problem, with $\widetilde{T} > T$ and which coincides with x on $[-\tau,T)$. So,

$$\text{graph } x \subseteq \text{graph } \widetilde{x}_{|[-\tau,T]}.$$

Since the mapping $t \mapsto (t,\widetilde{x}(t))$ is continuous on $[-\tau,T]$ and the latter is compact, it follows that graph $\widetilde{x}_{|[-\tau,T]}$ is compact.

Sufficiency. If $K_0 \subseteq \mathbb{R} \times \mathbb{R}^n$ is a compact set and the graph of x is included in K_0, then $\overline{\{x(t);\ t \in [-\tau,T)\}}$ is compact in $\mathbb{R}^n$. Hence the set $\{x_t;\ t \in [0,T)\}$ is uniformly bounded. From the hypothesis it follows that $t \mapsto f(t,x_t)$ is bounded on $[0,T)$. So, by virtue of Proposition 11.5.1, there exists $\lim_{t\uparrow T} x(t) = x^*$. Since $(T,x^*) \in K_0$, we can apply Theorem 11.4.1 to conclude that there exists $\widetilde{T} > T$ such that the problem

$$\begin{cases} v'(t) = f(t,v_t),\ t \in [T,\widetilde{T}], \\ v(t) = x(t), \qquad t \in [T-\tau,T] \end{cases}$$

has a solution $v : [T-\tau,\widetilde{T}] \to \mathbb{R}^n$. Finally, we have only to observe that the function $\widetilde{x} : [-\tau,\widetilde{T}) \to \mathbb{R}^n$, defined by

$$\widetilde{x}(t) = \begin{cases} x(t),\ t \in [-\tau,T), \\ v(t),\ t \in [T,\widetilde{T}], \end{cases}$$

is a solution of (11.5.1). The proof is complete. □

Theorem 11.5.2. *Let $f : \mathbb{R}_+ \times C([-\tau,0];\mathbb{R}^n) \to \mathbb{R}^n$ be a continuous function, let $\psi \in C([-\tau,0];\mathbb{R}^n)$ and let $x : [-\tau,T) \to \mathbb{R}^n$ be a solution of (11.5.1). Then, either x is non-continuable, or x can be continued up to a non-continuable one.*

Proof. We shall apply Zorn's Lemma. Let $x : [-\tau, T) \to \mathbb{R}^n$ be a solution of (11.5.1). If x is non-continuable we have nothing to prove. If x is continuable, it follows that the set $\mathcal{S}$ of all solutions of (11.5.1) extending x strictly at the right of T is nonempty. On this set, we define the partial order "$\preceq$" as follows: if $x_1 : [-\tau, T_1) \to \mathbb{R}^n$ and $x_2 : [-\tau, T_2) \to \mathbb{R}^n$ are in $\mathcal{S}$, we say that $x_1 \preceq x_2$ if $T_1 \le T_2$ and $x_1(t) = x_2(t)$ for each $t \in [-\tau, T_1)$. One may easily check that $(\mathcal{S}, \preceq)$ is an inductively ordered set. Accordingly there exists at least one maximal element $\widetilde{x}$ in $\mathcal{S}$. Finally, we have merely to observe that $\widetilde{x}$ is a non-continuable solution of (11.5.1) extending x. The proof is complete. □

From Theorems 11.2.1, 11.4.1 and 11.5.2 it follows:

Corollary 11.5.1. *Let $f : \mathbb{R}_+ \times C([-\tau, 0]; \mathbb{R}^n) \to \mathbb{R}^n$ be a continuous function. Then for each initial data $\psi \in C([-\tau, 0]; \mathbb{R}^n)$, the problem* (11.5.1) *has at least one non-continuable solution. If f is Lipschitz on bounded sets with respect to its second argument, then this solution is unique.*

We conclude this section with a sufficient condition on f ensuring the existence of global solutions of (11.5.1).

Theorem 11.5.3. *Let $f : \mathbb{R}_+ \times C([-\tau, 0]; \mathbb{R}^n) \to \mathbb{R}^n$ be a continuous function for which there exist two continuous functions $h, k : \mathbb{R}_+ \to \mathbb{R}_+$ such that*

$$\|f(t, v)\| \le k(t)\|v\|_{C([-\tau,0];\mathbb{R}^n)} + h(t), \tag{11.5.3}$$

for each $(t, v) \in \mathbb{R}_+ \times C([-\tau, 0]; \mathbb{R}^n)$. Then, for each $\psi \in C([-\tau, 0]; \mathbb{R}^n)$, the problem (11.5.1) *has at least one global solution, i.e. defined on $[-\tau, +\infty)$. If f is Lipschitz on bounded sets with respect to its second argument then this solution is unique.*

Proof. Let $\psi \in C([-\tau, 0]; \mathbb{R}^n)$ and let $x : [-\tau, T) \to \mathbb{R}^n$ be a non-continuable solution of (11.5.1) whose existence is ensured by Corollary 11.5.1. We will show that $T = +\infty$. Indeed, if we assume the contrary, we get

$$\|x(t)\| \le \|x(0)\| + \int_0^t k(s)\|x_s\|_{C([-\tau,0];\mathbb{R}^n)}\, ds + \int_0^t h(s)\, ds$$

for each $t \in [0, T)$. Set

$$\begin{cases} \alpha_0(t) = \|x(0)\| + \int_0^t h(s)\, ds, \ t \in \mathbb{R}_+, \\ \beta(t) = k(t), & t \in \mathbb{R}_+, \\ y(t) = \|x(t)\|, & t \in [-\tau, T). \end{cases}$$

Taking into account that both k and h are bounded on $[0, T]$, from Lemma 11.3.1, we conclude that $\|x\|$ is bounded on $[0, T)$ and thus it is bounded on $[-\tau, T)$. Again by (11.5.3), we deduce that the mapping $t \mapsto f(t, x_t)$ is bounded on $[0, T)$. Hence, by Proposition 11.5.1, there exists $\lim_{t \uparrow T} x(t) = x^*$. From Lemma 11.5.1, it follows that x is continuable – a contradiction. This contradiction can be eliminated only if $T = +\infty$ and this completes the proof. □

11.6 A Spring Mass System with Delay

In this section we analyze an illustrative example of ODE with delay which can be handled by the previously developed abstract existence theory.

Let $0 < \sigma \leq \tau$. If $\psi_1 \in C([-\tau, 0]; \mathbb{R})$ and $\psi_2 \in C([-\sigma, 0]; \mathbb{R})$, we define $\psi \in \begin{pmatrix} C([-\tau,0];\mathbb{R}) \\ \times \\ C([-\tau,0];\mathbb{R}) \end{pmatrix}$, by

$$\psi(t) = \begin{pmatrix} \widetilde{\psi}_1(t), \\ \widetilde{\psi}_2(t), \end{pmatrix}, \quad t \in [-\tau, 0],$$

where

$$\widetilde{\psi}_1(t) = \psi_1(t), \qquad t \in [-\tau, 0],$$

$$\widetilde{\psi}_2(t) = \begin{cases} \psi_2(t), & t \in [-\sigma, 0], \\ \psi_2(-\sigma), & t \in [-\tau, -\sigma). \end{cases}$$

The following mathematical model describing the evolution of a *spring mass system with delay*, inspired by McKibben [McKibben (2011)], was considered in [Burlică *et al.* (2016)]. Let $\beta > 0$, $\omega > 0$ and let us consider the system

$$\begin{cases} x''(t) + \beta x'(t) + \omega^2 x(t) = h(t, x(t-\tau), x'(t-\sigma)), & t \in [0, T], \\ x(t) = \psi_1(t), & t \in [-\tau, 0], \\ x'(t) = \psi_2(t), & t \in [-\sigma, 0], \end{cases} \tag{11.6.1}$$

where $h : [0,T] \times \begin{pmatrix} \mathbb{R} \\ \times \\ \mathbb{R} \end{pmatrix} \to \mathbb{R}$ is continuous. One may easily see that (11.6.1) can be rewritten as a first-order differential system with delay of the form

$$\begin{cases} x'(t) = y(t), & t \in [0,T], \\ y'(t) = -\omega^2 x(t) - \beta y(t) + h(t, x(t-\tau), y(t-\sigma)), & t \in [0,T], \\ x(t) = \psi_1(t), & t \in [-\tau, 0], \\ y(t) = \psi_2(t), & t \in [-\sigma, 0]. \end{cases} \tag{11.6.2}$$

In order to rewrite this system in a more convenient abstract form, let us observe that

$$\begin{cases} x(t) = x_t(0), & t \in [0,T], \\ y(t) = y_t(0), & t \in [0,T], \\ x(s-\tau) = x_s(-\tau), & s \in [0,T], \\ y(s-\sigma) = y_s(-\sigma), & s \in [0,T]. \end{cases}$$

Now, let us define the function $f : [0,T] \times \begin{pmatrix} C([-\tau,0];\mathbb{R}) \\ \times \\ C([-\tau,0];\mathbb{R}) \end{pmatrix} \to \begin{pmatrix} \mathbb{R} \\ \times \\ \mathbb{R} \end{pmatrix}$ by

$$f\left(t, \begin{pmatrix} z \\ w \end{pmatrix}\right) = \begin{pmatrix} w(0) \\ -\omega^2 z(0) - \beta w(0) + h(t, z(-\tau), w(-\sigma)) \end{pmatrix}$$

for each $t \in [0,T]$ and each $\begin{pmatrix} z \\ w \end{pmatrix} \in \begin{pmatrix} C([-\tau,0];\mathbb{R}) \\ \times \\ C([-\tau,0];\mathbb{R}) \end{pmatrix}$. Setting $u = \begin{pmatrix} x \\ y \end{pmatrix}$, the problem (11.6.2) and equivalently (11.6.1) can be rewritten as

$$\begin{cases} u'(t) = f(t, u_t), & t \in [0,T], \\ u(t) = \psi(t), & t \in [-\tau, 0]. \end{cases} \tag{11.6.3}$$

From Corollary 11.5.1 and Theorem 11.5.3, we deduce:

Theorem 11.6.1. *If the function* $h : [0,T] \times \begin{pmatrix} \mathbb{R} \\ \times \\ \mathbb{R} \end{pmatrix} \to \mathbb{R}$ *is continuous, then for each* $\begin{pmatrix} \psi_1 \\ \psi_2 \end{pmatrix} \in \begin{pmatrix} C([-\tau,0];\mathbb{R}) \\ \times \\ C([-\tau,0];\mathbb{R}) \end{pmatrix}$ *the problem* (11.6.3) *has at least one non-continuable solution* $u : [-\tau, T_m) \to \begin{pmatrix} \mathbb{R} \\ \times \\ \mathbb{R} \end{pmatrix}$, $u = \begin{pmatrix} \widetilde{x} \\ \widetilde{y} \end{pmatrix}$, *where* $x =$

$\widetilde{x}$ is a non-continuable solution of (11.6.1). *If, in addition, h is Lipschitz with respect to its last two arguments, then the solution u is unique and global.*

11.7 A Delay Glucose Level-Dependent Dosage

Also inspired from a model described by [McKibben (2011)], the system describing the evolution of the glucose level in the blood stream as well as in the gastrointestinal system in the presence of a feedback – called dosage – taking into account of the histories of both levels was discussed in [Burlică *et al.* (2016)]. Namely, let us consider

$$\begin{cases} y'(t) = -ay(t) + D_1(t, y(t-\tau), z(t-\sigma)), & t \in [0, T], \\ z'(t) = ay(t) - bz(t) + D_2(t, y(t-\tau), z(t-\sigma)), & t \in [0, T], \\ y(t) = \psi_1(t), & t \in [-\tau, 0], \\ z(t) = \psi_2(t), & t \in [-\sigma, 0]. \end{cases} \tag{11.7.1}$$

Here $0 < a \leq b$ and the dosages D_1, D_2 depend on the retarded value of the glucose level in the blood stream, $y(t-\tau)$, as well as on the glucose level in gastrointestinal system, $z(t-\tau)$. Let $0 \leq \sigma \leq \tau$ and let us define

$$\psi(t) = \begin{pmatrix} \widetilde{\psi}_1(t) \\ \widetilde{\psi}_2(t) \end{pmatrix}, \quad t \in [-\tau, 0],$$

as at the beginning of this section.

For each $y \in C([-\tau, 0]; \mathbb{R})$ and $z \in C([-\tau, 0]; \mathbb{R})$, we have

$$\begin{cases} y(t) = y_t(0), & t \in [0, T], \\ z(t) = z_t(0), & t \in [0, T], \\ y(s-\tau) = y_s(-\tau), & s \in [0, T], \\ z(s-\sigma) = z_s(-\sigma), & s \in [0, T]. \end{cases}$$

Now, let us define the function

$$f : [0, T] \times \begin{pmatrix} C([-\tau, 0]; \mathbb{R}) \\ \times \\ C([-\tau, 0]; \mathbb{R}) \end{pmatrix} \to \begin{pmatrix} \mathbb{R} \\ \times \\ \mathbb{R} \end{pmatrix}$$

by

$$f\left(t, \begin{pmatrix} v \\ w \end{pmatrix}\right) = \begin{pmatrix} -av_t(0) + D_1(t, v_t(-\tau), w_t(-\sigma)) \\ av_t(0) - bw_t(0) + D_2(t, v_t(-\tau), w_t(-\sigma)) \end{pmatrix}$$

for each $t \in [\,0, T\,]$ and each $\begin{pmatrix} v \\ w \end{pmatrix} \in \begin{pmatrix} C([\,-\tau, 0\,]; \mathbb{R}) \\ \times \\ C([\,-\tau, 0\,]; \mathbb{R}) \end{pmatrix}$. With the notations above, recalling that $u = \begin{pmatrix} y \\ z \end{pmatrix} \in \begin{pmatrix} \mathbb{R} \\ \times \\ \mathbb{R} \end{pmatrix}$, the problem (11.7.1) can be rewritten under the form

$$\begin{cases} u'(t) = f(t, u_t), \; t \in [\,0, T\,], \\ u(t) = \psi(t), \qquad t \in [\,-\tau, 0\,]. \end{cases} \tag{11.7.2}$$

From Corollary 11.5.1 and Theorem 11.5.3, we get:

Theorem 11.7.1. *If* $D_i : [0, T\,] \times \begin{pmatrix} \mathbb{R} \\ \times \\ \mathbb{R} \end{pmatrix} \to \mathbb{R}$, $i = 1, 2$, *are continuous then, for each* $\begin{pmatrix} \psi_1 \\ \psi_2 \end{pmatrix} \in \begin{pmatrix} C([\,-\tau, 0\,]; \mathbb{R}) \\ \times \\ C([\,-\tau, 0\,]; \mathbb{R}) \end{pmatrix}$, *the problem* (11.7.2) *has at least one non-continuable solution* $u : [-\tau, T_m) \to \begin{pmatrix} \mathbb{R} \\ \times \\ \mathbb{R} \end{pmatrix}$, $u = \begin{pmatrix} \widetilde{y} \\ \widetilde{z} \end{pmatrix}$, *where* $\begin{pmatrix} \widetilde{y} \\ z \end{pmatrix}$, *with* $z = \widetilde{z}_{|[\,-\sigma, T_m)}$, *is a non-continuable solution of* (11.7.1). *If, in addition,* D_i, $i = 1, 2$, *are Lipschitz with respect to the last two arguments, then the solution is unique and global, i.e.* $\widetilde{y} : [\,-\tau, T\,] \to \mathbb{R}$ *and* $z : [\,-\sigma, T\,] \to \mathbb{R}$.

11.8 Existence of Global Bounded Solutions

Let us consider the functional delay differential equation subjected to a nonlocal initial condition

$$\begin{cases} x'(t) = \mathcal{A}x(t) + h(t, x_t), & t \in \mathbb{R}_+, \\ x(t) = \psi(t), & t \in [\,-\tau, 0\,]. \end{cases} \tag{11.8.1}$$

Here $\mathcal{A} \in \mathcal{M}_{n \times n}(\mathbb{R})$, $\tau \geq 0$, $h : \mathbb{R}_+ \times C([\,-\tau, 0\,]; \mathbb{R}^n) \to \mathbb{R}^n$ is continuous and $\psi \in C([\,-\tau, 0\,]; \mathbb{R}^n)$. In the limiting case $\tau = 0$, i.e. when the delay is absent, $h : \mathbb{R}_+ \times \mathbb{R}^n \to \mathbb{R}^n$ and $\psi \in \mathbb{R}^n$.

We observe that the problem (11.8.1) is of the form (11.1.1) with the function $f : \mathbb{R}_+ \times C([\,-\tau, 0\,]; \mathbb{R}^n) \to \mathbb{R}^n$ appropriately defined, i.e. $f(t, v) := \mathcal{A}v(0) + h(t, v)$ for each $(t, v) \in \mathbb{R}_+ \times C([\,-\tau, 0\,]; \mathbb{R}^n)$. Hence all the results proved in the preceding sections are applicable to (11.8.1). See

Exercises 11.1 and 11.2. As seen in Theorem 11.5.3, under rather general assumptions on the function f defined as above, the problem (11.8.1) has global solutions. Now, we are interested in getting sufficient conditions for the global solutions of (11.8.1) to be bounded.

We begin with the assumptions we need in what follows.

(i') $\mathcal{A} \in \mathcal{M}_{n\times n}(\mathbb{R})$ and there exists $\omega > 0$ such that

$$\langle \mathcal{A}x, x\rangle \leq -\omega\|x\|^2$$

for each $x \in \mathbb{R}^n$;

(ii) $h : \mathbb{R}_+ \times C([-\tau, 0]; \mathbb{R}^n) \to \mathbb{R}^n$ is continuous and there exist ℓ and m such that

$$\|h(t, v)\| \leq \ell\|v\|_{C([-\tau,0];\mathbb{R}^n)} + m$$

for each $t \in \mathbb{R}_+$ and $v \in C([-\tau, 0]; \mathbb{R}^n)$;

(iii) ℓ and ω satisfy

$$\ell < \omega.$$

(iv) $\psi \in C([-\tau, 0]; \mathbb{R}^n)$.

The main boundedness result concerning (11.8.1) is:

Theorem 11.8.1. *If (i') $\sim$ (iv) are satisfied, then the unique solution, x, of the problem (11.8.1) whose existence is ensured by Theorem 11.5.3 is bounded on $[-\tau, +\infty)$, i.e. $x \in C_b([-\tau, +\infty); \mathbb{R}^n)$. More precisely, x satisfies*

$$\|x\|_{C_b([-\tau,+\infty);\mathbb{R}^n)} \leq \max\left\{\|\psi\|_{C([-\tau,0];\mathbb{R}^n)}, \frac{m}{\omega-\ell}\right\}, \tag{11.8.2}$$

where ω is given by (i'), while ℓ and m are given by (ii).

Proof. We begin by showing that x is bounded on $[0, +\infty)$. To this aim, let us observe first that x satisfies the variation of constants formula (4.3.4) in Remark 4.3.3. Namely

$$x(t) = e^{t\mathcal{A}}x(0) + \int_0^t e^{(t-s)\mathcal{A}}h(s, x_s)\,ds$$

for each $t \in \mathbb{R}_+$. From (i') and Remark 10.5.1, it follows that

$$\|e^{t\mathcal{A}}\|_{\mathcal{M}} \leq e^{-\omega t}$$

for each $t \in \mathbb{R}_+$. Next, let us observe that

$$\|x(t)\| \leq \|\psi\|_{C([-\tau,0];\mathbb{R}^n)} + \int_0^t e^{-(t-s)\omega} \left(\ell \|x_s\|_{C([-\tau,0];\mathbb{R}^n)} + m\right) ds \tag{11.8.3}$$

for each $t \in \mathbb{R}_+$. From (11.8.3), we get

$$\|x(t)\| \leq \|\psi\|_{C([-\tau,0];\mathbb{R}^n)} + \left(1 - e^{-t\omega}\right) \frac{\ell}{\omega} \left(\|x\|_{C([-\tau,t];\mathbb{R}^n)} + \frac{m}{\ell}\right) \tag{11.8.4}$$

for each $T > 0$ and each $t \in [0, T]$. We will show that x is bounded on $\mathbb{R}_+$. Let us assume for contradiction that x is unbounded on $\mathbb{R}_+$. Then there exists $(t_k)_{k \in \mathbb{N}}$ with

$$\lim_{k \to \infty} t_k = +\infty$$

such that

$$\|x(t_k)\| = \|x\|_{C([0,t_k];\mathbb{R}^n)} = \|x\|_{C([-\tau,t_k];\mathbb{R}^n)}$$

and

$$\lim_{k \to \infty} \|x(t_k)\| = \sup_{t \in \mathbb{R}_+} \|x(t)\| = +\infty.$$

Setting $t = t_k$ in (11.8.4) and taking into account the preceding equalities, we deduce

$$\|x\|_{C([0,t_k];\mathbb{R}^n)} \leq \|\psi\|_{C([-\tau,0];\mathbb{R}^n)} + (1 - e^{-\omega t_k}) \frac{\ell}{\omega} \left(\|x\|_{C([0,t_k];\mathbb{R}^n)} + \frac{m}{\ell}\right) \tag{11.8.5}$$

for each $k \in \mathbb{N}$. Dividing both sides in (11.8.5) by $\|x\|_{C([0,t_k];\mathbb{R}^n)}$, we get

$$1 \leq \frac{\|\psi\|_{C([-\tau,0];\mathbb{R}^n)}}{\|x\|_{C([0,t_k];\mathbb{R}^n)}} + \frac{\ell}{\omega}\left(1 - e^{-\omega t_k}\right)\left(1 + \frac{m}{\ell \|x\|_{C([0,t_k];\mathbb{R}^n)}}\right)$$

for each $k \in \mathbb{N}$. Passing to the limit for $k \to \infty$ in this inequality, we get

$$1 \leq \frac{\ell}{\omega} < 1$$

which is absurd. This contradiction can be eliminated only if x is bounded on $\mathbb{R}_+$. Since x is bounded on $[-\tau, 0]$, it follows that it is bounded on $[-\tau, +\infty)$.

To prove the estimate (11.8.2), let $x \in C([-\tau, +\infty); \mathbb{R}^n)$ be the unique solution of (11.8.1).

Since $\|x\|_{C([-\tau,0];\mathbb{R}^n)} = \|\psi\|_{C([-\tau,0];\mathbb{R}^n)}$, we get

$$\|x(t)\| \leq e^{-\omega t}\|x(0)\| + \int_0^t e^{-\omega(t-s)} \left(\ell \|x_s\|_{C([-\tau,0];\mathbb{R}^n)} + m\right) ds$$

$$\leq e^{-\omega t}\|x\|_{C([0,T];\mathbb{R}^n)} + \int_0^t e^{-\omega(t-s)}\left(\ell\|x\|_{C([-\tau,T];\mathbb{R}^n)} + m\right)ds$$

$$\leq e^{-\omega t}\|x\|_{C([0,T];\mathbb{R}^n)}$$

$$+\int_0^t e^{-\omega(t-s)}\left(\ell\max\{\|x\|_{C([0,T];\mathbb{R}^n)}, \|x\|_{C([-\tau,0];\mathbb{R}^n)}\} + m\right)ds,$$

for each $T > 0$ and $t \in [0,T]$. Hence

$$\|x(t)\| \leq e^{-\omega t}\|x\|_{C([0,T];\mathbb{R}^n)}$$

$$+\frac{\ell}{\omega}\left(1 - e^{-\omega t}\right)\left(\max\{\|x\|_{C([0,T];\mathbb{R}^n)}, \|\psi\|_{C([-\tau,0];\mathbb{R}^n)}\} + \frac{m}{\ell}\right) \quad (11.8.6)$$

for each $T > 0$ and $t \in [0,T]$.

If for each $T > 0$, $\|x\|_{C([0,T];\mathbb{R}^n)} \leq \|\psi\|_{C([-\tau,0];\mathbb{R}^n)}$, then the conclusion is obvious. Therefore, let us assume that there exists $T_0 > 0$ such that

$$\|x\|_{C([0,T_0];\mathbb{R}^n)} > \|\psi\|_{C([-\tau,0];\mathbb{R}^n)}.$$

Accordingly, for each $T \geq T_0$, from (11.8.6), we get

$$\|x(t)\| \leq e^{-\omega t}\|x\|_{C([0,T];\mathbb{R}^n)} + \frac{\ell}{\omega}\left(1 - e^{-\omega t}\right)\left(\|x\|_{C([0,T];\mathbb{R}^n)} + \frac{m}{\ell}\right) \quad (11.8.7)$$

for each $t \in [0,T]$. Now, let $\widetilde{t} > 0$ be such that

$$\|x(\widetilde{t})\| = \|x\|_{C([0,T];\mathbb{R}^n)}.$$

Setting $t = \widetilde{t}$ in (11.8.7) and observing that $\|x\|_{C([0,T];\mathbb{R}^n)} = \|x\|_{C([-\tau,T];\mathbb{R}^n)}$, we deduce

$$\|x\|_{C([-\tau,T];\mathbb{R}^n)} \leq \frac{m}{\omega - \ell}.$$

As $T \geq T_0$ is arbitrary, we get

$$\|x\|_{C_b([-\tau,+\infty);\mathbb{R}^n)} \leq \frac{m}{\omega - \ell}.$$

Thus x satisfies (11.8.2) and this completes the proof. □

11.9 Exercises and Problems

Exercise 11.1. *Let $\sigma \in \mathbb{R}$, $\tau > 0$ and let $h : [\sigma, +\infty) \times \mathbb{R}^n \to \mathbb{R}^n$ be a continuous function. Let $\theta \in [-\tau, 0]$ and let us define $f : [\sigma, +\infty) \times C([-\tau, 0]; \mathbb{R}^n) \to \mathbb{R}^n$ by*

$$f(t, v) := h(t, v(\theta))$$

for each $t \in [\sigma, +\infty)$ and each $v \in C([-\tau, 0]; \mathbb{R}^n)$. Show that f is continuous on $[\sigma, +\infty) \times C([-\tau, 0]; \mathbb{R}^n)$.

Exercise 11.2. *Let $\sigma \in \mathbb{R}$ and let $h : [\sigma, +\infty) \times \mathbb{R}^n \to \mathbb{R}^n$ be a continuous function satisfying the following condition:*

(c) for each $T > \sigma$ and $r > 0$ there exists $\ell = \ell(T, r) > 0$ such that

$$\|h(t, v) - h(t, w)\| \leq \ell \|v - w\|$$

for each $t \in [\sigma, T]$ and each $v, w \in \mathbb{R}^n$ satisfying

$$\begin{cases} \|v\| \leq r, \\ \|w\| \leq r. \end{cases}$$

Let $\tau > 0$, $\theta \in [-\tau, 0]$ and let us define $f : [\sigma, +\infty) \times C([-\tau, 0]; \mathbb{R}^n) \to \mathbb{R}^n$ by

$$f(t, v) := h(t, v(\theta))$$

for each $(t, v) \in [\sigma, +\infty) \times C([-\tau, 0]; \mathbb{R}^n)$. Show that f is continuous on $[\sigma, +\infty) \times C([-\tau, 0]; \mathbb{R}^n)$ and Lipschitz on bounded sets with respect to its second argument in the sense of Definition 11.2.1.

Exercise 11.3. *Let $h : \mathbb{R}_+ \times \mathbb{R}^n \to \mathbb{R}^n$ be continuous. Let $\tau > 0$, let $\theta \in [-\tau, 0]$ and let us define $f : \mathbb{R}_+ \times C([-\tau, 0]; \mathbb{R}^n) \to \mathbb{R}^n$ by*

$$f(t, v) := h(t, v(\theta))$$

for each $(t, v) \in \mathbb{R}_+ \times C([-\tau, 0]; \mathbb{R}^n)$. Show that f maps uniformly bounded subsets in $\mathbb{R}_+ \times C([-\tau, 0]; \mathbb{R}^n)$ into bounded subsets in $\mathbb{R}^n$ in the sense of Definition 11.5.2.

Exercise 11.4. *Let $h : \mathbb{R}_+ \times \mathbb{R}^n \to \mathbb{R}^n$ be continuous. Let $\tau > 0$, $\theta \in [-\tau, 0]$ and let us define $f : \mathbb{R}_+ \times C([-\tau, 0]; \mathbb{R}^n) \to \mathbb{R}^n$ by*

$$f(t, v) := h\left(t, \int_{-\tau}^{0} v(\theta)\, d\theta\right) \tag{11.9.1}$$

for each $(t, v) \in \mathbb{R}_+ \times C([-\tau, 0]; \mathbb{R}^n)$. Show that the function f, defined by (11.9.1)*, is continuous on $\mathbb{R}_+ \times C([-\tau, 0]; \mathbb{R}^n)$.*

Exercise 11.5. *Let $h : \mathbb{R}_+ \times \mathbb{R}^n \to \mathbb{R}^n$ be a continuous function satisfying condition (c) in Exercise* 11.2. *Let $\tau > 0$, $\theta \in [-\tau, 0]$ and let us define the function $f : \mathbb{R}_+ \times C([-\tau, 0]; \mathbb{R}^n) \to \mathbb{R}^n$ by* (11.9.1). *Show that f is continuous on $\mathbb{R}_+ \times C([-\tau, 0]; \mathbb{R}^n)$ and Lipschitz on bounded sets with respect to its second argument in the sense of Definition* 11.2.1.

Exercise 11.6. *Let $h : \mathbb{R}_+ \times \mathbb{R}^n \to \mathbb{R}^n$ be a continuous function. Let $\tau > 0$ and let us define the function $f : \mathbb{R}_+ \times C([-\tau, 0]; \mathbb{R}^n) \to \mathbb{R}^n$ by* (11.9.1). *Show that f maps uniformly bounded subsets in $\mathbb{R}_+ \times C([-\tau, 0]; \mathbb{R}^n)$ into bounded subsets in $\mathbb{R}^n$ in the sense of Definition* 11.5.2.

Problem 11.1. *Consider the Cauchy problem*

$$\begin{cases} x'(t) = h\left(t, \displaystyle\int_{-\tau}^{0} x(t+\theta)\, d\theta\right), & t \in [0, T], \\ x(t) = \psi(t), & t \in [-\tau, 0], \end{cases} \tag{11.9.2}$$

where $h : \mathbb{R}_+ \times \mathbb{R}^n \to \mathbb{R}^n$. Prove the following variant of Theorem 11.5.1 :

Theorem 11.9.1. *Let $h : \mathbb{R}_+ \times \mathbb{R}^n \to \mathbb{R}^n$ be a continuous function and let $\psi \in C([-\tau, 0]; \mathbb{R}^n)$. A necessary and sufficient condition for a solution $x : [-\tau, T) \to \mathbb{R}^n$ of* (11.9.2) *to be continuable is that the graph of x, i.e.*

$$\text{graph } x = \{(t, x(t)) \in [-\tau, +\infty) \times \mathbb{R}^n;\ t \in [-\tau, T)\}$$

is included in a compact subset of $\mathbb{R} \times \mathbb{R}^n$.

Chapter 12

Auxiliary Results

12.1 Elements of Vector Analysis

For $n \in \mathbb{N}^*$ we denote by $\mathbb{R}^n$ the set of all n-tuples $x = (x_1, x_2, \ldots, x_n)$ of real numbers which, with respect to the operations "+" (internal composition law) and "·" (external composition law) defined by

$$x + y = (x_1, x_2, \ldots, x_n) + (y_1, y_2, \ldots, y_n) = (x_1 + y_1, x_2 + y_2, \ldots, x_n + y_n)$$

for every $x, y \in \mathbb{R}^n$, and respectively by

$$\lambda \cdot x = \lambda(x_1, x_2, \ldots, x_n) = (\lambda x_1, \lambda x_2, \ldots, \lambda x_n)$$

for every $\lambda \in \mathbb{R}$ and every $x \in \mathbb{R}^n$, is an n-dimensional vector space over $\mathbb{R}$. In all that follows, $\langle \cdot, \cdot \rangle_n : \mathbb{R}^n \times \mathbb{R}^n \to \mathbb{R}$ is the standard inner product on $\mathbb{R}^n$, i.e.

$$\langle x, y \rangle_n = \sum_{i=1}^{n} x_i y_i$$

and $\| \cdot \|_n : \mathbb{R}^n \to \mathbb{R}_+$ is the induced Euclidean norm, i.e.

$$\|x\|_n = \sqrt{\langle x, x \rangle_n} = \left(\sum_{i=1}^{n} x_i^2 \right)^{1/2}$$

for every $x, y \in \mathbb{R}^n$. Whenever no confusion may occur, we will cancel the index n, writing $\langle x, y \rangle$ instead of $\langle x, y \rangle_n$ and $\|x\|$ instead of $\|x\|_n$. Also, we will cancel "·" by simply writing λx instead of $\lambda \cdot x$.

Let $\mathcal{M}_{n \times m}(\mathbb{R})$ be the set of all $n \times m$-matrices with real elements. In many situations we will identify an element $\mathcal{A} \in \mathcal{M}_{n \times m}(\mathbb{R})$ by a linear operator (denoted for simplicity by the same symbol) $\mathcal{A} : \mathbb{R}^m \to \mathbb{R}^n$, defined by

$$\mathcal{A}(x) = \mathcal{A}x$$

for every $x \in \mathbb{R}^m$, where x is a column vector.

On the set $\mathcal{M}_{n\times m}(\mathbb{R})$, which is clearly an $n \times m$-dimensional vector space over $\mathbb{R}$, we define the function $\|\cdot\|_{\mathcal{M}}$, by

$$\|\mathcal{A}\|_{\mathcal{M}} = \sup\{\|\mathcal{A}x\|_n\,;\ x \in \mathbb{R}^m,\ \|x\|_m \leq 1\}$$

for every $\mathcal{A} \in \mathcal{M}_{n\times m}(\mathbb{R})$. The next simple lemma is particularly useful in what follows.

Lemma 12.1.1. *The function $\|\cdot\|_{\mathcal{M}} : \mathcal{M}_{n\times m}(\mathbb{R}) \to \mathbb{R}_+$ is a norm on $\mathcal{M}_{n\times m}(\mathbb{R})$, i.e. it satisfies:*

(N_1) *$\|\mathcal{A}\|_{\mathcal{M}} = 0$ if and only if $\mathcal{A}$ is the null matrix;*
(N_2) *$\|\lambda\mathcal{A}\|_{\mathcal{M}} = |\lambda|\|\mathcal{A}\|_{\mathcal{M}}$ for every $\lambda \in \mathbb{R}$ and every $\mathcal{A} \in \mathcal{M}_{n\times m}(\mathbb{R})$;*
(N_3) *$\|\mathcal{A}+\mathcal{B}\|_{\mathcal{M}} \leq \|\mathcal{A}\|_{\mathcal{M}} + \|\mathcal{B}\|_{\mathcal{M}}$ for every $\mathcal{A}, \mathcal{B} \in \mathcal{M}_{n\times m}(\mathbb{R})$.*

Moreover, for every $x \in \mathbb{R}^m$ and every $\mathcal{A} \in \mathcal{M}_{n\times m}(\mathbb{R})$, we have:

(N_4) *$\|\mathcal{A}x\|_n \leq \|\mathcal{A}\|_{\mathcal{M}}\|x\|_m$.*

In addition, for every $\mathcal{A} \in \mathcal{M}_{n\times m}(\mathbb{R})$ and every $\mathcal{B} \in \mathcal{M}_{m\times p}(\mathbb{R})$, we have

(N_5) *$\|\mathcal{A}\mathcal{B}\|_{\mathcal{M}} \leq \|\mathcal{A}\|_{\mathcal{M}}\|\mathcal{B}\|_{\mathcal{M}}$.*

Proof. Since (N_1) and (N_2) are obvious, it remains to prove (N_3) $\sim$ (N_5). In order to check (N_3), let us observe that the operator $\mathcal{A}+\mathcal{B}$ is continuous from $\mathbb{R}^m$ to $\mathbb{R}^n$. Since $\|\cdot\|_n$ is continuous on $\mathbb{R}^n$, it follows that the function $x \mapsto \|(\mathcal{A}+\mathcal{B})x\|_n$ is continuous on $\mathbb{R}^m$. Furthermore, the set $B(0,1) = \{x \in \mathbb{R}^m\,;\ \|x\|_m \leq 1\}$ is compact and then, according to Weierstrass' theorem, it follows that the function above attains its supremum on $B(0,1)$. So, there exists $\xi \in B(0,1)$ such that

$$\|\mathcal{A}+\mathcal{B}\|_{\mathcal{M}} = \sup\{\|(\mathcal{A}+\mathcal{B})x\|_n\,;\ \|x\|_m \leq 1\} = \|(\mathcal{A}+\mathcal{B})\xi\|_n.$$

But

$$\|(\mathcal{A}+\mathcal{B})\xi\|_n \leq \|\mathcal{A}\xi\|_n + \|\mathcal{B}\xi\|_n \leq \|\mathcal{A}\|_{\mathcal{M}} + \|\mathcal{B}\|_{\mathcal{M}},$$

which completes the proof of item (N_3).

In order to prove (N_4), let us observe that, for $x = 0$, it is obviously satisfied. Let then $x \in \mathbb{R}^m$, $x \neq 0$. We have $\|x\|_m^{-1}x \in B(0,1)$, and therefore

$$\|\mathcal{A}(\|x\|_m^{-1}x)\|_n = \|x\|_m^{-1}\|\mathcal{A}x\|_n \leq \|\mathcal{A}\|_{\mathcal{M}},$$

which shows that (N_4) holds true for every $x \in \mathbb{R}^m$. Finally, from (N_4), we deduce that, for every $\mathcal{A} \in \mathcal{M}_{n\times m}(\mathbb{R})$, every $\mathcal{B} \in \mathcal{M}_{m\times p}(\mathbb{R})$ and every $x \in \mathbb{R}^p$, we have

$$\|\mathcal{A}\mathcal{B}x\|_n \leq \|\mathcal{A}\|_{\mathcal{M}}\|\mathcal{B}x\|_m \leq \|\mathcal{A}\|_{\mathcal{M}}\|\mathcal{B}\|_{\mathcal{M}}\|x\|_p.$$

Passing to the supremum for $x \in B(0,1)$ in the inequality above, we deduce (N_5). The proof is complete. □

Corollary 12.1.1. *For every $\mathcal{A} \in \mathcal{M}_{n\times n}(\mathbb{R})$ and every $k \in \mathbb{N}$ we have*

(N_6) $\|\mathcal{A}^k\|_{\mathcal{M}} \leq \|\mathcal{A}\|_{\mathcal{M}}^k$.[1]

Proof. The conclusion follows by a simple inductive argument reiterating the property (N_5). □

Remark 12.1.1. The norm $\|\cdot\|_{\mathcal{M}}$, defined on $\mathcal{M}_{n\times m}(\mathbb{R})$, is equivalent on the isomorphic space $\mathbb{R}^{n\times m}$ with the Euclidean norm. More precisely, there exist two constants $k_1 > 0$ and $k_2 > 0$, such that

$$k_1\|\mathcal{A}\|_{\mathcal{M}} \leq \|\mathcal{A}\|_e \leq k_2\|\mathcal{A}\|_{\mathcal{M}} \tag{12.1.1}$$

for every $\mathcal{A} \in \mathcal{M}_{n\times m}(\mathbb{R})$, where

$$\|\mathcal{A}\|_e^2 = \sum_{i=1}^{n}\sum_{k=1}^{m} a_{ik}^2.$$

Indeed, if $e_1, e_2, \ldots, e_m$ are the vectors of the canonical basis in $\mathbb{R}^m$, we have

$$\sum_{i=1}^{n} a_{ik}^2 = \|\mathcal{A}e_k\|_e^2 \leq \|\mathcal{A}\|_{\mathcal{M}}^2 \quad \text{for} \quad k = 1, 2, \ldots, m.$$

Adding side by side these inequalities, we deduce

$$\|\mathcal{A}\|_e^2 \leq m\|\mathcal{A}\|_{\mathcal{M}}^2.$$

Hence, for $k_2 = \sqrt{m}$, the second inequality in (12.1.1) is satisfied. On the other hand, as we have already seen in the proof of Lemma 12.1.1, there exists $\xi \in \mathbb{R}^m$, with $\|\xi\|_m \leq 1$, such that

$$\|\mathcal{A}\|_{\mathcal{M}}^2 = \|\mathcal{A}\xi\|_n^2 = \sum_{i=1}^{n}\left(\sum_{j=1}^{m} a_{ij}\xi_j\right)^2.$$

Using the Cauchy-Schwarz inequality[2] in order to majorize the sum over j, we deduce

$$\|\mathcal{A}\|_{\mathcal{M}}^2 \leq \sum_{i=1}^{n}\left(\sum_{j=1}^{m} a_{ij}^2\right)\left(\sum_{j=1}^{m}\xi_j^2\right) = \sum_{i=1}^{n}\sum_{j=1}^{m} a_{ij}^2\|\xi\|_m^2 \leq \|\mathcal{A}\|_e^2.$$

---

[1] $\mathcal{A}^k$ denotes the k-times product of the matrix $\mathcal{A}$ by itself. For $k = 0$, by definition $\mathcal{A}^0 = \mathcal{I}$, where $\mathcal{I}$ is the unit matrix in $\mathcal{M}_{n\times n}(\mathbb{R})$.

[2] We recall that the Cauchy-Schwarz inequality asserts that, for every system of real numbers $x_1, x_2, \ldots, x_m$ and $y_1, y_2, \ldots, y_m$, we have

$$\left(\sum_{j=1}^{m} x_j y_j\right)^2 \leq \left(\sum_{j=1}^{m} x_j^2\right)\left(\sum_{j=1}^{m} y_j^2\right).$$

This inequality shows that, for $k_1 = 1$, the first inequality in (12.1.1) also holds true.

We emphasize that (12.1.1) expresses the invariance of boundedness, continuity, differentiability of functions with values in $\mathcal{M}_{n\times m}(\mathbb{R})$, with respect to the two norms $\|\cdot\|_{\mathcal{M}}$ and $\|\cdot\|_e$.

Now let D be a nonempty subset of $\mathbb{R}$ and let $f : D \to \mathbb{R}^n$ be a function,

$$f(t) = (f_1(t), f_2(t), \ldots, f_n(t))$$

for every $t \in D$. In all that follows, we shall say that f has a certain property if all the partial functions $f_1, f_2, \ldots, f_n$ have that property. For instance, we will say that f is *differentiable* at $t \in D$ if all the functions f_i, with $i = 1, 2, \ldots, n$, are differentiable at t. If f is differentiable at $t \in D$, we denote by $f'(t)$ its *derivative* at t, i.e.

$$f'(t) = (f_1'(t), f_2'(t), \ldots, f_n'(t)).$$

By analogy, we shall say that $f : [\,a,b\,] \to \mathbb{R}^n$ is *Riemann integrable* on $[\,a,b\,]$ if all the partial functions f_i with $i = 1, 2, \ldots, n$ are Riemann integrable over $[\,a,b\,]$. In the case in which f has this property, we denote by

$$\int_a^b f(t)\,dt = \left(\int_a^b f_1(t)\,dt, \int_a^b f_2(t)\,dt, \ldots, \int_a^b f_n(t)\,dt\right)$$

its Riemann integral over $[\,a,b\,]$. The next lemma extends to arbitrary $n \in \mathbb{N}^*$ two well-known results for $n = 1$.

Lemma 12.1.2. *Let $f : [\,a,b\,] \to \mathbb{R}^n$ and $g : [\,a,b\,] \to \mathbb{R}^n$.*

(i) *If f and g are differentiable at $t_0 \in [\,a,b\,]$, then $\langle f, g\rangle : [\,a,b\,] \to \mathbb{R}$ is differentiable at t_0, and*

$$\frac{d}{dt}\left(\langle f, g\rangle\right)(t_0) = \langle f'(t_0), g(t_0)\rangle + \langle f(t_0), g'(t_0)\rangle. \tag{12.1.2}$$

In particular, if f is differentiable at $t_0 \in [\,a,b\,]$, then the function $\|f\|^2 : [\,a,b\,] \to \mathbb{R}_+$ is differentiable at t_0, and

$$\frac{d}{dt}\left(\|f\|^2\right)(t_0) = 2\langle f'(t_0), f(t_0)\rangle. \tag{12.1.3}$$

(ii) *If $f : [\,a,b\,] \to \mathbb{R}^n$ is Riemann integrable on $[\,a,b\,]$, then the function $\|f\| : [\,a,b\,] \to \mathbb{R}_+$ is Riemann integrable on $[\,a,b\,]$, and*

$$\left\|\int_a^b f(t)\,dt\right\| \leq \int_a^b \|f(t)\|\,dt. \tag{12.1.4}$$

Proof. In order to prove (i), let us recall that $\langle f, g\rangle : [a,b] \to \mathbb{R}$ is defined by

$$(\langle f, g\rangle)(t) = \langle f(t), g(t)\rangle = \sum_{i=1}^{n} f_i(t)g_i(t)$$

for every $t \in [a,b]$. Since all functions f_i and g_i, with $i = 1, 2, \ldots, n$, are differentiable at t_0, by the relation above, it follows that $\langle f, g\rangle$ is differentiable at t_0. In addition, we have

$$\begin{aligned}\frac{d}{dt}(\langle f, g\rangle)(t_0) &= \sum_{i=1}^{n}[f_i'(t_0)g_i(t_0) + f_i(t_0)g_i'(t_0)]\\ &= \langle f'(t_0), g(t_0)\rangle + \langle f(t_0), g'(t_0)\rangle,\end{aligned}$$

which proves (12.1.2). Clearly, (12.1.3) follows from (12.1.2) by taking $f = g$.

In order to prove (ii), let us observe that the function $\|f\| : [a,b] \to \mathbb{R}_+$ is defined by

$$\|f\|(t) = \|f(t)\| = \left(\sum_{i=1}^{n} f_i^2(t)\right)^{1/2}$$

for every $t \in [a,b]$. Since all functions f_i, with $i = 1, 2, \ldots, n$, are Riemann integrable, it follows that $\|f\|$ has the same property. Furthermore, let us consider $\Delta : a = t_0 < t_1 < \cdots < t_k = b$, a partition of the interval $[a,b]$, and let $\xi_i \in [t_i, t_{i+1})$, $i = 0, 1, \ldots, k-1$, be arbitrary intermediate points. We have

$$\|\sigma_\Delta(f, \xi_i)\| = \left\|\sum_{i=0}^{k-1}(t_{i+1} - t_i)f(\xi_i)\right\| \le \sum_{i=0}^{k-1}(t_{i+1} - t_i)\|f(\xi_i)\| = \sigma_\Delta(\|f\|, \xi_i).$$

Taking a sequence of partitions of the interval $[a,b]$, with the sequence of norms tending to zero, and a sequence of corresponding intermediate points, and passing to the limit in the inequality above, we get (12.1.4). The proof is complete. □

In the next lemma, all vectors considered are column vectors.

Lemma 12.1.3. *Let us consider $f : [a,b] \to \mathbb{R}^n$, $\mathcal{A} \in \mathcal{M}_{n\times n}(\mathbb{R})$ and $\mathcal{B} : [a,b] \to \mathcal{M}_{n\times n}(\mathbb{R})$.*

(i) *If f is Riemann integrable over $[a,b]$, then $\mathcal{A}f$ is Riemann integrable over $[a,b]$ and*

$$\mathcal{A}\left(\int_a^b f(t)\,dt\right) = \int_a^b \mathcal{A}f(t)\,dt.$$

(ii) *If $\mathcal{B}$ is Riemann integrable over $[a,b]$, then $\mathcal{B}^*$ is Riemann integrable over $[a,b]$ and*

$$\left(\int_a^b \mathcal{B}(t)\,dt\right)^* = \int_a^b \mathcal{B}^*(t)\,dt,$$

where $\mathcal{B}^*$ denotes the adjoint of the matrix $\mathcal{B}$.

(iii) *If $\mathcal{B}$ is Riemann integrable over $[a,b]$ and $x, y \in \mathbb{R}^n$, then $\langle \mathcal{B}(\cdot)x, y\rangle$ is Riemann integrable over $[a,b]$ and*

$$\left\langle \left(\int_a^b \mathcal{B}(t)\,dt\right) x, y\right\rangle = \int_a^b \langle \mathcal{B}(t)x, y\rangle dt.$$

Proof. Let $\Delta : a = t_0 < t_1 < \cdots < t_k = b$ be a partition of the interval $[a,b]$ and let $\xi_i \in [t_i, t_{i+1})$, $i = 0, 1, \ldots, k-1$, be arbitrary intermediate points. We have

$$\mathcal{A}(\sigma_\Delta(f, \xi_i)) = \sum_{i=0}^{k-1} (t_{i+1} - t_i)\mathcal{A}f(\xi_i) = \sigma_\Delta(\mathcal{A}f, \xi_i),$$

$$(\sigma_\Delta(\mathcal{B}, \xi_i))^* = \sum_{i=0}^{k-1} (t_{i+1} - t_i)\mathcal{B}^*(\xi_i) = \sigma_\Delta(\mathcal{B}^*, \xi_i)$$

and

$$\langle \sigma_\Delta(\mathcal{B}, \xi_i)x, y\rangle = \sum_{i=0}^{k-1} (t_{i+1} - t_i)\langle \mathcal{B}(\xi_i)x, y\rangle = \sigma_\Delta(\langle \mathcal{B}(\cdot)x, y\rangle, \xi_i).$$

Taking a sequence of partitions of the interval $[a,b]$, with the sequence of norms tending to zero, and a sequence of corresponding intermediate points, and passing to the limit in the equalities above, we get (i), (ii) and (iii). The proof is complete. □

12.2 Compactness in $C([a,b];\mathbb{R}^n)$

In this section we shall prove an analogue of Cesàro's lemma referring to sequences of continuous functions from $[a,b]$ to $\mathbb{R}^n$. We denote by $C([a,b];\mathbb{R}^n)$ the space of all continuous functions from $[a,b]$ to $\mathbb{R}^n$ and we endow this space with the *uniform convergence topology* on $[a,b]$. We recall that a sequence $(f_m)_{m\in\mathbb{N}^*}$ is convergent to f in the uniform convergence topology if $\lim_{m\to\infty} f_m(t) = f(t)$ uniformly for $t \in [a,b]$. In turn, this is equivalent to $\lim_{m\to\infty} \|f_m - f\|_\infty = 0$, where $\|\cdot\|_\infty : C([a,b];\mathbb{R}^n) \to \mathbb{R}_+$, called the *supremum norm*, is defined by $\|g\|_\infty = \sup_{t\in[\,a,b\,]} \|g(t)\|$ for each $g \in C([\,a,b\,];\mathbb{R}^n)$.

Definition 12.2.1. A family $\mathcal{F}$ in $C([\,a,b\,];\mathbb{R}^n)$ is *relatively compact* if every sequence in $\mathcal{F}$ has at least one subsequence which is uniformly convergent on $[\,a,b\,]$. If, in addition, the limit of each uniform convergent sequence in $\mathcal{F}$ belongs to $\mathcal{F}$, the family is called *compact.*

Definition 12.2.2. A family $\mathcal{F}$ in $C([\,a,b\,];\mathbb{R}^n)$ is *equicontinuous* at a point $t \in [\,a,b\,]$ if for every $\varepsilon > 0$ there exists $\delta(\varepsilon,t) > 0$ such that, for every $s \in [\,a,b\,]$ with $|t-s| < \delta(\varepsilon,t)$ we have

$$\|f(t) - f(s)\| \le \varepsilon,$$

for all $f \in \mathcal{F}$.

A family $\mathcal{F}$ is *equicontinuous* on $[\,a,b\,]$ if it is equicontinuous at each $t \in [\,a,b\,]$, in the sense described above.

A family $\mathcal{F}$ is *uniformly equicontinuous* on $[\,a,b\,]$ if it is equicontinuous on $[\,a,b\,]$ and $\delta(\varepsilon,t)$ can be chosen independent of $t \in [\,a,b\,]$.

Lemma 12.2.1. *A family $\mathcal{F}$ in $C([\,a,b\,];\mathbb{R}^n)$ is equicontinuous on $[\,a,b\,]$ if and only if it is uniformly equicontinuous on $[\,a,b\,]$.*

Proof. Clearly, each family which is uniformly equicontinuous on $[\,a,b\,]$ is equicontinuous on $[\,a,b\,]$. In order to prove the converse of this statement, we shall proceed for contradiction. So, let us assume that there exists a family $\mathcal{F}$ which is equicontinuous on $[\,a,b\,]$, but is not uniformly equicontinuous on $[\,a,b\,]$. This means that there exists $\varepsilon > 0$ such that, for every $\delta > 0$, there exist $t_\delta, s_\delta \in [\,a,b\,]$ and $f_\delta \in \mathcal{F}$, with $|t_\delta - s_\delta| \le \delta$ and $\|f_\delta(t_\delta) - f_\delta(s_\delta)\| \ge \varepsilon$. Taking $\delta = 1/m$ with $m \in \mathbb{N}^*$, and denoting by $t_m = t_\delta$, $s_m = s_\delta$, and $f_m = f_\delta$, we have

$$\begin{cases} |t_m - s_m| \le \dfrac{1}{m} \\ \|f_m(t_m) - f_m(s_m)\| \ge \varepsilon \end{cases}$$

for every $m \in \mathbb{N}^*$. Since the sequence $(t_m)_{m\in\mathbb{N}^*}$ is bounded, from Cesàro's lemma, it follows that it has one subsequence, convergent to an element $t \in [\,a,b\,]$. We denote by $(t_{m_p})_{p\in\mathbb{N}^*}$ this subsequence, and we observe that, from the first inequality above, it follows that $(s_{m_p})_{p\in\mathbb{N}^*}$ is convergent to t too. On the other hand, the family $\mathcal{F}$ is equicontinuous at t and therefore, for $\varepsilon > 0$ as above, there exists $\delta(\varepsilon,t) > 0$ such that, for every $s \in [\,a,b\,]$ with $|s-t| \le \delta(\varepsilon,t)$ and every $f \in \mathcal{F}$, we have $\|f(s) - f(t)\| \le \varepsilon/3$. Since both $(t_{m_p})_{p\in\mathbb{N}^*}$ and $(s_{m_p})_{p\in\mathbb{N}^*}$ converge to t, for $p \in \mathbb{N}^*$ large enough, we have $|t_{m_p} - t| \leq \delta(\varepsilon, t)$ and $|s_{m_p} - t| \leq \delta(\varepsilon, t)$. Then

$$\varepsilon \leq \|f_{m_p}(t_{m_p}) - f_{m_p}(s_{m_p})\|$$
$$\leq \|f_{m_p}(t_{m_p}) - f_{m_p}(t)\| + \|f_{m_p}(t) - f_{m_p}(s_{m_p})\| \leq \frac{2\varepsilon}{3}.$$

This contradiction can be eliminated only if $\mathcal{F}$ is uniformly equicontinuous on $[a, b]$. The proof is complete. □

Definition 12.2.3. A subset $\mathcal{F}$ is *uniformly bounded* on $[a, b]$ if there exists $M > 0$ such that for every $f \in \mathcal{F}$ and every $t \in [a, b]$, we have $\|f(t)\| \leq M$.

Theorem 12.2.1. (Ascoli-Arzelà) *A family $\mathcal{F}$ in $C([a, b]; \mathbb{R}^n)$ is relatively compact if and only if:*

(i) *$\mathcal{F}$ is equicontinuous on $[a, b]$;*
(ii) *$\mathcal{F}$ is uniformly bounded on $[a, b]$.*

Proof. We begin with the necessity of the conditions (i) and (ii). To this aim, let $\mathcal{F}$ be a relatively compact subset of $C([a, b]; \mathbb{R}^n)$ and let $\varepsilon > 0$. Let us assume for contradiction that $\mathcal{F}$ is not uniformly bounded. Then there would exist one sequence $(t_m)_{m\in\mathbb{N}}$ in $[a, b]$ and one sequence $(f_m)_{m\in\mathbb{N}}$ in $\mathcal{F}$ such that $\|f_m(t_m)\| \geq m$. Since $\mathcal{F}$ is relatively compact, the sequence $(f_m)_{m\in\mathbb{N}}$ has at least one subsequence $(f_{m_p})_{p\in\mathbb{N}}$ which is uniformly convergent to a continuous function f. Then, there exists $p_1 \in \mathbb{N}$ such that $\|f_{m_p}(t) - f(t)\| \leq 1$ for every $p \geq p_1$ and every $t \in [a, b]$. On the other hand, by virtue of Weierstrass' theorem, it follows that there exists $M > 0$ such that $\|f(t)\| \leq M$ for every $t \in [a, b]$. So, we have

$$m_p \leq \|f_{m_p}(t_{m_p})\| \leq \|f_{m_p}(t_{m_p}) - f(t_{m_p})\| + \|f(t_{m_p})\| \leq 1 + M$$

for every $p \geq p_1$. But this inequality contradicts $\lim_{p\to\infty} m_p = +\infty$. This contradiction can be eliminated only if $\mathcal{F}$ is uniformly bounded. Hence the condition (ii) is necessary for the relative compactness of the set $\mathcal{F}$.

Next, let us assume for contradiction that there exists a relatively compact set $\mathcal{F}$ which is not equicontinuous, or equivalently, is not uniformly equicontinuous on $[a, b]$. See Lemma 12.2.1. This means that there exists $\varepsilon > 0$ with the property that, for every $\delta > 0$ there exist $t_\delta, s_\delta \in [a, b]$ and $f_\delta \in \mathcal{F}$ with $|t_\delta - s_\delta| \leq \delta$ and $\|f_\delta(t_\delta) - f_\delta(s_\delta)\| \geq \varepsilon$. Taking $\delta = 1/m$ with $m \in \mathbb{N}^*$, and denoting by $t_m = t_\delta$, $s_m = s_\delta$, and $f_m = f_\delta$, we have

$$\begin{cases} |t_m - s_m| \leq \dfrac{1}{m} \\ \|f_m(t_m) - f_m(s_m)\| \geq \varepsilon \end{cases}$$

for every $m \in \mathbb{N}^*$. Since $\mathcal{F}$ is relatively compact there exist $f \in C([a,b];\mathbb{R}^n)$ and one subsequence, $(f_{m_p})_{p\in\mathbb{N}^*}$ of $(f_m)_{m\in\mathbb{N}^*}$ which is uniformly convergent on $[a,b]$ to a certain function f. It then follows that there exists $p_1(\varepsilon) \in \mathbb{N}^*$ such that, for every $p \geq p_1(\varepsilon)$, we have $\|f_{m_p}(t) - f(t)\| \leq \varepsilon/4$. Since f is uniformly continuous on $[a,b]$ and $\lim_{p\to\infty} |t_{m_p} - s_{m_p}| = 0$, there exists $p_2(\varepsilon) > 0$ such that, for every $p \geq p_2(\varepsilon)$, we have $\|f(t_{m_p}) - f(s_{m_p})\| \leq \varepsilon/4$. Accordingly, for every $p \geq \max\{p_1(\varepsilon), p_2(\varepsilon)\}$

$$\varepsilon \leq \|f_{m_p}(t_{m_p}) - f_{m_p}(s_{m_p})\| \leq \|f_{m_p}(t_{m_p}) - f(t_{m_p})\|$$

$$+\|f(t_{m_p}) - f(s_{m_p})\| + \|f(s_{m_p}) - f_{m_p}(s_{m_p})\| \leq \frac{3\varepsilon}{4}.$$

This contradiction can be eliminated only if $\mathcal{F}$ is uniformly equicontinuous on $[a,b]$. Hence (i) is also a necessary condition of relative compactness for the set $\mathcal{F}$. The proof of the necessity part is complete.

Sufficiency. The proof of the sufficiency is based on Cantor's diagonal procedure. More precisely, let $\mathcal{F}$ be a set in $C([a,b];\mathbb{R}^n)$ which satisfies (i) and (ii). Let $\{t_m;\ m \in \mathbb{N}^*\}$ be a countable dense subset of $[a,b]$. For instance, this could be the set of all rational numbers in $[a,b]$. Let $(f_m)_{m\in\mathbb{N}^*}$ be a sequence of elements in $\mathcal{F}$, and let us observe that the sequence $(f_m(t_1))_{m\in\mathbb{N}^*}$ is bounded in $\mathbb{R}^n$, because $\mathcal{F}$ satisfies (ii). From Cesàro's lemma, it follows that it has at least one subsequence which converges to a certain element $f(t_1) \in \mathbb{R}^n$. We fix and denote by $(f_{m_1}(t_1))_{m\in\mathbb{N}^*}$ such a subsequence. Let us consider now the sequence $(f_{m_1}(t_2))_{m\in\mathbb{N}^*}$, and let us observe that, again from Cesàro's lemma, this has at least one subsequence, $(f_{m_2}(t_2))_{m\in\mathbb{N}^*}$, convergent to an element $f(t_2) \in \mathbb{R}^n$. Moreover, as $(f_{m_2}(t_1))_{m\in\mathbb{N}^*}$ is a subsequence of $(f_{m_1}(t_1))_{m\in\mathbb{N}^*}$, we have $\lim_m f_{m_2}(t_1) = f(t_1)$. Repeating this procedure, we get a family of subsequences $(f_{m_p})_{m\in\mathbb{N}^*}$ of the sequence $(f_m)_{m\in\mathbb{N}^*}$ and one sequence $(f(t_p))_{p\in\mathbb{N}^*}$ in $\mathbb{R}^n$, with the property

$$\lim_m f_{m_p}(t_i) = f(t_i)$$

for every $p \in \mathbb{N}^*$ and $i = 1,2,\ldots,p$. Let us consider now the diagonal sequence $(f_{m_m})_{m\in\mathbb{N}^*}$ and let us observe that, in accordance with the choice of the subsequences above, we have

$$\lim_m f_{m_m}(t_p) = f(t_p)$$

for every $p \in \mathbb{N}^*$. In order to complete the proof, it suffices to show that $(f_{m_m})_{m\in\mathbb{N}^*}$ satisfies Cauchy's condition for the existence of the uniform limit on $[a,b]$. Hence, we will show that, for every $\varepsilon > 0$ there exists

$m = m(\varepsilon) \in \mathbb{N}^*$ such that, for every $m \geq m(\varepsilon)$, every $p \geq m(\varepsilon)$ and every $t \in [a, b]$, we have

$$\|f_{m_m}(t) - f_{p_p}(t)\| \leq \varepsilon.$$

Let $\varepsilon > 0$. Since the family $\mathcal{F}$ is equicontinuous on $[a, b]$, it is uniformly equicontinuous on $[a, b]$. See Lemma 12.2.1. Then, there exists $\delta(\varepsilon) > 0$ such that, for every $t, s \in [a, b]$ with $|t - s| \leq \delta(\varepsilon)$ and every $f \in \mathcal{F}$, we have

$$\|f(t) - f(s)\| \leq \frac{\varepsilon}{3}.$$

Since the interval $[a, b]$ is compact and the set $\{t_m;\ m \in \mathbb{N}\}$ is dense, it contains a finite subfamily $\{t_i;\ i = 1, 2, \ldots, p(\varepsilon)\}$ such that for every $t \in [a, b]$ there exists $i \in \{1, 2, \ldots, p(\varepsilon)\}$ with the property

$$|t - t_i| \leq \delta(\varepsilon).$$

As the family of convergent sequences $\{(f_{m_m}(t_i))_{m \in \mathbb{N}^*};\ i = 1, 2, \ldots, p(\varepsilon)\}$ is finite, there exists a natural number $m(\varepsilon)$ such that, for every $m \geq m(\varepsilon)$ and every $i = 1, 2, \ldots, p(\varepsilon)$, we have

$$\|f_{m_m}(t_i) - f(t_i)\| \leq \frac{\varepsilon}{3}.$$

It then follows

$$\|f_{m_m}(t) - f_{p_p}(t)\|$$

$$\leq \|f_{m_m}(t) - f_{m_m}(t_i)\| + \|f_{m_m}(t_i) - f_{p_p}(t_i)\| + \|f_{p_p}(t_i) - f_{p_p}(t)\| \leq \varepsilon$$

for every $m \geq m(\varepsilon)$ and every $p \geq m(\varepsilon)$. Hence $(f_{m_m})_{m \in \mathbb{N}^*}$ satisfies Cauchy's condition of uniform convergence on $[a, b]$. The proof is complete. □

Several useful consequences of Ascoli-Arzelà's theorem are listed below.

Corollary 12.2.1. *Let $\mathcal{F}$ be a relatively compact subset of $C([a, b]; \mathbb{R}^n)$. Then*

$$\mathcal{F}([a, b]) = \{f(t);\ f \in \mathcal{F},\ t \in [a, b]\} \tag{12.2.1}$$

is relatively compact in $\mathbb{R}^n$.

Proof. From *(ii)* in Theorem 12.2.1, it follows that $\mathcal{F}([a, b])$ is bounded. Since a subset of $\mathbb{R}^n$ is relatively compact if and only if it is bounded this completes the proof. □

Corollary 12.2.2. *Let U be nonempty and closed subset of $\mathbb{R}^n$, let $g : [a,b] \times U \to \mathbb{R}^n$ be a continuous function,*

$$\mathcal{U} = \{u; u \in C([a,b]; \mathbb{R}^n), u(t) \in U \text{ for } t \in [a,b]\}$$

and let $G : \mathcal{U} \to C([a,b]; \mathbb{R}^n)$ be the superposition operator associated to the function g, i.e.

$$G(u)(t) = g(t, u(t))$$

for every $u \in \mathcal{U}$ and $t \in [a,b]$. Then G is continuous from $\mathcal{U}$ in $C([a,b]; \mathbb{R}^n)$, both the domain and the range being endowed with the uniform convergence topology.

Proof. Let $(u_m)_{m \in \mathbb{N}}$ be a sequence in $\mathcal{U}$ which converges uniformly on $[a,b]$ to $u \in \mathcal{U}$. Obviously $\{u_m; m \in \mathbb{N}\}$ is relatively compact in $C([a,b]; \mathbb{R}^n)$. Then, according to Corollary 12.2.1, the set

$$K = \overline{\{u_m(t)\,;\ m \in \mathbb{N},\ t \in [a,b]\}} \subset U$$

is compact in $\mathbb{R}^n$. As a consequence, the restriction of g to $[a,b] \times K$ is uniformly continuous, i.e. for each $\varepsilon > 0$ there exists $\delta(\varepsilon) > 0$, such that, for each $(t,v), (s,w) \in [a,b] \times K$ with $|t-s| + \|v-w\| \leq \delta(\varepsilon)$, we have $\|g(t,v) - g(s,w)\| \leq \varepsilon$. Since $(u_m)_{m \in \mathbb{N}}$ converges uniformly on $[a,b]$ to u, there exists $m(\varepsilon) \in \mathbb{N}$ such that, for each $m \in \mathbb{N}$, $m \geq m(\varepsilon)$, and each $t \in [a,b]$, we have $\|u_m(t) - u(t)\| \leq \delta(\varepsilon)$. So, for each $m \in \mathbb{N}$, $m \geq m(\varepsilon)$, and each $t \in [a,b]$, we have $\|g(t, u_m(t)) - g(t, u(t))\| \leq \varepsilon$, thereby completing the proof. □

Some extension of the results in this sections to continuous functions with values in general Banach spaces can be found in [Vrabie (1995)] and [Vrabie (2003)].

12.3 Banach Spaces

Let X be a real vector space.

Definition 12.3.1. By definition, a *norm* on X is a function $\|\cdot\| : X \to \mathbb{R}_+$ satisfying:

(N_1) $\|x\| = 0$ if and only if $x = 0$;
(N_2) $\|\lambda x\| = |\lambda|\|x\|$ for every $\lambda \in \mathbb{R}$ and every $x \in X$;
(N_3) $\|x + y\| \leq \|x\| + \|y\|$ for every $x, y \in X$.

A pair $(X, \|\cdot\|)$, where X is real vector and $\|\cdot\|$ is a norm on X is called *normed real space.*

We introduce below the concept of convergent sequence and that of fundamental or Cauchy sequence in a normed real space $(X, \|\cdot\|)$.

Definition 12.3.2. A sequence $(x_k)_{k\in\mathbb{N}}$ is said to be *convergent* to $x \in X$ if $\lim_{k\to\infty} \|x_k - x\| = 0$ in $\mathbb{R}$.

One may easily verify that if $\lim_{k\to\infty} \|x_k - x\| = \lim_{k\to\infty} \|x_k - y\| = 0$ then $x = y$, i.e. the limit of a convergent sequence is unique.

Definition 12.3.3. A sequence $(x_k)_{k\in\mathbb{N}}$ is said to be *fundamental or Cauchy* if for each $\varepsilon > 0$ there exists $k_\varepsilon \in \mathbb{N}$ such that for each $k, p \in \mathbb{N}$ with $k \geq k_\varepsilon$ we have $\|x_{k+p} - x_k\| \leq \varepsilon$.

Definition 12.3.4. A normed real space $(X, \|\cdot\|)$ is called *real Banach space* or *complete real vector space* if every fundamental sequence $(x_k)_{k\in\mathbb{N}}$ in X is convergent to some element $x \in X$.

Remark 12.3.1. We recall that:

(*i*) The classical Cauchy theorem for real sequences stating that "*a real sequence is convergent if and only if it is fundamental*" shows that $(\mathbb{R}, |\cdot|)$, where $|\cdot|$ is the modulus function, is a real Banach space.

(*ii*) Consequently, the n-dimensional real vector space $\mathbb{R}^n$, endowed with the usual Euclidean norm – see Section 12.1 – is a real Banach space.

(*iii*) From Lemma 12.1.1, we know that $(\mathcal{M}_{n\times m}(\mathbb{R}), \|\cdot\|_{\mathcal{M}})$ is a real vector space. In fact, by observing that a sequence of matrices in $\mathcal{M}_{n\times m}(\mathbb{R})$ is convergent (fundamental) if and only if the corresponding sequences of real entries are convergent (fundamental), one may easily verify that $(\mathcal{M}_{n\times m}(\mathbb{R}), \|\cdot\|_{\mathcal{M}})$ is a real Banach space.

(*iv*) The vector space $(C([\,a, b\,]; \mathbb{R}), \|\cdot\|_\infty)$, where $\|\cdot\|_\infty$ is defined as in Section 12.2, is a real Banach space. This follows from the observation that a sequence in $C([\,a, b\,]; \mathbb{R})$ is convergent (fundamental) if and only if it is uniformly convergent (uniformly fundamental) on $[\,a, b\,]$.

(*v*) On the space $C_b([\,a, +\infty); \mathbb{R}^n)$ of all bounded continuous functions from $[\,a, +\infty)$ to $\mathbb{R}^n$, we define $\|\cdot\|_{C_b([\,a,+\infty);\mathbb{R}^n)} : C_b([\,a, +\infty); \mathbb{R}^n) \to \mathbb{R}_+$ by

$$\|x\|_{C_b([\,a,+\infty);\mathbb{R}^n)} := \sup\{\|x\|;\ t \in [\,a, +\infty)\},$$

for each $x \in C_b([\,a, +\infty); \mathbb{R}^n)$. Then $\|\cdot\|_{C_b([\,a,+\infty);\mathbb{R}^n)}$ is a norm. In addition, with respect to this norm, $\big(C_b([\,a, +\infty); \mathbb{R}^n), \|\cdot\|_{C_b([\,a,+\infty);\mathbb{R}^n)}\big)$ is a real Banach space.

Remark 12.3.2. In this terminology, Ascoli-Arzelà's Theorem 12.2.1 can be reformulated as:

Theorem 12.3.1. (Ascoli-Arzelà) *Let $\mathcal{F}$ be a family of functions in the Banach space $(C([a,b];\mathbb{R}^n), \|\cdot\|_\infty)$. A necessary and sufficient condition for each sequence in $\mathcal{F}$ to have at least one convergent subsequence in $C([a,b];\mathbb{R}^n)$ is that:*

(i) *$\mathcal{F}$ is equicontinuous on $[a,b]$;*
(ii) *$\mathcal{F}$ is uniformly bounded on $[a,b]$.*

12.4 The Projection of a Point on a Convex Set

Let K be a nonempty closed and convex subset of $\mathbb{R}^n$ and $x \in \mathbb{R}^n$. In what follows, we shall prove the existence of a unique point $\xi \in K$ with the property that $\|x - \xi\|$ equals the distance between x to K. More precisely we have:

Lemma 12.4.1. *Let K be a nonempty, closed and convex subset of $\mathbb{R}^n$. Then, for every $x \in \mathbb{R}^n$ there exists one and only one element $\xi \in K$ such that*

$$\operatorname{dist}(x, K) = \|x - \xi\|. \tag{12.4.1}$$

In addition, $\xi \in K$ enjoys the property (12.4.1) if and only if

$$\langle x - \xi, \xi - u \rangle \geq 0 \tag{12.4.2}$$

for every $u \in K$.

Proof. Since $\operatorname{dist}(x, K) = \inf\{\|x-y\|;\ y \in K\}$, there exists one sequence $(x_k)_{k\in\mathbb{N}^*}$, of elements in K, with the property

$$\operatorname{dist}(x, K) \leq \|x - x_k\| \leq \operatorname{dist}(x, K) + \frac{1}{k}$$

for every $k \in \mathbb{N}^*$. As $\|x_k\| \leq \|x_k - x\| + \|x\| \leq \operatorname{dist}(x, K) + \|x\| + \frac{1}{k}$ for every $k \in \mathbb{N}^*$, it follows that $(x_k)_{k\in\mathbb{N}}$ is bounded. Since K is closed, from Cesàro's lemma, it follows that there exists $\xi \in K$ such that, on one subsequence at least, we have

$$\lim_{k\to\infty} x_k = \xi.$$

Obviously $\operatorname{dist}(x, K) = \|x - \xi\|$, which proves the existence part. In order to prove the uniqueness of the point ξ, let $\eta \in K$ be another point with the property $\operatorname{dist}(x, K) = \|x - \eta\|$. Since K is convex, $\mu = \frac{1}{2}\xi + \frac{1}{2}\eta \in K$ and

$$\|x - \mu\| \leq \frac{1}{2}(\|x - \xi\| + \|x - \eta\|) = \operatorname{dist}(x, K).$$

We denote by $d = \operatorname{dist}(x, K)$. Let us observe that

$$\frac{1}{4}\|y + z\|^2 + \frac{1}{4}\|y - z\|^2 = \frac{1}{2}(\|y\|^2 + \|z\|^2)$$

for every $y, z \in \mathbb{R}^n$. Taking $y = \xi - x$ and $z = \eta - x$ in the equality above, we get

$$\|\mu - x\|^2 + \frac{1}{4}\|\xi - \eta\|^2 = \frac{1}{2}(\|\xi - x\|^2 + \|\eta - x\|^2) = d^2.$$

Since $\|\mu - x\|^2 = d^2$, by virtue of the preceding relation, we deduce that

$$d^2 + \frac{1}{4}\|\xi - \eta\|^2 = d^2,$$

which proves that $\xi = \eta$.

In what follows, we will show that (12.4.1) implies (12.4.2). From the definition of the point ξ, we have

$$\|x - \xi\|^2 \leq \|x - v\|^2$$

for every $v \in K$. This inequality is equivalent to

$$\langle \xi - v, -\xi + 2x - v\rangle \geq 0$$

for every $v \in K$. Let $u \in K$, $\lambda \in (0, 1)$ and $v = \lambda\xi + (1 - \lambda)u$. Since K is convex, it follows that $v \in K$, and therefore it satisfies the preceding inequality, i.e.

$$\langle (1 - \lambda)(\xi - u), -(1 + \lambda)\xi - (1 - \lambda)u + 2x\rangle \geq 0.$$

Dividing by $1 - \lambda$ and passing to the limit for λ tending to 1, we get

$$\langle \xi - u, -2\xi + 2x\rangle \geq 0,$$

inequality which proves (12.4.2). □

Definition 12.4.1. Let K be a nonempty, convex and closed subset of $\mathbb{R}^n$ and let $x \in \mathbb{R}^n$. The vector $\mathcal{P}_K(x) = \xi$ in K with the property (12.4.1) is called *the projection of the vector x on the set K*. The function $\mathcal{P}_K : \mathbb{R}^n \to K$, defined by $\mathcal{P}_K(x) = \xi$, where ξ is the projection of the vector x on the set K, is called *the projection operator on the set K*.

Lemma 12.4.2. *The projection operator on the set K is non-expansive, i.e.*

$$\|\mathcal{P}_K(x) - \mathcal{P}_K(y)\| \leq \|x - y\| \tag{12.4.3}$$

for every $x, y \in \mathbb{R}^n$. In addition, the operator $\mathcal{P}_K - \mathcal{I}$ is dissipative, i.e.

$$\langle (\mathcal{P}_K(x) - x) - (\mathcal{P}_K(y) - y), x - y \rangle \leq 0 \tag{12.4.4}$$

for every $x, y \in \mathbb{R}^n$.

Proof. Take $\xi = \mathcal{P}_K(x)$ and $u = \mathcal{P}_K(y)$ in (12.4.2). We have

$$\langle x - \mathcal{P}_K(x), \mathcal{P}_K(x) - \mathcal{P}_K(y) \rangle \geq 0.$$

Similarly, we get

$$\langle y - \mathcal{P}_K(y), \mathcal{P}_K(y) - \mathcal{P}_K(x) \rangle = \langle \mathcal{P}_K(y) - y, \mathcal{P}_K(x) - \mathcal{P}_K(y) \rangle \geq 0$$

for every $x, y \in \mathbb{R}^n$. Adding side by side the two inequalities, we deduce

$$\|\mathcal{P}_K(x) - \mathcal{P}_K(y)\|^2 \leq \langle x - y, \mathcal{P}_K(x) - \mathcal{P}_K(y) \rangle$$

for every $x, y \in \mathbb{R}^n$. From this relation and Cauchy-Schwarz inequality, it follows

$$\|\mathcal{P}_K(x) - \mathcal{P}_K(y)\|^2 \leq \|\mathcal{P}_K(x) - \mathcal{P}_K(y)\| \|x - y\|,$$

for every $x, y \in \mathbb{R}^n$, which proves (12.4.3).

Finally, let us observe that (12.4.4) is equivalent to

$$\langle \mathcal{P}_K(x) - \mathcal{P}_K(y), x - y \rangle \leq \|x - y\|^2$$

for every $x, y \in \mathbb{R}^n$. But the latter relation follows from the Cauchy-Schwarz inequality and (12.4.3), and this completes the proof. □

12.5 The Laplace Transform

The aim of this section is to present the fundamentals of a very important tool in solving Cauchy problems and Volterra integral equations, i.e. the Laplace Transform. The main idea is to introduce an isomorphism between two given function spaces which transforms the differentiation into the multiplication with the argument. So, a linear differential equation with constant coefficients is transformed into a polynomial equation which, once solved, produces the image of the solutions of the differential equation.

Let us denote by $\mathcal{O}$ the vector space of all functions $f : \mathbb{R} \to \mathbb{C}$ satisfying:

($\mathcal{O}_1$) $f(t) = 0$ for each $t \leq 0$;

($\mathcal{O}_2$) f is piecewise continuous and piecewise differentiable on $\mathbb{R}$;
($\mathcal{O}_3$) there exist $M > 0$ and $\omega \in \mathbb{R}$ such that $|f(t)| \leq Me^{\omega t}$ for each $t > 0$.

As we can easily see from the definition of the space $\mathcal{O}$, the value of f at $t = 0$ is of no interest and so, in many cases we will work with functions which are not defined at $t = 0$.

Definition 12.5.1. Let $f \in \mathcal{O}$. The number ω_f, defined by

$$\omega_f = \inf\{\omega \in \mathbb{R};\ \text{there exists } M > 0,\ |f(t)| \leq Me^{\omega t} \text{ for all } t > 0\},$$

is called the *growth index* of the function f.

Remark 12.5.1. For each $f \in \mathcal{O}$ and each $\omega > \omega_f$, there exists $M > 0$ such that

$$|f(t)| \leq Me^{\omega t}$$

for all $t > 0$.

We recall for easy reference that the Heaviside function or the *unit function*, $\theta : \mathbb{R} :\to \mathbb{R}$ is defined by

$$\theta(t) = \begin{cases} 0 \text{ for } t < 0 \\ 1 \text{ for } t \geq 0. \end{cases}$$

Remark 12.5.2. If $f : \mathbb{R} \to \mathbb{R}$ is any polynomial, exponential or any trigonometric function, then $\theta \cdot f \in \mathcal{O}$.

Throughout, we denote by $\mathcal{F}(\mathbb{C};\mathbb{C})$ the space of all functions $F : \mathbb{C} \to \mathbb{C}$. Also, if $p = \alpha + i\beta \in \mathbb{C}$, we denote by $\Re(p) = \alpha$ and by $\Im(p) = \beta$.

Proposition 12.5.1. *The set $\mathcal{O}$ defined above is an algebra with unit over $\mathbb{C}$.*

Proof. It is a simple exercise to show that for each $f, g \in \mathcal{O}$ and each $\lambda, \mu \in \mathbb{C}$, we have

(i) $\lambda f + \mu g \in \mathcal{O}$;
(ii) $f \cdot g \in \mathcal{O}$;
(iii) $\theta \cdot f = f$.

□

Proposition 12.5.2. *Let $f \in \mathcal{O}$. Then, for each $p = \alpha + i\beta \in \mathbb{C}$ with $\Re(p) > \omega_f$, the improper integral*

$$\int_0^\infty e^{-pt} f(t)\, dt$$

is absolutely convergent.

Proof. Let $0 < a < b$ and $p = \alpha + i\beta \in \mathbb{C}$ with $\alpha > \omega_f$. Let $\omega > \omega_f$ be such that $\alpha > \omega$. By Remark 12.5.1, we know there exists $M > 0$ such that

$$|f(t)| \leq Me^{\omega t}$$

for each $t \in \mathbb{R}$. Now, recalling that

$$e^{-pt} = e^{-\alpha t - i\beta t} = e^{-\alpha t} e^{-i\beta t} = e^{-\alpha t}(\cos \beta t - i \sin \beta t),$$

and $|\cos \beta t - i \sin \beta t| = \sqrt{\cos^2 \beta t + \sin^2 \beta t} = 1$, by virtue of $(\mathcal{O}_2)$, we get

$$\int_a^b |e^{-pt}||f(t)|\, dt \leq \int_a^b e^{-\alpha t} M e^{\omega t}\, dt = M \int_a^b e^{(-\alpha+\omega)t}\, dt.$$

Since, by hypothesis, $\alpha > \omega$, we have

$$\int_a^b e^{-\alpha t} M e^{\omega t}\, dt = \frac{M}{\alpha - \omega}\left[e^{(-\alpha+\omega)b} - e^{(-\alpha+\omega)a}\right].$$

But

$$\lim_{a \to \infty; a < b} \left[e^{(-\alpha+\omega)b} - e^{(-\alpha+\omega)a}\right] = 0$$

and thus, by virtue of the Cauchy test, it follows that the integral is convergent. This completes the proof. □

Let $\mathcal{P}(\mathbb{C})$ be the class of subsets of $\mathbb{C}$. If $D \in \mathcal{P}(\mathbb{C})$, we denote by $\mathcal{F}(D; \mathbb{C})$ the set of all functions $F : D \to \mathbb{C}$. If $F \in \cup_{D \in \mathcal{P}(\mathbb{C})} \mathcal{F}(D; \mathbb{C})$, we denote by $D(F)$ the domain of F.

Definition 12.5.2. An operator

$$\mathcal{T} : \mathcal{O} \to \bigcup_{D \in \mathcal{P}(\mathbb{C})} \mathcal{F}(D; \mathbb{C})$$

is called *almost additive* if for each $f, g \in \mathcal{O}$ and each $p \in D(\mathcal{T}[f]) \cap D(\mathcal{T}[f])$, we have

$$\mathcal{T}[f + g](p) = \mathcal{T}[f](p) + \mathcal{T}[g](p).$$

An almost additive operator $\mathcal{T}$ is called *almost linear* if, in addition, it is homogeneous, i.e. for each $(\lambda, f) \in \mathbb{C} \times \mathcal{O}$, we have $D(\mathcal{T}[f]) \subseteq D(\mathcal{T}[\lambda f])$ and for each $p \in D(\mathcal{T}[f])$,

$$\mathcal{T}[\lambda f](p) = \lambda \mathcal{T}[f](p).$$

Remark 12.5.3. If an operator $\mathcal{T}$ is almost additive, then $\mathcal{T}[0_{\mathcal{O}}] = 0_{\mathbb{C}}$, where $0_{\mathcal{O}}$ is the null function in $\mathcal{O}$, while $0_{\mathbb{C}}$ is the null function in $\mathcal{F}(\mathbb{C};\mathbb{C})$. This simply follows by observing that

$$\mathcal{T}[0_{\mathcal{O}}] = \mathcal{T}[0_{\mathcal{O}} + 0_{\mathcal{O}}] = 2\mathcal{T}[0_{\mathcal{O}}].$$

From this remark, we deduce that the condition: for each $(\lambda, f) \in \mathbb{C}\times\mathcal{O}$, we have $D(\mathcal{T}[f]) \subseteq D(\mathcal{T}[\lambda f])$, in Definition 12.5.2, can be substituted by: for each $(\lambda, f) \in \mathbb{C} \times \mathcal{O}$, with $\lambda \neq 0$, we have $D(\mathcal{T}[f]) = D(\mathcal{T}[\lambda f])$.

Definition 12.5.3. The operator

$$\mathcal{L} : \mathcal{O} \to \bigcup_{D\in\mathcal{P}(\mathbb{C})} \mathcal{F}(D;\mathbb{C}),$$

defined by

$$\begin{cases} D(\mathcal{L}[f(t)]) = \left\{p \in \mathbb{C};\ \displaystyle\int_0^\infty e^{-pt} f(t)\,dt \text{ is convergent}\right\} \\ \mathcal{L}[f(t)](p) = \displaystyle\int_0^\infty e^{-pt} f(t)\,dt, \quad \text{for } p \in D(\mathcal{L}[f(t)]), \end{cases}$$

is called *the Laplace Transform.* We denote by $F(p) = \mathcal{L}[f(t)](p)$ and we agree to say that f is the *original function* and that F is the *image function.*

Theorem 12.5.1. *Let $f \in \mathcal{O}$. Then, its Laplace Transform, $F(p) = \mathcal{L}[f(t)](p)$, is holomorphic in the half-plane $\Re(p) > \omega_f$. Moreover, its derivative is, in its turn, the Laplace transform of the function $t \mapsto -tf(t)$, i.e. for each $p \in \mathbb{C}$ with $\Re(p) > \omega_f$, we have*

$$F'(p) = \mathcal{L}[-tf(t)](p).$$

For the proof of Theorem 12.5.1, the interested reader is referred to Lemma 5, p. 57 in [Widder (1946)].

Theorem 12.5.2. *Let $f \in \mathcal{O}$ and let $F(p) = \mathcal{L}[f(t)](p)$. Then,*

$$\lim_{\Re(p)\to+\infty} F(p) = 0.$$

Proof. Evaluating the integral in the definition of the Laplace Transform, we get

$$|F(p)| \leq \frac{M}{(\Re(p) - \omega_f)},$$

thus the conclusion. □

We state, without proof, a sufficient condition for a complex function to be the image through the Laplace Transform of some function $f \in \mathcal{O}$.

Theorem 12.5.3. *Let $\sigma \geq 0$ and $F : \{p \in \mathbb{C};\ \Re(p) > \sigma\} \to \mathbb{C}$ be a holomorphic function and let us assume that for each $a > \sigma$*

$$\lim_{|p| \to +\infty} F(p) = 0,$$

uniformly with respect to the argument of the complex number p, when $p \to \infty$ in the half-plane $\{p \in \mathbb{C};\ \Re(p) > a\}$, and the integral

$$\int_{a-i\infty}^{a+i\infty} F(p)\, dp$$

is absolutely convergent. Then, there exists $f \in \mathcal{O}$ such that

$$F(p) = \mathcal{L}[f(t)](p),$$

for each $p \in \mathbb{C}$, $\Re(p) > \sigma$. More precisely, f is given by

$$f(t) = \frac{1}{2\pi i} \int_{a-i\infty}^{a+i\infty} e^{pt} F(p)\, dp,$$

for each $t \geq 0$.

For the proof, the reader is referred to [Lavrentiev and Sabat (1959)], p. 404.

Corollary 12.5.1. *Let $f \in \mathcal{O}$ be arbitrary, $F(p) = \mathcal{L}[f(t)](p)$, and let $n \in \mathbb{N}$, $n \geq 1$. Then $t \mapsto t^n f(t)$ belongs to $\mathcal{O}$ and*

$$\mathcal{L}[t^n f(t)](p) = (-1)^n \frac{d^n F}{dp^n}(p).$$

Proof. Clearly, $t \mapsto t^n f(t)$ belongs to $\mathcal{O}$. From Theorem 12.5.1, we know that F is holomorphic on the half-plane $\Re(p) > \omega_f$. Thus, it is infinitely many times differentiable on that half-plane. Theorem 12.5.1, combined with a mathematical induction argument, leads to the conclusion. □

Proposition 12.5.3. *The Laplace Transform is almost linear.*

Proof. Just look at Definition 12.5.2 and use the linearity of the integral involved in the definition of the Laplace Transform. □

Remark 12.5.4. Unlike stated in many textbooks, the Laplace transform is not linear in the usual sense. For instance, if $f(t) = \theta(t)e^t$ and $g(t) = -f(t)$, we have that

$$\mathcal{L}[f(t) + g(t)](p) = 0, \quad \text{for all} \quad p \in \mathbb{C},$$

while

$$\mathcal{L}[f(t)](p) + \mathcal{L}[g(t)](p) = 0, \quad \text{only for those} \quad p \in \mathbb{C}, \quad \text{with} \quad \Re(p) > 1.$$

Theorem 12.5.4. *For each $f \in \mathcal{O}$ and $a > 0$, the function $t \mapsto f(at)$, belongs to $\mathcal{O}$ and*

$$\mathcal{L}[f(at)](p) = \frac{1}{a}\mathcal{L}[f(t)]\left(\frac{p}{a}\right).$$

Proof. Clearly, $t \mapsto f(at)$ satisfies (i), (ii) and (iii) in the definition of the space $\mathcal{O}$. Further, the change of the argument $at = \theta$, in the integral defining the Laplace Transform of $t \mapsto f(at)$, yields

$$\mathcal{L}[f(at)](p) = \int_0^\infty e^{-pt} f(at)\,dt = \frac{1}{a}\int_0^\infty e^{-\frac{p}{a}\theta} f(\theta)\,d\theta = \frac{1}{a}\mathcal{L}[f(t)]\left(\frac{p}{a}\right),$$

as claimed. □

Theorem 12.5.5. *For each $f \in \mathcal{O}$ and each $\tau \in \mathbb{R}_+$, the function $t \mapsto f(t-\tau)$, belongs to $\mathcal{O}$ and*

$$\mathcal{L}[f(t-\tau)](p) = e^{-\tau p}\mathcal{L}[f(t)](p).$$

Proof. It is a simple exercise to show that $t \mapsto f(t-\tau)$ satisfies (i), (ii) and (iii) in the definition of the space $\mathcal{O}$. In addition, by changing the argument $t - \tau = s$, in the integral defining the Laplace Transform of $t \mapsto f(t-\tau)$, and observing that $f(t-\tau) = 0$ for each $t \in [0, \tau)$, we get

$$\mathcal{L}[f(t-\tau)](p) = \int_0^\infty e^{-pt} f(t-\tau)\,dt = \int_\tau^\infty e^{-pt} f(t-\tau)\,dt$$

$$= \int_0^\infty e^{-p(\tau+s)} f(s)\,ds = e^{-p\tau}\int_0^\infty e^{-ps} f(s)\,ds = e^{-\tau p}\mathcal{L}[f(t)](p),$$

as claimed. □

Theorem 12.5.6. *If $f \in \mathcal{O}$ is differentiable and $f' \in \mathcal{O}$, then*

$$\mathcal{L}[f'(t)](p) = p\mathcal{L}[f(t)](p) - f(0+0).$$

Proof. We have

$$\mathcal{L}[f'(t)](p) = \int_0^\infty e^{-pt} f'(t)\,dt = e^{-pt} f(t)\big|_{t=0}^{t=\infty} + p\int_0^\infty e^{-pt} f(t)\,dt$$

$$= p\mathcal{L}[f(t)](p) - f(0+0).$$ □

Corollary 12.5.2. *If $f \in \mathcal{O}$ is n-times differentiable and $f^{(k)} \in \mathcal{O}$, for $k = 1, 2, \ldots, n$, then*

$$\mathcal{L}[f^{(n)}(t)](p) = p^n\mathcal{L}[f(t)](p) - \sum_{k=0}^{n-1} p^{n-k-1} f^{(k)}(0+0).$$

Proof. The conclusion follows by mathematical induction with the help of Theorem 12.5.6. □

Theorem 12.5.7. *Let $f \in \mathcal{O}$. Then $t \mapsto \int_0^t f(s)\,ds$ belongs to $\mathcal{O}$ and*

$$\mathcal{L}\left[\int_0^t f(s)\,ds\right](p) = \frac{\mathcal{L}[f(t)](p)}{p}.$$

Proof. It is a simple exercise to show that $t \mapsto \int_0^t f(s)\,ds$ belongs to $\mathcal{O}$. By Theorem 12.5.6, we have

$$\mathcal{L}[f(t)](p) = \mathcal{L}\left[\frac{d}{dt}\int_0^t f(s)\,ds\right](p) = p\mathcal{L}\left[\int_0^t f(s)\,ds\right](p) - \int_0^{0+0} f(t)\,dt$$

$$= p\mathcal{L}\left[\int_0^t f(s)\,ds\right](p),$$

thus the conclusion. □

Theorem 12.5.8. *Let $f \in \mathcal{O}$ be such that $t \mapsto \frac{f(t)}{t}$ belongs to $\mathcal{O}$ and let $F(p) = \mathcal{L}[f(t)](p)$. If for each $p \in \mathbb{C}$, with $\Re(p) > \omega_f$, the integral*

$$\int_p^\infty F(q)\,dq$$

is convergent, then

$$\mathcal{L}\left[\frac{f(t)}{t}\right](p) = \int_p^\infty F(p)\,dp.$$

Proof. By virtue of Theorem 12.5.1, we know that

$$\frac{d}{dp}\{\mathcal{L}[f(t)]\}(p) = \mathcal{L}[-tf(t)](p).$$

Since F is holomorphic in the half-plane $\Re(p) > \omega_f$, it follows that F admits antiderivatives. Further, since the integral $G(p) = \int_p^\infty F(q)\,dq$ is convergent, then it follows that $G'(p) = -F(p)$. In view of Theorem 12.5.6, we have

$$\mathcal{L}[f(t)](p) = \mathcal{L}\left[t\frac{f(t)}{t}\right](p) = -\frac{d}{dp}\left\{\mathcal{L}\left[\frac{f(t)}{t}\right]\right\}(p).$$

Thus

$$\mathcal{L}\left[\frac{f(t)}{t}\right](p) = \int_p^\infty \mathcal{L}[f(t)](q)\,dq,$$

as claimed. □

Theorem 12.5.9. *Let $f \in \mathcal{O}$. Then, for each $a \in \mathbb{R}$, we have*
$$\mathcal{L}[e^{at}f(t)](p) = \mathcal{L}[f(t)](p-a).$$

Proof. Clearly, $t \mapsto e^{at}f(t)$ belongs to $\mathcal{O}$. We have
$$\mathcal{L}[e^{at}f(t)](p) = \int_0^\infty e^{-pt}e^{at}f(t)\,dt$$
$$= \int_0^\infty e^{-(p-a)t}f(t)\,dt = \mathcal{L}[f(t)](p-a),$$
as claimed. □

Theorem 12.5.10. (Borel) *For every $f, g \in \mathcal{O}$ the convolution product $f * g$, i.e.*
$$(f * g)(t) = \int_0^t f(t-s)g(s)\,ds$$
for each $t \in \mathbb{R}$, belongs to $\mathcal{O}$ and
$$\mathcal{L}[(f * g)(t)](p) = \mathcal{L}[f(t)](p)\mathcal{L}[g(t)](p).$$

Proof. Let $F(p) = \mathcal{L}[f(t)](p)$ and $G(p) = \mathcal{L}[f(t)](p)$. Multiplying both sides of the equality
$$F(p) = \int_0^\infty e^{-pt}f(t)\,dt,$$
by $G(p)$, we get
$$F(p)G(p) = \int_0^\infty e^{-p\tau}G(p)f(\tau)\,d\tau.$$
By virtue of Theorem 12.5.5, we have
$$e^{-p\tau}G(p) = \mathcal{L}[g(t-\tau)](p) = \int_0^\infty e^{-pt}g(t-\tau)\,dt.$$
Then
$$F(p)G(p) = \int_0^\infty f(\tau)\,d\tau \int_0^\infty e^{-pt}g(t-\tau)\,dt.$$
By changing the order of integration, we deduce
$$F(p)G(p) = \int_0^\infty e^{-pt}\,dt \int_0^\infty f(\tau)g(t-\tau)\,d\tau.$$
But $g \in \mathcal{O}$ and thus $g(t-\tau) = 0$ for $\tau > t$. Accordingly
$$\int_0^\infty f(\tau)g(t-\tau)\,d\tau = \int_0^t f(\tau)g(t-\tau)\,d\tau = (f * g)(t).$$
It follows that
$$F(p)G(p) = \mathcal{L}[f(t)](p)\mathcal{L}[g(t)](p) = \mathcal{L}[(f * g)(t)](p),$$
as claimed. The proof is complete. □

The next results are stated without proof. For details, the interested reader is referred to [Şabac (1981)], pp. 630 and 642.

Theorem 12.5.11. (The Inversion Formula of Mellin-Fourier) *Let $f \in \mathcal{O}$ and let $F(p) = \mathcal{L}[f(t)](p)$. Then, at each continuity point of f, we have*

$$f(t) = \frac{1}{2\pi i} \int_{a-i\infty}^{a+i\infty} e^{pt} F(p)\, dp,$$

where $a > \omega_f$.

Theorem 12.5.12. *Let $f, g \in \mathcal{O}$. Then $fg \in \mathcal{O}$ and*

$$\mathcal{L}[f(t)g(t)](p) = \int_{a-i\infty}^{a+i\infty} F(q)G(p-q)\, dq,$$

for each $p \in \mathbb{C}$ with $\Re(p) > \omega_f$.

12.5.1 *The Transforms of Some Elementary Functions*

We include below the list of the Laplace transforms of some usual functions. For a larger list see [Sneddon (1971)].

If $f : \mathbb{R} \to \mathbb{R}$ is such that $\theta f \in \mathcal{O}$, for the sake of simplicity, instead of $\theta f \in \mathcal{O}$ we will simply write $f \in \mathcal{O}$, understanding, of course, that for $t < 0$, f vanishes.

- $\mathcal{L}[\theta(t)](p) = \dfrac{1}{p}$ $\quad (\Re(p) > 0)$
- $\mathcal{L}[t](p) = \dfrac{1}{p^2}$ $\quad (\Re(p) > 0)$
- $\mathcal{L}[t^n](p) = \dfrac{n!}{p^{n+1}}$ $\quad (\Re(p) > 0)$
- $\mathcal{L}[e^{at}](p) = \dfrac{1}{p-a}$ $\quad (\Re(p) > a)$
- $\mathcal{L}[\sin at](p) = \dfrac{a}{p^2+a^2}$ $\quad (\Re(p) > 0)$
- $\mathcal{L}[\cos at](p) = \dfrac{p}{p^2+a^2}$ $\quad (\Re(p) > 0)$
- $\mathcal{L}[\sinh at](p) = \dfrac{a}{p^2-a^2}$ $\quad (\Re(p) > |a|)$
- $\mathcal{L}[\cosh at](p) = \dfrac{p}{p^2-a^2}$ $\quad (\Re(p) > |a|)$
- $\mathcal{L}[t^{-1/2}e^{-c/t}](p) = \sqrt{\dfrac{\pi}{p}}\, e^{-2\sqrt{cp}}$ $\quad (c > 0,\ \Re(p) > 0)$

- $\mathcal{L}[t^{-3/2}e^{-c/t}](p) = \sqrt{\dfrac{\pi}{c}}e^{-2\sqrt{cp}}$ $\qquad (c > 0,\ \Re(p) > 0)$
- $\mathcal{L}[e^{-c^2t^2/4}](p) = \dfrac{1}{c}\sqrt{\pi}e^{p^2/c^2}\,\mathrm{Erfc}\left(\dfrac{p}{c}\right)$ $\qquad (c > 0, p \in \mathbb{C})$[3]
- $\mathcal{L}[\mathrm{Erf}(t/2c)](p) = e^{c^2p^2}\,\mathrm{Erfc}(cp)/p$ $\qquad (c > 0, p \in \mathbb{C})$.

[3]Here we denoted by $\mathrm{Erfc}(x) = 1 - \mathrm{Erf}(x)$, where $\mathrm{Erf}(x) = \frac{2}{\sqrt{\pi}}\int_0^x e^{-s^2}\,ds$ is the *error function*.

Solutions

Chapter 1

Problem 1.1 Let $x : [a, b] \to \mathbb{R}$ be the curve we are looking for. The condition in the problem reads as

$$\frac{x(t)}{x(t)/x'(t)} = \frac{k}{x(t) - t},$$

or equivalently

$$x'(t) = \frac{k}{x(t) - t}$$

for every $t \in [a, b]$. This is a differential equation reducible to one with separable variables. The change of unknown function $y = x - t$ leads to

$$y'(t) = \frac{k - y(t)}{y(t)}$$

for every $t \in [a, b]$, whose general solution is defined by $y + \ln|k - y| + t + c = 0$, with c an arbitrary constant. It then follows that the family of curves with the desired property is implicitly defined by $x + \ln|k - x + t| + c = 0$, $c \in \mathbb{R}$.

Problem 1.2 Let $x : \mathbb{I}_x \to \mathbb{R}$ be the curve we are looking for with $3 \in \mathbb{I}_x$ and let $A(a, 0)$ and $B(0, b)$ be the intersection points of the tangent to the curve at the point $(t, x(t))$ with the coordinate axes. Since $(t, x(t))$ is the middle point of the segment AB, we have $a = 2t$ and $b = 2x$. See Figure S.1.1.

On the other hand, the slope of the tangent to the current point $(t, x(t))$ is $x'(t)$. The condition in the problem expresses by $x'(t) = -\frac{b}{a}$, or equivalently by $tx'(t) = -x(t)$. The equation above is with separable variables and has the general solution $tx = c$, with c real constant. Since $x(3) = 2$, we deduce $c = 6$. Consequently, the desired curve is the hyperbola of equation $tx = 6$.

Exercise 1.1 (1) This is an equation with separable variables having the general solution defined by $x(t) = \pm \arcsin \sqrt{\frac{\cos^2 t}{1+2c\cos^2 t}}$ for $t \in \mathbb{I}_x$, where the interval $\mathbb{I}_x \subset \left((2k-1)\frac{\pi}{2}, (2k+1)\frac{\pi}{2}\right)$ depends on the constant $c \in \mathbb{R}$.

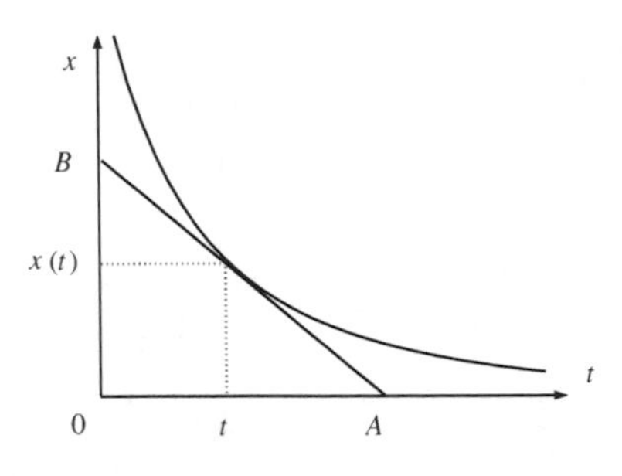

Fig. S.1.1

(2) This is a Bernoulli equation, but with separable variables too. The general solution is defined by $x(t) = ct(1-ct)^{-1}$ for $t \in \mathbb{I}_x$, where $\mathbb{I}_x$ depends on the integration constant $c \in \mathbb{R}$. The equation also admits the stationary solution $x(t) = -1$ for every $t \in \mathbb{R}$.

(3) This is an equation with separable variables having the general solution $x(t) = \pm\sqrt{2\ln|t| - t^2 + c}$ for every $t \in \mathbb{I}_x$, where $\mathbb{I}_x$ is an interval which does not contain 0 and depends on the integration constant $c \in \mathbb{R}$.

(4) The substitution $y = t + x$ leads to the equation with separable variables $y' = 1 + y^2$. Solving this equation and coming back to the function x, we get $x(t) = \tan(t+c) - t$ for every $t \in \left(-\frac{\pi}{2} - c, \frac{\pi}{2} - c\right)$, $c \in \mathbb{R}$.

(5) The substitution $y = 8t + 2x + 1$ leads to an equation with separable variables. The general solution of the initial equation is $x(t) = \tan(4t+c) - 4t - \frac{1}{2}$ for every $t \in \left(-\frac{\pi}{8} - \frac{c}{4}, \frac{\pi}{8} - \frac{c}{4}\right)$, $c \in \mathbb{R}$.

(6) The substitution $y = 2t + 3x + 1$ leads to an equation with separable variables. The general solution of the initial equation, after suitably denoting the constant of integration, is given in the implicit form by $t + 2x + 7\ln|2t + 3x - 13| = c$ with $c \in \mathbb{R}$.

(7) The substitution $y = 2t - x$ leads to an equation with separable variables. The general solution of the initial equation, after suitably denoting the constant of integration, is given in the implicit form by $5t + 10x - 3\ln|10t - 5x + 6| = c$ with $c \in \mathbb{R}$. The equation also has the solution $x : \mathbb{R} \to \mathbb{R}$ defined by $x(t) = 2t + \frac{6}{5}$, eliminated during the integration process of the equation with separable variables.

(8) This is an equation with separable variables having the general solution $x(t) = \pm\sqrt{\frac{c}{t^2-1} - 1}$ for every $t \in \mathbb{I}_x$, where $\mathbb{I}_x$ is an interval, depending on the constant $c \in \mathbb{R}$ and which does not contain ± 1.

Problem 1.3 As in the case of Problem 1.2, let $x : \mathbb{I}_x \to \mathbb{R}$ be the curve to be found with $1 \in \mathbb{I}_x$ and let $A(a, 0)$ and $B(0, b)$ be the intersection points of the normal to the curve at the point $(t, x(t))$ with the coordinate axes. Since $(t, x(t))$ is the middle of the segment AB, we have $a = 2t$ and $b = 2x$. See Figure S.1.2.

On the other hand, the slope of the normal to the curve at the current point $(t, x(t))$ is $-[x'(t)]^{-1}$. The condition in the problem expresses then by $-[x'(t)]^{-1} = -\frac{b}{a}$, or equivalently by $x(t)x'(t) = t$. The latter equation is with separable variables, and has the general solution $x^2 - t^2 = c$, with c constant.

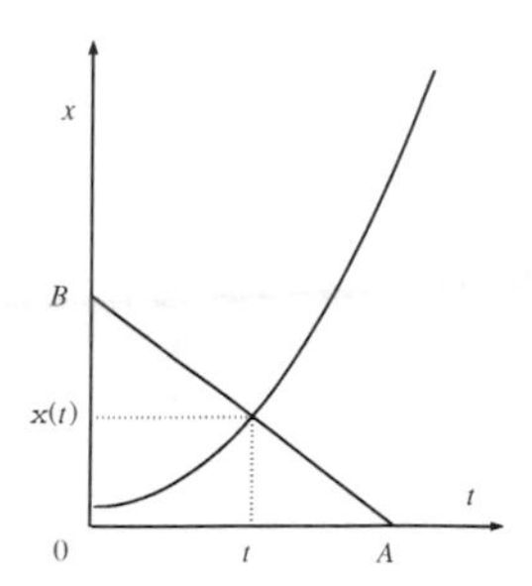

Fig. S.1.2

But $x(1) = 2$, and then $c = 3$. Consequently, the curve we are looking for is a hyperbola of equation $x^2 - t^2 = 3$.

Problem 1.4 Let $x : \mathbb{I}_x \to \mathbb{R}$ be the curve we are looking for. The required condition takes the equivalent form

$$x(t)/x'(t) = a,$$

equation which has the general solution $x(t) = ce^{t/a}$ for $t \in \mathbb{R}$, with $c \in \mathbb{R}$.

Problem 1.5 In this case we get the equation

$$x(t)/x'(t) = 2t,$$

for $t \geq 0$, which has the general solution $x(t) = c\sqrt{t}$ with $c > 0$.

Exercise 1.2 (1) Dividing by $t \neq 0$, the equation reduces to a homogeneous one having the general solution $x(t) = -t \ln |t| + ct$ for $t \in \mathbb{I}_x$, where $\mathbb{I}_x$ is an interval which does not contain 0, and $c \in \mathbb{R}$.

(2) Dividing by $t \neq 0$, the equation reduces to a homogeneous one whose general solution is $x(t) = \frac{c}{t} - \frac{t}{2}$ for $t \in \mathbb{I}_x$, where $\mathbb{I}_x$ is an interval which does not contain 0, and $c \in \mathbb{R}^*$. The equation also admits the solution $x(t) = -\frac{t}{2}$ for every $t \in \mathbb{R}$.

(3) Dividing by t^2, we get a homogeneous equation whose general solution is defined by $x(t) = t(\ln |t| + c)^{-1}$ for every $t \in \mathbb{I}_x$, where $\mathbb{I}_x$ is an interval which depends on the constant $c \in \mathbb{R}$ and does not contain 0. The equation also admits the solution $x(t) = 0$ for every $t \in \mathbb{R}$.

(4) Dividing by $2tx$, we get a homogeneous equation whose general solution, $x : \mathbb{I}_x \to \mathbb{R}$, is defined by

$$x(t) = \pm t\sqrt{\frac{t+c}{t}},$$

where $c \in \mathbb{R}$ and $\mathbb{I}_x$ depends on c and does not contain 0. At the same time the equation also admits the solution $x_{1,2}(t) = \pm t$ for every $t \in \mathbb{R}$.

(5) Obviously $x : \mathbb{R} \to \mathbb{R}$, $x(t) = 0$ for every $t \in \mathbb{R}$, is a solution of the equation. Dividing the equation by $t \neq 0$, we get a homogeneous equation whose solution

is given in the implicit form: $\ln|x| - \sqrt{\frac{t}{x}} = c$ for $t \in (-\infty, 0)$ and $\ln x + \sqrt{\frac{t}{x}} = c$ for $t \in (0, +\infty)$.

(6) Dividing by $t \neq 0$, we get a homogeneous equation. The general solution of the initial equation is $x : \mathbb{R} \to \mathbb{R}$, $x(t) = (c^2t^2 - 1)(2c)^{-1}$ for every $t \in \mathbb{R}$, where $c \in \mathbb{R}_+^*$.

(7) Dividing by $4x^2 + 3tx + t^2$, and simplifying the fraction thus obtained by $t^2 \neq 0$, we get a homogeneous equation. The general solution of the initial equation is implicitly defined by $(x^2 + t^2)^{3/2}(x + t) = c$, where $c \in \mathbb{R}$.

(8) Dividing by $2tx \neq 0$ the equation reduces to a homogeneous one. The general solution of the initial equation is $x : \mathbb{I}_x \to \mathbb{R}$, $x(t) = \pm t\sqrt{1 + ct}$ for every $t \in \mathbb{I}_x$, where $c \in \mathbb{R}$, and $\mathbb{I}_x$ depends on c.

Problem 1.6 Let $x : \mathbb{I}_x \to \mathbb{R}$ be the curve we are looking for, with $1 \in \mathbb{I}_x$. The required condition expresses as

$$\left| t - \frac{x}{x'} \right| = \sqrt{t^2 + x^2},$$

or equivalently as

$$t - \frac{x}{x'} = \pm\sqrt{t^2 + x^2}.$$

See Figure S.1.3.

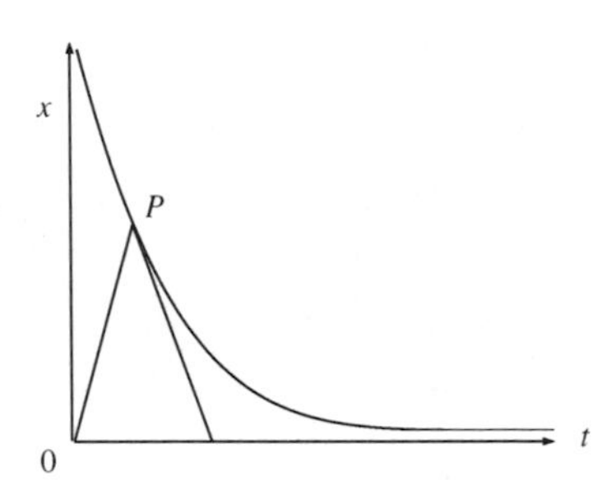

Fig. S.1.3

These equations are reducible to homogeneous equations. Analyzing the two cases, we deduce that only the equation $t - x/x' = \sqrt{t^2 + x^2}$ has a suitable solution ($x(1) = 0$), i.e. $x(t) = \pm 2\sqrt{1 - t}$.

Problem 1.7 Imposing the condition that $x = t^m y$ satisfy the equation, we deduce

$$mt^{m-1}y + t^m y' = f(t, t^m y) = t^{m-1} f(1, y),$$

or equivalently

$$y' = \frac{1}{t}(f(1, y) - my).$$

For the considered equation, we have $f(t, x) = x^2 - \frac{2}{t^2}$. Let us observe that

$f(\lambda t, \lambda^m x) = \lambda^{m-1} f(t,x)$ for every $(t,x) \in \mathbb{R}_+ \times \mathbb{R}_+$ and $\lambda \in \mathbb{R}_+$ if and only if

$$\lambda^{2m} x^2 - \frac{2}{\lambda^2 t^2} = \lambda^{m-1}\left(x^2 - \frac{2}{t^2}\right).$$

We observe that this condition is satisfied if and only if $m = -1$. Imposing $x = t^{-1}y$, we get $y' = \frac{1}{t}(y^2 + y - 2)$, equation which has the general solution $y(t) = \frac{c+2t^3}{c-t^3}$ with $c \in \mathbb{R}$. The general solution of the initial equation is then $x : \mathbb{I}_x \to \mathbb{R}_+$, $x(t) = \frac{c+2t^3}{ct-t^4}$, where $c \in \mathbb{R}$ and $\mathbb{I}_x$ is an interval which does not contain 0 and $\sqrt[3]{c}$.

Exercise 1.3 (1) The equation is reducible to a linear equation. Also, the equation is with separable variables. The general solution is $x : \mathbb{R} \to \mathbb{R}$, $x(t) = cte^t$, for every $t \in \mathbb{R}$, where $c \in \mathbb{R}$.

(2) The equation is reducible to a linear one. The solutions are $x : \mathbb{R} \to \mathbb{R}$, $x(t) = \frac{t^4}{6}$ for every $t \in \mathbb{R}$ and $x : \mathbb{I}_x \to \mathbb{R}$, $x(t) = \frac{t^4}{6} + \frac{c}{t^2}$ for every $t \in \mathbb{I}_x$, where $c \in \mathbb{R}^*$ and $\mathbb{I}_x = (0, +\infty)$ or $(-\infty, 0)$.

(3) This is an equation reducible to a linear one, with solutions $x_1 : \mathbb{R} \to \mathbb{R}$, defined by

$$x_1(t) = \begin{cases} \dfrac{e^t - 1}{t}, & t \in \mathbb{R}^* \\ 1 & t = 0, \end{cases}$$

and $x_2 : \mathbb{I} \to \mathbb{R}$, defined by $x_2(t) = (e^t + c)t^{-1}$ for every $t \in \mathbb{I}$, where $c \in \mathbb{R} \setminus \{-1\}$ and $\mathbb{I} = (0, +\infty)$ or $(-\infty, 0)$.

(4) The equation has the solution $x \equiv 0$. For $x \neq 0$ we will look for t as a function of x. We obtain the equation

$$\frac{dt}{dx} = \frac{3t}{2x} - \frac{x}{t}$$

which is both Bernoulli and homogeneous and which, by integration, leads to the implicit form of the general solution for the initial equation: $cx^3 + x^2 - t^2 = 0$, with $c \in \mathbb{R}_+^*$.

(5) The equation is reducible to a Bernoulli equation with $\alpha = 2$. The general solution is $x(t) = (t \ln|t| + ct)^{-1}$ for every $t \in \mathbb{I}_x$, where $\mathbb{I}_x$ depends on $c \in \mathbb{R}$. The equation also has the solution $x \equiv 0$.

(6) The substitution $x^2 = y$ leads to an equation reducible to a linear one. The general solution of the initial equation is $x : \mathbb{I}_x \to \mathbb{R}$, $x(t) = \pm\sqrt{t(c - \ln|t|)}$ for every $t \in \mathbb{I}_x$, where $\mathbb{I}_x$ does not contain 0 and depends on $c \in \mathbb{R}$.

(7) We observe that $x \equiv 0$ is a solution. For $x \neq 0$ we determine t as a function of x. We conclude that t satisfies the Bernoulli equation

$$\frac{dt}{dx} = -\frac{2}{x}t + t^2.$$

The general solution x of the initial equation is given in the implicit form by $(cx^2 + x)t = 1$, where $c \in \mathbb{R}$.

(8) The equation is reducible to a Bernoulli equation whose general solution, $x : \mathbb{I}_x \to \mathbb{R}$, is defined by $x(t) = \left(2t + ct^2\right)^{-1}$ for every $t \in \mathbb{I}_x$, where $\mathbb{I}_x$ depends on $c \in \mathbb{R}$. The equation also has the solution $x \equiv 0$.

Problem 1.8 We have

$$R'(t) = \frac{(x_2'(t) - x'(t))(x(t) - x_1(t)) - (x_2(t) - x(t))(x'(t) - x_1'(t))}{(x(t) - x_1(t))^2}$$

$$= \frac{a(t)(x_2(t) - x(t))(x(t) - x_1(t)) - (x_2(t) - x(t))a(t)(x(t) - x_1(t))}{(x(t) - x_1(t))^2} = 0$$

for every $t \in \mathbb{I}$. Hence R is a constant on $\mathbb{I}$. The geometric meaning of this result is the following: if x_1, x_2 are two distinct solutions of the linear equation $x'(t) = a(t)x(t) + b(t)$ and x is another solution, then $A(t, x(t))$ lies on the straight line passing through the points $A_1(t, x_1(t))$ and $A_2(t, x_2(t))$ and the ratio $\dfrac{AA_1}{AA_2}$ is constant.

Problem 1.9 We have

$$y'(t) = \frac{x_1'(t)x_2(t) - x_1(t)x_2'(t)}{x_2^2(t)}$$

$$= \frac{[a(t)x_1(t) + b(t)x_1^2(t)]x_2(t) - x_1(t)[a(t)x_2(t) + b(t)x_2^2(t)]}{x_2^2(t)}$$

$$= \frac{b(t)x_1(t)x_2(t)[x_1(t) - x_2(t)]}{x_2^2(t)} = b(t)[x_1(t) - x_2(t)]y(t)$$

which shows that $y'(t) = b(t)[x_1(t) - x_2(t)]y(t)$.

Problem 1.10 We denote by

$$A(t) = \frac{x_2(t) - x(t)}{x_2(t) - x_1(t)}$$

and let us observe that

$$A'(t) = \frac{(x_2'(t) - x'(t))(x_2(t) - x_1(t)) - (x_2'(t) - x_1'(t))(x_2(t) - x(t))}{(x_2(t) - x_1(t))^2}$$

$$= \frac{(x_2(t) - x_1(t))[a(t) + b(t)(x_2(t) + x(t))](x_2(t) - x(t))}{(x_2(t) - x_1(t))^2} -$$

$$- \frac{(x_2(t) - x_1(t))[a(t) + b(t)(x_2(t) + x_1(t))](x_2(t) - x(t))}{(x_2(t) - x_1(t))^2}$$

$$= \frac{b(t)(x_2(t) - x(t))(x(t) - x_1(t))}{x_2(t) - x_1(t)}.$$

Similarly,

$$C(t) = \frac{x_3(t) - x_1(t)}{x_3(t) - x(t)}$$

satisfies

$$C'(t) = \frac{b(t)(x_3(t) - x_1(t))(x_1(t) - x(t))}{x_3(t) - x(t)}.$$

But

$$\begin{aligned} B'(t) &= A'(t)C(t) + A(t)C'(t) \\ &= \frac{b(t)(x_2(t) - x(t))(x(t) - x_1(t))}{x_2(t) - x_1(t)} \cdot \frac{x_3(t) - x_1(t)}{x_3(t) - x(t)} \\ &+ \frac{x_2(t) - x(t)}{x_2(t) - x_1(t)} \cdot \frac{b(t)(x_3(t) - x_1(t))(x_1(t) - x(t))}{x_3(t) - x(t)} = 0 \end{aligned}$$

for every $t \in \mathbb{I}$. Hence B is a constant on $\mathbb{I}$.

Exercise 1.4 (1) This is an exact differential equation. The general solution is given in the implicit form by $t^2 + 2tx + 2x^2 = c$, where $c \geq 0$.

(2) This is an exact differential equation. The general solution is given in the implicit form by: $t^3 + 6tx + 3t^2 = c$, where $c \in \mathbb{R}$.

(3) This is an exact differential equation. The general solution is given in the implicit form by: $2t^3 - 9t^2x^2 + 12t + 2x^3 = c$, where $c \in \mathbb{R}$.

(4) This is an exact differential equation having the general solution given in the implicit form by $-t^4 + 2t^2x^2 + 4xt + x^4 = c$, with $c \in \mathbb{R}$.

(5) This is an equation reducible to an exact differential one by means of the integrating factor $\rho(x) = \frac{1}{x^4}$. The general solution is given in the implicit form by $t^2 - x^2 - cx^3 = 0$, where $c \in \mathbb{R}$. The equation also admits the solution $x \equiv 0$ eliminated during the reducing process of the initial equation to an exact differential one.

(6) This is an equation reducible to an exact differential one by means of the integrating factor $\rho(t) = \frac{1}{t^2}$. The general solution is given in the implicit form by $x^2 - t\ln|t| - ct = 0$, where $c \in \mathbb{R}$. From here, we deduce that $x : \mathbb{I}_x \to \mathbb{R}$ is defined by $x(t) = \pm\sqrt{t(c + \ln|t|)}$ for every $t \in \mathbb{I}_x$, where $\mathbb{I}_x$ depends on $c \in \mathbb{R}$.

(7) This is an equation reducible to an exact differential one by means of the integrating factor $\rho(x) = \frac{1}{x^2}$. We have the solutions $x \equiv 0$ and $x : \mathbb{I}_x \to \mathbb{R}$, defined by $x(t) = 2t(2c - t^2)^{-1}$ for every $t \in \mathbb{I}_x$, where $\mathbb{I}_x$ depends on $c \in \mathbb{R}$.

(8) This is an equation reducible to an exact differential one by means of the integrating factor $\rho(t) = \frac{1}{t}$. The general solution is given in the implicit form by $x\ln t + \frac{x^4}{4} = c$ for every $t > 0$, where $c \in \mathbb{R}$.

Exercise 1.5 (1) This is a Lagrange equation having the general solution in the parametric form:

$$\begin{cases} t(p) = 6p^2 + cp \\ x(p) = 4p^3 + \dfrac{1}{2}cp^2 \end{cases}, \; p \in \mathbb{R},$$

where $c \in \mathbb{R}$. The equation also admits the solution $x \equiv 0$.

(2) This is a Lagrange equation having the general solution in the parametric form:

$$\begin{cases} t(p) = \ln|p| - \arcsin p + c \\ x(p) = p + \sqrt{1 - p^2} \end{cases}, \; p \in (-1, 0) \;\text{ or }\; (0, 1),$$

where $c \in \mathbb{R}$. The equation also admits the solution $x \equiv 1$.

(3) This is a Lagrange equation having the general solution in the parametric form:

$$\begin{cases} t(p) = ce^{-p} - 2p + 2 \\ x(p) = c(1+p)e^{-p} - p^2 + 2 \end{cases}, \ p \in \mathbb{R},$$

where $c \in \mathbb{R}$.

(4) This is a Lagrange equation having the general solution in the parametric form:

$$\begin{cases} t(p) = -\dfrac{1}{3}p + \dfrac{c}{\sqrt{p}} \\ x(p) = -c\sqrt{p} - \dfrac{1}{6}p^2 \end{cases}, \ p > 0,$$

or

$$\begin{cases} t(p) = -\dfrac{1}{3}p + \dfrac{c}{\sqrt{-p}} \\ x(p) = c\sqrt{-p} - \dfrac{1}{6}p^2 \end{cases}, \ p < 0,$$

where $c \in \mathbb{R}$. The equation also has the solution $x \equiv 0$.

(5) Clairaut equation having the general solution $x : \mathbb{R} \to \mathbb{R}$, $x(t) = ct + t^2$ with $c \in \mathbb{R}$, and the singular solution in the parametric form:

$$\begin{cases} t(p) = -2p \\ x(p) = -p^2 \end{cases}, \ p \in \mathbb{R}.$$

Eliminating $p \in \mathbb{R}$, we get $x(t) = -\frac{t^2}{4}$ for every $t \in \mathbb{R}$.

(6) Clairaut equation, but with separable variables too. The general solution, $x : \mathbb{R} \to \mathbb{R}$, is given by $x(t) = ct + c$, with $c \in \mathbb{R}$. The equation does not admit singular solution.

(7) Clairaut equation. The general solution is $x : \mathbb{R} \to \mathbb{R}$, $x(t) = ct + \sqrt{1+c^2}$, with $c \in \mathbb{R}$. The singular solution is

$$\begin{cases} t(p) = -\dfrac{p}{\sqrt{1+p^2}} \\ x(p) = \dfrac{1}{\sqrt{1+p^2}} \end{cases}, \ p \in \mathbb{R}.$$

Eliminating p, we obtain $x : (-1, 1) \to \mathbb{R}$, $x(t) = \sqrt{1-t^2}$.

(8) Clairaut equation having the general solution $x : \mathbb{R} \to \mathbb{R}$, $x(t) = ct + \frac{1}{c}$, with $c \in \mathbb{R}^*$, and the singular solution

$$\begin{cases} t(p) = \dfrac{1}{p^2} \\ x(p) = \dfrac{2}{p} \end{cases}, \ p \in \mathbb{R}^*.$$

Eliminating the parameter p, we get $x : (0, +\infty) \to \mathbb{R}$, $x(t) = \pm 2\sqrt{t}$.

Problem 1.11 Let us choose a Cartesian system of coordinates with the origin at the fixed point. Let $x : \mathbb{I}_x \to \mathbb{R}$ be the function whose graph is the curve we are looking for. The equation of the tangent to the curve at the current point $(t, x(t))$ is $X - x(t) = x'(t)(T - t)$, while the distance from the origin to this tangent is constant if and only if there exists $c \in \mathbb{R}^*$ such that

$$\frac{tx'(t) - x(t)}{\sqrt{1 + x'^2(t)}} = c.$$

Solving this with respect to $x(t)$, we get a Clairaut equation having the general solution $x(t) = kt - c\sqrt{1 + k^2}$, with $k \in \mathbb{R}^*$, and the singular solution

$$\begin{cases} t(p) = \dfrac{cp}{\sqrt{1+p^2}} \\ x(p) = -\dfrac{c}{\sqrt{1+p^2}} \end{cases}, \quad p \in \mathbb{R}.$$

Eliminating p, we get the implicit equation of the curve: $x^2 + t^2 = c^2$, equation which represents a circle centered at the origin (at the fixed point considered) and of radius $|c|$, i.e. the distance from point to the tangent. Other solutions, of class C^1 only, can be obtained concatenating any arc of the circle with the two "semi-tangents" at the two endpoints of the arc. See Figure S.1.4.

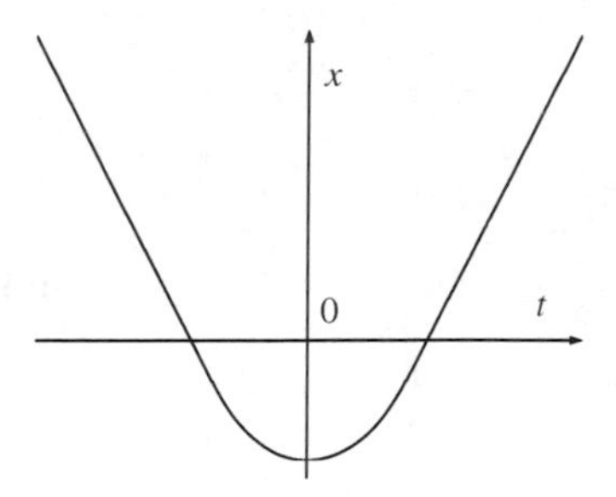

Fig. S.1.4

Problem 1.12 Let $x : \mathbb{I}_x \to \mathbb{R}$ be the function whose graph is the curve we are looking for. The equation of the tangent to the current point of the curve $(t, x(t))$ is $X - x(t) = x'(t)(T - t)$, while the intersection points of the tangent with the coordinate axes are $A\left(t - \frac{x(t)}{x'(t)}, 0\right)$, and $B(0, x(t) - tx'(t))$. See Figure S.1.5.

The condition imposed expresses analytically in the form

$$\left(t - \frac{x(t)}{x'(t)}\right)(x(t) - tx'(t)) = -c,$$

where $c \in \mathbb{R}^*$. Rearranging, we get a Clairaut equation with the general solution $x(t) = kt \pm \sqrt{ck}$, with $k \in \mathbb{R}^*$, $ck > 0$, and the singular solution $tx = -\dfrac{c}{4}$. We

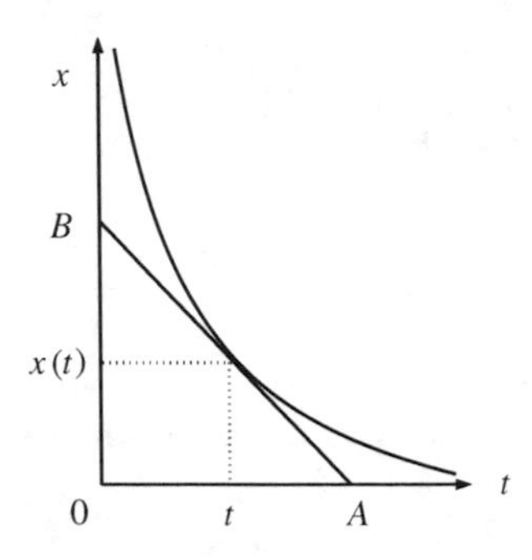

Fig. S.1.5

get other solutions, of class C^1 only, by concatenating an arc of hyperbola with the semi-tangent(s) at the end(s) of the arc.

Problem 1.13 We observe that the resultant of the two forces, gravitational and centrifugal, has the direction of the normal to the surface at the point considered. Taking Oy as rotational axis, and denoting by ω the angular speed, we get, for the axial plane section of the surface, the differential equation

$$g\frac{dy}{dx}(x) = \omega^2 x.$$

Problem 1.14 According to the Boyle–Mariotte law, the density is proportional with the pressure. So, the variation of the pressure from the altitude t to altitude $t+h$ is $p(t+h) - p(t) = -kp(t)h$. The equation obtained is $p'(t) = -kp(t)$. We deduce $p(t) = e^{-0.000167t}$.

Problem 1.15 The variation of the length on the portion x, $x+h$ is given by $s(x+h) - s(x) = kW(l-x)l^{-1}h$. We get $s'(x) = kW(l-x)l^{-1}$. It then follows that $s(l) = 0.5kWl$.

Problem 1.16 Let $y : [a,b] \to \mathbb{R}_+$ be the function defined by

$$y(t) = \int_a^t k(s)\, x(s)\, ds$$

for $t \in [a,b]$. Obviously y is differentiable on $[a,b]$ and $y'(t) = k(t)\, x(t)$ for every $t \in [a,b]$. Taking into account the inequality in the hypothesis and the fact that the function k is positive, we deduce

$$y'(s) \le k(s)\, y(s) + k(s)\, h(s)$$

for every $s \in [a,b]$. Multiplying both sides of the inequality above by

$$\exp\left(-\int_a^s k(\tau)\, d\tau\right),$$

we deduce

$$\frac{d}{ds}\left(y(s)\exp\left(-\int_a^s k(\tau)\,d\tau\right)\right) \le k(s)\,h(s)\exp\left(-\int_a^s k(\tau)\,d\tau\right).$$

Integrating from a to t, we get

$$y(t) \le \int_a^t k(s)\,h(s)\exp\left(\int_s^t k(\tau)\,d\tau\right) ds.$$

Since $x(t) \le h(t) + y(t)$ for every $t \in [\,a, b\,]$, this completes the proof.
Problem 1.17 From the Bellman's inequality, it follows

$$\begin{aligned}
x(t) &\le \xi + \int_a^t v(s)\,ds + \int_a^t k(s)\left(\xi + \int_a^s v(\tau)\,d\tau\right)\exp\left(\int_s^t k(\tau)\,d\tau\right) ds \\
&= \xi + \int_a^t v(s)\,ds - \int_a^t \left(\xi + \int_a^s v(\tau)\,d\tau\right)\frac{d}{ds}\exp\left(\int_s^t k(\tau)\,d\tau\right) ds \\
&= \xi + \int_a^t v(s)\,ds - \left(\xi + \int_a^s v(\tau)\,d\tau\right)\exp\left(\int_s^t k(\tau)\,d\tau\right)\Bigg|_a^t \\
&\quad + \int_a^t v(s)\exp\left(\int_s^t k(\tau)\,d\tau\right) ds \\
&= \xi\exp\left(\int_a^t k(s)\,ds\right) + \int_a^t v(s)\exp\left(\int_s^t k(\tau)\,d\tau\right) ds.
\end{aligned}$$

Problem 1.18 The proof follows, with minor modifications, the same way as that used for establishing Lemma 1.5.3.
Problem 1.19 Let us assume for contradiction that there exists $t_1 \in (0, T)$, such that $x(t_1) > y(t_1)$. Since x and y are continuous and $x(0) \le y(0)$, there exists $t_0 \in [\,0, T\,]$ with $t_0 < t_1$, such that $x(t_0) = y(t_0)$, and $x(t) \ge y(t)$ for every $t \in [\,t_0, t_1)$. Since f is nondecreasing, we have

$$\left(\frac{dx}{dt}(t) - \frac{dy}{dt}(t)\right)(x(t) - y(t)) \le 0$$

for every $t \in [\,t_0, t_1\,]$. Integrating this inequality over $[\,t_0, t_1\,]$, we get

$$(x(t_1) - y(t_1))^2 \le (x(t_0) - y(t_0))^2 = 0,$$

relation which contradicts the inequality $x(t_1) > y(t_1)$.

Chapter 2

Exercise 2.1 (a) $x(t) = t(2-t)^{-1}$ for $t \in [\,1, 2)$. (b) $x(t) = \frac{1}{2}te^{2-t^2/2}$ for $t \in [\,2, +\infty)$. (c) $x(t) = \tan 4t - 4t - \frac{1}{2}$ for $t \in [\,0, \frac{\pi}{8})$. (d) $x(t) = \sqrt{5(1-t^2)^{-1} - 1}$

for every $t \in [0,1)$. (e) $x(t) = -t\ln t + 2t$ for $t \in [1,+\infty)$. (f) $x(t) = (1-t^2)(2t)^{-1}$ for $t \in [1,+\infty)$. (g) $x(t) = t(\ln t + 1)^{-1}$ for $t \in [1,+\infty)$. (h) $x(t) = \sqrt{1+3t}$ for $t \in [1,+\infty)$. (i) $x(t) = te^t$ for $t \in [1,+\infty)$. (j) $x(t) = (t^6+11)(6t^2)^{-1}$ for $t \in [1,+\infty)$. (k) $x(t) = (e^t - e)t^{-1}$ for $t \in [1,+\infty)$. (l) $x(t) = (t\ln t - 1)^{-1}$ for $t \in [1,t^*)$, where t^* is the root of the transcendental equation $t\ln t - 1 = 0$. (m) $x(t) = \sqrt{t(4-\ln t)}$ for $t \in [1,e^4)$. (n) $x(t) = t^{-1}$ for $t \in [1,+\infty)$. (o) $x(t) = (2t - t^2)^{-1}$ for every $t \in [1,2)$. (p) x is implicitly defined by the equation $x^3 - x^2 + t^2 = 0$ for $t \in [1,+\infty)$.

Problem 2.1 One may easily state that the function z is continuous on $[a,c]$, differentiable on $[a,c]\setminus\{b\}$, and satisfies $z(a) = \xi$, and $z'(t) = f(t,z(t))$ for every $t \in [a,c]\setminus\{b\}$. From the continuity of the functions f and z and the last equality, we deduce that z' can be extended by continuity at the point b. But this means that z is of class C^1 on $[a,c]$ and, in addition, that it is a solution of $\mathcal{CP}(\mathbb{I},\Omega,f,a,\xi)$.

Problem 2.2 According to Lemma 2.2.2, for every $a \in \mathbb{I}$ and every $\delta > 0$ with $[a,a+\delta] \subset \mathbb{I}$, $\mathcal{CP}(\mathcal{D})$ has at least one solution $x : [a,a+\delta] \to \mathbb{R}^n$. Let $(\delta_k)_{k\in\mathbb{N}}$ be a sequence of positive numbers, with $(a+\delta_k)_{k\in\mathbb{N}}$, strictly increasing to $\sup\mathbb{I}$, and let $x_k : [a,a+\delta_k] \to \mathbb{R}^n$ be a fixed sequence of solutions of $\mathcal{CP}(\mathcal{D})$ satisfying $x_k(t) = x_{k+1}(t)$ for every $k \in \mathbb{N}$ and every $t \in [a,a+\delta_k]$. By Lemma 2.2.2 combined with Proposition 2.1.2, we can always find such a sequence. More precisely, let us take first the solution x_1 defined on $[a,a+\delta_1]$. Then the Cauchy problem

$$\begin{cases} y' = f(t,y) \\ y(a+\delta_1) = x_1(a+\delta_1) \end{cases}$$

has at least one solution $y_1 : [a+\delta_1,a+\delta_2] \to \mathbb{R}^n$. We define $x_2 : [a,a+\delta_2] \to \mathbb{R}^n$ by concatenating the functions x_1 and y_1. We continue the procedure inductively. Let us observe that the function $x : [a,\sup\mathbb{I}) \to \mathbb{R}^n$, defined by $x(t) = x_k(t)$ for $t \in [a,a+\delta_k]$, is well-defined. Furthermore, it is a global solution of $\mathcal{CP}(\mathcal{D})$.

Problem 2.3 If $\xi > 0$, then $x : [a,+\infty) \to \mathbb{R}$, $x(t) = \sqrt{t^2+\xi^2-a^2}$, is the unique global right solution of $\mathcal{CP}(\mathbb{R},\mathbb{R},f,a,\xi)$. Similarly, if $\xi < 0$ then the function $x : [a,+\infty) \to \mathbb{R}$, defined by $x(t) = -\sqrt{t^2+\xi^2-a^2}$, is the unique right global solution of $\mathcal{CP}(\mathbb{R},\mathbb{R},f,a,\xi)$. If $\xi = 0$, then $x : [a,+\infty) \to \mathbb{R}$, $x(t) = 0$, is the global solution we are looking for. Obviously the function f is not continuous at $(1,0)$ because $f(1,0) = 0$, while $\lim_{x\downarrow 0} f(1,x) = +\infty$.

Problem 2.4 Let $x : (c,0] \to \mathbb{R}$ be a saturated left solution of $\mathcal{CP}(\mathbb{R},\mathbb{R},0,0)$. Then we have $x'(t) = f(t,x(t))$ for every $t \in (c,0]$. Since $x'(0) = f(0,0) = -1$, and x is of class C^1, x' can only take the value -1. We recall that f has only the values ± 1. So $x(t) = -t + k$ with $k \in \mathbb{R}$. Since $x(0) = 0$, it follows that $k = 0$, and by consequence the unique saturated left solution of $\mathcal{CP}(\mathbb{R},\mathbb{R},0,0)$ is the function $x : (-\infty,0] \to \mathbb{R}$, defined by $x(t) = -t$.

Problem 2.5 Let us assume for contradiction that this would not be the case. Then there would exist a compact set $\mathcal{K} \subset \mathbb{I}\times\Omega$ such that for every $L > 0$ there exist $(t_L,x_L),(t_L,y_L) \in \mathcal{K}$ with $\|f(t_L,x_L) - f(t_L,y_L)\| > L\|x_L - y_L\|$. Taking $L = n$ with $n \in \mathbb{N}$, and denoting by $t_n = t_L$, $x_n = x_L$ and $y_n = y_L$, we have $\|f(t_n,x_n) - f(t_n,y_n)\| > n\|x_n - y_n\|$ for every $n \in \mathbb{N}$. Since $\mathcal{K}$ is compact and f is continuous on $\mathbb{I} \times \Omega$, it is bounded on $\mathcal{K}$. So there exists $M > 0$ such that $\|f(t,x)\| \leq M$ for every $(t,x) \in \mathcal{K}$. From this inequality and from the preceding one, we deduce that $n\|x_n - y_n\| \leq 2M$ for every $n \in \mathbb{R}$. It follows that $\lim_{n\to\infty}(x_n - y_n) = 0$. Using once again the compactness of $\mathcal{K}$, we may assume without loss of generality that there exists $(t^*, x^*) \in \mathcal{K}$ such that $\lim_{n\to\infty}(t_n, x_n, y_n) = (t^*, x^*, x^*)$. By hypothesis, we have that there exist a neighborhood $\mathcal{V}$ of (t^*, x^*), and $L = L(\mathcal{V}) > 0$, such that for every $(t,x), (t,y) \in \mathcal{V}$, we have $\|f(t,x) - f(t,y)\| \leq L\|x - y\|$. Since $\lim_{n\to\infty}(t_n, x_n, y_n) = (t^*, x^*, x^*)$, we deduce that there exists $n(\mathcal{V}) \in \mathbb{N}$ such that for every $n \geq n(\mathcal{V})$ we have $(t_n, x_n), (t_n, y_n) \in \mathcal{V}$. Consequently, for every $n \geq n(\mathcal{V})$, we necessarily have $n\|x_n - y_n\| < \|f(t_n, x_n) - f(t_n, y_n)\| \leq L\|x_n - y_n\|$ which, by virtue of the fact that $x_n \neq y_n$ for every $n \in \mathbb{N}$, leads to a contradiction: $(n < L)$ for every $n \geq n(\mathcal{V})$.

Problem 2.6 According to Problem 2.5, it suffices to prove that, for every (a, ξ) in $\mathbb{I} \times \Omega$, there exist a neighborhood $\mathcal{V}$ of (a, ξ) and $L > 0$, such that, for every $(t,x), (t,y) \in \mathcal{V}$, we have $\|f(t,x) - f(t,y)\| \leq L\|x - y\|$. Let $(a, \xi) \in \mathbb{I} \times \Omega$ and let $\mathcal{V}$ be a closed ball centered at (a, ξ) included in $\mathbb{I} \times \Omega$. Let $(t,x), (t,y) \in \mathcal{V}$, and let us observe that the function $\theta \mapsto (t, \theta y + (1-\theta)x)$ is continuous from $[\,0,1\,]$ and takes its values in $\mathcal{V}$, and this is because $\mathcal{V}$ is convex. Then the function

$$\theta \mapsto \frac{d}{d\theta}(f(t, \theta y + (1-\theta)x)) = \sum_{i=1}^{n} \frac{\partial f}{\partial x_j}(t, \theta y + (1-\theta)x)(y_j - x_j)$$

is continuous from $[\,0,1\,]$ in $\mathbb{R}^n$ too, and

$$\int_0^1 \frac{d}{d\theta}(f(t, \theta y + (1-\theta)x))\,d\theta = f(t,y) - f(t,x).$$

From the hypothesis, we know that $\partial f_i / \partial x_j$, $i,j = 1,2,\ldots,n$, are continuous on $\mathbb{I} \times \Omega$, and therefore they are bounded on the compact set $\mathcal{V}$. This means that there exists $M > 0$ such that

$$\left| \frac{\partial f_i}{\partial x_j}(s,z) \right| \leq M$$

for every $i,j = 1,2,\ldots,n$ and every $(s,z) \in \mathcal{V}$. Then, we have

$$\begin{aligned}
\|f(t,x) - f(t,y)\| &= \left\| \int_0^1 \frac{d}{d\theta}\left(f(t, \theta y + (1-\theta)x)\right) d\theta \right\| \\
&\leq \int_0^1 \left\| \frac{d}{d\theta}\left(f(t, \theta y + (1-\theta)x)\right) \right\| d\theta \\
&\leq \int_0^1 \left\| \sum_{j=1}^{n} \frac{\partial f}{\partial x_j}(t, \theta y + (1-\theta)x)(y_j - x_j) \right\| d\theta \leq \sqrt{n} M \|x - y\|,
\end{aligned}$$

and therefore $L = \sqrt{n} M$.

Problem 2.7 The substitution $x - t = y$ in the equation $x' = g(t, x)$ leads to the equation $y' = h(y)$, where $h : \mathbb{R} \to \mathbb{R}$ is defined by $h(y) = 1 + 2\sqrt[3]{y^2}$ for every $y \in \mathbb{R}$. This equation is with separable variables, while the associated Cauchy problem has the uniqueness property. Indeed, according to Theorem 1.3.1, the solution of $\mathcal{CP}(\mathbb{R}, \mathbb{R}, h, a, \xi)$ is

$$y(t, \xi) = H^{-1}(t - a)$$

for every $t \in \mathbb{R}$, where

$$H(z) = \int_{\xi}^{z} \frac{dy}{1 + 2\sqrt[3]{y^2}}.$$

In order to complete the proof, let us remark that the functions $x_1, x_2 : \mathbb{R} \to \mathbb{R}$, defined by $x_1(t) = t$ and $x_2(t) = \frac{1}{27}(t - a)^3 + t$ for $t \in \mathbb{R}$, are distinct solutions of $\mathcal{CP}(\mathbb{R}, \mathbb{R}, f, a, a)$.

Problem 2.8 We begin by observing that both functions $x \vee y$ and $x \wedge y$ satisfy the initial condition. Also, it is easy to see that $x \vee y$ and $x \wedge y$ are continuous on $\mathbb{J}$, differentiable on the open set $\{t;\ t \in \mathbb{J},\ x(t) \neq y(t)\}$, and satisfy the differential equation at every point in this set. This follows from the fact that the set above is at most a countable union of open intervals and, on each interval $\mathbb{J}_k$ from this union, we have either $x(t) < y(t)$ for every $t \in \mathbb{J}_k$, or $x(t) > y(t)$ for every $t \in \mathbb{J}_k$. In order to complete the proof, it suffices to show that $x \vee y$ and $x \wedge y$ are differentiable at every point $t \in \mathbb{J}$ at which $x(t) = y(t)$. Let t be such a point. Then, we have

$$\lim_{s \to t} \frac{x(s) - x(t)}{s - t} = x'(t) = f(t, x(t)) = f(t, y(t)) = y'(t) = \lim_{s \to t} \frac{y(s) - y(t)}{s - t}$$

which shows that $x \vee y$ and $x \wedge y$ are differentiable at t and both derivatives at this point are equal, i.e. $x'(t) = y'(t)$. From this simple remark, and from the fact that $(x \vee y)(t) = (x \wedge y)(t) = x(t) = y(t)$, it follows that both functions satisfy the differential equation at t, which completes the proof.

Problem 2.9 Let us assume for contradiction that there exists $t_0 \in [\,a, b_\xi) \cap [\,a, b_\eta)$ such that $x(t_0, \xi) > x(t_0, \eta)$. Since $(x(t_0, \xi) - x(t_0, \eta))(x(a, \xi) - x(a, \eta)) \leq 0$ and $t \mapsto x(t, \xi) - x(t, \eta)$ has the Darboux property being continuous, there exists $t_1 \in [\,a, t_0)$ such that $x(t_1, \xi) = x(t_1, \eta)$. From the uniqueness property, we deduce that $x(t, \xi) = x(t, \eta)$ for every $t \in [\,t_1, b_\xi) \cap [\,t_1, b_\eta)$. We get $x(t_0, \xi) = x(t_0, \eta)$ which is in contradiction with the supposition made. This contradiction can be eliminated only if $x(t, \xi) \leq x(t, \eta)$ for every $t \in [\,a, b_\xi) \cap [\,a, b_\eta)$.

Problem 2.10 First, we will analyze the particular case in which f is continuous on $\mathbb{I} \times \Omega$ and locally Lipschitz on Ω, and then, we will show that the general case reduces to the former one. In addition, let us observe that it suffices to prove the inequality only locally at the right of the point a. More precisely, if, in the hypotheses of the problem, there exists $c \in (a, b)$ such that $y(t) \leq x(t)$ for every $t \in [\,a, c\,]$, then $b = c^* = \sup\{c \in [\,a, b);\ y(t) \leq x(t) \text{ for } t \in [\,a, c\,]\}$. Indeed, if this is not the case, we must have $c^* < b$ and $y(c^*) \leq x(c^*)$. Now, from the fact that the inequality $y(t) \leq x(t)$ holds locally at the right of c^*, we get a contradiction, i.e. c^*, which is the supremum of a set, is strictly less than an element in that set. Let then $c \in (a, b)$ and let us define the function $z : [a, b) \to \Omega$ by $z(t) = y(t) - x(t)$ for every $t \in [a, c]$. Let us assume for contradiction that there exists $s \in [a, c]$ such that $z(s) > 0$. Since z is continuous and $z(a) \leq 0$, the set $\{t \in [a, s];\ z(t) = 0\}$ is nonempty, bounded and closed. So it has a last element τ. Obviously $z(\tau) = 0$. Also $z(t) > 0$ for every $t \in (\tau, s]$. Indeed, assuming for contradiction that this is not the case, it follows that there exists a point $t_1 \in (\tau, s)$ with $z(t_1) \leq 0$. Since z has the Darboux property, it follows that there exists $t_2 \in (t_1, s)$ with $z(t_2) = 0$, which is absurd because $\tau < t_2 < s$ and τ is the largest number in $[c, s]$ for which $z(\tau) = 0$. The contradiction we got can be eliminated only if $z(t) > 0$ for every $t \in (\tau, s]$. Let now $L > 0$ be the Lipschitz constant of the function f corresponding the compact set $K = \{(t, x(t));\ t \in [a, c]\} \cup \{(t, y(t));\ t \in [a, c]\}$. Then, we have

$$z'(t) \leq f(t, y(t)) - f(t, x(t)) \leq |f(t, x(t)) - f(t, y(t))| \leq L|y(t) - x(t)| = Lz(t)$$

for every $t \in [\tau, s]$. Since $z(\tau) = 0$, integrating the inequality $z'(\eta) \leq Lz(\eta)$ over $[\tau, t]$, we get

$$z(t) \leq \int_\tau^t Lz(\eta)\, d\eta$$

for every $t \in [\tau, s]$. From Gronwall's Lemma 1.5.2, it follows that $z(t) \leq 0$ for every $t \in [\tau, s]$, relation in contradiction with $z(t) > 0$ for every $t \in (\tau, s)$. This contradiction originates in the supposition that there exists $s \in [a, c]$ such that $z(s) > 0$. So, $z(t) \leq 0$ for every $t \in [a, c]$, which solves the problem in the particular case when f is locally Lipschitz on Ω.

We can now proceed to the general case. To this aim, let us fix $c \in (a, b)$ and let $r > 0$ be such that $B(\xi, r) \subset \Omega$. Let us also fix $M > 0$ such that

$$|f(t, x)| \leq M \tag{$*$}$$

for every $(t, x) \in [a, c]$, and let us observe that the set $C = [a, c] \times B(\xi, r)$ is compact and included in $\mathbb{I} \times \Omega$. From Weierstrass' approximation theorem, it follows that, for every $\varepsilon > 0$, there exists a polynomial $f_\varepsilon : \mathbb{I} \times \Omega \to \mathbb{R}$ with the property that

$$|f(t, x) - f_\varepsilon(t, x)| \leq \varepsilon$$

for every $(t, x) \in C$. Since f_ε is of class C^∞, it is Lipschitz on C. Let us consider the Cauchy problem

$$\begin{cases} x_\varepsilon'(t) = f_\varepsilon(t, x_\varepsilon(t)) + \varepsilon \\ x_\varepsilon(a) = \xi. \end{cases} \tag{$\mathcal{CP}$}_\varepsilon$$

Let $\delta = \min\{c - a, \frac{r}{M+2}\}$ and let us observe that, for every $\varepsilon \in (0, 1)$, $(\mathcal{CP})_\varepsilon$ has a unique saturated solution $x_\varepsilon : [a, b_\varepsilon) \to \overset{\circ}{B}(\xi, r)$ with the property that $b_\varepsilon \geq a + \delta$. Indeed, if we would assume by contradiction that $b_\varepsilon < a + \delta$, from $(*)$ it would follow that $f_\varepsilon(\cdot, x_\varepsilon(\cdot))$ is bounded on $[a, b_\varepsilon)$ by $M + 1$. This means that $|f_\varepsilon(t, x)| \leq |f_\varepsilon(t, x) - f(t, x)| + |f(t, x)| \leq \varepsilon + M \leq 1 + M$, and, by Proposition 2.4.1, there would exist $\lim_{t\uparrow b_\varepsilon} x_\varepsilon(t) = x^*$. In addition, in view of (ii) in Theorem 2.4.3, x^* must belong to the boundary of the set $B(\xi, r)$, i.e. $|x^* - \xi| = r$. On the other hand, we have

$$r = |x^* - \xi| \leq \int_a^{b_\varepsilon} |f_\varepsilon(s, x_\varepsilon(s))|\, ds + (b_\varepsilon - a)\varepsilon \leq (b_\varepsilon - a)[(M+1)+\varepsilon] < \delta(M+2) \leq r,$$

i.e. $r < r$, which is absurd. Consequently, x_ε is defined at least on $[\,a, a+\delta\,]$. Let us observe that $f(t,x) \leq f_\varepsilon(t,x) + \varepsilon$ for every $(t,x) \in [\,a, a+\delta\,]$, and therefore we have $y'(t) \leq f_\varepsilon(t, y(t))$, $x'_\varepsilon(t) = f_\varepsilon(t, x(t))$, and $y(a) \leq x_\varepsilon(a) = \xi$. According to the preceding item, we have that $y(t) \leq x_\varepsilon(t)$ for every $t \in [\,a, a+\delta\,]$. On the other hand, we can observe that the family $\{x_\varepsilon;\ \varepsilon \in (0,1)\}$ is uniformly bounded and equicontinuous on $[\,a, a+\delta\,]$. According to Theorem 12.2.1, there exists one sequence $(\varepsilon_k)_{k\in\mathbb{N}}$ tending to 0, such that the corresponding sequence of solutions of $(\mathcal{CP})_{\varepsilon_k}$, denoted for simplicity by $(x_k)_{k\in\mathbb{N}}$, be uniformly convergent on $[\,a, a+\delta\,]$ to a function $\tilde{x}$. Passing to the limit in $(\mathcal{CP})_{\varepsilon_k}$, we deduce that $\tilde{x}$ is a solution of the problem $x'(t) = f(t, x(t))$, $x(a) = \xi$. Since this problem has the uniqueness property, $\tilde{x} = x$. Passing to the limit in the inequality $y(t) \leq x_k(t)$ for every $t \in [\,a, a+\delta\,]$ we get the required inequality.

Problem 2.11 Let $x, y : \mathbb{J} \to \Omega$ be two solutions of $\mathcal{CP}(\mathbb{I}, \Omega, f, a, \xi)$. Taking the inner product on both sides of the equality $x'(t) - y'(t) = f(t, x(t)) - f(t, y(t))$ by $x(t) - y(t)$, using (i) in Lemma 12.1.2 and the condition in hypothesis, we deduce

$$\frac{1}{2}\frac{d}{dt}\|x(t) - y(t)\|^2 \leq \omega(t, \|x(t) - y(t)\|)\|x(t) - y(t)\|$$

for every $t \in \mathbb{J}$. Denoting by $z(t) = \frac{1}{2}\|x(t) - y(t)\|^2$, the inequality above may be rewritten in the form $z'(t) \leq \omega(t, \sqrt{2z(t)})\sqrt{2z(t)}$, or equivalently $(\sqrt{2z})'(t) \leq \omega(t, \sqrt{2z(t)})$ for every $t \in \mathbb{J}$. Since $\sqrt{2z(a)} = 0$ and the unique solution of $\mathcal{CP}(\mathbb{I}, \mathbb{R}_+, \omega, a, 0)$ is the identically null function, from Problem 2.10, it follows that $\sqrt{2z(t)} \leq 0$ for every $t \in \mathbb{J}$, which shows that $x(t) = y(t)$ for every $t \in \mathbb{J}$.

Theorem 2.3.1 follows from the previous result by taking $\omega : \mathbb{R} \to \mathbb{R}$, defined by $\omega(\eta) = L\eta$ for every $\eta \in \mathbb{R}$, where $L > 0$ is the Lipschitz constant corresponding to the function f on the set $[a, a+\delta\,] \times B(\xi, r)$. See the proof of Theorem 2.3.1. Theorem 2.3.3 follows from the preceding considerations by taking $\omega \equiv 0$.

Problem 2.12 We begin by observing that, for every $a > 0$, we have

$$\int_0^a \frac{d\eta}{\omega(\eta)} = +\infty.$$

Let us assume for contradiction that there exists a non-identically zero solution $x : [\,0, T) \to \mathbb{R}$. Since $\omega(r) \geq 0$, it follows that $x(t) \geq 0$. Consequently, there exists $t \in (0, T)$ such that $x(t) > 0$. From here, it follows that there exist $\alpha, \beta \in [\,0, T)$

with $\alpha < \beta$ and $x(\alpha) = 0 < x(t)$ for every $t \in (\alpha, \beta)$. Then, we have

$$1 = \frac{x'(t)}{\omega(x(t))}$$

for every $t \in (\alpha, \beta)$. Integrating this equality from α to β, we get

$$\beta - \alpha = \int_\alpha^\beta \frac{x'(t)\,dt}{\omega(x(t))} = \int_0^{x(\beta)} \frac{d\eta}{\omega(\eta)} = +\infty$$

which is absurd. The contradiction we got can be eliminated only if the unique saturated solution of the Cauchy problem considered is $x \equiv 0$.

Problem 2.13 Since f is Lipschitz and g is dissipative, we have

$$\langle f(t,x) + g(t,x) - f(t,y) - g(t,y), x - y\rangle \leq L\|x - y\|^2$$

for every $t \in \mathbb{I}$ and every $x, y \in \Omega$. We are in the hypotheses of Problem 2.11, with $\omega(r) = r$ for every $r \in \mathbb{R}_+$.

Problem 2.14 We have

$$\langle f(t,x) - f(t,y), x - y\rangle \leq \omega(\|x - y\|)\|x - y\|$$

for every $t \in \mathbb{I}$ and every $x, y \in \Omega$, and we are in the hypotheses of Problem 2.11.

Problem 2.15 Let $[\,a, a + \delta\,]$, $B(\xi, r)$ and $L > 0$ be chosen as in the proof of Theorem 2.3.1. We denote by $u(t) = e^{-L(t-a)}x(t)$ and by $v(t) = e^{-L(t-a)}y(t)$, and let us observe that u, v are solutions of $\mathcal{CP}(\mathbb{I}, \Omega_0, g, a, \xi)$, where the function $g(t,z) = e^{-L(t-a)}f(t, e^{L(t-a)}z) - Le^{-L(t-a)}z$ for every (t,z) in $\mathbb{I}\times\Omega_0$ with $\Omega_0 \subset \Omega$ suitably chosen. Since g satisfies the hypotheses of Theorem 2.3.3 (see the proof of Theorem 2.5.2), it follows that $u \equiv v$, or equivalently that $x \equiv y$ on $\mathbb{J}$.

Problem 2.16 The functions $x_1, x_2 : \mathbb{R} \to \mathbb{R}$, defined by $x_1(t) = 0$ for every $t \in \mathbb{R}$, and

$$x_2(t) = \begin{cases} \dfrac{(t+1)^3}{27} & \text{if } t < -1 \\ 0 & \text{if } t \in [\,-1, 0\,] \\ \dfrac{t^3}{27} & \text{if } t > 0 \end{cases}$$

are two distinct saturated solutions of $\mathcal{CP}(\mathbb{R}, \mathbb{R}, f, -1, 0)$.

Problem 2.17 Let $\mathbb{I} = \mathbb{R}$, $\Omega = \left(-\frac{\pi}{2}, \frac{\pi}{2}\right)$ and $f : \mathbb{I}\times\Omega \to \mathbb{R}$, $f(t,x) = \tan x$ for every $(t,x) \in \mathbb{I}\times\Omega$. One may easily see that f does not map bounded subsets in $\mathbb{I}\times\Omega$ into bounded subsets in $\mathbb{R}$.

Problem 2.18 The proof follows the same way as that of Theorem 2.4.5, except the phrase preceding the inequality (2.4.4), which in this case should read : "Since for every compact subset $\mathbb{J}$ in $\mathbb{I}$ and every bounded subset B in Ω, $f(\mathbb{J}\times\Omega)$ is bounded, as $[\,a, b\,]$ is compact and included in $\mathbb{I}$ and C is bounded, it follows that there exists $M > 0$ such that...". In addition, it is easy to see that the function $f : \left(-\frac{\pi}{2}, \frac{\pi}{2}\right) \times \left(-\frac{\pi}{2}, \frac{\pi}{2}\right) \to \mathbb{R}$, defined by $f(t,x) = \tan t \cdot \tan x$ for $(t,x) \in \left(-\frac{\pi}{2}, \frac{\pi}{2}\right) \times \left(-\frac{\pi}{2}, \frac{\pi}{2}\right)$ has the property in the problem, but it does not map bounded subsets in $\mathbb{I} \times \Omega$ in bounded subsets in $\mathbb{R}$. Hence, the class of functions with the property described in this problem is strictly broader than that of the functions f which map bounded subsets in $\mathbb{I} \times \Omega$ into bounded subsets in $\mathbb{R}^n$.

Problem 2.19 The answer is in the negative as we can see by taking the sets $\mathcal{K} = \{(x_1, x_2) \in \mathbb{R}^2;\ x_1 \geq 0,\ x_2 = 0\}$, $\mathcal{F} = \left\{(x_1, x_2) \in \mathbb{R}^2;\ x_1 > 0,\ x_2 = \dfrac{1}{x_1}\right\}$ which are closed in $\mathbb{R}^2$ and $\mathcal{K} \cap \mathcal{F} = \emptyset$, but $\operatorname{dist}(\mathcal{K}, \mathcal{F}) = 0$.

Problem 2.20 Let $x : [\,0, b) \to \mathbb{R}$ be a saturated solution of the Cauchy problem considered. This means that the vector-valued function $z : [\,0, b) \to \mathbb{R}^2$, defined by $z(t) = (x(t), x'(t))$ for every $t \in [\,0, b)$, is a saturated solution of the Cauchy problem

$$\begin{cases} x' = y \\ y' = -g(x) - f(y) \\ x(0) = \xi_1,\ y(0) = \xi_2. \end{cases} \tag{$\mathcal{CP}$}$$

Multiplying the equation $x'' + f(x') + g(x) = 0$ by x', integrating the equality thus obtained over $[\,0, t\,]$, and recalling that $G(x) \geq ax^2$ and $yf(y) \geq 0$, we get

$$\frac{1}{2}|x'(t)|^2 + a|x(t)|^2 \leq \frac{1}{2}|\xi_1|^2 + a|\xi_2|^2$$

for every $t \in [\,0, b)$. Since $a > 0$, it follows that the function z, defined as above, is a saturated bounded solution of $\mathcal{CP}(\mathcal{D})$. According to Corollary 2.4.3, $b = +\infty$, which completes the proof.

Problem 2.21 The uniqueness follows from Problem 2.13. We show that every saturated solution of $\mathcal{CP}(\mathbb{R}_+, \mathbb{R}^n, f+g, a, \xi)$ is global. By virtue of Corollary 2.4.3, in order to do this, it suffices to prove that, if $x : [\,a, b) \to \mathbb{R}^n$ is a solution of $\mathcal{CP}(\mathbb{R}_+, \mathbb{R}^n, f+g, a, \xi)$ with $b < +\infty$, then x is bounded on $[\,a, b)$. Let us observe that, since f is Lipschitz g is dissipative on $\mathbb{R}^n$, we have

$$\frac{1}{2}\frac{d}{ds}\|x(s)\|^2 \leq L\|x(s)\|^2 + [\|f(s,0)\| + \|g(s,0)\|]\|x(s)\|$$

for each $s \in [\,a, b)$. Integrating on both sides from a to t and using Lemma 1.5.3, we get

$$\|x(t)\| \leq \|\xi\| + \int_a^b [\|f(s,0)| + \|g(s,0)\|]\, ds + L\int_a^t \|x(s)\|\, ds$$

for every $t \in [\,a, b)$. From Gronwall's Lemma 1.5.2, it follows that

$$\|x(t)\| \leq \left(\|\xi\| + \int_a^b [\|f(s,0)\| + \|g(s,0)\|]\, ds\right) e^{L(b-a)}$$

for every $t \in [\,a, b)$, which completes the proof.

Problem 2.22 Since x is bounded on $[\,a, b)$, the set of its limit points for $t \uparrow b$ is nonempty and compact. In order to complete the proof, it suffices to show that this set contains exactly one element. To this aim, as x is saturated and

$b < t_2$, by (iii) in Theorem 2.4.3, we deduce that the set of these limit points is included in the boundary of the set (ω_1, ω_2) which is $\{\omega_1, \omega_2\}$. Assuming for contradiction that both ω_1 and ω_2 are limit points of x for $t \uparrow b$, it follows that there exist two sequences $(t_k)_{k\in\mathbb{N}}$ and $(s_k)_{k\in\mathbb{N}}$, both strictly increasing to b and such that $\lim_{k\to\infty} x(t_k) = \omega_1$ and $\lim_{k\to\infty} x(s_k) = \omega_2$. In addition, we may assume without loss of generality (taking two subsequences and relabelling, if necessary) that $t_k < s_k$ for every $k \in \mathbb{N}$. Let now $\omega \in (\omega_1, \omega_2)$. Then there exists $k_\omega \in \mathbb{N}$ such that $x(t_k) \in (\omega_1, \omega)$ and $x(s_k) \in (\omega, \omega_2)$ for every $k \geq k_\omega$. Since x is continuous, it has the Darboux property, and therefore, for every $k \geq k_\omega$, there exists $r_k \in (t_k, s_k)$ such that $x(r_k) = \omega$. Obviously $\lim_{k\to\infty} r_k = b$ and by consequence $\omega \in (\omega_1, \omega_2)$ is also a limit point of x for $t \uparrow b$. This contradiction can be eliminated only if the set of all limit points of x for $t \uparrow b$ is a singleton. The generalization to the n-dimensional case reads as follows: *if $\Omega \subset \mathbb{R}^n$ is an open set whose boundary contains only isolated points, $f : (t_1, t_2) \times \Omega \to \mathbb{R}^n$ is continuous, $a \in (t_1, t_2)$, $\xi \in \Omega$ and $x : [a, b) \to \Omega$ is a saturated solution of $\mathcal{CP}((t_1, t_2), \Omega, f, a, \xi)$ with $b < t_2$ and x is bounded on $[a, b)$, then there exists $\lim_{t\uparrow b} x(t) = x^*$.* The proof follows the same way as before, by observing that the line segment joining any two distinct points in the boundary of Ω contains the whole nontrivial subsegment (which does not reduce to a single point) included in Ω.

Problem 2.23 Multiplying the equation $x' = f(x)$ by x', integrating side by side over $[a, b]$, and taking into account that $x(a) = x(b)$, we get

$$\int_a^b x'^2(t)\,dt = \int_a^b f(x(t))x'(t)\,dt = \int_a^b \frac{d}{dt}(F(x(t))\,dt = F(x(b)) - F(x(a)) = 0,$$

where $F : \mathbb{R} \to \mathbb{R}$ is a primitive of the function f. Since x'^2 is continuous and non-negative, its integral over $[a, b]$ equals zero if and only if $x' \equiv 0$ on $[a, b]$. Hence x is constant on $[a, b]$. The result no longer holds in the case $f : \mathbb{R}^n \to \mathbb{R}^n$ for $n > 1$, as we can state by observing that the function $x : [0, 2\pi] \to \mathbb{R}^2$, defined by $x(t) = (x_1(t), x_2(t)) = (\sin t, \cos t)$ for $t \in [0, 2\pi]$, is a nonconstant solution of the problem

$$\begin{cases} x' = f(x) \\ x(a) = x(b), \end{cases} \tag{$\mathcal{P}$}$$

where $[a, b] = [0, 2\pi]$ and $f : \mathbb{R}^2 \to \mathbb{R}^2$ is given by $f(x_1, x_2) = (x_2, -x_1)$ for $(x_1, x_2) \in \mathbb{R}^2$. However, we can prove, by using very similar arguments, that if $f : \mathbb{R}^n \to \mathbb{R}^n$ is the gradient of a function of class C^1 $F : \mathbb{R}^n \to \mathbb{R}$, then every solution $x : [a, b] \to \mathbb{R}^n$, of class C^1, of the problem $(\mathcal{P})$ is constant[1]. In this case we have

$$\int_a^b \|x'(t)\|^2 dt = \int_a^b \langle f(x(t)), x'(t)\rangle dt = \int_a^b \left(\sum_{i=1}^n \frac{\partial F}{\partial x_i}(x(t))x_i'(t) \right) dt$$

[1] This condition is automatically satisfied for $n = 1$ because every continuous function $f : \mathbb{R} \to \mathbb{R}$ admits primitives, and thus it is the gradient of any of its primitives.

$$= F(x(b)) - F(x(a)) = 0.$$

Problem 2.24 Let $[a,b] \subset \mathbb{I}$. From Lemma 2.2.2 and the uniqueness assumption, it follows that, for every $\xi \in \mathbb{R}^n$, $\mathcal{CP}(\mathbb{I}, \mathbb{R}^n, f, a, \xi)$ has a unique solution $x(\cdot, \xi) : [a,b] \to \mathbb{R}^n$. Let $(\xi_k)_{k\in\mathbb{N}}$ be a sequence of points in $\mathbb{R}^n$ with $\lim_{k\to\infty} \xi_k = \xi$. Since the function f is bounded, it follows that there exists $M > 0$ such that $\|f(t,x)\| \leq M$ for every $(t,x) \in \mathbb{I} \times \mathbb{R}^n$. Also there exists $m > 0$ such that $\|\xi_k\| \leq m$ for every $k \in \mathbb{N}$. Then, we have

$$\|x(t,\xi_k)\| \leq \|\xi_k\| + \int_a^t \|f(s, x(s,\xi_k))\| \, ds \leq m + (b-a)M$$

and

$$\|x(t,\xi_k) - x(s,\xi_k)\| \leq \left| \int_s^t \|f(\tau, x(\tau, \xi_k))\| \, d\tau \right| \leq M|t-s|$$

for every $k \in \mathbb{N}$ and every $t, s \in [a,b]$. According to Theorem 12.2.1, it follows that the family of functions $\{x(\cdot,\xi_k);\ k \in \mathbb{N}\}$ is relatively compact in the space $C([a,b];\mathbb{R}^n)$ endowed with the uniform convergence topology. Therefore, in order to conclude the proof, it suffices to show that the only limit point of the sequence $(x(\cdot,\xi_k))_{k\in\mathbb{N}}$, in this topology, is $x(\cdot,\xi)$. Let then y be such a limit point. For the sake of simplicity, let us denote again by $(x(\cdot,\xi_k))_{k\in\mathbb{N}}$ the subsequence which is uniformly convergent to y. Then, according to Corollary 12.2.1, it follows that the set $U = \overline{\{x(t,\xi_k);\ k \in \mathbb{N},\ t \in [a,b]\}}$ is compact. Passing to the limit in the equality

$$x(t,\xi_k) = \xi_k + \int_a^t f(s, x(s,\xi_k)) \, ds$$

for k tending to $+\infty$, and using Corollary 12.2.2 with

$$U = \overline{\{x(t,\xi_k);\ k \in \mathbb{N},\ t \in [a,b]\}},$$

we deduce that y is a solution of $\mathcal{CP}(\mathbb{I}, \mathbb{R}^n, f, a, \xi)$. As, by hypothesis this problem has the uniqueness property, it follows that $y(t) = x(t,\xi)$ for every $t \in [a,b]$, which achieves the proof.

Problem 2.25 As we have seen in the proof of Problem 2.3, if $\xi > 0$, then the function $x(\cdot,\xi) : [0,+\infty) \to \mathbb{R}$, defined by $x(t,\xi) = \sqrt{t^2+\xi^2}$, is the unique global right solution for $\mathcal{CP}(\mathbb{R}, \mathbb{R}, f, 0, \xi)$. Then

$$|x(t,\xi) - t| \leq |\sqrt{t^2+\xi^2} - t| = \frac{\xi^2}{\sqrt{t^2+\xi^2}+t} \leq \xi$$

for every $t \geq 0$ and every $\xi > 0$. Hence $\lim_{\xi\downarrow 0} x(t,\xi) = t$ uniformly for $t \geq 0$. Nevertheless, the function $y(t) = t$ for $t \geq 0$ is not a solution of $\mathcal{CP}(\mathbb{R}, \mathbb{R}, f, 0, 0)$, because $y'(0) = 1 \neq f(0,0) = 0$. This discontinuity with respect to the initial data is a consequence of the discontinuity of the function f at the points of the form $(t, 0)$ with $t \in \mathbb{R}$.

Problem 2.26 For every $p > 0$, the unique solution of $\mathcal{CP}(\mathbb{R}, \mathbb{R}, f, 0, 0)_p$ is $x(\cdot, p) : [0, +\infty) \to \mathbb{R}$, defined by $x(t, p) = \sqrt{t^2 + p^2} - p$ for every $t \in [0, +\infty)$. We have

$$|x(t,p)| = \left| \frac{t^2}{\sqrt{t^2+p^2}} \right| \leq 1 \text{ and } |x'(t,p)| = \left| \frac{t}{\sqrt{t^2+p^2}} \right| \leq 1$$

for every $p \geq 0$ and every $t \in [0, 1]$. So, the family of functions $\{x(\cdot, p);\ p \geq 0\}$ is relatively compact in $C([0, 1]; \mathbb{R})$. From this observation and from the fact that $\lim_{p \downarrow 0} x(t, p) = t$ point-wise on $[0, 1]$, it follows that the convergence above is in fact uniform on $[0, 1]$. But the function $y : [0, 1] \to \mathbb{R}$, defined by $y(t) = t$, is not a solution of $\mathcal{CP}(\mathbb{R}, \mathbb{R}, f, 0, 0)_0$ because $y'(0) = 1 \neq f(0, 0, 0) = 0$. In this case too, the discontinuity of the solution with respect to the parameter p is caused by the discontinuity of the function f at the points of the form (t, x, p) with $x + p = 0$.
Problem 2.27 For every fixed $p \neq 0$, the function $x \mapsto 3\sqrt[3]{x^2 + p^2}$ is locally Lipschitz on $\mathbb{R}$, being of class C^1. Then, according to Theorem 2.3.1, it follows that, for every $p > 0$, $\mathcal{CP}(\mathcal{D})_p$ has the uniqueness property. On the other hand, as we have already seen in Example 2.3.1, $\mathcal{CP}(\mathcal{D})_0$ does not have the uniqueness property.
Problem 2.28 The idea of proof follows the very same lines as those in the proof of Theorem 2.8.1. So, if we assume that $x : [a, b] \to \mathbb{R}^n$ is a function implicitly defined by $F(t, x(t)) = F(a, \xi)$, it follows that $dF(t, x(t)) = 0$ for $t \in [a, b]$. This means that x is a solution of the system

$$\frac{\partial F_i}{\partial t}(t, x(t)) + \sum_{j=1}^{n} \frac{\partial F_i}{\partial x_j}(t, x(t)) \frac{dx_j}{dt}(t) = 0, \quad i = 1, 2, \ldots, n,$$

and $x(a) = \xi$. In other words, x is a solution of the Cauchy problem

$$\begin{cases} \begin{pmatrix} \dfrac{dx_1}{dt}(t) \\ \dfrac{dx_2}{dt}(t) \\ \vdots \\ \dfrac{dx_n}{dt}(t) \end{pmatrix} = - \begin{pmatrix} \dfrac{\partial F_1}{\partial x_1}(t,x) & \dfrac{\partial F_1}{\partial x_2}(t,x) & \dots & \dfrac{\partial F_1}{\partial x_n}(t,x) \\ \dfrac{\partial F_2}{\partial x_1}(t,x) & \dfrac{\partial F_2}{\partial x_2}(t,x) & \dots & \dfrac{\partial F_2}{\partial x_n}(t,x) \\ \vdots & & & \\ \dfrac{\partial F_n}{\partial x_1}(t,x) & \dfrac{\partial F_n}{\partial x_2}(t,x) & \dots & \dfrac{\partial F_n}{\partial x_n}(t,x) \end{pmatrix}^{-1} \begin{pmatrix} \dfrac{\partial F_1}{\partial t}(t,x) \\ \dfrac{\partial F_2}{\partial t}(t,x) \\ \vdots \\ \dfrac{\partial F_n}{\partial t}(t,x) \end{pmatrix} \\ x_1(a) = \xi_1, \\ x_2(a) = \xi_2, \\ \vdots \\ x_n(a) = \xi_n. \end{cases}$$

Conversely, each solution of this Cauchy problem is a function implicitly defined by $F(t, x(t)) = F(a, \xi)$. The existence part follows from Peano's Theorem 2.2.1, while the uniqueness is a consequence of the Local Inversion Theorem.

Chapter 3

Exercise 3.1 (1) We look for the solution as a power series of the form

$$x(t) = \sum_{k=0}^{\infty} c_k t^k.$$

Imposing the condition that x satisfies the equation, we deduce

$$\sum_{k=1}^{\infty} k c_k t^{k-1} - \sum_{k=1}^{\infty} k c_k t^k = 1 + t - \sum_{k=0}^{\infty} c_k t^k.$$

From the initial condition and identifying the coefficients, we get

$$\begin{cases} c_0 = 0 \\ c_1 = 1, \quad c_2 = 1/2, \\ c_{k+1} = c_k(k-1)/(k+1), \quad k = 2, 3, \dots \end{cases}$$

from where it follows

$$x(t) = t + \sum_{p=1}^{\infty} \frac{t^{p+1}}{p(p+1)},$$

the series being uniformly and absolutely convergent on $[-1, 1]$.

(2) Proceeding as in the preceding exercise, we find

$$x(t) = \sum_{p=0}^{\infty} (-1)^p \frac{t^{p+1}}{(p!)^2(p+1)},$$

the series being absolutely convergent on $\mathbb{R}$ and uniformly convergent on every compact interval.

(3) Similarly, we have

$$x(t) = \sum_{p=0}^{\infty} (-1)^p \frac{(t/2)^{2p}}{(p!)^2},$$

the series being absolutely convergent on $\mathbb{R}$ and uniformly convergent on every compact interval.

(4) The solution of the equation is

$$x(t) = \sum_{p=0}^{\infty} (-1)^p \frac{t^{2p}}{(2p+1)!}$$

for every $t \in \mathbb{R}$, i.e.

$$x(t) = \begin{cases} (\sin t)/t & \text{for } t \neq 0 \\ 1 & \text{for } t = 0. \end{cases}$$

(5) The solution of the equation is

$$x(t)=\sum_{p=0}^{\infty}\frac{t^p}{p!}$$

for every $t\in\mathbb{R}$, i.e. $x(t)=e^t$ for every $t\in\mathbb{R}$.

(6) The solution of the equation is

$$x(t)=\sum_{p=0}^{\infty}\frac{t^{2p}}{p!}$$

for every $t\in\mathbb{R}$, i.e. $x(t)=e^{t^2}$ for every $t\in\mathbb{R}$.

Problem 3.1 We look for x as $x(t)=\sum_{k=0}^{\infty}c_kt^k$. Asking that x satisfy the equation, we obtain $\sum_{k=2}^{\infty}k(k-1)c_kt^{k-2}-\sum_{k=1}^{\infty}2kc_kt^k+\sum_{k=0}^{\infty}2\lambda c_kt^k=0$. By identifying the coefficients, we get

$$\begin{cases}2\cdot 1c_2+2\lambda c_0=0\\ 3\cdot 2c_3-2c_1+2\lambda c_1=0\\ (k+2)(k+1)c_{k+2}-2kc_k+2\lambda c_k=0\end{cases}$$

for every $k\in\mathbb{N}$. From these equalities, we deduce

$$\begin{cases}c_{2p}=(-1)^p\dfrac{2^p\lambda(\lambda-2)\dots(\lambda-2p+2)}{(2p)!}c_0\\[2ex] c_{2p+1}=(-1)^p\dfrac{2^p(\lambda-1)(\lambda-3)\dots(\lambda-2p+1)}{(2p+1)!}c_1\end{cases}$$

for every $p\in\mathbb{N}$. Taking successively $(c_0,c_1)=(1,0)$ and $(c_0,c_1)=(0,1)$, we get the solutions

$$x_1(t)=1+\sum_{p=1}^{\infty}(-1)^p\frac{2^p\lambda(\lambda-2)\dots(\lambda-2p+2)}{(2p)!}t^{2p}$$

and

$$x_2(t)=t+\sum_{p=1}^{\infty}(-1)^p\frac{2^p(\lambda-1)(\lambda-3)\dots(\lambda-2p+1)}{(2p+1)!}t^{2p+1},$$

both series being absolutely convergent on $\mathbb{R}$, and uniformly convergent on every compact interval. Since the Wronskian of this system of solutions is nonzero at $t=0$, it follows that $\{x_1,x_2\}$ is a fundamental system of solutions for the Hermite equation. If $\lambda\in\mathbb{N}$, one can easily see that one of the two solutions is a polynomial. More precisely, if λ is even, x_1 is a polynomial of even degree, while if λ is odd, x_2 is a polynomial of odd degree. Conversely, if there exists a non-identically null polynomial which satisfies the equation, then this is a linear combination of x_1 and x_2. Since the first series contains only even powers of t, while the second one only odd powers, al least one of these must be a polynomial. From here it follows that there exists a first null coefficient in the series, which can happen only if $\lambda \in \mathbb{N}$.

Problem 3.2 For $\lambda = 0$ the equation takes the equivalent form $x'' + (1-t^2)x = 0$. Let us remark that a solution of this equation is $x(t) = e^{-t^2/2}$. For this reason it follows that, for every $\lambda \in \mathbb{R}$, the function $x(t) = y(t)e^{-t^2/2}$ satisfies the equation $-x'' + t^2 x = (2\lambda + 1)x$ if and only if y satisfies the Hermite equation $y'' - 2ty + 2\lambda y = 0$. Accordingly, the general solution of the equation is of the form $x(t) = [c_1 x_1(t) + c_2 x_2(t)]e^{-t^2/2}$, with $c_1, c_2 \in \mathbb{R}$, and x_1, x_2 determined in the solution of Problem 3.1. Then, if $\lambda \in \mathbb{N}$, again from Problem 3.1, we have that at least one of the two solutions is a polynomial. So, at least one of the solutions $x(t) = x_1(t)e^{-t^2/2}$, or $z(t) = x_2(t)e^{-t^2/2}$ is non-identically zero and bounded on $\mathbb{R}_+$.

Exercise 3.2 According to Theorem 3.1.1 the solution of the equation is an analytic function. Consequently $x(t) = \sum_{k=0}^{\infty} c_k t^k$. Then $x'(t) = \sum_{k=1}^{\infty} k c_k t^{k-1}$ and $x''(t) = \sum_{k=2}^{\infty} k(k-1) c_k t^{k-2}$. Substituting in the equation and identifying the coefficients, we get

$$\begin{cases} c_{3k} = \dfrac{1 \cdot 4 \ldots (3k-5) \cdot (3k-2)}{(3k)!} \cdot c_0 & \text{for every } k \in \mathbb{N}^* \\ c_{3k+1} = \dfrac{2 \cdot 5 \ldots (3k-4) \cdot (3k-1)}{(3k+1)!} \cdot c_1 & \text{for every } k \in \mathbb{N}^* \\ c_{3k+2} = 0 & \text{for every } k \in \mathbb{N}. \end{cases}$$

Hence

$$x(t) = c_0 \left(1 + \sum_{k=1}^{\infty} \frac{1 \cdot 4 \ldots (3k-5) \cdot (3k-2)}{(3k)!} \cdot t^{3k} \right)$$

$$+ c_1 \left(1 + \sum_{k=1}^{\infty} \frac{2 \cdot 5 \ldots (3k-4) \cdot (3k-1)}{(3k+1)!} \cdot t^{3k+1} \right)$$

for every $t \in \mathbb{R}$ (the two series converge on $\mathbb{R}$).

Problem 3.3 We look for $x(t) = t^{\alpha} \sum_{k=0}^{\infty} c_k t^k$. Termwise differentiation yields both $x'(t) = \sum_{k=0}^{\infty} (k+\alpha) c_k t^{\alpha+k-1}$ and $x''(t) = \sum_{k=0}^{\infty} (k+\alpha)(k+\alpha-1) c_k t^{\alpha+k-2}$. Substituting in the equation and identifying the free term, taking into account that $c_0 \neq 0$, it follows $\alpha = \pm n$. For $\alpha = n$, we get $c_{2k+1} = 0$ and

$$c_{2k} = (-1)^k \frac{n!}{2^{2k} \cdot k! \cdot (n+k)!} c_0$$

for every $k \in \mathbb{N}$. Taking $c_0 = 1/(n! \cdot 2^n)$, we deduce that a non-identically null solution, corresponding to $\alpha = n$, is given by

$$x_n(t) = \sum_{k=0}^{\infty} (-1)^k \frac{(t/2)^{n+2k}}{k! \cdot (n+k)!}$$

for every $t \in \mathbb{R}$ (the radius of convergence of the series above is $R = +\infty$). For $\alpha = -n$, we get $c_{2k+1} = 0$ for every $k \in \mathbb{N}$, $c_{2k} = 0$ for $k = 0, 1, \ldots, n-1$ and

$$c_{2k+2n} = (-1)^k \frac{n!}{2^{2k} \cdot k! \cdot (n+k)!} c_{2n}$$

for every $k \in \mathbb{N}$. Let us remark that, in this case, ($\alpha = -n$), every non-zero solution is of the form λx_n with $\lambda \in \mathbb{R}^*$. Indeed, taking $c_{2n} = 1/(n! \cdot 2^n)$, we deduce that the solution corresponding to this coefficient and to $\alpha = -n$ coincides with x_n.

Exercise 3.3 We look for the solution of the form $x(t) = \sum_{k=0}^{\infty} c_k t^k$. Substituting in the equation and identifying the coefficients, we get

$$c_{k+1} = \frac{(a+k)(b+k)}{(k+1)(c+k)} \cdot c_k$$

for every $k \in \mathbb{N}^*$. By virtue of the conditions $x(0) = 1$ and $x'(0) = (ab)/c$, it follows that $c_0 = 1$ and $c_1 = (ab)/c$. Then the solution is given by the sum of the Gauss hypergeometric series

$$x(t) = 1 + \sum_{k=1}^{\infty} \frac{(a+k-1)(a+k-2)\ldots a(b+k-1)(b+k-2)\ldots b}{n!(c+k-1)(c+k-2)\ldots(c+1)c} \cdot t^k$$

for every $t \in (-1, 1)$ (the radius of convergence of the series above is $R = 1$).

Problem 3.4 The solutions about 0 are of the form $x(t) = \sum_{k=0}^{\infty} c_k t^k$. Asking x to satisfy the equation and identifying the coefficients, we get

$$\begin{cases} c_2 = -\dfrac{\lambda(\lambda+1)}{2 \cdot 1} c_1 \\ c_3 = -\dfrac{(\lambda-1)(\lambda+2)}{3 \cdot 2} c_2 \\ c_{k+2} = -\dfrac{(\lambda-k)(\lambda+k+1)}{(k+2)(k+1)}. \end{cases}$$

Then we have

$$\begin{cases} c_{2k} = (-1)^k \dfrac{\lambda(\lambda-2)\ldots(\lambda-2k+2)(\lambda+1)(\lambda+3)\ldots(\lambda+2k-1)}{(2k)!} c_0 \\ c_{2k+1} = (-1)^k \dfrac{(\lambda-1)(\lambda-3)\ldots(\lambda-2k+1)(\lambda+2)(\lambda+4)\ldots(\lambda+2k)}{(2k+1)!} c_1, \end{cases}$$

for every $k = 1, 2, \ldots$. From here, it follows that a necessary and sufficient condition for a local solution about 0 to be a polynomial is that $\lambda \in \mathbb{N}$.

Problem 3.5 From Lemma 3.2.2, it follows that, for every "starting" continuous function $x_0 : [a, a+\delta] \to B(\xi, r)$, the sequence of successive approximations:

$$x_k(t) = \xi + \int_a^t f(s, x_{k-1}(s))\, ds$$

for every $k \in \mathbb{N}^*$ and $t \in [a, a+\delta]$ is well-defined. Let $x : [a, a+\delta] \to B(\xi, r)$ be the unique solution of $\mathcal{CP}(\mathcal{D})$ and let $m > 0$ such that $\|x_0(t) - x(t)\| \leq m$ for every $t \in [a, a+\delta]$. Using the fact that f is Lipschitz on $B(\xi, r)$, one proves by complete induction that

$$\|x_k(t) - x(t)\| \leq m \frac{L^k (t-a)^k}{k!}$$

for every $k \in \mathbb{N}$ and every $t \in [a, a+\delta]$. From here, we deduce that the error evaluation formula, in this general case, is

$$\|x_k(t) - x(t)\| \leq m \frac{L^k \delta^k}{k!}$$

for every $k \in \mathbb{N}$ and every $t \in [a, a+\delta]$.

Problem 3.6 Let $(x_k)_{k \in \mathbb{N}}$ be the sequence of successive approximations defined on $[a, a+\delta]$ with values in $B(\xi, r)$: $x_0(t) = \xi$ and

$$x_k(t) = \xi + \int_a^t f(s, x_{k-1}(s))\, ds$$

for every $k \in \mathbb{N}^*$ and $t \in [a, a+\delta]$. One shows by mathematical induction that

$$\|x_{k+1}(t) - x_k(t)\| \leq \frac{M}{L} \frac{[L(t-a)]^{k+1}}{(k+1)!} \tag{$*$}$$

for every $k \in \mathbb{N}$ and every $t \in [a, b]$. From the inequality above, we deduce that

$$\|x_{k+p}(t) - x_k(t)\| \leq \frac{M}{L} \sum_{i=1}^{p} \frac{[L(b-a)]^{k+i}}{(k+i)!}$$

for every $k, p \in \mathbb{N}$ and every $t \in [a, b]$. Since $\sum_{k=0}^{\infty} \frac{[L(b-a)]^k}{k!} = e^{L(b-a)}$, it follows that the sequence $(x_k)_{k \in \mathbb{N}}$ is uniformly Cauchy on $[a, b]$. So it is uniformly convergent on $[a, b]$ to a continuous function x. Passing to the limit in the recurrence relation which defines the sequence and taking into account Corollary 12.2.1, we deduce that x is the solution of the integral equation

$$x(t) = \xi + \int_a^t f(s, x(s))\, ds$$

and implicitly of $\mathcal{CP}(\mathcal{D})$. This achieves the proof of the existence part of Theorem 2.3.2. Since f is Lipschitz on $B(\xi, r)$, it follows that every two solutions $x, y : [a, a+\delta] \to B(\xi, r)$ of $\mathcal{CP}(\mathcal{D})$ satisfy

$$\|x(t) - y(t)\| \leq \int_a^t L \|x(s) - y(s)\|\, ds$$

for every $t \in [a, a+\delta]$. From Gronwall's Lemma 1.5.2, it follows that $x \equiv y$ which proves the uniqueness too.

Problem 3.7 Let us define the sequence of functions: $x_k : [a,b] \to \mathbb{R}^n$ by $x_0(t) = f(t)$ for every $t \in [a,b]$, and

$$x_k(t) = f(t) + \int_a^t g(t, \tau, x_{k-1}(\tau))\, d\tau$$

for every $k \in \mathbb{N}^*$ and every $t \in [a,b]$. Obviously all the terms of this sequence are continuous functions on $[a,b]$. Since $[a,b]$ is compact, there exists $M > 0$ such that $\|g(t,s,f(s))\| \leq M$ for every $(t,s) \in [a,b] \times [a,b]$. Then, we have $\|x_1(t) - x_0(t)\| \leq M(t-a)$ for every $t \in [a,b]$. Using the fact that the function g is Lipschitz on $\mathbb{R}^n$, one proves by mathematical induction that $(x_k)_{k\in\mathbb{N}}$ satisfies the inequality $(*)$ established in the solution of Problem 3.6. From this point, the proof follows that of Problem 3.6.

Problem 3.8 One shows for the beginning that $x : [a,b] \to \mathbb{R}$ is a solution of the Cauchy problem for the integro-differential equation if and only if x is continuous on $[a,b]$ and

$$x(t) = \xi + \int_a^t f(s, x(s))\, ds + \int_a^t \left(\int_a^s g(s, \tau, x(\tau))\, d\tau \right) ds$$

for every $t \in [a,b]$. We define the sequence $x_k : [a,b] \to \mathbb{R}^n$ of successive approximations: $x_0(t) = \xi$ and

$$x_k(t) = \xi + \int_a^t f(s, x_{k-1}(s))\, ds + \int_a^t \left(\int_a^s g(s, \tau, x_{k-1}(\tau))\, d\tau \right) ds$$

for every $k \in \mathbb{N}^*$ and every $t \in [a,b]$. Let us observe that all the terms of this sequence are continuous functions. Also, there exist $M_f > 0$ and $M_g > 0$ such that $\|f(t,\xi)\| \leq M_f$ and $\|g(t,s,\xi)\| \leq M_g$ for $(t,s) \in [a,b] \times [a,b]$. Let $M = M_f + (b-a)M_g$, and let us observe that $\|x_1(t) - x_0(t)\| \leq M(t-a)$ for every $t \in [a,b]$. Let $L_f > 0$ and $L_g > 0$ be the Lipschitz constants corresponding to the functions f and g, and let $L = L_f + (b-a)L_g$. Majorizing if necessary the double integral over $[a,t] \times [a,s]$ by the double integral over $[a,b] \times [a,t]$ and using the complete induction method, one shows that the sequence $(x_k)_{k\in\mathbb{N}}$ satisfies the inequality $(*)$ established in the solution of Problem 3.6. In what follows, one proceeds by analogy with the case of Problem 3.6.

Problem 3.9 Let $h : [0,T] \to \mathbb{R}^n$ be a continuous function and let $\xi \in \mathbb{R}^n$. According to Corollary 2.4.1, $\mathcal{CP}(\mathcal{D})$ has at least one saturated solution x defined either on $[0,T]$ or on $[0,T_m)$, with $T_m \leq T$. We will show in what follows that x is defined on $[0,T]$. To this aim, let us assume for contradiction that x is defined on $[0,T_m)$. Then, for every $s \in [0,T_m)$ and $\delta > 0$ with $s + \delta < T_m$ we have

$$x'(s+\delta) - x'(s) = \mathcal{A}x(s+\delta) - \mathcal{A}x(s) + h(s+\delta) - h(s).$$

Taking the inner product on both sides of this equality by $x(s+\delta)-x(s)$, using the dissipativity condition and (i) in Lemma 12.1.2, we deduce

$$\frac{1}{2}\frac{d}{ds}\left(\|x(s+\delta)-x(s)\|^2\right)\leq\langle h(s+\delta)-h(s),x(s+\delta)-x(s)\rangle.$$

Integrating this inequality over $[0,t]$ with $t+\delta<T_m$, we get

$$\|x(t+\delta)-x(t)\|^2\leq\|x(\delta)-\xi\|^2+2\int_0^t\langle h(s+\delta)-h(s),x(s+\delta)-x(s)\rangle ds.$$

From the Cauchy–Schwarz inequality, we have that

$$\langle h(s+\delta)-h(s),x(s+\delta)-x(s)\rangle\leq\|h(s+\delta)-h(s)\|\|x(s+\delta)-x(s)\|.$$

From this relation, the preceding one, and from Lemma 1.5.3, we deduce that

$$\|x(t+\delta)-x(t)\|\leq\|x(\delta)-\xi\|+\int_0^t\|h(s+\delta)-h(s)\|\,ds.$$

Since x is continuous at $t=0$, $x(0)=\xi$ and h is uniformly continuous on $[0,T]$, from this inequality, we conclude that x satisfies the Cauchy's condition of the existence of finite limit at the right at the point T_m. So, x can be extended to $[0,T_m]$, which is absurd. This contradiction can be eliminated only if x is defined on $[0,T]$. The uniqueness follows from the second inequality formulated in the problem, which we prove below. Let x_1, x_2 be two saturated solutions corresponding to the initial data ξ_i and to the functions h_i with $i=1,2$. Taking the inner product on both sides of $x_1'(t)-x_2'(t)=\mathcal{A}x_1(t)-\mathcal{A}x_2(t)+h_1(t)-h_2(t)$ by $x_1(t)-x_2(t)$, taking into account the dissipativity of the function $\mathcal{A}$ and using (i) in Lemma 12.1.2, we deduce

$$\frac{1}{2}\frac{d}{dt}\left(\|x_1(t)-x_2(t)\|^2\right)\leq\langle h_1(t)-h_2(t),x_1(t)-x_2(t)\rangle$$

for every $t\in[0,T]$. Integrating this inequality over $[0,t]$, we get

$$\|x_1(t)-x_2(t)\|^2\leq\|\xi_1-\xi_2\|^2+2\int_0^t\langle h_1(s)-h_2(s),x_1(s)-x_2\rangle ds$$

for every $t\in[0,T]$. From this inequality, the Cauchy–Schwarz inequality and from Lemma 1.5.3, it follows that

$$\|x_1(t)-x_2(t)\|\leq\|\xi_1-\xi_2\|+\int_0^t\|h_1(s)-h_2\|ds$$

for every $t\in[0,T]$, which completes the proof.

Problem 3.10 Since the function x_1 is continuous on $[0,T]$, it follows that there exists $M>0$ such that $\|x_1(t)-\xi\|\leq M$ for every $t\in[0,T]$. From the second inequality established in Problem 3.9, and from the fact that f is Lipschitz on $\mathbb{R}^n$ of constant $L > 0$, we deduce

$$\|x_{k+1}(t) - x_k(t)\| \leq \int_0^t L\|x_k(s) - x_{k-1}(s)\|\, ds$$

for every $k \in \mathbb{N}$ and every $t \in [\,0, T\,]$. From this inequality, and from the preceding one, using the method of complete induction, one shows that

$$\|x_{k+1}(t) - x_k(t)\| \leq M\frac{L^k t^k}{k!}$$

for every $k \in \mathbb{N}$ and every $t \in [\,0, T\,]$. So,

$$\|x_{k+p}(t) - x_k(t)\| \leq M\sum_{i=0}^{p-1} \frac{[L(b-a)]^{k+i}}{(k+i)!}$$

for every $k, p \in \mathbb{N}$ and every $t \in [\,a, b\,]$. Since $\sum_{k=0}^{\infty} \frac{[L(b-a)]^k}{k!} = e^{L(b-a)}$, it follows that the sequence $(x_k)_{k\in\mathbb{N}}$ is uniformly Cauchy on $[\,a, b\,]$, and hence uniformly convergent on this interval to a continuous function x. Passing to the limit in the recurrence relation (in the integral form) which defines the sequence and taking into account Corollary 12.2.1, we deduce that x is a solution of the integral equation

$$x(t) = \xi + \int_a^t [\mathcal{A}x(s) + f(s, x(s))]\, ds$$

and so, of $\mathcal{CP}(\mathcal{D})$. This completes the proof of the existence part of Problem 3.10. Since f is Lipschitz on $B(\xi, r)$, from the second inequality in Problem 3.9, it follows that every two solutions $x, y : [\,a, a+\delta\,] \to B(\xi, r)$ of $\mathcal{CP}(\mathcal{D})$ satisfy

$$\|x(t) - y(t)\| \leq \int_a^t L\|x(s) - y(s)\|\, ds$$

for every $t \in [\,a, a+\delta\,]$. From Gronwall's Lemma 1.5.2, it follows that $x \equiv y$ which completes the proof of the uniqueness part.

Problem 3.11 We begin by observing that, from the hypothesis imposed on the function $\mathcal{A}$, it follows that this is dissipative on $\mathbb{R}^n$. From Problem 3.9, we deduce that for every $\xi \in \mathbb{R}^n$ $\mathcal{CP}(\xi)$ has a unique global solution and therefore $\mathcal{P}$ is well-defined. Let $\xi, \eta \in \mathbb{R}^n$ and let us denote by x and y the two global solutions of $\mathcal{CP}(\xi)$ and $\mathcal{CP}(\eta)$ respectively. Taking the inner product on both sides of the equality $x'(t) - y'(t) = \mathcal{A}x(t) - \mathcal{A}y(t)$ by $x(t) - y(t)$, taking into account the dissipativity condition satisfied by $\mathcal{A}$ and using (i) in Lemma 12.1.2, we obtain

$$\frac{1}{2}\frac{d}{dt}\left(\|x(t) - y(t)\|^2\right) \leq -\omega^2\|x(t) - y(t)\|^2$$

for every $t \in \mathbb{R}_+$. Multiplying both sides of this inequality by the integrating factor $e^{2\omega^2 t}$, we deduce

$$\frac{d}{dt}\left(\frac{1}{2}e^{2\omega^2 t}\|x(t) - y(t)\|^2\right) \leq 0$$

for every $t \in \mathbb{R}_+$. From here, integrating over $[0, T]$, we get

$$\frac{1}{2}e^{2\omega^2 T}\|x(T) - y(T)\|^2 \leq \frac{1}{2}\|x(0) - y(0)\|^2.$$

Recalling that $x(0) = \xi$, $y(0) = \eta$, $x(T) = \mathcal{P}(\xi)$ and $y(T) = \mathcal{P}(\eta)$, the last inequality implies

$$\|\mathcal{P}(\xi) - \mathcal{P}(\eta)\| \leq q\|\xi - \eta\|$$

for every $\xi, \eta \in \mathbb{R}^n$, where $q = e^{-\omega^2 T}$. From this property it follows by mathematical induction that $\|\xi_{k+1} - \xi_k\| \leq q^k\|\xi_1 - \xi_0\|$ for every $k \in \mathbb{N}$ and therefore $\|\xi_{k+p} - \xi_k\| \leq \|\xi_1 - \xi_0\| \sum_{i=0}^{p-1} q^{k+i}$ for every $k, p \in \mathbb{N}^*$. Finally, by observing that the geometric series $\sum_{k=0}^{\infty} q^k$ is convergent because $q \in (0,1)$, it follows that the sequence $(\xi_k)_{k\in\mathbb{N}}$ is convergent to an element $\eta \in \mathbb{R}^n$. Passing to the limit in the recurrence relation $\xi_k = \mathcal{P}(x_{k-1})$ and taking into account the continuity of the function $\mathcal{P}$, we conclude that $\eta = \mathcal{P}(\eta)$, which is equivalent to $\eta = x(0, 0, \eta) = x(T, 0, \eta)$. The proof of (1) and (2) is complete. Finally, if f is T-periodic and x is a global solution of the equation $x'(t) = \mathcal{A}x(t) + f(t)$, then the function $x_T : \mathbb{R}_+ \to \mathbb{R}^n$, defined by $x_T(t) = x(t + T)$ is also a solution of the same equation. Since $x(T, 0, \eta) = \eta$, from the uniqueness property, it follows that $x(t + T, 0, \eta) = x(t, 0, \eta)$ for every $t \in \mathbb{R}_+$. This means that $x(\cdot, 0, \eta)$ is periodic of period T, which completes the proof of (3). In order to prove (4), let us observe that if $x : \mathbb{R}_+ \to \mathbb{R}^n$ is a T-periodic solution of the differential equation $x'(t) = \mathcal{A}x(t) + f(t)$ then $\xi = x(0)$ is a fixed point of the function $\mathcal{P}$, i.e. $\xi = \mathcal{P}(\xi)$.[2] Since $\mathcal{P}$ is a strict contraction ($\|\mathcal{P}(\xi) - \mathcal{P}(\eta)\| \leq q\|\xi - \eta\|$ for every $\xi, \eta \in \mathbb{R}^n$, where $q \in (0,1)$), it follows that $\mathcal{P}$ has at most one fixed point. The proof is complete.

Chapter 4

Problem 4.1 If x is bounded on $\mathbb{R}_+$, there exists $m > 0$ such that

$$|x(t)| \leq m$$

for every $t \in \mathbb{R}_+$. From the second equation in $(\mathcal{S})$, we deduce

$$|y(t) - y(s)| \leq m\left|\int_s^t |b(\tau)|\, d\tau\right|$$

[2]Under the extra-assumption that f is T-periodic, the converse of this assertion also holds true, as we have already seen.

for every $t, s \in \mathbb{R}_+$. Since b is absolutely integrable over $\mathbb{R}_+$, for every $\varepsilon > 0$ there exists $\delta(\varepsilon) > 0$ such that

$$\left|\int_s^t |b(\tau)|\, d\tau\right| \leq \varepsilon$$

for every $t, s \in \mathbb{R}_+$, with $t \geq \delta(\varepsilon)$ and $s \geq \delta(\varepsilon)$. From the inequality previously established, it follows that y satisfies the Cauchy's condition of the existence of the finite limit at $+\infty$. Let $\ell = \lim_{t\to+\infty} y(t)$. It follows then that $\lim_{t\to+\infty} x'(t) = \ell$. Assuming for contradiction that $\ell \neq 0$, we deduce that x is unbounded. Indeed, to fix the ideas, let us assume that $\ell > 0$. Then, there exists $t_0 > 0$ such that, for every $t \geq t_0$, we have $x(t) \in [\frac{\ell}{2}(t - t_0) + x(t_0), \frac{3\ell}{2}(t - t_0) + x(t_0)\,]$. Accordingly x is unbounded on $\mathbb{R}_+$. This contradiction can be eliminated only if $\ell = 0$, which proves (i).

In order to prove (ii), let us observe that the Wronskian of the system $(\mathcal{S})$ is constant. Let us consider then a fundamental system of solutions of $(\mathcal{S})$. Assuming that both solutions are bounded on $\mathbb{R}_+$, from what we have already proved, it follows that

$$c = \lim_{t\to+\infty} W(t) = 0$$

relation in contradiction with the fact that the system of solutions is fundamental. This contradiction can be eliminated only if at least one of the two solutions is unbounded on $\mathbb{R}_+$, which shows (ii).

Problem 4.2 By virtue of Theorem 2.6.1

$$\mathcal{X}(t) = \left(\frac{\partial S_i(t)x}{\partial x_j}\right)_{n\times n}$$

is the solution of the Cauchy problem

$$\begin{cases} \mathcal{X}'(t) = f_x(t, S(t)x)\mathcal{X}(t) \\ \mathcal{X}(a) = \mathcal{I}_n. \end{cases}$$

According to Liouville Theorem 4.1.5, we have

$$\det(\mathcal{X}(t)) = \det(\mathcal{X}(a))\exp\left(\int_a^t \operatorname{tr} f_x(s, S(s)\xi)\, ds\right)$$

$$= \det(\mathcal{X}(a))\exp\left[\int_a^t \left(\sum_{i=1}^n \frac{\partial f_i}{\partial x_i}(s, S(s)\xi)\right) ds\right] = \det(\mathcal{X}(a))$$

for every $t \in [\,a, b)$, which completes the proof.

Problem 4.3 Since H is of class C^2, from Schwarz theorem (on the equality of the second-order mixed partial derivatives), we conclude that the function $f : \mathbb{R}^{2n} \to \mathbb{R}^{2n}$, defined by

$$f(p, q) = \left(-\frac{\partial H}{\partial q_1}, \ldots, -\frac{\partial H}{\partial q_n}, \frac{\partial H}{\partial p_1}, \ldots, \frac{\partial H}{\partial p_n}\right),$$

where $p, q \in \mathbb{R}^n$, is divergence free. The conclusion follows from Problem 4.2.
Problem 4.4 From the definition of the matrix $e^{t\mathcal{A}}$ and from the continuity of the mapping $\mathcal{A} \mapsto \mathcal{A}^\tau$, it follows that

$$\left(e^{t\mathcal{A}}\right)^\tau = e^{t\mathcal{A}^\tau} = e^{-t\mathcal{A}} = \left(e^{t\mathcal{A}}\right)^{-1},$$

which shows that $e^{t\mathcal{A}}$ is orthogonal.
Problem 4.5 Let $\mathcal{X}$ be a fundamental matrix of the system $x' = \mathcal{A}x$ which is orthogonal at $t = 0$. Obviously $\mathcal{X}$ satisfies $\mathcal{X}' = \mathcal{A}\mathcal{X}$, and therefore

$$\frac{d}{dt}\left(\mathcal{X}^\tau(t)\right) = \mathcal{X}^\tau(t)\mathcal{A}^\tau.$$

Hence $\mathcal{X}^\tau$ is a solution of the Cauchy problem

$$\begin{cases} \mathcal{Y}' = \mathcal{Y}\mathcal{A}^\tau \\ \mathcal{Y}(0) = \mathcal{X}^\tau(0). \end{cases}$$

On the other hand $\mathcal{X}(t)\mathcal{X}^{-1}(t) = \mathcal{I}_n$, which implies $\left(\mathcal{X}(t)\mathcal{X}^{-1}(t)\right)' = 0$. Then we have

$$\mathcal{X}'(t)\mathcal{X}^{-1}(t) = -\mathcal{X}(t)\left(\mathcal{X}^{-1}\right)'(t)$$

or

$$\left(\mathcal{X}^{-1}\right)'(t) = -\mathcal{X}^{-1}(t)\mathcal{X}'(t)\mathcal{X}^{-1}(t) = \mathcal{X}^{-1}(t)\left(-\mathcal{A}\right) = \mathcal{X}^{-1}(t)\mathcal{A}^\tau.$$

It follows that $\mathcal{X}^{-1}$ is also a solution of the Cauchy problem above, while from the uniqueness part of Theorem 4.1.1, we deduce that $\mathcal{X}(t)^\tau = \mathcal{X}^{-1}(t)$ for every $t \in \mathbb{R}$.
Problem 4.6 The proof follows the same lines as those in the proof of the preceding problem.
Problem 4.7 We have

$$\mathcal{A}^k - \lambda^k\mathcal{I}_n = (\mathcal{A} - \lambda\mathcal{I}_n)(\mathcal{A}^{k-1} + \lambda\mathcal{A}^{k-2} + \cdots + \lambda^{k-1}\mathcal{I}_n)$$

for every $k \in \mathbb{N}^*$. From here, one observes that every root of the characteristic equation $\det(\mathcal{A} - \lambda\mathcal{I}_n) = 0$ is also a root of the equation

$$\det\left(\sum_{p=1}^{k}\frac{t^p\mathcal{A}^p}{p!} - \sum_{p=1}^{k}\frac{t^p\lambda^p}{p!}\mathcal{I}_n\right) = 0.$$

Since the function det is continuous, passing to the limit in the equality above for k tending to $+\infty$, we deduce that, if λ is a root of the equation $\det(\mathcal{A} - \lambda\mathcal{I}_n) = 0$, then, for every $t \in \mathbb{R}$, $e^{t\lambda}$ is a root of the equation $\det\left(e^{t\mathcal{A}} - \mu\mathcal{I}_n\right) = 0$.

Problem 4.8 The matrix $\mathcal{A}$ is symmetric if and only if $\langle \mathcal{A}x, y\rangle = \langle x, \mathcal{A}y\rangle$ for every $x, y \in \mathbb{R}^n$. So, if $\mathcal{A}$ is symmetric, we have

$$\left\langle \left(\sum_{p=1}^{k} \frac{t^p \mathcal{A}^p}{p!}\right) x, y \right\rangle = \left\langle x, \left(\sum_{p=1}^{k} \frac{t^p \mathcal{A}^p}{p!}\right) y \right\rangle$$

Passing to the limit for k tending to $+\infty$ in this equality and taking into account that the inner product is a continuous function of both variables, we deduce $\langle e^{t\mathcal{A}}x, y\rangle = \langle x, e^{t\mathcal{A}}y\rangle$ for every $x, y \in \mathbb{R}^n$ and $t \in \mathbb{R}$, which shows that $e^{t\mathcal{A}}$ is symmetric for every $t \in \mathbb{R}$.

Problem 4.9 Let $\mathcal{X} : \mathbb{R} \to \mathcal{M}_{n\times n}(\mathbb{R})$ be a fundamental matrix of the system with the property that $\mathcal{X}(0)$ is symmetric. Since the inverse of every self-adjoint matrix is symmetric, while $\mathcal{X}^{-1}(t) = \mathcal{X}(-t)$ for every $t \in \mathbb{R}$, it suffices to consider only the case $t > 0$. Let then $t > 0$, and let us choose $a > 0$ with the property that $t \in [0, a]$. Let $k \in \mathbb{N}^*$. Let us divide the interval $[0, a]$ in k equal parts $0 = t_0 < t_1 < \cdots < t_{k-1} < t_k = a$ and let us define $\mathcal{A}_k : [0, a] \to \mathcal{M}_{n\times n}(\mathbb{R})$, by $\mathcal{A}_k(t) = \mathcal{A}(t_i)$ for $t \in [t_i, t_{i+1})$, $i = 0, 1, \ldots, k-1$ and $\mathcal{A}_k(a) = \mathcal{A}(t_{k-1})$. Let us define the function $\mathcal{X}_k : [0, a] \to \mathcal{M}_{n\times n}(\mathbb{R})$ by $\mathcal{X}_k(t) = e^{(t-t_i)\mathcal{A}(t_i)}\mathcal{X}_k(t_i)$ for $t \in (t_i, t_{i+1}]$, $i = 0, 1, \ldots, k-1$ and $\mathcal{X}_k(0) = \mathcal{X}(0)$. One may easily see that $\mathcal{X}_k$ is continuous on $[0, a]$, differentiable on the set $[0, a] \setminus \{t_i;\ i = 1, 2, \ldots, k\}$, and satisfies

$$\mathcal{X}_k'(t) = \mathcal{A}_k(t)\mathcal{X}_k(t) \tag{$*$}$$

at every point of differentiability. Let us observe that $\mathcal{X}_k$ is obtained by the concatenation of the solutions of the Cauchy problems of the type

$$\begin{cases} \mathcal{Z}_i' = \mathcal{A}(t_{i-1})\mathcal{Z}_i \\ \mathcal{Z}_i(t_{i-1}) = \mathcal{Z}_{i-1}(t_{i-1}),\ \mathcal{Z}_0(0) = \mathcal{X}(0) \end{cases}$$

for $i = 1, 2, \ldots, k$. From the previous problem, we successively deduce that $\mathcal{Z}_i(t)$ is symmetric for every $t \in [t_{i-1}, t_i]$ and $i = 1, 2, \ldots, k$. So, $\mathcal{X}_k(t)$ has the same property for every $t \in [0, a]$. Finally, let us observe that the sequence of functions $(\mathcal{X}_k)_{k\in\mathbb{N}^*}$ is uniformly bounded and equicontinuous on $[0, a]$. This is an immediate consequence of the fact that $\mathcal{X}_k$ satisfies

$$\mathcal{X}_k(t) = \mathcal{X}(0) + \int_0^t \mathcal{A}_k(s)\mathcal{X}_k(s)\,ds$$

for every $k \in \mathbb{N}^*$ and every $t \in [0, a]$, and of the boundedness of the function $\mathcal{A}$ on the interval $[0, a]$. By virtue of Theorem 12.2.1, it follows that, at least one subsequence, $(\mathcal{X}_k)_{k\in\mathbb{N}^*}$ is uniformly convergent on $[0, a]$ to a function $\mathcal{Y}$. Since $\lim_{k\to\infty} \mathcal{A}_k = \mathcal{A}$ uniformly on $[0, a]$, passing to the limit in $(*)$, we deduce that $\mathcal{Y}(t) = \mathcal{X}(t)$ for every $t \in [0, a]$. We conclude the proof with the remark that $\mathcal{Y}(t)$ is symmetric for every $t \in [0, a]$ being the uniform limit of a sequence of functions having the same property.

Problem 4.10 Since $e^{t\mathcal{A}} = \mathcal{I}+t\mathcal{A}+\sum_{k=2}^{\infty}\frac{t^k\mathcal{A}^k}{k!}$ it follows that, for $t>0$ sufficiently small, all the elements of the matrix $e^{t\mathcal{A}}$, which are not on the diagonal, have the same sign with the corresponding ones of the matrix $t\mathcal{A}$. Hence the condition is necessary. In order to prove the sufficiency, let us observe that, by virtue of (ii) in Proposition 4.3.1, for every $t,s\in\mathbb{R}_+$, we have $e^{t\mathcal{A}} = e^{t(\mathcal{A}+s\mathcal{I})}e^{-st\mathcal{I}}$. Moreover, if s is large enough and $\mathcal{A}$ satisfies the condition of the problem, $t(\mathcal{A}+s\mathcal{I})$ has only positive elements. Then $e^{t(\mathcal{A}+s\mathcal{I})}$ has the same property too. Since $e^{-st\mathcal{I}} = e^{-st}\mathcal{I}$ has only positive elements and the product of two matrix with positive elements is a matrix with positive elements, this completes the proof.

Problem 4.11 Let us define $f:\mathcal{M}_{n\times n}(\mathcal{R})\to\mathcal{M}_{n\times n}(\mathcal{R})$ by $f(\mathcal{X}) = \mathcal{A}\mathcal{X}+\mathcal{X}\mathcal{B}$ for every $\mathcal{X}\in\mathcal{M}_{n\times n}(\mathcal{R})$. Following the same way as that used in the proof of Corollary 2.4.4, one concludes that f is globally Lipschitz and therefore the Cauchy problem considered has one unique global solution. To complete the proof, it suffices to show that $\mathcal{X}:\mathbb{R}\to\mathcal{M}_{n\times n}(\mathbb{R})$, defined by $\mathcal{X}(t) = e^{t\mathcal{A}}\mathcal{C}e^{t\mathcal{B}}$, is a solution of the Cauchy problem. We have $\mathcal{X}(0) = e^{0\mathcal{A}}\mathcal{C}e^{0\mathcal{B}} = \mathcal{C}$. From Theorem 4.3.1 it follows

$$\frac{d}{dt}(\mathcal{X})(t) = \frac{d}{dt}\left(e^{t\mathcal{A}}\right)\mathcal{C}e^{t\mathcal{B}} + e^{t\mathcal{A}}\frac{d}{dt}(\mathcal{C})e^{t\mathcal{B}} + e^{t\mathcal{A}}\mathcal{C}\frac{d}{dt}\left(e^{t\mathcal{B}}\right)$$

$$= \mathcal{A}e^{t\mathcal{A}}\mathcal{C}e^{t\mathcal{B}} + e^{t\mathcal{A}}\mathcal{C}e^{t\mathcal{B}}\mathcal{B} = \mathcal{A}\mathcal{X}(t)+\mathcal{X}(t)\mathcal{B}.$$

The proof is complete.

Problem 4.12 Let us observe that

$$\mathcal{A}\mathcal{X}+\mathcal{X}\mathcal{B} = -\int_0^{+\infty}\mathcal{A}e^{s\mathcal{A}}\mathcal{C}e^{s\mathcal{B}}\,ds - \int_0^{+\infty}e^{s\mathcal{A}}\mathcal{C}e^{s\mathcal{B}}\mathcal{B}\,ds$$

$$= -\int_0^{+\infty}\frac{d}{ds}\left(e^{s\mathcal{A}}\right)\mathcal{C}e^{s\mathcal{B}}\,ds - \int_0^{+\infty}e^{s\mathcal{A}}\mathcal{C}\frac{d}{ds}\left(e^{s\mathcal{B}}\right)ds$$

$$= -2e^{s\mathcal{A}}\mathcal{C}e^{s\mathcal{B}}\Big|_0^{+\infty} + \int_0^{+\infty}\mathcal{A}e^{s\mathcal{A}}\mathcal{C}e^{s\mathcal{B}}\,ds + \int_0^{+\infty}e^{s\mathcal{A}}\mathcal{C}e^{s\mathcal{B}}\mathcal{B}\,ds = 2\mathcal{C}-\mathcal{A}\mathcal{X}-\mathcal{X}\mathcal{B},$$

the last equality being satisfied if and only if $\lim_{s\to+\infty}e^{s\mathcal{A}}\mathcal{C}e^{s\mathcal{B}} = 0$. Since there exists $\lim_{s\to+\infty}e^{s\mathcal{A}}\mathcal{C}e^{s\mathcal{B}}$, in order to complete the proof it suffices to show that, the inf-limit, for s tending to $+\infty$, of each element of the matrix $e^{s\mathcal{A}}\mathcal{C}e^{s\mathcal{B}}$ is 0. To this aim let us observe that, by virtue of the convergence of the integral

$$\int_0^{+\infty}e^{s\mathcal{A}}\mathcal{C}e^{s\mathcal{B}}\,ds,$$

it follows that

$$\lim_{m\to+\infty}\int_m^{m+1}e^{s\mathcal{A}}\mathcal{C}e^{s\mathcal{B}}\,ds = 0.$$

Also, from the mean-value theorem, it follows that, for every element α_{ij} of the matrix $e^{t\mathcal{A}}\mathcal{C}e^{t\mathcal{B}}$ there exists $t_m(ij) \in [m, m+1]$ such that

$$\int_m^{m+1} \alpha_{ij}(s)\,ds = \alpha_{ij}(t_m(ij)).$$

From this relation and from the preceding one, it follows that

$$\lim_{m\to+\infty} \alpha_{ij}(t_m(ij)) = 0$$

for every $i, j = 1, 2, \ldots, n$, which completes the proof.

Problem 4.13 The power series which define both functions $t \mapsto \cos t\mathcal{A}$ and $t \mapsto \sin t\mathcal{A}$ are termwise differentiable. From this observation, we deduce that

$$\frac{d}{dt}(\cos t\mathcal{A}) = -\mathcal{A}\sin t\mathcal{A} = -(\sin t\mathcal{A})\mathcal{A} \;\text{ and }\; \frac{d}{dt}(\sin t\mathcal{A}) = \mathcal{A}\cos t\mathcal{A} = (\cos t\mathcal{A})\mathcal{A},$$

which proves (1). From here, it follows that

$$\begin{pmatrix} \cos t\mathcal{A} & \sin t\mathcal{A} \\ -\mathcal{A}\sin t\mathcal{A} & \mathcal{A}\cos t\mathcal{A} \end{pmatrix}' = \begin{pmatrix} 0 & \mathcal{I} \\ -\mathcal{A}^2 & 0 \end{pmatrix} \begin{pmatrix} \cos t\mathcal{A} & \sin t\mathcal{A} \\ -\mathcal{A}\sin t\mathcal{A} & \mathcal{A}\cos t\mathcal{A} \end{pmatrix}$$

which proves the first part of (2). The matrix $\mathcal{Z}(t)$ is fundamental for the system if and only if $\det \mathcal{Z}(0) \neq 0$, relation which holds true if and only if $\det \mathcal{A} \neq 0$.

Problem 4.14 Let us remark that, from the variation of constants formula (see Remark 4.3.3), it follows

$$x_k(t) = e^{(t-a)\mathcal{A}}\xi + \int_a^t e^{(t-s)\mathcal{A}}\left[f(s, x_{k-1}(s)) - \mathcal{A}x_{k-1}(s)\right] ds$$

for every $m \in \mathbb{N}^*$ and every $t \in [a, b]$. Let $L_1 > 0$ be the Lipschitz constant corresponding to the function f, let $M > 0$ be such that $\|x_1(t) - x_0(t)\| \leq M$ for every $t \in [a, b]$, and let us define $L = e^{(b-a)\|\mathcal{A}\|_{\mathcal{M}}}\left[L_1 + \|\mathcal{A}\|_{\mathcal{M}}\right]$. Using the fact that $\|\mathcal{A}\eta\| \leq \|\mathcal{A}\|_{\mathcal{M}}\|\eta\|$ for every $\eta \in \mathbb{R}^n$ (see (N_4), from Lemma 12.1.1), and by observing that $\|e^{(t-s)\mathcal{A}}\|_{\mathcal{M}} \leq e^{(b-a)\|\mathcal{A}\|_{\mathcal{M}}}$ for every $t, s \in [a, b]$ with $s \leq t$, one shows by mathematical induction that

$$\|x_{k+1}(t) - x_k(t)\| \leq \frac{ML^k(t-a)^k}{k!}$$

for every $k \in \mathbb{N}$ and every $t \in [a, b]$. From this point, the proof follows the same arguments as those used in the solution of Problem 3.10.

Exercise 4.1 The general solutions of the systems are:

(1) $\begin{cases} x_1(t) = c_1e^{-t} + c_2e^{5t} \\ x_2(t) = -c_1e^{-t} + 2c_2e^{5t}. \end{cases}$

(2) $\begin{cases} x_1(t) = c_1\cos t + c_2\sin t \\ x_2(t) = -c_1\sin t + c_2\cos t. \end{cases}$

(3) $\begin{cases} x_1(t) = c_1e^{-t}\cos t + c_2e^{-t}\sin t \\ x_2(t) = \frac{1}{5}(c_2 - 2c_1)e^{-t}\cos t - \frac{1}{5}(c_1 + 2c_2)e^{-t}\sin t. \end{cases}$

(4) $\begin{cases} x_1(t) = \frac{1}{8}(c_1e^{2t} + c_2 - 2t^2 - 2t - 1) \\ x_2(t) = \frac{1}{8}(c_1e^{2t} - c_2 + 2t^2 - 2t - 1). \end{cases}$

(5) $\begin{cases} x_1(t) = 2\sin t - (2c_1 + c_2)t + c_1 \\ x_2(t) = -2\cos t - 3\sin t + (4c_1 + 2c_2)t + c_2. \end{cases}$

(6) $\begin{cases} x_1(t) = (c_1 - 4c_2)e^{2t} + 4(c_1 + c_2)e^{-3t} + t^2 + t \\ x_2(t) = (-c_1 + 4c_2)e^{2t} + (c_1 + c_2)e^{-3t} - \frac{t^2}{2}. \end{cases}$

(7) $\begin{cases} x_1(t) = c_1e^t + c_2e^{-\frac{t}{2}}\cos\frac{\sqrt{3}}{2}t + c_3e^{-\frac{t}{2}}\sin\frac{\sqrt{3}}{2}t \\ x_2(t) = c_1e^t + \left(-\frac{1}{2}c_2 + \frac{\sqrt{3}}{2}c_3\right)e^{-\frac{t}{2}}\cos\frac{\sqrt{3}}{2}t \\ -\left(\frac{\sqrt{3}}{2}c_2 + \frac{1}{2}c_3\right)e^{-\frac{t}{2}}\sin\frac{\sqrt{3}}{2}t \\ x_3(t) = c_1e^t + \left(-\frac{1}{2}c_2 - \frac{\sqrt{3}}{2}c_3\right)e^{-\frac{t}{2}}\cos\frac{\sqrt{3}}{2}t \\ +\left(\frac{\sqrt{3}}{2}c_2 - \frac{1}{2}c_3\right)e^{-\frac{t}{2}}\sin\frac{\sqrt{3}}{2}t. \end{cases}$

(8) $\begin{cases} x_1(t) = -c_1e^{-t} + 2c_2e^{2t} \\ x_2(t) = -c_3e^{-t} + 2c_2e^{2t} \\ x_3(t) = (c_1 + c_3)e^{-t} + 2c_2e^{2t}. \end{cases}$

Here $c_1, c_2, c_3 \in \mathbb{R}$ and $t \in \mathbb{R}$.

Exercise 4.2 The general solutions of the equations are:

(1) $x(t) = c_1e^t + c_3e^{4t}$.

(2) $x(t) = c_1e^{-t} + c_2te^{-t}$.

(3) $x(t) = c_1\cos 2t + c_2\sin 2t$.

(4) $x(t) = c_1e^{-2t} + c_2e^{2t} + e^{2t}\left(\frac{t^3}{12} - \frac{t^2}{16} + \frac{t}{32} - \frac{1}{144}\right)$.

(5) $x(t) = c_1\cos 3t + c_2\sin 3t + \frac{1}{5}\cos 2t$.

(6) $x(t) = c_1\cos t + c_2\sin t + \sin t\,\ln|\sin t| - t\cos t$, for $t \in (k\pi, (k+1)\pi)$, $k \in \mathbb{Z}$.

(7) $x(t) = c_1\cos t + c_2\sin t + \frac{t^2}{4}\sin t + \frac{t}{8}(4\sin 2t\sin t + \cos 3t) - \frac{1}{32}\sin 3t + \frac{1}{4}\cos 2t\sin t$.

(8) $x(t) = c_1e^{2t} + c_2te^{2t} + \frac{t^3}{6}e^{2t}$.

(9) $x(t) = c_1e^{\sqrt{2}t} + c_2e^{-\sqrt{2}t} + e^{t^2}$.

Here $c_1, c_2, c_3 \in \mathbb{R}$ and, with the exception of item (6), $t \in \mathbb{R}$.

Exercise 4.3 The general solutions of the equations are:

(1) $x(t) = c_1 + c_2e^t + c_3e^{12t}$.

(2) $x(t) = c_1 + c_2e^{-t} + c_3e^t$.

(3) $x(t) = c_1e^{-t} + c_2e^{\frac{t}{2}}\cos\frac{\sqrt{3}}{2}t + c_3e^{\frac{t}{2}}\sin\frac{\sqrt{3}}{2}t$.

(4) $x(t) = c_1 e^t \sin t + c_2 e^t \cos t + c_3 e^{-t} \sin t + c_4 e^{-t} \cos t$.

(5) $x(t) = c_1 e^t + c_2 t e^t + c_3 t^2 e^t - t - 3$.

(6) $x(t) = c_1 \cos t + c_2 \sin t + c_3 t \cos t + c_4 t \sin t$.

(7) $x(t) = c_1 + c_2 t + c_3 e^t + c_4 t e^t + \frac{t^2}{2} e^t - 2te^t + 3e^t$.

(8) $x(t) = c_1 e^{-t} + c_2 \cos t + c_3 \sin t + \frac{2t-3}{8} e^t$.

(9) $x(t) = c_1 + c_2 e^{-3t} + c_3 t e^{-3t} + \frac{t^2}{18} - \frac{2t}{27}$.

Here $c_1, c_2, c_3 \in \mathbb{R}$ and $t \in \mathbb{R}$.

Exercise 4.4 (1) This is a Euler equation. The general solution, $x : \mathbb{I} \to \mathbb{R}$, is defined by $x(t) = c_1 t^{-1} + c_2 t^{-1} \ln |t|$ for every $t \in \mathbb{I}$, where $\mathbb{I} = (-\infty, 0)$ or $(0, +\infty)$ and $c_1, c_2 \in \mathbb{R}$, $c_1^2 + c_2^2 \neq 0$. The equation also admits the solution $x : \mathbb{R} \to \mathbb{R}$, $x \equiv 0$.

(2) Euler equation. The general solution is $x : \mathbb{I} \to \mathbb{R}$, $x(t) = c_1 t^{-1} + c_2 t^3$ for every $t \in \mathbb{I}$, where $\mathbb{I} = (-\infty, 0)$ or $(0, +\infty)$, where $c_1 \in \mathbb{R}^*$ and $c_2 \in \mathbb{R}$. The equation also admits the solution $x : \mathbb{R} \to \mathbb{R}$, defined by

$$x(t) = \begin{cases} c_1 t^3 \ , & t \geq 0 \\ c_2 t^3 \ , & t < 0 \end{cases} \text{ with } c_1, c_2 \in \mathbb{R}.$$

(3) This is a Euler equation whose general solution, $x : \mathbb{I} \to \mathbb{R}$, is defined by $x(t) = c_1 \cos(\ln t^2) + c_2 \sin(\ln t^2)$ for every $t \in \mathbb{I}$, where $\mathbb{I} = (-\infty, 0)$ or $(0, +\infty)$ and $c_1, c_2 \in \mathbb{R}$, $c_1^2 + c_2^2 \neq 0$. The equation also admits the solution $x : \mathbb{R} \to \mathbb{R}$, $x \equiv 0$.

(4) This is a Euler equation whose general solution, $x : \mathbb{R} \to \mathbb{R}$, is defined by $x(t) = c_1 t + c_2 t^2 + c_3 t^3$ for every $t \in \mathbb{R}$, where $c_1, c_2, c_3 \in \mathbb{R}$.

(5) The equation is reducible to one of Euler types by means of the substitution $3t + 2 = \tau$. The general solution is $x : \mathbb{I} \to \mathbb{R}$, defined by $x(t) = c_1 (3t+2)^{-\frac{4}{3}} + c_2$ for every $t \in \mathbb{I}$, where $\mathbb{I} = \left(-\infty, -\frac{3}{2}\right)$, or $\left(-\frac{3}{2}, +\infty\right)$, while $c_1 \in \mathbb{R}^*$, $c_2 \in \mathbb{R}$. The equation also admits the solution $x : \mathbb{R} \to \mathbb{R}$, defined by $x(t) = c$ for every $t \in \mathbb{R}$, where $c \in \mathbb{R}$.

(6) The equation is reducible to one of Euler types. The general solution, $x : \mathbb{I} \to \mathbb{R}$, of the initial equation is defined by $x(t) = c_1 t^{-1} + c_2 t^2$ for every $t \in \mathbb{I}$, where $\mathbb{I} = (-\infty, 0)$ or $(0, +\infty)$, while $c_1, c_2 \in \mathbb{R}$.

(7) This is an equation reducible to one of Euler types having the general solution $x : \mathbb{I} \to \mathbb{R}$, defined by $x(t) = c_1 \cos(\ln |t|) + c_2 \sin(\ln |t|)$ for every $t \in \mathbb{I}$, where $\mathbb{I} = (-\infty, 0)$ or $(0, +\infty)$ and $c_1, c_2 \in \mathbb{R}$.

(8) This is a non-homogeneous Euler equation. The general solution is

$$x(t) = \begin{cases} \frac{1}{2} t + c_1 t^2 + c_2 t^3 & \text{for } t \geq 0 \\ \frac{1}{2} t + c_1 t^2 + c_3 t^3 & \text{for } t < 0 \end{cases}, \text{ with } c_1, c_2, c_3 \in \mathbb{R}.$$

(9) The substitution $1+t = \tau$ leads to a non-homogeneous Euler equation with the general solution defined by $x(t) = c_1(1+t)^2 + c_2(1+t)^2 \ln |1+t| + (1+t)^3$ for every $t \in \mathbb{I}$, where $\mathbb{I} = (-\infty, -1)$ or $(-1, +\infty)$, $c_1 \in \mathbb{R}$ and $c_2 \in \mathbb{R}^*$. The equation

also has the solution $x : \mathbb{R} \to \mathbb{R}$, defined by $x(t) = c(1+t)^2 + (1+t)^3$ for every $t \in \mathbb{R}$, where $c \in \mathbb{R}$.

(10) The equation is of Euler type, non-homogeneous, having the general solution $x : \mathbb{I} \to \mathbb{R}$, defined by $x(t) = c_1 t + c_2 t \ln|t| + t \ln^2|t|$ for every $t \in \mathbb{I}$, where $\mathbb{I} = (-\infty, 0)$ or $(0, +\infty)$ and $c_1, c_2 \in \mathbb{R}$.

Chapter 5

Problem 5.1 Before proceeding to the proof of the four assertions, let us observe that, in the case of the equation considered, every fundamental matrix is of type 1×1 and of the form

$$\mathcal{X}(t) = \xi e^{\int_0^t a(s)\,ds}$$

for every $t \in \mathbb{R}$, where $\xi \in \mathbb{R}^*$.

(1) In view of Theorem 5.2.2, the null solution of the equation considered is stable if and only if there exists a fundamental matrix bounded on $\mathbb{R}_+$, or equivalently every fundamental matrix is bounded on $\mathbb{R}_+$. According to the remark from the beginning, this happens if and only if

$$x(t) = e^{\int_0^t a(s)\,ds} = e^{\int_0^{t_0} a(s)\,ds} e^{\int_{t_0}^t a(s)\,ds} \leq M \qquad (*)$$

for every $t, t_0 \in \mathbb{R}_+$, $t_0 \leq t$. If the mentioned inequality is satisfied, we have $x(t) \leq e^{K(0)}$ for every $t \in \mathbb{R}_+$, and therefore x is bounded on $\mathbb{R}_+$. Hence the null solution is stable. Conversely, if there exists $M > 0$ such that $x(t) \leq M$ for every $t \in \mathbb{R}_+$, then from $(*)$, one observes that the function $K : \mathbb{R}_+ \to \mathbb{R}$, which satisfies the inequality in question, can be taken

$$K(t_0) = \ln M - \int_0^{t_0} a(s)\,ds$$

for every $t_0 \in \mathbb{R}_+$.

(2) By virtue of Theorem 5.2.4, the null solution of the equation is uniformly stable if and only if there exists a fundamental matrix $\mathcal{X}(t)$ and there exists $M > 0$ such that $\|\mathcal{X}(t)\mathcal{X}^{-1}(t_0)\|_{\mathcal{M}} \leq M$ for every $t, t_0 \in \mathbb{R}_+$, $t_0 \leq t$. According to the initial remark, the null solution is uniformly stable if and only if

$$e^{\int_{t_0}^t a(s)\,ds} \leq M$$

for every $t, t_0 \in \mathbb{R}_+$, $t_0 \leq t$, or equivalently

$$\int_{t_0}^t a(s)\,ds \leq \ln M = K$$

for every $t, t_0 \in \mathbb{R}_+$, $t_0 \leq t$.

(3) According to Theorem 5.2.2, the null solution is asymptotically stable if and only if

$$\lim_{t\to+\infty} e^{\int_0^t a(s)\,ds} = 0,$$

which happens if and only if

$$\lim_{t\to+\infty} \int_0^t a(s)\,ds = -\infty.$$

(4) According to Theorem 5.2.5, the null solution of the equation is uniformly asymptotically stable if and only if

$$\lim_{t-s\to+\infty} e^{\int_s^t a(\tau)\,d\tau} = 0.$$

It is evident that the mentioned inequality implies the relation above and therefore the uniform asymptotic stability of the null solution. Conversely, let us assume that the null solution is uniformly asymptotically stable. Then, there exists $\mu > 0$, and for every $\varepsilon > 0$, there exists $T_\varepsilon \geq 0$, such that, for every $t_0 \geq 0$, every $t \geq t_0 + T_\varepsilon$ and every $\xi \in \mathbb{R}$ with $|\xi| \leq \mu$, we have

$$|\xi| e^{\int_{t_0}^t a(s)\,ds} \leq \varepsilon.$$

Let us fix $q \in (0,1)$, let us take $\xi = \mu$, $\varepsilon = q\mu$ and let us denote by $T = T_{k\mu}$. The preceding inequality rewrites, in this particular case, in the equivalent form

$$\int_{t_0}^t a(s)\,ds \leq \ln q \tag{**}$$

for every $t_0 \geq 0$ and every $t \geq t_0 + T$. On the other hand, the null solution is uniformly stable, being uniformly asymptotically stable. According to (2), there exists $K \geq 0$ such that

$$\int_{t_0}^t a(s)\,ds \leq K \tag{***}$$

for every $t_0 \geq 0$ and every $t \geq t_0$. Let $t \geq t_0$. Let us observe that there exists $m \in \mathbb{N}$ such that $t \in [t_0 + mT, t_0 + (m+1)T)$. We have

$$\int_{t_0}^t a(s)\,ds = \sum_{p=0}^{m-1} \int_{t_0+pT}^{t_0+(p+1)T} a(s)\,ds + \int_{t_0+mT}^t a(s)\,ds.$$

Let us remark that, by virtue of the inequality $(**)$, each of the first m terms in the sum above does not exceed $\ln q$, while the last term is bounded from above by K (see $(***)$). We deduce that

$$\int_{t_0}^t a(s)\,ds \leq K + m \ln q.$$

Since $t - t_0 \leq mT$, it follows that $m \geq \frac{1}{T}(t - t_0)$, and by consequence

$$\int_{t_0}^{t} a(s)\, ds \leq K + \frac{\ln q}{T}(t - t_0),$$

for every $t, t_0 \in \mathbb{R}$ with $t \geq t_0$. Hence, the mentioned inequality holds true for $K \geq 0$ and $\alpha = -\frac{\ln q}{T}$ determined as above.

Exercise 5.1 (1) The unique saturated solution $x(\cdot, a, \xi)$ of equation (1), satisfying the initial condition $x(a, a, \xi) = \xi$, is $x(\cdot, a, \xi) : [a, +\infty) \to \mathbb{R}$, defined by $x(t, a, \xi) = \xi e^{t-a}$ for $t \geq a$. So, $\lim_{t\uparrow+\infty} |x(t, \xi)| = +\infty$ for every $\xi \in \mathbb{R}^*$, and by consequence, the null solution is unstable.

(2) The unique saturated solution, $x(\cdot, a, \xi)$, of equation (2), which satisfies the condition $x(a, a, \xi) = \xi$, is $x(\cdot, a, \xi) : [a, +\infty) \to \mathbb{R}$, defined by $x(t, a, \xi) = \xi$ for $t \geq a$. Then, for every $\varepsilon > 0$ and every $a \geq 0$ there exists $\delta(\varepsilon, a) = \varepsilon > 0$ such that, for every $\xi \in \mathbb{R}$ with $|\xi| \leq \delta(\varepsilon, a)$ we have $|x(t, a, \xi)| \leq \varepsilon$ for every $t \geq a$. Hence, the null solution is uniformly stable.

(3) The unique saturated solution $x(\cdot, a, \xi)$ of equation (3) which satisfies the condition $x(a, a, \xi) = \xi$ is $x(\cdot, a, \xi) : [a, +\infty) \to \mathbb{R}$, $x(t, a, \xi) = \xi e^{-(t-a)}$ for $t \geq a$. So, $\lim_{t\uparrow+\infty} |x(t, \xi)| = 0$ for every $\xi \in \mathbb{R}^*$ and, by consequence, the null solution of equation (3) is globally and uniformly asymptotically stable.

(4) The function $f : \mathbb{R} \to \mathbb{R}$, defined by $f(x) = -2x + \sin x$, is of class C^1 and satisfies $f(0) = 0$ and $f'_x(0) = -1$. We are in the hypotheses of Theorem 5.3.3, and consequently the null solution of equation (4) is asymptotically stable.

(5) Let $a \geq 0$ and $\xi \in \mathbb{R}$. The unique saturated solution $x : \mathbb{I}_{a,\xi} \to \mathbb{R}$ of equation (5), which satisfies $x(a, a, \xi) = \xi$, is defined by

$$x(t, a, \xi) = \frac{\xi}{1 - \xi(t - a)}$$

for every $t \in \mathbb{I}_{a,\xi}$, where $\mathbb{I}_{a,\xi} = [a, +\infty)$ if $\xi \leq 0$ and $\mathbb{I}_{a,\xi} = [a, a + \frac{1}{\xi})$ if $\xi > 0$. Since $x(\cdot, a, \xi)$ is not global for $\xi > 0$, the null solution is not stable. We notice, however, that the continuity property (ii) required by Definition 5.1.5 is satisfied in this case "from the left" of $\xi = 0$. Indeed, this follows from the inequality

$$|x(t, a, \xi)| \leq \frac{|\xi|}{1 - \xi(t - a)} \leq |\xi|$$

for every $\xi \leq 0$ and every $t \geq a$.

(6) The unique saturated solution $x : \mathbb{I}_{a,\xi} \to \mathbb{R}$ of equation (6), which satisfies $x(a, a, \xi) = \xi$, is defined by

$$x(t, a, \xi) = \frac{\xi}{1 + \xi(t - a)}$$

for every $t \in \mathbb{I}_{a,\xi}$, where $\mathbb{I}_{a,\xi} = [a, a - \frac{1}{\xi})$ if $\xi < 0$ and $\mathbb{I}_{a,\xi} = [a, a + \infty)$ if $\xi \geq 0$. Since $x(\cdot, a, \xi)$ is not global for $\xi < 0$, the null solution is not stable. Again, the continuity property (ii) required by Definition 5.1.5 is satisfied in this case "from the left" of $\xi = 0$.

(7) The function $f : \left(-\frac{\pi}{2}, \frac{\pi}{2}\right) \to \mathbb{R}$ defined by $f(x) = -\tan x$ is of class C^1, $f(0) = 0$ and $f'_x(0) = -1$. We are in the hypotheses Theorem 5.3.3 and therefore the null solution of equation (7) is asymptotically stable.

(8) The function $f : \mathbb{R} \to \mathbb{R}$, defined by $f(x) = -\sin x$, is of class C^1, $f(0) = 0$ and $f'_x(0) = -1$. We are, also, in the hypotheses Theorem 5.3.3 and therefore the null solution of equation (7) is asymptotically stable.

(9) The same arguments used in the last two exercises lead to the conclusion that the null solution of equation (9) is asymptotically stable.

Exercise 5.2 (1) The roots of the characteristic equation $\det(\mathcal{A} - \lambda\mathcal{I}) = 0$ are $\lambda_{1,2} = -1 \pm \sqrt{2}$. Since $-1 + \sqrt{2} > 0$, the system (1) is unstable. See Theorem 5.2.7.

(2) The roots of the characteristic equation $\det(\mathcal{A} - \lambda\mathcal{I}) = 0$ are $\lambda_{1,2} = \pm i$. Since both these roots have the real part 0 and are simple, the system (2) is uniformly stable. See Theorem 5.2.7.

(3) The roots of the characteristic equation $\det(\mathcal{A} - \lambda\mathcal{I}) = 0$ are $\lambda_{1,2} = -1 \pm i$. Hence the matrix $\mathcal{A}$ is hurwitzian and therefore the system (3) is uniformly and globally asymptotically stable. See Theorem 5.2.6.

(4) Since the roots of the characteristic equation $\det(\mathcal{A} - \lambda\mathcal{I}) = 0$ are $\lambda_{1,2} = \frac{1}{2}(1 \pm \sqrt{13})$, and $\frac{1}{2}(1 + \sqrt{13}) > 0$, the system (4) is unstable. See Theorem 5.2.7.

(5) The matrix $\mathcal{A}$ is hurwitzian and, by consequence, the system (5) is uniformly and globally asymptotically stable. See Theorem 5.2.6.

(6) The roots of the characteristic equation $\det(\mathcal{A} - \lambda\mathcal{I}) = 0$ are $\lambda_1 = -4$ and $\lambda_2 = 0$. Since $\lambda_2 = 0$ is simple, the system (6) is uniformly stable. See Theorem 5.2.7.

(7) We may use Theorem 5.2.7, but we may also conclude directly, by observing that every global solution of the system (7), having equal components, is of the form $x(t) = c(e^t, e^t, e^t)$ for every $t \in \mathbb{R}$, where $c \in \mathbb{R}$. From this observation, it follows that the system is unstable.

(8) Let us observe that every global solution of the system (8) having equal components, is of the form $x(t) = c(e^{2t}, e^{2t}, e^{2t})$ for every $t \in \mathbb{R}$, where $c \in \mathbb{R}$. It follows that the system is unstable.

(9) The roots of the characteristic equation $\det(\mathcal{A} - \lambda\mathcal{I}) = 0$ are $\lambda_1 = 0$ and $\lambda_{2,3} = \pm i\sqrt{3}$. Since all these roots have the real part 0 and are simple, the system (9) is uniformly stable. See Theorem 5.2.6.

Problem 5.2 Using the variation of constants method — see Theorem 4.5.7 — we deduce that the general solution of the equation considered is

$$x(t, \xi_1, \xi_2) = \xi_1 \cos \omega t + \xi_2 \sin \omega t + \frac{1}{\omega} \int_0^t f(s) \sin \omega(t - s)\, ds,$$

where $\xi_1, \xi_2 \in \mathbb{R}$. Then we have

$$|x(t, \xi_1, \xi_2)| \le |\xi_1| + |\xi_2| + \frac{1}{\omega} \int_0^{+\infty} |f(s)|\, ds$$

for every $t \in \mathbb{R}_+$.

Problem 5.3 We begin by noticing that, in the case of the second-order equation, the uniform stability of the null solution is equivalent to the uniform stability of the null solution of the first-order system

$$\begin{cases} x'(t) = y(t) \\ y'(t) = -[\,\omega^2 + f(t)\,]x(t). \end{cases} \tag{$\mathcal{S}$}$$

Also, the equation being linear, every saturated solution is global. So, its unique saturated solution, $x(\cdot, a, \xi_1, \xi_2)$, with $x(a, a, \xi_1, \xi_2) = \xi_1$, $x'(a, a, \xi_1, \xi_2) = \xi_2$, is defined on $[\,a, +\infty)$. By the variation of constants method (see Theorem 4.5.7), we deduce that

$$x(t, a, \xi_1, \xi_2) = \xi_1 \cos \omega(t-a) + \frac{\xi_2}{\omega} \sin \omega(t-a) + \frac{1}{\omega} \int_a^t f(s)x(s, a, \xi_1, \xi_2) \sin \omega(t-s)\, ds$$

for every $t \geq a$. It follows that

$$|x(t, a, \xi_1, \xi_2)| \leq |\xi_1| + \frac{|\xi_2|}{\omega} + \frac{1}{\omega} \int_a^t |f(s)||x(s, a, \xi_1, \xi_2)|\, ds$$

and

$$|x'(t, a, \xi_1, \xi_2)| \leq \omega|\xi_1| + |\xi_2| + \int_a^t |f(s)||x(s, a, \xi_1, \xi_2)|\, ds$$

for every $t \geq a$. From Gronwall's Lemma 1.5.2, we get

$$|x(t, a, \xi_1, \xi_2)| \leq \left(|\xi_1| + \frac{|\xi_2|}{\omega}\right) \exp\left(\frac{1}{\omega} \int_a^t |f(s)|\, ds\right)$$

and

$$|x'(t, a, \xi_1, \xi_2)| \leq (\omega|\xi_1| + |\xi_2|) \exp\left(\int_a^t |f(s)|\, ds\right)$$

for every $\xi_1, \xi_2 \in \mathbb{R}$ and every $t \geq a$. Since f is absolutely integrable over $\mathbb{R}_+$, there exists $m > 0$ such that

$$\int_a^t |f(s)|\, ds \leq m$$

for every $a, t \in \mathbb{R}_+$, $t \geq a$. From the last three inequalities, recalling that $x' = y$, we deduce

$$x^2(t, a, \xi_1, \xi_2) + y^2(t, a, \xi_1, \xi_2) \leq M\left(\xi_1^2 + \xi_2^2\right)$$

for every $\xi_1, \xi_2 \in \mathbb{R}$ and every $a, t \in \mathbb{R}_+$, $t \geq a$, where $M > 0$ depends only on m and on ω, but not on a, t, ξ_1, ξ_2. This inequality shows that the null solution of the system $(\mathcal{S})$ is uniformly stable.

Problem 5.4 Since $\mathcal{A}$ is hurwitzian, according to Lemma 5.2.1, it follows that there exist $M > 0$ and $\omega > 0$ such that $\|e^{t\mathcal{A}}\|_{\mathcal{M}} \leq Me^{-\omega t}$ for every $t \geq 0$. Let us fix a number $L > 0$ with the property $ML - \omega < 0$. Using the fact that

$\lim_{t \to +\infty} \|\mathcal{B}(t)\|_{\mathcal{M}} = 0$, we conclude that there exists $c \geq 0$ such that $\|\mathcal{B}(t)\|_{\mathcal{M}} \leq L$ for every $t \geq c$, where $L > 0$ is fixed as above. So, $\|\mathcal{B}(t)x\| \leq L\|x\|$ for every $t \geq c$ and every $x \in \mathbb{R}^n$. The conclusion follows from a simple variant of Theorem 5.3.1 which instead of the hypothesis $\|F(t,x)\| \leq L\|x\|$ for every $(t,x) \in \mathbb{R}_+ \times \Omega$, uses the hypothesis $\|F(t,x)\| \leq L\|x\|$ for every $t \geq c$ and every $x \in \Omega$, where $c \geq 0$.

Problem 5.5 From the variation of constants formula — see Remark 4.3.3 — we have that the unique global solution $x(\cdot, \xi) : \mathbb{R}_+ \to \mathbb{R}^n$ of the system, which satisfies $x(0, \xi) = \xi$, also satisfies

$$x(t,\xi) = e^{t\mathcal{A}}\xi + \int_0^t e^{(t-s)\mathcal{A}}\mathcal{B}(s)x(s,\xi)\,ds$$

for every $t \in \mathbb{R}_+$. Since $\mathcal{A}$ is hurwitzian, according to Lemma 5.2.1, there exist $M > 0$ and $\omega > 0$ such that $\|e^{t\mathcal{A}}\|_{\mathcal{M}} \leq Me^{-\omega t}$ for every $t \in \mathbb{R}_+$. Then, we have

$$\|x(t,\xi)\| \leq Me^{-\omega t}\|\xi\| + Me^{-\omega t}\int_0^t e^{\omega s}\|\mathcal{B}(s)\|_{\mathcal{M}}\|x(s,\xi)\|\,ds$$

for every $t \in \mathbb{R}_+$. Multiplying both sides of this inequality by $e^{\omega t}$, and denoting by $y(t) = \|x(t)\|e^{\omega t}$, we deduce

$$y(t) \leq M\|\xi\| + M\int_0^t \|\mathcal{B}(s)\|_{\mathcal{M}}y(s)\,ds$$

for every $t \in \mathbb{R}_+$. From Gronwall's Lemma 1.5.2, it follows that

$$y(t) \leq M\|\xi\|\exp\left(M\int_0^t \|\mathcal{B}(s)\|_{\mathcal{M}}ds\right)$$

for every $t \in \mathbb{R}_+$. Since

$$\int_0^{+\infty} \|\mathcal{B}(s)\|_{\mathcal{M}}ds = m < +\infty,$$

from the preceding inequality, we deduce that $y(t) \leq k\|\xi\|$ for every $t \in \mathbb{R}_+$, where $k = Me^{Mm}$. Multiplying both sides of this inequality with $e^{-\omega t}$, and recalling the definition of $y(t)$, we get the conclusion.

Problem 5.6 The change of variable $s = \frac{t^{m+1}}{m+1}$ leads to a system of the type considered in Problem 5.4 except that, in this case, the interval of definition of the function $\mathcal{B}$ is $(0, +\infty)$ and not $[\,0, +\infty)$. Indeed, putting $x(t) = y(s)$, we have

$$\frac{dx}{dt}(t) = \frac{dy}{ds}\frac{ds}{dt}(t) = \frac{dy}{ds}(s)[(m+1)s]^{\frac{m}{m+1}},$$

and the initial system is equivalent to

$$\frac{dy}{ds}(s) = [\,\mathcal{A} + \mathcal{B}(s)\,]y(s),$$

$\mathcal{A} = \mathcal{A}_0$, $\mathcal{B}(s) = [(m+1)s]^{-\frac{1}{m+1}}\mathcal{A}_1 + [(m+1)s]^{-\frac{2}{m+1}}\mathcal{A}_2 + \cdots + [(m+1)s]^{-\frac{m}{m+1}}\mathcal{A}_m$. From the hypothesis, we know that $\mathcal{A}$ is hurwitzian. Also, we may easily see that $\lim_{s\to+\infty} \|\mathcal{B}(s)\|_{\mathcal{M}} = 0$. From this point, following the same way as that used for solving Problem 5.4, one shows that for every $a > 0$ there exists $\delta(a) > 0$ such that $\lim_{s\to+\infty} y(s,a,\xi) = 0$ for every $\xi \in B(0,\delta(a))$. We conclude the proof by recalling the relationship between $y(s)$ and $x(t)$.

Problem 5.7 Since $\xi_1 < \xi_2$, from Problem 2.9, we deduce that $x(t,\xi_1) < x(t,\xi_2)$ for every $t \geq 0$ and also that, for every $a \geq 0$ and every $\xi \in (x(a,\xi_1), x(a,\xi_2))$, the unique saturated solution $x(\cdot,a,\xi)$ of the equation considered, which satisfies $x(a,a,\xi) = \xi$, also satisfies $x(t,a,\xi) \in (x(t,\xi_1), x(t,\xi_2))$ for every t in the interval of definition. Assuming that this solution is not global, it follows that it is bounded on the interval of existence $[a, T_m)$. In view of Corollary 2.4.3, it follows that it is continuable. This contradiction can be eliminated only if $x(\cdot,\xi)$ is global. From the condition in the statement and the preceding inequality, it follows that, for every $a \geq 0$ and every $\varepsilon > 0$, there exists $\delta(\varepsilon,a) > 0$ such that for every $a \geq a_\varepsilon$, every $t \geq a_\varepsilon$ and every $\eta \in (x(a,\xi_1), x(a,\xi_2))$ we have $|x(t,a,\eta) - x^*| \leq \varepsilon$. We distinguish between two cases: $a \geq a_\varepsilon$ and $a < a_\varepsilon$. If $a \geq a_\varepsilon$, taking $\delta(\varepsilon,a) = \min\{x(a,\xi) - x(a,\xi_1), x(a,\xi_2) - x(a,\xi)\}$, we get that $|x(t,a,\eta) - x(t,\xi)| \leq \varepsilon$ for every $t \geq a$, which is nothing but the condition of stability. If $a < a_\varepsilon$, then, from Theorem 2.5.2, it follows that there exists $\delta(\varepsilon,a) > 0$ such that, for every $\eta \in \mathbb{R}$ with $|x(a,\xi) - \eta| \leq \delta(\varepsilon,a)$, we have $|x(t,a,\xi) - x(t,a,\eta)| \leq \varepsilon$ for every $t \in [a, a_\varepsilon]$. Let us observe that, from the definition of both a_ε and $\delta(\varepsilon,a)$, we have $|x(t,a,\eta) - x(t,\xi)| \leq \varepsilon$ for every $\eta \in \mathbb{R}$ with $|x(a,\xi) - \eta| \leq \delta(\varepsilon,a)$ and every $t \geq a$. Consequently, the solution $x(\cdot,\xi)$ is stable. Finally, as $\lim_{t\to+\infty} x(t,a,\eta) = x^*$ for every $\eta \in (x(a,\xi_1), x(a,\xi_2))$, we deduce that $x(\cdot,\xi)$ is asymptotically stable.

Problem 5.8 The answer is in the negative. Indeed, the function $f : (-1,1) \to \mathbb{R}$, defined by

$$f(x) = \begin{cases} x(1-x) & \text{if } x \in [0,1) \\ x(1+x) & \text{if } x \in (-1,0] \end{cases}$$

for each $x \in (-1,1)$, is locally Lipschitz, $f(0) = 0$, all saturated solutions of the Cauchy problem are global and bounded, but the null solution is not stable. To see that this is the case, we observe that for each $\xi \in (0,1)$ the unique global solution, $x(\cdot,\xi)$, of the Cauchy problem is given by

$$x(t,\xi) = \frac{\xi e^t}{1 + \xi(e^t - 1)}.$$

Problem 5.9 From the Lagrange formula, we have $f(x) = \lambda x + g(x)$, where

$$\lim_{x\downarrow 0} \frac{g(x)}{x} = 0. \qquad (*)$$

Multiplying the equation by x, we get $\frac{1}{2}\frac{d}{dt}\left(x^2\right) = \lambda x^2 + g(x)x = 0$. From $(*)$, it follows that there exists $r > 0$ such that $\lambda x + g(x) > 0$ for every $x \in \mathbb{R}$ with $|x| \leq r$. Accordingly, every solution, which "enters or is" in the interval $[-r, r]$, tends to leave this interval. From here, it follows that the null solution of the equation cannot be asymptotically stable.

Exercise 5.3 (1) The function on the right-hand side of the system $f : \mathbb{R}^2 \to \mathbb{R}^2$ is defined by $f(x) = (f_1(x), f_2(x)) = (-x_1 + x_2^2, -x_1^3 - 2x_2)$ for $x = (x_1, x_2) \in \mathbb{R}^2$. We can easily see that the matrix

$$f_x(0) = \mathcal{A} = \begin{pmatrix} -1 & 0 \\ 0 & -2 \end{pmatrix}$$

has both characteristic roots real and strictly negative. By Theorem 5.3.3, the null solution is asymptotically stable.

(2) With the notations in the preceding exercise, we have

$$f_x(0) = \mathcal{A} = \begin{pmatrix} 1 & 0 \\ 0 & -4 \end{pmatrix}.$$

Since the matrix above has one strictly positive characteristic root, by virtute of Theorem 5.3.4, the null solution is unstable.

(3) The matrix

$$f_x(0) = \mathcal{A} = \begin{pmatrix} -1 & 5 \\ 0 & -1 \end{pmatrix}$$

has both characteristic roots real and strictly negative. In view of Theorem 5.3.3, the null solution is asymptotically stable.

(4) We have

$$f_x(0) = \mathcal{A} = \begin{pmatrix} 2 & 0 \\ 0 & -1 \end{pmatrix}.$$

Since this matrix has a strictly positive characteristic root, by Theorem 5.3.4, it follows that the null solution is unstable.

(5) The matrix

$$f_x(0) = \mathcal{A} = \begin{pmatrix} -1 & 0 \\ -4 & -5 \end{pmatrix}$$

has both characteristic roots strictly negative. According to Theorem 5.3.3, the null solution is asymptotically stable.

(6) The matrix

$$f_x(0) = \mathcal{A} = \begin{pmatrix} 0 & 2 \\ 0 & -3 \end{pmatrix}$$

has one null characteristic root. In this case, none of the theorems proved in Section 5.3 can help with respect to stability.

Exercise 5.4 We begin with the remark that, for all systems considered in this exercise, we will look for Lyapunov functions which are independent of t, and this because all these systems are autonomous.

(1) The system is of the form $x' = f(x)$, where the function on the right-hand side, $f : \mathbb{R}^2 \to \mathbb{R}^2$, is defined by $f(x) = (f_1(x), f_2(x)) = (-x_1^3 + x_2, -x_1 - 2x_2^3)$ for

every $x = (x_1, x_2) \in \mathbb{R}^2$. One observes that $V : \mathbb{R}^2 \to \mathbb{R}$, defined by $V(x) = \frac{1}{2}\|x\|^2$ for every $x \in \mathbb{R}^2$, is of class C^1, $V(x) = 0$ if and only if $x = 0$ and satisfies

$$f_1(x)\frac{\partial V}{\partial x_1}(x) + f_2(x)\frac{\partial V}{\partial x_2}(x) = -x_1^4 - 2x_2^4 \leq 0$$

for every $x \in \mathbb{R}^2$. Accordingly, V is a Lyapunov function for the system. From Theorem 5.4.1, it follows that the null solution is stable. Let us remark that the function V also has the extra-properties required by Theorem 5.4.2. More precisely, V satisfies $V(x) \leq \lambda(\|x\|) = \frac{1}{2}\|x\|^2$ and

$$f_1(x)\frac{\partial V}{\partial x_1}(x) + f_2(x)\frac{\partial V}{\partial x_2}(x) = -x_1^4 - 2x_2^4 \leq -\frac{1}{2}\|x\|^4 = -\eta(\|x\|)$$

for every $x \in \mathbb{R}^2$. Obviously the functions $\lambda, \eta : \mathbb{R}_+ \to \mathbb{R}_+$, defined by $\lambda(r) = \frac{1}{2}r^2$ and $\eta(r) = \frac{1}{2}r^4$ are continuous, nondecreasing and satisfy $\lambda(r) = \eta(s) = 0$ if and only if $r = s = 0$. According to Theorem 5.4.2, it follows that the null solution is asymptotically stable. Moreover, because for every $x \in \mathbb{R}^2$, we have $V(x) = \frac{1}{2}\|x\|^2 = \omega(\|x\|)$ and $\lim_{r\to+\infty} \omega(r) = +\infty$, in Theorem 5.4.3, we conclude that the system considered is globally asymptotically stable.

(2) In the case of this system the function on the right-hand side, $f : \mathbb{R}^2 \to \mathbb{R}^2$, is $f(x) = (f_1(x), f_2(x)) = (-x_1^5 - 3x_2, 3x_1 - 4x_2^3)$ for every $x = (x_1, x_2) \in \mathbb{R}^2$. We remark that $V : \mathbb{R}^2 \to \mathbb{R}$, defined by $V(x) = \frac{1}{2}\|x\|^2$ for every $x \in \mathbb{R}^2$, is a Lyapunov function. Indeed, V is of class C^1, $V(x) = 0$ if and only if $x = 0$ and satisfies

$$f_1(x)\frac{\partial V}{\partial x_1}(x) + f_2(x)\frac{\partial V}{\partial x_2}(x) = -x_1^6 - 4x_2^4 \leq 0$$

for every $x \in \mathbb{R}^2$. Let us observe that the restriction of this function to the set $\Omega_0 = \{(x_1, x_2);\ |x_1| < 1,\ |x_2| < 1\}$ (which obviously is an open neighborhood of the origin), satisfies all the hypotheses of Theorem 5.4.3. This follows from the remark that, on this set, we have $\frac{1}{2}(x_1^2+x_2^2)^4 \leq x_1^8 + x_2^8 \leq x_1^6 + 4x_2^4$, which implies

$$f_1(x)\frac{\partial V}{\partial x_1}(x) + f_2(x)\frac{\partial V}{\partial x_2}(x) \leq -\frac{1}{2}\|x\|^8$$

for every $x \in \mathbb{R}^2$. In order to complete the proof we only have to observe that the functions $\lambda, \eta : \mathbb{R}_+ \to \mathbb{R}_+$, defined by $\lambda(r) = \frac{1}{2}r^2$ and $\eta(r) = \frac{1}{2}r^8$ are continuous, nondecreasing and satisfy $\lambda(r) = \eta(s) = 0$ if and only if $r = s = 0$. According to Theorem 5.4.2, it follows that the null solution is asymptotically stable.

(3) The function $f : \mathbb{R}^2 \to \mathbb{R}^2$ on the right-hand side of the system considered is defined by $f(x) = (f_1(x), f_2(x)) = (-x_1 + 5x_2^3, -x_1^3 - 3x_2)$ for $x = (x_1, x_2) \in \mathbb{R}^2$. We observe that $V : \mathbb{R}^2 \to \mathbb{R}$, defined by $V(x) = \frac{1}{4}(x_1^4 + 5x_2^4)$ for every $x \in \mathbb{R}^2$, is a Lyapunov function. Indeed, V is of class C^1, $V(x) = 0$ if and only if $x = 0$ and satisfies

$$f_1(x)\frac{\partial V}{\partial x_1}(x) + f_2(x)\frac{\partial V}{\partial x_2}(x) = -x_1^4 - 15x_2^4 \leq 0$$

for every $x \in \mathbb{R}^2$. Let us observe that V satisfies all the hypotheses of both Theorems 5.4.2 and 5.4.3. Indeed, V is bounded from below by the function $\omega : \mathbb{R}_+ \to \mathbb{R}_+$, defined by $\omega(r) = \frac{1}{8}r^4$ for every $r \in \mathbb{R}_+$, and which satisfies the condition $\lim_{r\to+\infty} \omega(r) = +\infty$. Finally, let us observe that a possible choice of the functions $\lambda, \eta : \mathbb{R}_+ \to \mathbb{R}_+$ in Theorem 5.4.2 is $\lambda(r) = \frac{5}{4}r^4$ and $\eta(r) = \frac{1}{2}r^4$. According to Theorem 5.4.2, it follows that the null solution of the system is asymptotically stable, while from Theorem 5.4.3, we deduce that the system is globally asymptotically stable.

(4) Let us observe that the unique global solution of the system considered, $x(\cdot, 0, (\xi, 0))$, which satisfies $x(0, 0, (\xi, 0)) = (\xi, 0)$, is $x(t, 0, (\xi, 0)) = \xi(e^t, 0)$ for every $t \in \mathbb{R}_+$. So, the null solution of the system is unstable.

(5) The function $f : \mathbb{R}^2 \to \mathbb{R}^2$ on the right-hand side of the system considered is defined by $f(x) = (f_1(x), f_2(x)) = (-\sin x_1 + x_2, -4x_1 - 3\tan x_2)$ for every $x = (x_1, x_2) \in \mathbb{R} \times \left(-\frac{\pi}{2}, \frac{\pi}{2}\right)$. One observes that $V : \mathbb{R}^2 \to \mathbb{R}$, $V(x) = 2x_1^2 + \frac{1}{2}x_2^2$ for every $x \in \mathbb{R}^2$ is a Lyapunov function for the system on $\Omega_0 = \left(-\frac{\pi}{2}, \frac{\pi}{2}\right) \times \left(-\frac{\pi}{2}, \frac{\pi}{2}\right)$. Indeed, V is of class C^1, $V(x) = 0$ if and only if $x = 0$ and satisfies

$$f_1(x)\frac{\partial V}{\partial x_1}(x) + f_2(x)\frac{\partial V}{\partial x_2}(x) = -x_1 \sin x_1 - 3x_2 \tan x_2 \leq 0$$

for every $x \in \left(-\frac{\pi}{2}, \frac{\pi}{2}\right) \times \left(-\frac{\pi}{2}, \frac{\pi}{2}\right)$. According to Theorem 5.4.1, the null solution is stable. Observing that, on a sufficiently small neighborhood of the origin, $(-\delta, \delta)$, we have $y \sin y \geq \frac{y^2}{2}$ and $y \tan y \geq \frac{y^2}{2}$, we deduce that the system satisfies the hypotheses Theorem 5.4.2 too, with $\lambda(r) = 2r^2$ and $\eta(r) = \frac{1}{2}r^2$ for every $r \in \mathbb{R}_+$. Hence the null solution is asymptotically stable.

(6) The function $f : \mathbb{R}^2 \to \mathbb{R}^2$ on the right-hand side of the system considered is defined by $f(x) = (f_1(x), f_2(x)) = (-2\text{sh}\, x_1 + 4x_2^3, -x_1^3 - 2x_2)$. One observes that $V : \mathbb{R}^2 \to \mathbb{R}$, defined by $V(x) = x_1^4 + 4x_2^4$ for every $x \in \mathbb{R}^2$, is a Lyapunov function for the system on $\mathbb{R}^2$. On $\Omega = \{x \in \mathbb{R}^2;\ \|x\| < 1\}$, V satisfies the hypotheses of Theorem 5.4.2. Hence the null solution is asymptotically stable.

Chapter 6

Exercise 6.1 (1) Adding side by side the three equations, we get $x_1' + x_2' + x_3' = 0$. Hence every solution of the system satisfies $x_1 + x_2 + x_3 = c_1$. So, one prime integral is the function $U_1 : \mathbb{R}^3 \to \mathbb{R}$, defined by $U_1(x_1, x_2, x_3) = x_1 + x_2 + x_3$. Multiplying the equation of rank i with x_i, $i = 1, 2, 3$, and adding the equalities thus obtained, we deduce $x_1x_1' + x_2x_2' + x_3x_3' = 0$. Hence the function $U_2 : \mathbb{R}^3 \to \mathbb{R}$ defined by $U_2(x_1, x_2, x_3) = x_1^2 + x_2^2 + x_3^2$ is also a prime integral. Since

$$\left(\frac{\partial U_i}{\partial x_j}\right)_{2\times 3}(x_1, x_2, x_3) = \begin{pmatrix} 1 & 1 & 1 \\ 2x_1 & 2x_2 & 2x_3 \end{pmatrix},$$

it follows that U_1, U_2 are independent about any non-stationary point. Indeed, let us observe that (x_1, x_2, x_3) is non-stationary if and only if $x_1 \neq x_2$, or $x_1 \neq x_3$, or $x_2 \neq x_3$, situations in which the rank of the matrix above is 2.

(2) From the system, we deduce $x_1' - x_2' + x_3' = 0$ and $x_1x_1' - x_2x_2' = 0$. So, the functions $U_i : \mathbb{R}^3 \to \mathbb{R}$, $i = 1, 2$, defined by $U_1(x_1, x_2, x_3) = x_1 - x_2 + x_3$, and $U_2(x_1, x_2, x_3) = x_1^2 - x_2^2$ respectively, are prime integrals for the system. The only stationary points of the system are of the form $(0, 0, x_3)$. Since the rank of the matrix

$$\left(\frac{\partial U_i}{\partial x_j}\right)_{2\times 3}(x_1, x_2, x_3) = \begin{pmatrix} 1 & -1 & 1 \\ 2x_1 & -2x_2 & 0 \end{pmatrix}$$

is 2 at every point (x_1, x_2, x_3) with $x_1 \neq x_2$ or $x_1 = x_2 \neq 0$, it follows that the two prime integrals are independent at any non-stationary point.

(3) From the first two equations, we deduce $x_1x_1' + x_2x_2' = 0$, while from the first and the last equations, we obtain $x_1'/x_1 = x_3'/x_3$. Then, two prime integrals are $U_1(x_1, x_2, x_3) = x_1^2 + x_2^2$ and $U_2(x_1, x_2, x_3) = x_3/x_1$ defined on the set $\Omega = \{(x_1, x_2, x_3) \in \mathbb{R}^3 \,;\; x_1 \neq 0\}$. We have

$$\left(\frac{\partial U_i}{\partial x_j}\right)_{2\times 3}(x_1, x_2, x_3) = \begin{pmatrix} 2x_1 & 2x_2 & 0 \\ \\ \dfrac{-x_3}{x_1^2} & 0 & \dfrac{1}{x_1} \end{pmatrix}.$$

Obviously, the rank of this matrix is 2 at every non-stationary point and therefore the two prime integrals are independent at all non-stationary points.

(4) Subtracting the first two equations, we deduce $x_1' - x_2' = (x_1 - x_2)(x_3 - 1)$ equality, which along with the third equation, leads to

$$\frac{x_1' - x_2'}{x_1 - x_2} = \frac{x_3'}{x_3 + 1}.$$

Hence, the function U_1, defined on $\Omega_1 = \{(x_1, x_2, x_3) \,;\; x_3 \neq -1\}$ by

$$U_1(x_1, x_2, x_3) = \frac{x_1 - x_2}{x_3 + 1},$$

is a prime integral for the system. Adding the first two equations and repeating the manipulations above, we deduce that the function U_2, defined on $\Omega_2 = \{(x_1, x_2, x_3) \,;\; x_3 \neq 1\}$ by

$$U_2(x_1, x_2, x_3) = \frac{x_1 + x_2}{x_3 - 1},$$

is also a prime integral for the system. A point (x_1, x_2, x_3) is stationary for the system if and only if $x_1 = -x_2$ and $x_3 = 1$, or $x_1 = x_2$ and $x_3 = -1$. One may easily state that the rank of the matrix

$$\left(\frac{\partial U_i}{\partial x_j}\right)_{2\times 3}(x_1,x_2,x_3)=\begin{pmatrix} \dfrac{1}{x_3+1} & -\dfrac{1}{x_3+1} & -\dfrac{x_1-x_2}{(x_3+1)^2} \\ \\ \dfrac{1}{x_3-1} & \dfrac{1}{x_3-1} & -\dfrac{x_1+x_2}{x_3-1} \end{pmatrix}$$

is 2 at all non-stationary points.

(5) All solutions satisfy $x_1'x_2 - x_1x_2' = 0$ and $x_1'x_2 + x_1x_2' + x_3' = 0$. Two prime integrals, on $\mathbb{R}^3$, are $U_1(x_1,x_2,x_3) = x_1/x_2$ and $U_2(x_1,x_2,x_3) = x_1x_2+x_3$. The stationary points of the system are of the form $(0,0,x_3)$. So, the rank of the matrix

$$\left(\frac{\partial U_i}{\partial x_j}\right)_{2\times 3}(x_1,x_2,x_3)=\begin{pmatrix} \dfrac{1}{x_2} & -\dfrac{x_1}{x_2^2} & 0 \\ \\ x_2 & x_1 & 1 \end{pmatrix}$$

is 2 at every non-stationary point of the system.

(6) From the first two equations, we deduce $x_1'x_2+x_1x_2'=0$ which shows that the function $U_1:\mathbb{R}^3\to\mathbb{R}$, defined by $U_1(x_1,x_2,x_3)=x_1x_2$, is a prime integral for the system. Since "along the solutions" of the system we have $x_1x_2=c$, from the first and the last equations, it follows that $-x_1(1+x_1^2)x_1' = cx_3'$. So, we have $cx_3+x_1^2/2+x_1^4/4=c_2$. Then, "along the solutions" of the system, we have $x_1x_2x_3+x_1^2/2+x_1^4/4=c_2$. Another prime integral is $U_2:\mathbb{R}^3\to\mathbb{R}$, defined by $U_2(x_2,x_2,x_3)=x_1x_2x_3+x_1^2/2+x_1^4/4$. The stationary points of the system are of the form $(0,0,x_3)$, while the rank of the matrix

$$\left(\frac{\partial U_i}{\partial x_j}\right)_{2\times 3}(x_1,x_2,x_3)=\begin{pmatrix} x_2 & x_1 & 0 \\ \\ x_2x_3+x_1+x_1^3 & x_1x_3 & x_1x_2 \end{pmatrix}$$

is 2 at every non-stationary point of the system for which $x_1\neq 0$.

(7) From the first two equations, we deduce $x_1'x_2-x_1x_2'=0$. So, the function $U_1(x_1,x_2,x_3)=x_1/x_2$, whose domain is $\Omega_1=\{(x_1,x_2,x_3);\ \mathbb{R}^3,\ x_2\neq 0\}$, is a prime integral for the system. Let us observe that $x_1'/x_1+x_2'/x_2=x_3'/x_3$, which shows that the function $U_2(x_1,x_2,x_3)=(x_1x_2)/x_3$, whose domain is given by $\Omega_2=\{(x_1,x_2,x_3);\ \mathbb{R}^3,\ x_3\neq 0\}$, is also a prime integral for the system. The stationary points of the system are of the form $(x_1,0,0)$ or $(0,x_2,0)$. At every non-stationary point of the system for which $x_1x_2x_3\neq 0$, the rank of the matrix

$$\left(\frac{\partial U_i}{\partial x_j}\right)_{2\times 3}(x_1,x_2,x_3)=\begin{pmatrix} \dfrac{1}{x_2} & -\dfrac{x_1}{x_2^2} & 0 \\ \\ \dfrac{x_2}{x_3} & \dfrac{x_1}{x_3} & -x_1x_2x_3^{-2} \end{pmatrix}$$

is 2.

(8) We have both $\dfrac{x_1'}{2-x_1}+2\dfrac{x_3'}{x_3}=0$ and $\dfrac{x_1x_1'+x_2x_2'+x_3x_3'}{x_1^2+x_2^2+x_3^2}=-\dfrac{x_3'}{2x_3}$. From these relations, it follows that $U_1, U_2 : \{(x_1,x_2,x_3)\in\mathbb{R}^3\,;\ x_3\neq 0\}\to\mathbb{R}$, defined by $U_1(x_1,x_2,x_3)=(2-x_1)/x_3^2$, and $U_2(x_1,x_2,x_3)=(x_1^2+x_2^2+x_3^2)/x_3$, are prime integrals for the system. The stationary points of the system are all the points of the circle of equations $x_2=0$ and $(x_1-2)^2+x_3^2=4$. The rank of the matrix

$$\left(\frac{\partial U_i}{\partial x_j}\right)_{2\times 3}(x_1,x_2,x_3)=\begin{pmatrix} -\dfrac{1}{x_3^2} & 0 & -\dfrac{4-2x_1}{x_3^3} \\ \\ \dfrac{2x_1}{x_3} & \dfrac{2x_2}{x_3} & -\dfrac{x_1^2+x_2^2-x_3^2}{x_3^2}\end{pmatrix}$$

is 2 at all non-stationary points, except for $(2,0,2)$ and $(2,0,-2)$.

Problem 6.1 Let us observe that the function U is a prime integral for the system if and only if the function $V:(0,+\infty)\times(0,+\infty)\to\mathbb{R}$, defined by $V=\ln(U)$ has the same property. Let us also observe that V is nonconstant, of class C^1, and satisfies

$$\frac{\partial V}{\partial x}(x,y)(a-ky)x+\frac{\partial V}{\partial y}(x,y)(-1)(b-hx)y$$

$$=\left(h-\frac{b}{x}\right)(a-ky)x-\left(k-\frac{a}{y}\right)(b-hx)y=0.$$

According to Theorem 6.1.1, it follows that V, or equivalently U, is a prime integral.

Problem 6.2 One observes that $U_1, U_2:\mathbb{R}^3\to\mathbb{R}$, $U_1(x_1,x_2,x_3)=x_1+x_2+x_3$ and $U_2(x_1,x_2,x_3)=x_1^2+x_2^2+x_3^2$ for every $(x_1,x_2,x_3)\in\mathbb{R}^3$, are prime integrals for the system. So, every trajectory of the system is included in a circle, i.e. the intersection of the plane of equation $x_1+x_2+x_3=c_1$ and the sphere of equation $x_1^2+x_2^2+x_3^2=c_2$. We complete the proof by observing that every saturated solution is global.

Problem 6.3 Let us observe that the graph of the prime integral V, defined in the solution of Problem 6.1, is a "paraboloid-like" surface whose vertex has the coordinates $(b/h, a/k, U(b/h,a/k))$. So, the intersection of this graph with every plane, parallel with the xOy-plane, is a simple closed curve. See Figure S.6.1. Since the trajectory of any solution of the system is the projection of such a curve on the xOy-plane, this, in its turn, is a simple closed curve.

Problem 6.4 Dividing the first equation by x, we deduce $\frac{x'}{x}=a-ky$. Integrating this equality on $[\,t,t+T\,]$ and taking into account that x is periodic of period T, we deduce

$$aT-k\int_t^{t+T}y(s)ds=0,$$

or equivalently $y_m=a/k$. Analogously, we obtain $x_m=b/h$.

Problem 6.5 Let us remark that, if U is a prime integral of the system which has a strict local minimum at ξ, then $V(x)=U(x)-U(\xi)$ is also a prime integral. By a simple translation argument, we may assume without any loss of generality,

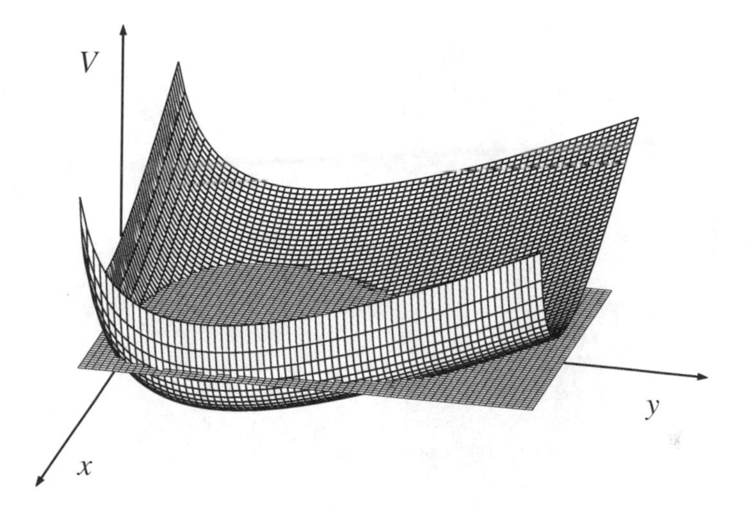

Fig. S.6.1

that $\xi = 0$. Obviously, V is positive defined (see Lemma 5.4.1), $V(0) = 0$, and satisfies

$$\sum_{i=1}^{n} f_i(x)\frac{\partial V}{\partial x_i}(x) = 0.$$

So, V is a Lyapunov function for the system. We are then in the hypotheses of Theorem 5.4.1 from where we get the conclusion.

Problem 6.6 The conclusion follows from Problem 6.5, by observing that the prime integral, defined in the solution of Problem 6.1 by $V(x,y) = \ln(U(x,y))$, has local strict minimum at $(b/h, a/k)$.

Problem 6.7 If the autonomous system $x' = f(x)$ admits an injective prime integral, it follows that all the solutions of the system are constants. Hence, the system is of the form $x' = 0$. Let us assume for contradiction that there exists a non-autonomous system $x' = f(t,x)$ which admits an injective prime integral U. As in the autonomous case, it follows that the graph of each solution reduces to one point which is absurd as long as there exist solutions defined on intervals containing at least two points. So, under minimal continuity assumptions on f, the non-autonomous system $x' = f(t,x)$ cannot admit injective prime integrals

Problem 6.8 Let $x : [\,a, T_m) \to \mathbb{R}^n$ be a solution of the system and let us assume that $T_m < +\infty$. Since $U(x(\cdot))$ is constant on $[\,a, T_m)$, while U is coercive, it follows that x is bounded on $[\,a, T_m)$. According to Corollary 2.4.3, it follows that x is not saturated. Hence every saturated solution is global. If $\lim_{\|x\|\to+\infty} U(x) = -\infty$ then $-U$ is a coercive prime integral for the system. Hence, the result still holds true even in this case.

Problem 6.9 It follows that

$$\sum_{i=1}^{n}\left[-\frac{\partial H}{\partial q_i}(p,q)\frac{\partial H}{\partial p_i}(p,q)+\frac{\partial H}{\partial p_i}(p,q)\frac{\partial H}{\partial q_i}(p,q)\right]=0.$$

According to Theorem 6.1.1, H is a prime integral for the system.

Problem 6.10 We observe that, a function of class C^1, $U:\Omega_0\subset\Omega\to\mathbb{R}$, satisfies the condition (6.1.2) in Theorem 6.1.1 with respect to the function f if and only if it satisfies the same condition with respect to the function λf.

Problem 6.11 Let us assume for contradiction that there exists a prime integral for the system, $U:\mathbb{R}^2\to\mathbb{R}$. Since its general solution is $x_1(t)=\xi e^{2t}, x_2(t)=\eta e^t$ for $t\in\mathbb{R}$, we have that $U(\xi e^{2t},\eta e^t)=U(\xi,\eta)$ for every $(\xi,\eta)\in\mathbb{R}^2$ and every $t\in\mathbb{R}$. Letting t approach $-\infty$, we conclude that $U(\xi,\eta)=U(0,0)$ for every $(\xi,\eta)\in\mathbb{R}^2$, i.e. U is a constant function. This contradiction can be eliminated only if there is no prime integral of the system considered, defined on the whole $\mathbb{R}^2$. On the other hand, the function $U:\{(x_1,x_2)\in\mathbb{R}^2;\ x_1>0\}\to\mathbb{R}$, defined by $U(x_1,x_2)=x_2^2/x_1$, is a prime integral for the system.

Problem 6.12 The first part of the problem follows from the fact that the function $U:\mathbb{R}^n\to\mathbb{R}$, defined by $U(x)=\|x\|^2$, is a prime integral for the system. Hence, the trajectory of any solution lies on a sphere centered at the origin and with radius depending on the solution. In addition, if $\mathbb{I}=[0,+\infty)$, as every saturated solution of the system is bounded, according to Theorem 5.2.2, it follows that the system is stable.

Exercise 6.2 (1) The characteristic system in the symmetric form is

$$\frac{dx_1}{x_2^2-x_3^2}=\frac{dx_2}{x_3^2-x_1^2}=\frac{dx_3}{x_1^2-x_2^2}.$$

We have $dx_1+dx_2+dx_3=0$ and $x_1^2dx_1+x_2^2dx_2+x_3^2dx_3=0$. Hence, the functions $U_1,U_2:\mathbb{R}^3\to\mathbb{R}$, defined by $U_1(x_1,x_2,x_3)=x_1+x_2+x_3$ and respectively by $U_2(x_1,x_2,x_3)=x_1^3+x_2^3+x_3^3$, are prime integrals for this system. The stationary points of the system are of the form (x_1,x_2,x_3) with $x_1=\pm x_2=\pm x_3$. One may easily see that the prime integrals above are independent about any non-stationary point. Therefore, the general solution of the equation is defined by $z(x_1,x_2,x_3)=F(x_1+x_2+x_3,x_1^3+x_2^3+x_3^3)$, where $F:\mathbb{R}^2\to\mathbb{R}$ is a function of class C^1.

(2) The characteristic system is

$$\frac{dx_1}{-x_1e^{x_2}}=\frac{dx_2}{1}=\frac{dx_3}{x_3e^{x_2}},$$

and so, $x_3dx_1+x_1dx_3=0$ and $x_1^{-1}dx_1+e^{x_2}dx_2=0$. We have the prime integrals $U_1(x_1,x_2,x_3)=x_1x_3$ and $U_2(x_1,x_2,x_3)=x_1e^{e^{x_2}}$, defined on $\mathbb{R}^3$. The system has no stationary points, and the two prime integrals are independent at every point. The general solution of the equation is $z(x_1,x_2,x_3)=F(x_1x_3,x_1e^{e^{x_2}})$, where $F:\mathbb{R}^2\to\mathbb{R}$ varies in the set of functions of class C^1.

(3) The characteristic system is

$$\frac{dx_1}{x_1(x_2 - x_3)} = \frac{dx_2}{x_2(x_3 - x_1)} = \frac{dx_3}{x_3(x_1 - x_2)}.$$

We have

$$dx_1 + dx_2 + dx_3 = 0 \text{ and } \frac{dx_1}{x_1} + \frac{dx_2}{x_2} + \frac{dx_3}{x_3} = 0.$$

Consequently, $U_1(x_1, x_2, x_3) = x_1 + x_2 + x_3$ and $U_2(x_1, x_2, x_3) = x_1x_2x_3$, defined on $\mathbb{R}^3$, are prime integrals for the system. A point (x_1, x_2, x_3) is stationary for the system if and only if $x_1 = x_2 = x_3$ or $x_i = x_j = 0$ for $i, j = 1, 2, 3$, $i \neq j$. The two prime integrals are independent at any non-stationary point and therefore the general solution is $z(x_1, x_2, x_3) = F(x_1 + x_2 + x_3, x_1x_2x_3)$, where F ranges over the set of all real functions, of class C^1, defined on $\mathbb{R}^2$.

(4) The equation is quasi-linear. So, we are looking for the solution as a function x_3 implicitly defined by a relation of the form $\phi(x_1, x_2, x_3(x_1, x_2)) = c$, where the function ϕ is the solution of the first-order linear partial differential equation

$$(x_1 - x_3)\frac{\partial \phi}{\partial x_1} + (x_2 - x_3)\frac{\partial \phi}{\partial x_2} + 2x_3\frac{\partial \phi}{\partial x_3} = 0.$$

The characteristic system is

$$\frac{dx_1}{x_1 - x_3} = \frac{dx_2}{x_2 - x_3} = \frac{dx_3}{2x_3}.$$

We have

$$\frac{dx_1 + dx_2}{x_1 + x_2} = \frac{dx_3}{2x_3} \text{ and } \frac{dx_2}{dx_3} = \frac{x_2}{2x_3} - \frac{1}{2}.$$

Thus, the functions $U_1(x_1, x_2, x_3) = \dfrac{x_1 + x_2}{2x_3}$ and $U_2(x_1, x_2, x_3) = \dfrac{(x_2 + x_3)^2}{x_3}$, defined on the set $\{(x_1, x_2, x_3) \in \mathbb{R}^3;\ x_3 \neq 0\}$, are prime integrals for the characteristic system. The only stationary point of the system is the origin. One observes that the prime integrals above are independent at any non-stationary point for which $x_3 \neq 0$ and $x_2 \neq -x_3$. The general solution is implicitly defined by a relation of the form $F\left(\dfrac{x_1 + x_2}{2x_3}, \dfrac{(x_2 + x_3)^2}{x_3}\right) = c$, where F ranges over the set of all real functions of class C^1, defined on $\mathbb{R}^2$, and $c \in \mathbb{R}$.

(5) The equation is quasi-linear. So, we are looking for the solution as a function x_3, implicitly defined by a relation of the form $\phi(x_1, x_2, x_3(x_1, x_2)) = c$, where the function ϕ satisfies the first-order linear partial differential equation

$$x_3\frac{\partial \phi}{\partial x_1} - x_3\frac{\partial \phi}{\partial x_2} + (x_2 - x_1)\frac{\partial \phi}{\partial x_3} = 0.$$

The characteristic system is

$$\frac{dx_1}{x_3} = \frac{dx_2}{-x_3} = \frac{dx_3}{x_2 - x_1}.$$

The solutions of the system satisfy $x_1 - x_2 = c_1$ and $x_1^2 - x_2^2 + x_3^2 = c_2$, and therefore the functions $U_1, U_2 : \mathbb{R}^3 \to \mathbb{R}$, defined by $U_1(x_1, x_2, x_3) = x_1 - x_2$ and $U_2(x_1, x_2, x_3) = x_1^2 - x_2^2 + x_3^2$, are prime integrals for the characteristic system. The stationary points of the system are of the form $(a, a, 0)$, while the two prime integrals are independent at every non-stationary point except for the origin. Hence the general solution of the initial equation is implicitly defined by a relation of the form $F(x_1 - x_2, x_1^2 - x_2^2 + x_3^2) = c$, where $F : \mathbb{R}^2 \to \mathbb{R}$ ranges over the set of all real functions of class C^1 and $c \in \mathbb{R}$.

(6) We look for the solution as a function, implicitly defined by $\phi(x_1, x_2, x_3(x_1, x_2)) = c$, where ϕ is the solution of the first-order linear partial differential equation

$$x_1 \frac{\partial \phi}{\partial x_1} + x_2 \frac{\partial \phi}{\partial x_2} + \left(x_3 + \frac{x_1 x_2}{x_3} \right) \frac{\partial \phi}{\partial x_3} = 0.$$

The characteristic system attached is

$$\frac{dx_1}{x_1} = \frac{dx_2}{x_2} = \frac{x_3 dx_3}{x_3^2 + x_1 x_2}.$$

Along the solutions of the system we have $\dfrac{x_1}{x_2} = c_1$. Therefore, a prime integral of the characteristic system is $U_1 : \{(x_1, x_2, x_3) \in \mathbb{R}^3;\ x_2 \neq 0\} \to \mathbb{R}$, defined by $U_1(x_1, x_2, x_3) = \dfrac{x_1}{x_2}$. We also have

$$\frac{d(x_1 x_2)}{x_1 x_2} = \frac{d(x_3^2)}{x_3^2 + x_1 x_2}.$$

Denoting by

$$\begin{cases} x_1 x_2 = u \\ x_3^2 = v, \end{cases}$$

the equation above may be rewritten in the form $\frac{dv}{du} = \frac{v}{u} + 1$. From this equation, we deduce $\dfrac{v}{u} - \ln |u| = c_2$. So, $U_2 : \{(x_1, x_2, x_3) \in \mathbb{R}^3;\ x_1 x_2 \neq 0\} \to \mathbb{R}$, defined by $U_2(x_1, x_2, x_3) = \dfrac{x_3^2}{x_1 x_2} - \ln |x_1 x_2|$ is a second prime integral. The system has no stationary points, and the two prime integrals are independent at every point (x_1, x_2, x_3) for which $x_1 x_2 x_3 \neq 0$. The general solution of the initial equation is implicitly defined by a relation of the form $F\left(\dfrac{x_1}{x_2}, \dfrac{x_3^2}{x_1 x_2} - \ln |x_1 x_2| \right) = c$, where $F : \mathbb{R}^2 \to \mathbb{R}$ ranges over the set of all functions of class C^1, while $c \in \mathbb{R}$.

(7) The general solution is implicitly defined by $F(x_1^2 - x_2^2, 2x_2^2 - x_3^2) = c$, where $F : \mathbb{R}^2 \to \mathbb{R}$ ranges over the set of all functions of class C^1, and $c \in \mathbb{R}$.

(8) We look for the solution z, defined by $\phi(x_1, x_2, x_3, z(x_1, x_2, x_3)) = c$, where

ϕ is the solution of the first-order linear partial differential equation

$$(1+\sqrt{z-a_1x_1-a_2x_2-a_3x_3})\frac{\partial\phi}{\partial x_1}+\frac{\partial\phi}{\partial x_2}+\frac{\partial\phi}{\partial x_3}+(a_1+a_2+a_3)\frac{\partial\phi}{\partial z}=0.$$

The attached characteristic system is

$$\frac{dx_1}{1+\sqrt{z-a_1x_1-a_2x_2-a_3x_3}}=\frac{dx_2}{1}=\frac{dx_3}{1}=\frac{dz}{a_1+a_2+a_3}.$$

From the equality of the last three ratios, we deduce that $U_1, U_2 : \mathbb{R}^4 \to \mathbb{R}$, defined by $U_1(x_1,x_2,x_4,z)=x_2-x_3$ and $U_2(x_1,x_2,x_3,z)=x_3-\dfrac{z}{a_1+a_2+a_3}$, are prime integrals for the system. From the derived proportion

$$\frac{dx_2}{1}=\frac{a_1dx_1+a_2dx_2+a_3dx_3-dz}{a_1\sqrt{z-a_1x_1-a_2x_2-a_3x_3}},$$

we deduce that $U_3 : \{(x_1,x_2,x_3,z)\in\mathbb{R}^4;\ z>a_1x_1+u_2x_2+a_3x_3\}\to\mathbb{R}$, defined by $U_3(x_1,x_2,x_3,z)=a_1x_2+2\sqrt{z-a_1x_1-a_2x_2-a_3x_3}$, is also a prime integral. The system has no stationary points. The three prime integrals are independent at all points in the common part of the domains of definition. The general solution of the initial equation is implicitly defined by

$$F\left(x_2-x_3, x_3-\frac{z}{a_1+a_2+a_3}, a_1x_2+\sqrt{z-a_1x_1-a_2x_2-a_3x_3}\right)=c,$$

where $F:\mathbb{R}^3\to\mathbb{R}$ ranges over the set of all functions of class C^1, and $c\in\mathbb{R}$. The equation also admits the "special" solution $z=a_1x_1+a_2x_2+a_3x_3$, eliminated during the determination of the prime integral U_3.

Exercise 6.3 (1) The prime integral of the characteristic system is $x^2+y^2=c$. So, the general solution is $z=\varphi(x^2+y^2)$, and the solution of the Cauchy problem is $z=\cos\sqrt{x^2+y^2}$.

(2) The characteristic system is

$$\frac{dx}{x}=-\frac{dy}{z}=\frac{dz}{0}.$$

Two independent prime integrals are

$$\begin{cases} z=c_1 \\ \ln|x|+\dfrac{y}{z}=c_2. \end{cases}$$

Substituting $x=s$, $y=s^2$ and $z=s^3$ in the system above and eliminating s, we get $\frac{1}{3}\ln|c_1|+c_1^{-1/3}=c_2$. Finally, substituting c_1 and c_2 from the system, we get the implicit equation of the solution $\frac{1}{3}\ln|z|+z^{-1/3}=\ln|x|+\frac{y}{z}$.

(3) The characteristic system is

$$\frac{dx}{x} = -\frac{dy}{y} = -\frac{dz}{2z}$$

and so, two independent prime integrals are

$$\begin{cases} xy = c_1 \\ xz^2 = c^2. \end{cases}$$

The general solution of the equation is $u = \varphi(xy, xz^2)$. From the initial condition, we deduce $\varphi(y, z^2) = \sin(y + z)$, and consequently, the solution of the Cauchy problem is $u(x, y, z) = \sin(xy + z\sqrt{x})$.

(4) The characteristic system is

$$\frac{dx}{x} = \frac{dy}{y + x^2} = \frac{dz}{z}.$$

Two independent prime integrals are

$$\begin{cases} \dfrac{x}{z} = c_1 \\ \\ \dfrac{y}{x} - x = c_2. \end{cases}$$

Hence, the general solution of the equation is given implicitly by $F\left(\frac{z}{x}, \frac{y-x^2}{x}\right) = c$, with F of class C^1. By the implicit function theorem, we get $z = xf\left(\frac{y-x^2}{x}\right)$. From the initial condition we deduce that $2f\left(\frac{y-4}{2}\right) = (y-4)^3$. Consequently, $f(s) = 4s^3$ and $z(x, y) = 4\frac{(y-x^2)^3}{x^2}$.

Problem 6.13 We consider for the beginning the attached homogeneous problem

$$\begin{cases} \dfrac{\partial z}{\partial t} + a\dfrac{\partial z}{\partial x} = 0 \\ \\ z(0, x) = \varphi(x) \end{cases}$$

whose characteristic system is

$$\frac{dt}{1} = \frac{dx}{a}.$$

A prime integral for this system is $U(t, x) = x - at$ for every $(t, x) \in \mathbb{R} \times \mathbb{R}$. The general solution of the homogeneous equation is then $z(t, x) = F(x - at)$, where $F : \mathbb{R} \to \mathbb{R}$ is of class C^1. From the Cauchy condition, it follows that the unique solution of the homogeneous Cauchy problem is $z(t, x) = \varphi(x - at)$ for $(t, x) \in \mathbb{R} \times \mathbb{R}$. In order to determine the solution of the non-homogeneous equation, we shall use a variation of constants-like method. More precisely, we will look for the solution in the form $z(t, x) = \psi(t, x - at)$, where $\psi : \mathbb{R} \times \mathbb{R} \to \mathbb{R}$

is a function of class C^1 which will be determined by imposing the condition that z be the solution of the non-homogeneous problem. We have

$$\begin{cases} \dfrac{\partial z}{\partial t}(t,x) = \dfrac{\partial \psi}{\partial t}(t, x-at) - a\dfrac{\partial \psi}{\partial y}(t, x-at) \\ \\ \dfrac{\partial z}{\partial x}(t,x) = \dfrac{\partial \psi}{\partial y}(t, x-at), \end{cases}$$

and consequently

$$\frac{\partial \psi}{\partial t}(t,y) = f(t, y+at).$$

From this equation, we deduce

$$\psi(t,y) = \psi(0,y) + \int_0^t f(s, y+as)\, ds$$

for every $t \in \mathbb{R}$. Finally, the last equality and the initial condition yield

$$z(t,x) = \varphi(x-at) + \int_0^t f(s, x-a(t-s))\, ds$$

for every $t \in \mathbb{R}$.

Problem 6.14 First, let us consider the attached homogeneous problem

$$\begin{cases} \dfrac{\partial z}{\partial t} + a(t)\dfrac{\partial z}{\partial x} = 0 \\ \\ z(0,x) = \varphi(x), \end{cases}$$

whose characteristic system is

$$\frac{dt}{1} = \frac{dx}{a(t)}.$$

A prime integral for this system is

$$U(t,x) = x - \int_0^t a(s)\, ds$$

for $(t,x) \in \mathbb{R} \times \mathbb{R}$. The general solution of the homogeneous equation is given by $z(t,x) = F\left(x - \int_0^t a(s)\, ds\right)$, where $F : \mathbb{R} \to \mathbb{R}$ is of class C^1. From the Cauchy condition it follows that the unique solution of the homogeneous Cauchy problem is

$$z(t,x) = \varphi\left(x - \int_0^t a(s)\, ds\right)$$

for $(t,x) \in \mathbb{R} \times \mathbb{R}$. We look for the solution to the non-homogeneous equation in the form

$$z(t,x) = \psi\left(t, x - \int_0^t a(s)\,ds\right),$$

where $\psi : \mathbb{R} \times \mathbb{R} \to \mathbb{R}$ is a function of class C^1 which satisfies

$$\begin{cases} \dfrac{\partial z}{\partial t}(t,x) = \dfrac{\partial \psi}{\partial t}\left(t, x - \displaystyle\int_0^t a(s)\,ds\right) - a(t)\dfrac{\partial \psi}{\partial y}\left(t, x - \displaystyle\int_0^t a(s)\,ds\right) \\[2ex] \dfrac{\partial z}{\partial x}(t,x) = \dfrac{\partial \psi}{\partial y}\left(t, x - \displaystyle\int_0^t a(s)\,ds\right), \end{cases}$$

from where

$$\frac{\partial \psi}{\partial t}(t,y) = f\left(t, y + \int_0^t a(s)\,ds\right).$$

From this equation, we deduce

$$\psi(t,y) = \psi(0,x) + \int_0^t f\left(s, y + \int_0^s a(\tau)\,d\tau\right) ds$$

for every $t \in \mathbb{R}$. Finally, from the last equality and the initial condition, it follows

$$z(t,x) = \varphi\left(x - \int_0^t a(\tau)\,d\tau\right) + \int_0^t f\left(s, x - \int_s^t a(\tau)\,d\tau\right) ds$$

for every $t \in \mathbb{R}$.

Problem 6.15 The characteristic system attached to the homogeneous equation is

$$\frac{dt}{1} = \frac{dx_1}{a_1} = \frac{dx_2}{a_2} = \cdots = \frac{dx_n}{a_n}.$$

Its general solution is $z(t,x) = U(x - ta)$, where $U : \mathbb{R}^n \to \mathbb{R}$ ranges over the set of all functions of class C^1 and $x - ta = (x_1 - a_1 t, x_2 - a_2 t, \ldots, x_n - a_n t)$. The solution of the homogeneous equation which satisfies the corresponding initial condition is $z(t,x) = \varphi(x - ta)$ for every $(t,x) \in \mathbb{R} \times \mathbb{R}^n$. We seek for the solution of the non-homogeneous equation in the form $z(t,x) = \psi(t, x - ta)$ for every $(t,x) \in \mathbb{R} \times \mathbb{R}^n$, and we deduce

$$z(t,x) = \varphi(x - ta) + \int_0^t f(s, x - (t-s)a)\,ds$$

for every $(t,x) \in \mathbb{R} \times \mathbb{R}^n$.

Problem 6.16 By analogy with the solution of Problem 6.14, we deduce that

$$z(t,x) = \varphi\left(x - \int_0^t a(\tau)\,d\tau\right) + \int_0^t f\left(s, x - \int_s^t a(\tau)\,d\tau\right) ds$$

for every $(t,x) \in \mathbb{R} \times \mathbb{R}^n$.

Problem 6.17 First, we consider the attached homogeneous problem

$$\begin{cases} \dfrac{\partial z}{\partial t} + ax\dfrac{\partial z}{\partial x} = 0 \\ z(0,x) = \varphi(x) \end{cases}$$

whose characteristic system is

$$\frac{dt}{1} = \frac{dx}{ax}.$$

A prime integral for this system is $U(t,x) = xe^{-at}$ for every $(t,x) \in \mathbb{R} \times \mathbb{R}$. The general solution of the homogeneous equation is $z(t,x) = F(xe^{-at})$, where $F : \mathbb{R} \to \mathbb{R}$ is of class C^1. From the Cauchy condition, it follows that the unique solution of the homogeneous Cauchy problem is $z(t,x) = \varphi(xe^{-at})$ for $(t,x) \in \mathbb{R} \times \mathbb{R}$. In order to find the solution of the non-homogeneous equation, we shall use the variation of constants-like method. More precisely, we will look for the solution in the form $z(t,x) = \psi(t, xe^{-at})$, where $\psi : \mathbb{R} \times \mathbb{R} \to \mathbb{R}$ is a function of class C^1, which will be determined by letting z to be a solution of the non-homogeneous problem. We have

$$\begin{cases} \dfrac{\partial z}{\partial t}(t,x) = \dfrac{\partial \psi}{\partial t}(t, xe^{-at}) - axe^{-at}\dfrac{\partial \psi}{\partial y}(t, xe^{-at}) \\ \dfrac{\partial z}{\partial x}(t,x) = e^{-at}\dfrac{\partial \psi}{\partial y}(t, xe^{-at}), \end{cases}$$

from where

$$\frac{\partial \psi}{\partial t}(t,y) = f(t, ye^{at}).$$

From this equation, we deduce

$$\psi(t,y) = \psi(0,y) + \int_0^t f(s, ye^{as})\, ds$$

for every $t \in \mathbb{R}$. Finally, from the last equality and the initial condition, it follows

$$z(t,x) = \varphi(e^{-at}x) + \int_0^t f(s, xe^{-a(t-s)})\, ds$$

for every $t \in \mathbb{R}$.

Problem 6.18 The solution is given by

$$z(t,x) = \varphi\left(e^{-\int_0^t a(\tau)\, d\tau} x\right) + \int_0^t f\left(s, e^{-\int_s^t a(\tau)\, d\tau} x\right) ds$$

for every $(t,x) \in \mathbb{R} \times \mathbb{R}$.

Problem 6.19 We have

$$z(t,x) = \varphi\left(e^{-t\mathcal{A}}x\right) + \int_0^t f\left(s, e^{-(t-s)\mathcal{A}}x\right) ds$$

for every $(t,x) \in \mathbb{R} \times \mathbb{R}^n$, where $e^{-t\mathcal{A}}$ is the exponential of the matrix $-t\mathcal{A}$.
Problem 6.20 The solution is

$$z(t,x) = \varphi\left(\mathcal{X}^{-1}(t)x\right) + \int_0^t f\left(s, \mathcal{X}^{-1}(t)\mathcal{X}(s)x\right) ds$$

for every $(t,x) \in \mathbb{R} \times \mathbb{R}^n$, where $\mathcal{X}(t)$ is a fundamental matrix of the first-order linear differential system $x' = \mathcal{A}(t)x$.
Problem 6.21 The normal to the family of cones is $(x_2, x_1, -2\alpha x_3) = \left(x_2, x_1, -2\frac{x_1 x_2}{x_3}\right)$. The implicit equation of the surface we are looking for is $\varphi(x_1, x_2, x_3) = 0$. From the orthogonality condition, we get the first-order linear partial differential equation for φ, i.e.

$$x_2 \frac{\partial \varphi}{\partial x_1} + x_1 \frac{\partial \varphi}{\partial x_2} - 2\frac{x_1 x_2}{x_3}\frac{\partial \varphi}{\partial x_2} = 0.$$

Solving the attached characteristic system

$$\frac{dx_1}{x_2} = \frac{dx_2}{x_1} = -\frac{x_3 dx_3}{2x_1 x_2},$$

we get the prime integrals $x_1^2 - x_2^2 = c_1$ and $x_1^2 + x_2^2 + x_3^2 = c_2$ which are independent at each non-stationary point of $f(x_1, x_2, x_3) = \left(x_2, x_1, -\frac{2x_1 x_2}{x_3}\right)$. Using the equations of the circle, we get the compatibility condition: $c_2 = 5$. Thus, the equation of the surface is $x_1^2 + x_2^2 + x_3^2 = 5$, which is nothing but the equation a sphere centered at the origin and having radius $\sqrt{5}$.

Chapter 7

Exercise 7.1 (1) One observes that, for any choice of the function φ, the sequence satisfies all the conditions in Definition 7.1.2, and therefore it converges to 0 in $\mathcal{D}(\mathbb{R})$. (2) If φ is non-identically zero, the sequence, although uniformly convergent to 0, is not convergent in $\mathcal{D}(\mathbb{R})$ because it does not satisfy the condition (ii) in Definition 7.1.2. Indeed, in this case, there exists at least one $t \in \mathbb{R}$ such that $\varphi'(t) \neq 0$. Let $t_k = t/k$ for $k \in \mathbb{N}^*$, and let us observe that $\varphi_k'(t_k) = \varphi'(t) \neq 0$ for every $k \in \mathbb{N}^*$, which shows that the sequence of the first-order derivatives does not converge uniformly to 0 on $\mathbb{R}$. (3) If φ is non-identically zero, the sequence, although uniformly convergent to 0, is not convergent in $\mathcal{D}(\mathbb{R})$ because it does not satisfy the condition (i) in Definition 7.1.2. Indeed, in this case, there exists at least one $t \in \mathbb{R}^*$ such that $\varphi(t) \neq 0$. Let us observe that the term of rank k

of the unbounded sequence $(kt)_{k\in\mathbb{N}^*}$ lies in the support of the term of the same rank of the sequence of functions considered.

Exercise 7.2 (1) $\dot{x}(t) = 2\delta(t)$. (2) $\dot{x}(t) = \cos t\, 2\delta(t) - \sin t \operatorname{sgn}(t) = 2\delta(t) - \sin t \operatorname{sgn}(t)$. (3) $\dot{x}(t) = t\delta(t) + \operatorname{sgn}(t) = \operatorname{sgn}(t)$. (4) $\dot{x}(t) = t\delta(t-1) + \operatorname{sgn}(t-1) = \delta(t-1) + \operatorname{sgn}(t-1)$. (5) $\dot{x}(t) = \sin t\, \delta(t) + \cos t\, \theta(t) = \cos t\, \theta t$. (6) $\dot{x}(t) = e^t\delta(t) + e^t\theta t = \delta(t) + e^t\theta t$.

Problem 7.1 For every test function $\varphi \in \mathcal{D}(\mathbb{R})$, we have

$$(\eta(t)\dot{\delta}(t), \varphi(t)) = (\dot{\delta}(t), \eta(t)\varphi(t)) = -(\delta(t), \eta'(t)\varphi(t) + \eta(t)\varphi'(t))$$

$$= -\eta'(0)(\delta(t), \varphi(t)) - (\eta(t)\delta(t), \varphi'(t)$$

$$= (-\eta'(0)\delta(t), \varphi(t)) + (-\eta(0)\delta(t), \varphi'(t)) = (-\eta'(0)\delta(t) + \eta(0)\delta'(t), \varphi(t)),$$

which proves the equality.

(1) Let $\varphi \in \mathcal{D}(\mathbb{R})$. We have

$$(t\delta^{(m)}(t), \varphi(t)) = (\delta^{(m)}(t), t\varphi(t)) = (-1)^m \left(\delta(t), \sum_{k=0}^{m} C_m^k t^{(k)} \varphi^{(m-k)}(t)\right),$$

where $C_m^k = \frac{k!(m-k)!}{m!}$. Since $t^{(k)} = 0$ for $k = 2, 3, \ldots, m$ and $(\delta(t), t\varphi^{(m)}(t)) = 0$, we deduce that

$$(t\delta^{(m)}(t), \varphi(t)) = (-1)^m (\delta(t), C_m^1 \varphi^{(m-1)}(t))$$

$$= (-1)^m (-1)^{(m-1)} (m-1)(\delta^{(m-1)}(t), \varphi(t)) = -(m-1)(\delta^{(m-1)}(t), \varphi(t)),$$

which completes the proof.

(2) Let $\varphi \in \mathcal{D}(\mathbb{R})$. We have

$$(t^m\delta^{(m)}(t), \varphi(t)) = (-1)^m \left(\delta(t), \sum_{k=0}^{m} C_m^k (t^m)^{(k)} \varphi^{(m-k)}(t)\right).$$

Since $(\delta(t), (t^m)^{(k)}\varphi^{(m-k)}(t)) = 0$ for $k = 1, 2, \ldots, m-1$, it follows

$$(t^m\delta^{(m)}(t), \varphi(t)) = (-1)^m m!(\delta(t), \varphi^{(m)}(t)),$$

which completes the proof.

(3) As before, we have

$$(t^k\delta^{(m)}(t), \varphi(t)) = (-1)^m \left(\delta(t), \sum_{p=0}^{m} C_m^p (t^k)^{(p)} \varphi^{(m-p)}(t)\right).$$

Since $k \leq m-1$, it follows that $(\delta(t), (t^k)^{(p)}\varphi^{(m-p)}(t)) = 0$ for $k = 1, 2, \ldots, m$. The proof is complete.

Problem 7.2 We denote by $c = \int_{\mathbb{R}} x(t)\, dt$ and we observe that $\int_{\mathbb{R}} x_n(t)\, dt = c$ for every $n \in \mathbb{N}$. Let $\phi \in \mathcal{D}(\mathbb{R})$. We have

$$\int_{\mathbb{R}} x_n(t)\phi(t)\, dt = \int_{-1/n}^{1/n} nx(nt)\phi(t)\, dt = \int_{-1}^{1} x(s)\phi(s/n)\, dt \xrightarrow{n} c\phi(t).$$

Hence

$$\lim_{n\to\infty} x_n = c\delta(t)$$

on $\mathcal{D}(\mathbb{R})$.

Exercise 7.3 (1) $x(t) = (e^{-1} - e^{-2t})\theta(t)$. (2) $x(t) = te^{-t}\theta(t)$. (3) $x(t) = \frac{1}{2}\sin 2t\,\theta(t)$. (4) $x(t) = \frac{1}{4}(1 - e^{-2t} - 2t2^{-2t})\theta(t)$. (5) $x(t) = \frac{1}{2}(e^t + e^{-t} - 2)\theta(t)$. (6) $x(t) = \frac{1}{2}(1 - 2e^t + e^{2t})\theta(t)$.

Exercise 7.4 (1) $x(t) = (e^t\,\theta(t)) * (e^{-1} - e^{-2t})\theta(t)$. (2) $x(t) = (t\,\theta(t)) * (te^{-t}\theta(t))$. (3) $x(t) = (\cos 2t\,\theta(t)) * (\frac{1}{2}\sin 2t\,\theta(t))$. (4) $x(t) = (\sin t\,\theta(t)) * (\frac{1}{4}(1 - e^{-2t} - 2te^{-2t})\theta(t))$. (5) $x(t) = (t\,\theta(t)) * (\frac{1}{2}(e^t + e^{-t} - 2)\theta(t))$. (6) $x(t) = (te^t\,\theta(t)) * (\frac{1}{2}(1 - 2e^t + e^{2t})\theta(t))$ where, according to Example 7.2.1,

$$(f * g)(t) = \int_0^t f(\tau)g(t - \tau)\, d\tau.$$

Exercise 7.5 The solutions are:

(1)
$$x(t) = \begin{cases} e^{-t} & \text{for } t \in [0, 1] \\ e^{-t} - e^{-(t-1)} + 1 & \text{for } t > 1. \end{cases}$$

(2)
$$x(t) = \begin{cases} 0 & \text{for } t \in [1, 3] \\ \frac{7}{4}e^{2(t-3)} - \frac{t}{2} - \frac{1}{4} & \text{for } t > 3. \end{cases}$$

(3)
$$x(t) = \begin{cases} -t & \text{for } t \in [1, 2] \\ -t + t\ln\frac{t}{2} & \text{for } t > 2. \end{cases}$$

(4)
$$x(t) = \begin{cases} 0 & \text{for } t \in [0, \frac{\pi}{2}] \\ \tan(-\cos t) & \text{for } t > \frac{\pi}{2}. \end{cases}$$

(5)
$$x(t) = \begin{cases} 1 & \text{for } t \in [0, \frac{\pi}{4}] \\ \arctan t & \text{for } t > \frac{\pi}{4}. \end{cases}$$

(6)
$$x(t) = \begin{cases} 1 & \text{for } t \in [0, 1] \\ \sqrt{t} & \text{for } t > 1. \end{cases}$$

Problem 7.3 Using Definition 7.5.1, one observes that the multi-valued functions in (1), (4) and (5) are upper semi-continuous on $\mathbb{R}$, while the others, only on $\mathbb{R}^*$.

Problem 7.4 We denote by graph (F) the graph of the multi-function F. Let $((x_k, y_k))_{k\in\mathbb{N}}$ be a sequence in graph (F), convergent to (x, y) and let us assume for contradiction that (x, y) does not belong to the graph of F. Since $x \in K$ (because K is closed), from the assumption made, we have that y does not belong to $F(x)$. But $F(x)$ is closed and $\{y\}$ is compact. According to Lemma 2.5.1, we know that $\text{dist}\,(y, F(x)) = \delta > 0$. Then, the set $D = \{z \in \mathbb{R}^n;\ \text{dist}\,(z, K) < \delta/2\}$ is open, includes $F(x)$, while $\overline{D}$ does not contain y. Since F is upper semi-continuous, for k large enough, we have $y_k \in F(x_k) \subset D$, which shows that $y \in \overline{D}$. This contradiction can be eliminated only if graph (F) is a closed set.

Problem 7.5 Let us assume for contradiction that there exists $x \in K$ with the property that F is not upper semi-continuous at x. Then, there exists $D \subset \mathbb{R}^n$ with $F(x) \subset D$, such that, for every neighborhood V of x, there exists $x_V \in V$ such that $F(x_V)$ is not included in D. Let $k \in \mathbb{N}^*$. Take $V_k = B(x, \frac{1}{k})$ and let us denote by $x_{V_k} = x_k$. From the assumption made, it follows that there exists $y_k \in F(x_k)$ with $y_k \in \mathbb{R}^n \setminus D$. Since F takes values in a compact set, $(y_k)_{k\in\mathbb{N}^*}$ is convergent (at least on a subsequence) to an element y. On the other hand, $(x_k)_{k\in\mathbb{N}^*}$ is convergent to x, and therefore (x, y) belongs to the graph of F. This means that $x \in K$ and $y \in F(x)$. At the same time, because for every $k \in \mathbb{N}^*$ y_k belongs to the closed set $\mathbb{R}^n \setminus D$, it follows that $y \in \mathbb{R}^n \setminus D$, relation in contradiction with $y \in F(x) \subset D$. The contradiction can be eliminated only if F is upper semi-continuous on K.

Problem 7.6 Let $x \in K$ and let D be an open set with $\overline{\text{conv}\, F(x)} \subset D$. Since $F(x)$ is compact, $\overline{\text{conv}\, F(x)}$ is compact too, and $\overline{\text{conv}\, F(x)} \cap \partial D = \emptyset$. According to Lemma 2.5.1, we have that $\text{dist}\,(\overline{\text{conv}\, F(x)}, \partial D) = \delta > 0$. Then, the set D_δ, defined by $D_\delta = \{y \in \mathbb{R}^n;\ \text{dist}\,(y, \overline{\text{conv}\, F(x)}) < \delta/2\}$, is convex, open and satisfies $F(x) \subset D_\delta \subset \overline{D_\delta} \subset D$. Since F is upper semi-continuous at x, there exists a neighborhood $V \subset K$ of x such that $F(z) \subset D_\delta$ for every $z \in V$. Since D_δ is convex, we have $\overline{\text{conv}\, F(z)} \subset \overline{D_\delta} \subset D$ for every $z \in V$. Hence, $\overline{\text{conv}\, F}$ is upper semi-continuous at x. Since x is arbitrary in K, this completes the proof.

Problem 7.7 Since f is bounded, from Cesàro's lemma, it follows that F is nonempty and closed valued. Moreover, F takes the values in a compact set. According to Problem 7.5, in order to complete the proof, it suffices to show that graph (F) is closed. So, let $((x_p, y_p))_{p\in\mathbb{N}^*}$ be a sequence in graph (F) convergent to (x, y). From the definition of F, we have that, for every $p \in \mathbb{N}^*$, there exists a sequence $(x_{p,k})_{k\in\mathbb{N}^*}$ with $\lim_{k\to\infty} x_{p,k} = x_p$ and $\lim_{k\to\infty} f(x_{p,k}) = y_p$. Then, for every $p \in \mathbb{N}^*$, there exists $k_p \geq p$ such that we have both $\|x_{p,k_p} - x_p\| \leq 1/p$ and $\|f(x_{p,k_p}) - y_p\| \leq 1/p$. Then $\lim_{p\to\infty} x_{p,k_p} = x$ and $\lim_{p\to\infty} f(x_{p,k_p}) = y$, which shows that $y \in F(x)$. The proof is complete.

Problem 7.8 Let $\Omega = \mathbb{R}$, $\Sigma = \{0\}$ and $f : \omega \to \mathbb{R}$ be defined by $f(x) = 3\sqrt[3]{x^2}$ for every $x \in \mathbb{R}$. Then Σ is viable for $x' = f_{|\Sigma}(x)$ but Σ is not invariant for $x' = f(x)$, because the latter equation has at least two solutions which satisfy $x(0) = 0$. See Example 2.3.1.

Problem 7.9 Let $\xi \in \Sigma$. According to Definition 7.7.3, we have $\eta \in \mathcal{T}_\Sigma(\xi)$ if $\liminf_{t\downarrow 0} \frac{1}{t}\, \text{dist}\,(\xi + t\eta, \Sigma) = 0$. Let $s > 0$, and let us observe that

$$\liminf_{t\downarrow 0} \frac{1}{t}\, \text{dist}\,(\xi + ts\eta, \Sigma) = s \liminf_{t\downarrow 0} \frac{1}{ts}\, \text{dist}\,(\xi + ts\eta, \Sigma)$$

$$= s \liminf_{\tau\downarrow 0} \frac{1}{\tau}\, \text{dist}\,(\xi + \tau\eta, \Sigma) = 0.$$

Hence $s\eta \in \mathcal{T}_\Sigma(\xi)$. In order to complete the proof, it remains to show that $\mathcal{T}_\Sigma(\xi)$ is a closed set. To this aim, let $(\eta_k)_{k\in\mathbb{N}^*}$ be a sequence of elements in $\mathcal{T}_\Sigma(\xi)$, convergent to η. We have

$$\frac{1}{t}\,\mathrm{dist}\,(\xi+t\eta,\Sigma)\leq\frac{1}{t}\|t(\eta-\eta_k)\|+\frac{1}{t}\,\mathrm{dist}\,(\xi+t\eta_k,\Sigma)$$

$$=\|\eta-\eta_k\|+\frac{1}{t}\,\mathrm{dist}\,(\xi+t\eta_k,\Sigma)$$

for every $k\in\mathbb{N}^*$. So, $\limsup_{t\downarrow 0}\frac{1}{t}\,\mathrm{dist}\,(\xi+t\eta,\Sigma)\leq\|\eta-\eta_k\|$ for every $k\in\mathbb{N}^*$. Since $\lim_{k\to\infty}\|\eta-\eta_k\|=0$, it follows that $\liminf_{t\downarrow 0}\frac{1}{t}\,\mathrm{dist}\,(\xi+t\eta,\Sigma)=0$, which completes the proof.

Problem 7.10 Let us observe that a vector $\eta\in\mathcal{T}_\Sigma(\xi)$ if and only if there exists a function $x:[\,0,1\,]\to\Sigma$ with $x(0)=\xi$, $\displaystyle\lim_{h_k\downarrow 0}\frac{1}{h_k}(x(h_k)-\xi)=\eta$, for some sequence $h_k\downarrow 0$. But, in the particular case of the set Σ in the problem, this relation holds if and only if $\langle\eta,\nabla U(\xi)\rangle=0$, which completes the proof.

Problem 7.11 Let us consider $f:\mathbb{R}^3\to\mathbb{R}^3$ and $U:\mathbb{R}^3\to\mathbb{R}$, defined by $f(x_1,x_2,x_3)=(-x_2+x_3^2,x_1,-x_1x_3)$ and $U(x_1,x_2,x_3)=x_1^2+x_2^2+x_3^2$ for every $(x_1,x_2,x_3)\in\mathbb{R}^3$. It is easy to see that f is locally Lipschitz being of class C^∞. Therefore, the uniqueness hypothesis in Theorem 7.9.1 is satisfied. In order to check the tangency condition: $f(\xi)\in\mathcal{T}_\Sigma(\xi)$ for every $\xi\in\Sigma$, according to Problem 7.10, it suffices to show that $\langle f(\xi),\nabla U(\xi)\rangle=0$ for every $\xi\in\Sigma$. But this condition is satisfied because $\langle f(\xi),\nabla U(\xi)\rangle=-2\xi_1\xi_2+2\xi_1\xi_3^2+2\xi_1\xi_2-2\xi_1\xi_3^2=0$.

Problem 7.12 We will analyze only the case of the Oy-axis. So, let $f:\mathbb{R}^2\to\mathbb{R}^2$ and $U:\mathbb{R}^2\to\mathbb{R}$ be defined by $f(x,y)=((a-ky)x,-(b-hx)y)$ and respectively by $U(x,y)=x$, for every $(x,y)\in\mathbb{R}^2$. Since f is locally Lipschitz, the uniqueness hypothesis in Theorem 7.9.1 is satisfied. We have $\Sigma=\{(x,y)\in\mathbb{R}^2;\;U(x,y)=0\}$, while in order to check the tangency condition in Theorem 7.9.1, in accordance with Problem 7.10, we must show that $\langle f(\xi_1,\xi_2),\nabla U(\xi_1,\xi_2)\rangle=0$ for every $(\xi_1,\xi_2)\in\Sigma$. This condition is satisfied because $f(\xi_1,\xi_2)=f(0,\xi_2)=(0,-b\xi_2)$ and $\nabla U(\xi_1,\xi_2)=(1,0)$, for every $(\xi_1,\xi_2)\in\Sigma$. Similarly, one proves that the Ox-axis is invariant for the Lotka–Volterra system. By virtue of the remarks above, in order to show that the first quadrant is an invariant set for the system, it suffices to observe that every solution, issued at a point of the boundary of this set, remains there as long as it exists. But this amounts to proving that every solution, which reaches the origin remains there, assertion which is obvious in view of the uniqueness property.

Problem 7.13 By virtue of Definition 7.7.3, we deduce that for every $\xi\in\partial\Sigma$ we have $\mathcal{T}_{\partial\Sigma}(\xi)\subset\mathcal{T}_\Sigma(\xi)$ and therefore the condition in the statement implies: for every $\xi\in\partial\Sigma$ we have $f(\xi)\in\mathcal{T}_\Sigma(\xi)$. Since for every $\xi\in\overset{\circ}{\Sigma}$, $\mathcal{T}_\Sigma(\xi)=\mathbb{R}^n$, it follows that f satisfies the sufficiency part in Theorem 7.7.2.

The condition is not necessary, as we can state from the following simple example. Let $\Sigma=\{(x,y)\in\mathbb{R}^2;\;x^2+y^2\leq 1\}$, and let us consider the system

$$\begin{cases}x'=-x-y\\ y'=x-y.\end{cases}$$

Multiplying the first equality by x and the second by y, we deduce that

$$\frac{1}{2}\frac{d}{dt}\left(x^2(t)+y^2(t)\right)=-\left(x^2(t)+y^2(t)\right)$$

for every $t\in[0,T]$. Hence $x^2(t)+y^2(t)=e^{-2t}(x^2(0)+y^2(0))<1$ for every $(x(0),y(0))\in\partial\Sigma$. So, although the set Σ is invariant for the system, its boundary $\partial\Sigma$ is not. According to the necessity part of Theorem 7.7.2, it follows that one cannot have $f(\xi)\in\mathcal{T}_{\partial\Sigma}(\xi)$ for every $\xi\in\partial\Sigma$. Hence, the condition in Theorem A is not necessary.

Chapter 8

Exercise 8.1 Let us denote by $x(t)=y''(t)$ for $t\geq 0$. We successively have $\int_0^t x(s)\,ds=y'(t)-1$, and $\int_0^t\int_0^s x(\theta)\,d\theta\,ds=\int_0^t\frac{ds}{ds}\int_0^s x(\theta)\,d\theta\,ds=s\int_0^s x(\theta)\,d\theta\big|_{s=0}^{s=t}-\int_0^t sx(s)\,ds=\int_0^t(t-s)x(s)\,ds$. But $\int_0^t\int_0^s x(\theta)\,d\theta\,ds=y(t)-t=-y''(t)-t=-x(t)-t$, and accordingly $x(t)=-t-\int_0^t(t-s)x(s)\,ds$.

Exercise 8.2 As in the preceding exercise, we denote by $x(t)=y''(t)$ for $t\geq 0$. Then $\int_0^t x(s)\,ds=y'(t)$, and $\int_0^t\int_0^s x(\theta)\,d\theta\,ds=\int_0^t\frac{ds}{ds}\int_0^s x(\theta)\,d\theta\,ds=s\int_0^s x(\theta)\,d\theta\big|_{s=0}^{s=t}-\int_0^t sx(s)\,ds=\int_0^t(t-s)x(s)\,ds$. But $\int_0^t\int_0^s x(\theta)\,d\theta\,ds=y(t)-1=-y''(t)-ty'(t)-1=-x(t)-t\int_0^t x(s)\,ds-1$, and accordingly $x(t)=-1-\int_0^t(2t-s)x(s)\,ds$.

Exercise 8.3 Let us denote by $x(t)=y''(t)$ for $t\geq 0$. We have $\int_0^t x(s)\,ds=y'(t)+1$, and $\int_0^t\int_0^s x(\theta)\,d\theta\,ds=\int_0^t\frac{ds}{ds}\int_0^s x(\theta)\,d\theta\,ds=s\int_0^s x(\theta)\,d\theta\big|_{s=0}^{s=t}-\int_0^t sx(s)\,ds=\int_0^t(t-s)x(s)\,ds$. But $\int_0^t\int_0^s x(\theta)\,d\theta\,ds=y(t)+t-1=te^{-t}+(\sin t)e^{-t}y'(t)-e^{-t}y''(t)+t-1=te^{-t}+(\sin t)e^{-t}\left[\int_0^t x(s)\,ds-1\right]-e^{-t}x(t)+t-1$, and so $x(t)=t-\sin t+e^t(t-1)+\int_0^t\left[\sin t-e^t(t-s)\right]x(s)\,ds$.

Exercise 8.4 Let us denote by $x(t)=y''(t)$ for $t\geq 0$. We have $\int_0^t x(s)\,ds=y'(t)$, and $\int_0^t\int_0^s x(\theta)\,d\theta\,ds=\int_0^t\frac{ds}{ds}\int_0^s x(\theta)\,d\theta\,ds=s\int_0^s x(\theta)\,d\theta\big|_{s=0}^{s=t}-\int_0^t sx(s)\,ds=\int_0^t(t-s)x(s)\,ds$. But $\int_0^t\int_0^s x(\theta)\,d\theta\,ds=y(t)=-y''(t)+\cos t=-x(t)+\cos t$, and so $x(t) = \cos t - \int_0^t (t-s)x(s)\,ds$.

Problem 8.1 Let us consider the equation

$$\begin{cases} y''(t) + ay'(t) + by(t) = f(t) \\ y(a) = y_a, \quad y'(a) = y'_a. \end{cases}$$

Let $x(t) = y''(t)$ for $t \in [a,b]$. We have $\int_a^t x(s)\,ds = y'(t) - y'_a$ and $y(t) - (t-a)y'_a - y_a = \int_a^t \int_a^s x(\theta)\,d\theta\,ds = \int_a^t \frac{ds}{ds} \int_a^s x(\theta)\,d\theta\,ds = s\int_a^t x(\theta)\,d\theta|_{s=a}^{s=t} - \int_a^t sx(s)\,ds = \int_a^t (t-s)x(s)\,ds$. On the other hand, $y(t) = b^{-1}f(t) - ab^{-1}y'(t) - b^{-1}y''(t) = b^{-1}f(t) - ab^{-1}\left[\int_a^t x(s)\,ds + y'_a\right] - b^{-1}x(t)$. So, $b^{-1}f(t) - ab^{-1}\left[\int_a^t x(s)\,ds + y'_a\right] - b^{-1}x(t) = (t-a)y'_a - y_a + \int_a^t (t-s)x(s)\,ds$. Finally, we get $x(t) = f(t) - ay'_a + b(a-t)y'_a + by_a + \int_a^t [-b(t-s) - a]x(s)\,ds$. Thus, $k(t,s) = \tilde{k}(t-s)$, where $\tilde{k}(t) = -bt - a$, for $t \in \mathbb{R}$, as claimed.

Problem 8.2 Let us denote by $y(t) = \int_a^t k_2(s)x(s)\,ds$. Differentiating both sides, we get $y'(t) = k_2(t)x(t)$. The integral equation can be rewritten as $x(t) = h(t) + k_1(t)y(t)$, or equivalently $y'(t) = k_2(t)h(t) + k_1(t)k_2(t)y(t)$. The solution of this first-order nonhomogeneous linear equation is $y(t) = e^{\int_a^t k_1(s)k_2(s)\,ds}\xi + \int_a^t e^{\int_s^t k_1(\theta)k_2(\theta)\,d\theta} k_2(s)h(s)\,ds\,ds$. But $y(a) = 0$ and so $\xi = 0$. We finally get $x(t) = h(t) + k_1(t)\int_a^t e^{\int_s^t k_1(\theta)k_2(\theta)\,d\theta} k_2(s)h(s)\,ds$.

Exercise 8.5 (1) Multiplying both sides of the equation by e^{-t}, we get $e^{-t}x(t) = te^{-t} - \int_0^t e^{-s}x(s)\,ds$. Let $y(t) = e^{-t}x(t)$ for $t \in \mathbb{R}$. Then $y(t) = te^{-t} - \int_0^t y(s)\,ds$ which shows that

$$\begin{cases} y'(t) = -y(t) + e^{-t} - te^{-t} \\ y(0) = 0. \end{cases}$$

Thus $y(t) = \int_0^t e^{-(t-s)}\left[e^{-s} - se^{-s}\right]\,ds = e^{-t}\int_0^t (1-s)ds = e^{-t}\left(t - \frac{t^2}{2}\right)$. So, $x(t) = t - \frac{t^2}{2}$.

(2) Differentiating both sides, we obtain

$$\begin{cases} x'(t) = x(t) + e^t \\ x(0) = 1. \end{cases}$$

Then

$$x(t) = e^t + \int_0^t e^{t-s}e^s\,ds = e^t + te^t.$$

(3) Multiplying both sides by e^{-t}, and denoting $e^{-t}x(t) = y(t)$, we get

$$y(t) = 1 + \int_0^t y(s)\, ds.$$

So,

$$\begin{cases} y'(t) = y(t) \\ y(0) = 1. \end{cases}$$

Thus $y(t) = e^t$ and, accordingly, $x(t) = e^{2t}$.

(4) Applying the Laplace Transform on both sides and denoting by $X(p) = \mathcal{L}(x(t))$, we get

$$X(p) = \frac{1}{p^2+1} + 2X(p)\frac{p}{p^2+1}$$

which implies that

$$X(p) = \frac{1}{(p-1)^2}.$$

So, $x(t) = \mathcal{L}^{-1}(\frac{1}{(p-1)^2}) = te^t$.

(5) Differentiating both sides, we get

$$\begin{cases} x'(t) = x(t) - \sin t \\ x(0) = 1. \end{cases}$$

Thus,

$$\begin{aligned} x(t) = e^t - \int_0^t e^{t-s} \sin s\, ds &= e^t \left(1 - \int_0^t e^{-s} \sin s\, ds\right) \\ &= \frac{1}{2}(e^t + \sin t + \cos t). \end{aligned}$$

(6) Multiplying both sides by e^{-t} and denoting $e^{-t}x(t) = y(t)$, we obtain

$$y(t) = e^{-t}\sin t + 2\int_0^t y(s)\, ds.$$

So,

$$\begin{cases} y'(t) = 2y(t) + e^{-t}\cos t - e^{-t}\sin t \\ y(0) = 0. \end{cases}$$

Hence

$$\begin{aligned} y(t) &= \int_0^t e^{2(t-s)}[e^{-s}\cos s - e^{-s}\sin s]\, ds \\ &= e^{2t}\int_0^t e^{-3s}[\cos s - \sin s]\, ds \\ &= \frac{1}{5}\left[e^{2t} - e^{-t}\cos t + 2e^{-t}\sin t\right]. \end{aligned}$$

Finally,

$$x(t) = e^t y(t) = \frac{1}{5}\left[e^{3t} - \cos t + 2\sin t\right].$$

Exercise 8.6

(1) Let us denote $e^t x(t) = y(t)$. We then have $e^t \int_0^t y(s)\,ds = 2t$, which implies that $y(t) = 2e^{-t}(1-t)$. Thus, $x(t) = 2e^{-2t}(1-t)$.

(2) Let us denote by $e^{-t}x(t) = y(t)$. We get $e^t \int_0^t y(s)\,ds = t$. So, $y(t) = e^{-t}(1-t)$ and thus $x(t) = 1-t$.

(3) This is a Volterra equation of the first kind. The problem is that, in our case, the condition (iii) in Theorem 8.3.1, i.e. $k(s,s) \neq 0$ for each $s \in [0,b]$, is violated exactly in $t = 0$. We will see however that the equation has infinite-many solutions. Indeed, differentiating both sides of the equation, yields $-tx(t) + \int_0^t x(s)\,ds = -\frac{t^2}{2}$. Differentiating once again, we get $x'(t) = 1$ and thus $x(t) = t + c$ with $c \in \mathbb{R}$. It is a simple exercise to show that, for each $c \in \mathbb{R}$, $x(t) = t + c$ is a solution of the initial problem. So, the initial Volterra integral equation has infinite many solutions.

(4) As in the preceding case, this is a Volterra equation of the first kind whose kernel does not satisfy the condition (iii) in Theorem 8.3.1. So, it cannot be reduced to a Volterra equation of the second kind satisfying the standard sufficient conditions for existence and uniqueness in Theorem 8.1.1. Differentiating both sides of the equation, yields $\int_0^t (2t-4s)x(s)\,ds = t^3$. Further, differentiating two times the equation thus obtained, we get successively $-2tx(t)+2\int_0^t x(s)\,ds = 3t^2$ and $-2tx'(t)-2x(t)+2x(t) = 6t$. So, $x'(t) = -3$ which implies that $x(t) = -3t+c$, with $c \in \mathbb{R}$. A simple calculation shows that, for each $c \in \mathbb{R}$, $x(t) = -3t + c$ is a solution of the initial equation. Thus, in this case too, we have existence but not uniqueness.

(5) Let us assume first that x is a solution of the equation. Then,

$$\int_0^\theta \frac{x(s)}{(\theta - s)^\alpha}\,ds = h(\theta)$$

for each $\theta \geq 0$, and therefore

$$\int_0^t \frac{1}{(t-\theta)^{1-\alpha}} \int_0^\theta \frac{x(s)}{(\theta - s)^\alpha}\,ds\,d\theta = \int_0^t \frac{h(\theta)}{(t-\theta)^{1-\alpha}}\,d\theta,$$

for $t > 0$. Changing the order of integration, yields

$$\int_0^t x(s)\,ds \int_s^t \frac{d\theta}{(t-\theta)^{1-\alpha}(\theta - s)^\alpha} = \int_0^t \frac{h(\theta)}{(t-\theta)^{1-\alpha}}\,d\theta.$$

The substitution $\theta = s + \tau(t-s)$ leads to

$$\int_s^t \frac{d\theta}{(t-\theta)^{1-\alpha}(\theta-s)^\alpha} = \int_0^1 \tau^{-\alpha}(1-\tau)^{\alpha-1}\,d\tau$$

$$= B(1-\alpha,\alpha) = \Gamma(1-\alpha)\Gamma(\alpha) = \frac{\pi}{\sin\alpha\pi},$$

where B and Γ are the *Beta*, and respectively the *Gamma* function of Euler. Hence

$$\int_0^t x(s)\,ds = \frac{\sin\alpha\pi}{\pi}\int_0^t \frac{h(s)}{(t-s)^{1-\alpha}}\,ds,$$

which, by differentiation, yields

$$x(t) = \frac{\sin\alpha\pi}{\pi}\frac{d}{dt}\left(\int_0^t \frac{h(s)}{(t-s)^{1-\alpha}}\,ds\right) = \frac{\sin\alpha\pi}{\pi}\frac{d}{dt}\left(\int_0^t \frac{h(t-s)}{s^{1-\alpha}}\,ds\right)$$

$$= \frac{\sin\alpha\pi}{\pi}\left(\frac{h(0)}{t^{1-\alpha}} + \int_0^t \frac{h'(s)}{(t-s)^{1-\alpha}}\,ds\right).$$

We notice that x, given by the formula above, is not defined at $t = 0$ unless $h(0) = 0$. If $h(0) \neq 0$, x can be viewed as a solution of the Generalized Abel Equation in a broad sense, i.e. by admitting that, although it satisfies the singular equation, it may have a singularity at $t = 0$. Of course, if we are looking for a function x which is continuous and well-defined at $t = 0$, we have to impose the necessary condition $h(0) = 0$ in Theorem 8.3.1 as well as that h is differentiable at $t = 0$. Then, in this case, we get $x(t) = \dfrac{\sin\alpha\pi}{\pi}\displaystyle\int_0^t \frac{h'(s)}{(t-s)^{1-\alpha}}\,ds$.

Exercise 8.7 Let us assume that there exists $b > 0$ such that the Volterra equation of the first kind has a solution, $x : [0,b] \to \mathbb{R}$. Differentiating both sides, yields

$$\int_0^t x(s)\,ds = 1,$$

which implies that $x(t) = 0$ for each $t \in [0,b]$. But the null function does not satisfy the equation. Thus, the equation in question has no solution.

Alternatively, taking the Laplace Transform on both sides of the equation, and using Theorem 12.5.10, we get $\mathcal{L}[t * x(t)](p) = \dfrac{1}{p}\mathcal{L}[x(t)](p) = \dfrac{1}{p}$, for each $p \in \mathbb{C}$ with $\Re(p) > \omega_x$. Thus, $\mathcal{L}[x(t)](p) = 1$, which is impossible because, by virtue of Theorem 12.5.2, we necessarily have $\displaystyle\lim_{\Re(p)\to+\infty} \mathcal{L}[x(t)](p) = 0$.

Problem 8.3 We shall use the successive approximation method. Namely, let

$$\begin{cases} y_0(t) = x(t) & \text{for } t \in [a,b] \\ y_{n+1}(t) = h(t) + \lambda\displaystyle\int_a^t k(t,s)y_n(s)\,ds & \text{for } n = 0,1,2,\dots, \text{ and } t \in [a,b]. \end{cases}$$

Repeating the proof of Theorem 8.1.1, we deduce that $(y_n)_n$ is uniformly convergent to $[a, b]$ to the unique solution of the integral equation

$$y(t) = h(t) + \lambda \int_a^t k(t,s)y(s)\,ds.$$

Further, by (8.2.1) in Theorem 8.2.1, we know that

$$y(t) = h(t) + \lambda \int_a^t R(t,s,\lambda)y(s)\,ds, \tag{S.8.1}$$

where R is the resolvent kernel of the equation. Next, let us observe that, by hypothesis, we have $x(t) = y_0(t) \leq y_1(t)$ for each $t \in [a,b]$. Since both λ and k are positive, a simple induction argument shows that $y_n(t) \leq y_{n+1}(t)$ for each $n \in \mathbb{N}$ and each $t \in [a,b]$. Indeed, if we assume that $y_{n-1}(t) \leq y_n(t)$, it readily follows that

$$y_n(t) = h(t) + \lambda \int_a^t k(t,s)y_{n-1}(s)\,ds \leq h(t) + \lambda \int_a^t k(t,s)y_n(s)\,ds = y_{n+1}(t)$$

for each $n \in \mathbb{N}$ and each $t \in [a,b]$. Hence $x(t) \leq y_n(t)$, for each $n \in \mathbb{R}$ and each $t \in [a,b]$. Passing to the limit for $n \to \infty$ in the inequality above and taking into account that y satisfies (S.8.1), we conclude the proof.
Problem 8.4 We show first that the resolvent kernel of the integral equation

$$x(t) = h(t) + \int_a^t k(s)x(s)\,ds$$

is given by $R(t,s,\lambda) = k(s)e^{\lambda \int_s^t k(\tau)\,d\tau}$ and then to apply Problem 8.3. To compute the resolvent kernel, let us observe that

$$\begin{cases} k_1(t,s) = k(s) \\ k_2(t,s) = \displaystyle\int_s^t k(\theta)k(s)\,d\theta = k(s)\int_s^t k(\tau)\,d\tau \\ \vdots \\ k_m(t,s) = \dfrac{k(s)}{(m-1)!}\left[\displaystyle\int_s^t k(\tau)\,d\tau\right]^{m-1} \qquad \text{for } m = 1,2,\ldots, \end{cases}$$

the general formula for k_m being obtained by mathematical induction. So,

$$\begin{aligned} R(t,s,\lambda) &= \sum_{m=1}^{\infty} \lambda^{m-1}k_m(t,s) = k(s)\sum_{m=1}^{\infty} \frac{\lambda^{m-1}}{(m-1)!}\left[\int_s^t k(\tau)\,d\tau\right]^{m-1} \\ &= k(s)e^{\lambda \int_s^t k(\tau)\,d\tau}. \end{aligned}$$

Let $h \geq 0$ be constant. From Problem 8.3 we know that

$$x(t) \leq h + h\int_a^t k(s)e^{\int_s^t k(\tau)\,d\tau}ds = h - h\int_a^t \frac{d}{ds}\left[e^{\int_s^t k(\tau)\,d\tau}\right] ds = he^{\int_a^t k(\tau)\,d\tau}.$$

Problem 8.5 Since $R(t,s,\lambda) = \sum_{n=1}^{\infty} \lambda^n k_n(t,s)$, to complete the proof, it suffices to show that, for each $n = 1, 2, \ldots$, there exists a function $\tilde{k}_n : \mathbb{R} \to \mathbb{C}$, such that $k_n(t,s) = \tilde{k}_n(t-s)$, for each $a \leq s \leq t \leq b$. To this aim, we proceed by induction. By hypothesis, we know that $k_1(t,s) = k(t-s)$. So, let us assume that for some $n = 2, 3, \ldots$ there exists $\tilde{k}_n : \mathbb{R} \to \mathbb{C}$ such that $k_n(t,s) = \tilde{k}_n(t-s)$, for each $a \leq s \leq t \leq b$. We will prove that, there exists $\tilde{k}_{n+1} : \mathbb{R} \to \mathbb{C}$ such that $k_{n+1}(t,s) = \tilde{k}_{n+1}(t-s)$, for each $a \leq s \leq t \leq b$. From the inductive assumption and the definition of k_{n+1}, we get $k_{n+1}(t,s) = \int_s^t k(t-u)\tilde{k}_n(u-s)\,du$, for each $a \leq s \leq t \leq b$. The change of the argument $u - s = \theta$, in the integral above, yields $k_{n+1}(t,s) = \int_0^{t-s} k(t-s-\theta)\tilde{k}_n(\theta)\,d\theta$, which shows that $k_{n+1}(t,s) = \tilde{k}_{n+1}(t-s)$, where $\tilde{k}_{n+1}(r) = \int_0^r k(r-\theta)\tilde{k}_n(\theta)\,d\theta$.

Exercise 8.8

(1) By mathematical induction, we deduce that the iterated kernels are

$$\begin{cases} k_1(t,s) = e^{t-s} \\ k_2(t,s) = \int_s^t e^{t-\theta}e^{\theta-s}\,d\theta = (t-s)e^{t-s} \\ \vdots \\ k_m(t,s) = \dfrac{(t-s)^{m-1}}{(m-1)!}e^{t-s}. \end{cases}$$

So, in view of Definition 8.2.1, the resolvent kernel is given by

$$R(t,s,\lambda) = \sum_{m=1}^{\infty} \lambda^{m-1}\frac{(t-s)^{m-1}}{(m-1)!}e^{t-s} = e^{(1+\lambda)(t-s)}.$$

(2) By mathematical induction, we obtain that the iterated kernels are

$$\begin{cases} k_1(t,s) = t-s \\ k_2(t,s) = \int_s^t (t-\theta)(\theta-s)\,d\theta = \dfrac{(t-s)^3}{3!} \\ \vdots \\ k_m(t,s) = \dfrac{(t-s)^{2m-1}}{(2m-1)!}. \end{cases}$$

So,

$$R(t,s,\lambda) = \sum_{m=1}^{\infty} \lambda^{m-1} \frac{(t-s)^{2m-1}}{(2m-1)!}$$

$$= \frac{1}{\sqrt{\lambda}} \sum_{m=1}^{\infty} \frac{[\sqrt{\lambda}(t-s)]^{2m-1}}{(2m-1)!} = \frac{1}{\sqrt{\lambda}} \sinh \sqrt{\lambda}(t-s).$$

(3) By mathematical induction, we obtain that the iterated kernels are

$$\begin{cases} k_1(t,s) = e^{t^2-s^2} \\ k_2(t,s) = \displaystyle\int_s^t e^{t^2-\theta^2} e^{\theta^2-s^2}\, d\theta = (t-s)e^{t^2-s^2} \\ \vdots \\ k_m(t,s) = \dfrac{(t-s)^{2m-3}}{(2m-3)!} e^{t^2-s^2}, \qquad m = 2,3,\ldots. \end{cases}$$

So,

$$R(t,s,\lambda) = e^{t^2-s^2} \left[1 + \sum_{m=2}^{\infty} \lambda^{m-1} \frac{(t-s)^{2m-3}}{(2m-3)!} \right]$$

$$= e^{t^2-s^2} \left[1 + \sqrt{\lambda} \sum_{m=1}^{\infty} \frac{[\sqrt{\lambda}(t-s)]^{2m-1}}{(2m-1)!} \right] = \left[1 + \sqrt{\lambda} \sinh \sqrt{\lambda}(t-s) \right] e^{t^2-s^2}.$$

(4) By mathematical induction, we obtain that the iterated kernels are

$$\begin{cases} k_1(t,s) = \dfrac{\cosh t}{\cosh s} \\ k_2(t,s) = \displaystyle\int_s^t \frac{\cosh t}{\cosh\theta} \frac{\cosh\theta}{\cosh s}\, d\theta = (t-s)\frac{\cosh t}{\cosh s} \\ \vdots \\ k_m(t,s) = \dfrac{(t-s)^{2m-3}}{(2m-3)!} \dfrac{\cosh t}{\cosh s}, \qquad m = 2,3,\ldots. \end{cases}$$

So,

$$R(t,s,\lambda) = \frac{\cosh t}{\cosh s} \left[1 + \sum_{m=2}^{\infty} \lambda^{m-1} \frac{(t-s)^{2m-3}}{(2m-3)!} \right]$$

$$= \frac{\cosh t}{\cosh s} \left[1 + \sqrt{\lambda} \sum_{m=1}^{\infty} \frac{[\sqrt{\lambda}(t-s)]^{2m-1}}{(2m-1)!} \right] = \left[1 + \sqrt{\lambda} \sinh \sqrt{\lambda}(t-s) \right] \frac{\cosh t}{\cosh s}.$$

Exercise 8.9

(1) By virtue of Theorem 8.2.1 combined with (1) in Exercise 8.8, we have

$$x(t) = t + \int_0^t R(t,s,1)\, ds = t + \int_0^t e^{(1+1)(t-s)}\, ds$$

$$= t + \int_0^t e^{2(t-s)}\,ds = \frac{e^{2t} + 2t - 1}{2}.$$

(2) By virtue of Theorem 8.2.1 combined with (2) in Exercise 8.8,

$$x(t) = 1 + \int_0^t \sinh(t-s)\,ds = \cosh t.$$

(3) By virtue of Theorem 8.2.1 combined with (3) in Exercise 8.8,

$$x(t) = e^{t^2} + \int_0^t [1 + \sinh(t-s)]\, e^{t^2 - s^2}\,ds.$$

Exercise 8.10

(1) Applying the Laplace Transform on both sides, denoting $X(p) = \mathcal{L}[x(t)](p)$, and using Theorem 12.5.10, we get $X(p) = \frac{1}{p-1} - \frac{1}{p-1}X(p)$. Thus, $X(p) = \frac{1}{p}$, which implies that $x(t) = 1$, for $t \geq 0$.

(2) Arguing as before, we get $X(p) = \frac{1}{p^2+1} + \frac{p}{p^2+1}X(p)$, and thus $X(p) = \frac{1}{p^2-p+1} = \frac{1}{(p-\frac{1}{2})^2 + [\frac{\sqrt{3}}{2}]^2}$. But, by virtue of Theorems 12.5.9 and 12.5.4, we have $\mathcal{L}[e^{\lambda t}f(t)] = \mathcal{L}[f(t)](p-\lambda)$ and $\mathcal{L}[f(at) = \frac{1}{a}\mathcal{L}[f(t)]\left(\frac{p}{a}\right)$. Therefore, $\mathcal{L}[e^{\lambda t}\sin(at)](p) = \frac{a}{(p-\lambda)^2+a^2}$. In our case $a = \frac{\sqrt{3}}{2}$ and $\lambda = \frac{1}{2}$. So, $x(t) = \frac{2\sqrt{3}}{3}e^{t2}\sin\frac{\sqrt{3}}{2}t$.

(3) Applying the Laplace Transform on both sides and denoting by $X(p) = \mathcal{L}[x(t)](p)$, we get $X(p) = \frac{1}{p^2-1} - \frac{1}{p^2-1}X(p)$. Hence $X(p) = \frac{1}{p^2}$ which implies that $x(t) = t$, for each $t \geq 0$.

(4) Arguing as before, we get $X(p) = \frac{1}{p^2+1} + \frac{1}{p^2}X(p)$, wherefrom we deduce that $X(p) = \frac{p}{p^2+1} \cdot \frac{p}{p^2-1}$. From Theorem 12.5.10, we conclude that $x(t) = \cos t * \cosh t$, for each $t \geq 0$.

Exercise 8.11

(1) Applying the Laplace Transform on both sides of the equation, denoting by $X(p) = \mathcal{L}[x(t)](p)$, and using Theorem 12.5.9, we get $\frac{1}{p^2-1}X(p) = \frac{6}{(p+1)^4}$. So, after some elementary calculations, we deduce $X(p) = -\frac{12}{(p+1)^3} + \frac{6}{(p+1)^2}$. Thanks to Theorem 12.5.9, we conclude that $x(t) = -6t^2e^{-t} + 6te^{-t}$, for $t \geq 0$.

(2) Applying the Laplace Transform on both sides of the equation, we get $\frac{1}{p-1}X(p) = \frac{2}{p^2}$. Hence $X(p) = \frac{2}{p^2} - \frac{2}{p^3}$, which implies that $x(t) = 2t - t^2$, for each $t \geq 0$.

(3) Arguing as before, we get $\frac{p}{p^2+1}X(p) = -\frac{d}{dp}\left(\frac{1}{p^2+1}\right) = \frac{2p}{(p^2+1)^2}$. Thus, $X(p) = \frac{2}{p^2+1}$, which implies that $x(t) = 2\sin t$, for $t \geq 0$.

Chapter 9

Exercise 9.1

(1) Let $q \in \mathcal{D}^1(0,1)$. We have $\int_0^1 q'(t)\,dt = q(1) - q(0) = 1$, because q' is piecewise continuous. So, the functional is constant on $\mathcal{D}^1(0,1)$.

(2) We have $\int_0^1 q(t)q'(t)\,dt = \frac{1}{2}[q^2(1) - q^2(0)] = \frac{1}{2}$, and so, the functional is constant on $\mathcal{D}^1(0,1)$.

(3) We have $\int_0^1 tq(t)q'(t)\,dt = \frac{1}{2}\int_0^1 t\frac{d}{dt}(q^2)(t)\,dt = \frac{1}{2} - \frac{1}{2}\int_0^1 q^2(t)\,dt$. From this relation, it follows that $\inf\left\{\int_0^1 tq(t)q'(t)\,dt;\ q \in \mathcal{D}^1(0,1)\right\} = -\infty$, simply because there are positive functions $p \in \mathcal{D}^1(0,1)$ such that $\int_0^1 p(t)\,dt$ is as big as we wish. Thus, the minimum problem has no solution. In fact, in this case, the Euler–Lagrange Equation is degenerate, i.e. it reduces to $q(t) = 0$, and thus there are no extremals verifying the boundary conditions $q(0) = 0$ and $q(1) = 1$.

Exercise 9.2

(1)The Euler–Lagrange Equation is $q'' + q = \sin t$, whose solutions are

$$q(t) = c_1 \sin t + c_2 \cos t - \frac{t}{2}\cos t.$$

(2) The Euler–Lagrange Equation is $q'' - q = \cosh t$, whose solutions are

$$q(t) = c_1 e^t + c_2 e^{-t} + \frac{t}{2}\cosh t + \frac{1}{2}\sinh t.$$

(3) The Euler–Lagrange Equation is $q'' - q = \sin t$, whose solutions are

$$q(t) = c_1 e^t + c_2 e^{-t} - \frac{1}{2}\sin t.$$

Exercise 9.3

(1) Since the functional is nonnegative and $q^*(t) = t$ for $t \in [0,1]$, satisfies the boundary conditions, and the value of the functional at q^* is 0, it follows that this is a global minimum.

(2) The Euler–Lagrange Equation is degenerate, i.e. $2(t - q) = 0$, which has the solution $q(t) = t$ for $t \in [0,1]$. Since this extremal does not satisfy the boundary conditions, it follows that the functional has no local weak minimum.

(3) Since the Lagrangian does not depend explicitly on q, instead of the Euler–Lagrange Equation, we can use the prime integral $\frac{\partial L}{\partial q'} = c$ which, in our case, has the form $\frac{q'}{t\sqrt{1+q'^2}} = c$. Thus, $q'^2 = \frac{c^2t^2}{1-c^2t^2}$. The general solution of this equation is $q(t) = -\frac{1}{c}\sqrt{1-c^2t^2} + d$. So, $(q-d)^2 + t^2 = \frac{1}{c^2}$ and thus the extremals are circles with center on the q-axis. From the boundary conditions $q(1) = 0$ and $q(2) = 1$, we get $c = \frac{1}{\sqrt{5}}$ and $d = 2$. It follows that the equation of the extremals is $(q-2)^2 + t^2 = 5$. In fact there are two extremals: $q_{1,2}(t) = 2 \pm \sqrt{5-t^2}$ and we may easily see that $J[q_1] = J[q_2]$, simply because $\sqrt{1 + {q_1'}^2(t)} = \sqrt{1 + {q_2'}^2(t)}$. Since for $t \in [1,2]$, along both extremals the second variation of the Lagrangian, i.e. $\delta^2 L[q(t)][\,\cdot\,] = \frac{1}{t\sqrt{1+q'^2}(1+q'^2)}$ is strongly positive, it follows that both extremals are local weak minima for the problem.

Problem 9.1 Let $q : [a,b] \to \mathbb{R}$ be a function in $\mathcal{D}^1(q_a, q_b)$. The length of the curve is given by $J[q] = \int_a^b \sqrt{1 + q'^2(t)}dt$. The Lagrangian of the problem is

$L(t,q,q') = \sqrt{1+q'^2}$. Since L does not depend explicitly on t, we may use the energy integral, i.e. $L(q(t),q'(t)) - q'(t)\frac{\partial L}{\partial q'}(q(t),q'(t)) = c$, for each $t \in [a,b]$, which in our case has the form $\sqrt{1+q'^2(t)} - \frac{q'^2(t)}{\sqrt{1+q'^2(t)}} = c$, for each $t \in [a,b]$. Equivalently, we have $\frac{1}{\sqrt{1+q'^2(t)}} = c$, for each $t \in [a,b]$, which shows that there exists a constant k such that $q'(t) = k$, for each $t \in [a,b]$. Taking into account that $q(a) = q_a$ and $q(b) = q_b$, we conclude that $q(t) = \frac{q_b - q_a}{b-a}t + \frac{bq_a - aq_b}{b-a}$, for $t \in [a,b]$ which is nothing but the line segment connecting (a,q_a) with (b,q_b). This extremal is in fact a local weak minimum point for J because $\delta^2 L[q]$ is strongly positive. In fact, q is a global minimum because the Lagrangian is strictly convex.

Problem 9.2 The functional in this case is $J[q] = 2\pi\int_a^b q(t)\sqrt{1+q'^2(t)}\,dt$. Clearly, each minimum point of J is a minimum point for the functional $K(q) = \int_a^b q(t)\sqrt{1+q'^2(t)}\,dt$. The Lagrangian is $L(t,q,q') = q\sqrt{1+q'^2}$, and, since it does not depend explicitly on t, we can use the energy integral. We have $q(t)\sqrt{1+q'^2(t)} - q(t)\frac{q'^2(t)}{\sqrt{1+q'^2(t)}} = c$, which implies that $q(t) = c\sqrt{1+q'^2(t)}$. Setting $q' = \sinh\theta$ we get $q = c\cosh\theta$. But $dt = \frac{dq}{q'} = \frac{c\sinh\theta\, d\theta}{\sinh\theta} = c\,d\theta$, and so, $t = c\theta + d$. Thus the extremals of the minimum problem consists of a family of catenaries $q(t) = c\cosh\frac{t-d}{c}$, where c and d are constants. Depending on the ratio ℓ/m, where ℓ is the length of the line segment joining the points (a,q_a) and (b,q_b), and $r = \max\{q_a,q_b\}$, there are three possible cases:

(1) if the boundary conditions $q(a) = q_a$ and $q(b) = q_b$ uniquely determine the function q, then q is a local weak minimum.

(2) if the boundary conditions $q(a) = q_a$ and $q(b) = q_b$ determine two functions q_1 q_2, then one and only one of them is the local weak minimum of the problem.

(3) if there is no extremal satisfying the boundary conditions, there is no local weak minimum. This case has the following geometrical interpretation. If the distance between the points (a,q_a) and (b,q_b) is very large comparing with the distance of these points to the t-axis, then the curve q degenerate to a multi-valued one which is the union of the three segments joining $(a,0)$ with (a,q_a), $(a,0)$ with $(b,0)$ and $(b,0)$ with (b,q_b). So, in this case, the surface of revolution degenerates in a union of two circles of radius q_a and q_b respectively and the line segment of endpoints (a,q_a) and (b,q_b).

Problem 9.3 As in the case of Theorem 9.2.2, it is easy to show that a necessary condition for a function, $q \in \mathcal{D}^1$, to be a local weak $\mathcal{D}^1$-minimum[3] for $(\mathcal{P}_0)$ is that q satisfy $\int_a^b \left[\frac{\partial L}{\partial q}(t,q(t),q'(t))\eta(t) + \frac{\partial L}{\partial q'}(t,q(t),q'(t))\eta'(t)\right] dt = 0$, for each $\eta \in \mathcal{D}^1$. We note that here there is no need to assume that $\eta(a) = \eta(b) = 0$. So, if the Lagrangian is of class C^2, integrating by parts, we get

$$\int_a^b \left\{\frac{\partial L}{\partial q}(t,q(t),q'(t)) - \frac{d}{dt}\left[\frac{\partial L}{\partial q'}(t,q(t),q'(t))\right]\right\}\eta(t)\,dt$$
$$+\eta(b)\frac{\partial L}{\partial q'}(b,q(b),q'(b)) - \eta(a)\frac{\partial L}{\partial q'}(a,q(a),q'(a)) = 0,$$

[3] We say that q is a local weak $\mathcal{D}^1$-minimum for $(\mathcal{P}_0)$, if for each $\eta \in \mathcal{D}^1$, the function $g(\varepsilon) = J[q + \varepsilon\eta]$ has a local minimum at $\varepsilon = 0$.

for each $\eta \in \mathcal{D}^1$. Since $\eta(a)$ and $\eta(b)$ are arbitrary in $\mathbb{R}$, we necessarily have

$$\frac{\partial L}{\partial q'}(a, q(a), q'(a)) = 0 \quad \text{and} \quad \frac{\partial L}{\partial q'}(b, q(b), q'(b)) = 0.$$

So, from Du Bois Raymond's Lemma 9.2.1, we have

$$\frac{\partial L}{\partial q}(t, q(t), q'(t)) - \frac{d}{dt}\left[\frac{\partial L}{\partial q'}(t, q(t), q'(t))\right] = 0.$$

Problem 9.4 Reasoning as in Problem 9.3, we deduce that the necessary conditions in order that q is a local $\mathcal{D}(q_a, \ell_{t=b})$ minimum[4] for the problem $(\mathcal{P}_1)$ is

$$\begin{cases} \dfrac{\partial L}{\partial q}(t, q(t), q'(t)) - \dfrac{d}{dt}\left[\dfrac{\partial L}{\partial q'}(t, q(t), q'(t))\right] = 0 \\ q(a) = q_a, \quad \dfrac{\partial L}{\partial q'}(b, q(b), q'(b)) = 0. \end{cases}$$

Problem 9.5 The functional is in this case (up to a multiplicative positive constant, i.e. $1/\sqrt{2g}$) $J[q] = \int_0^1 \sqrt{\frac{1+q'^2(t)}{q(t)}}dt$. In this case, we have chosen the coordinate system, tOq, with the t-axis (the axis of the abscises) oriented as usual, i.e. from the left to the right, but the q-axis (the axis of the ordinates) oriented in the sense of the gravitational force, i.e. downward. A prime integral for the Euler–Lagrange Equation is $\sqrt{\frac{1+q'^2(t)}{q(t)}} - \frac{q'^2}{\sqrt{q(t)[1+q'^2(t)]}} = \frac{1}{c}$, which can be equivalently rewritten as $\sqrt{\frac{q}{c^2-q}}\, dq = dt$. The solution of this equation is given by the family of cycloids, depending on two constants $\frac{c^2}{2}$ and k, i.e. $\begin{cases} t + k = \frac{c^2}{2}(\theta - \sin\theta) \\ q = \frac{c^2}{2}(1 - \cos\theta). \end{cases}$

From Problem 9.4, we know that $q(0) = 0$ and $\frac{q'(1)}{\sqrt{q(1)[1+q'^2(1)]}} = 0$. So, $q(0) = 0$ and $q(1) = c^2$, and hence we deduce that $k = 0$ and $c = \frac{\sqrt{2}}{\pi}$. So, the equations of the extremal are $\begin{cases} t = (1/\pi)(\sin\theta - \theta) \\ q = (1/\pi)(\cos\theta - 1) \end{cases}$, $\quad \theta \in [0, \pi]$.

Problem 9.6 In the case considered, the Lagrangian is defined by

$$L(t, x_1, y_1, z_1, x_2, y_2, z_2, \ldots, x_n, y_n, z_n, x_1', y_1', z_1', x_2', y_2', z_2', \ldots, x_n', y_n', z_n')$$

$$= \frac{1}{2}\sum_{i=1}^{n} m_i[{x_i'}^2(t) + {y_i'}^2(t) + {z_i'}^2(t)] - U(t, x_1, y_1, z_1, x_2, y_2, z_2, \ldots, x_n, y_n, z_n).$$

So, the Euler–Lagrange System is

$$-\frac{\partial U}{\partial x_i} - \frac{d}{dt}[m_i x_i'] = 0, \quad -\frac{\partial U}{\partial y_i} - \frac{d}{dt}[m_i y_i'] = 0, \quad -\frac{\partial U}{\partial z_i} - \frac{d}{dt}[m_i z_i'] = 0,$$

[4]We say that q is a local weak $\mathcal{D}^1(q_a, \ell_{t=b})$-minimum for $(\mathcal{P}_1)$, if for each $\eta \in \mathcal{D}^1$ with $\eta(a) = 0$, the function $g(\varepsilon) = J[q + \varepsilon\eta]$ has a local minimum at $\varepsilon = 0$.

and $F_i = -\left(\frac{\partial U}{\partial x_i}, \frac{\partial U}{\partial y_i}, \frac{\partial U}{\partial z_i}\right)$ is the force acting on the ith particle, we finally deduce

$$\begin{cases} m_i x_i'' = F_{ix} \\ m_i y_i'' = F_{iy} \\ m_i z_i'' = F_{iz}, \end{cases}$$

which are nothing but the Newton's Equations of Motion.
Problem 9.7 We only have to observe that the canonical variables, corresponding to integral

$$\int_{t_0}^{t_1} L\,dt = \int_{t_0}^{t_1} (T - U)\,dt,$$

are

$$\begin{cases} p_{ix} = \dfrac{\partial L}{\partial x_i'} = m_i x_i' \\ p_{iy} = \dfrac{\partial L}{\partial y_i'} = m_i y_i' \\ p_{iz} = \dfrac{\partial L}{\partial z_i'} = m_i z_i', \end{cases}$$

and thus the Hamiltonian H is given by

$$H = \sum_{i=1}^{n} [x_i' p_{ix} + y_i' p_{iy} + z_i' p_{iz}] - L = 2T - (T - U) = T + U,$$

which is the total energy of the system. If the system is conservative, i.e. if U does not depend explicitly on t, it follows that L does not depend explicitly on t and thus, by Theorem 9.8.1, it follows that H is a prime integral for the canonical Euler–Lagrange System. But this means that H, which is nothing but that the total energy of the system, is constant along the trajectories of the system.

Chapter 10

Exercise 10.1 We have

$$\|g(u) - g(v)\| = \left\|\sum_{i=1}^{\infty} \alpha_i [u(t_i) - v(t_i)]\right\| \leq \sum_{i=1}^{\infty} \alpha_i \|u(t_i) - v(t_i)\|$$

$$\leq \left(\sum_{i=1}^{\infty} \alpha_i\right) \sup_{t \in [t_1, +\infty)} \|u(t) - v(t)\| \leq \|u - v\|_{C_b([a, +\infty); \mathbb{R}^n)}$$

for each $u, v \in C_b(\mathbb{R}_+; \mathbb{R}^n)$, where $a = t_1 > 0$. Consequently g satisfies the hypothesis (iv) in Theorem 10.2.1, as claimed.
Exercise 10.2 We have

$$\|g(u) - g(v)\| = \left\|\int_a^{\infty} k(s)[u(s) - v(s)]\,ds\right\| \leq \int_a^{\infty} k(s)\|u(s) - v(s)\|\,ds$$

$$\leq \left(\int_a^\infty k(s)\,ds\right) \|u-v\|_{C_b([\,a,+\infty);\mathbb{R}^n)} \leq \|u-v\|_{C_b([\,a,+\infty);\mathbb{R}^n)}$$

for each $u, v \in C_b(\mathbb{R}_+;\mathbb{R}^n)$. Hence g satisfies (iv) in Theorem 10.2.1, as claimed.

Problem 10.1 Consider the sequence of successive approximations given by (10.5.1). We recall that x_{k+1} is the unique solution of the problem

$$\begin{cases} x'_{k+1}(t) = \mathcal{A}x_{k+1}(t) + f(t, x_{k+1}(t)),\ t \in \mathbb{R}_+, \\ x_{k+1}(0) = x_k(T),\ k \in \mathbb{N}, \end{cases}$$

where the first term $x_0 \in C_b(\mathbb{R}_+;\mathbb{R}^n)$ is arbitrary but fixed. From (i'') and (ii''), we deduce

$$\left\langle x'_{k+1}(t), x_{k+1}(t)\right\rangle \leq \langle \mathcal{A}x_{k+1}(t), x_{k+1}(t)\rangle$$

$$+ \langle f(t, x_{k+1}(t) - f(t,0), x_{k+1}(t)\rangle + \langle f(t,0), x_{k+1}(t)\rangle$$

$$\leq -\omega\|x_{k+1}(t)\|^2 + m\|x_{k+1}(t)\|$$

for each $k \in \mathbb{N}$ and $t \in \mathbb{R}_+$. Thus

$$\frac{1}{2}\frac{d}{dt}\left(\|x_{k+1}(t)\|^2\right) \leq -\omega\|x_{k+1}(t)\|^2 + m\|x_{k+1}(t)\|$$

for each $k \in \mathbb{N}$ and $t \in \mathbb{R}_+$. This inequality shows that

$$\|x_{k+1}(t)\| \leq \max\left\{\|x_k(T)\|, \frac{m}{\omega}\right\}$$

for each $k \in \mathbb{N}$ and $t \in \mathbb{R}_+$. It follows that $x_{k+1} \in C_b(\mathbb{R}_+;\mathbb{R}^n)$ whenever $x_k \in C_b(\mathbb{R}_+;\mathbb{R}^n)$ and thus $(x_k)_{k\in\mathbb{N}}$ is well-defined. Next, we will prove that the telescopic series

$$x_0(t) + \sum_{k=0}^\infty [x_{k+1}(t) - x_k(t)]$$

satisfies the hypotheses of Lemma 10.3.1 and thus the sequence $(x_k)_{k\in\mathbb{N}}$ is uniformly convergent on $\mathbb{R}_+$ to some function $x \in C_b(\mathbb{R}_+;\mathbb{R}^n)$. Since

$$x'_{k+1}(t) - x'_k(t) = \mathcal{A}x_{k+1}(t) - \mathcal{A}x_k(t) + f(t, x_{k+1}(t)) - f(t, x_k(t))$$

for each $k \in \mathbb{N}$ and $t \in \mathbb{R}_+$, we deduce that

$$\langle x'_{k+1}(t) - x'_k(t), x_{k+1}(t) - x_k(t)\rangle \leq -\omega\|x_{k+1}(t) - x_k(t)\|^2$$

for each $k \in \mathbb{N}^*$ and $t \in \mathbb{R}_+$. From (12.1.3) in Lemma 12.1.2, we conclude that

$$\frac{d}{dt}\left(\|x_{k+1}(t) - x_k(t)\|^2\right) \leq -2\omega\|x_{k+1}(t) - x_k(t)\|^2$$

for each $k \in \mathbb{N}$ and $t \in \mathbb{R}_+$. This implies that

$$\|x_{k+1}(t) - x_k(t)\|^2 \leq e^{-2\omega t}\|x_{k+1}(0) - x_k(0)\|^2$$

for each $k \in \mathbb{N}$ and $t \in \mathbb{R}_+$ and so

$$\|x_{k+1}(t) - x_k(t)\| \leq e^{-\omega t}\|x_{k+1}(0) - x_k(0)\|$$

for each $k \in \mathbb{N}$ and $t \in \mathbb{R}_+$. Consequently

$$\|x_{k+1}(t) - x_k(t)\| \leq e^{-\omega t}\|x_k(T) - x_{k-1}(T)\| \leq e^{-\omega t}\|x_k - x_{k-1}\|_{C_b([T,+\infty);\mathbb{R}^n)}$$

for each $k \in \mathbb{N}^*$ and $t \in \mathbb{R}_+$. Now, let us observe that

$$\|x_{k+1} - x_k\|_{C_b([T,+\infty);\mathbb{R}^n)} \leq e^{-\omega T}\|x_k - x_{k-1}\|_{C_b([T,+\infty);\mathbb{R}^n)}$$

$$\leq \cdots \leq \left(e^{-\omega T}\right)^k \|x_1 - x_0\|_{C_b([T,+\infty);\mathbb{R}^n)}.$$

The preceding inequalities show that we are in the hypotheses of Lemma 10.3.1 and thus the sequence $(x_k)_{k\in\mathbb{N}}$ is uniformly convergent on $[T, +\infty)$. Moreover, from the initial condition and (iv), we get

$$\|x_{k+1}(t) - x_k(t)\| \leq e^{-\omega t}\|x_{k+1} - x_k\|_{C_b([T,+\infty);\mathbb{R}^n)}$$

for each $t \in \mathbb{R}_+$ which shows that $(x_k)_{k\in\mathbb{N}}$ converges even on $\mathbb{R}_+$ to some function $x \in C_b([T, +\infty); \mathbb{R}^n)$. Passing to the limit in the recurrence relation defining $(x_k)_{k\in\mathbb{N}}$, we conclude that x is a solution of (10.1.1).

As concerns the uniqueness, reasoning as above, we note that if x and y are solutions of (10.1.1) with $g(u) = u(T)$ then

$$\|x(t) - y(t)\| \leq e^{-\omega t}\|x - y\|_{C_b([T,+\infty);\mathbb{R}^n)}$$

for each $t \in \mathbb{R}_+$. We deduce

$$\|x - y\|_{C_b([T,+\infty);\mathbb{R}^n)} \leq e^{-\omega T}\|x - y\|_{C_b([T,+\infty);\mathbb{R}^n)}.$$

Since $0 < e^{-\omega T} < 1$, it follows that $x(t) = y(t)$ for each $t \in [T, +\infty)$. But $x(0) = x(T) = y(T) = y(0)$ and so x coincides with y on $\mathbb{R}_+$. Since the global asymptotic stability follows from similar arguments, the proof is complete.

Problem 10.2 Let $T > 0$ and $r > 0$ be arbitrary but fixed and let $K = B(\xi, r)$, i.e. the closed ball with center ξ and radius r in $C([0, T]; \mathbb{R}^n)$. Let us define the function $g : C([0, T]; \mathbb{R}^n) \to \mathbb{R}^n$ by

$$g(u) = \xi + h\left(\int_0^T u(s)\, ds\right)$$

for each $u \in C([0,T];\mathbb{R}^n)$, let $x_0 \in K$ and let us consider the sequence of successive approximations given by

$$x_{k+1}(t) = g(x_k) + \int_0^t f(s, x_{k+1}(s))\,ds,$$

for each $k \in \mathbb{N}$ and $t \in [0,T]$. We will show first that, if $T > 0$ is sufficiently small then the sequence $(x_k)_{k\in\mathbb{N}}$ is well-defined, i.e. $x_{k+1} \in K$ whenever $x_k \in K$. Let $x_k \in K$ be arbitrary. Denote by

$$M = \sup\{\|f(t,x)\|;\ (t,x) \in [0,T] \times B(\xi, r)\}$$

and let us observe that

$$\|x_{k+1}(t) - \xi\| \leq \left\|h\left(\int_0^T x_k(s)\,ds\right)\right\| + \int_0^T \|f(s, x_k(s))\|\,ds$$

$$\leq \ell_2 \left\|\int_0^T x_k(s)ds\right\| + TM \leq T(\ell_2 r + M)$$

for each $t \in [0,T]$. Keeping $r > 0$ fixed and choosing a sufficiently small $T > 0$ such that $T(\ell_2 r + M) \leq r$, we get

$$\|x_{k+1}(t) - \xi\| \leq r$$

for each $t \in [0,T]$. But this shows that $x_{k+1} \in K$.

Next, with $T > 0$ and $r > 0$ fixed as above, let $\ell_1 = \ell_1(T,K)$ be the Lipschitz constant of the restriction of f to $[0,T] \times K$ – see (jj) in Theorem 10.8.2. We have

$$\|x_{k+1}(t) - x_k(t)\| \leq \left\|h\left(\int_0^T x_k(s)\,ds\right) - h\left(\int_0^T x_{k-1}(s)\,ds\right)\right\|$$

$$+ \int_0^t \|f(s, x_{k+1}(s)) - f(s, x_k(s))\|\,ds$$

which yields

$$\|x_{k+1}(t) - x_k(t)\| \leq T\ell_2 \|x_k - x_{k-1}\|_{C([0,T];\mathbb{R}^n)} + \int_0^t \ell_1 \|x_{k+1}(s) - x_k(s)\|\,ds$$

for each $k \in \mathbb{N}^*$ and $t \in [0,T]$. From Gronwall's Lemma 1.5.2, we deduce

$$\|x_{k+1} - x_k\|_{C([0,T];\mathbb{R}^n)} \leq T\ell_2 e^{\ell_1 T} \|x_k - x_{k-1}\|_{C([0,T];\mathbb{R}^n)}$$

for each $k \in \mathbb{N}^*$. Now, let us observe that if $T > 0$ is sufficiently small, we have

$$T\ell_2 e^{\ell_1 T} < 1.$$

Consequently, the telescopic series

$$x_0(t) + \sum_{k=0}^{\infty}(x_{k+1}(t) - x_k(t))$$

satisfies the hypotheses of Lemma 10.3.1 with $\alpha_k = (\ell_2 e^{\ell_1 T})^k \|x_1 - x_0\|_{C([0,T];\mathbb{R}^n)}$ for $k \in \mathbb{N}$. Hence $(x_k)_{k\in\mathbb{N}}$ is uniformly convergent on $[0,T]$ to a function x which is a solution of (10.8.1). Finally, we only have to observe that, by the procedure above, T can be chosen independent of ξ.

Problem 10.3 Let us observe that (10.8.2) can be written as

$$\begin{cases} u'(t) = f(u(t)), \ t \in [0,T], \\ u(0) = g(u), \end{cases}$$

where $f : \mathbb{R}^2 \to \mathbb{R}^2$ and $g : C([0,T];\mathbb{R}^2) \to \mathbb{R}^2$ are defined by

$$f(v) = \begin{pmatrix} ax - bxy \\ -cy + dxy \end{pmatrix}$$

for each $v = \begin{pmatrix} x \\ y \end{pmatrix} \in \mathbb{R}^2$ and

$$g(u) = \xi + h\left(\int_0^T u(s)\,ds\right) = \begin{pmatrix} \xi_1 \\ \xi_2 \end{pmatrix} + \begin{pmatrix} h_1\left(\int_0^T y(s)\,ds\right) \\ h_2\left(\int_0^T x(s)\,ds\right) \end{pmatrix}.$$

for each $u \in C([0,T];\mathbb{R}^2)$, $u = \begin{pmatrix} x(\cdot) \\ y(\cdot) \end{pmatrix}$. Clearly f is of class C^1 and so it is locally Lipschitz on $\mathbb{R}^2$ and g satisfies

$$\|g(u) - g(v)\| \leq T\ell_2 \|u - v\|_{C([0,T];\mathbb{R}^2)}$$

for each $u, v \in C([0,T];\mathbb{R}^2)$. Consequently, all the hypotheses of Theorem 10.8.2 are satisfied, wherefrom we get the conclusion.

As concerns the contribution of the history functions h_1 and h_2 to the behavior of the solution of (10.8.2), one can easily see that h_1 diminishes the initial datum $x(0)$ whenever the mean over $[0,T]$ of the population of predators y increases, while h_2 augments the initial datum $y(0)$ according to the decrease of the mean over $[0,T]$ of the population of preys. Thus the nonlocal initial conditions are, in some sense, some feedback laws which try to keep a "good balance" between the two populations of prey and predators, respectively.

Chapter 11

Exercise 11.1 Let $(t, v) \in [\sigma, +\infty) \times C([-\tau, 0]; \mathbb{R}^n)$ be arbitrary but fixed and let $((t_k, v_k))_{k\in\mathbb{N}}$ be a sequence in $[\sigma, +\infty) \times C([-\tau, 0]; \mathbb{R}^n)$ such that $(t_k)_{k\in\mathbb{N}}$ converges to t and $(v_k)_{k\in\mathbb{N}}$ is uniformly convergent to v on $[-\tau, 0]$. In particular, $(v_k(\theta))_{k\in\mathbb{N}}$ is convergent to $v(\theta)$ in $\mathbb{R}^n$. Since h is continuous, we conclude that

$$\lim_{k\to\infty} f(t_k, v_k) = \lim_{k\to\infty} h(t_k, v_k(\theta)) = h(t, v(\theta)) = f(t, v)$$

which shows that f is continuous at (t, v). Since $(t, v) \in [\sigma, +\infty) \times C([-\tau, 0]; \mathbb{R}^n)$ is arbitrary, it follows that f is continuous on $[\sigma, +\infty) \times C([-\tau, 0]; \mathbb{R}^n)$, as claimed.

Exercise 11.2 Let us observe that f is continuous on $[\sigma, +\infty) \times C([-\tau, 0]; \mathbb{R}^n)$. See Exercise 11.1. Now, let $T > \sigma$, let $t \in [\sigma, T]$ and $v, w \in C([-\tau, 0]; \mathbb{R}^n)$ be such that

$$\begin{cases} \|v\|_{C([-\tau,0];\mathbb{R}^n)} \leq r, \\ \|w\|_{C([-\tau,0];\mathbb{R}^n)} \leq r. \end{cases}$$

Clearly

$$\begin{cases} \|v(\theta)\| \leq r, \\ \|w(\theta)\| \leq r. \end{cases}$$

Hence we have

$$\|f(t, v) - f(t, w)\| = \|h(t, v(\theta)) - h(t, w(\theta))\|$$

$$\leq \ell\|v(\theta) - w(\theta)\| \leq \ell\|v - w\|_{C([-\tau,0];\mathbb{R}^n)}$$

which shows that f Lipschitz on bounded sets with respect to its second argument in the sense of Definition 11.2.1.

Exercise 11.3 Let $T > 0$ be arbitrary and let U be a uniformly bounded subset of $C([-\tau, 0]; \mathbb{R}^n)$. Then there exists $r > 0$ such that

$$\|v(t)\| \leq r$$

for each $v \in U$ and each $t \in [-\tau, 0]$. Since h is continuous and $[0, T] \times B(0, r)$ is compact, there exists $M > 0$ such that

$$\|h(t, u)\| \leq M$$

for each $(t, u) \in [0, T] \times B(0, r)$. Consequently

$$\|f(t, v)\| \leq \|h(t, v(\theta))\| \leq M$$

for each $(t, v) \in [0, T] \times U$, which proves that f maps uniformly bounded subsets in $\mathbb{R}_+ \times C([-\tau, 0]; \mathbb{R}^n)$ into bounded subsets in $\mathbb{R}^n$ in the sense of Definition 11.5.2.

Exercise 11.4 Let $(t,v)\in\mathbb{R}_+\times C([-\tau,0];\mathbb{R}^n)$ be arbitrary but fixed and let $((t_k,v_k))_{k\in\mathbb{N}}$ be a sequence in $\mathbb{R}_+\times C([-\tau,0];\mathbb{R}^n)$ such that $(t_k)_{k\in\mathbb{N}}$ converges to t and $(v_k)_{k\in\mathbb{N}}$ converges uniformly to v on $[-\tau,0]$. Then

$$\lim_{k\to\infty}\int_{-\tau}^{0}v_k(\theta)\,d\theta=\int_{-\tau}^{0}v(\theta)\,d\theta$$

and consequently

$$\lim_{k\to\infty}h\left(t_k,\int_{-\tau}^{0}v_k(\theta)\,d\theta\right)=h\left(t,\int_{-\tau}^{0}v(\theta)\,d\theta\right).$$

Hence f is continuous at (t,v). Since $(t,v)\in\mathbb{R}_+\times C([-\tau,0];\mathbb{R}^n)$ is arbitrary, we conclude that f is continuous on $\mathbb{R}_+\times C([-\tau,0];\mathbb{R}^n)$, as claimed.

Exercise 11.5 The continuity of f follows from Exercise 11.4. To prove that f is Lipschitz on bounded sets with respect to its second argument in the sense of Definition 11.2.1, let $T>0$ and $r>0$ and let $\ell=\ell(T,r)$ be as in condition (c) in Exercise 11.2. Let $v,w\in C([-\tau,0];\mathbb{R}^n)$ be such that

$$\begin{cases}\|v\|_{C([-\tau,0];\mathbb{R}^n)}\leq\dfrac{r}{\tau},\\ \|w\|_{C([-\tau,0];\mathbb{R}^n)}\leq\dfrac{r}{\tau}.\end{cases}$$

Then

$$\begin{cases}\left\|\displaystyle\int_{-\tau}^{0}v(\theta)\,d\theta\right\|\leq r,\\ \left\|\displaystyle\int_{-\tau}^{0}w(\theta)\,d\theta\right\|\leq r.\end{cases}$$

Hence

$$\|f(t,v)-f(t,w)\|\leq\left\|h\left(t,\int_{-\tau}^{0}v(\theta)\,d\theta\right)-h\left(t,\int_{-\tau}^{0}w(\theta)\,d\theta\right)\right\|$$

$$\leq\ell\left\|\int_{-\tau}^{0}v(\theta)\,d\theta-\int_{-\tau}^{0}w(\theta)\,d\theta\right\|\leq\ell\tau\|v-w\|_{C([-\tau,0];\mathbb{R}^n)}.$$

But this inequality shows that f is Lipschitz on bounded sets with respect to its second argument in the sense of Definition 11.2.1.

Exercise 11.6 Let $T>0$ be arbitrary and let U be a uniformly bounded subset of $C([-\tau,0];\mathbb{R}^n)$. Then there exists $r>0$ such that

$$\|v(t)\|\leq r$$

for each $v\in U$ and each $t\in[-\tau,0]$. Then

$$\left\|\int_{-\tau}^{0}v(\theta)\,d\theta\right\|\leq\int_{-\tau}^{0}\|v(\theta)\|\,d\theta\leq\tau r.$$

Since h is continuous and $[0,T] \times B(0,\tau r)$ is compact, there exists $M > 0$ such that

$$\|f(t,v)\| = \left\|h\left(t, \int_{-\tau}^{0} v(\theta)\,d\theta\right)\right\| \leq M$$

for each $(t,v) \in [0,T] \times B(0,\tau r)$. Hence f maps uniformly bounded subsets in $\mathbb{R}_+ \times C([-\tau,0];\mathbb{R}^n)$ into bounded subsets in $\mathbb{R}^n$ in the sense of Definition 11.5.2, as claimed.

Problem 11.1 Let us observe that the problem (11.9.2) can be written as

$$\begin{cases} u'(t) = f(t,u_t), & t \in [0,T], \\ u(t) = \psi(t), & t \in [-\tau,0], \end{cases}$$

where f is the continuous function defined as in Exercise 11.4. In order to get the conclusion, it suffices to observe that, from Exercise 11.6, it follows that f maps uniformly bounded subsets in $\mathbb{R}_+ \times C([-\tau,0];\mathbb{R}^n)$ into bounded subsets in $\mathbb{R}^n$ in the sense of Definition 11.5.2. Hence Theorem 11.5.1 applies and this completes the proof.

Bibliography

Arnold, V., Avez, A. (1967). *Problèmes Ergodiques de la Mécanique Classique*, Gauthier-Villars Paris.

Arnold, V. (1974). *Équations Différentielles Ordinaires*, Editions MIR-Moscow.

Arnold, V. (1978). *Mathematical Methods of Classical Mechanics*, Graduate Texts in Mathematics **60**, Springer Verlag, New York, Heidelberg, Berlin.

Arecchi, F. T., Lisi, F. (1982). Hopping Mechanism Generating $1/f$ Noise in Nonlinear Systems *Phys. Rev. Lett.* **49**, 94–98.

Arrowsmith, D. K., Place, C. M. (1982). *Ordinary Differential Equations*, Chapman and Hall, London-New York.

Aubin, J. P., Cellina, A. (1984). *Differential Inclusions*, Springer Verlag, Berlin, Heidelberg, New York, Tokyo.

Aubin, J. P. (1991). *Viability Theory*, Birkhäuser, Boston, Basel, Berlin.

Barbu, V. (1976). *Nonlinear Semigroups and Differential Equations in Banach Spaces*, Noordhoff International Publishing, Leyden, The Netherlands.

Barbu, V. (1985). *Ecuaţii Diferenţiale*, Editura Junimea, Iaşi.

Barbu, V. (2010). *Nonlinear Differential Equations of Monotone Type in Banach Spaces*, Springer Monographs in Mathematics.

Bliss, G. A. (1946). *Lectures on the Calculus of Variations*, University of Chicago Press, Chicago–Illinois.

Bouligand, H. (1930). Sur les surfaces dépourvues de points hyperlimités, *Ann. Soc. Polon. Math.*, **9**, 32–41.

Braun, M. (1983). *Differential Equations and Their Applications*, Applied Mathematical Sciences **15**, 3rd Edition, Springer Verlag, New York, Heildelberg, Berlin.

Brezis, H. (1975). Monotone operators, nonlinear semigroups and applications, Proceedings of the International Congress of Mathematicians (Vancouver, B.C. 1974), *Canadian Math. Congress Montreal, Que.* Vol. **2**, 249–255.

Burlică, M., Necula, M., Roşu, D., Vrabie, I. I. (2016). *Delay Differential Evolutions Subjected to Nonlocal Initial Conditions*, CRC Research Monograph Series.

Byszewski, L. (1991). Theorems about the existence and uniqueness of solutions of semilinear evolution nonlocal Cauchy problems, *J. Math. Anal. Appl.*,

162, 494–505.

Cârjă, O. (2003). *Unele Metode de Analiză Funcţională Neliniară*, MATRIX ROM, Bucureşti.

Cârjă, O., Necula, M., Vrabie, I. I. (2004). Local invariance via comparison functions, *Electronic Journal of Differential Equations*, **50**, 1–14.

Cârjă, O., Necula, M., Vrabie, I. I. (2007). *Viability, Invariance and Applications*, North-Holland Mathematics Studies **207**, Elsevier North-Holland.

Cârjă, O., Ursescu, C. (1993). The characteristics method for a first-order partial differential equation, *An. Ştiin. Univ. "Al. I. Cuza" Iaşi, Secţ. I a Mat.*, **39**, 367–396.

Cârjă, O., Vrabie, I. I. (2005). *Differential Equations on Closed Sets*, Volume 2 of Handbook of Differential Equations: Ordinary Differential Equations, Cañada, A., Drabek, P. and Fonda, A. Editors, Mathematics & Computer Science, Elsevier B.V.

Coddington, E. A., Levinson, N.(1955). *Theory of Differential Equations*, McGraw-Hill Book Company Inc. New York, Toronto, London.

Corduneanu, C. (1977). *Principles of Differential and Integral Equations*, Chelsea Publishing Company, The Bronx New York, Second Edition.

Courant, R., Hilbert, D. (1962). *Methods of Mathematical Physics. Vol. II. Partial Differential Equations*, Interscience Publishers, Inc., New York.

Craiu, M., Roşculeţ, M. (1971). *Ecuaţii Diferenţiale Aplicative. Probleme de Ecuaţii cu Derivate Parţiale de Ordinul Întîi*, Editura didactică şi pedagogică, Bucureşti.

Cronin, J. (1980). *Differential Equations. Introduction and Qualitative Theory*, Pure and Applied Mathematics, Marcel Dekker Inc. New York and Basel.

Demidovich, B. (1973). Editor, *Problems in Mathematical Analysis*, MIR Publishers.

Dunford, N., Schwartz, J. T. (1958). *Linear Operators Part I: General Theory*, Interscience Publishers, Inc. New York.

Elsgolts, L. (1980). *Differential Equations and the Calculus of Variations*, 3rd printing, MIR Publishers.

Gantmacher, F. (1959). *Applications of the Theory of Matrices*, New York: Interscience. Translated from Russian by Brenner, J.L.

Gelfand, I. M., Fomin, S. V. (1963). *Calculus of Variations*, Prentice-Hall, Inc. Englewood Cliffs, New Jersey.

Giusti, E. (2003). *Direct Methods in Calculus of Variations*, World Scientific, New Jersey - London - Singapore - Beijing - Hong Kong - Taipei - Chennai.

Glăvan, V., Guţu, V., Stahi, A. (1993). *Ecuaţii Diferenţiale prin Probleme*, Editura Universitas, Chişinău.

Halanay, A. (1972). *Ecuaţii Diferenţiale*, Editura didactică şi pedagogică, Bucureşti.

Hale, J. K. (1969). *Ordinary Differential Equations*, Pure and Applied Mathematics, Volume XXI, Wiley-Interscience A division of John Wiley & Sons, New York, London, Sydney, Toronto.

Hartman, P. (1964). *Ordinary Differential Equations*, Wiley, New York.

Havârneanu, T. (2007). *Ecuaţii Integrale*, Editura Alexandru Myller Iaşi.

Hirsch, M. W. (1984). The dynamical system approach to differential equations, *Bull. Amer. Math. Soc.*, **11**, 1–64.

Hirsch, M. W., Smale, S. (1974). *Differential Equations, Dynamical Systems, and Linear Algebra*, Academic Press, New York and London.

Hubbard, J. H., West, B. H. (1995). *Differential Equations: a Dynamical System Approach. Part I Ordinary Differential Equations*, Corrected reprint of the 1991 edition, Text in Applied Mathematics, **5**, Springer Verlag, New York.

Ionescu, D. V. (1972). *Ecuaţii Diferenţiale şi Integrale*, Editura didactică şi pedagogică, Bucureşti.

Kelley, W. G., Peterson, A. C. (2010), *The Theory of Differential Equations Classical and Qualitative*, Second Edition, Universitext, Springer.

Krasnov, M., Kissélev, A., Makarenko, G. (1977). *Équations Integrales*, Éditions MIR Moscou.

Lavrentiev, M. A., Sabat, B. V. (1959). *Metody teorii funktsii complecsnogo peremennogo*, Moscva – Leningrad.

Lefter, C.-G. (2006a). *Calculul Variaţiilor şi Controlul Ecuaţiilor Diferenţiale*, Editura Alexandru Myller Iaşi.

Lefter, C.-G. (2006b). *Ecuaţii Difcrenţiale şi Sisteme Dinamice*, Editura Alexandru Myller Iaşi.

Lorenz, E. N. (1963). Deterministic nonperiodic flow, *J. Atmospheric Sci.*, **20**, 130–141.

Lotka, A. J. (1920a). Undamped oscillations derived from law of mass action, *J. Am. Chem. Soc.*, **42**, 1595–1599.

Lotka, A. J. (1920b). Analytical note on certain rhythmic relations in inorganic systems, *Proc. Natl. Acad. Sci. U.S.A.*, **6**, 410–415.

Lotka, A. J. (1925). *Elements of Physical Biology*, New York.

Lu, Yunguang (2003). *Hyperbolic Conservation Laws and the Compensated Compactness Method*, Chapman & Hall/CRC Monographs and Surveys in Pure and Applied Mathematics **128**, Boca Raton, London, New York, Washington D.C.

Malkin, I. G. (1952). *Teoriya ustoychivosti dvizhenia*, Gostekhizdat.

McKibben, M. (2011) *Discovering Evolution Equations with Applications.* Vol. I *Deterministic Models*, Chapman & Hall/CRC Appl. Math. Nonlinear Sci. Ser.

Murray, J. D. (1977). *Lecture on Nonlinear-Differential-Equation Models in Biology*, Clarendon Press, Oxford.

Neuhauser, C. (2001). Mathematical challenges in spatial ecology, *Notices of the Amer. Math. Soc.*, **48**, 1304–1314.

Nicolescu, M., Dinculeanu, N., Marcus, S. (1971a). *Analiză Matematică*, Vol.I, Fourth Edition, Editura didactică şi pedagogică, Bucureşti.

Nicolescu, M., Dinculeanu, N., Marcus, S. (1971b). *Analiză Matematică*, Vol.II, Second Edition, Editura didactică şi pedagogică, Bucureşti.

Nicolis, G. (1995). *Introduction to Nonlinear Science*, Cambridge University Press.

Nistor, S., Tofan, I. (1997). *Introducere în Teoria Funcţiilor Complexe*, Editura Universităţii "Al. I. Cuza" Iaşi.

Pavel, N. H. (1984). *Differential Equations, Flow Invariance and Applications*, Research Notes in Mathematics 113, Pitman, Boston–London–Melbourne.

Picard, E. (1890). Memoire sur la théorie des équations aux dérivées partielles et la méthode des approximations successives, Chap. V (*Journal de Mathémathique et et Journal de l'École Polytechnique.*

Piccinini, L. C., Stampacchia, G., Vidossich, G. (1984). *Ordinary Differential Equations in* $\mathbb{R}^n$, Applied Mathematical Sciences 39, Springer Verlag, New York, Berlin, Heildelberg, Tokyo.

Pontriaghin, L. (1969). *Equations Différentielles Ordinaires*, Éditions MIR, Moscow.

Putzer, E. J. (1966). Avoiding the Jordan canonical form in the discussion of linear systems with constant coefficients. Amer. Math. Monthly, **73**, 2–7.

Rudin, W. (1976). *Principles of Mathematical Analysis*, Third Edition. McGraw–Hill Book Company, New York.

Şabac, I., G. (1981). *Matematici Speciale*, Editura Didactică şi Pedagogică, Bucureşti.

Severi, F. (1930). Su alcune questioni di topologia infinitesimale, *Ann. Polon. Soc. Math.*, **9**, 97–108.

Sneddon, I. H. (1971). *The Use of Integral Transforms*, McGraw-Hill, New York.

Thirring, W. (1978). *A Course in Mathematical Physics* 1. *Classical Dynamical Systems*, Springer Verlag, New York, Wien.

Vladimirov, V. S., Mihailov, V. P., Varaşin, A. A., Karimova, H. H., Sidorov, Iu. V., Şabunin, M. I. (1981). *Culegere de Probleme de Ecuaţiile Fizicii Matematice*, Editura ştiinţifică şi enciclopedică, Bucureşti.

Volterra, V. (1926). Variazzioni e fluttuazioni del numero d'individui in specie animali conviventi, *Mem. Acad. Lincei*, **2**, 31–113.

Vrabie, I. I. (1995). *Compactness Methods for Nonlinear Evolutions*, Second Edition, Addison-Wesley and Longman, 75.

Vrabie, I. I. (2003). C_0*-semigroups and Applications*, North-Holland Mathematics Studies, **191**, North-Holland, Elsevier.

Widder, D. V. (1946). *The Laplace Transform*, Princeton University Press.

Wieleitner, H. (1964). *Istoria Matematicii. De la Déscartes până în Mijlocul Secolului al XIX-lea*, Editura ştiinţifică, Bucureşti.

Wrede, R., Spiegel, M. R. (2002). *Theory and Problems of Advanced Calculus* Second Edition, Schaum's Outline Series McGraw-Hill.

Ye, Yan-Qian, Cai, Sui-Lin, Huang, Ke-Cheng, Luo, Ding-Jun, Ma, Zhi-En, Wang, Er-Nian, Wang, Ming-Shu, Yan, Xin-An (1986). *Theory of Limit Cycles*, AMS Translation Math. Monographs 66.

Index

algorithm
Putzer's, 138

behavior
chaotic, 193
ergodic, 7

capacity, 42
Cauchy problem, 50
autonomous, 54
chaos, 193
concatenation principle, 53
condition
Frobenius' integrability, 299
Hugoniot–Rankine, 230
Jacobi necessary, 337
Nagumo tangency, 281
conditions
first compatibility
of Erdmann-Weierstrass, 332
cone, 281, 302
Bouligand–Severi tangent, 281
configuration coordinates, 234
conjugate
pairs, 337
points, 337
conservative
system, 348
contraction principle, 7
controller, 188
curve
characteristic, 218

derivative, 392
classical, 246
Dini, 288
distributional, 246
right directional, 288
differential inclusion, 264
disintegration constant, 30
distribution, 242
derivative of order k, 246
Dirac delta, 243
Dirac delta concentrated at a, 245
homothety, 245
of type function, 242

energy integral, 331
entropy, 191
growth, 232
equation
Abel, 316
Airy, 118
autonomous, 13, 54
Bernoulli, 22
Bessel, 118
characteristic, 146
Clairaut, 26
singular solution, 26
delay functional differential, 366
eikonal, 14
Euler, 148
Euler Second, 337
Euler–Lagrange, 327
exact, 23

first-order differential in normal form, 12
first-order linear partial differential, 214
first-order quasi-linear partial differential, 213
first-order vector differential, 12
fluxional, 3
Fredholm integral, 103
Gauss, 119
generalized Abel, 316
Hermite, 118
homogeneous, 142
integro-differential, 103, 119
Lagrange, 25
parametric equations of the general solution, 26
Legendre, 119
Liénard, 42, 172
linear, 20
linear homogeneous, 20
linear non-homogeneous, 20
logistic, 34
n^{th}-order scalar differential incomplete, 27
n^{th}-order scalar differential, 11
n^{th}-order scalar differential in normal form, 12
neutral, 103
non-homogeneous, 142
of big waves in long rivers, 223
of forced oscillations, 147
of the gravitational pendulum, 32
of the harmonic oscillator, 31
of the mathematical pendulum, 32
of the small oscillations of the pendulum, 32
order of an, 11
ordinary differential, 11
partial differential, 11, 103
pendulum, 28
quasi-homogeneous, 46
Riccati, 23
scalar differential, 11
solvable by quadratures, 18
traffic, 221
Van der Pol, 42, 172
Volterra
of the first kind, 310
of the second kind, 305
with convolution kernel, 316
Volterra integral, 103, 119
with separable variables, 18
error function, 412
evolutor, 127
exterior tangency condition, 289
extremal
for the problem $(\mathcal{P})$, 324

family
compact, 395
equicontinuous, 395
of straight lines
envelope of, 26
relatively compact in $C([a,b];\mathbb{R}^n)$, 395
uniformly bounded, 396
uniformly equicontinuous, 395
first variation
of a functional, 324
forced harmonic oscillations, 147
formula
exponential, 9, 117
variation of constants, 20, 121, 132
function
T-anti-periodic, 361
absolutely continuous, 260
beta of Euler, 481
Carathéodory, 260
coercive, 234
comparison, 289
concatenate, 53
continuous at (t,ψ), 366
continuous on C, 366
control, 188
cylindrical, 3
differentiable, 392
dissipative, 62
gamma of Euler, 481
Hamilton, 234
Heaviside, 252
implicitly defined, 91

input, 188
Lipschitz on bounded sets with respect to its second argument, 367
locally Lebesgue integrable, 225, 242, 261
locally Lipschitz, 60
locally Lipschitz on bounded sets with respect to its second argument, 363
lower semi-continuous (l.s.c), 75
Lyapunov, 173
autonomous, 177
maps uniformly bounded subsets in $\mathbb{R}_+ \times C([-\tau, 0]; \mathbb{R}^n)$ into bounded subsets in $\mathbb{R}^n$, 376
matrix-valued, 72
modulus of continuity, 194
multi-valued, 264, 266
multi-valued upper semi-continuous, 267
negative definite, 173
output, 188
perturbing, 169
piecewise continuous, 319
positive definite, 173
Riemann integrable, 392
sensitivity, 86
set-valued, 266
state, 188
unit, 252, 404
unknown, 11

generalized momenta, 234
growth index
of a function, 404

Heaviside function, 404
homogeneous equation, 21

image function, 406
inductance, 42
inequality
Bellman, 47
regulating, 189
initial
condition, 17
state, 17
time, 17
initial-value problem, 50
integral
first, 204
general, 17
independent prime, 206
Kurzweil–Henstock, 8
Lebesgue integral, 8
prime, 204, 211
interval
nontrivial, 15
iterated kernels, 309

Laplace Transform, 406
law
Boyle–Mariotte, 422
conservation, 223
Hooke, 47
Malthus, 33
Newton's second, 31
of conservation of the energy, 234
of mass action, 40
of radioactive disintegration, 29
lemma
Bihari, 43
Brezis, 44
Gronwall, 44
Zorn, 66
Leray–Schauder topological degree, 7
Lie–Jacobi bracket, 296
limit cycle, 8, 182, 195
limit point, 66

mapping
nonexpansive, 75
Poincaré, 120
matrix
associated, 124, 143
fundamental, 124, 144
Hurwitz, 168
Hurwitzian, 165
sensitivity, 86
mean ergodic, 196
method

Carbon dating, 30
Cauchy, 219
Cauchy–Lipschitz–Peano, 4
characteristic, 219
comparison, 8
of integrating factor, 2, 24
of majorant series, 4
of polygonal lines, 106
of successive approximations, 4, 7, 103
of tangents, 4
parameter, 25
polygonal lines, 3
small parameter, 87
stability by the first approximation, 171
the Euler explicit, 97
the Euler implicit, 97, 112
variation of constants, 2, 145

minimum
local $\mathcal{D}^n(0,0)$, 323
local weak, 323

model
Levins, 35
Lotka, 39
pray-predator, 35
pursuit-evasion, 264
Verhulst, 34

multifunction, 266

norm, 399
supremum, 395

observed output, 188

operator
almost additive, 405
almost linear, 405
evolution, 127
feedback, 188
n^{th}-order differential, 255
observation, 188
synthesis, 188

original function, 406

Poincaré's Inequality, 340

point
equilibrium, 159, 206
interior of K
in the direction η, 323
stationary, 159, 206

Poisson bracket, 345

polynomial
characteristic, 146

predator, 35

prey, 35

principal period, 359

principle
Hamilton's least action, 323

problem
Cauchy, 3, 17
Cauchy for a quasi-linear first-order partial differential equation, 218
initial-value, 3, 17
Lurie-Postnikov, 189
nonlocal, 349
of inverse tangents, 1
variational, 320

product
convolution, 249
of a distribution by a function of class C^∞, 245

projection
of a vector on a set, 402
operator on a set, 402
subordinated to $\mathbb{V}$, 290

property
comparison, 291
global uniqueness, 59
local existence and uniqueness, 295
local uniqueness, 59
semigroup, 112
uniqueness, 90

proximal neighborhood, 290

quadrature, 18

quasi-polynomial, 141

rate constant of a reaction, 40

regular
Lagrangian, 332

regulating parameters, 188

resolvent kernel
of a Volterra equation, 310
resonance, 147
rule
Leibniz, 247

second principle of thermodynamics, 232
second variation
of a functional, 324
semigroup, 112
of non-expansive operators, 182
sequence
Cauchy, 400
convergent, 400
convergent in $\mathcal{D}(\mathbb{R})$, 241
fundamental, 400
of successive approximations, 105
series
of matrices, absolutely convergent, 133
of matrices, convergent, 133
of matrices, uniformly convergent, 133
set
convex, 182
globally invariant, 287
invariant, 287
locally closed, 280
normal cone, 273
ω-limit, 182
viable, 280
solution, 366
a.e., 268, 269
asymptotically stable, 157, 160
bilateral, 50
blowing up in finite time, 68
Carathéodory, 260
classical, 225, 240
continuable, 375
continuable at the left, 50, 63
continuable at the right, 50, 63
distribution, 252
ε-approximate, 283
elementary, 256
equilibrium, 36
fundamental system of, 124, 144
general, 17, 214
general of Clairaut equation, 26
generalized, 240, 252, 256
global, 50, 294, 375
at the left, 50
to the right, 50
globally uniformly stable, 357
left, 50
local, 50, 218
non-continuable, 375
of $\mathcal{CP}(\mathcal{D})$, 50
of a first-order linear partial differential equation, 214
of a first-order quasi-linear partial differential equation, 214
of a first-order vector differential equation, 15
of a gradient system, 294
of a system of first-order differential equations, 15
of an nth-order scalar differential equation, 15
of an nth-order scalar differential equation in the normal form, 15
persistently stable, 194
piecewise, 326, 329
right, 50
robust, 194
saturated, 375
saturated at the left, 50
saturated to the right, 50
singular, 2
stable, 37, 156, 160
stationary, 36, 159
the orbit, 182
the trajectory, 15, 16, 182
uniformly asymptotically stable, 157, 160
uniformly stable, 157, 160
unstable, 37
viscosity, 8
with finite speed of propagation, 224
space

complete real vector, 400
normed real, 400
of testing, or test functions, 241
real Banach, 400
strongly positive
second variation, 342
support of a function, 241
system
action of the, 348
autonomous, 13
characteristic, 214
closed loop, 188
control, 188
Euler Second, 337
Euler–Lagrange, 330
canonical form, 344
first-order of linear differential equations, 122
gradient, 294
Hamiltonian, 150, 234, 344
homogeneous, 122
homogeneous attached to, 122
homogeneous with exact differentials, 295
in symmetric form, 214
in variations, 85
Lorenz, 192
Lotka, 40
Lotka–Volterra, 36, 40, 233
Navier–Stokes, 10
non-homogeneous, 122
of n first-order differential equations, 12
perturbed, 169
prey-predator, 40
robust, 86
spring mass with delay, 379
Volterra, 36

theorem
Banach's fixed point, 7
Cauchy–Lindelöff, 99
Cayley–Hamilton, 137
Crandall–Liggett generation, 9
Hille–Yosida, 9
implicit function, 12
Liouville, 128, 144, 150
Nagumo, 281
of structure of the matrix e^{tA}, 140
Peano, 58
Picard, 105
Poincaré–Lyapunov, 169
Putzer, 138
Routh–Hurwitz, 168
theory
bifurcation, 6
of distributions, 8
of dynamical systems, 7
topology
uniform convergence, 394
translation
of a distribution, 245

variable
observed, 188
variational inequality, 273
vector
tangent in the sense of Bouligand–Severi, 280
tangent in the sense of Federer, 282
vector fields
commuting, 296

Wronskian, 125, 144

Printed in the United States
By Bookmasters